Simon Snyder Rathvon

The Lancaster Farmer

Simon Snyder Rathvon

The Lancaster Farmer

ISBN/EAN: 9783744735339

Printed in Europe, USA, Canada, Australia, Japan

Cover: Foto ©ninafisch / pixelio.de

More available books at **www.hansebooks.com**

"THE FARMER IS THE FOUNDER OF CIVILIZATION."—WEBSTER.

A MONTHLY NEWSPAPER:

DEVOTED TO

AGRICULTURE AND HORTICULTURE, PRACTICAL ENTOMOLOGY, DOMESTIC ECONOMY AND GENERAL MISCELLANY.

EDITD BY PROF. S. S. RATHVON.

VOLUME XIV.—1882.

INDEX TO SCIENTIFIC NAMES.

Alce Americanus, 1
ANOCILIDÆ, 2
Anguilla bostoniensis, 2
Agaraoini, 5
Agaricus campestris, 8
Aocularia mellous, 6
Agaricus tridens, 6
Asclepius cornuti, 80
Agrotis vulgaris, 100
Annuals, 120
Articulata, 125
Aphidiphaga, 132
Arctomys monax, 156
Boletus luteus, 6
" versipella, 6
" scaber, 6
" granulata, 6
" bovinus, 6
Batrachus tan, 30
Butalis cerealis, 93
Belostoma americana, 116
Botrysis bassiana, 119
Bivoltius, 120
Belosoma tessolatum, 124
Cervus virginianus, 1
" canadensis, 1
Coprinus comatus, 5, 13
Clytocybe dealbatus, 6
" geotropus, 6
" nebularis, 6
" odorus, 6
Collybia esculenta, 6
" fucipes, 6
" longipus, 6
" radicatus, 6
Clitopilus orcella, 6
" soula, 6
Coprinus atromentarius, 6
Cantharillus cibarius, 6
Clavaria cristata, 6
" fastigiata, 6
" flava, 6
Clytus robinea, 18
" pictus, 18
Cicada septendecim, 31
Canthon laevis, 116
Calosoma, 124

Chrysocbus, 124
Caloptinus femur rubrum, 130
Copris, 132
COCCINELLIDÆ, 132
Ceratocampa regalis, 133
Dermocybe cunameus, 6
Danais, 89
Dactylis glomerata, 109
Danais archippis, 130
Dryocampa imperialis, 156
Dalana ministra, 161
Empusa musce, 119
Etheostomidæ, 124
Eriphus suturalis, 124
Elater, 132
Erythre comissima, 156
" pulchella, 156
" muhienburgia, 156
Fiber zibethicus, 1
Formica rufa, 13
Gyromatia esculenta, 6
Galeruca xanthomalæna, 17
Gossypium herbaceum, 31
Gymnetus nitida, 116
Gordius equaticus, 145
Hygrophores cbuineus, 6
Hydnum repanda, 6
Helvilla californica, 6
" crispa, 6
Helianthus globosus, 90
" festuosus, 90
HEMIPTERA, 116
HYMENOPTERA, 131
HOMOPTERA, 161
ICHNEUMONIDÆ, 131
Lepiota excoriatus, 6
" illanitus, 6
" procerus, 6
" rachodes, 6
Lactarius deliciosus, 6
" luculens, 6
" piperatus, 6
" volemus, 6
Lycoperdon giganteum, 6
LONGICORNIA, 18
Lumbricus terrestris, 108
Lepidoptera, 118, 131

Languria mozardi, 124
Locusta migratoria, 129
Lencania unipuncta, 132
Mus decumanus, 1
" rattus, 1
Morchella esculenta, 5, 6
" conica, 6
Marasmius oreades, 6
Melangaster variegatus, 6
Myrmecocystus hortus deorum, 7
Mollusca, 125
Macrosila carolina, 130, 156
" 5-maculata, 130
Microgaster congregata, 131
Mydas fata, 132
Meloe angustiocllis, 156
MYRIAPODA, 177
Necrophorus marginatus, 146
Omphalla oniscus, 6
Orthosoma cylindrica, 124
Orgya leucostigma, 132
PTERONTZONIDÆ, 2
Pteromyzon americanus, 2
Pleurotus ostreatus, 6
" pomelia, 6
" ulmacium, 6
Psaliota arvensis, 6
" campestris, 6
" cretaceus, 6
" pratensis, 6
" silvatica, 6
Pholiata mutabilis, 6
" aquinoscus, 6
Paxillus giganteus, 6
Polyporus sulphureus, 6
Pogonomyrmex occidentalis, 5
Passer domesticus, 17
Pyrethrum roseum, 65, 72, 74, 82, 83
" cincrariæfolcum, ibid.
" wilimeti, 72
" caucasum, 72
Phryganea cinera, 84
" semifasciata, 84
Prodenia lineatella, 98
Pyralis farinalis, 98
Poa pratensis, 109
" compressa, 109

Pembrina, 120
Panlays tophyton, 120
PRONOSPERNIÆ, 120
Prionus laticonia, 124
PENCIDÆ, 124
Pteromalus puparum, 131
Pieris rapa, 131
Papilio asterias, 132
Pelidiuta punctata, 132
Percica lævis, 145
Phylloxera vastata, 147
Politiscus festivus, 156
" erythacus, 156
Polydesmus, 178
Rhus vernix, 2
" glabrum, 2
Russula adjusta, 6
" alutacea, 6
" heterophyla, 6
" lepida, 6
Radiata, 125
Sliex purpurea, 2
Solklago, 18, 124
Spongia, 31
Strongulus syngamus, 68
Sitophilus granarius, 99
Sericaria mori, 118
Strougulidæ, 145
Silpha americana, 146
Scolopendra heros, 177
Spirobolus marginatus, 177
Tremella mesenterica, 6
Turdus fuscocus, 31
Trichius spiralis, 57
Triton jeffersoni, 60
Theridion trigonum, 60
" globosum, 60
Tetraopes tomator, 89, 124
Trevoltius, 120
Trogus fulvus, 132
Tettigonia vitis, 161
Volvaria Combyebuss, 6
Vanadium, 124
Vanadiate of lead, 124

CONTENTS OF VOLUME FOURTEEN.

EDITORIAL.

Our Fourteenth Volume, 1
The Moose-Deer 100 Years Ago, 1
Killehinic, 2
The Value of Snow, 2
Kitchen Garden for January, 2
Winter Blooming, 2
"Aid and Comfort," 3
How Do Eels Breed? 3
Excerpta, 3, 20, 37, 52, 67, 84, 100, 114, 132, 163, 180
Our Responsibility, 5
Edible Fungi, 5
The English Sparrow, 17
Ourselves, 18
February Snows, 18
Wood-worm, 18
Planting Trees on Railway Embankments, 18
The Largest Tree in the World, 19
Shifty, Thrifty France, 19
Kitchen Garden for February, 19
Poultry Exhibition, 20
Rules and Exceptions, 20
Writing for the Farmer, 20
Our Apology, 33
"Our Winged Friends," 33
Kitchen Garden for March, 33
Why not Write for the Farmer ? 33
The Bane and Antidote, 34
"Revised Fruit List," 34
Eating Before Sleeping, 35
How Long are We to Live? 36
The Will and the Deed, 37
Ensilage, 49
April Meeting, 49
Snails in Gardens, 50
Kitchen Garden for April, 50
Phenomenal 51,
Eating Between Meals, 51
Pyrethrum, 65
Gapes vs. Entomology, 65
A New Industry of Lancaster County, 67
Lime in Soil, 67
Queries and Answers, 69
The Proposed New Department of Agriculture, 81
Increase of our Crops, 81
Potash in Plants, 81
Kitchen Garden for June, 81
Exports of Cheese, 83
The Conestoga Flying Fish, 83
Pyrethrum Roseum, 83
Vonnor Predicts a Bad Summer, 83
Caddice Flies, 84
Eggs, 84
Our Crops, 84
Egg Culture in France, 97
Gapes in Chickens, 97
Entomological Notes—Directions for Seeding Insects, 87
Kitchen Garden for July—Quality and Vitality of Seeds, 98
How to Kill Wheat Moth, 98
Our Local Crops, 98
Destroying Weevil, 99
Effects of Baking on Flour, 99
Phosphoric Acid in Plants, 99
A Mare's Nest, 100
Three Wonders, 100
A Chosen People, 113
Green Corn Pudding, 113
Kitchen Garden for August, 113
Good Husbandry, 113
How to Preserve Stable Manure, 114
Gapes and Eels, 114
A Big Bug, 116
Tomato Horn Worm, 116
Goldsmith Beetle, 116
The English Sparrows, 116
State and County Fairs of 1882, 129
Kitchen Garden for September (Seedpurchasing a Matter of Confidence), 129
Insect Migrations, 139
The Wheat Crop of 1882, 130
Tobacco Worms—Curious Facts Concerning Them, 130
The Royal Horned Caterpillar, 133
The Stanwich Nectarine, 145
Luscious Grapes, 145
Something about "Hair-Worms" and Eels, 145
Kitchen Garden for October, 146
Necrophore, 146
Seedling Peach, 146
The History of the Tomato, 161
"Leaves," 161
Kitchen Garden for November, 161
Insects Injurious to Forests and Shade trees, 161
A Plea for Trees, 162
The Farmer's Creed, 162
Volume Fourteen, 177
Myriapods, 177
The Tariff and Free Trade, 178
The Turkey, 179
Kitchen Garden for December, 179

CONTRIBUTIONS.

Hybridizing Fruits and Flowers, 6
Persimmons, 6
The Egg—Its Contents, and How it is Made, 22
Fruit Belts, 23
Chinese Fruit Pear, 24
Commercial Fertilizers, 23
Forestry, 38
Strawberries, 39
Practical Poultry Notes, 39
Domestic Hints, 39
Practical Recipes, 39
Comparative Value of Farms between Now and Fifty Years Ago, 83
Our Wheat Crops, 86
The Uses of Pruning, 102
Balance of Trade, 102, 164, 182
Gapes in Poultry, 117
Lime, 117
Tariffs and Their Effects, 118
Gapes in Poultry, 133
Shallow Cultivation, 133
Not the Tariff Question, 133
The Eel—Its Habit and Growth, 133
The Value of Clover on Land, 147
The Leaves, 147
Save the Peach-stones, 147
"The Farmer's Friend," 181
A Sure Preventive of Chicken Cholera, 182
The Balance-of-Trade Delusion, 182

ESSAYS.

The Growth and Consumption of Timber Trees in America, 40
"Our Winged Friends," 41
Seedling Fruits, 44
Fruits and Vegetables—Their Culture, 53
The Bright Side of Horticulture, 54
Horticultural Fertilizers, 55
Some Practical Points in Peach Culture, 69
The Management of an Orchard, 71
Insects and Some of Their Relations to the Vegetable Kingdom, 86

SELECTIONS.

Farming about the Rocky Mountains, 6
"Go to the Ant," 7
A Great Southern Farmer, 7
Lime as a Preservative, 8
On Square Acre, 8
Yards in a Mile, 8
Wheat Crop of the United States, 8
A Plain and Easy Way of Curing Hams, 8
The Part which Worms Play in Nature, 9
Spare the Trees, 10
Let the Frost Help You, 10
Tobacco Review—The Old Year and the New One, 11
Berks County Agricultural Society, 11
Poultry Show, 24
White Vein—Cause of the Disease in Tobacco — The Early Cutting Theory—Convincing Experiences, 26
Tobacco Growing—Profits Realized by some Experts—Early Buying in the Field—Result of Careful Handling—A Growing Tobacco—Cost of Growing Tobacco—Another Paying Crop—Still Another—In Conclusion, 26
American Silks Good, 27
Coal Tar and Alkali in Peach Culture, 28
Points in Cows, 28
The New Wheat Region, 55
How to Deodorize Stables, 56
Utilizing Rough Ground, 56
The Building of Houses, 56
When to Cut Grass, 57
Feeding Poultry and Raising Chicks, 57
Vegetable Condiments, 57
Trichinosis, 57
Testing Cream, 58
Application of Liquid Manure, 58
Early Price of Pennsylvania Lands, 58
A Home Fruit Canning Factory, 59
History of Pyrethrum, 72
Quince Culture, 74
Poultry Farming, 75
Poultry Abundant, but Dear, 75
Notes on French Agriculture, 76
The Benevolent Sunflower, 89
Our Timber Lands, 90
Roots and How to Grow Them, 91
Green Manures, 92
U. S. Department of Agriculture, 103
The Happy Grainger, 103
Underdraining, 104
Education for Farmers, 104
Success in Farming, 104
The Department of Agriculture, 105
Fancy Butter, 106
All about Poultry, 106
Talks about Fruit, 107
Silk Culture, 118
Minerals at the Exposition, 121
Diversified Farming in the South, 121
The Mosquito, 122
A Grand Harvest, 134
Occupation and Longevity, 134
The War in Egypt, 135
The Climate in Different Parts of the Union, 135
Pure and Wholesome, 135
Temperature and Rainfall, 136
Barn Yard Manures, 136
Preserving Fence Posts, 136
Some Wheat Statistics, 136
Importance of Having a Good Queen, 137
Draining of Land, 137
The Practical and the Scientific in Agriculture, 138
Fighting the Phylloxera in Europe, 147
Protecting Plants During Winter, 148
Self Dependence, 148
The Preservation of Forests from Wanton Destruction, and Tree Planting, 148
Cultivation of Peppers, 150
How to Bottle Wine, 150
Practical Forestry Illustrated, 151
Summer, 152
How to Keep Houses Healthy, 152
The Coming Horse, 152
The Trade in Nuts, 152
Work and Leisure, 153
Stable Cleaning, 153
Worthless Dogs, 153
The Black Walnut, 153
Trees, Climate, and Soil, 164
Heavy Manuring, and How ? 165
Artificial Incubation, 166
Indian Corn in Kansas—Its Value and Importance, 166
The Effect of a Good Silo, 167
Agricultural Prosperity Should Benefit the Farmer, 167
Tree-Planting in Streets and Grounds, 168
The Fair Season, 168
Italian Bees and How Italianize Common Black Bees, 168
Preventable Losses on the Farm, 169
Yield and Condition of Crops, 169
The Virtues of Coffee, 183
Feeding Stock in Winter, 183
The Rational Method of Tree-Pruning, 184
Letter from the Mother of Bayard Taylor to Prof. E. V. Riley, 185
Soiling Cattle, 185
Smoke House at Small Expense, 186
The Sugar Beet, 186

OUR LOCAL ORGANIZATIONS.

Lancaster County Agricultural and Horticultural Society, 11, 28, 45, 76, 92, 107, 122, 138, 154, 170, 187
The State Grange, 12
Poultry Association, 13, 30, 47, 77, 93, 108, 123, 140, 155, 171, 187
Fulton Farmers' Club, 30, 47, 59, 77, 124, 140, 163, 171
Linnæan Society, 13, 30, 47, 60, 93, 109, 124, 156, 171, 188
State Board of Agriculture, 16, 140

ENTOMOLOGICAL.

Swarming Ants and Allied Phenomena, 60
Curculio in Plum Culture, 61
Birds and Canker Worms, 61

AGRICULTURE.

Look after the Implements, 14
Do Your Own Repairing, 14
Ensilage Silos, 14
Bad Seed, 14
Planting Tobacco, 31
Improved Grasses, 31
Rotation of Crops, 31
Sowing the Seed, 61
Clover and Grass, 61
Clover, 61
Ploughing, 61
Potatoes, 61
Onions, 61
French Farming, 78
Sand Farming, 78
Crop Prospects, 78
Fence Posts, 78
Rotation of Crops, 94
Manure Made under Cover, 94
Exports of Breadstuffs, 94
Corn Culture in Gardens, 94
Green Crops, 108
Loading Hay, 108
Manure under Cover, 108
Plaster, 108
The Largest Land Owner on the Continent, 108
Best Pasture Grass, 109
Pacific Coast Wheat Items, 109
Lying in Fallows, 123
A Short-sighted View, 125
Select Your Own Seed Wheat, 125
A Talk about Grasses, 125
Pasture Grasses, 141
Experiments with Green Manuring, 141
Wheat Raising, 142
What of the Future as Regards Grain, 142
What Manure Loses by Heating, 142
Good Crops in Alabama, 142
Magnesia for Wheat, 142
Wheat Growing, 156
An Excellent Fertilizer, 156
How to Remove Stumps, 156
The Telephone on the Farm, 156
Octagonal Barns, 156
The Use of the Roller, 172
Progressive Farmers, 172
Effect of Draining, 172
Fall Plowing, 172
Ivory Wheat and Milo Maize, 188
Economy on the Farm, 188
Rules Adopted by the Hay Trade, 188
Effects of Broom Corn on the Soil, 188
The Agricultural Interests of the Country, 188
Small Potatoes, 189

HORTICULTURE.

Rosebushes, 14
Pears, 14
Notes on Orchard and Garden Work, 14
Making Butter, 14
How to Make Tea and Coffee, 14
Butter Easily Spoiled, 15
The Rhubarb Plant, 61
Mulberry Trees, 61
An Excellent Old Apple, 62
An Experiment in Potato Planting, 62

INDEX.

Apples for Medicine, 78
Greenhouse and Window Plants, 78
Profit to Onions, 78
Celery Culture, 78
How the Chinese Make Dwarf Trees, 78
An Abundant Apple Crop, 94
What Kills Fruit Trees, 94
Early Turnips, 94
Summer Grape Pruning, 109
The Care and Pruning of Peach Trees, 109
The Delaware Peach Crop, 109
Strawberry Beds, 109
Quince Culture, 109
The Peach Crop, 125
Value of Fruit, 125
Shallow Cultivation for Fruits, 125
The Vegetable Garden, 125
Fig Culture, 126
Keeping Grapes Fresh, 142
Beneficial Effect of Mulching on Berries, 142
Taking in Fall Flowers, 142
Saves the Peachstones, 143
A Hint for Winter Gardening, 143
York Imperial Apple, 157
Keeping Apples, 157
Apple Notes, 157
Root Pruning, 157
The Cherry and Apple, 157
Pear Raising, 172
The Effect of Dry Weather on Apples, 172
Saving Cabbage till Spring, 173
The Fruit Supply, 173
Bananas and Plaintains, 173
Winter Flowers in the Window, 189
Preserving Garden Flowers, 189

HOUSEHOLD RECIPES.
Light Gingerbread, 15
Cocoanut Cake, 15
Chocolate Cake, 15
Rock Cake, 15
Gingerbread, 15
English Buns, 15
Almond Cake, 15
Milk Biscuit, 15
Soft Gingerbread, 15
Doughnuts, 15
Raisin Pie, 15
Corn Bread, 15
Cocoanut Pudding, 15
Baked Soup for Invalids, 15
Baked Indian Pudding, 15
Orange Pie, 31
New England Baked Indian Pudding, 31
Chicken Pie, 31
Prune Pudding, 31
A Nice Way of Cooking Cold Meats, 31
Chocolate Cake, 31
Breakfast Husks, 31
Preparing Carrots, 31
Barley Soup, 71
Cornstarch Cakes, 31
French Tapioca Pudding, 31
Sweet Macaroni, 32
Oatmeal Pudding, 32
Wholesale Pie Crust, 32
Stewed Apples and Rice, 33
To Make a Cheap Wash or Paint, 62
Rice, Milanaise Style, 62
Macaroni and Ham, 62
Poor Man's Plum Pudding, 62
Fig Pudding, 62
Yorkshire Pudding, 62
Warm Slaw, 62
Cold Slaw, 62
Lincoln Cake, 62
Pastry, 62
To Clean Marble, 62
Valuable Hints, 62
Cocoanut Cookies, 62
To Renovate Black Grenadine, 62
To Wash Silk Stockings, 62
Cornstarch Bake, 62
Black Bean Soup, 62
To Clean Musty Barrels, 62
Cottage Gingerbread, 62
Household Weights and Measures, 62
Scotch Butter Candy, 62
Tapioca Pudding, 78
Bread Pudding, 78
Chili Sauce, 78
Clam Chowder, 78
Saddle of Lamb, 78
Tomato Soup, 78
Oyster Soup, 78
Chicken Salad, 78
White Sauce, 79
Sugar Kisses, 79
Queen of Puddings, 79
Lemon Pudding Sauce, 79

Bird's Nest Pudding, 79
Orange Pudding, 79
Green Corn Patties, 79
Boston Cream Cake, 79
Flake Pie Crust, 79
Superior Doughnuts, 79
Cookies, 79
Custard Pie, 79
Graham Rolls, 79
Rice Waffles, 79
Steamed Indian Loaf, 79
Muffins, 79
Lemon Pie, 79
Pumpkin Pie, 79
Graham Muffins, 79
Turkey Soup, 79
Fish Sauce, 79
Cabbage Salad, 79
Cottage Pudding, 79
Suet Pudding, 79
Boiled Bread Pudding, 79
Lowell Pudding, 79
Hominy Muffins, 79
Potato Cakes, 79
Oyster Fritters, 79
Corn Oysters, 79
Boiled Leg of Lamb, 79
Tapioca Pudding, 79
Snow Pudding, 79
Beefsteak Rolls, 95
Deviled Ham, 95
Yankee Plum Pudding, 95
French Beefsteaks, 95
Squash Pie, 95
Delightful Pudding, 95
To Make Tough Meat Tender, 95
Cabbage Salad, 95
Scalloped Oysters, 96
Roast Shoulder of Veal, 95
Western Cookies, 95
Fairy Apple, 95
Deep Apple Pie, 110
Pan Dowdy, 110
Fried Apple, 110
Apple Toast, 110
Apple and Bread Pudding, 110
Racket Club Pudding, 110
Jelly Pudding, 110
Cheese Crust, 110
Pumpkin Pie, 110
Plain Mince Pie, 110
Welsh Rare-Bit, 110
Omelette, 110
Chicken and Green Peas, 110
Bean Soup, 110
Codfish, 110
Broiled Birds, 110
Sago and Wine, 110
Beef Juice, 110
Wine Jelly, 110
Toast, 110
Barley Water, 110
Egg and Wine, 110
Milk Punch, 110
Cucumber Mangos, 126
Peach Mangos, 126
Veal a la Mode, 126
Breast of Veal Baked with Tomatoes, 126
Breast of Veal Braised, 126
White Sauce, 126
Boiled Tongue, 126
Boiled Corned Beef, 126
Boiled Ham, 126
Pork Chops, Spanish Style, 126
Roast Pork, 126
Pork Tenderloins, 126
Irish Stew, 126
Persillade of Mutton, 126
Fried Breast of Mutton, 126
Breading 126
Ragout of Cold Beef and Vegetables, 126
Roast Leg of Lamb or Mutton, 126
Garlic Cloves, 126
Fig Pudding, 143
To Whiten Scorched Linen, 143
To Cook Turnips, 143
Almond Cake, 143
Pan Dowdy, 143
Smothered Chicken, 143
Pumpkin Pie, 143
Sheep's-head Soup, 143
Pickled Onions, 143
Lemon Pudding, 143
Ready-made Glue, 143
Apple Jelly, 143
A Remedy for Diptheria, 143
Household Hints, 143
Dry Curing Pork and Beef, 143
Stewed Corn, 144
Brown Sauce, 144
Boiled Sweet Corn, 144
Stewed Corn and Tomatoes, 144
Chow-Chow, 157

Stuffed Tomatoes, 157
Pancakes, 157
Rissole Soup, 157
Lamb Chops, 157
Potato Mound, 157
Ladies' Cabbage, 157
Damson Tart, 158
Potato Porridge, 158
Roasted Sweetbreads, 158
Boil and Blanch the Sweetbreads, 158
Potato Croquettes, 158
Rice Pudding Cold, 158
Breakfast Cakes, 158
Cream Nectar, 158
Potatoes au Maitre d'Hotel, 158
Stewed Tomatoes and Onion, 158
Stewed Pears with Rice, 158
Ox-Cheek Soup, 158
Stewed Calf's Hearts, 158
Apple Soufflé Pudding, 158
Graham Bread, 173
Indian Cake, 173
Crullers, 173
Doughnuts, 173
Buns, 173
Roast Mutton, 173
Mashed Potatoes, 173
Mashed Turnips, 173
Baked Potatoes, 173
Apple Pudding, 173
Spanish Cream, 173
Boiled Flank of Beef, 173
Meat Hash, 173
Veal Loaf, 173
Tomato Sauce, 173
Steamed Beef Steak Pudding, 173
Stewed Lobster, 174
Boiled Rice, 174
Boiled Cider, 174
Steamed Pudding, 174
Nice Griddle Cakes, 174
Cottage Pudding, 174
Griddle and Indian Cakes, 174
Escalloped Mutton, 174
Mocked Oyster Soup, 174
Excellent Gold Cake, 174
Lemon Cake, 174
Fried Chicken, 174
Plain Fruit Cake, 174
Boiled Rice Pudding, 174
Okra Soup Equal to Turtle Soup, 174
Steamed Brown Bread, 174
Rhubarb Pies, 174
Roast Turkey Garnished with Sausage, 189
Mashed Turnips, 189
Canned Corn Pudding, 189
Cranberry Sauce, 189
Orange Snow and Snowdrift Cake, 189
Oyster Soup, 189
Boiled Chicken, 189
Browned Potatoes, 189
Baked Sweet Potatoes, 189
Scalloped Squash, 189
Baked Custards, 189
Simple White Soup, 189
Stewed Fillet of Veal, 190
Spinach, 190
Boiled Beans, 190
Mashed Potatoes, 190
Queen's Toast, 190
Brown Gilet Soup, 190
Minced Turkey and Eggs, 190
Stewed Potatoes, 190
Celery, 190
A Plain Rice Pudding, 190

LIVE STOCK.
The Care of Cows, 15
Charcoal for Sick Animals, 15
Hints about Horses, 15
Hay for Swine, 15
Warts on Horses, 15
The Horse Shoe and Its Application, 15
Sawdust for Bedding, 62
Salting Stock, 63
Floors for Horse Stables, 63
Charcoal for Sick Animals, 63
The Hog Crop, 63
Tying up Calves, 63
Man's Treatment of the Horse, 63
Advantages of Small Flocks, 63
"Loss of Cud," 63
Training Heifers to Milk, 63
Bedding for Cows, 63
Inoculation of Animals, 63
Care of Horses' Legs, 79
Care of Sheep, 79
Watering Horses, 79
Save and Care for the Pigs, 79
How to Grow a Pig, 79
A Nevada Stock Raiser, 80
Improving the Stock on the Farm, 95

Keep up the Flow of Milk, 95
Care of Dairy Vessels, 95
Raise the Good Cow's Heifer Calf, 95
Spoiling a Young Horse, 110
The Pig in Agriculture, 110
Sheep Raising in Dakota, 111
Treatment of the Cow, 111
Advice of a Lancaster County Blacksmith on How to Shoe Horses, 126
Training Horses, 126
The Best Farm Horses, 127
Draught Horses, 127
Is Horseshoeing Useless, 127
Keep the Stable Clear of Flies, 127
Remedy for Sale Hole in Cow's Teat, 127
Care of Horses, 127
The Stock, 127
Improved Sheep, 144
Management of Pigs, 144
A New Cattle Disease, 144
Raising a Colt, 158
Hints on Raising Stock, 158
Swine Raising—A Different System Desirable, 158
More Frequent Milking, 158
Jersey Cows and their Records, 158
Facts about Horses, 159
Overloading Cows' Stomachs, 159
Quarantined Cattle, 159
Cattle-Raising in Montana, 174
To Utilize Jersey Bulls, 174
The Shropshire Sheep, 174
Rearing Sheep for Their Milk, 174
Making Good Pork, 175
The Coming Sheep, 175
Cotton-seed Meal for Live Stock, 190
Dry Food For Hogs, 190
Lincoln Sheep, 190
Pasturing and Soiling Hogs, 190
Growth of Colts, 190
Sheep, 191
Training Horns, 191
Cattle Range of Wyoming, 191

POULTRY.
Sunflower Seed for Poultry, 64
Grain in Vegetables, 64
Poultry Upon the Farm, 64
Dressing and Keeping Poultry, 64
Common Sense in the Poultry Yard, 64
The Roup in Fowls, 64
Poultry, 64
A Writer in the Poultry Monthly, 80
A House for 200 Fowls, 80
Questions about Eggs and Fowls, 80
Raising Sunflowers for Hens, 80
Care of Young Turkeys, 80
How Chickens are Born, 80
A Cheap Chicken Coop, 80
Hawaiian Geese, 80
One Variety, 95
Treatment of Young Ducks, 96
A Profitable Hennery, 70
Floors for Poultry Houses, 111
Fowl Fattening, 111
Onions for Chicken Cholera, 111
Cramming Poultry, 111
Wild Chickens, 111
Good Hatching,
Poultry Gossip, 127
Feather and Egg Eating, 127
Geese, 128
The Wonders of Incubation, 128
A Meat Diet, 128
Food for Laying Hens, 128
Guinea Hens, 160
Care of Fowls, 160
Ducks, 160
Which is the More Profitable ?, 160
Fattening Turkeys, 160
Farm and Workshop Notes, 160
Moulting, 175
How to be Rid of Them, 175
Poultry Nonsense, 191
Poultry, 191
Women as Poultry Raisers, 191
To Fatten Fowls or Chickens in Four or Five Days, 191
Winter Rations for Hens, 191
Pekin Ducks, 192

APIARY.
Some Information About the Queen Bee, 159
Twelve Facts for Beginners, 159
A System for Wintering 159
Preparing for Winter, 159

LITERARY.
Literary and Personal, 16, 32, 48, 64, 80, 111, 128, 144, 160, 176, 192

ONE DOLLAR PER ANNUM.—SINGLE COPIES 10 CENTS.

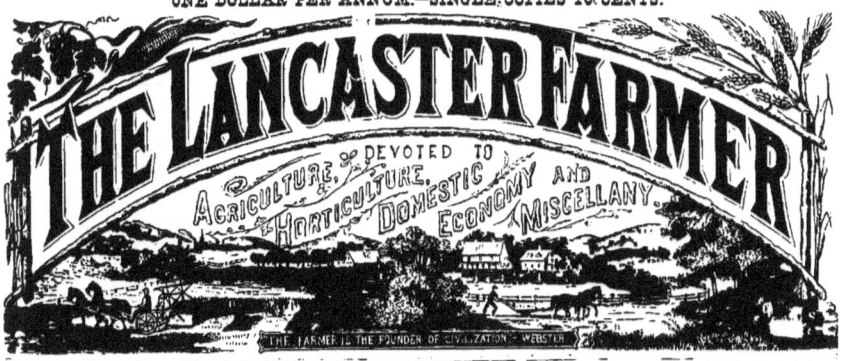

THE LANCASTER FARMER

DEVOTED TO Agriculture, Horticulture, Domestic Economy and Miscellany.

THE FARMER IS THE FOUNDER OF CIVILIZATION.—WEBSTER.

Dr. S. S. RATHVON, Editor.   LANCASTER, PA. JANUARY, 1882.   JOHN A. HIESTAND, Publisher.

Entered at the Post Office at Lancaster as Second Class Matter.

CONTENTS OF THIS NUMBER.

EDITORIAL.

Our Fourteenth Volume,	1
The Moose-Deer 100 Years Ago,	1
Killeknic,	2
The Value of Snow,	2
Kitchen Garden for January,	2
Winter Blooming,	3
"Aid and Comfort,"	3
How Do Eels Breed ?	3
Excerpts,	
Miscellaneous—Economical—Moral Economy—Domestic Economy.	
Our Responsibility,	5
Edible Fungi,	5

CONTRIBUTIONS.

Hybridising Fruits and Flowers,	6
Persimmons,	6

SELECTIONS.

Farming About the Rocky Mountains,	6
"Go to the Ant,"	7
A Great Southern Farmer,	7
Lime as a Preservative,	8
One Square Acre,	8
Yards in a Mile,	8
Wheat Crop of the United States,	8
A Plain and Easy Way of Curing Hams,	9
The Part which Worms Play in Nature,	9
Spare the Tree,	10
Let the Frost Help You,	10
Tobacco Review—The Old Year and the New One,	11
Consumption in 1881—Stock on Hand on January 1, 1882—The Crop of 1881 and Visible Supply—Receipts in 1881—Sales Each Month—Remarks Prices—Quotations January 1, 1881.	
Berks County Agricultural Society,	11

OUR LOCAL ORGANIZATIONS.

Lancaster County Agricultural and Horticultural Society,	11
Crop Reports—Election of Officers—"Can the Grain Grower Dispense with Nitrogenous Fertilizers?"	
The State Grange,	12
Wednesday's Proceedings—The Proceedings on Thursday.	
Poultry Association,	13
Treasurers' Report—Election of Officers—Miscellaneous Business.	
Linnæan Society,	13
Library—Papers Read—Elections—New Business.	

AGRICULTURE.

Look After the Implements,	14
Do Your Own Repairing,	14
Ensilage Silos,	14
Bad Seed,	14

HORTICULTURE.

Rosebushes,	14
Pears,	14
Notes on Orchard and Garden Work,	14

Making Butter,	14
How to Make Tea and Coffee,	14
Butter Easily Spoiled,	15

HOUSEHOLD RECIPES

Light Gingerbread,	15
Cocoanut Cake,	15
Chocolate Cake,	15
Rock Cake,	15
Gingerbread,	15
English Buns,	15
Almond Cake,	15
Milk Biscuit,	15
Soft Gingerbread,	15
Doughnuts,	15
Raisin Pie,	15
Corn Bread,	15
Cocoanut Pudding,	15
Baked Soup for Invalids,	15
Baked Indian Pudding,	15

LIVE STOCK.

The Care of Cows,	15
Charcoal for Sick Animals,	15
Hints About Horses,	15
Hay for Swine,	15
Warts on Horses,	15
The Horse Shoe and its Application,	15
Literary and Personal,	15
Board of Agriculture,	16

SEND IN YOUR SUBSCRIPTIONS
—FOR—

THE FARMER
FOR 1882.

The cheapest and one of the best Agricultural papers in the country.

Only $1.00 per year.

JOHN A. HIESTAND, Publisher,
No. 9 North Queen st., Lancaster, Pa.

SEEDS, BULBS, PLANTS.

J. LEWIS CHILDS, QUEENS, N. Y.

Jan-3m

WE WANT OLD BOOKS.
WE WANT GERMAN BOOKS.
WE WANT BOOKS PRINTED IN LANCASTER CO.
We Want All Kinds of Old Books.
LIBRARIES, ENGLISH OR GERMAN BOUGHT.
Cash paid for Books in any quantity. Send your address and we will call.

REES WELSH & CO.,
23 South Ninth Street, Philadelphia.

D. M. FERRY & CO'S ILLUSTRATED, DESCRIPTIVE, PRICED SEED ANNUAL FOR 1882

Will be mailed FREE to all applicants, and to customers without ordering it. It contains five colored plates, 600 engravings, about 200 pages, and full descriptions, prices and directions for planting 1500 varieties of Vegetable and Flower Seeds, Plants, Fruit Trees, etc. Invaluable to all. Send for it. Address,
D. M. FERRY & CO., Detroit, Mich.

Jan-4m

$66 A week in your own town. Terms and $5 outfit free Address H. HALLETT & Co., Portland, Maine.
jun-1yr*

PENSIONS For SOLDIERS, widows, fathers, mothers or children. Thousands yet entitled. Pensions given for loss of finger, toe, eye or rupture, varicose veins or any Disease. Thousands of pensioners and soldiers entitled to INCREASE and BOUNTY. PATENTS procured for Inventors. Soldiers land warrants procured, bought and sold. Soldiers and heirs apply for your rights at once. Send Stamps for "The Citizen-Soldier," and Pension, Bounty and Patent Laws, blanks and instructions. We can not refer to thousands of Pensioners and Clients. Address N. W. Fitzgerald & Co., Claim Attys. Lock Box 585, Washington, D. C.
Jan

LIGHT BRAHMA EGGS

For hatching, now ready—from the best strain in the county—at the moderate price of
$1.50 for a setting of **13 Eggs.**
L. RATHVON,
No. 9 North Queen st., Examiner Office, Lancaster, Pa.

WANTED.—CANVASSERS for the LANCASTER WEEKLY EXAMINER In Every Township in the County. Good Wages can be made. Inquire at
THE EXAMINER OFFICE,
No. 9 North Queen Street, Lancaster, Pa.

$72 A WEEK. $12 a day at home easily made. Costly Outfit free. Address TRUE & Co., Augusta, Maine.
jun-1yr*

SEND FOR SPECIAL PRICES

On Concord Grape-vines, Transplanted Evergreens, Tulip, Poplar, Linden Maple, etc., Tree Seedlings and Trees for timber plantations by the 100,000.
J. JENKINS' NURSERY,
3-2-79
WINONA, COLUMBIANA CO., OHIO.

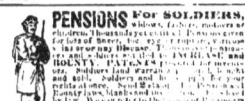

PENSIONS For SOLDIERS, children, Thousands entitled. Pensions given for any Disease. Increase in Pensions obtained. PATENTS procured for Inventors. Soldiers' land warrants, bought and sold. Soldiers apply for your rights at once. Send stamps for blanks and instructions. We can refer to thousands of Pensioners and Clients. Address N. W. FITZGERALD & Co.,
PATENT ATT'YS, Lock Box 585, Washington, D. C.
dec-1t

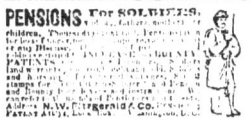

WELL-AUGER. Ours is guaranteed to be the cheapest and best in the world. Also nothing can beat our SAWING MACHINE. No getting out of repair in 5 minutes. Pictorial books free. W. GILES, Chicago, Ill.
-6m)

PENNSYLVANIA RAILROAD SCHEDULE.
Trains leave the Depot in this city, as follows:

WESTWARD.	Leave Lancaster.	Arrive Harrisburg.
Pacific Express	2:40 a. m.	4:05 a. m.
Way Passenger	5:00 a. m.	7:30 a. m.
Niagara Express	11:00 a. m.	11:20 a. m.
Hanover Accommodation	11:05 a. m.	
Mail train via Mt. Joy	10:20 a. m.	12:40 p. m.
No. 2 via Columbia	11:25 a. m.	12:55 p. m.
Sunday Mail	10:50 a. m.	12:40 p. m.
Fast Line	2:30 p. m.	3:25 p. m.
Frederick Accommodation	2:55 p. m.	Col. 2:45 p. m.
Harrisburg Accom	5:45 p. m.	7:40 p. m.
Columbia Accommodation	7:20 p. m.	Col. 8:30 p. m.
Harrisburg Express	7:30 p. m.	9:40 p. m.
Pittsburg Express	9:50 p. m.	10:10 p. m.
Cincinnati Express	11:00 p. m.	12:15 a. m.
EASTWARD.	Lancaster.	Philadelphia
Cincinnati Express	2:55 a. m.	5:00 a. m.
Fast Line	5:58 a. m.	7:40 a. m.
Harrisburg Express	8:05 a. m.	10:00 a. m.
Columbia Accommodation	9:10 p. m.	12:00 p. m.
Pacific Express	1:00 p. m.	3:40 p. m.
Sunday Mail	2:00 p. m.	5:00 p. m.
Johnstown Express	3:05 p. m.	5:30 p. m.
Day Express	5:55 p. m.	7:20 p. m.
Harrisburg Accom	6:25 p. m.	9:30 p. m.

The Hanover Accommodation, west, connects at Lancaster with Niagara Express, west, at 9:35 a. m., and will run through to Hanover.
The Frederick Accommodation, west, connects at Lancaster with Fast Line, west, at 2:10 p. m., and runs to Frederick.
The Pacific Express, east, on Sunday, when flagged, will stop at Middletown, Elizabethtown, Mount Joy and Landisville.
The only trains which run daily. Others daily, except Monday.

NORBECK & MILEY,

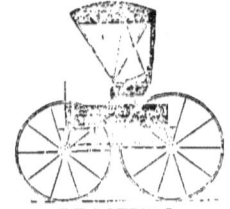

PRACTICAL
Carriage Builders,

COX & CO'S OLD STAND.

Corner of Duke and Vine Streets,
LANCASTER, PA.

THE LATEST IMPROVED

SIDE-BAR BUGGIES,

PHÆTONS,

Carriages, Etc.

THE LARGEST ASSORTMENT IN THE CITY.

Prices to Suit the Times.

REPAIRING promptly attended to. All work guaranteed.
79-3-

S. B. COX,
LANCASTER, PA.
Manufacturer of
Carriages, Buggies, Phaetons, etc.
CHURCH ST., NEAR DUKE, LANCASTER, PA.

Large Stock of New and Second-hand Work on hand very cheap. Carriages Made to Order Work Warranted or one year. [7r-4-12

EDW. J. ZAHM,
DEALER IN
AMERICAN AND FOREIGN
WATCHES,
SOLID SILVER & SILVER PLATED WARE,
CLOCKS,
JEWELRY & TABLE CUTLERY.
Sole Agent for the Arundel Tinted
SPECTACLES.
Repairing strictly attended to.

ZAHM'S CORNER,
North Queen-st. and Centre Square, Lancaster, Pa.
79-1-12

E. F. BOWMAN,

AT LOWEST POSSIBLE PRICES,
Fully guaranteed.
No. 106 EAST KING STREET,
79-1-12] *Opposite Leopard Hotel.*

ESTABLISHED 1832.

G. SENER & SONS,
Manufacturers and dealers in all kinds of rough and finished
LUMBER,
The best Sawed SHINGLES in the country. Also Sash, Doors, Blinds, Mouldings, &c.
PATENT O. G. WEATHERBOARDING
and PATENT BLINDS, which are far superior to any other. Also best COAL constantly on hand.

OFFICE AND YARD:
Northeast Corner of Prince and Walnut-sts.,
LANCASTER, PA.
79-1-12]

PRACTICAL ESSAYS ON ENTOMOLOGY,
Embracing the history and habits of
NOXIOUS AND INNOXIOUS
INSECTS,
and the best remedies for their expulsion or extermination.

By S. S. RATHVON, Ph. D.
LANCASTER, PA.

This work will be highly illustrated, and will be put to press (as soon after a sufficient number of subscribers can be obtained to cover the cost) as the work can possibly be accomplished.
79-2-

$77 a month and expenses guaranteed to Agents. Outfit free. SHAW & CO., Augusta, Maine.

Fruit, Shade and Ornamental Trees.
Plant Trees raised in this county and suited to the climate. Write for prices to
LOUIS C. LYTE,
Bird-in-Hand P. O., Lancaster co., Pa.
Nursery at Smoketown, six miles east of Lancaster.
79-1-12

WIDMYER & RICKSECKER,
UPHOLSTERERS,
And Manufacturers of
FURNITURE AND CHAIRS.

WAREROOMS:
102 East King St., Cor. of Duke St.,
LANCASTER, PA.
79-1-12]

Special Inducements at the
NEW FURNITURE STORE
OF
W. A. HEINITSH,
No. 15 1-2 E. KING STREET
(over Bursk's Grocery Store), Lancaster, Pa.
A general assortment of furniture of all kinds constantly on hand. Don't forget the number.
15 1-2 East King Street,
Nov-1y] (over Bursk's Grocery Store.)

For Good and Cheap Work go to
F. VOLLMER'S
FURNITURE WARE ROOMS,
No 309 NORTH QUEEN ST.,
(Opposite Northern Market),
Lancaster, Pa.
Also, all kinds of picture frames. nov-1y

GREAT BARGAINS.
A large assortment of all kinds of Carpets are still sold at lower rates than ever at the
CARPET HALL OF H. S. SHIRK,
No. 202 West King St.
Call and examine our stock and satisfy yourself that we can show the largest assortment of these Brussels, three ply, and ingrain at all prices—at the lowest Philadelphia prices.
Also on hand a large and complete assortment of Rag Carpet.
Satisfaction guaranteed both as to price and quality.
You are invited to call and see my goods. No trouble in showing them even if you do not want to purchase.
Don't forget this notice. You can save money here if you want to buy.
Particular attention given to customer work.
Also on hand a full assortment of Counterpanes, Oil Cloths and Blankets of every variety. [nov-.yr.

PHILIP SCHUM, SON & CO.,
38 and 40 West King Street.
We keep on hand of our own manufacture,
QUILTS, COVERLETS,
COUNTERPANES, CARPETS,
Bureau and Tidy Covers, Ladies' Furnishing Goods, Notions, &c.
Particular attention paid to customer Rag Carpet, and scouring and dyeing of all kinds.
PHILIP SCHUM, SON & CO.,
Nov-1y Lancaster, Pa.

THE HOLMAN LIVER PAD;
Cures by absorption without medicine.
Now is the time to apply these remedies. They will do for you what nothing else on earth can. Hundreds of citizens of Lancaster say so. Get the genuine at
LANCASTER OFFICE AND SALESROOM,
22 East Orange Street.
Nov-1yr

C. R. KLINE,
ATTORNEY-AT-LAW,
OFFICE: 15 NORTH DUKE STREET,
LANCASTER, PA.
Nov-1y

The Lancaster Farmer.

Dr. S. S. RATHVON, Editor.   LANCASTER, PA., JANUARY, 1882.   Vol. XIV. No. 1.

EDITORIAL.

OUR FOURTEENTH VOLUME.

Like the "Ghost of Banquo," here we are again, in response to many friends who have greeted us with the significant greeting—"May your shadow never grow less." This would be a dreadful greeting to a *fat* man; but then dear reader, *we are not fat*—never have been fat, and, without any double meaning whatever, we never *expect* to be fat at our present rate of feeding. Cæsar is said to have said, "Let me have men about me that are *fat*." Mark Antony was fat and was popular—Cassius was lean and was unpopular, if not hated by Cæsar. Now we want to get out of the category of leanness, since leanness works such disparagement to its possessor—or rather its victims; and, it seems to us the shortest and surest road out, would be *five thousand new* and *true* subscribers to our volume for 1882. *Two hundred* contributors, *one hundred* correspondents and *one dozen* reporters, or communicators. We are not very particular *where* these subscribers, correspondents, contributors, &c., are from, so that they are not representatives of districts in which a "Killdeer" could not live; for, we don't care to know how *poor* soil may be, and how *lean* its cultivators may become and yet have the power to breathe; but, how *rich* the one may be, and how *fat* the other. But this is not all. We want to know how *how* they have become rich and fat, that we may be able to instruct others to "go and do likewise." These we consider very moderate wants, in such a magnificent "kingdom" as Lancaster county, and where people are reputed to live on "the *fat* of the land."

It would be a most laudable ambition for the farmers of Lancaster county, just to see how *fat* they can make us and their representative journal, within the year 1882. It would be something to be jolly over next Christmas, and enhance the pleasures of the occasion. Dear public, try it "for once."

We have nothing in the form of premiums or bribes to offer, but would rather that every one should be bribed by his own feelings of right, of justice, and of humanity, when he contemplates his duties toward the institutions and enterprises of his county, his state, or his entire country. It may require a greater struggle, a larger *quantum* of self-denial and self-compulsion, to obey the dictates of his understanding—

"Unbought by compact
And unbribed by gain,"

but then, after the deed is accomplished, the doer of it will feel just so much the better than he would have felt had he yielded to the *flattering inducement* to do otherwise.

We are not so exceedingly selfish as to admonish our patrons and readers to patronize no journal but ours. That is not at all our meaning; for, we would have them patronize all they can afford to, and ours too *especially* ours. "Eat them all yourself, and give me

some," was the "small boy's" advice to his "chum," when other "small boys" were begging his sugar plums. So we advise, when other publishers are trying to obtain your patronage through a *premium*, take them all if you can, but don't forget to take the FARMER; for it is "bone of your bone and flesh of your flesh," and like you, "to the manor-born"—a local anchor, mooring you to local homes, wherever you may be.

There is no love more ennobling than pure unselfish domestic, or local love. Men often in the heyday or prime of life indulge in the glittering sensation of foreign loves, foreign scenes and foreign festivities; only to return in maturer life, or in life's decline, to the loves, the homes, and the affiliations of their earlier days. It is very much the same in regard to home literature, home publications. Publications that contain a record of the local doings and sentiments of those who represent the local industries and enterprises of a district where we have, or once had, our local homes, possess a local value far beyond that of mere dollars and cents. And, if perchance, we should become possessed of an old volume, or even a single paper, that recorded the events of our early days, and the names and occupations of our cotemporaries, we are apt to peruse it from "end to end" and gaze upon it with all the fond affection that a grandparent does upon the tiny shoes of the first born. And why? Because they are *ours*—part and parcel of our local history, local experience, and local memories, however common-place and humble they may appear to others.

By the generous assistance of local contributors and local correspondents we desire to make the LANCASTER FARMER for 1882 a local hand-book of Agriculture, Horticulture, Floriculture, gardening, domestic employment, and scientific miscellany, that will always be referred to with pleasure and with profit, long years after its projectors, its editors and its publishers are "gathered to their fathers."

At no period in the history of agriculture, is a publication of its development and progress considered of more vital importance than the present period. Agricultural and Domestic journals are springing up almost every where in our broad land, and by a reference to our "Literary and Personal" columns, it will be perceived that our own Pennsylvania furnishes several new enterprises of the kind. We sincerely hope they may all realize their most sanguine expectations; and, that they may all become *"fat,"* and—if they prefer it—also "ragged and sassy."

We cannot indulge in any special promises for 1882. We hope to be all that you see fit to make us; but, under any circumstances, we think we can with confidence point you to the past as a guarantee for the future.

Of course, the opening year is still one of undeveloped anticipation, and as the tide of time flows on, it will become manifest who is to be carried successfully on its flow, and who buried beneath its flood. But, whatever woes betide us, both religion and philosophy admonish us to reconcile ourselves to "the things that be," as the best condition in which to learn the lessons which experience teaches.

In conclusion, may one and all be blest with a prosperous and HAPPY NEW YEAR.

THE MOOSE-DEER 100 YEARS AGO.

"Captain Harrison," stated to Judge Henry in 1775, that the moose-deer reigned the master of the forest at that period, about Fort Halifax, at the junction of the Sabasticoog and Kennebec rivers; but that when he first settled there, about 1745, the common deer (*Cervus virginianus*) which now inhabits our more southern climate, was the only animal of the deer kind found in all those regions, to their knowledge, unless it was the Elks (*Cervus canadensis*), and those only occasionally. In a short space of time after his location in the country the Moose-deer (*Alce americanus*) appeared in small numbers, but increased annually afterwards, and as the one species became more numerous, the other diminished, so that the common, or Virginia deer, at the time of this information (1775), according to Captain Harrison, was totally driven from that quarter.

This, in the mind of the narrator seemed to imply that animals, like human beings, whether forced by necessity, or from choice, do emigrate.

Perhaps the most notable instance of one species of animal displacing another, is to be found in the Rat, immediately around us, but that was not a matter of choice, for the intruder and usurper was brought here, perhaps against his will.

The Rat that now predominates Lancaster county, and perhaps the entire State of Pennsylvania, if not the whole country, is the "Norway Rat" (*Mus decumanus*) and he has almost entirely displaced the "Black Rat" (*Mus rattus*) which is our native species—indeed we do not remember to have seen but one (dead and partly decayed) specimen in all our life, and that was forty years ago. The reader will please not infer (from a similarity of sound in name) that we refer to the "Musk-Rat" (*Fiber zibethicus*) for that animal is now almost as common as ever it was.

It may seem singular that an animal apparently so slow and stupid as the Moose, should have supplanted one so bright and swift as the Virginia deer. It has been said that the dull, slow "Gray Fox," which was once very abundant in Pennsylvania, has retired further north on account of the invasion of the swift, cunning and sprightly "Red Fox;" but the cases, in regard to special characterics, are here reversed. It occurs also that one species of plant will displace another. Permit "Canada Thistle" to domiciliate itself in the soil, and soon nothing will be found growing but the thistle.

KILLEKINIC.

"The 'Red Willow,' (*Salix purpurea*) which is a native of the United States, is spread throughout our climate. The outer bark is of a deep red color, peels off in a very thin scale, the inner is scraped off with a knife, and is dried either in the sun or over the fire. The scent, when burning, is delightful. To increase the flavor, the Indians pluck the current year's branches of the "Upland Sumac," and dry it in bunches over the smoke of a fire. An equal part of the Red willow bark added to as much of the dryed Sumac forms the Killekenic of the Indians. One third part of leaf tobacco added to the aforenamed ingredients, and the mass rubbed finely together in the palms of the hands, makes that delicious fume, so fascinating to the red, and also to the white men. Great care, however, must be taken, not to use the "Swamp Sumac" (*Rhus vernix*) instead of the Upland (*Rhus glabrum*) as the former is most poisonous, and resembles the latter in the bark and leaf so much that an inexperienced eye might be deceived. The difference may be distinctly marked by observing that the bunch of berries of the Upland Sumac, is a cone closely attached to each other, and when ripe of a reddish color. The berries of the Swamp Sumac hang loosly pendent from a lengthy footstalk, and when ripe are a greenish grey. On the authority of Natanis and "Corn Planter," distinguished Indian chiefs, it is stated, that the person who should smoke the Swamp Sumac would forfeit his eyesight. The Vanilla of South America has been applied by the Spanish manfacinrers of tobacco in various ways; it is strange that we have never assayed *Killekenic*."

The above, from Judge Henry's "Campaign against Quebec" in 1775, we reproduced, merely to admonish the lovers of the weed of a resource, should the tobacco crop at any time totally fail, or be so "cornered" that poor people could not afford to use it. Surely the white man ought to be as good a botanist as the Indian, and not make the mistake of getting the *Swamp* instead of the *Upland* Sumac.

We hazard the suggestion that the above compound might be superior to much of the "stuff" now sold as smoking tobacco; although so far as concerns *ourselves personally*, we prefer the pure, simple tobacco to any compound, whatever it may be scented with. To us, any outside ingredient added to tobacco to give it quality, indicates that it is not good tobacco.

THE VALUE OF SNOW.

If snow possesses no properties that are valuable to the soil as a fertilizer, such an opinion has at least long existed among men of intelligence. We have just finished reading "an interesting account of the hardships and sufferings of a *Band of Heroes* who traversed the wilderness in the *Campaign against Quebec* in 1775," written by Judge John Joseph Henry, of Lancaster, and published by William Greer, in 1812. In speaking of the immense and long continued snows of Canada, where he was held a prisoner for over seven months in 1776, the writer remarks as follows:

"An observation may be made in this place with propriety, that is, that in the climates of all high southern or northern regions, the soil is very rich and prolific. The beneficial operation of nature is, in all likelihood attributable to the nitrous qualities which the snow deposits. Of the fact, that nitre is the principal ingredient which causes fertility in the earth, no man of observation, can at this day, reasonably doubt. The earth is replete of it. Wherever earth and shade unite, it is engendered and becomes apparent. This idea is proved by the circumstance, that nitre may be procured from caves, the earth of cellars, outhouses, and even from common earth, if kept under cover. During the Revolution, when powder was so necessary, we everywhere experienced the good effects of this mineralogical discovery; it gives me pleasure to say, that it is most fairly ascribable to our German ancestors. The snows which usually fall in Canada about the middle of November, and generally cover the ground until the end of April, in my opinion, fill the soil with those negative salts, which forward the growth of plants. This idea was evinced to my vague and inconsiderate mind, from observations then made, and which were more firmly established by assurances from Captain Prentis, that muck or manure which we employ in southern climates is *there* (in Canada) never used. In that country, the moment the ground is freed from snow, the grass and every species of plant, springs forward in the most luxuriant manner."

These observations were made over a hundred years ago, and although Canada may have, in the mean time, learned to recuperate her soil by "muck or manure," it does not obliterate the fact that a good bed of snow during a long, cold winter, is of immense benefit to the soil and winter crops.

KITCHEN GARDEN FOR JANUARY.

In the Middle States, January is unfavorable to out-door labor; in the garden especially, little can be usually done. The forcing-beds and green-houses will of course, require particular attention, and the active man may find something to do in preparing for a more congenial season.

Poles and *rods* for beans and peas may be made ready to be used when needed; and compost heaps formed. Compost is beyond all comparison the very best form in which to apply fertilizers to most vegetable crops, and ample supplies may be readily made by proper attention, as the materials present themselves from time to time during the year.

Fruit trees may be pruned; hedges clipped —those formed of evergreens not till after frost has disappeared—asparagus-beds top-dressed, preparatory to being dug when frost has ceased. When new ones are to be made, plant the colossal, *Hot-beds* for early forcing may be made, and other 'jobs will present themselves in anticipation of spring. Where there exists the will to work, the opportunity for the useful disposition of time is ever present.—*Landreth Ru. Reg.*

These suggestions are applicable to any kind of garden, whether a kitchen garden, flower garden, or large market garden; and yet there are many gardens that receive no attention whatever until the planting time is immediately at hand; in the mean time they are the common depositories of boxes, weeds, old cans, broken crockery, coal ashes, oyster-shells, sticks and stones, and any and every kind of rubbish that people wish to expel from their sheds, yards and houses; unless they may choose to erect a pyramid of such trash convenient to their kitchen doors. Doubtless some will allege that all this rubbish contains fertilizing substances that will be imparted to the soil during the winter, and hence they take that slovenly way of recuperating their gardens. Even admitting that, to some extent, such is the case, it must be evident that such fertilization must be very irregularly distributed, and may not be supplied where it is most needed. The January work on either a farm or a garden will depend a great deal on the kind of weather we have. If there is three feet of snow on the ground during the entire month, much of the work of *order* must be deferred to a more "convenient season."

WINTER BLOOMING.

Up to the incoming of the new year the season has been a remarkably mild one, although not by any means a unique one. Apples, pears, peaches and cherries have bloomed, and in some instances have borne fruit, in the months of October, November and December. Dandelions and other flowering plants have also bloomed in the open air. Snakes and other reptiles have gone abroad, ants have swarmed and large fungi have sprung up as late as the third of November. This illustrates that the whole living world is always ready for vital action, as soon as heat, light and air supervene, no matter what period in the year it may be. The old theory that the sap descends to the roots of all perennial vegetation, and the plant world indulges in a long winter sleep, has no existence in fact, as a universal condition. It seems as if all nature was always in a state of watchfulness for the opportunities engendered by heat, light and air. We have seen caterpillars revive in January and feeding as voraciously as they do in July and August, when food was obtainable, and perhaps within three days thereafter they have retired and relapsed into their winter hibernations. Trees seem to be always full of sap, which is ready to swell or break forth from their leaf and flower buds, whenever the atmospheric condition is favorable to their inflorescence and fructification, either in late autumn, midwinter or early spring. Nevertheless the intervention of winter in our climate is necessary to that repose and recuperation which plants require, to enable them to bear a prolific and perfect crop of flowers or fruit. Even if summer was prolonged during the whole year, it is doubtful whether a second crop would pay for gathering or possess the requisite quality when gathered. Second blooming, and second crops are usually abnormal and abortive. Vegetation attempts something by such phenomena that it cannot successfully carry out in practical results. Rest and recreation is something that is needed. The winter bloomer may not be worth much next summer.

"AID AND COMFORT."

A distinguished editorial contemporary, in reviewing the present status and the past progress of his journal, very significantly remarks: "That our labors have been duly appreciated is shown by the assistance we have had from the able and progressive men and women who have contributed their best thoughts to its pages, and by their kindly and zealous efforts continuously put forth for the extension of its circulation."

Any publication that is fortunate enough to realize the foregoing text, comes within the possibility of ultimate success; for, single handed and alone, either a publisher or an editor, "can do nothing." "Variety, is the spice of life," which adds interest to a journal, especially when that variety consists of the "best thoughts" of zealous and intelligent men and women contributors. The LANCASTER FARMER has been long enough before the public to entitle it to the *medianship* of the best thoughts of the agriculturists of Lancaster county, and that it is, not so is one of the things that is incomprehensible to those who reside beyond its borders. With perhaps a single exception, the LANCASTER FARMER has lived longer than any other agricultural paper ever published in Pennsylvania. There is not a more convenient, a safer, or a more permanent *recorder* of the best thoughts of the people of the county and the State, than is to be found between the covers of the FARMER; and the citizens of the county had "better believe it." There is as much agricultural, mechanical, professional and scientific thought—and as much of the substances upon which thought exercises itself—in Lancaster county, as in any other county in the State, and our local journal is the *book* in which they should be *recorded* and transmitted to posterity. The text we quote above, is the language of an octogenarian, who has occupied the chair editorial for more than half a century, and he virtually acknowledges that his success is due to the *assistance* which he has received from contributors and canvassers of both sexes. It is the same with every periodical publication in the land. The greatest flow of thought must come from other sources than the minds of their editors only, or they will lack that flavor which renders them agreeable to the public.

HOW DO EELS BREED?

In the *New Era* of December 10, I saw a communication, signed by Mr. William Neal, of Port Deposit, in reference to the *eel-breeding* question, and as we can develop facts oft-times by agitating such subjects on which we are not thoroughly familiar, I would respectfully ask to be admitted to your columns on this subject. His theory accords perfectly with mine, that eels breed like other fish, but I cannot believe the lamprey to be the female eel—that is, the eel which is caught so extensively in the Susquehanna. I have spent my early boyhood on the banks of the Susquehanna, and have played at its water's edge many a summer's day, gathering shells of the fresh water mollusca and watching the tiny fishes disport themselves in its limpid waters.

One morning some time in May, I cannot, unfortunately, fix the date, I noticed a black line along the shore, which proved to be a host of small eels migrating up the river. A few years after that I noticed them again, but they did not continue so long as the first time, when they occupied a full day and night in passing, some straggling along the next day at eight o'clock. This, then, proves to me that they migrate, and do not breed in the upper waters of the river.

Now, then, an experiment which was made in my presence by my mother proved to me that they have *eggs!* We took of that substance which an eel contains, and is usually called fat by fishermen, and fried it, and it yielded no oil as the fat of other fishes does, which proved to my mind that it is the *ovarium* and not fat, but the eggs are so small that they are not recognized by the naked eye.

From the above, I infer that the eels breed in the deep waters of the bay in the mud, and in the spring ascend the river to grow, and in the fall they descend again to deposit their eggs in the mud of the bay. It seems to me that nature itself proves that the lamprey is not the female, as but very few were caught, possibly no more than three, during the sixteen years I spent at the river, while as many as 500 eels were caught by us in one night in pots in the Turkey Hill falls, which, according to Mr. Neal's theory, would have been all males! In all other forms of life the sexes are nearly equal in the numbers of the males and females, but in the eels the disparity would be too great. I would refer Mr. Neal to an able article by Dr. Rathvon in the December number of the *Lancaster Farmer*, in which the eel question is thoroughly ventilated.—*E. K. Hershey, Creswell, Dec.* 10, 1881.

In the *New Era* of December 21, 1881, I noticed a very sensible article from E. K. Hershey on the subject of eels, their migrations and breeding habits, etc., in which he alludes to a communication from Mr. William Neal, of Port Deposit, which, it appears, was published in the *New Era* of December 10. Somehow that communication entirely escaped my observation, nor could I find it in that issue, although I looked for it after my attention was called to it by Mr. Hershey's article. It appears to me that nothing could possibly be more absurd than that the lamprey is the female of the common eel, and that through her the race is perpetuated. The lamprey is far removed from the common eel. Between the *Anguillidæ*, or eel family, and the *Petromyzonidæ*, or lamprey family, there is a very wide difference both in structure and habit. The pipe fish, the sea horse, the puffers, the sun fishes, the trunk fishes, the sturgeons, the dog fish, the sharks, the threshers, the hammer-heads, the saw-fishes, the rays and the torpedoes, are all families and genera intermediate between the eels and the lampreys. The common eel of the United States is the *Anguilla bostoniensis*, and has the dorsal and anal fins continuous around the end of the tail, forming by their union a pointed caudal extremity. They have also very conspicuously a pair of pectoral fins, which are entirely absent in the lamprey. Although the latter has an irregular dorsal fin, it is destitute of the anal fin, and the caudal termination is not pointed but broad. The common lamprey is the *Petromyzon Americanus*. This species has a maxillary ring armed with strong teeth, and they attach themselves to other fishes. Many years ago I saw one nearly two feet long attached to a large sucker that had been caught in a shad seine in the Susquehanna. I have also known hundreds of the smaller ones to be dug out of the mud flats of the Susquehanna, opposite Marietta, and used as bait. The female lamprey is no more the mother of the common eel than she is the mother of the blacksnake, or the boa constrictor, and it is a matter of surprise that any one living near the bays and rivers of our country should for a single moment entertain such a fallacy. The question of eel breeding must be decided without the aid of the lamprey.

EXCERPTS.

MISCELLANEOUS.

THE great gray slug has a supply of 28,000 teeth.

THERE are forty-six species of the English cuckoo.

DUSTERS were at first made of the tails of oxen or foxes.

The earliest mention of parks is among the Persians.

THE leech has eight or ten eyes set in its back near the head.

GAUZE is said to receive its name from Gaza, in Palestine, where it was first made.

THE part of the human body which shows the greatest variety of color is the iris of the eye.

THE moon was pronounced by Anaxagoras, 500 B. C., to be an earth having mountains and valleys.

KING ALFRED used to measure time by a device of twelve candles, graduated so as to burn two hours each.

THE cochineal insect is very small, a pound of cochineal being calculated to contain not less than 70,000 in a dried state.

A PECULIAR violet odor is emitted from the males of some species of Brazilian butterflies, the female being not at all fragrant.

To neutralize the sting of gnats and mosquitos, English sportsmen rub the part affected with cerumen, or the wax of the ear.

THE fashion in men's hats changes far more often in England, France, and America than in other countries. The sombrero worn in Don Quixote's time is in fashion in Spain to-day.

WITHIN the past sixty years the value of gold has fluctuated from $15\frac{1}{2}$ to $15\frac{3}{4}$ times that of silver, averaging about $15\frac{1}{2}$ time and never falling so low as that of fifteen times such value.

AMONG the early Romans a kind of festi, or annals, was kept by driving nails into the wall of the Temple of Minerva; and in public calamities, in time of pestilence, etc., a nail was fastened in the Temple of Jupiter.

CHRISTIAN names are so called by having been given to converts in baptism as substitutes for their former pagan appellations, many of which were borrowed from the names of their gods, and were therefore rejected as profane.

THE institution of the "Order of the Bath" originated in the custom of the Franks, who, when they conferred knighthood, bathed before they performed the ceremony, and from this habit came the title Knight of the Bath.

A well-fed frog is more susceptible to poison than one which has been fasted for weeks.

CORALS often permanently change color, when subjected to different conditions of living.

THE Japanese bronze brass by boiling it in a solution of sulphate of copper, alum and verdigris.

A concentrated beam of electric light carried seven miles has furnished sufficient light to read by.

THE solar atmosphere contains sodium iron, calcium, magnesium, nickel, barium, copper and zinc.

ACCORDING to Ehrenberg a cubic inch of water may contain more than 800,000,000,000 of animalcule.

PLATINUM when alloyed with silver becomes soluble in nitric acid, which, does not affect it while unalloyed.

PIG iron contains from ninety-five to ninety-seven parts of pure iron, and three to five of carbon, with small quantities of sulphur, phosphorus and silicon.

HAY, like most vegetable products, contains much material which is soluble in water. On this material its nutritive value depends, and its removal by dampness seriously injures the crop.

THE floods and droughts of the present time will probably lead farmers and others to a careful reconsideration of the question regarding the proportion which wooded ought to bear to cleared land.

LAST year the German wire mills supplied England with 30,000 tons of wire, and Russia with 40,000 tons. France received from Germany from 12,000 to 15,000 tons of steel wire for sofa springs, and America not less than 30,000 from the same source.

FROM surveys taken in the province of Ufa, Russia, it appears that the former forest area of 17,577,000 acres has now been diminished by more than 3,500,000 acres, and yet the population is only three to the square mile.

EDWARD RICHARDSON, of Mississippi, is the largest cotton raiser in the world, the Khedive of Egypt coming second. Mr. Richardson owns some 52,000 acres of cotton land, from which he raised last year more than 12,-000 bales. He gins, spins, and weaves it, and has oil mills as well. Mr. Richardson has amassed a fortune variously estimated at from $15,000,000 to $20,000,000.

THERE are 700,000 Masons in the United States.

THE length of the East river suspension bridge is 5,993 feet.

THE quantity of soda imported into the United States from England in 1847 was 8,000 tons.

IN an edition of Ptolemy's geography, 1540, a double-tailed mermaid figures in one of the plates.

THERE are seventy-two national cemeteries for the burial of the Union and Confederate dead.

AMONG the natives of India white quartz, boiled in milk, is used as a remedy for sick children.

A wire 400 feet long can be made from one grain of silver. Such a wire is finer than human hair.

THE ancient Chinese used hydropathy as a cure for certain diseases, among others chronic rheumatism.

STEEL needles were invented by the Spanish Moors, before which thorns or fish bones, with a hole pierced for an eye, were used. The first needles made in London were made in the reign of Henry VIII, by a Moor.

THE first book published in the North American colonies was, it is supposed, an "Almanac calculated for New England, by Mr. William Pierce," which appeared in Cambridge in 1639. It was printed by Stephen Daye, but not a copy of it now exists.

THOSE of us who in winter complain that the sun has not power of warmth should bear in mind Professor Young's recent remark, that if we could build up a solid column of ice from the earth to the sun, two miles and a quarter in diameter, spanning the inconceivable abyss of 93,000,000 miles, and if then the sun should concentrate its power upon it, it would dissolve and melt, not in an hour nor in a minute, but in a single second ; one swing of the pendulum and it would be water seven more and it would be dissipated in vapor.

ECONOMICAL.

A teaspoonful of saltpeter to a pail of water will kill worms in the roots of squashes.

PIGS are able to consume far more food in proportion to their weight than either sheep or oxen.

SHEEP provided with cotton-seed meal as an auxiliary feed are the best restorers of worn-out pastures.

COMBS and wattles of fowls may be prevented from freezing by oiling them so as to prevent their getting wet.

SWEET apples are an excellent feed for cows, if supplied in moderate quantities and under favorable circumstances.

GREEN manuring, or the plowing in of green crops, is especially adapted for light sandy soils, which need humus to increase their retentive power.

AT some time, during the fall or winter give the thin spots in meadows and pastures an even coat of manure. Harrow in spring and sow grass seed.

A farm can be stocked with sheep cheaper than with any other animals. Sheep will come nearer to utilizing everything which grows on the farm.

JOSEPH HARRIS says that we can make our lands poor by growing clover and selling it, or we can make them rich by growing clover and feeding it out on the farm.

ADD a little glycerine to the grease applied to harness, and it will be kept in a soft and pliable state, in spite of the ammoniacal exhalations of the stable, which tend to make it brittle.

ALL noxious weeds, such as dock, skunk cabbage and others may be killed by pouring a small quantity of kerosene oil over the young plants. They may also be cut off with a hoe several inches below the surface and salt dropped on the cut-off root.

SOME people think that grapevines will grow anywhere because in their boyhood they found strong, luxuriant wild vines growing in damp places. Do not make a mistake. Vines on low lands suffer much by the winter. Hillsides and lean soils are good for grapes for hard winters. For manure that of the cowyard is good.

GREASE, says a writer in the *Rural New Yorker*, is fatal to all insect life. Insects breathe by means of small pores on their sides. Grease or oil that comes in contact with the insects closes the pores and stops the breathing. Mercurial ointment kills as much by the lard in it as by the mercury—that is, so far as the vermin are concerned, but not as to the animals that lick it off from their bodies, so that almost any oily or greasy application will be destructive to insect vermin that infest animals if it is applied where it will do the most good.

THE importation of opium by this country which in 1861 was 109,536 pounds, in 1871 had grown to 315,121, and in 1880 amounted to 533,451 pounds. These figures indicate an immense increase in opium eating. In 1876 it was estimated that the number of people having the habit was 225,000, and now it is thought to be fully 500,000. Some persons become so accustomed to the drug as to take immense doses. A Missouri farmer took forty grains of morphin at once without apparent injury, and there are several cases reported in which sixty grains a day were taken regularly.

A statistician has been figuring upon the annual consumption by American manufacturers of the precious metals, which he estimates as $13,000,000 gold and $3,000,000 silver. Two-thirds of the latter is used in making plate. Of the gold, the greater part goes for rings and watch cases. It is estimated that there are about 350,000 wedding rings given in this country every year, averaging $2 each in cost. There are 100,000 more rings given as *gages d'amour* and a still larger number bestowed in holiday presents.

MORAL ECONOMY.

PEOPLE'S intentions can only be decided by their conduct.

STRIVE for the best, and provide against the worst.

BE graceful if you can ; but if you can't be graceful, be true.

HE who throws out suspicion should at once be suspected himself.

AN effort made for the happiness of others lifts us above ourselves.

THERE is always room for a man of force, and he makes room for many.

TIME once passed never returns; the moment which is lost is lost forever.

PRIDE breakfasted with Plenty, dined with Poverty and supped with Infamy.

THERE is a past which is gone forever. But there is a future which is still our own.

THERE are few occasions when ceremony may not be dispensed with ; kindness never.

A father's blessing builds houses for his children, but a mother's curse tears them down.

READING, study, thinking, observation and sensible conversation makes the mind grow.

ARGUMENT in company is generally the worst sort of conversation, and in books the worst of reading.

THE discovery of what is true and the practice of what is good are the two most important objects of life.

WE can't be too much on our guard against reactions, lest we rush from one fault into another contrary fault.

NOTHING so adorns the face as cheerfulness. When the heart is in flower, its bloom and beauty pass to the features.

A wise man in the company of those who are ignorant has been compared by the sages to a beautiful girl in the company of blind men.

A person that would secure to himself great deference will, perhaps, gain his point by silence as effectually as by anything he can say.

No school is more necessary to childr n than patience, because either the will must be broken in childhood, or the heart in old age.

DOMESTIC ECONOMY.

ADD all refuse matter to the compost heap.

ENGLISH farmers use bone dust on pastures, but prefer superphosphates for sown crops.

THE addition of charcoal in the soil deepens the tint of dahlias, hyacinths and petunias.

SCIONS, it is claimed, carry with them the bearing year of the tree from which they were taken.

SHELTER and warmth, with regularity in feeding, are essential to success in the management of cattle.

CLEAR the ground now on which you expect to put small fruit plants next year. Do it thoroughly, too.

ALL that you wish to know of any new breeds of fowls will not be learned from those who are anxious to sell them.

SCRUB sheep are dear even for no price at all. On a good farm they are as bad as rusty nails on a new house.

FEED windbroken horses frequently and little at a time. Grind the food. Give plenty of salt and little water at a time.

TEACH your children not to annoy or maltreat the toad. Try rather to coax him to your garden. He will destroy many insects.

A LITTLE grease or kerosene on the legs of fowls will remove scabs in a short time. Two applications are sometimes needed.

AS A partial antidote for drought, keep the land rich, plow deeply, and cultivate as often as possible. Cultivation always tells.

ON many farms there are some old cattle and old sheep that can only be kept at a loss. It is economy to fatten them for the butcher.

THOSE who have been feeding the surplus fruit to hogs say that their stock are in excellent and healthy condition. Fruit makes fine sweet pork.

WHERE the ground is infested with white grubs it would be advisable to mix salt sparingly with the soil before setting out strawberry plants.

AN offensive odor from decaying vegetables will be absorbed by milk. A pair of old shoes or a pair of barnyard overalls in a cellar where there is milk are likely to contaminate it.

PUT your stock in a good condition to stand the winter by giving a little fodder of some kind early in the morning. A slight breakfast of cornstalks or some other food will be greatly relished while the air is cold and the grass frozen. Colts, calves and lambs need particular attention at this season.

A CONNECTICUT farmer says that the butt ends of potatoes and the kernels of corn from the butt ends of the ears, each produced crops that were materially better than where the opposite course was pursued. In the case of potatoes the stalks from the butt end were much the larger and more forward at the first hoeing. The increase in corn was some twenty per cent. in favor of the butt end kernels.

THE most profitable way to raise beef cattle is to keep them constantly in a thrifty and improving condition. It is not necessary to keep very young stock rolling in fat, but there should always be an abundance of nutritious food to help nature in its development. To allow stock to run down in flesh and become ill conditioned, simply because it is not designed for market for some time, is the height of folly.

OUR RESPONSIBILITY.

We are not at all responsible for the non-appearance of papers read before the Agricultural and Horticultural Society, in the columns of the LANCASTER FARMER, unless immediately directed to us. Under present circumstances, it is almost impossible for us to attend a meeting, and very unfortunately for us we cannot hear what is said and done when we do attend, but we nevertheless feel, and ever have felt, a deep interest in its welfare; hence, we never have said or done anything, or intended to do anything prejudicial to its standing and its usefulness in the community. The publisher of the FARMER sends his reporter to the meetings of the society, to report its proceedings for his daily and weekly papers, and when that part of the FARMER (which we do not profess to control) is made up, his foreman very naturally selects the proceedings found in the paper issued from the same office. These are details that do not belong to our specialty as editor. If any member of the society discovers that the papers he has read before it are omitted, neglected, or suppressed, he must hold the publisher, or reporter, or both, responsible, and not the editor, for we can under no circumstances be held accountable for that which has never come into our possession. Any intelligent person who reads the prospectus of the FARMER will find that all communications, contributions and essays should be sent to the editor, and all subscriptions, advertisements and business matters to the publisher, in order to insure due attention.

The charge against us in the proceedings of the last meeting, under the caption of "INSECTS," (which, by the way, is like the play of HAMLET, with the part of Hamlet left out) is a disingenuous one, and entirely foreign to the character we thought we had been endeavoring to cultivate; and the author of the charge is consciously or unconsciously exercising himself unnecessarily on our account. So far as the matter relates to ourself individually, we attach little or no importance to it; but the attribution to us of motives which we never for a single moment entertained, and the feeble attempt to create the impression that we have been acting prejudicially to the interests, the edification, and the dignity of the society, imposes upon us the necessity of making this explanation—especially as the association is one of those with which we have been identified from its very origin, and which we have always esteemed.

EDIBLE FUNGI.

The *Book* on the *Fungi* of the United States has not yet been written, it seems—at least, it has not yet been published—and especially the *book* on the *edible fungi*. Such a book is needed, but possibly the enterprise of publishing it "would not pay." On the night of the 3d of November, 1881, a fungus sprung up in our garden nearly twelve inches in height, belonging to the order AGARACINI, which differed from any we had before noticed. The *pileus*, or hood, was tall bell-shaped—nearly a cone—flaring a little at the bottom, which was surrounded with a broad fringe, and was nearly six inches in height. The gills, inside, were of a purplish rown in color, and rather finer than the common species. The stock or stem was over ten inches high and tapered to the top, and the pileus was so delicately poised on its upper point, that the least draft of air caused an active oscillation. The color, externally, was a tarnished white, and the surface of the pileus was covered with fine fibers, gathered in tufts, giving it the appearance of disordered plush. The diameter at open mouth of the pileus was 3¼ inches. After we first discovered it, it did not increase any in size, and we let it remain three days before we took it up, and then only to prevent it from freezing, as the weather had suddenly changed to cold. We referred a drawing of it to Professor Farlow, of Cambridge, Mass., and he kindly determined it for us as *Coprinus comatus*, and further informed us that it was excellent when cooked. And *there* is just the "rub," for doubtless we have many edible fungi in our country, if we only could tell t'other from which."

Many long years ago, when large districts in our county (that are now cultivated meadows and fields) were woodlands, we were familiar with a white species of fungus, which was commonly called "pipe-stems," because they grew in clusters, or bunches, from six to a dozen grouped together, and bent at top like a common white clay pipe. These were gathered by a few knowing families, cooked, and much relished. But the people generally did not trust them, although the common "Mushroom" (*Agaricus compestris*) and the common "Morell" or "Mauricle" (*Morchella esculanti*) were in very common use.

We have now before us a catalogue of the "Pacific Coast Fungi," giving a list of the species systematically arranged, their localities, their authorities, and the simple fact of the edible species, together with their technical names, without any descriptions whatever. This catalogue is published under the auspices of the California Academy of Sciences, and is, perhaps, the first attempt of the kind in this country.

The catalogue includes 750 species, only 61 of which are edible; but even this is an immense number, compared with the popular knowledge on the subject. It is not claimed however, that this contains *all* that are indigenous to the Pacific Coast, as there are as many more which have not yet been "worked up." We append a list of the edible species, and suggest that a descriptive work on these, accurately illustrated, is what is wanting on the subject at the present time.

Lepiota excoriatus,
" illanitus,
" procerus,
" rachodes,
Armillaria melleus,
Clitocybe dealbatus,
" gentropus,
" nebularis,
" odorus,
Collybia esculentus,
" fusipes,
" longipus,
" radicatus,
Omphalea onicus,
Pleurotus ostreatus,
" pometis,
" ulmacium,
Volvaria Icoulageanus,
Clitopilus orcella,
" soulo,
Pholiota mutabilis,
" squamosus,
*Tralliota arvensis,
" campestris,
" arctaceus,
" pratensis,
" silvaticus,
Coprinus atramentarius,
" 1 comatus,
Dermocyle cinnamoneus,

Pazillus giganteus,
Hygrophones chrimens,
Lactarius deliciosus,
" insulsus,
" ? piperatus,
" volumus,
Russula adjusta,
" alutacea,
" heterophyla,
" lepida,
Cantharilius ciburius,
Marasmius oreades,
Polyporus sulphureus,
Hydnorum repandum,
Clavaria cristata,
" fastigiata,
" flava,
Tremella mesenterica,
Meanguster variagatus,
? Lycoperdon giganteum,
4 Morchella esculenta,
" conica,
Gyrometia esculenta,
Helvella californica,
" crispa,
5 Bohtus luteus,
" versipelles,
" scaber,
" granulates,
" bovinus.

Of course, it is presumed that the edible species must generally attain such a size as to make their possession an object ; but many of those of the general list must be very small, or found in the form of *moulds*, *smuts*, *rusts*, etc., but even the smaller species, microscopically considered, are very pretty and interesting objects; and, as the cause, or the result, of animal and vegetable disease, they occupy a prominent position in their relations to the interests of the human family. As the population of a country increases, and advances in taste and culture, its culinary preparations will be correspondingly developed as a domestic necessity, and many of the products of nature will become objects of cultivation that had been hitherto considered useless. The cultivation of *Mushrooms*, *Morells*, *Truffles*, and other species of *fungi*, has long been a source of considerable revenue to European nations, and may become so in America.

*In an addenda to the list we find the single species of *Agaricus tridens*, and that is a new one. What has usually been included in this genus (the various Mushrooms) will be found in the genus *Psalliota*.
1. This is the species to which we refer in the above description, and we may mention here that it was the only individual we have ever noticed on our premises, or elsewhere. Of course, it may be looking for it, we probably might have found it long ago.
2. On this be the species in which we have referred as the "pipe stem?" We have not seen one for at least five and forty years.
3. This species has been frequently found in the county of Lancaster. A specimen is now in the Museum of the Linnaean Society that measures fifteen inches in diameter, and developed in one night in the city of Lancaster. They commonly take the name of "puff balls."
4. This is the popular "Meurlele" of Lancaster county, and is more or less abundant in many localities every summer, being frequently to be found in our markets.
5. Species of this genus are abundant in Lancaster county, but we are not aware that any of them are edible. They are usually found on trees, logs, stumps, etc., and some of them are very pretty.

CONTRIBUTIONS.

For THE LANCASTER FARMER.
HYBRIDISING FRUITS AND FLOWERS.
DECEMBER 28, 1881.

Mr. Editor—Dear Sir: If I were a young man, as you know I am not—I would go strongly into this interesting operation. What a number of new and superior grapes, pears, peaches and flowers have been already produced by this truly interesting process ! But great improvements are yet to be made. I well remember Van Mon's experiment in producing many new pears, yet he only made progress by raising seedlings, and grafting the seedlings on older trees, thus causing them to bear in advance of the seedlings—then again planting the seeds of these and going through the same operation, until the sixth and seventh generation, each generation an improvement on the original—thus producing many superior pears. But we do not know that Van Mons practiced crossing his fruits, and so far as we know, Mr. Rogers, of Salem, Mass., was the first who practically proved that the grape could by thus crossing the pollen from one flower to another—and in this way he has *originated* over fifty new varieties ! Yet many good botanists, at the time, denied that the Rogers grapes were crosses. But though they were all seedlings of a Fox grape and crossed with a pollen from the exotic, or vinefera species—these grapes all lost their foxiness, and are now among our best grapes—vigorous, healthy, hardy and great bearers of delicious grapes, half native and half foreign.

Since Mr. Rogers' successful experiments, many others have tried the same, with more or less success. Notably Mr. Campbell of Ohio, Mr. Ricketts of Newburgh, New York and Dr. Wylie, of Chester, South Carolina.

There is yet plenty of room and time for our young horticulturists to "go and do likewise !" All fruits and flowers may be changed in this way, and if carefully performed, the seedlings will be different and some very superior varieties may be thus produced.

As stated above, if I yet had the hope of living many years (which of course I have not) what an interest and pleasure it would be, to thus spend my leisure time producing new fruits, vegetables and flowers. J. B. G.

For THE LANCASTER FARMER.
PERSIMMONS.
DECEMBER 29th, 1881.

Editor of the Farmer—Dear Sir : In the December No. of the FARMER, page 184. L. S. R. speaks highly of our native Persimmons and says "he has some very superior varieties," which is interesting to lovers of that fruit. I fully agree with him, so far as our natives are concerned. But when he says "it is useless to waste our time, trying the Japanese varieties, it reminds me of the story of the "Fox and the grapes." Poor Reynard saw some very fine grapes on a high tree, and as he could not get them, he passed along, saying "they were only *sour grapes*," so he did not want them.

I am fully of the opinion, that if L. S. R. would once get a taste of these Japan varieties, he would change his opinion. 'Tis true, they are not in the northwest what they call *Iron clads*, but even our natives are frequently injured by severe cold winters. My own trees, some half a dozen—*apparently* not injured by the last severe winter, yet they, the trees, must have suffered some, as none of my trees bore any fruit this last season. The same happened with all my Chinese and hybrid pears—as none of the trees had a single fruit—while last year, two trees had over a bushel; yet these Chinese varieties are remarkable for bearing large crops every year; the trees appear all right, but the flowerbuds must have suffered. I have had the "Shalea," or Chinese sand pear near fifty years, off and on. Sometimes the trees were killed by severe winters, but I always got grafts again from friends to whom I had given grafts. These varieties of pears always produce heavy crops of large and showey pears, as do also the Kiefer hybrid, and several others no doubt would also be crossed by bees and insects. No blight on these pears.

But the Japan persimmons are very different from our natives; even before fully ripe, they have none of that astringency so peculiar to our natives. The fruit is larger and I believe they will in time be acclimated in our middle States. A friend tells me he had several varieties, and the trees froze down to the snow line, but all sprouted up again and a graft on top of a native was not injured.

By planting the trees on high ground, they might live; but as many people grow orange and lemon trees in tubs, or boxes, these Japan persimmon trees can easily be grown in the same way. J. B. G.

SELECTIONS.

FARMING ABOUT THE ROCKY MOUNTAINS.

Those of us who have become habituated to green fields and shady woods—who have been helped by the rainfall and have done little of our own to water the crops—could not easily bring ourselves to think much of those dry regions where little but cactus and other succulent plants grow naturally ; where all is gray and cheerless, and artificial watering alone produces all a human being is to eat. Yet these apparently inhospitable places are paradises for some people, and in many respects have advantages which we do not enjoy.

In the ages of the past we look to Egypt as the pioneer in work of civilization ; and yet her vast agriculture was solely artificial. There was little rain, and the mighty Nile river, as the poet says, had to bleed through a thousand pores in order to make the grain and the grass to grow. Dependent solely on their own resources, they always had bread to eat ; while the countries supposed to be more favored of Nature often left their people to starve. Joseph's brethren heard the good news that there was corn in Egypt, when famine stalked all over their own fate land.

There is no danger in these days of railroads and electric telegraphs that the stories of ancient famines will ever be repeated, as "history repeats itself" in our lands. The day after the Mississippi overflows, Massachusets ships food for the inundated ones; and if the grasshoppers eat up the crops of Nebraska, the loss is made within a few days by the sympathies of Eastern brethren. But if ever a general Eastern destruction of crops should occur, who knows but these despised arid western plains would not be fully able to come to our rescue?

People often suppose that where crops are raised by irrigation, the land under culture must necessarily be limited; but this is not the case. At the very base of the Rocky Mountains most of the farmers work forty acre lots; many one hundred and fifty; while some are reported as having over three hundred acres in wheat. Of course this is nothing in comparison with what many Western people have in the more nature-favored regions; but it is very large for artificial work, and quite large enough.

As we have said, the natural charms of nature-watered lands will ever have the greatest charms for the average man; but it is a matter of great interest to watch what other places can do and are doing, and this Colorado illustration gives a new one of a point we now and then make, that, whatever may be local ills, every part of the world has its own advantages.

"GO TO THE ANT."

Rev. Dr. H. C. McCook, a Presbyterian minister of Philadelphia who was entertained by Dr. J. A. Ehler during the meeting of synod in Lancaster, has for years made close study of the ant a specialty. It has been known a long while that an ant exists in New Mexico which secretes honey after some fashion. Travelers have told of Indian feasts in which the ant was served up "in her own honey" as a species of animated honey-cell. But there was need of a careful examination of the habits of these ants on the part of same one who had the scientific spirit and some training in the observation of insect life. Dr. McCook undertook the long journey from Philadelphia to New Mexico for the sole purpose of playing Paul Pry on the interiors of the honey-ants—the interiors in two senses, for his purpose was not alone accomplished by observing them at work in their underground burrows, or rather in the singular galleries which they drive through soft sandstone rock; it was also necessary to examine their anatomy and find out how and by what organs they secrete the limpid honey. All of which Dr. McCook has done, and curious enough are the habits of these little favorites of Æsop. The sluggard would hardly profit were he enjoined to go to the honey-ant of the garden of the Gods (*Myrmecocystus hortusdeorum*). Could he see the galleries made specially for those ants which secrete the honey, and note the care taken of them by the worker ants, and witness the absolute quiet in which these honey makers loaf away the entire day and night, the moral would not be what it was intended. He would regard with envy the swollen crop of the honey-maker, the assiduity of its servants and attendants both to keep it neat and to feed it with fresh honey from the neighboring oaks, and the laziness with which, when it does move at all it pushes itself or is dragged by the busy workers from one gallery to another. And even the slender worker might not seem to the sluggard so bad an ant, for none goes out by daylight, and it is only when the sun sets that these peculiar creatures, turning night into day, sally out for food. Hereafter the revised reading will be: "Go to the ant, thou sluggard, but not to the *Myrmecocystus hortusdeorum*."

Ants have been astonishing us now for a century, and yet there seems no end to the variety of their tricks and performances. Till found in the Garden of the Gods, it was not supposed that the honey ants existed further north than New Mexico. They have been found at Brownsville, Santa Fe. Matamoras and the City of Mexico. Dr. McCook found their nests on the tops of dry ridges in the picturesque section on the Eau qui Bouille, Colorado, called the Garden of the Gods. He followed them at night, lantern in hand, for several evenings in succession before discovering what they fed on. The long train of workers was easily traced to thickets of scrub oaks. Finally, on the third night, they were seen on the oak twigs running from one oak gall to another and sucking a juice secreted by the gall. Each active gall had the larva of the gall fly within; the ants passed by those from which the mature insect had escaped. Nests were then laid bare with pick and shovel and the workers caught in the act of feeding the honey bearers. These apparently were of the same breed, even the same cast, as the workers, and only different in the monstrous swelling of an anterior stomach, which Dr. McCook calls a crop. Like a crop, this part does not digest the honey, it merely distills and purifies it: and worker ants when hungry will go up to a honey ant and ask for honey from its crop, just as a young pigeon is fed from the crop of its mother. Among the many plates in this volume, which show the habits and dwellings of the ants so clearly that the story hardly needs the aid of text, we see workers feeding the honey bearers with the contents of their own little crops on returning from a midnight foray, and others taking toll both from the raiders and from the distended honey-bearers. These latter are seen hanging from the rough ceilings of the larger galleries in a half-torpid state, for all the world like single Delaware grapes. Dr. McCook describes them as very light of color, shining and transparent. The honey is singularly pure and liquid. In summer it has a slight tartness that is very refreshing, but in winter even this is not tasted. The Indians serve them up as a delicacy exactly like very tender fruit. The Mexicans are said to press the honey bearers exactly as if they were grapes, and even to make a sort of wine or liquor from them. Dr. McCook dissents from another observer who recommends that attention should be given to the ants as honey-producers for the market. He is of the opinion that the number of honey-bearers is too small in each community. A large colony would not have more than 600, which would yield not more than half a pound. But it is likely that any one who should experiment with them would devise means of doubling the number of honey-makers. Dr. McCook's other argument is stronger, namely, that the destruction of insect life involved in obtaining the honey will be likely to prejudice people against it. He might have also remembered to mention the natural disgust which most people have towards insects like the ant, which are never associated in their minds with food otherwise than as corrupters and pilferers. To many people the smell of ants is intensely hateful.

It would take too much space to follow Dr. McCook in his discoveries of the intimate life of these ants; their care for each other and their occasional utter indifference; their sloth and activity; their ferocity and apparent good temper under provocation. On the whole, the report is extremely in their favor. They are hard-working, stubborn, long-suffering when other ants run their mines among their galleries, and so prudent in laying up stores of food for a bad day that they actually store it in living kegs, which move, indeed, with difficulty, but still can drag themselves out of the way of immediate danger. Notwithstanding all the doctor has done, there is yet more to study. Which of the workers are they that begin to get swelled crops and finally take to the honey room? What do the honey bearers look like after several months during which the colony has not stirred abroad? Do they find other honey food beside the galls on the oak? Do the Southern colonies secrete more honey or less? How much of the honey habit is voluntary in the individual? How much chance? There is no end to the problems before the students of this singular little creature. The second part of the book relates to another Western ant, *Pogonomyrmex occidentalis*, whose fortresses and cleared spaces on the prairies might have formed the models on which some of the earthworks of the mound-builders of the Mississippi valley were arranged. Every night these ants close their gates with large pebbles, thus reversing the habit of the *Myrmecocystus*. They are continually attacked by a very minute ant called the Erratic, which fastens on like bull-terriers to an ox, and are greatly dreaded by the large ant. Ants of other species run their burrows into those of the *Occidentalis*, "jumping their claims," but the latter do not mind. Even the eggs and larvæ of two kinds have been found in one gallery. Dr. McCook has issued a prospectus for a large work on American spiders, to be printed if sufficient subscribers send in their names.

A GREAT SOUTHERN FARMER.

Brains will find or make a pathway to success under any conditions, and brains have been the wealth-creating factor in the case of the large planters. It is by business shrewdness and the economy of wholesale dealings that E. F. Bailey, of Jefferson county, Fla., succeeds in making money, though he has never improved upon the old methods of cultivating his 6,000 acres; it is by brains that the managers of the Capehart plantations on Albemarle sound are able to add constantly to the number of their acres, the land added last year being valued at $52,000; and it is by brains, and not by the mere vastness of his farming operations, that Edward Richardson, of Mississippi, the greatest cotton-raiser in the world, has amassed his immense fortune, now estimated at from $15,000,000 to $20,000,000. The means by which Mr. Richardson has achieved phenomenal success as a planter are worthy of a moment's study for the lessons they convey. His business is a comprehensive one, including everything relating to cotton. He not only raises cotton, but gins, spins and weaves it, is a large dealer and has oil mills as well. He was clear-sighted enough to perceive that there is a special profit in each process and operation through which cotton passes from the field to the consumer of cotton goods, and he had the capital and ability to organize a business which makes all these profits his own. He owns some 32,000 acres of land, and last year raised over 12,000 bales of cotton—a greater number than the Khedive of Egypt, who is the next largest cotton raiser in the world. Mr. Richardson is not a "high" farmer, a bale to three acres being the average production of his land, which is largely tilled by tenants on the share system. The 36,000 pounds of seed cotton which he annually gets from his land are ginned by his own gins—which do public ginning also—and pressed, baled and compressed, so much as is shipped as raw material, on his own plantation. The seed, which is ordinarily worth $6 a ton, and is to a great extent wasted by other planters, is ground and pressed for the oil. The hulls are used for fuel in this process, and the ashes sold and used for fertilizers. From a ton of seed he obtains 35 gallons of oil worth 35 cents a gallon—$12.25. The cake remaining after the oil is pressed out is worth rather more for fuel than the seed itself, selling readily for home use or shipment to England at $6 to $7 a ton. Each ton of cotton-seed, therefore, nets rather more than $20—the bulk used as fuel being taken into account. Mr. Richardson's mill at Corinth receives and manufactures a large part of his crop, and another profit is added on the sales of yarns and sheetings, drillings, cottonades, etc., a profit which is considerably enhanced by the elimination of shipping charges, insurance, broker's commission, and other tolls levied on cotton shipped to distant mills.—*Letter to New York Times.*

LIME AS A PRESERVATIVE.

It would be interesting to record the many evidences of the value of lime in arresting decay. As long as 1760 a Mr. Jackson, a chemist, obtained permission to prepare timber for the ship yards, by immersing it in a solution of salt water, lime, muriate of soda, etc.; another practical experimentalist suggested slaked lime, thinned with a solution of glue, for mopping the timbers of a ship. The preservation of timber has been attempted by surrounding it with pounded lime; several attempts have been made to preserve timber by the use of lime. Mr. Britten, in his work on "Dry Rot," mentions a number of cases where lime has been of service. He says, "quick-lime with damp has been found to accelerate putrefaction in consequence of its extracting carbon; but when dry and in such large quantities as to absorb all moisture from the wood, the *wood is preserved* and the *sap hardened.*" "Vessels long in the lime trade have afforded proof of this fact, also examples in plastering laths which are generally found sound where they have been found dry." The joists and sleepers of the basement floors are rendered less subject to decay by a coating of limewhite; and this might be renewed at intervals. The same writer adds, "it does not appear practicable to use limewater to any extent for preserving timber, because water holds in solution only about 1-500 part of lime, which quantity would be too inconsiderable; it, however, renders timber more durable, but at the same time very hard and difficult to be worked." These facts are instructive; they show, at least, that lime in a sufficient quantity kept dry is a valuable preservative agent, and some practical chemist might earn a deserved repute if he could prepare a lime solution that would be capable of rendering so substantial a service to all builders. Such a solution would be at least sufficiently remunerative to make it worth while to try a few experiments in this direction. It is stated on good authority that the white ant in India costs the government £100,000 a year for repairing woodwork bridges, etc., caused by its depredations. Concrete basements have been found to resist the encroachments of the ant. Dr. Darwin proposed a process of timber preservation some yars ago, in which an absorption of limewater was effected, and after that had dried, a weak solution of sulphuric acid, so as to form sulphate of lime in the pores of the wood. The growth of dry-rot or fungus on timber has been prevented by limewater, and many instances have been mentioned of its value. The cleansing and sanitary virtues of lime are more generally known. The painter uses limewater to kill the grease upon his work instead of turpentine; and soot stains on the outside of flues have been removed by the agency of thick warm limewash. The value of limewhite as a wash for walls, as a purifier of the air in sheds, stables, and other buildings is unquestionable, though all limewashed roof-timbers have rather a rough and penurious look. As a preservative coating to the joists of floors and other timbers not exposed to damp, it seems worthy of a more extended trial.

ONE SQUARE ACRE.

The number of square feet in an acre is 43,560. In order to have this area the piece of land must be of such a length and breadth that the two multiplied together will produce the above number. Thus an acre of land might be 43,560 feet long by one foot broad; 21,780 feet long by two feet broad; 12,250 feet long by three broad and so on. If the acre of land is to be exactly square, each side must be as nearly as possible 280 feet 1-2 inches. The nearest you can come to an exactly square acre with an even number of feet on the sides is to make it 220 feet long by 180 broad.

YARDS IN A MILE.

Mile in England or America, 1,760 yards.
Mile in Russia, 1,100 yards.
Mile in Italy, 2,407 yards.
Mile in Scotland and Ireland, 2,200 yards.
Mile in Poland, 4,100 yards.
Mile in Spain, 5,028 yards.
Mile in Germany, 5,866 yards.
Mile in Sweden and Denmark, 7,233 yards.
Mile in Hungary, 8,800 yards.
A league in England and America, 5,280 yards.

WHEAT CROP OF THE UNITED STATES.

The following is the estimated wheat crop of the United States for 1881, according to the figures furnished by the department of agriculture at Washington. The figures for 1879 are from the census returns:

	1879. Bushels.	1881. Bushels.
Maine	666,301	524,800
New Hampshire	164,729	165,00
Vermont	341,416	399,000
Massachusetts	5,791	
Rhode Island	240	
Connecticut	38,525	36,000
New York	11,298,067	13,26 ,000
New Jersey	1,891,852	2,444,600
Pennsylvania	19,512,051	19,829,000
Delaware	1,181,776	127,600
Maryland	8,251,792	6,612,5 0
Virginia	7,847,021	7,150,000
North Carolina	3,128,194	4,442,100
South Carolina	937,074	1,600,800
Georgia	3,139,514	2,487,450
Florida	421	
Alabama	1,596,852	1,111,800
Mississippi	217,626	203,200
Louisiana	1,053	
Texas	2,977,921	3,247,800
Arkansas	1,205,612	921,4 0
Tennessee	7,299,84	6,752,0 0
West Virginia	3,580,680	4,2 0,000
Kentucky	11,359,087	8,970,000
Ohio	46,016,452	30,177,680
Michigan	31,632,234	21,765,000
Indiana	47,415,5 0	20,780,800
Illinois	51,171,818	28,544,680
Wisconsin	21,506,418	16,150,500
Minnesota	34,708,258	3,19 ,000
Iowa	1 ,02,728	4,671,350
Missouri	24,966,724	23,877,600
Kansas	17,311,638	19,148,800
Nebraska	3,816,721	13,628,600
California	29,802,378	20,367,800
Oregon	7,477,294	13,490,000
Nevada	20,438	
Colorado	420,356	1,300,00 0
The Territories	7,738,590	14,588,600
Total United States	458,105,747	381,479,580

A PLAIN AND EASY WAY OF CURING HAMS.

The principle thing in curing hams is to get them just salt enough to keep them and not so salt as to injure the flavor and cause them to become hard. Hams should be neatly trimmed and cut rounding, to imitate as closely as may be the hams of commerce. Trim closely, so there shall be no masses of fat left at the lowest extremity of the hams. The shoulders may be cut in shape convenient for packing, and they should be salted in separate packages from the hams.

Hams are cured by both dry salting and brine. When dry salting is employed the hams are rubbed often with salt and sugar. Between each rubbing they are bunched up on platforms or tables, the surface of which is spread with a layer of salt, and each ham is also covered with salt. When taken up to rub, which is usually done five or six times, a shallow box is at hand in which to do the work.

When brine is used, prepare a pickle strong enough to float an egg and stir into it a sufficient amount of sugar and molasses to give it a sweetened taste. Some add a little saltpeter to color the meat. In moderate quantity it is commonly accepted as beneficial. Cover the hams with the pickel and place the packages where the temperature is uniform and above freezing. For hams of twelve pounds, four weeks will be sufficient; larger hams must remain in the brine a longer time. In general, three to seven weeks embraces the extreme of time required for domestic curing of hams, varying as to the size of the hams, temperature and time when they will be required for use. When it is designed to preserve hams through the summer they must not be removed from the pickle too soon.

Shoulders require much the same treatment

as do hams, and both should be carefully smoked. The preservative principle of smoke is known as creosote. Smoke made by burning corn-cobs is highly esteemed, but those engaged in curing meats on a large scale prefer the smoke obtained from dry hickory that has been stripped of its bark. The smoking process must not be too much hurried or the creosote will not have time to penetrate the entire substance of the meat. Ten days' smoking is usually sufficient, unless the pieces are very large and thick.

A process in ham-curing practiced by some of the leading packing-houses consists in creating the smoke in an oven outside of the smoke-house and passed through underground pipes into it. The smoke, rising from the floor to the top of the house, encounters two opposite currents of air drawn from the outside. These currents cause the smoke to form into a rapidly revolving horizontal column which passes among the hams. The smoke is not warm, and there is no heat to melt the hams or hot air to blacken them. The hams under this process are smoked in very much less time than by the old method.

While canvassing hams has nothing to do with their flavor, it is a protection from insects, and will pay the farmer for the extra labor. It should be done before warm weather. Wrap each ham in coarse brown paper and then sew it up in cotton cloth cut to suit the size, following the shape of the ham. When covered as described, dip them in a wash made of lime-water and colored with yellow ochre. Hang up in a cool place to dry. The wash closes the interstices of the muslin, and the whole forms a perfect protection against insects. The room in which any kind of cured meat is stored should be dry and cool, and the darker the better.

THE PART WHICH WORMS PLAY IN NATURE.

The latest fruit of Charles Darwin's labors in the field of physical research is presented in a volume treating of *The Formation of Vegetable Mould Through the Action of Worms*. The term vegetable mould is commonly applied to that superficial layer of soil, generally of a blackish color and a few inches in thickness, which covers the whole surface of the land in every moderately humid country. The uniform fineness of the particles of which it is composed is one of its chief characteristic features, and this may be well observed in any recently ploughed field, where the top layer is exposed on the sides of a furrow. It is the object of this book to show that the fine earth composing this superficial layer has been brought up to the surface by worms in the form of castings or excrement. We are thus led to conclude that all the so-called vegetable mould which is strewn over the surface of the ground has passed many times through the intestinal canals of worms, and hence the the term "animal mould" would be in some respects more appropriate than the term in common use.

Some of the conclusions reached in this volume were suggested in a paper published by Mr. Darwin many years ago, but many scientists rejected his conclusions with respect to the part played by worms in the formation of the mould, on account of their assumed incapacity to do so much work. This seems to have been an instance of that inability to sum up the effects of a continually recurrent cause which has often retarded the progress of science. In order to meet the objection raised, Mr. Darwin resolved to make more observations of the same kind as those previously published, and to attack the problem on another side by weighing all the castings thrown up within a given time in a measured space, as well as by ascertaining the rate at which objects left on the surface are buried by worms. It appears that near Maer Hall, in Staffordshire, quick lime had been spread, about the year 1827, thickly over a field of good pasture land which had not since been ploughed. Some square holes were dug in this field in the beginning of October, 1837, and the sections showed a layer of turf formed by the matted roots of the grasses, half an inch in thickness, beneath which, at a depth of three inches from the surface (the 2½ inches intervening being vegetable mould), a layer of the lime in powder or in small lumps could be distinctly seen running all round the vertical sides of the holes. Coal cinders had been spread over a part of this same field in the year 1834, and when the holes mentioned were dug—that is, after an interval of three years—the cinders formed a line of black spots round the holes at a depth of one inch beneath the surface, parallel to and above the white layer of lime. Over another part of this field cinders had been strewn only about half a year before, and these either lay on the surface or were entangled among the roots of the grasses. Here Mr. Darwin saw the commencement of the burying process, for worm castings had been heaped on several of the smaller fragments. After an interval of 4½ years this field was re-examined and now the two layers of lime and cinders were found almost everywhere at a greater depth than before by nearly one inch. It follows that mould to an average thickness of one-fifth of an inch had been annually brought up by the worms and spread over the surface of the field. Mr. Darwin cites a number of instances in which he was able to compute the rate of mould formation by worms, which of course, must vary according to the nature of the subsoils. The rate, for example, must become very much slower after a bed of mould several inches in thickness has been formed; for the worms then live chiefly near the surface and burrow down to a greater depth so as to bring up fresh earth from below only during the winter, when the weather is very cold, or during midsummer, when the earth is very dry. Of course, too, relatively few worms would be found in stony ground, and their production of mould would be comparatively slow. The effect, however, of their action, even in such cases, is astonishing when extended periods of time are considered, as the following example shows. We are told that a field near Mr. Darwin's house was last ploughed in 1841, then harrowed, and left to become pasture land. For several years it was clothed with an extremely scant vegetation, and was so thickly covered with small and large flints (some of them half as large as a child's head,) that it came to be known as "the stony field." Mr. Darwin says he can remember doubting whether he should live to see these larger flints covered with vegetable mould and turf. But the smaller stones disappeared before many years had elapsed, as did every one of the larger ones after time; so that after thirty years a horse could gallop over the compact turf from one end of the field to the other and not strike a single stone with his shoes. This was certainly the work of worms, for though castings were not frequent for several years, yet some were thrown up month after month, and these gradually increased in numbers as the pasture improved. Still more striking was the burying of a path paved with flagstones, which in 1843 ran across Mr. Darwin's farm. The worms threw up many castings in the interstices of these stones, and although during several years the path was weeded and swept, yet ultimately the weeds and worms prevailed, the path became covered up, and after several years no traces of it was left. On removing in 1877 the thin overlaying layer of turf, the small flagstones, all in their proper places, were found covered by an inch of fine mould. It will surprise most readers to learn how large an amount of mould may be formed by worms on the surface of a field in a single year. Mr. Darwin calculates that the castings ejected annually by each earthworm weigh, on an average, more than twenty ounces. It has been estimated by other observers that 53,767 worms exist in an acre of land ; but this estimate is based on the number found in gardens. Assuming that only half the number named, or about 27,000 worms to the acre, live on pasture land, and that each worm annually ejects twenty ounces, we should have fifteen tons as the weight of the castings annually thrown up on an acre of land, and helping to form the layer of vegetable mould.

Archæologists are probably not aware how much they owe to worms for the preservation of many ancient objects ; coins, gold ornaments, stone implements, etc., if dropped on the surface of the ground will infallibly be buried by the castings of worms in a few years, and will thus be safely preserved. For instance, some years ago a grass field not far from Shrewsbury was ploughed up, and a surprising number of iron arrow heads were found at the bottom of the furrows, which no doubt had been left strewn on the battle-field of Shrewsbury in the year 1403. In Abinger, Surrey, on a trench being dug in 1876, the concrete floor of the atrium or reception room belonging to a Roman villa was disclosed at a depth of two or two and one-half feet. At first sight it appeared impossible that the vegetable mound covering the pavement could have been brought up by worms, but upon close inspection the concrete was found decayed and completely permeated with worm burrows. Through these channels in the softened mortar the worms have been throwing up their castings from the ground beneath, and heaping on the concrete pavement a layer of fine earth, during many centuries and perhaps for a thousand years. The coins discovered in this place dated from 133 to 375 A. D. The pavement of Beaulieu Abbey in Hampshire now lies at a depth of from 6½ to 11½ inches beneath the surrounding turf-covered surface. A part of this pavement has been uncovered, but requires continual sweeping to remove the worm castings, which otherwise would soon rebury it. A large number of analogous excavations described in this volume demonstrates how considerable a part worms have played in the concealment of Roman and other old buildings in England, although no doubt, the washing down of soil from neighboring higher lands and the deposition of dust have largely co-operated in the work of burial.

It is plain enough, from the data collected in this book, that worms have played a more important part in the history of the world than most persons would imagine. Few of us, indeed, when we behold a wide, turf-colored expanse, are aware that its smoothness, on which so much of its beauty depends, is mainly due to all the inequalities having been slowly leveled by worms. It is a marvelous reflection that the whole of the superficial mould over any such expanse has passed and will again pass every few years, through the bodies of worms. The creatures which exercise so important a function in the physical economy are poorly provided with sense organs, for they cannot be said to see, although they can just distinguish between light and darkness ; they are completely deaf, and have only a feeble power of smell ; the sense of touch alone is well developed. It may well be questioned whether there are many other animals which have played a more considerable part in the history of the earth than have these lowly organized beings. Some other animals, however, still more lowly organized—namely, corals, have done even more conspicuous work by constructing innumerable reefs and islands in the great ocean, but these are almost wholly confined to the tropical zones.

SPARE THE TREE.

No subject is of graver import to the future of this continent than the protection and preservation of its forests. Sir Samuel Baker, who recently returned from a hunting expedition in the Big Horn country of Wyoming, said that the extensive and wanton burning of the Rocky mountain woodlands was an evil of such magnitude that he was astonished to find hundreds of square miles in a blaze, carrying on the march of devastation until quenched by a heavy rain-fall or arrested by the high mountain tops above the timber line. The reckless miners and thoughtless hunters, traders and travelers, who are responsible for this prodigious waste, bid fair to convert fertile valleys and copious river sources into arid deserts and dried-up gulches.

It is a well-known law that forest destruction of a wholesale character diminishes rainfall, and eventually banishes it altogether. Hence the anxiety of the more enlightened governments to save their native and primeval timber intact, knowing that its reproduction and preservation are the life's blood of the country itself. What will be the ultimate result, judging from evidences of the East ? Our rich Western regions will become gradually parched ; brooks and streams will die out forever ; important feeders of a great river system will become extinct, lowering the level perhaps of such a channel as the Mississippi river, and one word will be written across the face of the country—desolation.

That this is no exaggeration may be understood from the fact that it was recently reported at the annual meeting of the Geographical society of Vienna by Councilor Wex, that the Volga is decreasing in volume, owing to the destruction of wood in its valley, so as to materially affect the level of the Caspian Sea and the Sea of Aral. It is apparent therefore, that the most vital question in connection with that wonderful domain beyond the Rocky mountains is the preservation of its forests. As long as it is possible for one adventurer to build his camp fire in the wood and leave it to the mercy of the winds, thus laying waste what would be a respectable county in our commonwealth, this destruction and consequent physical disorder will go on. Appropriate legislation sternly executed is only a partial remedy. The science of forestry, as studied and applied in the older countries of Europe, must be introduced and cultivated here.

In nearly all of the countries of the Old World forestry, in connection with climatology, geology and kindred branches, is taught in nearly all the universities, and the several governments take an especial interest in expert graduates in this branch. Particularly is this true, curiously enough, in countries where is the largest proportion of woodland, as in Russia, Sweden, Germany and Austria. The lowest occurs in Great Britain, Denmark, Spain and Holland. Over forty-two per cent. of the acreage of Russia is forest, while Britain has but a little over three per cent. In Germany more attention is given to arboriculture than in any other western power.

America, of all quarters of the world, is the most thickly wooded with the primeval forest, and was of vast extent and contained a great variety of species, covering, with insignificant exceptions, all that portion of our continent which was occupied by the colonists ; but now it is doubtful, according to the very best authorities, if any State of the Union, save Oregon, has more woodland than it ought permanently to preserve. Our Eastern and Middle States were at one time dense forests, while now Pennsylvania alone has preserved her timber. The other States are compelled to send to Canada and the West to supply their market. Our government, however, began early to perceive the danger of indiscriminate forest felling.

In 1817, and again in 1831, statutes were passed to restrict spoliation. Yet it may be judged that the woodland is largely suffering when we remember that there are over 30,000 saw mills in the United States, nearly all doing a flourishing business. In some States special legislation provides for adequate protection, and in California, a State forester has been appointed. The devastation in that State has been enormous, and in Texas also, where the supply of trees is totally inadequate and where destructive tornadoes prevail, together with extensive fires.

In view of the facts stated, it is plain that intelligent and prompt action should be taken by Congress to prevent further spoliation. The absolute necessity is apparent in the not encouraging fact that already over two-fifths of the entire area of the United States is so arid that even artificial irrigation cannot now redeem it ; indeed, west of the Mississippi, owing to the forest fires largely, one-sixth of the entire territory alone is susceptible to cultivation. In Colorado, New Mexico, Arizona, Nevada, Utah, Wyoming, Idaho and Montana, not one-fifth of the area can ever be rendered available, and it is doubtful without expedients now unknown, if any of these territories will support more than 300,000 people at a time ; and in Wyoming not over 5,000 square miles in the 100,000 square miles of area can be termed arable land.

The question then arises : What is the best method of achieving practical results for the preservation of whatever physical advantages we possess in our national domain, and no inquiry of greater magnitude can be addressed to the Forty-seventh Congress.—*New York Sun.*

LET THE FROST HELP YOU.

Few fully appreciate how much a freezing of the ground does to set at liberty the plant-food locked up in almost all soils. Water, in freezing, expends about one-eighth of its bulk, and with tremendous force. Water, if confined in the strongest rock and frozen, will burst it assunder. The smallest particles of soil, which are in fact only minute bits of rock, as the microscope will show, if frozen while moist are broken still finer. This will go on all winter in every part of the field or garden reached by the frost; and as most soils contain more or less elements that all growing plants are crops need, a good freezing is equivalent to adding manure or fertilizers. Hence it is desirable to expose as much of the soil as possible to frost action, and the deeper the better, for the lower soil has been less drawn upon and is richer in plant food. Turn up the soil this month wherever practicable. If thrown into ridges and hollows, in field and garden, the frost will penetrate so much deeper. Further, plowing or spading the soil now exposes insects and weed roots to killing by freezing. Still further, soils thrown up loosely will dry out earlier in spring, and admit earlier working, which is often a great gain when a day or two may decide in favor of a successful crop.—*American Agriculturist.*

The great wheat exporters of Russia are becoming alarmed at the tremendous competition they have to encounter. Hungary and the Danubian principalities were the first to appear in the Western markets, but the construction of a railway to Odessa restored the the equilibrium. Then the American competition commenced, and has ruined the inhabitants of the wheat-producing districts of the Muscovite empire. Wheat is abundant in the interior—more so than for many years past—but there is scarcely any communication with the seaboard. The great military railways run right through the country, but there are few feeding lines. The roads and canals and the core of the wheat in transport are in as primitive state as when Russia had no competitor in the field. If a prompt move is not made by the government—which is scarcely to be expected at present—Russian wheat will soon be driven out of the Western markets by United States enterprise and the new field opening up in India.

TOBACCO REVIEW—THE OLD YEAR AND THE NEW ONE.

We give the following excellent review of the seed leaf trade during the past year from the *Tobacco Leaf*. It goes over the ground very fairly, as we think, and will be found to be of unusual interest to all persons interested either in growing or manufacturing seed leaf tobaccos :

The year 1880 opened with an estimated stock of seed leaf of 327,000 cases, consisting of 52,000 cases of old of all kinds and 275,000 cases of the growth of 1880. The product of 1880 was estimated as follows : New England, 40,000 cases, Pennsylvania, 110,000; New York, 20,000; Ohio and Indiana, 50,000; Wisconsin and other Western States, 55,000. Total, 275,000.

The sales in the New York market during the year were 130,000 cases, of which 13,228 were for export.

Consumption in 1881.

According to the returns to the office of Internal Revenue at Washington, there were consumed in the making of 2,642,528,130 cigars in the fiscal year ending June 30, 1881, 50,012,669 pounds of leaf tobacco, which, at 350 pounds per case, are equal to 168,608 cases. This allows 22½ pounds of leaf to a thousand cigars. The case is here reckoned at 350 instead of 400 pounds, both because the various packings may average that, and because the revenue calculations are based upon net weight.

From the aggregate of cases must be deducted Havana, Sumatra, and other varieties of leaf used in making cigars. Substitutes for Havana are latterly used with freedom by manufacturers, and we subtract for surrogates of all kinds the equivalent of one-ninth: in other words, 18,734 cases, or about 65,000 bales, leaving about 150,000 cases of seed leaf converted into cigars; in precise figures, 149,974 cases.

The fiscal year equally divides the calendar year, and it will be a modest assumption to say that, if 75,000 of 150,000 case, were used in the first half of 1881, the last half, just ended, certainly appropriated as many more. It is well known that the manufacturing trade was more active in the latter than in the former period. Besides the requirements for cigars, not less than 25,000 cases of seed leaf were embraced in the production of cigarets and smoking tobacco in the past year.

Stock on Hand on January 1, 1882.

From New York there were exported in 1881 36,504 cases, and from Baltimore 3,968—total, 40,553 cases. Tabulating the disappearances, the exhibit is as follows :

Home consumption	175,000 cases.
Export	40,552 "
Total	215,552 "

Accepting the estimate at the beginning of the year, namely, 3.7,000 case, the above total indicates a remainder of old stock on the 1st of January, 1882, amounting to 111, 478 cases—not an inconvenient quantity, though large. Pennsylvania, it will be noticed is credited with a crop of 110,000 cases in 1880, and there are tradesmen who assert that several thousand cases might properly be taken from that figure. Those so inclined may do so.

The Crop of 1881 and Visible Supply.

Among experienced packers and samplers opinion differs widely respecting the quantity of seed leaf harvested in 1881. Maximum estimates place it at 260,000 cases; minimum at 220,000, the majority agreeing on the latter, which sums up as annexed :

New England,	40,000 cases.
Pennsylvania,	75,000 "
New York,	25 000 "
Ohio,	30,000 "
Wisconsin and other Western States,	50,000 "
Total,	220,000 "

Low as this total may seem: it is probably not greatly at variance with the actual fact. Assuming that it is an approximate, the visible supply appears to be as follows :

Old stock,	- - -	111,478 cases.
New stock,	- - -	220,000 "
Total old and new,	- -	331,478

Receipts in 1881.

Of seed leaf tobacco there were received in New York in

1881,	- - -	90,301 cases.
1880,	- - -	79,792 "

Sales Each Month.

		Cases.
January,	- - -	7,800
February,	- - -	10,000
March,	- - -	9,501
April,	- - -	6,950
May,	- - -	7,923
June,	- - -	17,130
July,	- - -	10,479
August,	- - -	11,400
September,	- - -	22,100
October,	- - -	17,000
November,	- - -	5,481
December,	- - -	4,550
Total,	- - -	130,006

The total sales of seed leaf in this market in 1880 were 92,457 cases, showing an increase in 1881 of 38,539 cases.

Comparative exhibit of the export of seed leaf and cuttings in New York since

		Cases.
January, 1881,	- - -	36,594
Same time in 1880,	- - -	31,847
Same time in 1879,	- - -	23,942

Remarks.

The year has closed with the largest volume of sales on record. The highest previous figure was reached in 178, when 124, 502 cases were sold. There is reason to believe that dealers in this staple have, as a rule, enjoyed a prosperous trade, and it is to be hoped and expected that they will have similar good fortune in the year now entered upon. Apparently this year commences with a little larger stock than last year did, but it must be borne in mind that crop estimates are not based upon positive data, and the figures set down for the several producing sections mentioned above may be too high in some instances, as well as too low. Possibly Pennsylvania is credited with 15,000 cases too much. When the writer saw the 1881 crop in the field in the latter part of August, he would have been reluctant to believe that it would yield 60,000 cases, the drouth seeming to have dwarfed beyond salvation much of that which was then standing. Succeeding rains and a late growing season helped to improve the situation very materially. This fact, and the circumstance that a larger acreage than ever before was planted, incline many to the belief that the product of the State will not be far from 75,000 cases. Some estimate it at ¥0,000.

The above estimates for the other States are certainly not in excess. If the writer were to express his own opinion, he would credit New England with 45,000 rather than 40,000 cases. For when he saw the New England crop, which, also, was late in August, it had the promise of undiminished fullness and excellence.

It is not necessary to refer here to the characteristics of the new crop as a whole. Some of it will be good and some indifferent, as is always the case. By and bye it will come forward for sale, when its merits and demerits will be made manifest. The growing season began and ended well, but its perfection was marred by the want of rain when rain was most needed. It need surprise no one, consequently, if some of the crop shall hereafter be found defective. Good tobacco is grown only in good conditions, and these were wanting in the hot and dry month of August in most of our tobacco growing regions.

The premature buying and high prices paid for some of 1881 crop in the field are likely to be obstacles in the way of a completely satisfactory trade in 1882. They are surely going to impede the export trade, which is to be regretted for commercial reasons. This year it is to be hoped there will be less haste than there was last year. Yet, when early and extravagant buying commenced, the situation seemed to justify the movement. With the vast manufacturing requirement, there is good reason for anticipating a large trade.

Prices.

Prices continued steady throughout 1881. Except the slight advance effected on some grades in August when only a half crop was anticipated, no change is perceptible in the year's tables of quotations. The market commenced and closed strong. Ohio shipping sorts are a trifle lower than they were, but all other kinds rule at the long prevailing rates.

Quotations January 1, 1881.

New England—Crop 1880, wrappers :

Common	15@17
Medium	18@20
Fine	25@35
Selections	40@50
Seconds	11@12½
Havana seed	20 to 25

Pennsylvania—Crop 1880, assorted lots :

Low	10@12
Fair	13@15
Fine	18@50
Wrappers	18 to 50
Fillers	6 to 7

New York—Crop 1880, assorted lots :

Common	8 to 10
Medium	12 to 14
Good	15 to 18

Ohio—

Crop 1880, assorted lots	11@12½
Wrappers	11 to 20

Wisconsin—

Crop 1880, assorted lots	6@10
Wrappers	12 to 20
Havana seed	12½@16

Berks County Agriculture Society.

At the annual meeting of the Berks County Agricultural Society, held at Reading on Saturday afternoon in the Courthouse, the old officers were re-elected, to wit : President, Jacob G. Zerr; Treasurer, William S. Ritter; Secretary, Cyrus T. Fox. The society, for the first time in fifteen years, is out of debt, and with a balance of $1,506 in its treasury. A resolution was adopted authorizing an agreement with the Park Commissioners to open the fair ground for the purpose of a public park, provided the consent of the County Commissioners be obtained, and that the city councils appropriate the moneys necessary to the improvement of the premises.

OUR LOCAL ORGANIZATIONS.

LANCASTER COUNTY AGRICULTURAL AND HORTICULTURAL SOCIETY.

The regular monthly meeting of the Lancaster county Agricultural Society convened in this city, Monday afternoon, January 2. The following named persons were present : M. D. Kendig, Creswell; Daniel Smeych, city; H. M. Engle, Marietta; Dr. C. A. Greene, city; Casper Hiller, Conestoga; Henry Kurtz, Mount Joy; W. W. Griest, city; Hebron Herr, West Lampeter; Enos. Weaver, Strasburg; F. R. Diffenderfer, city; Calvin Cooper, Bird-in-Handt Johnson Miller, Lititz; William H. Brosius, Liberty square, John H. Landis, Millersville; J. H. Hershey, Salunga; S. A. Hershey, Salunga; S. P. Eby, city; C. L. Hunsecker, Manheim township; Wash L. Hershey, Chickies; E. H. Hoover, Manheim township, J. M. Johnston, city.

President J. F. Witmer being absent, Vice President Henry M. Engle swung the gavel.

Enos H. Weaver of Strasburg, and Hebron Herr, of west Lampeter, were elected members of the society.

Crop Reports.

Calvin Cooper reported the grain fields in fine condition; abundant rain has fallen and everything promises well.

Henry Kurtz, of Mount Joy, never saw better wheat and seldom saw the grain as promising as at present; especially is this the case with wheat sown on tobacco land; from which fact Mr. Kurtz concluded that tobacco does not injure the land. There is considerable short leaf about his neighborhood and much of the tobacco does not color much.

John H. Landis, of Millersville, saw dandelions in bloom in Bucks county, last week, and saw good wheat on his own native Manor. Half the tobacco is stripped.

Martin D. Kendig remarked that a neighbor seeded rye to the latter part of November, and it came up well.

In Donegal wheat is good, said H. M. Engle, but much freezing and thawing during December may have endangered it more or less; and as young clover is pastured closely Mr. Engle feared the effects in spring. Rain fall for December was 5 5-10 inches; for the year 38¾ inches.

Election of Officers.

On motion of Johnson Miller, the regular business was now suspended and the society proceeded to nominate and elect officers for the ensuing year.

For president, Joseph F. Witmer, of Paradise was renominated.

For vice presidents, Henry M. Engle, of Marietta, and Jacob B. Garber,of Columbia, were renominated.

For recording secretary, M. D. Kendig, John H. Landis, Johnson Miller and Calvin Cooper positively and peremptorily declined nomination, and finally the honor was cast upon John C. Linville, of the Gap, who was absent.

For corresponding Secretary, Calvin Cooper of Bird-in-Hand, was nominated.

For treasurer, M. D. Kendig, of Creswell, was re-nominated.

There being no more than the constitutional number placed in nomination for the above offices the nominees were declared elected:

For managers the following were nominated and the figures attached indicate the number of votes each received: Wm. H. Brosius, 11; John H. Landis, 8; Casper Hiller, 7; Calvin Cooper, 6; Enos H. Weaver, 6; Hebron Herr, 5; Daniel Smeych, 4; Johnson Miller, 4; E. S. Hoover; 4. The first five named were declared the duly elected managers.

Casper Hiller, of Conestoga, read an essay entitled

"Can the Grain Grower Dispense with Nitrogenous Fertilizers?"

In order to have a proper understanding of the subject it may not be amiss to give the analysis and cost of several of the principal manures in the market.

A ton of well prepared bone contains about 400 pounds of phosphoric acid, valued at $30, and about 80 pounds of nitrogen, valued at $16.

A ton of acidulated South Carolina rock contains about 340 pounds of phosphoric acid, which can be bought for $25.

A ton of nitrate of soda costs about $80 and contains about 20 per cent. of nitrogen.

These figures show that nitrogen adds one-third to the price in the bone manure, and in the nitrate of soda the nitrogen makes up the whole cost, showing that nitrogen is an expensive ingredient, and for that reason the question put to me is worthy of consideration.

My experiments have been on too limited a scale to be of much value, but as far as they went (on corn only) the indications are that nitrogenous fertilizers are non-paying on my soil.

In my experiments, I have used phosphate rock; raw bone, dissolved bone, and a nitrogenous flesh fertilizer, and have come to the conclusion that phosphoric acid is the paying ingredient in those manures.

Extensive experiments have been made at the Eastern Experimental farm, that prove that nitrogenous fertilizers are not profitable on that farm.

The application of nitrate of soda and sulphate of ammonia did on no occasion yield sufficient increase of grain to pay for the fertilizers, while phosphate rock, a purely non-nitrogenous fertilizer, gave more increase of grain than stable manure, or ground bone, or bone superphosphate.

[See report of John L. Carter, to State Board of Agriculture for 1877 and 1878].

Chemical analysis shows that nitrogen is an important element in all our grain crops, but these experiments would show that there is a bountiful supply of it in our soil, or that the atmosphere furnishes all that is needed to perfect the crop. There are good authorities who contend that the later is the case.

From the foregoing you can perceive that my answer is, the grain grower can dispense with nitrogenous fertilizers.

The Board of Managers made the following appointments for the ensuing year:

Entomologist and Botanist—S. S. Rathvon.
he inist—Jno. C. Linville.
Mineralogist—B. K. Hershey.
Librarian—S. P. Eby.

The following questions were continued until next ... the referees being absent: "Can dairy cows be kept in good condition by the soiling process, and is the butter as good?" J. Frank Landis.

"What is the best time to plow land for spring crops?" John C. Linville.

"Ought rank growing wheat to be pastured?" Ja

John H. Landis offered the following resolution: "Resolved, that the thanks of this society be extended to Jos. F. Witmer for the fair and impartial manner in which he has presided over our deliberations." Unanimously adopted.

Henry M. Engle called the attention of the society to the fact that the State Fruit Growers' Association would meet in Harrisburg on the third Wednesday in January, and on motion of Mr. Cooper a committee of three was appointed to represent the Lancaster society at the fruit growers' meeting. That committee consists of Calvin Cooper, M. D. Kendig and Wm. H. Brosius.

Messrs. F. R. Diffenderffer and C. L. Hunsecker were appointed to audit the treasurer's account, which they did, and reported it correct, and a balance of $79.31 in the hands of treasurer Kendig.

On motion of Calvin Cooper the bounty due from the county to the society for the years 1880 and 1881 was ordered to be ascertained and the bill presented to the commissioners.

Dr. C. A Greene read an essay on insects, and said, Some weeks ago my attention was called to the fact that Mr. Rathvon has left out for some months my name from the various accounts of the proceedings of our society, both from the LANCASTER FARMER and the daily *Examiner*, and I have been unable to answer the inquiries of my friends, why it was done? Whether my various questions asked of our tailor friend had become a nut so hard to crack that he has become rancorous, or whether from jealousy, I know not. I do know that it is rather an insult to our organization, and although personally I care nothing for it and presume it will not shorten my life one day, yet as directed against the society, I bring it to your notice.

[The subject matter of the essay gets as close to the point as the doctor's communications generally do, but it also contains a few inaccuracies which we desire to correct. In the first place he accuses Dr. Rathvon of intentionally omitting his (Dr. Greene's) name from the reports. The FARMER takes its reports from the *Examiner*, and whatever blame is attached to the omission of Dr. Greene's valuable essays from the reports, must therefore be borne by this journal. We tried to give a faithful report of the proceedings, and only determined to exclude Dr. Greene's name after he had taken us to task for maliciously misrepresenting him by publishing a stricture upon the tobacco buyers in this city, which was delivered from his own lips at a meeting of the society. When the doctor found that his words had got him into hot water, he attempted to throw the blame from his shoulders upon ours, and we then came to the conclusion that we would in the future give him no cause to complain. Had the doctor not blamed the wrong person for what he deems an "insult to our organization," this reply to his essay on "insects" would not have been written.—REPORTER DAIL YEXPRESS.]

A discussion here ensued as to the best means of gaining a better attendance of members and increasing the interest in the society's proceedings. Calvin Cooper moved that each member bring his wife to the February meeting, and if he chances not to be mated, let him bring some other congenial companion. The motion was not pressed to a vote, but it was favorably regarded by the members present.

Mr. Engle suggested competitive essays as one means by which to awaken some activity.

C. L. Hunsecker thought the strictures of some of the reporters on the political discussion at the last meeting both severe and unkind. He had yet to learn that farmers are not allowed to express their opinions in any place. He thought Dr. Greene's essays could well be dispensed with, and was favorable to Mr. Engle's suggestion in regard to competitive essays.

Calvin Cooper suggested that the chair appoint an essayist at every meeting and accept no excuses whatever.

Ephraim S. Hoover thought that none but agricultural and horticultural subjects should be introduced for discussion—no politics; he also thought if an essayist was appointed, the appointee would feel it more of a duty to respond.

Mr. Hunsecker then moved that the chair be empowered to appoint an essayist at each meeting, the appointee to chose his own subject; provided, however, it is germain to agriculture and horticulture, not politics, for then the "reporters would catch us by the ear."

Calvin Cooper moved to amend by imposing a fine of fifty cents for failure to perform the duty assigned, and Ephraim Hoover, by proposed amendment, increased this sum to $1.00. Both of these amendments were voted down, and the main question was passed.

C. L. Hunsecker was appointed essayist for the February meeting.

THE STATE GRANGE.

The ninth annual session of the State Grange of Pennsylvania was held in the parlor of the Park Hotel, in the city of Williamsport, Pa., during the week beginning at 1:30 o'clock p. m., on Tuesday, December 18th, 1881. About four hundred Patrons were in attendance during the session, representing one hundred and twenty-seven Granges, located in fifty counties of the State.

On Tuesday evening an address of welcome, was delivered by Hon. C. D. Eldred of Grange No. 71, Lycoming county, which was responded to on the part of the State Grange by W. T. Everson, of Erie county. The annual address of State Master L. Rhone was delivered the same evening.

Wednesday's Proceedings.

On Wednesday morning the various committees were announced by the Master, and reports were heard from the different officers, the latter showing the order in the State to be increasing in membership and improving in efficiency of grange work.

W. A. Armstrong, Master of New York State Grange, addressed the afternoon meeting. Also Dr. Calder, of Dauphin county, State lecturer.

A public meeting was held in the Court House on Wednesday evening, with Hon. Wm J. Wood, of Lycoming county, as chairman. Worthy Master Rhone, the first speaker, referred to the importance of protecting the agricultural interests, showing that all other business is greatly dependent upon the prosperity of the farmers, and stated briefly the objects of the organization there represented.

Governor Hoyt said, being in Williamsport accidentally, he had submitted to the hospitality and persistence of the Grangers. He did not propose to discuss the technicology of the farmer's occupation. He hoped to see the Pennsylvania farmer put on an equal footing with the Western farmer. Improved methods here would soon make Pennsylvania soil as productive as the thin exhausted soil of the West.

He congratulated those present for the manifest indication of a revival of the farming interest, (referring to the large audience of farmers before him.) Dr. Calder spoke of the isolated condition of farm life in America and contrasted it with the East, where the farmers live in villages, thus affording better opportunities for social intercourse and mental improvement. He claimed that the Patrons of Husbandry had done much toward making up this deficiency here. He referred to the great want of information amongst farmers, citing numerous instances where this want was most noticeable.

He believed there was no better way of inducing

THE LANCASTER FARMER.

the acquisition of knowledge and of retaining it than in imparting it to our associates, and that grange meetings afforded an excellent opportunity for doing this.

Amos Holstein, of Montgomery county, read an essay on "Woman's Work in the Grange."

Past State Master V. E. Piollett discussed the relations of the Grange to corporations. He said the Grange grew out of a necessity for some association by which the interests of the agricultural class might be preserved. Twenty-five out of the fifty millions of people in our country are directly interested in agriculture, yet we have almost no voice in legislative bodies, where corporations are regulated. He thought the corporations had too much power. They must be regulated by law more effectually. Transportation companies should only be allowed to charge what would be a fair compensation, and not "what the traffic will bear." Effective laws have been established in Illinois and even in Georgia, where the railroad companies are required to put up their rates of freight and fare in their station houses. He wished it understood that the patrons made no war on associated capital, but demanded their rates without unjust discrimination.

The courthouse was crowded with citizens of Williamsport and farmer from the vicinity.

The Proceedings on Thursday.

On Thursday the Secretary made his report, showing over 350 active Granges in the State and about 12,000 members.

The Park Hotel has ample accommodations for entertaining the entire State Grange, as well as affording a suitable place for meeting, and nearly all the members availed themselves of its hospitality.

The citizens of Williamsport have shown a lively interest in making our sojourn here as pleasant as possible, and through their instrumentality and the kindness of Superintendent Neilson, of the Elmira Division of the Northern Central Railway, a free excursion was tendered members to the dairy farm of Judge Smith, several miles south of the city, where the Cooly system of setting milk is being tested in connection with the coiling method of keeping cows.

The representatives in attendance from Lancaster county are John H. Eplar, of Conoy Grange, No. 697, and W. P. Bolton and wife, of Fulton Grange, No. 66. The State Grange adjourned at noon on Friday.

THE POULTRY SOCIETY.

The regular monthly meeting of the Lancaster County Poultry Association, not having been held on the first Monday of the month, as is customary, was held Monday morning, January 9, 1781.

The meeting was called to order by President Tshudy.

The following membership were present: H. H. Tshudy, Lititz; J. B. Lichty, city; George A. Geyer, Silver Spring; T. Frank Evans, Lititz; F. R. Diffenderfer, Charles Lippold, Charles E. Long, city; J. A. Stober, Schoeneck; John A. Schum, W. A. Schoenberger, city; M. L. Grider, Mount Joy; J. B. Long, city; T. D. Martin, Lititz; Dr. E. H. Witmer, Neffsville; J. A. Garman, Leacock.

The minutes of the last meeting were read and approved.

J. B. Lichty gave notice that he would offer an amendment to the Constitution providing for the annual election of the officers of the society in February of each year, instead of in January as now.

Treasurer's Report.

T. Frank Evans read his annual report, by which it was shown there is at present in his hands the sum of $3.23.

The President appointed Messrs. Stober and Long to audit the Treasurer's accounts. This was done and they were found to be correct. This report was received and the committee discharged.

Election of Officers.

H. H. Tshudy withdrew from the candidacy for President. An election being had, the result stood

as follows: President, George A. Geyer; Vice Presidents, M. L. Grider, Charles Lippold; Recording Secretary, J. B. Lichty; Corresponding Secretary, Joseph R. Trissler; Treasurer, T. Frank Evans; Executive Committee, Dr. E. H. Witmer, John A. Schum, J. B. Long, Wm. A. Schoenberger, J. A. Stober.

Miscellaneous Business.

Charles E. Long offered a resolution that exhibitors from a distance drawing premiums be paid first. Carried.

A resolution was offered instructing the Secretary to notify members who have not paid their annual dues to do so by March first. Carried.

The new president, George A. Geyer, assumed the duties of his position.

On motion, the old Executive Committee were instructed to hold over until after the exhibition.

The Secretary stated that 350 entries have already been made, and a good many more are expected. The Lancaster county exhibitors have not come out so strongly. There will be 200 entries of pigeons. The variety of birds is much larger than last year. Rhode Island, Maryland, New Jersey, New York, Canada, Ohio and Delaware, will be represented. One exhibitor has made 35 entries. On the whole, the prospects for a successful exhibition are very good.

On motion, the meeting adjourned.

THE LINNÆAN SOCIETY.

The annual meeting of the society was held at the residence of Mr. Chas A. Heinitsh, East King Street, on Thursday evening, December 29, 1881, and was well attended. The president, Prof. Stahr and the secretary, Dr. M. L. Davis, in their chairs.

As the proceedings of the Society are generally published in three different newspapers, they are usually not read, unless to correct errors. After the customary opening and collection of dues, the following additions and donations were made.

Library.

International Scientists' Directory for 1881-2, 434 pages, demi octavo.

Annual Report of Commissioner of Patents, for 1880, 450 pages quarto.

Alphabetical list of Patentees and Inventors, January to June, 1881, 238 pages quarto.

Proceedings of Academy of Natural Sciences, for June and July, 1881.

Catalogue of the Fungi of the Pacific Coast, 46 pages, med. octavo, from California Academy of Natural Science.

Nos. 21, 22, 23 and 24, *Official Gazette of United States Patent Office*.

Lancaster Farmer, for December, 1881.

Four Book Catalogues and sundry Circulars.

One Envelope, containing 19 Historical and Biographical Scraps.

No additions or donations were made to *Museum*.

The *Curators* reported 3,500 plants added to the *Herbarium* of the Society during the year 1881; also, 1,000 specimens of minerals; 100 Indian relics; 50 Historical specimens; 150 Entomological; 300 in Paleontology, and 101 in Mamemology, Ornithology, Ichthyology, Reptilia and miscellaneous.

Total over 4,100 added to the museum.

The *Librarian* reported 160 books, pamphlets and serials to the Library during 1881, besides a large number of catalogues and circulars; also, 37 envelopes containing 500 Historical and Biographical Scraps. 10 original papers were read during the year. Since the organization of the society, 500 original papers were read before it, only 26 of which were published.

The *Treasurer* reported the receipts, including the balance on hand last January, for the year $41.00, and the expenditures $27.14 leaving a balance in the Treasury of $18.75. The whole amount of cash received by the society during 20 years was $1268.86, and the expenditures the same, less the balance now in the Treasury.

Papers Read.

Prof. Stahr read an interesting paper on the swarming of the "Brown Ant," (*Formica Rufa*) early in the month of November last. This was some weeks later than the usual period of swarming, but their past season has been rather extraordinary for its mildness—causing many trees to re-blossom, and in some instances to bear a second crop of fruit. As there was no weather during which ants could not have swarmed at their usual period (August and September), the question might well arise, "Did the same colony swarm a second time, as the apples, pears, cherries, &c., blooms.

Prof. Rathvon read an illustrated paper on a species of *fungus* (*Coprinus Comatus*) which sprung up in his garden on the night of November 3d, 1881. This was another illustration of a retarded warm season, and the effect of such weather, among the subjects of the animal and vegetable worlds. The *fungus* alluded to was one of the edible species, and was fully ten inches in height.

Dr. M. L. Davis, the chairman of the Committee on the state of the Society," appointed at the last meeting, read an interesting paper, full of good practical suggestions relating to the welfare, the progress, and a greater efficiency of the Society, and the manner in which this could most probably be effected.

The secretary read a paper from Mrs. Gibbons on some of the peculiarities in voting on questions, brought before societies in Ireland, and on the continent of Europe. She observed that at a conference held last summer at Cologne, the president "generally or always" put the *affirmative* only. At a meeting of Teachers in Ireland, the chairman put the *negative* only, and if no one voted "no" he would declare the motion "passed unanimously."

Elections.

Mr. H. M. Herr was balloted for and unanimously elected an active member of the Society.

The annual election of officers resulted as follows: *President*, Prof. J. S. Stahr; *Vice Presidents*, Profs. T. R. Baker and J. H. Dubbs: *Cor. Secretary*, Dr. Knight; *Rec. Secretary*, Dr. M. L. Davis; *Treasurer*, Prof. S. S. Rathvon; *Librarian*, Mrs. L. M. Zell; *Curators*, S. S. Rathvon, C. A. Heinitsh, Jno. B. Kevinski and Wm. L. Gill.

New Business.

Being the annual meeting, and reports and elections, occupying the time, no business other than the ordering of bills reported to be paid, was brought before the Society.

Extract from a Report on General Finance.

"We have tabulated these financial statistics of the Society merely to show by comparison with other associated enterprises in the city and county of Lancaster, what a little wheelbarrow we have been pushing forward during the last twenty years, when according to the magnitude of the subject, we should have been enabled to drive a "six-horse Conestoga Team.

"It would be quite safe to say, that such a collection as the society possesses, could not now be made for two times the amount it has cost us: and this *fact* should stimulate a desire for its preservation and perpetuation, among the intelligent and moneyed citizens of Lancaster: for, extinguish this institution and its museum, and such another nucleus could not be formed again for fifty years to come. No future scientists would feel encouraged to begin such a work again. The public seems to have very little comprehension of its magnitude, especially since more than one-half of it is necessarily packed away in drawers and boxes. Indeed, there are people who seem to think, that we are in some way, peculiarly enriching ourselves."

After a very pleasant meeting, and a general interchange of sentiment, the society adjourned to meet on the last Saturday in January, 1882.

Now is the time to subscribe for THE FARMER for 1882. Subscription price only $1 per year, the cheapest Agricultural Journal in the country.

AGRICULTURE.

Look After the Implements.

As winter approaches we cannot refrain from saying that the careful, thoughtful farmer never allows his plows, harrows, cultivators, mowing and reaping machines, hay tedders and other implements and machinery, to be exposed to the weather, or where they can be damaged by fowls or stock. He provides a covered place for them all where the rains and snows cannot penetrate, with either board flooring or placed upon scantling to raise them from the ground. Such portions of the iron likely to rust should be painted over slightly with any cheap oil paint, and it will add to the preservation and appearance of all implements and machinery, especially if the woodwork is also painted. When this is inconvenient the iron should be cleaned of dirt and greased with pieces of fat pork. They should also be put in good repair during the winter, in order to be ready for operating when needed in the spring. Leaving this repairing until another season opens frequently causes damaging delays which should always be provided against. There is nothing like being always ready with these things for any emergency.—*Germantown Telegraph.*

Do Your Own Repairing.

We think that almost every farmer will agree with us that every farm should have its own workshop, and every cultivator of the land should understand how to use it. He may not do so when he first enters upon farming on coming of age; but after a year or so of what we call apprenticeship, when he finds that to "know how to do things" is absolutely indispensable he will rapidly learn to attend to most of his own repairing of the ordinary implements and machines upon his premises, instead of incurring delay, expense and uncertainty by depending upon professionals at a distance. Rather than to be without a workshop and the necessary tools, one should be erected expressly for the purpose, in a convenient spot and daily warmed in winter, so as to be ready at all times for use, in which many odd jobs can be done also not immediately connected with the farm.

All ordinary wooden repairing ought to be done by the farmer and his hands during rainy days and in winter, when there is plenty of time on hand for that purpose. Every part of a wheelbarrow, except the wheel, ought to be made on the premises; new forks and handles of iron rakes, repairing even some portions of the farm machinery, building of garden and yard fences, repairing roofs, building of corncribs, hog-pens, wagon and cart shelvings, making of the frames of hotbeds, and all the many jobs constantly requiring to be done about a well-conducted place too numerous too mention. A person becomes very handy in the use of good tools after a short experience, and saves many a dollar without consuming any time necessary for the usual demands of the farm.

Ensilage Solus.

There has been of late considerable falling off in the talk about silos and the value of ensilage as a separate food. At the beginning of the mania the preserved cornfodder in its perfectly fresh, green state was to accomplish everything unassisted. Milk, butter and cheese were to be produced, condition of the cattle maintained, and health secured solely by the feeding of ensilage; and, altogether, it was to be effected at a rate of economy that must satisfy every one at short notice that this newly-discovered method of making the most out of the products of the earth at the least expense must commend itself to the favorable attention of every agriculturist.

But has it done so? We need hardly say that it has not. Ensilage by itself, as a food for even milch cows, is not recommended by those who seem to be mostly experienced in the use of it. Almost all extensive feeders employ at the same time other feed, which takes away from the fresh fodder its distinctive features or qualities as a separate food. One farmer says the fodder comes out of the silo in good condition and is eaten up clean by the cattle; but, he "mixes with good cut hay," which is given in two feeds per day; but to secure proper results "some concentrated feed must be added," such as cakemeal bran, etc. And this is the way the question is now treated. We don't pretend to say that this combined food is not very good—excellent—and that cattle will not give plenty milk and thrive upon it, but we beg to be allowed to say, without being much abused for it, that we doubt the economy involved, or that any labor is saved, or that any profits are obtained over the system in vogue before a silo was ever built.

Bad Seed.

It should be remembered that it is easier to deteriorate a crop by choosing bad seed, or even by carelessly neglecting the selection of good seed, than it is to improve upon a variety already acknowledged to be good. The down hill road is the easiest traveled.

HORT:CL TURE.

Rosebushes.

A correspondent of the New York *Observer* says: Never give up a choice but decaying rosebush till you have tried watering it two or three times with soot tea. Take soot from a chimney or stove with which wood is burned, and make tea of it. When cold, water the rosebush with it. When all is used, pour boiling water a second time on the soot. The shrub will quickly send out thrifty shoots, the eaves will become large and thick, and the blossoms will be larger and more richly tinted than before. To keep the plants clear of insects syringe them with quassia tea. Quassia chips can be obtained at the apothecary's.

Pears.

The pear as a fruit stands next in popularity to the apple, and has, like it, been known and cultivated from time immemorial. It is mentioned by the earliest writers as a fruit growing abundantly in Syria, Egypt, as well as Greece, and it appears to have been brought into Italy from these places about the time that Sylla made himself master of the latter country (68 B. C.,) and from thence it spread over Europe to Britain. Homer mentions the "pendant pear" as one of the fruits of the orchard of Laertes (Odys. 24 C. 2891). Theophrastus often speaks in praise of them and of the great productiveness of old pear trees in his works. That learned physician of ancient times, Galen, considered pears as containing in a greater degree more strengthening and astringent virtues than apples. The Greeks and Romans have several kinds of pears whose names included their taste and form. Pliny describes about forty varieties cultivated in Italy. Of all pears, he says, the Crustumine is the most delicate and agreeable; this fruit Columella places first in his catalogue. Then there was the Falernian pear, which was esteemed for its abundant juice, which Pliny compares to wine.

The Tiberian pears were so named because they were the sort Tiberius, the emperor, preferred, and they grew to a larger size than most pears; others were named after the persons who had introduced or cultivated them. Some, Pliny tells us, are reproached with the name of proud pears, because they ripened early and would not keep. There were also winter pears, pears for baking, etc., as in the present day. Nevertheless, Pliny did not consider this fruit, in an uncooked state, good for the constitution, for he states all pears whatever are but a heavy meat, even to those in good health, unless boiled or baked with honey, when they become extremely wholesome to the stomach. Some pears were used as counter-poison against venomous mushrooms; the ashes of pear trees were also used for the same medical purpose. The ancients appear to have had a curious notion respecting the effect of this fruit on beasts of burden, for Pliny tells us a load of apples or pears, however small, is singularly fatiguing to them. The best way to counteract this, they say, is to give the animal some to eat, or at least show them the fruit before starting. Virgil speaks of pears which he had from Cato.—*Science Gossip.*

Notes on Orchard and Garden Work.

One who depends upon the garden and orchard for his living will be very apt to know which products bring him the best returns. With the farmer the orchard and the garden are often looked upon as of little importance, if not regarded as necessary evils. Both manure and labor are grudgingly supplied and then at a time too late for the best results. In the general summing up of the business of the year, let the farmer take into account the return from the garden and orchard or fruit garden. We do not refer to the supplies of vegetables and fruit consumed at home, for health and comfort cannot be expressed in dollars and cents, but the actual money returns throughout the year. Much, of course, will depend upon the location in reference to market, but we are sure that in the majority of cases a carefully kept account, in which all the odd quarters and dollars are presented, will result in a determination to enlarge and improve the ground devoted to fruit, vegetables and flowers. The time has passed when choice fruits were regarded as a luxury; and the farmer who cannot afford to provide his table with a large variety of garden vegetables is living behind the age. The man who sees only the market value of any product of the soil may not care for a handsome lawn and a flower garden filled with choice plants; but he only half lives who is blind to the beauty of these things.

Making Butter.

The following method of making butter was pursued by the Farmington Creamery Company, Farmington, Conn., in the production of a premium lot: The milk was cooled and aerated before it came to the creamery, was received once a day, was mixed at once in a receiving vat, thence drawn into deep, open coolers, 8 by 20 inches, and set floating in a pool of cold spring water. It was skimmed in twenty-four hours, the cream again mixed in a vat and allowed to stand twenty-four hours and become slightly acid. It was then churned in a barrel-churn, and dashed running about forty-eight strokes to the minute, and the butter coming in about forty five minutes. The butter was worked by a lever worker and salted one ounce to the pound. After standing twenty hours it was again worked over and packed in tubs. If our butter has any particular merit we ascribe it largely to the cows, which are mostly Guernseys and Jerseys and their grades; the Guernseys giving the color, the Jerseys the dryness, and both solidity.

How to Make Tea and Coffee.

The Scotch do not say "to make tea," but "to *infuse* the tea," which is more correct in every respect. Good tea is an infusion, not a decoction. By *boiling* the tea leaves, you get from them a bitter principle, and you drive off the delicate perfume of the tea. For this reason, the tea-pot should never be kept hot by letting it stand on the top of a cooking stove, over a lamp, or where it is likely to be made to boil. Excessively bad tea is made in some parts of the continent by people who do not know better, by putting a small pinch of tea into a large kettle of water, and letting it boil till they have extracted all its *coloring* matter, in which they think the goodness of tea consists. A metal teapot is better than an earthenware one, and the brighter it is kept the better the tea. Rinse the teapot with boiling water. Put in a heaping spoonful of tea for each person, and one for the pot. Pour over it just enough boiling water to soak the tea. Let it stand a few minutes, and then fill up the pot with boiling water. Do not put in carbonate of soda to soften the water and make the tea draw better—*i. e.*, to make a wretched saving of tea, unless you are in absolute poverty. The water, in fact, is softened by boiling, which causes it to deposit some of the matters it held in solution; witness the "fur" in long used tea kettles, and the lime which settles at the bottom of many waters after boiling. A cup of tea is an excellent thing after any fatigue, and its refreshing effects may then be followed up by more substantial nutriment.

Coffee in English middle class houses is often badly served. It should not be *boiled*, nor made in quantity twice a week, to be heated up when wanted. The kernels should be sufficiently and equally roasted. As it is the roasting which develops the aroma, under roasted coffee is so much lost, whilst over roasted is much driven off and wasted or lost in another direction. Of the two faults the former is the worst. Unroasted coffee is useless. Circumstances very often compel the buying of coffee ready ground, almost always ready roasted; but the more quickly coffee is used after both roasting and grinding the better. It is only a healthy amusement to give a coffee mill a few turns. Coffee is easily roasted at home (it should be done in the open air) in an iron cylinder or barrel of small diameter, standing on two feet, over a coke and cinder, or, better, a charcoal fire, turned by a handle like that of a grindstone.

If you make the coffee in a biggin, put into the filter a good dessert spoonful for every person, and first of all only pour a few spoonfuls of boiling water sufficient to soak it, and after letting it stand in a warm place for a quarter of an hour, then pour on the rest of the boiling water, and let it percolate. The time to take coffee is either in the morning (with milk mixed in due proportion) or after lunch or an early dinner. In the evening it is to be avoided, unless you intend, like Lady Macbeth, to "murder sleep;" for which you are sure to be punished next morning.

Butter Easily Spoiled.

A farmer's wife writes: Of all the products of the farm butter is the most liable to be tainted by noxious odors floating in the atmosphere. Our people laid some veal in the cellar, from which a little blood flowed out, and was neglected until it had commenced to smell. The result was that a jar of butter we were packing smelled and tasted like spoiled beef. We know of an instance where there was a pond of filthy, stagnant water a few hundred feet from the house, from which an offensive effluvium would be borne on the breeze directly to the milk-room when the wind was in a certain direction, the result of which was that the cream and butter would taste like the disagreeable odor coming from the pond. As soon as the pond was drained there was no more damaged butter. It is remarkable how easily butter is spoiled.

HOUSEHOLD RECIPES.

LIGHT GINGERBREAD.—Three cups of flour, one of sugar, one of butter and one of molasses; three eggs beaten light, one tablespoonful of ginger, one teaspoonful of pearlash and some cloves. Beat the butter in sugar as for pound cake, then add other ingredients, putting in the pearlash last. Bake them in cake tins.

COCOANUT CAKE.—Take the whites of eight eggs beaten to a froth, one half cup of butter stirred to a cream, half cup sugar, half cup sweet milk, two and a half cups sifted flour, teaspoonful cream of tartar, half a teaspoonful of soda. Make of this three flat cakes, bake on pie tins and while warm spread with icing, and grate on cocoanut between each cake.

CHOCOLATE CAKE.—Take the yolks of ten eggs, and use just the same quantity of everything as you did for the cocoanut cake, grating chocolate upon the icing between each cake. The whites of two eggs beaten till they will not slide from the plate, and enough pulverized white sugar to make it very thick, will make enough icing for one cake.

ROCK CAKE.—The whites of four eggs beaten very light, one pound of loaf sugar added to them, three-fourths of a pound of sweet almonds slightly bruised. Bake on paper in tins.

GINGERBREAD.—Three pounds of flour, one pound of butter, half pound sugar, quart of molasses, two ounces of ginger, one ounce of cinnamon, ounce of allspice, an ounce and a half of cloves. Wash before baking with molasses and water.

ENGLISH BUNS.—One pound of flour, half pound of sugar, quarter pound butter, same of cinnamon, half pint of raisins; rub them all together and mix with milk and four or five drops of pearlash. Wash them after they are baked with sugar and water.

ALMOND CAKE.—One pound of sugar, half pound of flour, ten eggs, ounce of bitter almonds, a glass of rose water; beat the yolks till they are quite a batter, then add the sugar and beat it well; having previously pounded the almonds fine in the rose water, add them to the yolks; the whites must be beaten very light and then add the flour just stirred into the other ingredients. Bake an hour and ten minutes in rather a quick oven.

MILK BISCUIT.—One quart of milk, pound of butter, enough flour to thicken it, and a small teacupful of yeast; set them to rise early in the morning.

SOFT GINGERBREAD.—Six cups of flour, two of sugar, two of butter, two of molasses, two of milk, four eggs, a tablespoonful of ginger and a little allspice; beat the butter, sugar and eggs light, then stir in the other ingredients. Add a teaspoonful of pearlash dissolved in vinegar.

DOUGHNUTS.—Three pounds of flour, 1½ pounds of sugar, one pound of butter, six eggs, two wine-glasses good yeast, mix them with milk to a paste, set it to rise, shape them and fry in lard.

RAISIN PIE.—Ingredients: Raisins, one pound; lemon, one; white sugar, one cup; flour, two tablespoonfuls. Boil the raisins covered with water an hour; add the lemon, sugar and flour. Will make three pies.

CORN BREAD.—Three cups of cornmeal, one and one half cups of flour, one and one half cups of sweet milk, five eggs, four teaspoonfuls of baking powder, a little sugar. *Another Way:* One cup of cornmeal, two cups of flour, one half cup of sugar, three fourths of melted butter, one cup of milk, three eggs, three teaspoonfuls of baking powder.

COCOANUT PUDDING.—Ingredients: Milk, three pints; fine bread crumb, one teacup; cocoanut, one teacup; eggs, six; sugar, one teacupful; rind of lemon, one. Soak the bread crumbs for two hours in a pint and a half of the milk, and the grated meat of the cocoanut also; then add the well beaten eggs and the lemon rind grated, the sugar the rest of the milk. Stir well and bake. Do not let it remain long enough in the oven to become watery.

BAKED SOUP FOR INVALIDS.—This recipe is of use for invalids; it is easy to make, and cooks cannot well blunder. Take a pound of juicy steak from which all the fat has been removed, cut it up in pieces of about an inch square; salt and pepper it slightly, take a stone jar to hold two pints; pour into it a pint and a half of cold water, a teaspoonful of whole rice; cover the jar with a saucer and let it bake slowly for four hours; remove any fat present.

BAKED INDIAN PUDDING.—One quart of milk, one cup of molasses (best) one teaspoonful of salt, one quarter pound suet chopped fine, half teaspoonful powdered cloves and allspice together. Let milk come to a boil and stir in cornmeal enough to make it the consistency of thin batter, add suet and salt, stirring constantly to prevent its becoming lumpy; remove from the fire and let it become partially cool then stir in the molasses and cloves and allspice. Pour into into an earthen baking dish and bake moderately three fourths of an hour.

LIVE STOCK.

The Care of Cows.

The comfort of the cow has much to do with the quality of her milk. In hot weather the annoyance produced by flies and excitement caused by fighting them make the night's milk still poorer than it otherwise would be. Chemical analysis has shown a great falling off of fat of the milk in the same cow when chased by a dog. Any unusual excitement of the cow affects the fat in her milk. Extremes of heat and cold also affect the milk. In a case where cows went into a stream in hot weather, and stood several hours in the water above the knees, there was a falling off of the butter product from the same quantity of milk. This is accounted for by the extra food required to keep up the animal heat being carried off by the water. When we consider the fact that milk is secreted from the blood, we can readily see the effect that must be produced by excitement on the nervous system of the cow. In a case occurring in the city of Albany, N. Y., where a nervous cow was milked by a passionate man, who whipped and otherwise ill-treated her at milking, the milk was given to a child who had been healthy, but after using the milk, became ill and suffered from intestinal irritation, followed by a fever which seemed to affect the brain and nervous system. This illness was placed directly to the milk of the ill-treated cow. —*National Live Stock Journal.*

Charcoal for Sick Animals.

In nine cases out of ten when an animal is sick, the digestion is wrong. Charcoal is the most efficient and rapid corrective. The hired man came in with the intelligence that one of the finest cows was very sick, and a kind neighbor proposed the usual drugs and poisons. The owner being ill and unable to examine the cow, concluded that the trouble came from overeating, and ordered a teaspoonful of pulverized charcoal to be given in water. It was mixed placed in a junk bottle, the head turned upward and the water turned downward. In five minutes improvement was visible, and in a few hours the animal was in the pasture quietly grazing. Another instance of equal success occurred with a young heifer which had become badly bloated by eating green apples after a hard wind. The bloat was so severe that the sides were as hard as a barrel. The old remedy, saleratus, was tried for correcting the acidity. But the attempt to put it down always raised coughing and it did little good. Half a teacupful of fresh powdered charcoal was given. In six hours all appearance of the bloat had gone, and the heifer was well.

Hints About Horses.

Oats should be bruised for an old horse, but not for a young one, because the former, through age and defective teeth, cannot chew them properly. The young horse can do so, and they are thus properly mixed with saliva and turned into wholesome nutriment. There is no nourishment in bad hay, and cheapness should never tempt you to use it. Damaged corn is also exceedingly injurious. Sprinkle hay with salted water. It is more easily digested. For a saddle or coach horse half a peck of sound oats and eighteen pounds of good hay are sufficient. If the hay is not good add a quarter of a peck more oats. A horse which works harder may have rather more of each; one that works less should have less. Rack feeding is wasteful. The better plan is to feed with chopped hay from a manger, because the food is not then thrown about, and is more easily chewed and digested.

Hay for Swine.

In the opinion of an exchange hay is very beneficial to swine. Swine need rough food as well as horses, cattle or the human race. To prepare it you should have a cutting box or hay cutter, and the greener the better. Cut the hay as short as oats, or shorter, and mix with bran shorts or middlings and feed as other food. Hogs soon learn to like it, and if soaked in swill or other slop food it is highly relished by them. In winter use for the hogs the same hay you feed for horses and you will find that, while it saves bran or other food, it puts on flesh as rapidly as anything that can be given them. In summer the use of hay can be commenced as early as the grass will do to cut, and when run through the cutting box can be used to advantage by simply soaking in fresh water until it sours.

Warts on Horses.

These fungous growths appear in the horse most frequently about the mouth, nose and lips, but they are occasionally found upon other parts of the body. They are sometimes found in large numbers about the lips of colts, and are generally rubbed off, or drop off. If, however, they grow large and become deeply rooted, they may be cut off by passing a needle through the center, armed with double thread and tied tightly around the neck on each side. This prevents the possibility of the ligatures being rubbed off. Or they may be painted over with the permanganate of potash, a few applications of which will entirely destroy warts of a large size, or they may be removed with a knife.—*Jennings.*

The Horse Shoe and its Application.

The number and disposition of the nails depend upon the kind of shoe. For speed the light draft, from five to seven may be employed, while for heavy horses and for heavy draft the number may be increased. Where few nails are used they should be more widely distributed than is usually the custom. When it is remembered that the introduction of every nail is so much injury to the structures of the foot, it will readily be seen that the smaller the number requisite for the purpose the better for the animal. In driving the nails, it is essential that a thick short hold of the crust should be had rather than a long thin one. Not only is the shoe thus held more firmly, but there is a probability that the nail holes may, by the downward growth of the horn, be removed at the next shoeing, which in most cases should not exceed an interval of four or five weeks. The points of the nails should be shortened to just that length which will permit them to be turned over and hammered down smoothly, with perhaps the least possible rasping. The common method of rasping notches for the extremities of the nails is not advisable. In fact, as I have already said, the rasp should never be used upon the external walls of the hoof except in cases of absolute necessity to prevent striking of the opposite limb. Its use destroys the natural polish, exposes parts beneath, which are not fitted for such exposure, renders the horn brittle, and liable at any moment to quarter cracks and other maladies.

LITERARY AND PERSONAL.

THE GUARDIAN, a monthly magazine for young men and women, Sunday-schools and families. Edited by Rev. J. H. Dubbs, D.D. It is a long time since we have seen the face of this "old familiar," which, if we recollect rightly, originated here in the city of Lancaster, under the editorial auspices of the late Rev. Henry Harbaugh, more than thirty years ago, and for a time was also printed here.

That the *Guardian* should have been permitted to exercise its vigilant functions for such a long period without break or interruption, evinces that it has been faithful to its trust, or has had a cordon of sustaining and indulgent friends. No. 1, Vol. 33, (January 1882) of this excellent publication has found its way to our table; and we scan it with more than ordinary interest, not only on account of old memories, (for some years it was on our exchange list) but also for its healthy tone, its undoubted moral and intellectual attitude, and its continued editorial ability. It is an octavo of 36 pages, in tinted covers, and is issued by the *Reformed Church*

Publication Board, 907 Arch Street, Philadelphia, Pa., at the very low price of $1.25 per year in advance.

We congratulate the board in securing an editor so worthy of being the successor of such distinguished predecessors as Doctors Harbaugh and Bausman. We feel confident that the *Guardian* will lose none of its "*Life, Light, Love,*" under the editorial management of Dr. Dubbs; and therefore we heartily commend it to the favorable consideration of our readers, whatever their religious faith may be.

Faithful to its motto, it can inculcate nothing that will be detrimental to that spiritual rest for which our frail humanity is yearning, in the eternal world.

ADDRESS OF HON. GEO. B. LORING, Commissioner of Agriculture, and other proceedings of the COTTON CONVENTION held at Atlanta, Georgia, November 2, 1881. Uniform in size and mechanical execution with the Serial Bulletins of the Department of Agriculture—pp. 36.

We are under obligations to Prof. C. V. Riley, for a copy of this valuable contribution to the Agricultural and Entomolgical Literature of the Country, as developed through cotton culture and its protection from the ravages of noxious insects.

Practical or applied entomology certainly means *something* in its relations to the general crops of our diversified country, however insignificant it may seem in favored localities. Prof. Riley says—"Whenever we begin to carefully estimate the losses which, as a nation, we sustain from insect ravages, the figures always startle, and you will doubtless be surprised to learn that they reached in a single year nearly $400,000,000." This *estimate* is just as likely to fall far below the real amount of damage, as it is to reach beyond it, but under any circumstances, who among our readers can practically comprehend this amount in detail. *Ten hundred thousand dollars*—or one million—seems to be a vast amount, dissipated annually through the instrumentality of insects, and even this amount, to a man accustomed to labor at 75 cents a day, cannot be fully comprehended. Prof. Riley's remarks before the convention aforesaid includes among other things—methods of counteracting injurious insects—the cotton worm—natural history of the cotton worm—improved appliances—poisoning from below, &c., exemplified by a detailed context, and only requires a vigilant and intelligent co-operation to produce the desired effect. It is not sufficient that we *know* what to do, but that we *do* it. All success lies in that.

REPORT OF THE COMMISSIONER OF AGRICULTURE, for the year 1881, 58 pp.—uniform with the above. This report contains concise statements of work in the various divisions of the Agricultural Department, including—Division of Garden and Ground—Botanical Division—Microscopical Division—Chemical Division—Entomological Division—Seed Distribution—(1,325,922 papers of Vegetable Seed were distributed from July 1, 1880 to June 30, 1881). Statistical Division—Forestry—Artesian Wells—Agriculture on the Pacific Slope—Examination of Wools and Animal Fibers—Grape Culture and Wine-Making—Manufacture of Sugar from Sorghum—Tea Culture—Contagious Diseases of Domesticated Animals—Sugar from Beets, and the Operations of the Disbursing Office. It appears that Congress has only appropriated $195,300, for all the divisions of the department for the year ending June 30, 1881, which seems small, compared with the subject of Agriculture—the basis of all the other interests in the country. A copious appendix is attached to this report, containing communications from competent authorities, on contagious Pleuro-pneumonia, and Foot-and-mouth disease, and matters relating thereto.

THE AMERICAN FARMER, No. 1, Vol. 1, series 9, comes to us a four-columned royal quarto of 16 pages, and henceforth is to be published semi-monthly, at $1.50 a year, by Samuel Sands & Son, W. Baltimore street, Baltimore, Md. The *American Farmer*—heretofore published in octavo form—never occupied an inferior position in the world of agricultural literature, and its patrons and the public may rest fully assured that it takes no retrogressive step in its "new departure;" but, on the contrary, if we are at all competent to judge from the clean, neat, and mechanically executed journal before us, a very perceptible step forward. The *Farmer* is a veteran in the journalistic enterprises of the country, and its senior editor is one of that distinguished band of patriarchs who have devoted their energies to the dissemination of useful knowledge. The *Farmer* is one of the oldest—if not the very oldest—agricultural journals published in the country, and, therefore, as a guarantee of the future, it can refer with confidence to the past, for no journal could have sustained itself so long without possessing unquestionable merit. We tender our holiday greetings, and wish it a *happy New Year*, commending it to its patrons and an appreciative public.

LANDRETH'S RURAL REGISTER AND ALMANAC, published annually for gratuitous distribution, 1882. This is a royal octavo of 82 pages, in colored paper covers, and amply illustrated with accurate figures of the vegetable productions, the raising the seeds of which the publishers make a specialty. The bird's eye view of the central portion of Bloomsdale farm will give a tolerable idea of the magnitude of the concern, from which it will be perceived that twenty-one buildings are required for its successful operation, and additional ones are projected. Send for the *Register* by all means.

THE PENNSYLVANIA FARMER, a demi-folio of 16 pages, good paper and fair print. Published monthly at Mercer, Pa., at $1 a year in advance by F. H. Umholtz, editor and proprietor. No. 1, Vol. 1 of this excellent journal is now before us, and is an able representative of the interests it specializes—"Farm, Field, Garden and Home."

THE SCIENTIFIC TIMES, a weekly record of American progress in science, art, finance, commerce and manufactures. This is an old caterer for the farmers and artizans of the country. It is finely illustrated, and no doubt it is crowned with merited success.

ST. LOUIS MILLER, a semi-monthly journal, devoted to the interests of the milling trade. In the absence of any other evidence, this journal alone implies that St. Louis is a very large village, and has a very large grain and flour trade. A copy of No. 3, Vol. 7, (January 6, 1882) has found the desk of our *sanctum*, and remembering that just forty-seven years ago we sojourned a month in the city of St. Louis, the presence, the magnitude and the general make up of the journal before us, impresses us with the immense progress the city must have made since 1836 when her population was ten thousand less than Lancaster is to-day. But the *Miller*—it may be called a demi-folio (17 by 13) of 16 pages and has five closely printed columns to the page—printed on calendered paper, and profusely illustrated with all kinds of new and improved machinery pertaining to mills and milling. Its pictorial advertisements alone cannot but be interesting to any one having "half an eye," or half an idea, on the subject of mechanics.

Its 80 columns of reading and advertising matter relate almost exclusively to the grain and flour trade, and collaterals appertaining to that trade, (only in one little corner do we find the "humbug" artificial ear-drums advertised, which had better been filled with "beans.") Nobody, certainly, ought to starve in St. Louis for the want of bread, at least.

The receipts of flour for the year 1881 bbls	1,559,691
Shipments of flour for the same time .	2,619,529
Receipts of wheat in bushels was .	11,619,741
Shipments of wheat for the same time .	6,891,080
Receipts of corn for the same time .	20,219,310
Shipments of corn for the same time .	14,451,900
Receipts of oats for the same time .	5,950,900
Shipments of oats for the same time .	3,108,386
Receipts of rye for the same time .	419,914
Shipments of rye for the same time .	803,490
Receipts of barley for the same time .	2,327,632
Shipments of barley for the same time .	182,219

The last two items may illustrate a large consumption of *Barley* in St. Louis. If that is a *fact*, it has its redeeming quality in the significant other *fact*, that the quantum of *Rye* was comparatively small. The *St. Louis Miller* is published by Thomas & Stone, and as above indicated, is a semi-monthly, at $2.00 a year, or $1.25 for six months; and every intelligent and progressive miller ought to be a subscriber.

ANNUAL REVIEW OF THE APPLETON POST—*Appleton, Wisconsin, Thursday, December 29th*, 1881.—This is a folio (15 by 21) of 24 pages, and 6 columns to the page, abounding in interesting historical, statistical, geographical, and local matter. The quality and make-up of the paper are excellent, and the numerous illustrations up to the modern standard. Accompanying the whole is an extra sheet 25 inches square, containing on one side a map of Ledyard, Wisconsin, scale 200 ft. to 1 inch, and on the other side, a map of Outagamie county, Wis. The illustrations mainly relate to the city of Appleton, and are, Second Ward High School; St. Joseph's Church and school Buildings; Lawrence University; the Ravine looking west from foot of Prospect street; Fourth Ward Ravine, near the upper dam; a double page map of the city of Appleton; Memorial Presbyterian Church; College Avenue, looking east; College Avenue, looking west from Durkee street; First National Bank Building; full page bird's-eye view of the city of Appleton; Appleton Water Powers, Nos. 1, 2 and 3: Upper Dam; New Courthouse; a view of the city from University Dome; Marston & Beveridge's Hub and Spoke Factory; the new Brewery; besides sundry smaller and personal illustrations. Of course, we know this is all to give the city of Appleton "a lift" in her competitive progress with other progressive towns in the "great west," and no one can find fault with this; for, if people who own a town and live in it, do not put their hands to the wheel and help it forward, they can hardly deserve success, as things now go in this nether world. The water power of Appleton from this showing must be immense, for it seems to be literally a dammed—oh, a city of dams; which, in these days of fire and explosion, is a matter of vast importance. We are indebted to Mr. Mike K. Gochenaur, formerly of this county, for a complimentary copy of this annual number of the *Post*, and we commend the enterprising manifestations of that far off town to the favorable consideration of our patrons and readers. In looking over the 144 columns of the choice reading matter of this lively journal, we feel our local Old-fogyism the more impressive, notwithstanding our efforts, in later years, to move onward.

BOARD OF AGRICULTURE.

The annual meeting of the State Board of Agriculture will be held in Harrisburg, commencing Wednesday, January 25, at 2 p. m. The following is a full list of subjects of essays and discussions, furnished by Secretary Thomas J. Edge, from whom all information may be had:

Treatment and management of Dairy Cows, Hon. C. C. Musselman, of Somerset.

Agriculture of the Old and New World, F. Jackel, of Blair.

The Common Law and Statutes of Pennsylvania Regulating Surface and Underground Water Courses between Land Owners, Hon. M. C. Beebe, of Venango.

Lessons of 1881, and the Outlook for 1882, E. Reeder, Bucks.

Associated Dairying, John I. Carter, of Chester co.

Farmers' Gardens and Truck Patches, Rev. J. Calder, Harrisburg.

Preparation of the Ground for Wheat, J. D. Lyle, Butler.

Is the Importation of Foreign Live Stock an Advantage to the Pennsylvania Farmer? A. D. Shimer of Northampton.

Scenography in Agriculture, H. C. Demming, of Harrisburg.

The Best method of Fire Insurance for Farm Buildings and their Contents, Henry C. Tyler, of Susquehanna.

Weeds and their Eradication, Col. D. H. Wallace, of Lawrence.

Production and Preservation of Apples, J. Miles, of Erie.

During the evening sessions or at other times, at the option of the Board, addresses will be delivered on the following topics:

The Relation of the Soil and Crops to Heat and Moisture, Prof. W. H. Jordan, Pennsylvania State College.

Agricultural Education, Prof. S. B. Heiges, of York.

On a subject not assigned, by Col. Frank Mantor, of Crawford.

The above programme will not be strictly adhered to, as other topics will probably be introduced by members of the Board. Any question of a proper nature, if handed to the secretary, will be referred to a suitable person for answer.

Important to Grocers, Packers, Hucksters, and the General Public.

THE KING FORTUNE-MAKER.

OZONE

A New Process for Preserving all Perishable Articles, Animal and Vegetable from Fermentation and Putrefaction, Retaining their Odor and Flavor.

"**OZONE—Purified air, active state of Oxygen.**"—*Webster.*

This preservative is not a liquid pickle, or any of the old and exploded processes, but is simply and purely OZONE, as produced and applied by an entirely new process. Ozone is the antiseptic principle of every substance, and possesses the power to preserve animal and vegetable structures from decay.

There is nothing on the face of the earth liable to decay or spoil which Ozone, the new Preservative, will not preserve for all time in a perfectly fresh and palatable condition.

The value of Ozone as a natural preserver has been known to our abler chemists for years, but, until now, no means of producing it in a practical, inexpensive, and simple manner have been discovered. Microscopic observations prove that decay is due to septic matter or minute germs, that develop and feed upon animal and vegetable structures. Ozone, applied by the Prentiss method, seizes and destroys these germs at once, and thus preserves. At our office in Cincinnati can be seen almost every article that can be thought of, preserved by this process, and every visitor is welcomed to come in, taste, smell, take away with him, and test in every way the merits of Ozone as a preservative. We will also preserve, free of charge, any article that is brought to our office to us, and return it to the sender, for him to keep and test.

FRESH MEATS, such as beef, mutton, veal, pork, poultry, game, fish, &c., preserved by this method, can be shipped to Europe, subjected to atmospheric changes and return to this country in a state of perfect preservation.

EGGS can be tied at a cost of less than one dollar a thousand dozen, and be kept in ordinary rooms six months or more, thoroughly preserved, the yolk held in its normal condition, and the eggs as fresh and perfect as on the day they were treated, and will sell as strictly "choice." The advantage in preserving eggs is readily seen; there are seasons when they can be bought for 8 or 10 cents a dozen, and by holding them, can be sold for an advance of from one hundred to three hundred per cent. One man, with this method, can preserve 5,000 dozen in a day.

FRUITS may be permitted to ripen in their native climate, and can be transported to any part of the world. The juice expressed from fruits can be held for an indefinite period without fermentation—hence the great value of this process for producing a temperance beverage. Cider can be held perfectly sweet for any length of time.

VEGETABLES can be kept for an indefinite period in their natural condition, retaining their color and flavor, treated in their original packages at a small expense. All grains, flour, meal, etc., are held in their normal condition.

BUTTER, after being treated by this process, will not become rancid.

Dead animal bodies, treated before decomposition sets in, can be held in a natural condition for weeks, without puncturing the skin or mutilating the body in any way. Hence the great value of Ozone to undertakers.

There is no change in the chemical particular in the article thus preserved, and no trace of any foreign or unnatural odor or taste.

The process is so simple that a child can operate as well and as successfully as a man. There is no expensive apparatus or machinery required.

A room filled with different articles, such as eggs, meat, fish, etc., can be treated at one time, without additional trouble or expense.

In fact, there is nothing that Ozone will not preserve. Think of everything you can that is liable to sour, decay, or spoil, and then remember that we guarantee that Ozone will preserve it in exactly the condition you want it for any length of time. If you will remember this it will save asking questions as to whether Ozone will preserve this or that article—**it will preserve anything and everything you can think of.**

There is not a township in the United States in which a live man can not make any amount of money, from $1,000 to $10,000 a year, that he pleases. We desire to get a live man interested in each county in the United States, in whose hands we can place this Preservative, and through his agency the business which every county ought to produce.

A FORTUNE

Awaits any Man who Secures Control of OZONE in any Township or County.

A. C. Bowen, Marion, Ohio, has cleared $2,000 in two months. $2 for a test package was his first investment. Woods Brothers, Lebanon, Warren County, Ohio, made $6,000 on eggs purchased in August and sold November 1st. $2 for a test package was their first investment.

F. K. Raymond, Morgantown, Belmont Co., Ohio, is clearing $2,000 a month in handling and selling Ozone. $2 for a test package was his first investment.

D. F. Wilds r, Charlotte, Eaton Co., Mich., has cleared $1,000 a month since August. $2 for a test package was his first investment.

J. B. Gaybord, 80 La Salle St., Chicago, is preserving eggs, fruit, etc., for the commission men of Chicago, charging 1½c per dozen for eggs, and other articles in proportion. He is preserving 3,000 dozen eggs per day, and on his business is making, besides all his own salary, $2 for a test package was his first investment.

The Cincinnati Feed Co., West 4th St. south street, is making $5,000 a month in handling brewers' malt, preserving and shipping it as feed to all parts of the country. Mail unpreserved sours in 24 hours. Preserved by Ozone it keeps perfectly sweet for months.

These are instances where we have asked for the privilege of publishing. There are scores of others. Write to any of the above parties and get the evidence direct.

Now, to prove the absolute truth of every thing we have said in this paper, **we propose to place in your hands the means of proving for yourself that we have not claimed half enough.** The man who has no doubts any of these statements, and who is interested sufficiently to make the trip, we will pay all traveling and hotel expenses for a visit to this office, if he fail to prove any statement that we have made.

How to Secure a Fortune with Ozone.

A test package of Ozone, containing a sufficient quantity to preserve one thousand dozen eggs, or other articles in proportion, will be sent to any applicant on receipt of $2. This package will enable the applicant to pursue any line of tests and experiments he desires, and satisfy himself as to the extraordinary merits of Ozone as a Preservative. After having thus satisfied himself, and had time to look the field over to determine what he wishes to do in the future—whether to sell the article to others or to confine it to his own use, or any other line of policy which is best suited to him and to his township or county—we will enter into an arrangement with him that will make a fortune to him and give us good profits. We will give exclusive township or county privileges to the first responsible applicant who orders a test package and desires to control the business in his locality. The man who secures control of Ozone for any special territory, will enjoy a monopoly which will surely enrich him.

Don't let a day pass until you have ordered a Test Package, and if you desire to secure an exclusive privilege we assure you that delay may deprive you of it, for the applications come in to us by letter and mail—many by telegraph. "First come first served" is our rule.

If you do not care to engage in advance for the test package, we will send it (C. O. D.), but this will put you to the expense of charges for return money. Our correspondence is very large; we have all we can do to attend to the shipping of orders and giving attention to our working agents. Therefore we can not give any attention to letters which do not order Ozone. If you think of any article that you are doubtful about Ozone preserving remember we *guarantee that it will preserve it, no matter what it is.*

REFERENCES.

We desire to call your attention to a few references which no enterprise or firm based on any thing but the soundest business success and highest commercial merit could procure.

We refer, by permission, to the following gentlemen: Edward C. Hoye, Member Board of Public Works; E. O. Eshelby, City Comptroller; Amor Smith, Jr., Collector Internal Revenue; Wulsin & Worthington, Attorneys; Martin H. Harrell and B. F. Hopkins, County Commissioners; W. S. Cappellier, County Auditor; all of Cincinnati, Hamilton County, Ohio. These gentlemen are each familiar with the merits of Ozone Ozone and know from actual observation that we have without question

The Most Valuable Article in the World.

The $2 you invest in a test package, will surely lead you to secure a township or county, and then your way is absolutely clear to make from $2,000 to $10,000 a year.

Give your full address in every letter, and send your letter to

PRENTISS PRESERVING COMPANY, (Limited.)
S. E. Cor. Ninth & Race Sts., Cincinnati, O.

Nov-3m

THE LANCASTER EXAMINER

OFFICE

No, 9 North Queen Street,

LANCASTER, PA.

THE OLDEST AND BEST.

THE WEEKLY

LANCASTER EXAMINER

One of the largest Weekly Papers in the State.

Published Every Wednesday Morning,

Is an old, well-established newspaper, and contains just the news desirable to make it an interesting and valuable Family Newspaper. The postage to subscribers residing outside of Lancaster county is paid by the publisher.

Send for a specimen copy.

SUBSCRIPTION:

Two Dollars per Annum.

THE DAILY

LANCASTER EXAMINER

The Largest Daily Paper in the county.

Published Daily Except Sunday.

The daily is published every evening during the week. It is delivered in the City and to surrounding Towns accessible by railroad and daily stage lines, for **10 cents a week.**

Mail Subscription, free of postage—One month, 50 cents; one year, **$5.00.**

THE JOB ROOMS.

The Job rooms of THE LANCASTER EXAMINER are filled with the latest styles of presses, material, etc., and we are prepared to do all kinds of Book and Job Printing at as low rates and short notice as any establishment in the State.

SALE BILLS A SPECIALTY.

With a full assortment of new cuts that we have just purchased, we are prepared to print the finest and most attractive sale bills in the State.

JOHN A. HIESTAND, Proprietor,

No. 9 North Queen St.,
LANCASTER, PA.

WHERE TO BUY GOODS
IN
LANCASTER.

BOOTS AND SHOES.

MARSHALL & SON, No. 12 Centre Square, Lancaster, Dealers in Boots, Shoes and Rubbers. Repairing promptly attended to.

M. LEVY, No. 3 East King street. For the best Dollar Shoes in Lancaster go to M. Levy, No. 3 East King street.

BOOKS AND STATIONERY.

JOHN BAER'S SON'S, Nos. 15 and 17 North Queen Street, have the largest and best assorted Book and Paper Store in the City.

FURNITURE.

HEINITSH'S, No. 15½ East King st., (over China Hall) is the cheapest place in Lancaster to buy Furniture. Picture Frames a specialty.

CHINA AND GLASSWARE.

HIGH & MARTIN, No. 15 East King st., dealers in China, Glass and Queensware, Fancy Goods, Lamps, Burners, Chimneys, etc.

CLOTHING.

MYERS & RATHFON, Centre Hall, No. 12 East King St. Largest Clothing House in Pennsylvania outside of Philadelphia

DRUGS AND MEDICINES.

G. W. HULL, Dealer in Pure Drugs and Medicines Chemicals, Patent Medicines, Trusses, Shoulder Braces, Supporters, &c., 15 West King st., Lancaster, Pa.

JOHN F. LONG & SON, Druggists, No. 13 North Queen St. Drugs, Medicines, Perfumery, Spices, Dye Stuffs, Etc. Prescriptions carefully compounded.

DRY GOODS.

GIVLER, BOWERS & HURST, No. 25 E. King St., Lancaster, Pa., Dealers in Dry Goods, Carpets and Merchant Tailoring. Prices as low as the lowest.

HATS AND CAPS.

C. H. AMER, No. 39 West King Street, Dealer in Hats, Caps, Furs, Robes, etc. Assortment Large. Prices Low.

JEWELRY AND WATCHES.

H. Z. RHOADS & BRO, No. 4 West King St. Watches, Clock and Musical Boxes. Watches and Jewelry Manufactured to order.

PRINTING.

JOHN A. HIESTAND, 9 North Queen st., Sole Bills, Circulars, Posters, Cards, Invitations, Letter and Bill Heads and Envelopes neatly printed. Prices low.

Thirty-six Varieties of Cabbage, 26 of Corn, 28 of Cucumbers, 41 of Melon, 33 of Peas, 28 of Beans, 17 of Squash, 25 of Beet and 40 of Tomato, with other varieties in proportion, a large portion of which were grown on my five seed farms, will be found in my **Vegetable and Flower Seed Catalogue for 1882**. Sent FREE to all who apply. Customers of last Season need not write for it. All seed sold from my establishment warranted to be fresh and true to name, so far, that should it prove otherwise, I will refill the order gratis. The original introducer of **Early Ohio** and **Burbank Potatoes, Marblehead, Early Corn, the Hubbard Squash, Marblehead Cabbage, Phinney's Melon**, and a score of other New Vegetables, I invite the patronage of the public. New Vegetables a specialty.

JAMES J. H. GREGORY,
Marblehead, Mass.
Nov-3mo]

EVAPORATE YOUR FRUIT.
ILLUSTRATED CATALOGUE
FREE TO ALL.
AMERICAN DRIER COMPANY,
Chambersburg, Pa.
Ap1-tf

FARMING FOR PROFIT.

It is conceded that this large and comprehensive book, (advertised in another column by J. C. McCurdy & Co., of Philadelphia, the well-known publishers of Standard works,) is not only the newest and handsomest, but altogether the BEST work of the kind which has ever been published. Thoroughly treating the great subjects of general Agriculture, Live-Stock, Fruit-Growing, Business Principles, and Home Life; telling just what the farmer and the farmer's boys want to know, combining Science and Practice, stimulating thought, awakening inquiry, and interesting every member of the family, this book must exert a mighty influence for good. It is highly recommended by the best agricultural writers and the leading papers, and is destined to have an extensive sale. Agents are wanted everywhere. jan-tf

CIDER MILLS!
Wine Presses!

Fruit Presses, Apple Slicers,
Fodder and Ensillage Cutters,
GRAIN FANS,
Grain and Fertilizer Drills,
Broad-cast Seed Sowers,
Corn Shellers, Corn Mills,
Grain Mills, etc., etc.

FOR SALE BY
D. LANDRETH & SONS,
AGRICULTURAL AND HORTICULTURAL IMPLEMENT
AND
SEED WAREHOUSE,
Nos. 21 and 23 South Sixth Street,
BETWEEN MARKET AND CHESTNUT STS.,
— and —
No. 4 ARCH STREET,
apr-6m   PHILADELPHIA.

MERCHANT TAILORING.

1848 (The Oldest of All.) 1881
RATHVON & FISHER,
MERCHANT TAILORS AND DRAPERS,

respectfully inform the public that having disposed of their entire stock of Ready-Made Clothing, they now do, and for the future shall, devote their whole attention to the CUSTOM TRADE.

All the desirable styles of CLOTHS, CASSIMERES, WORSTEDS, COATINGS, SUITINGS and VESTINGS constantly on hand, and made to order in plain or fashionable style promptly, and warranted satisfactory.

All-Wool Suit from $10.00 to $30.00.
All-Wool Pants from 3.00 to 10.00.
All-Wool Vests from 2.00 to 6.00.

Union and Cotton Goods proportionately less. Cutting, Repairing, Trimming and Making, at reasonable prices.

Goods retailed by the yard to those who desire to have them made elsewhere.

A full supply of Spring and Summer Goods just opened and on hand.

Thankful to a generous public for past patronage they hope to merit its continued recognition in their "new departure."

RATHVON & FISHER,
PRATT HALL TAILOR'S,
No. 101 North Queen Street,
LANCASTER, PA.
1848   1881

GLOVES, SHIRTS, UNDERWEAR.
SHIRTS MADE TO ORDER,
AND WARRANTED TO FIT.
E. J. ERISMAN,
56 North Queen St., Lancaster, Pa.
75-1-12]

A HOME ORGAN FOR FARMERS.

THE LANCASTER FARMER,

A MONTHLY JOURNAL,

Devoted to Agriculture, Horticulture, Domestic Economy and Miscellany.

Founded Under the Auspices of the Lancaster County Agricultural and Horticultural Society.

EDITED BY DR. S. S. RATHVON.

TERMS OF SUBSCRIPTION:
ONE DOLLAR PER ANNUM,
POSTAGE PREPAID BY THE PROPRIETOR.

All subscriptions will commence with the January number, unless otherwise ordered.

Dr. S. S. Rathvon, who has so ably managed the editorial department in the past, will continue in the position of editor. His contributions on subjects connected with the science of farming, and particularly that specialty of which he is so thoroughly a master—entomological science—some knowledge of which has become a necessity to the successful farmer, are alone worth much more than the price of this publication. He is determined to make "The Farmer" a necessity to all households.

A county that has so wide a reputation as Lancaster county for its agricultural products should certainly be able to support an agricultural paper of its own, for the exchange of the opinions of farmers interested in this matter. We ask the co-operation of all farmers interested in this matter. Work among your friends. The "Farmer" is only one dollar per year. Show them your copy. Try and induce them to subscribe. It is not much for each subscriber to do but it will greatly assist us.

All communications in regard to the editorial management should be addressed to Dr. S. S. Rathvon, Lancaster, Pa., and all business letters in regard to subscriptions and advertising should be addressed to the publisher. Rates of advertising can be had on application at the office.

JOHN A. HIESTAND,
No. 9 North Queen St., Lancaster, Pa.

$5 TO $20 per day at home. Samples worth $5 free. Address STINSON & Co., Portland, Maine.
jun-1yr*

ONE DOLLAR PER ANNUM.—SINGLE COPIES 10 CENTS.

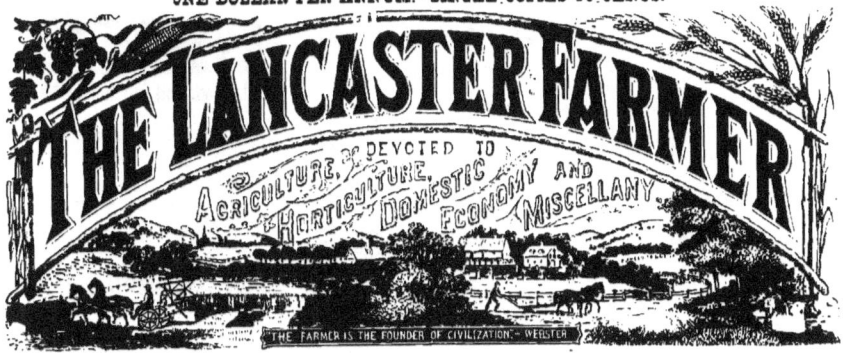

Dr. S. S. RATHVON, Editor.   LANCASTER, PA. FEBRUARY, 1882.   JOHN A. HIESTAND, Publisher.

Entered at the Post Office at Lancaster as Second Class Matter.

CONTENTS OF THIS NUMBER.

EDITORIAL.

The English Sparrow, ... 17
Ourselves, ... 18
February Snows, ... 18
Wood-worms, ... 18
Planting Trees on Railway Embankments, . 18
The Largest Tree in the World, . . . 19
Shifty, Thrifty, France, ... 19
Kitchen Garden for February, ... 20
Poultry Exhibition, ... 20
Rules and Exceptions, ... 20
  Hybrids Not Always Barren—A Breeding Mare Mule.
Writing for the Farmer, ... 20
Excerpts, ... 20
  Miscellaneous — Domestic — Scientific — Moral Economy — Historical-Statistical.

CONTRIBUTIONS.

The Egg—Its Contents and How it is Made, 22
Fruit Belts, ... 23
Chinese Fruit Pear, ... 23
Commercial Fertilizers, ... 24

SELECTIONS.

Poultry Show, ... 24
White Vein—Cause of the Disease in Tobacco, 26
  The Early Cutting Theory—Convincing Experiments.
Tobacco Growing—Profits Realized by Some Experts, ... 26
  Early Buying in the Field—Result of Careful Handling—An Excellent Crop—Cost of Growing Tobacco—Another Paying Crop—Still Another—In Conclusion.
American Silk Goods, ... 27
Coal Tar and Alkali in Peach Culture, . 28
Points in Cows, ... 28

OUR LOCAL ORGANIZATIONS.

Lancaster County Agricultural and Horticultural Society, ... 28
  Crop Reports—Growth and Consumption of Timber Trees in America—What Causes White Vein in Tobacco—Hard Growing Wheat—Dairy Cows and the Soiling System.
Poultry Association, ... 30
Fulton Farmers' Club, ... 30
  Exhibits and Answers to Questions—Examining the Hog's Farm—Literary Exercises.
Linnæan Society, ... 31
  Museum—Library—Distress—Locust Reach New Form.

AGRICULTURE

Planting Tobacco, ... 31
Improved Grasses, ... 31
Rotation of Crops, ... 31

HOUSEHOLD RECIPES.

Orange Pie, ... 31
New England Baked Indian Pudding, . 31
Chicken Pie, ... 31
Prune Pudding, ... 31

A Nice Way of Cooking Cold Meats, . 31
Chocolate Cake, ... 31
Breakfast Rusks, ... 31
Preparing Carrots, ... 31
Barley Soup, ... 31
Cornstarch Cakes, ... 31
French Tapioca Pudding, ... 32
Sweet Macaroni, ... 32
Oatmeal Pudding, ... 32
Wholesale Pie Crust, ... 32
Stewed Apples and Rice, ... 32
Literary and Personal, ... 32

REID'S CREAMERY
SIMPLEST & BEST
Agents Wanted

BUTTER WORKER
Most Effective and Convenient
Power Workers.
Capacity 10,000 lbs. per DAY
Price $6, $10, Stripping, Cooler, etc.
Send for Illustrated Circular.
A. H. REID,
S. 17th Street, Phila., Pa.

Eggs! Eggs!

From all the leading varieties of pure bred Poultry Bramahs, Cochins, Hamburgs, Polish Game, Dorking and French Fowls, Plymouth Rocks and Bantams, Rouen and Pekins Ducks. Send for Illustrated Circular.
T. SMITH, P. M., Fresh Pond, L. I., N. Y.
feb-3m

SEEDS BULBS, PLANTS.
Beautiful Illustrated Catalogue Free.

J. LEWIS CHILDS, QUEENS, N. Y.
Jan-3m

WE WANT OLD BOOKS.

WE WANT GERMAN BOOKS.
WE WANT BOOKS PRINTED IN LANCASTER CO.
We Want All Kinds of Old Books.
LIBRARIES, ENGLISH OR GERMAN BOUGHT.
Cash paid for Books in any quantity. Send your address and we will call.
REES WELSH & CO.,
23 South Ninth Street, Philadelphia.

D. M. FERRY & CO'S
ILLUSTRATED, DESCRIPTIVE AND PRICED
SEED ANNUAL
FOR 1882
Will be mailed FREE to all applicants, and to customers without ordering it. It contains four colored plates, 600 engravings, about 200 pages, and full descriptions, prices and directions for planting 1500 varieties of Vegetable and Flower Seeds, Plants, Fruit Trees, etc. Invaluable to all. Send for it. Address,
D. M. FERRY & CO., Detroit, Mich.
Jan-3m

$66 a week in your own town. Terms and $5 outfit free. Address H. HALLETT & CO., Portland, Maine.
Jan-1y†

PENSIONS For SOLDIERS
widows, fathers, mothers or children. Thousands yet entitled. Pensions given for loss of finger, toe, eye or rupture, varicose veins or any Disease. Thousands of Pensioners and soldiers entitled to INCREASE and BOUNTY PATENTS procured for Inventors. Soldiers, land warrants procured, bought and sold. Soldiers and heirs apply for your rights at once. Send stamps for "The Citizen-Soldier," and Pension and Bounty laws, blanks and instructions. We can refer to thousands of Pensioners and Clients. Address,
N. W. Fitzgerald & Co., PENSION & PATENT ATT'YS, Lock Box 585, Washington, D. C.
Ja-3t

LIGHT BRAHMA EGGS

For hatching, now ready—from the best strain in the country—at the moderate price of
$1.50 for a setting of 13 Eggs.
L. RATHVON
No. 9 North Queen St., Examiner Office, Lancaster, Pa.

WANTED. CANVASSERS for the
LANCASTER WEEKLY EXAMINER
In Every Town-ship in the County. Good Wages can be made. Inquire at
THE EXAMINER OFFICE.
No. 9 North Queen Street, Lancaster, Pa.

$72 A WEEK. $12 a day at home easily made. Costly outfit free. Address TRUE & CO., Augusta, Maine.
ju-0-1y†

SEND FOR SPECIAL PRICES

On Concord Grapevines, Transplanted Evergreens, Tulip, Poplar, Larches, Maples, etc. Tree seedlings and Tree seed bulbs, planted by the 1000.
J. JENKINS' NURSERY,
WINONA, COLUMBIANA CO., OHIO.

PENSIONS For SOLDIERS

children. Thousands yet entitled. Pensions given for any Disease. INCREASE FOR ANY PATENTS procured for inventors. Soldiers, land warrants bought and sold. Soldiers and heirs apply at once for your rights. Send stamps for Instructions. Address,
N. W. Fitzgerald & Co.
PENSION & PATENT ATT'YS, Lock Box 585 Washington D. C.

WELL-AUGER

One is guaranteed to be the cheapest and best in the world. Also can bore one a SAWING MACHINE. It saws off a log in 2 minutes. Pictorial books free. W. GILES, Chicago, Ill.

PENNSYLVANIA RAILROAD SCHEDULE

Trains leave the Depot in this city, as follows:

WESTWARD	Leave Lancaster	Arrive Harrisburg
Pacific Express	2:30 a.m.	4:05 a.m.
Way Passenger	5:00 a.m.	7:50 a.m.
Niagara Express	11:00 a.m.	11:40 a.m.
Hanover Accommodation	11:05 p.m.	Col. 10:00 a.m.
Mail train via Mt. Joy	10:20 a.m.	12:50 p.m.
No. 2 via Columbia	11:35 a.m.	12:55 p.m.
Sunday Mail	10:50 a.m.	1:20 p.m.
Fast Line	2:30 p.m.	3:05 p.m.
Frederick Accommodation	2:55 p.m.	Col. 2:15 p.m.
Harrisburg Accom.	5:15 p.m.	7:10 p.m.
Columbia Accommodation	7:20 p.m.	Col. 8:20 p.m.
Harrisburg Express	7:30 p.m.	9:30 p.m.
Pittsburg Express	8:30 p.m.	10:10 p.m.
Cincinnati Express	11:50 p.m.	12:45 a.m.

EASTWARD	Lancaster	Philadelphia
Cincinnati Express	2:55 a.m.	7:00 a.m.
Fast Line	5:08 a.m.	7:40 a.m.
Harrisburg Express	8:05 a.m.	10:00 a.m.
Columbia Accommodation	9:10 p.m.	12:30 p.m.
Pacific Express	1:00 p.m.	3:30 p.m.
Sunday Mail	2:00 p.m.	5:00 p.m.
Johnstown Express	3:05 p.m.	5:30 p.m.
Day Express	5:35 p.m.	8:00 p.m.
Harrisburg Accom.	6:25 p.m.	9:30 p.m.

The Hanover Accommodation, west, connects at Lancaster with Fast Line, west, at 2:10 p.m., and runs to Frederick. The Frederick Accommodation, west, connects at Lancaster with Fast Line, west, at 2:55 p.m., and runs to Frederick. The Pacific Express, east, on Sunday, when flagged, will stop at Middletown, Elizabethtown, Mount Joy and Landisville.

The only trains which run daily. Runs daily, except Monday.

NORBECK & MILEY,

PRACTICAL
Carriage Builders,

COX & CO'S OLD STAND.

Corner of Duke and Vine Streets,
LANCASTER, PA.

THE LATEST IMPROVED

SIDE-BAR BUGGIES,
PHÆTONS,
Carriages, Etc.

THE LARGEST ASSORTMENT IN THE CITY.

Prices to Suit the Times.

REPAIRING promptly attended to. All work guaranteed.

S. B. COX,
Manufacturer of

Carriages, Buggies, Phaetons, etc.

CHURCH ST., NEAR DUKE, LANCASTER, PA.

Large Stock of New and Second-hand Work on hand very cheap. Carriages Made to Order. Work Warranted for one year.

THE LANCASTER FARMER.

EDW. J. ZAHM,

DEALER IN

AMERICAN AND FOREIGN
WATCHES,

SOLID SILVER & SILVER PLATED WARE.

CLOCKS.
JEWELRY & TABLE CUTLERY.

Sole Agent for the Arundel Tinted

SPECTACLES.

Repairing strictly attended to

ZAHM'S CORNER,
North Queen-st. and Centre Square, Lancaster, Pa.

E. F. BOWMAN.

Watches & Clocks

AT LOWEST POSSIBLE PRICES,
Fully guaranteed.

No. 106 EAST KING STREET,
Opposite Leopard Hotel.

ESTABLISHED 1832.

G. SENER & SONS,
Manufacturers and dealers in all kinds of rough and finished

LUMBER,

The best sawed SHINGLES in the country. Also Doors, Blinds, Mouldings, &c.

PATENT O. G. WEATHERBOARDING

and PATENT BLINDS, which are far superior to any other. Also best COAL constantly on hand.

OFFICE AND YARD:
Northeast Corner of Prince and Walnut-sts.,
LANCASTER, PA.

PRACTICAL ESSAYS ON ENTOMOLOGY,

Embracing the history and habits of

NOXIOUS AND INNOXIOUS
INSECTS,

and the best remedies for their expulsion or extermination.

By S. S. RATHVON, Ph. D.
LANCASTER, PA.

This work will be highly illustrated, and will be put in press as soon after a sufficient number of subscribers can be obtained to cover the cost, as the work can possibly be accomplished.

$77 a month and expenses guaranteed to Agents. Outfit free. SHAW & CO., Augusta, Maine.

TREES

Fruit, Shade and Ornamental Trees.

LOUIS C. LYTE.
Bird-in-Hand P. O., Lancaster co., Pa.

Plant Trees raised in this county and suited to this climate. Write for prices to

Nursery at Smoketown, six miles east of Lancaster.

WIDMYER & RICKSECKER,
UPHOLSTERERS,
And Manufacturers of

FURNITURE AND CHAIRS.
WAREROOMS,

102 East King St., Cor. of Duke St.
LANCASTER, PA.

Special Inducements at the
NEW FURNITURE STORE
W. A. HEINITSH,
No. 15 1-2 E. KING STREET

(over Bursk's Grocery Store, Lancaster, Pa.
A general assortment of furniture of all kinds constantly on hand. Don't forget the number,

15 1-2 East King Street,
(over Bursk's Grocery Store.)

For Good and Cheap Work go to
F. VOLLMER'S
FURNITURE WARE ROOMS,
No. 309 NORTH QUEEN ST.
(Opposite Northern Market,)
Lancaster, Pa.

Also, a limited of picture frames.

GREAT BARGAINS.

A large assortment of all kinds of carpets are still sold at lower rates than ever at the

CARPET HALL OF H. S. SHIRK,
No. 202 West King St.

Call and examine our stock and satisfy yourself that we can show the largest assortment of these Brussels, three ply, and ingrain at all prices—at the lowest Philadelphia prices. Also on hand a large and complete assortment of Rag Carpet. Satisfaction guaranteed both as to price and quality. You are invited to call and see my goods. No trouble to show them, even if you do not want to purchase. Don't forget this notice. You can save money here if you want to buy. Particular attention given to customer work. Also on hand a fine assortment of Counterpanes, Oil Cloths and Blankets of every variety.

PHILIP SCHUM, SON & CO.,
38 and 40 West King Street.
We keep on hand of our own manufacture,
QUILTS, COVERLETS,
COUNTERPANES, CARPETS,
Bureau and Toilet Covers, Ladies' Furnishing Goods, Notions, etc.
Particular attention paid to customer Rag Carpet, and scouring and dyeing of all kinds.

PHILIP SCHUM, SON & CO.,
Lancaster, Pa.

THE HOLMAN LIVER PAD!
Cures by absorption without medicine.

Now is the time to apply these remedies. They will do for you what nothing else on earth can. Hundreds of evidences of Lancaster say so. Get the genuine at
LANCASTER OFFICE AND SALESROOM,
22 East Orange Street.

C. R. KLINE,
ATTORNEY-AT-LAW,

OFFICE: 15 NORTH DUKE STREET,
LANCASTER, PA.

The Lancaster Farmer.

Dr. S. S. RATHVON, Editor. LANCASTER, PA., FEBRUARY, 1882. Vol. XIV. No. 2.

THE ENGLISH SPARROW.

"Australia imported English sparrows to kill worms, but it found that the birds are the worst pest of the two, and bounties are offered for their destruction."

It appears that Australia has repeated the blunder of America, in importing the "English sparrows to kill worms," and now both countries are "down on the sparrow," because he cannot be forced to habitually do violence to the instincts of his nature. The "English sparrow" (*Passer domesticus*) is a *Finch*, belonging to the family FRINGILLIDÆ, and therefore, by nature, is a granivorous bird, and not strictly speaking, an insectivorous one. It is not the fault of the sparrow that it don't eat worms, any more than it is the fault of the lion, because he don't "eat straw like an ox." There was therefore no more wisdom in importing the English sparrow for the purpose of destroying insects, than there would have been in importing an English pigeon to destroy mice. It is nothing to the purpose to allege that they *do* occasionally, or under certain extraneous circumstances, eat insects, for that is only a *negative* quality at best. If birds *must* be imported to destroy insects, those of a *positive* character should have been imported.

The sparrow however, is sufficiently *positive* in the direction his instincts lead him, although as a uniform, or exclusive feeder on insects he is *negative*, and it is almost, if not quite, impossible for him to be otherwise, without doing violence to his own physical organization. If he were purely insectivorous he would not be eking out a precarious living in our cold uncongenial clime during winter, when insects are not obtainable by birds of his mandibular conformation; he would instinctively migrate with other members of the "feathered realm," who habitually feed on insects. It seems to us that this *fact* alone ought to be *prima facie* evidence that no confidence can be placed on the English sparrow as a reliable remedy against the multiplication of noxious insects. It is very probable that he may, and perhaps *does*, appropriate some insects when he can get nothing else, or when he is providing a repast for his young family, in common with many other birds that are not strictly insectivorous. Until young birds are fully competent to provide for themselves, the parent birds usually furnish them such food as is best adapted to their juvenile condition, and in that respect they perhaps do less violence to the laws of physiology than human parents do, in the rearing their own progeny. A tender juicy worm is easier to digest than a hard dry seed, especially during the period of helpless inactivity. The sparrow then, being a granivorous bird, its normal food is grain, or seeds, and when these are not obtainable it will appropriate that which nearest approximates to grain, or seeds, and hence the English sparrows beget themselves to cities, towns, villages and hamlets, where they can obtain bread-crumbs, undigested grains in the droppings of animals—especially

those of the horse—and in the absence of these, the young buds of trees and shrubbery. The streets of Lancaster city are full of them both Summer and Winter, whilst in the surrounding country there are few or none of them. Two summers ago a creeping vine on a gable in East Orange street, contained fifty or one hundred sparrows nests, and each one of them was the cradle of a brood or two during the season. Within, perhaps, a hundred yards of it stood several elm trees, badly infested by the "elm leaf beetle," (*Galeruca xanthomelæna*.) There were tens of thousands of these insects in the larva, the *pupa*, and the *imago* states; but no one that ever watched those birds ever saw them fly in the direction of those infested elm trees, and this seemed the more remarkable, inasmuch as the leaves, the branches, the trunks, and the pavement under the trees, were literally swarming with the insects in all their stages of development; moreover, the birds were rearing their young, and, from their appearance alone, one would suppose that if ever there was an insect that might be expected to excite the appetite of a small bird, it would have been these Elm-leaf Beetles. But no, they were totally ignored. Now, notwithstanding all this, it is not our intention to disparage the English sparrow; for, as we before intimated, he has his *place* in the economy of nature, and those who have forced him out of it must take the consequences. He is doing all he can, under the circumstances in which he is placed, and the highest reasoning creature can do no more. He doubtless is doing some good in his own peculiar way. If he does not destroy the number of insects we think he should, he may be gobbling up the seeds of many noxious weeds, when he can get access to them, and that is surely *something*.

But, in dealing with the sparrow we do not think it would be wise to follow the example of Australia, by offering bounties for their destruction; for this might furnish the other horn of the dilemma, as it did on another notable occasion in that same Australia. The government of New South Wales offered bounties for the destruction of the owls and hawks of that colony some years ago, on account of the depredations they committed upon the poultry of the farmers, through which they were finally exterminated. But then the more destructive rodents increased most fearfully, soon the country was overrun by rats, mice, rabbits, &c., and so great was the destruction of the pasture fields, that a single woolgrower or a single district lost fifteen hundred sheep by starvation. A similar event occurred many years ago in Scotland. There seem to be certain balances in the economy of nature, the equilibrium of which, if destroyed, or undue preponderance be given in either direction, results in disaster to the interests of the aggressor; and often too, in a manner that was wholly uncontemplated.

We cannot therefore say what effect the total destruction of the sparrows would have

upon the vegetable world, but there is room for rational inference that it would not be a favorable one; because, *indirect* influences may be so intensified as to produce more injury than those that are *direct*. Although the sparrows belong to the Finch family, yet within that family there are groups, some of which are more decidedly granivorous than others. Although their natural proclivities may lead in that direction, still they are not so exclusively seed-eating as the true *finches*, of which the canaries may be regarded as a familiar type. The sparrows have the conical bills of pure granivorous birds, but they are more decidedly notched than most of the other groups of the family. We therefore not only may *infer* that they capture insects—especially during the breeding season—but they have often been *seen* in the very act of doing so; and, if each bird captures and kills a single female insect in the spring before she has deposited her eggs, the benefit resulting from it may be incalculable. The destruction of a few buds of fruit trees in early spring, is surely not to be compared with the general interest of the crop; and perhaps such a contingency could be obviated by furnishing the birds with necessary food.

We have been portraying the English sparrow as he actually is, and not as people may think he *ought* to be. From his *status* in the classified arrangement of the feathered tribes, we freely confess that we are not at all disappointed in him. When the wag, dressed in an ox's hide appeared before Baron Cuvier, in order to frighten him, he enquired what he wanted; and when the wag replied in a sepulchral tone—"I want to eat you," the Baron significantly replied, in an unconcerned manner—"Hoofs, horns, *Herbivorous*. You can't do it." We knew of a cow that ate the frill of a woman's sunbonnet; and we also knew of a cat that ate pickles, but those articles were not their normal food of course. Nor can it be said that they habitually fed on that kind of provender. Under similar circumstances, aided by domestication, sparrows are occasionally seen capturing and eating insects, but it can hardly be regarded as a normal characteristic.

On the whole then, from what we *know* of the English sparrow through local experience, and the general tone of the public press, we have committed a blunder in introducing it into the United States for the purpose of destroying our surplusage of insects; and, whatever we intend to ultimately "do about it," it is perhaps well that the masses of the people should have some knowledge of what he is and how he lives. He is a shrewd, pugnacious and persevering little elf, and 'tis a pity he should have gained such a "disreputable reputation."

We would not recommend, therefore, that a government bounty should be offered for his head, when a simple repeal or suspension of the law for his protection, and his elevation to the status of a "game bird," would as effectually, and more cheaply accomplish the desired end. It is true, he is not a very large bird, but then it would only take the more of them to make a "mess," and the process of extinction would be more rapid. There are "four and twenty" of them that infest our premises, that we would sooner see "baked in a pie," than to endure their spattering attempts at whitewashing, especially when anything valuable happens to stand beneath their winter rookeries; and yet, we rather like their social presence.

OURSELVES.

"The fourteenth volume of the LANCASTER FARMER begins with the January number. The industrious editor, Dr. Rathvon, has for years given his time and talents to pushing the agricultural interests of this grand agricultural county ahead through its columns, and although ill requited for his laborious work will still keep his hands on the helm. He cordially invites contributions and correspondents to render him what aid they can in making the FARMER a still more valuable vehicle of progressive agriculture and pomology, and there should be a generous response from all sides to his invitation. We observe that he alludes to the ungenerous and unjust charge made against him by a member of the Agricultural Society, that he had suppressed, out of jealousy, or for some other cause, the essays of the offended member from the society proceedings as published in the FARMER. It was hardly worth his while to go to the trouble. No one gave the silly accusation a moment's thought. The charge that Dr. Rathvon was unable to answer the entomological queries of his captious critic brought smiles to the face of every member of the society present. All that individual ever knew about insects is not a tenth part of what Dr. Rathvon has forgotten about them."

The foregoing from the columns of a recent issue of the DAILY NEW ERA, needs no comment from us, save the expression of a grateful sentiment for the kindly recognition of our labors, our integrity, and the status of the journal we seem called upon to edit.

FEBRUARY SNOWS

The morning of February 1st, 1882, was ushered in by an eight inch snow, followed by one of about twelve inches on the 4th, but it cannot be said that the temperature was uncomfortably cold. On the morning of the 2nd (candlemas) the clouds had entirely dispersed, and the sun shone out bright and clear. This according to an ancient tradition (perhaps confined to Pennsylvania alone) was ominous of a prolonged winter, and a late spring. It was ground-hog day, and the tradition is something like the following :

If the ground-hog comes out of his hole on candlemas and sees his shadow reflected by the sun, he immediately returns to his winter lair, and resumes his state of hibernation for six weeks longer. But if the sky is clouded and he is unable to see his shadow, he remains out, and the spring will be an early one. We are unwilling to say anything calculated to undermine this ancient conceit, but really the groundhog is not much of a prophet after all. Six weeks from candlemas would carry us to about the 17th of March, which is "St. Patrick's day in the morning." Now, we are nearly "three-score and ten," and yet we never knew Spring to commence much before the 17th of March, but have known it to be "bitter cold" after that date on frequent occasions. Besides, there is that qualifying "if,"

which places the subject in the company of probabilities, or inferentialities. If we live, perhaps we may be able to say more about it six weeks later in the season, for which we are content to wait.

WOOD-WORMS.

"AN old experienced farmer says that hickory cut in July or August will not become worm-eaten. Oak, chestnut, walnut or other timber cut from the middle of July to the last of August will last twice as long as when cut in winter. When oak is cut at this season, if kept off the ground, it will season through if two feet in diameter, and remain perfectly sound for many years; whereas if cut in winter or spring it will become sap-rotten in a few years."

Perhaps the most common worm that infests hickory timber is the larva of the "Painted Clytus," (Clytus pictus) a longhorned Beetle (Longicornia) of a dark mottled greenish color, striped obliquely with yellow on the wing-covers, and transversely on the thorax. The "Locust-tree borer" (Clytus robinia) similarly marked, is very nearly like the first named ; so much so indeed as to be easily confounded with it, and some entomologists are of opinion that the species are identical, or at most, only varieties. Be that as it may, the painted clytus is usually found in early spring—even as early as the beginning of April, or of May—whilst the locust-tree clytus is usually found in early autumn. Many years ago, before coal was used as a household fuel as universally as it is now, we laid in our supply of hickory wood in autumn. This we had sawed in convenient lengths to suit the size of the stove. Invariably, almost every spring these beetles would evolve by hundreds, and issue through the cellar grates. After a week or ten days the insects would entirely disappear, and no clytes would be seen until about the month of September, or early in October, when the different species of solidago would be fairly swarming with them. This, together with a difference in the length of their horns (antennae) and other minor characters has been deemed sufficient to establish two species of these. Now, if hickory wood is cut before these insect deposite their eggs in it, it is not likely to be infested by these worms, and the same may be said of oaks, chestnut, walnut and other timber. More respect must be paid to the season in which the mature insects are abroad, pair and oviposit, whatever the month may be. Cutting timber in July and August might elude the attacks of the locust clytus, because the sexes are usually found in coitu on the bloom of the solidago in September, and as late as October ; but we think it would have to be cut in this latitude, before July to elude the attacks of the painted clytus, or whatever the woodboring insect may be. The whole success of eluding the attacks of woodboring insects hinges upon their ovipositing periods. If the substances which form niduses for their eggs are removed before the eggs are deposited they are likely to escape; always provided the insects are indifferent whether it is fallen or standing timber. These periods the farmer has better opportunities to observe than the closet entomologist. Whenever he finds beetles in coitu he may infer that the next act after that, will be the deposition of eggs. Separate from these circumstances, days, weeks, months and signs mean

nothing. The instincts of insects may lead them to avoid fallen timber because of its liability to be used before their progeny could be developed therein, or because the eggs require some moisture to facilitate incubation ; or because both the eggs and the newly excluded embryo might be sun-killed before the latter could penetrate hard or dry timber. Nothing is more fatal to the young larvæ and eggs of some insects than a hot sun. In perfect freedom insects would hardly deposit their eggs on stone or iron.

The foregoing has no relation to those woodboring insects that manifest a preference for dead timber, whether standing or fallen, decayed parts of living timber, or that which is very much rotted; but even many of these choose such parts of it as contain some moisture, whilst others may be found in timber almost as hard and dry as old bones.

PLANTING TREES ON RAILWAY EMBANKMENTS.

In our sylvian enthusiasms, our theories of tree replenishment may not be borne out by practice. A writer in the Journal of Forestry, for December, 1881, discourses on "some objections or restrictions which apply to planting and rearing timber on railway embankments;" and briefly indicated, they are, "First, the risk of windfall; second, the risk of fire; third, the lodgment of leaves against the rails; and fourth, the hindrance of view over the adjacent country."

When we reflect that the great storm which passed over Leicestershire, England, last October, uprooted or ruinously damaged, on the estates of Belvoir Castle alone, 319 oaks, 165 spruce, 266 larch, 102 elms, 124 ash, 70 spanish chestnut, 13 linn, 18 sycamore, 19 beech, 15 poplar, 16 birch, 6 cherry, 2 each of silver fur and Turkey oak, 6 of Scotch fir, and 1 each of mountain-ash, bird-cherry, maple and horse-chestnut trees, and that usually the United States is more stormy than England, we must admit a very serious obstacle to the enterprise.

Again, when we reflect upon the sorrowful devastations of the forest fires of Michigan a few months ago, the effects of which her people are still suffering, we are compelled to acknowledge an other source of danger, especially if any of the resinous pines should be planted. (The leaves of pines burn readily in consequence of the turpentine they contain, even when quite green). Although the third objection might be obviated by "sweepers" in front of the wheels, and at any rates would only continue for a brief season each year, yet the lodgment of these leaves in excavations might become a source of danger for an indefinite period, and moreover, would always be in danger of being ignited in dry weather.

The hindrance of a view of the adjacent country would be a serious objection to those who travel for pleasure, and desire to see the country they are passing through. However beautiful the trees may become, to have them on either side of a road for hundreds of miles — or perhaps thousands — would be like passing through a long forest—or a deep cut or tunnel—and hence it would become monotonous if not a dark and gloomy avenue of transit.

The risks from windfalls would doubtless, in time, become a very considerable one, and perhaps the main one. In the dim long ago, it seemed to have been tacitly understood that the poor people of the towns and villages of our State, were privileged to gather the windfalls of contiguous forests and appropriate them to their own use. At all events they did gather them and carry them home, and no owner of forest premises ever interfered with their right to do so, and perhaps would have been looked upon as a very mean man if he had.

Now, we well remember how thickly the ground of those forests was strewn with windfalls after every storm, and how eagerly the poor hied themselves thither to get the first choice of them, and this was particularly the case when the trees were large and old. Think of a train of cars entering such a dangerous avenue on a dark and stormy night, feeling its way cautiously for a whole night in momentary fear of encountering these windfalls on the track.

Again, where would railroad companies find room to plant their trees? Where the roads ran at grade there might be little difficulty, but where they penetrated hills by deep excavations, or where elevated on high embankments, they evidently would be compelled to plant them on the sides of those excavations and embankments, and thus the projecting limbs would, in time, spread and interfere with the transit of cars, or with the efficiency of the telegraph wires. They could not plant them along the margins of their grants, for there they would trespass upon the contiguous properly; moreover, railroad companies are granted the right of way to transport passengers and merchandise, and not to convert them into timber forests, jeopardizing the lives and property of their patrons.

We give these views for what they may be worth, and not as an unqualified endorsement of them, nor yet as a positive dissent. There is evidently two sides to the question, and before we commit ourselves to either side, we should contemplate the subject both pro and con. It doubtless would be pleasant to ride through a long shaded avenue in summer time, but if this should add to the present burden of danger, much might be lost and nothing gained.

THE LARGEST TREE IN THE WORLD.

"The biggest tree in the world is not in California, as every one supposed, but in Australia. The Champion of the Yosemite Valley must give way to the "Peppermint" trees, on the Dandenong range of hills in Australia. Baron Von Muller who is a great authority on botanical subjects, asserts that he has seen one of these trees of the enormous height of St. Paul's Cathedral," 480 feet.

The above paragraph is credited to Land and Water, by the Journal of Forestry. This will never do. We must find a tree 480 feet and one inch in height, even if we have to splice it. Perhaps the great Santa Barbara grape vine might be trellised up to a greater height than the Australian peppermint, but then it would hardly pay, as they would be sure to "beat it"—perhaps with a pumpkin vine—in Australia. The race in "big things" seems to lie between California and Australia, and the remainder of the world are mere spectators.

On a "second thought," perhaps, it would be better just now to "divide the honors," for the same authority nominally accords to New York State—"the largest orchard in the world," namely, that belonging to Mr. Kinstry, on the banks of the Hudson, containing 24,000 apple-trees. 4,000 cherry-trees, 1,600 pear-trees, 500 peach-trees, 500 chestnut-trees, 200 plum-trees, 15,000 grape-vines, and 6,000 raspberry-trees. Of apples alone Mr. K. sold over 30,000 barrels, and a proportionate quantity of other kinds of fruit last year. Taken as a whole, Mr. Kinstry's may be the largest orchard in the world; but, if the number of peach trees was thousands instead of hundreds, it would not equal some of the peach orchards of the little State of Delaware, and the plums can be outnumbered, we think, by orchards in Michigan.

We are not sure that big trees, big orchards and big farms, are the best things for the general welfare and equity of a country—and this might also apply to big establishments of any kind, unless it were such as could not be conducted on small or medium scales—as railroads and canals for instance, and which cannot accomplish their objects, or accommodate the demands of the public, without traversing hundreds or thousands of miles.

SHIFTY, THRIFTY, FRANCE.

"The Montezuma (N. Y.) marshes are likely to become as valuable as a coal mine. The marl is being shipped to France by the thousands of tons, being used there as a deodorizer and entering into the manufacture of artificial fertilizers. Seeing that we are exporting so much fertilizing matter in the shape of cereals, and beef, pork and mutton, we really ought to keep all crude fertilizing matter at home to replenish those fields whose fertility has been shipped to Europe."

The French seem to know "whats what" on subjects of "fragrance and fertility;" something which we Americans are slow in learning. Marl, the qualities of which should be familiar to American farmers, is absolutely unknown in many districts of our country, not very far either from localities where thousands of tons of it have been imbedded these many hundreds—perhaps thousands—of years.

The next step may be for France to send our marl back to us again, in the form of expensive fertilizers. Perhaps then it would become popular, for it would be French, and 'you know' we are a very "Frenchy people in tastes. Here we sweat and whorize, and France sends over and takes our deodorizers. They are a wonderfully prolific people any how, in scientific, domestic, manufacturing and social expedients and economies; and although comparatively poor in area and virgin fertility, yet they are rich in resources, and could live sumptuously on what we waste, or willfully throw away.

"A market for the sale of toads to gardeners is held every week in Paris. A hundred good toads bring from $15 to $17. They are brought packed in damp moss in well-ventilated casks."

There it is again: the next step will be to import American toads, as companions to our marl. As we are said to be a money getting people, it would not be surprising, if some of our enterprising experts would engage in the exportation of toads—especially if it "pays" —seeing that we have such a low appreciation of them as domestic auxiliaries on this side of the Atlantic.

"The highest mountain on the north American continent, is Mount St. Elias in Alaska, whose elevation is 17,780 feet. Next to it come the volcano of Popocatapetl, in Mexico, 17,700 feet, and Orizaba, also in Mexico, 17,370 feet.

We own the highest mountain: that surely is some compensation. France can't deprive us of that "any way." But should the ice crop fail, we may find her hankering after our Alaska ice. She would profit by it, although it, it seems, cannot. Well, let her take it, and the marl and toads along with it. She can't take our highest mountain, nor our noxious insects (indeed these she won't take, she is satisfied with our phylloxera) nor our stenches.

Irony apart, these paragraphs carefully culled, are significant. Before another century in our history transpires, our country may have a more practical illustration of the use of marl and toads, and perhaps of high mountains too, than it has now. These are the bountiful provisions of nature, that have been permitted by a power outside of nature, for our utilization when the proper time and circumstances harmonize in their discovery and development.

KITCHEN GARDEN FOR FEBRUARY.

In the Middle States, frost usually prevents out-door efforts in the way of gardening. Next month however, will bring its labors, and we can now only prepare to follow them. It is presumed all persons into whose hands t..is Journal is likely to fall are provided with that cheap and simple means of enjoyment, a HOTBED, for forwarding tender vegetables. We do not mean the more expensive structure under which delicacies are provided ready for the table, but a plain box, of suitable size and figure, with sash and shutter to fit, under which plants of cabbage, tomato, egg-plant, &c., may be raised in anticipation of spring, and on its arrival, to be transplanted in the open air. If there be one who has a garden still unfurnished with what we have just described, let him take our word for it he will, on trial, thank us for urging its immediate provision; no country family can half enjoy the comforts within reach who are unprovided with such a structure; a glance at one in use will give the necessary information as to the construction. Towards the close of this month (if the weather be very severe it may be prudent to defer it awhile), the seeds just named may be planted under glass; watch them lest they suffer from frost, or, as is not unfrequently the case, from want of sufficient air as the weather becomes milder, when they all need increased water. If the remarks under the head of January are referred to, perhaps something may be found which will apply with equal force to the present month.

We can only speak in general terms of the work which may be advantageously done now, preparatory to the active season which approaches. The thoughtful man will study out the subject for himself and leave nothing undone which may expedite the varied and pressing labors of spring. If tools and implements are likely to be needed, he will provide them in due season; repair the old ones, examine and realize, if need be, the sashes of the forcing frames, long before they are actually required; overhaul his stock of seeds, and make out a list of those which may be needed, so that they may be in hand before the time of sowing: thus not only his interest, but his personal comfort will be advanced, and those little trifles which perplex the careless and improvident, may be made sources of enjoyment. With each duty discharged at the proper time, with "a place for every thing and everything in its place," many rough spots in life's journey may be made smoother.—*Landreth's Rur. Reg.*

Comment on the above is unnecessary. It speaks for itself, and contains the essence of the admonition—"He that is forewarned is forearmed,"—or at least he *may* be, if he will.

POULTRY EXHIBITION.

This rare pageant, which finally closed on Wednesday evening at 10 o'clock. January 18, 1882, was the distinguishing feature of the month, in the domestic history of Lancaster county; and, independent of its pecuniary outcome, it may be "scored" as an unqualified *success*, as it deserved to be. Whatever indifference may be manifested in quarters from which we would naturally look for active sympathy, one thing seems certain; namely, that there is considerable of a "chicken-fever" in Lancaster city and county, and any one who visited the exhibition must have been impressed with the evidence that the *chicken* was determined to be "seen and heard."

The "birds" themselves, very graphically represented the different nationalities of the human family, and their vociferations, attitudinizings, genuflections and gyrations may fitly have represented the babel and the movements upon some foreign quay, where diverse nations are wont to meet in promiscuous intercourse, although limited by a ruling power apart from themselves. The awards of the premiums will be found on another page in this number of the FARMER, where it finds a permanent record and may be referred to by those who participated in the exhibition, when other records have perished. And that is not all. It stands as a *living* record, creditable to the energy and perseverance of its originators and conductors.

RULES AND EXCEPTIONS.

Hybrids Not Always Barren.

The general sterility of the mule has given rise to an impression that hybrids are generally sterile, and indeed the term mule and hybrid have become almost synonymous. Scientific agriculturists and other philosophers have even built theories on this supposed universal sterility, and we are not sure but some theories in the popular general science of the day are founded on those supposed facts. But as one swallow does not make a summer, so does not this one great fact about the mule make a general law. Facts opposing this general application of the principle are numerous and must be familiar to most observing persons. One of the most interesting that we have seen recently relates to the progeny of the common buffalo and the domestic cow. The progeny breed freely and are said to be good milkers, and there is even some prospect that the fact may be utilized in the production of a very hardy and valuable race.

It is to be regretted that the race of observers is so limited, while students evrywhere abound. Though the fact that hybrids are not necessarily sterile is sustained by numerous instances if people will only look about them, few know of it who are studying up these questions, not because they do not exist, but because they are not in the books."—*Germantown Telegraph.*

Time was, within our remembrance, when the mule was booked and discussed without an exception, as a perfectly sterile animal, but, subsequently a voice came up from Alabama that a female mule had foaled, just as any other female animal of the horse kind would. This then was an exception to the rule, in the minds of those who credited the story. Then came a similar report from Florida, from Kentucky and elsewhere, but nothing authenticated, save by newspaper paragraphs. Perhaps none of these cases were sufficiently authentic to break the theoretical rule in the minds of many scientists. Now however, according to our extract below, taken from the *National Live Stock Journal*, the French savants at least, have been compelled to admit the fact. This, of course totally destroys the rule, because that can hardly be considered a rule to which there are so many exceptions; and yet there are admitted rules to which, it is said, there are more exceptions than cases that are covered by the rule.

But, then, single isolated facts, however well attested, do not entirely exhaust the subject, for there are phases of the question suggested by both of our quotations, that would seem to need a more definite exemplification, especially in regard to the fertilizing animal—whether an *asinus* or a *caballus*. Moreover, the fertile or non-fertile character of the progeny involves a question of some interest. If the *cause* of this departure from a general rule can be determined, the matter might be turned to additional profit in mule-culture. As the *Telegraph* suggests, it would be well for those "who are studying up these questions," to investigate, and if they can find nothing "in the books," to see that it is duly placed on record there.

A Breeding Mare Mule.

A breeding mare mule was recently exhibited at the Jardin d'Acclimatation in Paris, which has produced three colts. As the French savants have hitherto been very incredulous as to reports of mule breeding, it is stated that they carefully inquired into this case, and became satisfied that it was true. We have heard of mare mules occasionally breeding in America, but we do not recollect the year and locality of this, or whether the sire was a jack or a stallion, and shall be obliged to any of our readers who can furnish these particulars; also, what sort of an animal the produce turned out to be. In the above instance of mule breeding in France, the sire was a stud-horse.—*Chicago Live-Stock Journal.*

WRITING FOR THE FARMER.

FRIEND RATHVON: "Why is it that our people of Lancaster county will not write more for the FARMER? Surely there are many who could give valuable information. More original matter would make the paper more interesting."—*J. B. G.*

"That's so"—eminently and absolutely so —and yet the desirable thing is not done; but we can conscientiously say it is not through any example, any unwillingness, or any refusal of ours. We have however erected no tribunal before which we arraign any one for delinquencies of this kind. Contributions of this kind, like church contributions, should be voluntary. There is no power except *self-compulsion* that can be legitimately exercised in such a matter. If those who are able to write, *choose* to "pass over Jordan" without having left a record for the benefit of posterity, they are not accountable to us. It would be a great relief to us, if we had more intelligent contributors, and would greatly add to the interest of the journal, the editorial labors of which, have devolved upon us these many years; but we must reconcile ourselves to the situation. It cannot continue forever, and we do not believe that our condition in the "forever" will be in the least damaged through the labors we have endured here; because, when we go hence, we do not expect, or even *desire*, to go to a land of apathy and idleness, but to one of *use* and perpetual progress; and our capacity of enjoyment *there* will be proportioned to our efforts to labor usefully *here*. . . . We wish some one would answer our aged friend's query: we confess we cannot. If the ship can be saved by throwing us overboard, like Jonah of old, we will cheerfully submit to the sacrifice. All that we have borne in conducting the FARMER thus far down the stream of time, may never be known until our "book of life" is opened. Perhaps, if a local journal were established to advance the interests of our secular craft, we might be as remiss in our contributions to its columns as those are who ought to "write for the FARMER;" but we *think* we would not. A *love* for writing however, must be cultivated, founded on *use*, before men will become habitual and voluntary writers; unless they write for emolument, and then it becomes a task.

In reply to our venerable friend's financial inquiry, we would say, that his remittance was duly received and placed to the credit of those for whom it was intended; and the acknowledgment will be found on the labels of the different papers.

As pertinent to this subject, but without claiming that we fill the measure of the following from the columns of a cotemperary journal, we quote it as a morally wholesome admonition to all. "Thousands of men breathe, move and live; pass off the stage of life, and are heard of no more. Why? They did not a particle of good in the world, and none were blessed by them; none could point to them as the instruments of their redemption: not a line they wrote, not a word they spoke, could be recalled, and they perished—their light went out in darkness and they were remembered no more than the insects of yesterday. Will you thus live and die? Live for *some thing*. Do good, and leave behind you a monument of virtue that the storms of time can never destroy. Good deeds will shine as bright on earth as the stars of heaven."

EXCERPTS.

MISCELLANEOUS.

GEORGIA has fifty cotton mills in operation and others in course of erection.

THE last census return place the "defective" list of persons in the United States at over 500,000. The list comprises the deaf dumb, blind, insane, idiotic and pauper.

TWENTY-FIVE acres of tobacco have been grown this year at Putney, Vt., which therefore claims to be the banner town in the State in this respect.

"LYING figures," says Mr. Dunlap, of The Chicago Tribune, "are those in a current newspaper article to the effect that with $2,300 a person can go to Dakota and realize a net profit in wheat culture of $19,000 the first year."

THE statement that the Canada thistle is spreading over a large part of the Northern or Middle States is not creditable to the enterprise of farmers. It should be eradicated by eternal vigilance.

THE Philadelphia Farmer has already predicted that there will not be even a fair crop of peaches next year, should the coming winter and spring be favorable. The freezing last winter and the hot, dry fall, told severely. Blossom buds, usually prominent before frost, are shriveled and show but little strength.

IT is estimated that the Barton drovers handled 3,000 cattle this season, and two Craftbury men have sold 7,000.

IF the novice would stick to one or two kinds of fowls in the beginning, less losses and disappointments would be the result.

OIL OF TURPENTINE is recommended to keep harness free from mold.

ALL manner of decaying vegetable matter should be added to the compost heap instead of being left to accumulate about the dooryard, where it will prove a fruitful source of malaria. Turn the heap occasionally and keep it moist to prevent fire fang.

GREEN manuring, or the plowing in of green crops, is especially adapted for light, sandy soils, which need humus to increase their retentive power.

A. B. GROFF, of Michigan, is said to have exhibited an onion seventeen inches in circumference, weighing upward of two pounds.

A MIXTURE of muriate of potash, fish guano or sulphate of ammoniate and superphosphate of lime, is an excellent fertilizer for corn.

OREGON had 100,000 tons of wheat for export, this year.

TENCH, a French food-fish, have been introduced in the Central Park pond, in New York.

WHEN artificial teeth were made of ivory, the canine teeth of hippopotamus were highly valued by dentists for that purpose, on account of keeping color better than any other kind of ivory.

IN the construction of the tubular bridge over Menai strait, England, there were used 2,000,000 bolts, averaging seven-eighths of an inch in diameter, four inches in length. The quantity of iron consumed for the purpose amounted in length to 126 miles, and in weight 900 tons.

PROBABLY 10,000 is an underestimate of the number of eggs shed annually by the herring.

DOMESTIC.

COMBS and wattles of fowls may be prevented from freezing by oiling them so as to prevent their getting wet.

PIGS are able to consume far more food in proportion to their weight than either sheep or oxen.

THE Italian bee was first imported into America in September, 1859, and ever since importation and home breeding of queens has been constantly gaining, until at present the supply rather exceeds the demand, and importers are opening a new field by introducing other races of bees.

LIKE the blackberry, the raspberry bears the fruit upon the cane of the previous year's growth, which, after fruitage, dies, the new cane coming forward for the next year's crop.

IN the orchard the thumb and forefinger are a better pruning instrument than the knife, and the latter than the shears or the saw; but the former must be used in the nick of time.

TWO cows well sheltered in winter, will produce more milk and butter than three unsheltered animals, though no more than half the feed required for the three should be given to the two.

IF the cucumber which grows nearest the root be saved for seed for a number of years in succession, the result will be a smaller and earlier variety. If the fruit on the extremity be saved it will produce a larger and later variety.

WHATEVER you undertake in the poultry line be sure to cultivate a thorough knowlege of its details before launching out with full steam in a haphazard way.

SHEEP should be tagged regularly, and kept clean. They should be culled every year, and those in any manner deficient in form or age should be put in a separate pasture and fattened for the butcher.

EGGS from hens partake in a great degree of the flavor and quality of the food, proving that they should be fed on clean wholesome food. One may get onious instead of eggs by feeding hens on onions.

TOMATO SOUP.—One pint of milk, one quart of water, one pint of tomatoes; two crackers powdered, and one and a half teaspoonfuls soda. Boil twenty minutes.

To break up setting hens have seven pens, one for each day of the week, then all hens found wanting to set on any day of the week should be put in the pen corresponding to that day. Keep them in five days. By this arrangement it is easily told how long each hen or pen of hens have been in.

SAVE the middle grains of the fine ears of corn for seed.

HOGS should be allowed to have a heap of coal ashes. They will be all the healthier for it.

BEEF and mutton are not flavored by feeding turnips to the animals—at least this is the statement of some who have tried it.

AN orchard should never be planted in a clay soil unless the latter is underdrained, after which it becomes one of the best soils for apples and pears.

LET every farmer keep all the stock he can possibly afford to—and generally he can afford to keep more than he does. The dependence of farming for all time must be mainly on stock.

YOUNG cows do not give as rich milk as do those of mature age. A lean cow gives poor milk and a fat one rich milk.

SCIENTIFIC.

THE latitude of England is the same as that of Labrador, and the former country is only saved from the coldness and desolation of the latter by the warmth of the gulf stream.

INSECTS are often attracted from a distance by artificial flowers, but they never light on them, leading us to believe that they are guided by some other sense than that of sight.

IT is recommended that, as the common ailanthus tree is diœcious, only the female trees should be propagated for shade in towns, the male having the disagreeable odor.

THE assertion that iron and platinum when raised to incandescence, are transparent to light, has been proved false by a series of experiments.

SOME engineers of Dundee, Scotland, have tried with success a new gun for throwing a line to a wrecked vessel. The gun is about two feet in length.

THE impression that flowers are never found double in a wild state is an incorrect one, the fact being that this is frequently one of nature's variations.

HERR HANSEN has found that the blue color in milk is due to the presence of peculiar microscopical organisms—known as bacteria—which multiply very rapidly, and in so doing produce a blue matter resembling aniline. These organisms render the milk unfit for food, especially for persons of weak digestive power.

M. H. F. BLANFORD reports that he has observed white ants in the act of emitting rythmical sounds. Another observer, Mr. F. P. Pascoe, has heard a peculiar sound, in fields of Southern Europe, which was found to be the song of a small lizard. It is generally believed that these creatures have no power of producing vocal sounds.

As we ascend from the earth the air grows thinner and thinner. From this fact astronomers believe that the limit of the atmosphere is 200 miles from the earth's surface.

COAL consists of from eighty to ninety-five per cent. of carbon mixed with a small proportion of mineral substances, which, after it is burned, remain as ashes, and of an inflammable gas contained in its interstices.

IN Alpine regions there are more narrow, partly-closed flowers than elsewhere, and a greater proportion of long-tongued insects, the flora seeming to be exactly adapted to the insects feeding on its honey.

The roes of various kinds of fish contain from about 30,000 to over 3,086,000 eggs.

The lion's teeth seem formed rather for destruction than for the chewing of his food.

A FOUR-FINGERED monkey, in its native state, has been seen to go down to the edge of a stream, rinse its mouth and then clean its teeth with one of its fingers.

IN Bavaria medical men are shorter lived than any other class. Out of every 100 individuals, 53 Protestants clergymen, 41 professors, 39 lawyers or magistrates, 34 Catholic priests, but only 26 doctors reach the age of 50.

THE octopus has a gland which secretes an inky fluid, and this he squirts out, making a thick, dark cloud behind him which baffles his pursuer at the same time that it helps himself to dart away. Mr. Darwin asserts that the octopus often takes deliberate aim at an enemy when it squirts out this unpleasant fountain.

OSTRICHES, when the full number of eggs have been laid, invariably place one of them outside the nest—the nest consisting naturally of a hollow scooped out of the land by the action of the wings and legs of the birds. It has been found that these eggs are reserved as food for the chicks, which are often reared in a natural stall, miles away from a blade of grass or other food.

MORAL ECONOMY.

INDUSTRY need not wish.

TRUTH is the basis of every virtue.

AVARICE is the mother of many vices.

THE path of truth is a plain and safe path.

OLD injuries are seldom canceled by new benefits.

HE that cannot live well to-day cannot to-morrow.

THE fountain of content must spring up in the mind.

FALSEHOOD sinks us into contempt with God and man.

THE road to home and happiness lies over small stepping stones.

THE touchstone by which men try us is most often their own vanity.

THERE is a long and wearisome step between admiration and imitation.

A MAN explodes with indignation when a woman ceases to love him, yet he soon finds consolation ; a woman is less demonstrative when deserted, and remains longer inconsolable.

IT is hard to personate and act a part long, for where truth is not at the bottom nature will always be endeavoring to return, and will peep out and betray itself one time or another.

HISTORICAL.

LIBRARIES existed in Egypt contemporaneously with the Trojan war.

THE earliest account of a diving bell in Europe is at Nuremburg, 1664.

CHAUCER cecived a pitcher of wine every day from the cellar of Edward III.

THE fine Syrian sponge is usually employed for the toilet, owing to its texture.

ON account of the scarcity of wood in India the people burn manure for fuel.

THE first normal school in America was established in Concord, Vt., in 1823.

CLOVES have been brought into the European market for more than 2,000 years.

THE Egyptians placed a mummy at their festal boards to remind them of immortality.

STATISTICAL.

THE value of property, as assessed, for purposes of taxation, in the United States, is $16,897,135,567, or $336.80 per capita for a population of 50,155,783. The New England States, with 4,010,529 of the population, hold $2,652,076,586 of the property, or $661.27 per capita; that is to say, with considerably less than one-twelfth of the population they have about two-thirteenths of the wealth of the country. The Middle States have $5,567,973,818 of property to 11,756,055 inhabitants, or $473.55 per capita; the Western States have $6,180,524,614 to 18,524,080 people, or $333.63 per capita; and the South, with 15,257,393 people, assesses its own property at $2,360,246,800, or only $155.20 for each person. The States which have the most wealth have also the heaviest debts. In New England the state, county and town indebtedness amounts to $44.54 per capita; in the Middle States, $41.57; in the West, $13.17, and in the South, $13.43. The difference does not exactly correspond with the difference in wealth, but it does approximately.

CONTRIBUTIONS.

For THE LANCASTER FARMER.

THE EGG—ITS CONTENTS AND HOW IT IS MADE.

My friend, as you are a close observer of nature, I should like you to explain to me the contents of this egg, and how it is made. It is composed of the ova, or yolk; and the albumen, or white, and a thin skin covering the same, and a shell enclosing the whole. What is the yolk composed of? It is composed of blood, assimilated through the working power of the hen; it also contains a portion of oil, derived from the grain that she may eat. What is the white composed of? It is a thick mucilage, made from any green substance that she may eat; young growing grass is preferable. Hens do not lay so well in winter, as the material for this purpose is in its dry state; the milk is made from the refuse of the woody, fibrous substance of the grass. The shell is composed of lime, or any hard substance easy to decompose; oyster shell, broken in small bits, is the best. Where are those ovas or yolks first formed ? They grow in a cluster on the spine, coming through a tuft of soft skin, perforated with small holes, and between the lungs and the kidney (fowls having but one), there is one forming every twenty-four or thirty-six hours, so long as they are in the laying mood. How long after the first appearance of the ova, before the egg is laid? From fifteen to twenty days; the ova, or yolk, is enclosed in a thin skin; as it grows the skin stretches; and when matured, the skin breaks, and it drops out into the mouth of the ova duct, which is somewhat of a funnel shape. The mouth then closes, and the yolk is swallowed into the first division of the duct ; it then opens again, ready for the next, always on the 'alert. When two drop at the same time, it forms a double yolk ; this is only a freak of nature, and the good condition of the hen. The first division of the duct is about five inches in length ; and in passing this division it makes three revolutions, and the white is put on in three separate layers. The next division is of the same length, and passing in a rotary motion, turning to the left with the small end first, opening the way as it passes, the same as swallowing. In this division is where the skinning process is performed ; and also in this is where it gets its shape, depending on the freeness of the duct to yield to its passage. The next division is six inches long; in this it receives the shell, which is a thin fluid, in color to suit the breed that is laying it, as it is the color of the egg that proves the genuineness of a thorough-bred fowl. At the terminus of the third division the duct is of a globe shape ; here the egg turns over, and passes big end first, which is head first, according to nature. How long is the ova duct ? It is from fifteen to twenty inches. This ova duct must be a curiously constructed affair. It is. At the terminus of each division there is an elbow, and the inner side is very soft, with a silk-like feel, and is composed of folds, each one lapping partially over the other, and soft and pliable ; the first division being the coarser, and increasing in fineness of folds, and more numerous ; and as the egg passes each division, it presses from beneath them the amount necessary for the make-up of the same, and no more. How is this egg fertilized, and when? Through the influence of the male bird, which passes through a small tube or duct, lying along the spine and making a connection with the cluster of small undeveloped ovas. How long will this egg keep, that I may rely upon its hatching, providing I turn it over every day ? You can't turn it over; you may turn the shell, but not the inner portion of the egg, as it is hung in the centre by two spiral cords, one being attached to each end of the yolk made fast to a thin net-work covering the yolk, and passes through the white and is fastened to the membrane or skin lining the shell. Each one of those cords is twisted the contrary way from the other, holding it the heavy side down all the time. This proves that the egg is growing and forming into its proper state, whilst passing the duct, as well as taking on its outward coating at the same time. Why is the head of the chick in the larger end of the egg? Because, when it is ready to extricate itself it has a greater distance to draw back its head and propel forward again with a heavy stroke, until the shell is cracked to admit air. This is its first breathing. How is it that it strikes its place every time? Because its head and neck is under the left wing ; therefore it is supported by the same, and kept on a level. By this means it strikes the same place every time ; it soon gains strength and knocks a hole through the shell. What is its mode of growth in the shell? It is made up entirely of the albumen or white; the first coating, or layer, forms the bone and sinews; the second the flesh, the third the skin; the first formation are two black specks, which are the eyes, one on each side of the spiral cord at the larger end ; next the skull bone between, the neck and spine, legs and wings attaching; at nine days there is life; at the end of two weeks the white is consumed; the two spiral cords make a connection in the stomach and protrude from the navel; now being formed into blood veins, and enclosing the yolk in a network of small ones; through these the chick derives its nourishment from the yolk; transforming back to its former substance, blood, after cracking the shell, it gains strength very fast, and those two blood veins commence drawing into the belly, and lifting what remains of the yolk, and draws it in also; it now has strength to stretch out its tiny legs the yolk being out of the way of its toenails, there is no danger; the navel being closed, and with its feet at the bottom and head and shoulders at the top, the shell divides in two halves and the chick rolls out. What have we that comes into the world, I may say, on a more scientific principle than the fowl, take it from the first formation of the ova. Such is nature; the Almighty has made all things in wisdom, and for our benefit, and there are so many ways to cook the egg, also the chick, and every way of each it is calculated to tickle the palate. Take care of your poultry.—*W. I. P.*

For The Lancaster Farmer.

FRUIT BELTS.

Close observers could hardly have failed to notice that for a number of years past certain sections of Lancaster county have produced better apples than others. The section lying east of a line drawn northward from Christiana through Leacock, West Earl and the western parts of Ephrata and Clay townships, grows finer apples than the section west of that line. In the southern part of the county is another small fruit belt. This includes part of Martic and Drumore, Fulton and Little Britain townships. Any one desirous of verifying these assertions need only compare the fruit brought to our market from New Holland, Ephrata and other points in the several townships named, with that brought from Conestoga, Manor, Hempfield and other places in the western section.

How can we account for this difference? The cause cannot be in the soil, as that is not materially different in the several locations. Latitude and longitude cannot have a marked effect on so small a scope as a single county.

Our hot and dry summers for a number of years, no doubt, have been the great hindrance that we had to contend with in successful apple culture. Can it be that those eastern and southern belts have more rain?

Our main supply of rain during the summer season, comes from thunder showers, and these, as is well known, are more or less subject to attraction by mountain ranges and large streams of water.

In severe drought, for many years past, the observation has been made from a certain point in the county, that a thunder shower to reach that point must almost invariably arise north-east from that point. If it rises only a few degrees north of that point it will go in an eastern direction and discharge itself through the north-eastern part of the county. If the shower forms a few degrees further south, it will discharge itself obliquely across the southern part of the county. Very often thunder showers starting at the point named, divide, one part thereof going east, while the other takes the southeastern course.

The attractive points, no doubt, are the Conewago hills on the north and the Susquehanna river on the south.

In most seasons these attractive points lose much of their force, and then thunder showers appear to be able to move in all directions.

The past season thunder showers were rare, the drought was very severe over the greater part of the country, but the few thunder showers we had almost invariably followed the Susquehanna, and the southern belt suffered but little from drought, as the crop of apples, potatoes and corn there raised fully shows.—*Casper Hiller.*

CHINESE FRUIT PEAR.

Columbia, Pa., February 10, 1882.

"Sha lea," or Chinese Sand Pears.
"Suet lea," or Chinese Snow Pears.

Friend S. S. Rathvon:

During 1852, I got a tree from the late Wm. R. Prince, of Flushing, Long Island, New York. It grew vigorously, and in a few years produced a heavy crop of its large and beautiful fruit. The pears are large and showy, but they never become soft or eatable, unless cooked. We did not know what they were good for, and we let them rot on the ground, but we have since discovered that for canning, for preserves, for applebutter, these pears can't be excelled.

If you wish to boil applebutter, and use them with the cider instead of apple, you will have an article that any person would prefer to all apples—all I can say is—it "tastes different!" At one time I raised a lot of seedlings; these grew from two to five feet high the first season. Other seedling pear trees rarely grew as many inches the first year. Of course I thought these would make excellent stocks to work on other pears; but I soon found that other pears, though growing freely on this stock for a year or two, did not continue their vigor, but stopped growing, became stunted, mossy, and bore poor, knotty fruit, and would not make thrifty trees. Yet these Chinese and crosses all take kindly on other pear stocks or trees.

A friend in Columbia, to whom I gave grafts many years ago, set them on top of a large pear tree; and this has never been injured by cold, blight, or any thing else, but bears lots of its large fruit every year.

The last severe winter, (22° below zero), apparently did not injure any of my trees, yet the flower buds must have been injured, as none of my trees had a flower, or bore a single specimen of fruit the last season.

In 1880, the "Sha lea" and my seedling both bore heavy crops, though quite small trees. My seedling on a limb four feet long and an inch in diameter where it branches out from the main trunk, had thirty seven (37) large pears. One morning I went out with a basket intending to take them off—but lo! a person who probably had a better right to them than I had, cleaned them all of!

I have never seen or heard of a well authenticated case of blight on any of this class of pears. I have had other pear trees killed by blight that stood only twenty feet from them. There have been rumors of them blighting, but these rumors want confirmation.

At Rochester, New York, they have what they call the Japan Pear. This may be what Mr. Prince called the "Suet lea," or Chinese Snow Pear, or a cross of it, as it is of the same class as all the other of the Chinese varieties—the Kieffer, Le Conte and the rest. This Japan variety is certainly a most excellent eating pear, as I can fully acknowledge from a specimen sent me last fall by Charles Downing, of Newburg, New York. This specimen was as round as a ball, with stem and eye a little depressed, twelve inches in circumference and of excellent quality.

Now where any trees of these Chinese species, such as have already been mixed or crossed with good pears, are growing and bearing fruit, the probability is, that their seeds being planted, these seedlings will still retain their peculiar growth and health, and the prospect of still further improvement is very promising. However, it would be better not to depend on bees to carry the pollen as they may take pollen from the poorest pears, but by opening the flowers on a Chinese, and carefully removing the pistles before the pollen is ripe, and then with a camel's hair brush take the pollen off of a flower of a superior variety, and apply to the stigma of the one you wish to impregnate, you can hardly fail of success, and a new and superior class of pears will be the result. J. B. Garber.

COMMERCIAL FERTILIZERS.

The question of the comparative values of the various kinds of fertilizers manufactured or sold in this section of the State, appears to be a matter of special interest with the farmers and others in the fast-improving agricultural district of which Oxford (Chester county) is the centre. According to an act of the Legislature of the 28th of June, 1879, every package of commercial fertilizer offered for sale is required to have stamped upon it the name of the manufacturer, the place where manufactured, the weight, and an analysis stating the percentage therein contained of nitrogen, or its equivalent of ammonia in an available form, of potash soluble in water, of phosphoric acid, &c., every manufacturer or importer of such fertilizers being required to pay a license to the State varying from ten to thirty dollars, according to quantity sold, and to the with the Secretary of the Board of Agriculture a copy of the analysis above referred to. Any person selling or offering for sale any commercial fertilizer without the required analysis, or stating that it contains more of the specified constituents than it really does, it shall be liable to a fine, ranging in amount from twenty-five to one hundred dollars for the first offence, and not less than two hundred dollars for each subsequent offence—one half to go to the informer, provided the informer is a purchaser for his own use. It is made the duty of the board of agriculture to analyze such specimens of fertilizers as may be furnished by its agents, said samples to be accompanied with proper proof that they were fairly drawn, and the money paid for licenses is to constitute a special fund out of which the expenses of analysis are to be paid.

In pursuance of this act, Prof. F. A. Genth, "Chemist of the Pennsylvania Board of Agriculture" has made a tabular statement, giving the chemical analysis of more than one hundred different kinds of fertilizers which are sold in the State, most of which are also manufactured here, but some are imported from Maryland, New Jersey, New York, Ohio, Illinois, and even from Missouri. The money value of the different manurial ingredients is rated by Prof. Genth as follows, viz; "soluble and reverted phosphoric acid 10 cts. per pound; insoluble phosphoric acid from bone, 6 cts.; from South Carolina Rock, 5 cts.;

potash, 6 cts.: ammonia 17½ cts. per pound." On the basis of these rates and of the analysis of the different samples tested, the Professor gives the estimated value per ton of each kind embraced in the table, as also the selling price of the same at the place of selection. If his figures are in the least to be depended on, a great deal of money is wasted by farmers in the purchase of fertilizers whose value is much less than the cost. Of the whole number of samples given in the table the sell-price of more than two-thirds of them is greater than the calculated value, and in some cases very much greater. For instance, the "Complete Bone Phosphate" from the Allentown Manufacturing Company, worth only $25.21 is sold at $35; "Plant Food" from Frederick, Md., selling at $40, is worth but $30.78; the "Economical Fertilizer" of Baugh & Sons, Philadelphia, worth $25.33, they sell at $33, &c.; the "Ammoniated Bone Phosphate" of Josiah Cope & Co., near Oxford, worth $27.16, is sold at $35; the "Fossil Alkalite" of Reeve & Co., selling for $15, is only worth $2.05; and the "Ammoniated Bone Phosphate" of the Susquehanna Fertilizing Company, at Oxford, selling at $34, is given as worth only $10.46. On the other hand, the fertilizers produced by a considerable number of manufacturers, according to Prof. Genth's table, are worth much more than they are sold for—as the "Raw Bone Phosphate" of Job Pugh, Oxford, selling for $35, is given as worth $38.08; the "Superior Acid Phosphate" of the Susquehanna Fertilizing Company, at Oxford, selling at $25, is worth $28.40; the "High Grade S. C. Rock" of the Waring Manufacturing Company, at Colora, Md., selling at $25 per ton, is set down as worth $32.78; and "Waring's Q. and L. Bone," by the same company, selling at $35, is rated by Prof. Genth to be worth $48.49.

The publication in the *Oxford Press* of the table from which the above figures are taken has raised some excitement among the manufacturers and dealers in fertilizers in that neighborhood, and as one of the first results, the following advertisement appeared in the *Oxford Press* of last week:

A PUBLIC MEETING of Farmers and Manufacturers of Fertilizers will be held in Grange Room at Lincoln Station, (near Oxford,) on Wednesday, Feb. 8th, at 1 P. M., to consider the following question, viz.: Should farmers in buying fertilizers be guided by their estimated value as determined by analyses and published by the State Board of Agriculture? The purpose of the meeting is to hear from the manufacturers on this subject.

What conclusion was arrived at by the meeting, if any, we have not heard. The manufacturers and dealers in the fertilizers pronounced to be of comparatively little value will naturally feel dissatisfied, and probably will endeavor to have Prof. Genth's verdict set aside, while those whose productions are declared by him to be of high value, will pretty certainly maintain the correctness of his calculations and conclusions. It will be for the farmers to decide between them.—J. P.

[We never supposed, that either the act of the Legislature or the analysis of the Chemist of the State Board of Agriculture, were to be regarded as an arbitrary and unchangeable *ultimatum*, governing the manufacture and price of Chemical Fertilizers. We rather regarded it as a preliminary experiment under the sanctions of law, to prevent present pos sible impositions and frauds in the manufacture and sale of these manurial substances. The law, however, should not be wiped out, merely because in its execution it happens to discriminate in favor of one set of fertilizers and against another set. Whatever inequalities may exist, should be corrected by a readjustment of the scale of prices, after a fair and calm investigation of the subject. Manufacturers, regarding the matter from a merely selfish standpoint, will avail themselves of the endorsement of the State chemist, no matter how worthless their goods may be: and those who deem their goods discredited will, of course, be dissatisfied, perhaps, like the Irishman in Court, who *feared* that justice would be done him. After all, it is possible that the chemical composition stamped upon the outside of a sack of manure, may not prove a protection against the fraud inside.]

SELECTIONS.

POULTRY-SHOW.

Premiums Awarded—Some Special Points.

The third annual exhibition of the Lancaster Poultry association, which opened, in Excelsior Hall, on Thursday, January 12th, 1882, closed on Wednesday evening the 18th. It was, in all respects, the best and most successful exhibition of poultry ever given in this county, if not in the state, and was attended by a far larger number of visitors than either of the preceding shows given under the auspices of the society.

Below will be found the list of premiums awarded by the judges, and paid by the so society.

Class 1.—Asiatics.

Light Brahma—Fowls, Dr. D. F. Royer, 1st and four specials; Wm. F. McLean, 2d; H. H. Hewitt, 3d. Chicks, Hon. C. S. Cooper and Dr. D. F. Royer, tied for 1st and 2d premiums and special, and divided them; T. M. Nelson 3d.

Dark Brahma—Fowls, Dr. D. Royer 1st. Chicks, Dr. D. F. Royer 1st and 2d, and tied Zimmerman and Hoffer for 3d, beside taking two specials.

White Cochin—Fowls, A. S. Flowers, 1st and special; J. F. Shaffer 2nd and 3d. Chicks, A. S. Flowers, 1st 2d and 3d, and several specials.

Black Cochin—Fowls, Samuel G. Engle 1st and 2d and several specials: J. F. Shaffer 3d. Chicks, T. Frank Evans 1st and 2d and specials; Dr. E. H. Witmer 3d.

Buff Cochin—Fowls, L. K. Bennett 1st; Zimmerman & Hoffer 2d; M. B. Weidler 3d. Chicks, no 1st premium; J. B. Long 2d; A. B. Hostetter, 3d.

Partridge Cochin—Fowls, H. S. Garber 1st and 2d and five specials; C. E. Long 3d. Chicks, H. S. Garber 1st and 2d and tied Dr. D. Royer for 3d and special.

Lanshan—Fowls, Dr. D. F. Royer 1st; T. Frank Evans 2d. Chicks, A. H. Sharpless 1st.; Dr. D. F. Royer 2d.

Games.

Black Breasted Red—Fowls, Dr. D. F. Royer, 1st and 3d; E. N. Denman, 2d. Chick. Dr. D. F. Royer, 1st and specials; E. F. Denman, 2d.

Brown Breasted Reds—T. B. Dorsey, 1st; T. K. Bennett 2d and a tie between them for special. Chicks, T. K. Bennett 1st and two specials.

Ginger Red—Fowls, T. K. Bennett 1st; no competition.

Yellow Duckwing—T. K. Bennett 1st—no competition. Chicks, T. K. Bennett 1st.

White Game—Fowls, T. B. Dorsey 1st, 2d and special.

Black Game—Fowls and chicks, no 1st prem., T. K. Bennett 2d.

Gray—Chicks, T. B. Dorsey 1st.

B. B. Red Malay—Fowls, D. M. Brosey 1st. Chicks, D. M. Brosey 1st and 2d.

Class 3—Game Bantams.

Black Breasted Red—Fowls, T. K. Bennett 1st, 3d and two specials; Charles E. Long 2d. Chicks, T. B. Brosey 1st and special; Frank Selak 2d; George Snyder 3d.

Brown Breasted Red—Fowls and chicks, T. B. Dorsey 1st—no competition.

Ginger Red—Chicks, J. L. Otto, 1st and special—no competition.

Yellow Duckwing—Fowls, Dr. J. C. Maple 1st, 2d and special; Chas. E. Long 3d. Chicks, T. K. Bennett 1st, Chas. E. Long, 2d.

Silver Duckwing—Fowls, Dr. J. C. Maple 1st and special, and tied T. K. Bennett for 2d and 3d. Chicks, George Snyder 1st; Dr. J. C. Maple 2d; Aug. L. Wentzel 3d.

Red Pyle—Fowls, T. B. Dorsey 1st; Geo. Snyder 2d; Chas. E. Long, 3d. Chicks, T. K. Bennett 1st, and ties Dr. Maple for 2d and 3d.

White Pyle—Fowls, George Snyder 1st; Dr. J. C. Maple 2d. Chicks, J. B. Lichty 1st; Dr. Maple 2d.

Black—Fowls, T. K. Bennett 1st—no competition. Chicks, T. B. Dorsey 1st and special—no competition.

White—Fowls, T. B. Dorsey 1st. Chicks, J. L. Otto 1st; T. B. Dorsey 2d.

Gray—Chicks, J. L. Otto 1st—no competition.

Class 4—Hamburgs.

Black—Fowls, Snyder & Hartman, 1st and 2d. Chicks, Geo. C. Liller, 1st and four specials; T. K. Bennett, 2d; T. B. Dorsey, 3d.

Silver Penciled—Fowls, S. M. Nelson, 1st—no competition. Chicks, Mrs. Kate Yearsley Ash, 1st and 2d.

Golden Penciled—Fowls, J. W. Bruckhart, 1st; Snyder & Hartman, 2d; T. K. Bennett, 3d. Chicks, T. B. Dorsey, 1st; J. W. Bruckhart, 2d and 3d.

Silver Spangled—Fowls, no 1st premium; Wm. F. McLean, 2d; Hon. J. A. Stober, 3d. Chicks, T. B. Dorsey, 1st and special; Hon. J. A. Stober, 2d and 3d.

Class 5—Spanish.

Black Spanish—Chicks, John Grosh, 1st—no competition.

White Leghorn—Fowls, Henry Neater, 1st and 3d; Dr. D. F. Royer, 2d. Chicks, Robert R. Morris, 1st and six specials; John B. Trissler, 2d and 3d.

Brown Leghorn—Fowls, Dr. D. F. Royer, 1st—no competition. Chicks, Jos. H. Trissler, 1st, 2d and five specials; M. L. Grelder, 3d.

Class 5—American.

Plymouth Rock—Fowls, Dr. D. F. Royer, 1st and five specials; Aug. L. Wentzel, 2d; Lount Lattin, 3d. Chicks, Dr. D. F. Royer,

1st and 2d and five specials; Aug. L. Wentzel, 3d.

Dominique—Chicks, John Wilcox, 1st and special; M. L. Greider, 2d; T. K. Bennett, 3d.

American Sebright—Fowls, Mrs. Kate Yearsley Ash, 1st. Chicks, G. C. Morris, 1st and 2d; Mrs. Kate Yearsley Ash, 3d.

Black Java—Chicks, M. L. Greider, 1st and special; Lount Lattin, 2d and 3d.

Erminettes—Fowls, Kate Yearsley Ash, 1st—no competition.

Class 7—Polish, Plain or Bearded.

White Crested White—Fowls, J. W. Carroll, 1st; T. B. Dorsey, 2d. Chicks, Wm. A. Schoenberger, 1st.

White Crested Black—Fowls Dr. D. F. Royer 1st—no competition. Chicks, Dr. D. F. Royer, 1st and 2d and four specials; J. W. Bruckhart, 3d.

Golden Bearded—Fowls, T. B. Dorsey, 1st; T. K. Bennett, 2d; Wm. A. Schoenberger, 3d. Chicks, T. B. Dorsey, 1st; f. K. Bennett, 2d; J. W. Carroll, 3d.

Silver-Bearded—Fowls and chicks, T. B. Dorsey, 1st—no competition.

Class 8—French.

Houdan—Fowls, B. S. Koons, 1st; Richard Preusser, 2d. Chicks, T. W. Wyman, 1st and four specials; T. M. Nelson, 2d; H. H. Hewitt, 3d.

Class 9—Dorkings.

White—Fowls, no 1st premium; W. J. Kirby, 2d and special—no competition.

Colored—Chicks, H. H. Tshudy, 1st—no competition.

Class 10—Bantams.

Golden Seabright—Fowls, no 1st premium; Dr. J. Maple, 2d and two specials; B. S. Koons, 3d.

Silver Sebright—Fowls, no 1st or 2d premiums; Dr. D. F. Royer, 3d and two specials.

Rose Comb White—Fowls, J. L. Otto, 1st, 2d and two specials. Chicks, J. L. Otto, 1st —no competition.

Japanese—Fowls, T. B. Dorsey, 1st—no competition. Chicks, B. S. Koons, 1st—no competition.

Black African—Fowls, Chas. Lippold, 1st; T. B. Dorsey, 2d; Dr. J. C. Maple, 3d.

Class 11—Miscellaneous.

Silky—Fowls and chicks, Wm. M. McLean, 1st—no competition.

Class 12—Turkeys.

Bronze—Fowls, B. L. Wood, 1st and three specials; Samuel G. Engle, 2d; T. M. Nelson, 3d. Chicks, B. L. Wood, 1st.

White—Fowls and chicks, J. W. Bruckhart, 1st and two specials.

Narragansett—M. L. Greider, 1st and special—no competition.

Class 13—Ducks.

Pekin—Dr. D. F. Royer, 1st and special; Geo. A. Geyer, 2d; J. W. Bruckhart, 3d.

Rouen—Geo. A. Geyer, 1st.

Colored Muscovy—Dr. D. F. Royer, 1st.

Cayuga—Dr. D. F. Royer, 1st.

Class 14—Geese.

Toulouse—George A. Geyer, 1st and special.

Class 15—Ornamental.

White Guineas—J. B. Garman, 1st—no competition.

Pearl Guineas—John M. Hagans, 1st—no competition.

Breeding Pens.

The following premiums were awarded the exhibitors of breeding pens consisting of one cock and four hens in the classes named:

White Cochins—A. S. Flowers, 1st and one special.

Partridge Cochins—H. S. Garber, 1st and one special.

Games—J. B. Lichty, 1st and one special.

W. F. Bantams—George Snyder 1st premium.

Silver Penciled Hamburgs—J. W. Bruckhart, 1st premium.

White Leghorns—R. R. Morris, 1st premium.

Brown Leghorns—M. L. Greider, 1st premium.

Plymouth Rocks—James Black, 1st.

Brahmas—J. B. Long, 1st.

Class 16—Pigeons.

Carriers—Black, John E. Schum, 1st and 2d; J. M. Skiles, Jr., 3d. Blue, John E. Schum, 1st. Dun, John E. Schum, 1st. White, Chas. Lippold, 2d.

Pouters—Checkered, Henry Neater, 1st. Red Pied, Chas. Lippold, 1st; Geo. C. Liller, 2d.

Barbs—Henry Neater, 2d; Geo. C. Liller, 3d. Red, J. M. Hagans, 2d. Yellow, J. M. Hagans, 1st. White, John E. Schum, 2d.

Fantails—Crested, J. M. Skiles 1st; J. M. Hagans, 2d. White, Chas. Lippold, 1st; J. M. Hagans, 2d; F. A. Pennington, 3d. Plain white, Chas. Lippold, 1st; J. M. Hagans, 2d; F. A. Pennington, 3d. Black, J. M. Skiles, Jr., 1st. Blue, J. M. Hagans, 1st. Yellow, C. S. Greider, 1st; Chas. Lippold, 2d; J. M. Skiles, Jr., 3d. Dun, J. M. Skiles, Jr., 2d.

White Calcutta—Crested, F. A. Pennington, 1st.

Jacobins—Black, John E. Schum, 1st; F. A. Pennington, 2d; J. M. Hagans, 3d. Red, Charles Lippold, 2d; F. A. Pennington, 3d. Yellow, Charles Lippold, 2d; J. M. Hagans, 3d. White, Henry Neater, 3d.

Tumblers—Baldhead, black, Charles Lippold, 1st; J. M. Hagans, 2d. Blue, C. S. Greider, 1st. Bearded, yellow, Charles Lippold, 1st. Red, Charles Lippold, 1st. Short-Faced, Charles Lippold, 1st; John E. Schum, 2d. Inside Tumblers, black, red and mottled, Charles Lippold, 1st.

Turbits—Solid colors, black, Henry Neater, 1st; Charles Lippold, 2d; J. M. Hagans, 3d. Red, J. M. Skiles, Jr., 2d. White, John E. Schum, 1st; Henry Neater, 2d; Joseph Eibel, 3d. Yellow, Henry Neater, 1st; Charles Lippold, 2d; John E. Schum. 3d. Turbits, winged—Red, Charles Lippold, 1st; John E. Schum, 2d; J. M. Hagans, 3d. Yellow, John E. Schum, 1st; J. M. Hagans, 2d; George C. Liller, 3d. Black, Charles Lippold, 2d. Silver, J. M. Hagans, 2d. Tailed Turbits—Black, John E. Schum, 1st and 2d; Thomas Humphreyville, 3d. Blue; Charles Lippold, 1st. Dun, J. M. Skiles, Jr., 1st; Charles Lippold, 2d.

Trumpeters—Black, John E. Schum, 1st and 3d; Charles Lippold, 2d. Yellow, John E. Schum, 2d. White, John E. Schum, 1st J. M. Skiles, 2d; Annie May Raymond, 3d. Mottled, Charles Lippold, 1st; John E. Schum, 2d; C. S. Greider, 3d.

Antwerps—Blue, Charles Homan, 1st and 2d; Joseph Eibel, 3d. Silver, Joseph Eibel, 1st; Charles Lippold, 2d; Christ. E. Barr, 3d. Red checkered, J. M. Hagans, 1st; Charles Lippold, 2d; John E. Schum, 3d. Blue checkered, John E. Schum, 1st; Jos. Eibel, 2d and 3d.

African Owls—White, John E. Schum, 1st; Chas. Lippold, 2d. Blue, Chas. Lippold, 1st; John E. Schum, 2d.

English Owls—Blues, Thos. Humphreyville, 1st. Silver, Chas. Lippold, 1st; C. S. Greider, 2d. Yellow, J. M. Skiles, 2d; Chas. Lippold, 3d.

Nunflowers—Red, J. M. Hagans, 1st; John E. Schum, 2d. Blue, J. M. Skiles, 1st; John E. Schum, 2d; J. M. Hagans, 3d. Black, Chas. Lippold st; John E. Schum, 2d and 3d. Yellow, John E. Schum, 1st and 2d.

Magpies—Red, J. M. Hagans, 1st. Yellow, Henry Neater, 1st; J. M. Hagans, 2d and 3d. Black, J. M Hagans, 1st.

Nuns—Black, J. M. Skiles, 1st; Chas. Lippold 2d. Yellow, John E. Schum, 1st.

Snells—Black, Chas. Lippold, 1st. Red, J. M. Skiles, 3d.

Mooorheads—J. M. Hagans, 1st and 2d; Geo. C. Liller, 3d.

Quakers—Blue, Chas. Lippold, 2d.

Frill-Backs—John E. Schum, 1st.

Birmingham Rollers—Charles Lippold, 1st; John E. Schum, 2d.

Archangels—J. M. Hagans, 1st; John E. Schum, 2d.

Priests—John E. Schum, 1st.

Ice Pigeons—John E. Schum 1st and 2d.

Class 17—Cage and Ornamental Birds.

Belgium Canary—Charles Lippold, 1st and 2d; Zachariah M. Weaver, 3d.

Cardinal—Chas. Lippold, 2d—no competition.

Mocking Bird—Wm. Killinger, 1st—no competition.

Gold Finch—Chas. Lippold, 1st—no competition.

Ring Dove—J. B. Garman, 1st—no competition.

Bull Finch—Chas. Lippold, 1st no competition.

Pertinent Poultry Points.

T. B. Dorsey, of St. Denis, Md., was very properly awarded, in addition to numerous other premiums, the special premium of $10 for the best collection of fowls on exhibition. Fine, as is his collection, he added largely to it while in Lancaster by purchasing a large number of the finest birds exhibited by other breeders.

H. S. Garber, of Mount Joy, enjoyed the distinction of exhibiting the fowl scoring the highest number of points. It is a Partridge Cochin hen and scored 99 points out of a possible 100. It was awarded, as a special premium, a folding exhibition coop, and in connection with its mate, a fine cock scoring 95½ points, took first premium and a year's subscription to the weekly *Intelligencer*. Mr. Garber's breeding pen of Partridge Cochins was awarded the silver cup valued at $10, given by M. L. Greider for the breeding pen of White, Black, Buff or Partridge Cochins scoring the greatest number of points.

A. S. Flowers, of Mount Joy, carried off almost all the premiums, regular and special on White Cochins, none of his birds scoring less than 92 points and some of them as high as 98. His exhibits attracted much attention.

T. Frank Evans, of Lititz, and S. G. Engle, of Marietta, divide the principal honors and profits on exhibits of Black Cochins, their birds scoring well up in the nineties.

Of Games, T. K. Bennett, of Phillipsburg, had the largest number, the greatest variety and took the greatest number of premiums. One of his Br. B. Red pullets scored 97¼ points, was awarded $3 as the best of her class and a special valued at $10.

The display of Bantams was very large and fine, and the bulk of the premiums were awarded Messrs. Dorsey and Bennett, and Dr. Maple, of Trenton, N.J.

There were many fine exhibits of Hamburgs, of all colors; but George C. Liller, of this city took the lead. He entered but a single pair of Black Hamburg chicks and with them he was awarded the following; for best pair Hamburg chicks 1st premium and special; for best Hamburg pullet of her class special premium; for best Hamburg cockerel, special premium; for best pair of Hamburgs of any variety, special premium. It is not often that a "single pair" wins so big a pot.

Jos. R. Trissler, of this city, roped in a majority of the premiums offered for Brown Leghorns, and Robert R. Morris, of Pottsville, for White Leghorn chicks; though Henry Neater, of York, took first premium for White Leghorn fowls.

So far as Plymouth Rocks were concerned, Dr. D. F. Royer, of Shady Grove, Franklin county, sat down on everybody else. In addition to the regular cash premiums awarded him he carried off ten specials, valued at $35.50.

B. L. Wood, of Doe Run, Chester county, was awarded first premium for best pair of Bronze Turkeys, the cock scoring 97 and the hen 99 points. S. G. Engle had on exhibition a heavier pair, but the cock had accidentally hurt his wing which reduced his score.

J. W. Brackhart, of Salunga, took first and two special premiums for a fine pair of white turkeys. H. H. Tshudy, of Lititz, showed a henvier pair, but the cock was "disqualified" because he had a small black feather in his tail.

There was a warm contest between John E. Schum and Chas. Lippold, both of Lancaster, for the pigeon championship. Each of these fanciers took some thirty premiums, but according to a close calculation made by the executive committee Schum came out one point ahead, and was awarded an additional premium of $10 for the best collection, Lippold taking a $5 premium for second best.

WHITE VEIN—CAUSE OF THIS DISEASE IN TOBACCO.

There are a few things connected with tobacco growing more aggravating to the grower than to find on stripping his tobacco that the small ribs or veins are not colored like the rest of the leaf, causing it to present a streaked appearance. The farmer, very naturally, asks himself the cause, and soon has some theory to account for it.

A number of these theories have come under the writer's observation, and some have been tested by him, and a record of his experiences, it is thought, will be of interest to others and stimulate them to test the matter more fully, both by experiment and observation.

The Early Cutting Theory.

The first theory, as near as I can recollect, was given me six or seven years ago, and was the too early cutting of the crop. We cut several hundred stalks quite green, in order to give a road through the field. This was colored so nicely, while the balance of the crop that year, which had ripened, contained so much white vein that it refuted this theory at once, and I began to inquire for another.

The next year a friend cut his crop rather over-ripe, had plenty of white vein, and he jumped to the conclusion that over-ripeness caused it. The next year several neighbors cut early and still had white vein.

Another gentleman proposed that old worn lands caused the vein, and that on new lands it would cure all right. This also proved incorrect, as I have had white vein on the best of new ground.

Convincing Experiences.

Thus I continued groping in the dark until the summer of 1879, in which I had experiences which convinced me I had at last reached the right solution of the problem. That season I had a variety of tobacco known in our neighborhood as the "Hanging Leaf Hoover," which is of slow growth. After topping, it received but a slight rain until it was cut off. On stripping it, I found the tobacco all nicely cured except that the first five or six branch ribs or veins from the tips of the leaves are white!

I reasoned as follows: The phenomenon is often witnessed of the human heart becoming so weakened by disease that it is not able to propel the life power, the blood, into the extremities, the feet and hands, thus causing them to die first, often as long as several days before the heart ceases to beat; so the plant, by continued drouth, became so weakened that the sap did not circulate to the veins at the extreme points of the leaves, and they died before the tobacco was cut and could not possibly cure brown, as could those which were nearer the life centre of the plant—the stalk—and therefore grew more perfect.

This, then, I think is the cause of white vein; either from drouth or some other cause the plant becomes stunted before cutting, and the veinlets are no longer vitalized and cannot cure as do those of stalks which continue growing vigorously from the time of planting until it is cut, and in this new land has the advantage, as it pushes the plant to perfection quicker than old soils.

I do not think, as some do, that white vein is under our control, but that it depends entirely on the weather after topping, and I think if farmers will but reflect how the growing season was when they had much white vein, they will invariably have found it dry and hot.—E. K. B., in New Era.

TOBACCO GROWING—PROFITS REALIZED BY SOME EXPERTS.

The past year was a remarkable one in several ways for the tobacco growers of Lancaster county. The planting season opened very auspiciously, and the young plants were, perhaps, never set out under more favorable circumstances. For a time all went well and the crop came along famously. But at the season when rains were most needed by the maturing plants, a long-continued drouth set in, which continued without intermission until the crop was harvested. What promised to be the largest crop ever grown in this county proved the smallest we have had in recent years.

Early Buying in the Field.

But the early planted fields had advanced so far towards maturity when the dry spell came that they suffered comparatively little from want of rain. The belief that there would be a very short crop woke up the buyers to a study of the situation, and as the previous year's crop had been very defective, each buyer became very desirous of securing some of the choice lots of the present season. The result was that about the middle of August buyers by the dozen came pouring into the county, overrunning every portion of it in their search of choice lots, which, when found, they at once purchased while still standing in the field, paying unprecedented prices for them. Nothing to match this scramble for the weed had before been seen among us, and perhaps nowhere else in the United States. Perhaps one-half the entire product of the county was purchased in this way, and even after the furore had spent its greatest force, the buying continued steadily until nearly the whole product of the county was secured by the eager buyers.

Result of Careful Handling.

Purchasers, however, by the terms of their contracts, bound the farmers to an unusually careful handling of their crops and the latter, fearful that the high prices paid by the former might induce them to find fault with the purpose of breaking their contracts, were careful to manipulate their crops with even more than their usual care. The result has been that much of the present crop is in some particulars the best and most carefully handled we have ever seen, and has proved unusually profitable to the growers, as we hope and believe it will also be to the liberal men who have bought it. Tobacco has been delivered at the packing houses in this city during the present month equal in quality to any ever grown in Pennsylvania, and although the weight per acre is considerably below the average of some other years, the greatly increased prices received for the crop have run the value per acre realized by some farmers fully up, if not beyond, that of any previous year. Several instances of this kind have come to our notice during the present week, and we have deemed the matter of sufficient interest to give the figures here.

An Excellent Crop.

The first crop to which we call attention was that grown by Mr. Moses Snavely, of Pequea township, purchased by Messrs. Skiles & Frey, of this city, and received by them on last Wednesday. It was not a large crop, consisting of only 10,400 plants, grown on something less than two acres of ground. It was planted in rows four feet apart, and 28 inches apart in the rows. The crop was sold in the early fall at 33 cents through, and when delivered was found to consist, after careful assorting by the grower, of 1,640 pounds of wrappers over 24 inches long, 764 pounds of wrappers under 24 inches, 490 pounds of seconds, and 377 pounds of fillers, making a total of 3,271 pounds, by no means a large yield so far as pounds are concerned, but the great price of 33 cents brought the value of the crop to $1,079.43, for which sum the fortunate grower received a check.

Cost of Growing Tobacco.

It is needless for us to say this lot of tobacco is a superb one. The leaves are long, silky, soft and tough, and the butts of the "hands" are as even as if they had been planed off. It has been well handled, as it deserved to be. To show how much labor and expense was incurred in the production of this lot of tobacco, the grower, at our request, made a detailed estimate, which will show not only what figures can be realized from tobacco growing, but what care and and attention are required to raise a first-class crop. A year ago, at the request of the Census Department, we procured from a number of well known growers careful estimates of the cost of growing an acre of tobacco; we have often wished to give them in these columns, but as they are to appear in the government report we have not felt at liberty to use them until then. The following estimate will, however, serve to show growers elsewhere something of the cost of growing fine tobacco here:

Interest on value of land ($250 per acre)	$ 30 00
Marking and care of seed bed	5 00
Plowing two acres one time	5 00
Harrowing ground three times	8 00
Making out rows	1 50
Setting out plants	8 00
Cultivating with shovel-harrow five times	10 00
Hoeing three times, eighteen days	18 00
Worming, topping and suckering	35 00
Cutting and hanging in barn	8 00
Interest on cost of barn, lath, etc	10 00
Stripping and preparing for market	40 00
Bringing to market	8 00
Value of manure used	25 00
Total cost	$212 00

Here we have as the total cost of the crop $212.00; the field was less than two acres, but to avoid fractions, we will call it two full acres, and we therefore find that the cost per acre was $106.00. This leaves the grower a net profit of $433.71 per acre, which, all things considered, is truly a wonderful result. The field was so much less than two acres that, strictly speaking, the profits may fairly be set down at $450 per acre. There was not one day during the entire growing season that hands were not at work in the field. The worming was not done once or twice a week, but every day; nor was this task left to children. In short, the labor steady throughout the season, and nothing was left undone to secure success. The sum realized shows that it pays to give the tobacco crop careful attention.

Another Paying Crop.

Messrs. Skiles & Frey received the crop grown on 3½ acres, grown by Mr. Jacob Stehman, of Manor township. The yield, in weight, was much greater in this case than in the preceding one, having been 7,737 pounds, or 2,210 per acre, but the price paid was only 24½ cents through; this netted the grower $1,895.56 for his crop, or at the rate of $541.58 per acre. If we allow for cost of cultivation at the same rates as estimated in the crop mentioned above, we have as the net profit per acre $435.58, which nearly equals the results secured by Mr. Snavely. Let us suppose, for a moment, that Mr. Stehman had received the same price for his crop per pound that Mr. Snavely did, the result would have been that his 3½ acres would have yielded him a gross sum of $2,553.21, or at the rate of $729.46 per acre, and deducting $106 as the cost per acre for cultivation and expenses, we get the net sum of $623.46 as profit realized from a single acre grown in tobacco.

Still Another.

But we have still another case we shall lay before our readers. Mr. John J. Long, of Drumore township, on last Monday, delivered at the packing house of Mr. Daniel Mayer, in this city, his crop grown on 1½ acres of ground, weighing 3,059 pounds, and for which he was paid the sum of $978-88, or at the rate of 32 cents per pound through. This is a yield of 2,038 pounds per acre, which at the price paid, would amount to $653.16 per acre. Deducting Mr. Snavely's allowance of $106 as the cost per acre, we have a net profit of $446.16 realized from a single acre of Lancaster county grown tobacco.

The above figures, be it remembered, are not ideal ones. They are actual facts. They are from the books of the purchasers and the checks received by the sellers. They represent three transactions consummated during the present week. They are not isolated cases, either. We have no doubt others like them have occurred of which we have not heard, and that still others, and not a few of them either, will transpire before the present crop is delivered.

In Conclusion.

A few papers in neighboring counties, whose ignorance far outruns their sense and discretion, have from time to time, been proclaiming that their tobacco farmers are as skillful as our own and their crops as good or even better. We have been content to let these sheets blow their penny trumpets uncontradicted. We now confront them with facts. If they have others that equal or exceed them, we will gracefully acknowledge that their tobacco growers are more skillful and their product superior; but nothing short of actual facts will answer—bare assertions will not serve the purpose. We have no desire to belittle the product of our neighbors; there is no occasion to do so. But when we can get such prices as are given above and realize sums per acre that exceed those received by the growers of seedleaf anywhere in the United States, we think our claim to be the champion tobacco growers of the country is pretty well founded.

AMERICAN SILK GOODS.

The silk trade of America and the subject of sericulture generally cannot be said to have enlisted that attention outside of the circle directly interested to which they are reasonably entitled. Any one reading the volume published under the direction of the Silk Association of America, by Mr. William C. Wyckoff, of New York, will certainly be interested and very probably be surprised.

"Everybody," he says in his preface, "knows that silk goods, both domestic and foreign, are cheaper now than formerly, but comparatively few persons are aware that the American goodsare, as a rule, better as well as cheaper. That there is much general ignorance on this subject may be shown in many ways—perhaps the most striking illustration is presented by the fact that nearly the entire product of some of our silk mills is still represented as of European make in the final sales of the retailer to the customer. In fact, our manufacturers are obliged to make better fabrics than their foreign rivals in order to attain the market where imported articles held a long established reputation." Census bulletin No. 69, prepared by Mr. Wycoff as a specialt agent of the Census office, gave 31,440 as the greatest number of hands employed at any one time during the year ending June 30, 1880, in the various factories, to whom $9,-107,835 were paid in wages. The total number of factories reporting was 383, representing a capital investment of $18,899,500, and employing 8,467 looms. The total net value of finished goods produced was set at $34,-410,463, the gross value of materials and supplies being $22,371,300. The principal articles of production were: Machine twist, $6,-000,275; ribbons, $5,935,005; fringes and dress trimmings, $1,950,275; dress goods, $4,115,-205; handkerchiefs, $3,862,550; cords, tassels and millinery trimmings, $1,392,355; upholstering and millinery trimmings; 1,392,355; satins, $1,101,875. It is not very easy to make comparisons with imported silks, as the invoice value of these latter is said to be on an average twenty-five per cent. under the real figure, while the duty and dealers' profits have to to be added, but is estimated that rather more than a third of the silk goods used in the United States were of American manufacture. About ninety-five per cent. of the imports come through New York. This amount in 1880 was $33,105,460, or about $7,500,000 more than in the preceding year, a showing without parallel since 1871, when the imports amounted to $33,899,719. In 1877 and 1878 a figure of barely $20,000,000 was attained, the figures being eloquent as to the financial condition of the community. Silk, it may be said, stands fourth in the list of duty-paying imports, contributing $18,556,400 to the Treasury, and so ranking after sugar, wool and iron. The imports of raw silk in the last fiscal year amounted to $9,138 bales, valued at $10,683,167, a falling off from 21,741 bales, valued at $11,949,743 in the previous year, but far in advance of former seasons.

Perhaps the craze which most frequently agitates the agricultural community is that of producing silk for home manufacture. There is no difficulty in breeding and rearing silk-worms if one has time, patience and mulberry trees at command, but there is no market for the cocoons, the manufacturer wanting reeled silk—not cocoons. The manufacture of silk thread, Mr. Wycoff tells us, though it has now outgrown foreign competition, was a long while "in the wilderness." American housewives had a prejudice in favor of Italian sewing silk, and Massachusetts manufacturers had to humor them by affecting foreign packages and wrappers, and compounding "Italian" trade names. The sewing-machine has completely revolutionized the business and brought about the invention of tube-twist. American sewing silk has an extremely high standard of purity—a fact which has naturally helped to drive out English goods which, by the addition of dye, are made to yield from eighteen to twenty-five pounds for each pound of raw silk. Thousands of cords of white birch from Maine are annually converted into spools, and many English makers come to the United States for these little articles, which an ingenious machine centres and prints—printing on the wood is preferred to labeling—at the rate of 100 a minute. The cabinets given by manufacturers to new customers with the first purchase cost about 1½ per cent. of the total sales; one firm has spent $150,000 in this sort of advertising. A $50 cabinet is nothing out of the way, and at times their value will reach $300 or $450. In dress goods, plain black fabrics are the hardest to make, as every defect in them is perceptible, and until a very recent period their successful manufacture was scarcely expected in the United States, principally on account of the costliness of the skilled labor required. Now nearly a third of the plain silks are made here, and the industry is making steady progress, thanks especially to the care given to the quality of the article, while European manufactures are only too apt to load theirs with dye. A simple test is to burn a small quantity of the threads, pure silk will immediately crisp and leave a pure charcoal; heavily-dyed silk will smoulder and leave a yellow, greasy ash. Very few velvets are manufactured in the United States, but the production of figured dress silks, grenadines, satins and the like, is large and growing; American linings have a high reputation, and American ingenuity has proved equal to the task of producing a satisfactory and lasting silk for umbrellas. Silk handkerchiefs have come into vogue during the last eight years and especially since the Centennial Exhibition. The manufacture of ribbons begun in 1861 as an experiment, there being a demand for particular shades, which, it was thought, could be more speedily met by making than by importing. Now the business has grown to great proportions. Curiously enough, nearly all the designs for American ribbons originate in American factories, frequently months in advance of the introduction of the goods into the market.

COAL TAR AND ALKALI IN PEACH CULTURE.

Apropos to what has been said about protecting peach stems from borers and from yellows by scalding the trees, and putting gas tar-impregnated sawdust or sand about the collar, Mr. Storm, of East Tyrone, Pa., reports that having some peach trees about as far gone as they could be to retain any life, a nurseryman whom he consulted about tarring the stems, told him that it was not much difference what he used, for trees so far gone did not recover, but tarring the whole stem would be sure to finish them. Mr. S., thinking that desperate cases need desperate remedies, and wishing to experiment further with tar, cleaned out a basin round the base of the trees and poured in a pool of tar, entirely surrounding and soaking the collar, as he had done with advantage before to some borer-infected apple trees.

The result was that the peach trees threw out strong, healthy shoots the next season, and have maintained vigorous growth during two seasons since. Other cases within the experience of the writer have proved that tar, or even coal oil, can be applied to the bark of young trees with impunity during the winter, but a coating of it in the summer is speedily fatal. A workman, however, once mistaking directions, added a quantity of tar to a wash of soap and sulphur, which he was directed to apply to the stems of some young orchard trees in June. The tar, not mixing well, showed itself in daubs and streaks here and there on the stems, covering them nearly or quite half, and, being irremovable, was an eyesore for years. None of the trees suffered seriously, however, excepting in a few cases, where they had a heavy coat; these took on the appearance of being bark-bound and impeded in their growth.

In Mr. Rutter's late excellent work on the peach, he shows that the free use of alkaline washes and manures, especially potash and lime, will preserve a peach orchard from the yellows, as well as from other destroyers of its fertility. Mr. Rutter has had thirty-five years of very extensive and varied experience, and his reliability is beyond question. So far as regards the borers, the carbolic acid of coal tar is most convenient and effective. It mixes in water well by stirring it first into hot, strong soapsuds. A pint of the crude acid, costing 25 cents, is recommended to four or five gallons of soft soap; which, diluted, will make twenty gallons of wash, to be applied in June, and again in August for assured effectiveness, although the June application usually suffices.

POINTS IN COWS.

Points in stock are the badges of purity. What are known as "points" are certain conformations, outlines of shapes and marks of color which specify that the animal possessing them is truly and distinctly a member of the class demanding the specifications possessed. The average farmer gives but little attention to the finer points, but with his experience, and habit of association, judges very critically at times. While farmers are seemingly anxious to improve, they endeavor to do so without knowing in which direction to benefit themselves. Nearly every farmer claims to be an expert at selecting milch cows, yet in breeding his stock he does not consider first what he is to breed for. Does he stop to consider whether he wishes the offspring of his favorite cow to be a superior milker or a great butter producer? The influence of the sire is to be considered above all others in such a matter. Jersey bulls are scattered far and wide now, and are within the reach of all, and yet the dairyman who sends his milk to market, and cares not to make butter, is foolish in patronizing Jersey bulls. The Jerseys are for butter-producing only, and are not heavy milkers. The milk such cows give is very rich; it is almost pure cream; but it does not come up in quantity. The farmer who desires large yields of milk from cows should seek to have transmitted to his young stock the blood of the Holstein or Ayrshire; for, although the milk from cows of these breeds is not as rich in quality as that from the Jerseys or Guernseys, they greatly excel them in quantity. Thus, those farmers living within reach of cheese factories can better promote their interests by selecting Holsteins or Ayrshires for improving their stock; while those who send butter to market should have nothing but the butter-producers.

A great milker shows her qualities in her looks and make-up. The eyes and hair also give good indications. The first point for a farmer's observation, and the principal one, is to observe that she does not show a tendency to become "beefy," or rounding with points that denote good fattening qualities. A first-class cow does not take on fat as a rule, but is rather bony and ugly-looking. The shape of the Jersey should be deer-like, with a large, mild-looking eye and soft feeling of hide to the touch. The udder should be full, reaching far up at the rear. One of the most prominent points is the large milk ducts (sometimes as large as a person's arm) running from the udder to the middle of the stomach. They are sure indications of good milking qualities. Jerseys have black nozzles and tongues, the udder being usually smoother than in other breeds, and velvet-like when examined by touch. The Holsteins are a very large breed of cows, equalling the Shorthorns in size, but largely excelling them in milking qualities. The young male calves from such cows can be kept with profit, as the Holsteins, when fed for the purpose, make not only good beef but equal to the best. Oxen from this stock are nearly equal to the Devons. Their color is usually black and white.

But in endeavoring to breed for milk it should not be forgotten that two excellent characteristics are rarely found in a single breed. Thus we must not expect to find good milkers among the Shorthorns, nor have choice beef from the milch cows. A cow cannot make milk and beef at the same time. If her tendencies are toward milk she will be hard to fatten; if she keeps extra fat it means that she is a better flesh-former than a milk-producer. A great deal depends on the feed, as a matter of course; but the breed must first be taken into consideration, if an increase in the herd is contemplated.—*Philadelphia Record.*

OUR LOCAL ORGANIZATIONS.

AGRICULTURAL AND HORTICULTURAL SOCIETY.

The horny-fisted sons of toil turned out fairly well at the meeting on Monday afternoon, February 6th, 1882, but the horny-fisted daughter was conspicuous by her absence. She did not respond to the invitation extended to her at the last meeting, although Mr. Calvin Cooper, the mover of the invitation, explained that he had induced Mrs. C. to come far enough to ascertain that no ladies were present, and—well when a woman won't she won't, you know. It was whispered around, however, that there was a little more punctiliousness observable in the appearance of some of the bachelor and younger members by reason of the anticipated influx of farmeresses, so to speak.

The following were present: President, Joseph F. Witmer, of Paradise; James Wood, Oak Hill; Calvin Cooper, Bird-in-Hand; M. D. Kendig, Creswell; F.

R. Diffenderffer, city; W. W. Griest, city; Enos H. Weaver, Strasburg; Hebron Herr, Lampeter; J. F. Landis, East Lampeter; D. W. Graybill, Petersburg; J. M. Johnston, city; C. L. Hunsecker, Manheim township; Wash L. Hershey, Chickies; Levi S. Reist, Oregon; H. G. Rush, West Willow; Jno. H. Landis, Millersville; Cyrus Neff, Mountville, H. K. Myers, Millersville; Eph. S. Hoover, Manheim township; John Huber, Pequea.

In the absence of Secretary Linville, ex Secretary Keudig was recalled.

Crop Reports.

James Wood, Little Britain, reported a good crop of snow and nothing else visible.

C. L. Hunsecker noticed before the snow fell that the wheat and grass looked well.

J. F. Landis reported the water fall for January in East Lampeter to be 4¼ inches.

Wash. L. Hershey regarded the crops generally in a favorable condition but did not particularize.

"Growth and Consumption of Timber Trees in America"

was the subject of an essay by C. L. Hunsecker. It was substantially thus:

Although a dense forest, almost untrodden by civilized man, yet as early as 1720 John Bartram planted on the banks of the Schuylkill below Philadelphia a garden containing many forest trees of North America, and in the reign of Queen Anne, 1702–14, an act of Parliament was passed for the protection of the colonial forests. In 1750 the felling of white pine was prohibited. About the same time iron furnaces were established in Virginia, Maryland, Pennsylvania, New York and New Jersey, but the apprehension of the scarcity of fuel was not realized, for coal came to the relief. And no doubt when the necessity again arrives for a new departure, a substitute will doubtless be found for the wood now used in ties, telegraph poles, fences, etc. To instance the enormous consumption of timber Mr. Hunsecker said that during the year 1880, 1,500,000,000 feet were cut in Minnesota, Mississippi, Alabama, Florida and Texas. On the other hand, an immense amount of timber stands in this country—700,000 square miles of it; besides Iowa, Kansas, Utah, etc., are becoming wooded by the planting of trees. Favorable legislation has and is doing much for the western prairie, in some of the States, "Arbor Day" being a regular holiday for the planting of trees. It may be said, however, that the cultivation of forests is greatly neglected in most countries, and in many a very sensible want of wood is felt. How shall we remedy this? Who shall plant trees? In the Old World governments can coerce, but in our country it must depend upon the will of the citizen. Among some of the public benefactors in this line are Mr. Fisher, the red cedar pencil manufacturer, who planted a large tract of land in cedars; the Landreths, of Philadelphia, who have been planting large areas in Virginia with catalpa, ailanthus, white oak, hickory, etc., and some Scotch immigrants, who are building up forests in Missouri.

Mr. Hunsecker's essay was quite lengthy, and contained much interesting statistical information pertaining to the timber interests of America.

What Causes White Vein in Tobacco.

Mr. Hebron Herr read the following essay on the above subject. It was as follows:

This is a very important question, and one that should elicit the attention and consideration of all growers of the weed. Numerous arguments have been advanced on various occasions, but have not proven satisfactory to the public in general.

In my little experience I have discovered that tobacco which has been grown perfectly, nothing interfering with its growth from the small and tender plant up to perfectly matured stalk, will invariably cure with the veins the desired color. Therefore, the cause which produces white veins in our tobacco is attributable first to a diseased condition of the plant in one or another stage of its growth. Tobacco may assume this diseased condition at various stages of its growth. It may become diseased in our plant beds, or when being transplanted from our plant beds into the field, or when half matured, or after it is fully matured. At any of these stages it may become diseased, and never grow healthy thereafter. Fully matured tobacco may become diseased by being permitted to stand in our fields in the hot and dry sun after it has ceased to grow. It may be left remaining on the field after maturity without any disastrous effect when the soil is in a moist and growing condition. By permitting tobacco to remain standing on our fields in the hot and dry sun after ceasing to grow it becomes subject to changes by the influence of the sun's rays robbing it of its vitality and retarding the copious flow of the nourishing elements which impart the life and vigor to the plants, evidently leaving the plant in a diseased condition when harvested. We should be very careful when growing tobacco in seasons as the last two were to harvest our crop immediately on the plant arriving at maturity or before it ceases to grow. It is better to have our tobacco an inch or so shorter and harvest it in a healthy condition, than a few inches longer and harvest it in an unhealthy condition. The next point to be taken into consideration, and one also pertaining to the cause of white veins is that of curing the tobacco after it has been grown and harvested. Experience being the best teacher and guide has taught me that the more we retard the curing of our tobacco, subjecting it to undergo a number of changes while curing, the better the color will be and the leaf will possess more of that fine silken condition, and with a much less frequency of white veins. Moisture, I claim, is one of the prime essentials in the curing of tobacco. Our curing houses, therefore, should be so constructed in the first place not to have them built so high, and invariably have a ground floor or so arranged, if having another floor, that it could be opened to permit the moisture and dampness to draw up through the tobacco to assist in retarding the rapid curing. Tobacco which I cured in my tobacco cellar, when coming to shipping I discovered no white veins; also that which I cured in another building possessed very little; it also was subject to the influence of a ground floor, while that which was cured in my tobacco house possessed white veins, and the higher up in my shed the more numerous they became. Now, the cause or reasons which I give for this is that the tobacco which I cured in my cellar cured slowly, undergoing numerous chances of becoming moist, and when too moist was by ventilation caused to become dry, reviving the vitality and vigor of the plants which had become dormant or inactive when harvested; while that cured in the house above, having not possessed this advantage, cured very rapidly, and the more rapid the more frequent would white veins appear. Houses being covered with slate are often spoken of as not being beneficial to curing tobacco, owing no doubt to the heated condition in which it becomes during the day, causing the tobacco to cure too rapidly. Tobacco harvested while in a green state seldom cures white veins. This is owing to the green and sappy condition of the leaf, causing it to cure more slowly; and its being in a green state is subjected to more numerous changes. We should aid our tobacco as much as possible while curing by closing tightly our houses during the day, and opening them at night. Therefore, in conclusion, the cause of white veins in our tobacco is attributable, first, to a diseased condition of plants while growing, and secondly, to an improper method of curing; and as a preventive we should endeavor to raise healthy plants, transplant properly, cultivate frequently and trust in Him who is the giver of all good gifts to send us copious showers to assist in a rapid growth, and we will be enabled to grow tobacco possessing very few white veins.

Mr. Cooper thought Mr. Herr had struck the nail on the head in saying that when cut in a green and succulent condition the weed was in little danger of white vein. Many of us cut the weed too ripe; we wait for too much length. Better have less length and also less white vein.

Mr. Graybill wanted to know why on a good healthy stalk one or two leaves will sometimes have white vein; to which Mr. Herr responded that part of the stalk was diseased, just as one finger on a man's hand may be diseased and the other digits perfectly healthy.

Mr. Graybill also wanted to know why you can sometimes sweat out white veins; and in response, Mr. Herr thought it was a poor rule that would not work both ways; therefore if you can sweat in white veins you can also sweat them out.

Enos H. Weaver read an article from the *Country Gentleman*, taking substantially the same views expressed by Mr. Herr, which were also in accord with Mr. Weaver's experience.

Mr. Cyrus Neff knew that some varieties of tobacco were subject to white vein.

President Witmer had been told that two well known growers in the eastern end leave their tobacco stand until very ripe, and they never have white vein. Two years ago Mr. Witmer cut some tobacco when only reasonably ripe, cured it properly, and out of two acres, only 300 pounds were marketable.

Mr. J. F. Landis said the best tobacco he ever grew stood at least two weeks after it was ripe.

Mr. J. H. Landis, through other growers, had learned that many concur that when a drought is followed by a wet spell and a consequent growth of the weed the white vein is very numerous.

This was just directly opposite to President Witmer's experiences. Under just these circumstances two years ago he had a very fine crop and very little white vein.

In Mr. Graybill's section, three years ago, heavy rains and hail occasioned a re-setting; then came a dry spell, then a heavy rain, and it was the best crop Mr. G. ever had.

Mr. Hoover also had a theory, to wit: In '79, about the second week of August, heavy rains succeeded the long drought, and the result was a magnificent crop. About the time tobacco needs rain the most is the topping time; if the plant then lack nourishment the leaf will probably show a defect in the shape of white veins.

Rank Growing Wheat.

"Ought Rank Wheat to be Pastured?" was answered by James Wood. It depends upon the winter that follows. If we have an open winter it might be no disadvantage to pasture it out; but if the winter was severe, pasturing would be less desirable and quite risky. On the whole, he thought that rank wheat should not be pastured.

Mr. Hunsecker thought pasturing would have a tendency to check the rankness of wheat and thus make a better crop.

Dairy Cows and the Soiling System.

Can dairy cows be kept in as healthy condition by the soiling system, and is the butter as good? was answered by J. F. Landis.

"Can dairy cows be kept in as healthy condition by the soiling system, and is their butter as sweet?"

If by the soiling system we mean the feeding of cows through the summer months in small inclosures or stables, and only take the parts into consideration touched upon by my question, I am decidedly opposed to the system. In order to have healthy cows it is essential that we have good food, pure air, pure water, light and comfort. I claim this cannot be had in a small lot or stable to so full an extent as in the field. The first part of the question, as answered, answers the second. In order to have sweet butter we must practice cleanliness from the time the milk leaves the cow until the butter is on the bread. There are few things so absorbent of surrounding odors as butter. I have seen good butter condemned here on our market because the persons making it placed it in a kettle in which cheese was placed, or anything else having an odor, which, in itself, may not be objectionable. When that butter is put upon the table it has lost its sweetness. I claim that the soiling system, to some extent, affects the health of cows as well as the sweetness of the butter.

President Witmer did not agree with the referee. He thought there were many advantages in the soiling system and only one objection, viz: the disadvantage of the labor attending it. Mr. W.'s cows never did as well as under the soiling system.

Mr. Neff is trying to feed his cattle entirely in the stable. Last year from December to February he never took out the cows even for water, and they never did better. Mr. Neff saw no reason why cattle could not be kept as well and in as healthy condition in as out of the stable. He cleaned his stables twice a day.

Mr. Cooper, one of the committee at the State Fruit Growers' Society, reported one of the most interesting meetings the society ever held. The room was crowded, and all the essays and discussions evinced great interest on the part of the members.

The following questions are on the programme for next meeting:

"Should patent fertilizers be applied to tobacco, and if so at what time?" Referred to D. W. Graybill.

"Can we not dispense with the division fences with profit?" Eph. S. Hoover.

"Should we encourage the introduction of new varieties of apples?" L. S. Reist.

"What is the best time for sowing cloverseed?" Enos H. Weaver.

"Is sub soiling beneficial?" John C. Linville.

Adjourned.

POULTRY ASSOCIATION.

The Association held their meeting on Monday morning, February 6. The following were present: President, G. A. Geyer, of Springville; J. B. Lichty, Secretary, city; M. L. Greider, Mount Joy; H. H. Tshudy, Lititz; J. B. Long, city; C. E. Long, city; J. F. Witmer, Paradise; Charles Lippold, city; John K. Schum, city; W. W. Griest, city; Washington Hershey, Chickies; F. R. Diffenderfer, city; J. M. Johnston, city.

Secretary Lichty in his report for the year 1881, stated that there were forty-five members in good standing, the average attendance was thirteen, and during the year twenty-five members had been elected, of which number only six paid the membership fee. Members are in the arrears to the extent of $150. The Secretary suggested that those in arrears prior to January 1, 1882, be notified that upon the payment of their dues to that date, the association will place their names upon the honorary list of membership. The total number of entries at the late show was 551, but only 520 birds were exhibited, of which 327 were poultry, 185 pigeons and 7 cage birds; also 33 breeding pens were exhibited. Cash premiums paid amount to $473.50; other special premiums swelled the total value of premiums awarded to $539.75; the only class in which the entrance fees exceeded the premiums paid was the Spanish, consisting principally of Leghorns, although there was a loss of fifty cents on the Asiatics. The varieties on which the entrance fees more than reimbursed the society for premiums are: Light Brahmas, Dark Brahmas, Black and Partridge Cochins, B. B. R. Game, B. B. R. Game Bantams, White and Brown Leghorns, Plymouth Rocks and S. S. Bantams. In the pigeon list but five varieties paid, viz: White Crested Fantails, White Trumpeters, Blue and Blue Checkered Antwerps, and Blue English Owls. There were sixty-two varieties of poultry which averaged over five; seventy-one of pigeons and six cage birds. Of the seventy-one varieties of pigeons exhibited but five paid the association. All premiums to foreign and local exhibitors have been paid, greatly to the credit of the society.

Treasurer J. B. Long then submitted his annual report, showing that $5,253.13 had come into his hands from the receipts of the show and other sources. $821.74 had been paid out, leaving in the treasury at present thirty-nine cents.

On motion of Mr. C. E. Long, the Secretary, was authorized to employ some one to collect the outstanding dues.

John Schlomridge, of Ephrata, was elected a member of the society.

The Secretary was authorized to place on the honorary list of membership such names as in his discretion he might see fit.

By a mistake express charges amounting to $5.40 had been charged to certain exhibitors. As the society had offered to pay this itself it promptly agreed to shoulder the debt.

Adjourned.

FULTON FARMERS' CLUB.

The February meeting of the club was held at the residence of Joseph R. Blackburn. Members present: E. H. Haines, Wm. P. Haines, Montillion Brown, J. R. Blackburn, S. L. Gregg and Wm. King. The family of Josiah Brown was represented by his wife and daughter, that of Grace A. King by her son Joel, and that of Lindley King by S. Lindley Jackson. Visitors: Neal Hambleton and wife, Layman C. Blackburn and wife, Edw. Stubbs and wife, and Samuel J. Kirk and wife. The attendance would no doubt have been considerably larger but for the driving snow storm.

Exhibits and Answers to Questions.

Joseph R. Blackburn exhibited a large plate of fine winesap apples, which, judging by the way they vanished, the Club thought to be very fine.

E. H. Haines asked if winesap apple trees grow as fast and come into bearing as soon as other varieties? There were several present who had winesap trees. They all spoke of as growing as well as other varieties, and being good bearers.

Montillion Brown had been reading lately in an agricultural paper of a kind of winter oats, or oats that could be sowed in the fall, and wished to know if any one present had any knowledge of it.

E. H. Haines said that he believed that there was such a variety of oats. Some time ago there was a kind of oats mix'd with the wheat that the winter failed to kill.

It was the general opinion of the club that such a variety would not be at all desirable, as it would be too late to sow after the corn had been harvested, and it would be liable to give trouble by getting mixed with the wheat.

Neal Hambleton asked the proper time for trimming an Osage Orange hedge.

E. H. Haines said that he had experience with hedges, and thought that he knew. Whenever there is a growth of six or eight inches long cut it off. It is easily done, and all that is cut off dries up and gives no further trouble. It is easier to trim three or four times a year than once. If the shoots are left to grow for a year, pruning breaks up the hedge and makes it look badly and leaves a lot of brush to burn.

Layman C. Blackburn had noticed an article in the New York Tribune, by J. F. Wade, giving an account of the writer's experience with an old orchard which he had taken to hand when it was in a plight that it would take pages to describe, and a number of the trees so nearly dead that the owner said they could not be saved. By pruning, scraping off the old bark and the filth in wet times, when they could easily be removed, and plowing five times in a season, and dragging in proportion, he had completely renewed the trees and brought them into profitable condition. He (L. C. Blackburn) wished to know the opinion of the club as to probable success in renovating old orchards la general by such treatment.

E. H. Haines : If the trees get into bad condition by neglect, there might be some hope of success, but if the trees were old, we might as well try to rejuvenate an old man.

Ed. Stubbs could not agree with the writer that trees might be made to bear every year by thorough cultivation. He had seen trees that were cultivated that did not bear well.

Neal Hambleton thought that the reason that trees bore only every other year was that they needed more rest than they got through the winter.

Montillion Brown had two Queen apple trees, one of which did not miss a crop for ten years in succession. It was favorably located, and the bugs ran around it and kept the ground loose. The other trees were not in so favorable a location, and bore every other year. There was not much difference in their vitality.

S. L. Gregg had a similar experience with two trees. The hogs ran around one of them, and it bore every year, but it wore out sooner than the other, which only bore every other year?

William King : Is it advisable to trim old trees ?
Ed. Stubbs: Don't think it is. They will die sooner than if let alone.

L. C. Blackburn : If trees are well trimmed when they are young they will not need much after they grow old. It not attended to when young they will have to be trimmed when they are old.

E. H. Haines said that his father once let some Yankee graft some old trees. The grafts nearly all grew and bore fruit, but it finally killed the trees.

Mary Ann Brown said that they once had an old orchard trimmed and it gradually died off. It never did any good after, Mr. Brown had some old trees on which large limbs were dying out off, leaving several feet remaining on the tree.

S. L. Gregg had an article read from the Oxford Press entitled "High Farming," giving an account of the extraordinary productions of a farm in Lower Oxford township, Chester county. Some of the members thought it a very clever advertisement. The owner is a manufacturer of fertilizers.

Examining the Host's Farm.

The forenoon session was now adjourned, both members and visitors retiring to the dining room, where for some time they diligently occupied their time in putting themselves outside of the good things of this world. After exhausting all their powers in this line of business they plunged out through the whirling storm to look at the condition of the live stock of the host. After again convening in the house, criticisms were called for.

Montillion Brown : He has some very nice fat steers. Did not go to see his wheat field. Suppose it is good.

William P. Haines had noticed some very thrifty pigs.

Neal Hambleton spoke of the fine condition of the pigs, and also of their pen, which was well arranged and everything about in neat order.

E. H. Haines, would like to see a good hog pen. He had been in search of one for some time and had come to the conclusion that a perfect hog pen or chicken house were things that had not yet been invented. This led to quite a discussion on the construction of hog houses.

Literary Exercises.

The literary exercises of the club were next taken up, when Carrie Blackburn recited "The Boy Convict's Story," an account of a young man who, in his boyhood, had been kept in strict surveillance by his pious parents, who made the atmosphere of his home fright with propriety, until at last he left the parental roof to seek abroad the recreations and pleasures for which he was longing and which were denied him at home. But he strayed too far, and finally was betrayed into the commission of crime.

Neal Hambleton congratulated the little girl on her choice of a selection for recitation. He thought that the young should not be restrained in civil amusement. It is a mistake that we do not mingle more with the young. Games and other amusements should not be discouraged, but we should use every means to make home cheerful to the children.

E. H. Haines was afraid that there was too much truth in the boy's story. We should mingle freely with our children and let them see society as it is. Young people who are kept in restraint and isolated for fear of their being contaminated do not know what value to put on what they see and hear. They are liable to be taken in.

Montillion Brown thought one reason why people in this country did not mingle more freely with the children was owing to the secluded nature of the farmer's life. This little Club gave an opportunity for a more social feeling between old and young. We should take the young with us and have them take an interest in it.

S. L. Gregg thought that the young of the present day had many advantages that we were deprived of when we were young, and they should be encouraged to avail themselves of them.

Mabel King recited the Wayside Inn. Lauretta A. Kirk recited The Two Dimes, and Ella Brown, The Independent Farmer.

The next meeting of the Club will be held at the residence of Joseph Griest, Fulton township, March 4th.

JANUARY MEETING OF THE LINNÆAN.

The Society convened on Saturday afternoon, January 28, in the hall of the Y. M. C. A., the President, Prof. Stahr, and the Secretary, Dr. Davis, occupying their respective chairs. After the formal opening and the collection of dues, the following donations to the museum and library were announced by the curators:

Museum.

A fine specimen of the "Frog Fish" (*Batrachus tau*), six inches in length, from Mr. Frank Moffett, was donated through Mr. Daniel Heitshu. This fish was received from Baltimore in a cargo of oysters, and was still alive when it arrived in Lancaster, although it must have been out of the water several days.

Eight specimens of the "seventeen-year cicada" (*Cicada septendecim*), in the pupa form, donated by Mr. Geo. Heusel, florist, East Orange street. These insects were dug out of the ground, eighteen inches below the surface, and about four feet from the outside wall of Mr. H.'s greenhouse, on the 16th of January, and were alive when received by the curators. Prof. Riley is of the opinion that they belong to his brood No. 8 and will appear the "coming summer."

A beautiful specimen of Wilson's thrush (*Turdus fuscescens*), found in a dying condition in East Orange street, in November last, and donated by W. De C. Rathvon. As this bird usually migrates in September or the beginning of October, it must have been deceived by the pleasant autumn weather, and was suddenly overcome by cold.

A very perfect Indian implement, found on a small island in the Susquehanna river, near Safe Harbor, was donated by Mrs. A. H. Reist, of No. 119 South Queen street, through Mr. William Roehm. This relic of the Red Man is of an unusual form, and was probably used in "barking" trees, or in skinning large animals, or both.

Specimens of "Georgia cotton" (*Gossypium herbaceum*) were donated by Mr. J. J. Sprenger, of Rome, Georgia. Height of plant, four feet six inches; spread, three feet, and contained one hundred and twenty bolls. This was represented as an average of the plants in the field where it grew.

Fine specimens of "Sponge" (*Spongia— ?*) gathered on Nantucket Beach, Mass., last Summer, and donated by Mr. D. Maxwell, of Baltimore. One specimen is of a very delicate and uncommon form.

Library.

Nos. 25 and 26, vol. 20, of the Official Gazette of the United States Patent Office, from the Department of the Interior. The LANCASTER FARMER, 1882.

Historical.

Three envelopes, containing forty seven local and foreign, historical, biographical and scientific scraps, by the Curators.

A twenty dollar bill of the Lancaster Bank, dated July 1, 1852, by Mr. J. G. Thackara.

Papers Read.

Dr. Rathvon read three papers containing notes on the history and habits of the "Frogfish," the "Thrush," and the "Cicadas" donated.

Mrs. Gibbons read an interesting letter from a friend in Nebraska, relating to some phenomenal peculiarities of the wells in the district where that friend resides.

New Bus ness.

The President announced the chairman of the standing committee for the year 1882, said chairman, under the provision of the constitution, being authorized to appoint each two colleagues from among the members and correspondents of the society. Mammology, Dr. M. L. Davis; Ornithology, William L. Gill; Herpetology, W. T. Bolton; Ichthyology, C A. Heinitsh; Entomology, S. S Rathvon; Botany, Mrs. L. D. Zell; Geology, Prof. I. S. Geist, Paleontology, Prof. T. R. Baker; Microscopy, Dr. Knight; Mineralogy, J. B. Kevinski; Archæology, Prof. J. H. Dubbs; Natural Historical Miscellany, Mrs. P. E. Gibbons.

After passing upon bills presented, and the usual social and scientific intercourse, the society adjourned to meet on the last Friday evening in February, of which the hour and place will be announced by the secretary in due time.

AGRICULTURE.

PLANTING TOBACCO.

M. Quad describes the method of planting tobacco in Virginia as follows:

The tobacco crop in Virginia has long been a source of great revenue, and there was a time when any agriculture outside of tobacco raising was supposed to be a losing business. Tobacco land must be prepared as carefully as the average farmer would prepare a garden. The beds for the plants are generally prepared on a piece of new land and in localities sheltered from winds and having a southern exposure. The ashes from the burned shrubs, leaves and limbs are carefully worked into the bed. The seed is then sown as we sow for cabbage, and the plants come up the same. When they are large enough to transplant they are set about three feet apart, and about 4,200 plants is the average for an acre of ground. At a certain age the plants must be "pruned," which consists of breaking off the shoots and suckers and pinching off the head, and again the tobacco worms must be hunted off the plants. Tobacco growers generally put in corn and other crops as well, so that hands can be shifted from one growing product to another as necessity requires or the state of the crop permits. A fair average per acre is 700 pounds. This must be sun-dried on scaffolds in the field, and afterward hung in barns and smoked. The average price for this heavy tobacco is seven cents per pound. An acre of land is thus made to yield about $500. Growers estimate 1,000 pounds to every hand employed, and the care of the tobacco crop is only one half their labor.

Improved Grasses.

In many respects grass-culture has not kept pace with improvements in other branches. We are continually getting new plants, new trees, new fruits, new vegetables, new grains, but a new grass is never thought of. We have the same orchard-grass, the same redtop, and the same timothy, that we have over a hundred years ago; and so far as the drift of thought goes, we shall have the same grasses for a hundred years to come. And yet there is no reason that we can see why there should not be improved grasses, as well as improvements in any other thing; and there doubtless would be if public attention was drawn to the matter as it should be.

We have to be sure, during the past twenty years or so, been treated to Hungarian grass or millet, a harsh, coarse thing, of little merit except for the very heavy crops it produces; and Lucerne is no better. There surely must be others which it would be of advantage to introduce. We see in foreign agricultural journals that some attention is being given to a species of grass called Tussock-grass, from its growing in large bunches, and which from its description appears to be closely allied to orchard grass. It does not appear to be considered very hardy; but there are no doubt a number of places on this continent where it would find itself entirely at home. It is said to grow five or six feet in height, and to produce vegetation of great fineness of quality and exceedingly nutritious. When once a field is set with it, it is seldom killed out by other vegetation, and goes on producing good crops for a great many years. It appears to be hardy in Hungary, and if so it ought to stand considerable frost. We wish all this was just so as stated, but we are afraid that in this as in so many other new and reputed valuable products of the soil, we shall not hear much of it in the future.

But it is not only the introduction of new species that grass crops and grass lands may be improved—there might be selected good varieties of kinds we now grow, just as we have selected good kinds of other things. There seems to be here a good chance for somebody.—*Germantown Telegraph.*

Rotation of Crops.

In a well planned system of farming, the subject of crop rotations should be carefully considered, as one of the essential elements of success in its highest and best sense. It seems to be the prevailing opinion that the alternation of crops, in sytematic order, is a modern invention that was gradually developed as a direct result of the applications of science to the art of agriculture. The early writers on agriculture, even from the times of the Romans, have, however, quite uniformly urged the advantages of a succession of crops from the teachings of experience. They were satisfied that a variety of crops grown in succession, all other conditions being equal, would give a greater aggregate yield than could otherwise be obtained. The reasons for the success of the system could not, it is true, be given, but practical men were fully agreed in urging its importance, and many systems of rotation, more or less perfect, were planned, some of which became the prevailing rule of farm practice in particular localities. That these practical rules of alternating crops of different habits and modes of growth are based on correct, but not explained, principles, has been shown by direct experiment.—*Dr. Manly Miles in American Agriculturist.*

HOUSEHOLD RECIPES.

ORANGE PIE.—Grate the rinds of two oranges and squeeze the juice. Cream a quarter of a pound of butter and by degrees add half a pound of sugar. Beat in the yolks of six eggs (already well beaten), then the rinds and juice of the oranges. Beat the whites of the egg to a stiff froth and mix them lightly in the other ingredients. Bake in paste lined tin pie plates.

NEW ENGLAND BAKED INDIAN PUDDING.—One quart of milk, three quarters cup of molasses, two teaspoons ginger, one-half teaspoon cinnamon, a bit of salt. Stir these thoroughly together and let come to a boil. Have ready three dessert spoonfuls of Indian meal wet in a little cold milk; put into the hot milk and after stirring thoroughly let it boil five minutes. This should be made early in the morning and set away to cool. When needed for dinner take two eggs well beaten, two tablespoonfulls melted butter, half a teacup cold milk, stir this into the first mixture and let it bake two hours.

CHICKEN PIE.—Choose a rather tender fowl, pluck off the pen feathers, singe off the hairs with a piece of burning paper, then wipe the fowl with a clean damp cloth, draw it carefully by slitting the skin at the back of the neck and taking out the crop without tearing the skin of the breast; loosen the heart, liver and lungs by introducing the forefinger at the neck; and then draw them, with the entrails, from the vent. Unless you have broken the gall, or the entrails in drawing the bird, *do not wash it*; for this greatly impairs the flavor, and partly destroys the nourishing qualities of the flesh. Cut it in joints and put it in a hot frying pan with an ounce of butter and two ounces salt pork cut in dice, and fry it brown. When it is brown stir an ounce of flour with it, and let the flour brown; season it with a teaspoonful of salt, a level teaspoonful of pepper and a tablespoonful of chopped parsley; cover it with boiling water and let it simmer gently for an hour, or until the chicken is tender.

PRUNE PUDDING.—One half pound of prunes boiled; soft and thick, remove the stones and sweeten well; then add the whites of six eggs beaten stiff; chop the prunes fine, then stir in the eggs; put into a dish and bake a light brown. Serve with sweetened cream.

A NICE WAY OF COOKING COLD MEATS.—Chop the meat fine; season with salt, pepper, onion or else tomato catsup. Fill a tin breadpan two-thirds full; covered it over with mashed potato, which has been salted and has milk in it; lay bits of butter over the top and set into a Dutch or stove oven for fifteen or twenty minutes.

CHOCOLATE CAKE.—One cup of sugar, tablespoonful of butter, one heaping cup of flour, one teaspoonful of cream tartar sifted in flour, and half a teaspoonful of soda dissolved in a tablespoonful of sweet milk. Filling—whites of three eggs beaten to a stiff froth, one cup of sugar (pulverized), and three tablespoonfuls of grated chocolate, and vanilla to taste. Bake the cake in jelly-cake tins in three layers, and spread the mixture between and on top. Eat within thirty-six hours after baking.

BREAKFAST RUSKS.—Two cupfuls of sweet milk, two eggs, two teaspoonfuls of cream tartar, one teaspoonful of soda, half a cupful of white sugar, about four small cupfuls of flour. Beat the eggs very light; put the cream of tartar in the flour, and add the soda the last thing. Bake in a long pan in a quick oven.

PREPARING CARROTS.—Carrots prepared in this way make a good side dish or entree: Scrape and wash them: boil until they are tender, in as little water as will serve to keep them covered; put in a large pinch of salt: when the carrots can be easily pierced with a broom splint drain off the water and roll the carrots in flour: put a lump of butter in a saucepan and set on the stove; when hot put the carrots in and fry until brown; the carrots may be cut in two parts or cooked whole; turn them so that they will brown on all sides.

BARLEY SOUP.—Two or three pounds of beef from the skin, two pounds of cracked bones, an onion, four stalks of celery, four potatoes, a gallon of water, pepper and salt. Put all into the soup pot and boil very gently three hours. Wash a cup of barley and boil in a very little clear watea twenty minutes. Strain the soup, pressing hard, boil up, skim, add the barley and simmer thirty minutes.

CORNSTARCH CAKES.—Take the whites of three eggs, one cupful of sugar, two-thirds of a cupful of sweet milk, two-thirds of a cupful of melted butter, one teaspoonful of cream tartar, half teaspoonful of soda, half a cupful of cornstarch, one small teaspoonful of lemon extract, one and one-half cupfuls of flour. Mix the cornstarch, flour and cream tartar together and sift all through a sieve. For yellow cakes take the yolks of the eggs and make the same with these exceptions: Leave out the cornstarch and but a cupful of butter instead of two-thirds soda buttermilk and soda instead of cream tartar and sweet milk.—*Country Gentleman.*

FRENCH TAPIOCA PUDDING.—Take two ounces of tapioca and boil it in a half a pint of milk by degrees, and boil until the tapioca becomes very thick; add a well-beaten egg, sugar and flavoring to taste, and bake three quarters of an hour. This preparation of tapioca is superior to any other, is nourishing, and suitable for delicate children.

SWEET MACARONI.—Break up a quarter of a pound of the best macaroni into small lengths, and boil it in two quarts of water with a large pinch of salt, until perfectly tender; drain away the water, add to the macaroni into the stew pan a cupful of milk and a quarter of a pound of sifted lump sugar, and keep shaking over the fire until the milk is absorbed; add any flavoring, serve. Stewed fruit may be served with the macaroni.

OATMEAL PUDDING.—Mix two ounces of fine Scotch oatmeal in a quarter of a pint of milk; add to it a pint of boiling milk; sweeten to taste, and stir over the fire for ten minutes; then put in two ounces of sifted bread crumbs; stir until the mixture is stiff, then add one ounce of shred suet and one or two well-beaten eggs; add a little lemon flavoring or greated nutmeg. Put the pudding into a buttered dish and bake s owly for an hour.

WHOLESALE CONGRESS.—Healthy piecrust is made of thin, sweet cream and flour, with a little salt. Don't knead, Bake in a quick oven. Another way is, sift a quart or two of flour in a pan. Stir in the centre a little salt and half a teaspoonful of soda well pulverized. Put in the hole a cup of soft (not liquid) lard, or butter and lard mixed; stir it thoroughly with the flour; next add two scant cups of good sour milk or buttermilk. Stir all quickly with the flour in such a way that you need hardly touch it with your hands till you can roll it out. Bake quick. This will make three or four pies.

STEWED APPLES AND RICE.—Peel good baking apples, take out the cores with a scoop, so as not to injure the shape of the apples; put them in a deep baking-dish and pour over them a syrup made by boiling sugar in the proportion of one pound to a pint of water; put a little piece of shred lemon inside of each apple and let them bake very slowly until done, but not in the least broken. If the syruf is thin, boil it until it is thick enough; takeout the lemon peel, and put a little jam inside each apple, and between them little heaps of well boiled rice. This dish may be served either hot or cold.

LITERARY AND PERSONAL.

THE AMERICAN BEE JOURNAL.—This oldest and ablest paper devoted exclusively to progressive bee culture, published in the country, is now issued in a royal octavo form, weekly, at $2.00 a year, by Thomas C. Newman, editor and proprietor, No. 974 West Madison street, Chicago, Illinois. This is a far better and more convenient form than that of a quarto, in which it was issued in 1881. Indeed, having been published as an octavo for seventeen years, the wisdom of changing to a quarto for a single year, seems to have been questionable, and now returning to nearly the original form is a concession that the departure was not a wise one. But, whatever its *form* may have been, its *substance* is, and always has been, of the highest apicultural order, and we don't see how any one who makes bee keeping a specialty, can afford to do without it.

THE WESTERN PLOWMAN, (not "ploughman" but *plowman*). A brand new agricultural royal quarto of 16 pages, published by John H. Porter—J. W. Ware, Editor—monthly, at Moline, Illinois, at the very low price of *fifty* cents a year, (with a premium worth a dollar), No. 1, vol. 1, of this "baby elephant," has found its way to our *sanctum*, and we confess we are prepossessed in its favor; for, being "devoted to the interests of the home, the farm and the family," it makes place for healthy literature in general, as well as farming and domestic affairs. The material is of good quality and the imprint especially, agreeable to the infirm of sight, being bold, plain, and easily read. Its very title insinuates economy, for it saves two letters in spelling and obviates the likelihood of any foreigner pronouncing it *Pluffman*. The very paper to interest the household, and help to while away the weary hours of a long and lonesome winter's day. It deserves to prosper, and we think it will.

THE HOME ECONOMIST, devoted to the interests of social economy. "A guide to every department of practical life," a beautiful folio, published monthly, by F. S. Blanchard and Company—Luke Goodwin, Editor—at Worcester, Mass., at 60 cts. a year, with many *premium* inducements No. 1, vol. 1, for January, 1882, received. It would be almost impossible to determine the literary progress of the country, from the rapid increase of publications alone. Low subscriptions, conspicuous advertisements, and showy premiums, are prominent factors in their material success, whatever their real merits may be.

THE AMERICAN POULTRY YARD.—A weekly illustrated journal; devoted specially to the interests of fowl breeders, fanciers, farmers, marketers and dealers. H. H. Stoddard, publisher, Hartford, Conn. $1.50 a year. We can add nothing to the merited reputation this journal has already attained, and that reputation is built on character. A six columned folio that *ought* to be patronized by all intelligent poultrymen of the country. Its illustrations are beautiful and significant, and its literary matter unexceptionable.

A GENERAL INDEX to the contents of fourteen popular treatises on natural philosophy, for the use of students, teachers, and artizans, by a Massachusetts teacher. Published by Ivison, Blakeman, Taylor and Co., Chicago, Ill. 108 royal 8 vo. Such a work, extended to other subjects, would be invaluable to those residing in the vicinity of a good public library, as it would obviate the necessity of owning a large library themselves.

WARD'S NATURAL SCIENCE BULLETIN, published at Ward's Natural Science Establishment, Rochester, New York. Price 50 cents per annum. This is an illustrated quarto quarterly of 16 pages, and being the presentation of Ward's Natural History Establishment, it occupies an entirely new field in Scientific literature. It is mainly devoted to the exposition and advertisement of Ward's Commercal Museum. No. 1, vol. 2, of this rare journal is now before us and in its leading editorial its publisher says—"It will mainly contain original matter: articles from various contributors on subjects connected with their particular departments: observations on specimens received at the aforesaid establishment, and interesting notes from collecting naturalists in the field." "There will be given from time to time many useful hints on the collecting, preparation and care of natural history specimens, and valuable receipts for compounds necessary for the collector, and practical working naturalist." Perhaps many of our readers are entirely ignorant of such an establishment as Ward's, where they may obtain at all times any thing in the "line" of natural history, from a "needle to an anchor," or from a tiny chinchbug, up to a gigantic elephant. Ward also deals extensively in models of extinct animals, and in stuffed specimens, in skeletons, craniums, &c., &c. To any one having the least taste for natural history and practical Taxidermy, this journal is invaluable. Prof. Ward himself has just recently returned from Australia and other foreign climes, "bringing with him mines of wealth;" much of which is entirely new to the scientific world. Mammals, birds—their nests and eggs—reptiles, crustaceans, mollusks, shells, corals, radiates, minerals, fossils, fishes, weapons and implements, utensils, plants, &c., &c. Prof. Ward also publishes a series of 17 catalogues of his specimens from 12 to 144 pp., ranging in price from 10 cents to $1.25, in which are enumerated what he has for sale, and the prices of the same. H. A. Ward, No. 2 College Avenue, Rochester, New York.

THE SOUTHERN CULTIVATOR.—We have received the January number of *The Southern Cultivator and Dixie Farmer*, the oldest, as it is the best, agricultural journal in the Southern States. It is now published by Jas. P. Harrison & Co., of Atlanta. Dr. W. L. Jones, for years the editor of this popular journal, retains his position; Dr. J. S. Lawton is the associate. Under this management, *The Southern Cultivator* will not only maintain its former high standard, but, with the assistance of ample capital and increased facilities, and contributions from the most eminent and popular writers on agriculture in this country, will attain a higher standing than ever. The number before us is a gem. No journal of its kind can excel it in the value of its reading matter, the beauty of its illustrations, and its adaptation to the demands of Progressive Southern agriculture. The illustrated title page is the finest of the kind we have ever seen. *The Southern Cultivator and Dixie Farmer* should be read and studied by every farmer and planter in the South. The terms, $1.50 a year, with special rates for clubs, are remarkably low. We advise our farmer friends to subscribe for it.

THE ARKANSAS FARMER.—"Non-partisan, non-political, but devoted to the real interests of our farmers." Little Rock, Arkansas, January 15, 1882. This is a seven columned folio (18 by 24) issued at $1.50 per year, weekly. The copy before us is No. 3, Vol. 1, and is, therefore, brand new; and, if it continues as it has begun, and does not prove a success, there must be something agriculturally—if not financially or socially—very "crooked" among the farmers and artizans of Arkansas. Its general make-up will average with the country folios of the North, and its editorials, contributions and selections are solid and instructive. We rejoice in its advent, for it seems to presage "better days a coming" for *Arkansaw*

REPORT of the "Pennsylvania Fruit-Growers' Society," prepared by its officers, 1881. An octavo of 69 pages, and contains the Constitution and By-Laws of the Society, lists of officers, committees, life members, annual members, and proceedings of the meeting held in Gettysburg in January of last year. The report contains two splendid full-page colored illustrations of the "Miner plum" and the "Cumberland triumph strawberry," with "Hersh's seedling," Striuestown pippin," apples, and the Maxatawney grape, concluded by an index of contents. Peculiarly situated as the society is, with the State as its printer and publisher, "more is the pity" that its reports only get into circulation about one year after the meeting of the society has adjourned.

THE SOUTHERN PLANTER, devoted to agriculture, horticulture, live stock and the household; a semi-monthly quarto of 15 pages, in tinted covers, published by Rolfe S. Saunders, Richmond, Va., at $3 a year. No. 1 of the 43d volume of this journal is before us, and although it has arrived at a patriarchal age among the literary institutions of the South, it seems to have lost none of the vigor of its youth, for it announces its intention to change, in the near future, to a weekly, and ought to be sustained.

JOURNAL of the *American Agricultural Association* for July and October, 1881, published quarterly at $2.00 per year, single copies, seventy-five cents. This is a Royal octavo, in tinted and embellished paper covers, containing 26: pages with 44 pages of advertisements. This multiplied by two would swell the volume to 524 pages annually of choice agricultural literature, contributed by some of the most distinguished agricultural writers of the country. Profusely illustrated with fine engravings, diagrams and "black-line charts, together with a multitude of statistics relating to the agricultural interests and resources of the country.

As corelating to the agricultural interests of the country are those of the transportation of agricultural products; hence the question of "THE RAILROAD AND THE FARMER" is discussed in lengthy articles by the *Hon. L. E. Chittenden* and the editor, Joseph H. Reall, in which the latter criticises the former in his paper, replying to a former paper by Mr. Atkinson on the same subject. Not having seen Mr. Atkinson's paper, and not having carefully read either Mr. Chittenden's or the Editor's, we refrain from expressing any sentiment at this time, any further than to say that our sympathies are with the FARMER in all the rights which legitimately belong to him, and especially to those in which he is the victim of unjust discrimination by Railroad companies.

THE SEED ANNUAL, for 1882, of D. M. FERRY & Co., Detroit, Mich., has been laid upon our table, and it is a perfect beauty in its line of operation. Its illustrated and descriptive space is equivalent to at least 180 pages, and, including the embellished covers, it has ten full pages colored lithograpic illustrations, embracing 72 figures of fruits, vegetables and flowers. It has also ten full page wood cuts, illustrating their seed stores in Detroit and Windsor, Can.; views on their seed farm, packing house, mailing department, box factory, &c., &c., besides *five hundred and forty* finely executed woodcuts, illustrating fruits, flowers, vines, ornamental plants, trees, shrubbery, vegetable, garden implements, &c., &c., and is perhaps as good a work on practical "Garden Botany," as any amateur needs.

INTERNATIONAL SCIENTIST'S DIRECTORY, for 1881-2, by S. E. CASSINO, Boston, Mass., containing the names, special departments of science, &c., &c., of amateur and professional naturalists, chemists, physicists, astronomers, &c., &c., in America, Europe, Asia, Africa and Oceanica. 12 mo. Over 400 pages. Paper $2.00. Cloth $2.50. Published December 1, 1881. This is undoubtedly the best work of its kind ever issued from the press of the United States, or perhaps any where else in the world, and no scientist should be without it. No one can form any conception of the labor it must have required to collect the information required in compiling such a work, without carefully examining its pages. Address, S. E. Cassino, No. 32 Hawley st.

THE ORIENTAL CASKET, a repository of literary gems; comprising poetry, tales, sketches, essays, wit, wisdom, humor, &c, from the world of literature, science and art. Edited by Emerson Bennett, and published by L. Lum Smith, 912 Arch street, Philadelphia, Pa, at $2.00 per annum: issued monthly. This is truly "a paper for all times, all people, and all places," and our chief regret is, that *we*, individually, have so little time to read it. The February number (Vol. 1, No. 2), of this magnificent journal has honored our table, and we find it all that its title claims it to be; truly "a casket of gems," contributed by a score of distinguished writers, both American and foreign. The material and the typographical execution is equal to any published in the Union at least, and its "orient pearls at random strung" will be found appreciable by a diversity of readers. It may be called (in size) a demi-folio, of 16 pages, or about the size of the *Scientific American* (12 by $16\frac{1}{2}$), clean and solid; no advertisement, no gaudy type, and no illustrations; but is enveloped in an embellished, tinted paper cover; and contains sufficient literary matter to feed a whole household, from Grandparents down to little Jo and Susey. If there is no "vacancy" for it in the realm of literature, it looks vigorous enough to make one. Our readers will observe that it is an entirely "new broom;" and perhaps they could not do better than help to make it an old one; but if they are unable to make up their minds, the 33 flattering editorial notices on the second page of the cover, ought to convey the necessary assurance that they cannot go wrong.

Important to Grocers, Packers, Hucksters, and the General Public.

THE KING FORTUNE-MAKER,
OZONE

A New Process for Preserving all Perishable Articles, Animal and Vegetable from Fermentation and Putrefaction, Retaining their Odor and Flavor.

"OZONE—Purified air, active state of Oxygen."—*Webster.*

This preservative is not a liquid pickle, or any of the old and exploded processes, but is simply and purely OZONE, as produced and applied by an entirely new process. Ozone is the antiseptic principle of every substance, and possesses the power to preserve animal and vegetable structures from decay.

There is nothing on the face of the earth liable to decay or spoil which Ozone, the new Preservative, will not preserve for all time in a perfectly fresh and palatable condition.

The value of Ozone as a natural preserver has been known to our older chemists for years lost, until now, no means of producing it in a practical, inexpensive, and simple manner have been discovered.

Microscopic observations prove that decay is due to septic matter or minute germs, that develop and feed upon animal and vegetable structures. Ozone, applied by the Prentiss method, seizes and destroys these germs at once and thus preserves. At our office in Cincinnati can be seen almost every article that can be thought of, preserved by us, and every visitor is welcomed to come in, taste, smell, take away with him, and test in every way the merits of Ozone as a preservative. We will also preserve, free of charge, any article that is brought or sent prepaid to us and return it to the sender, for him to keep and test.

FRESH MEATS, such as beef, mutton, veal, pork, poultry, game, fish, &c., preserved by this method, can be shipped to Europe, subjected to atmospheric changes and return to this country in a state of perfect preservation.

EGGS can be treated at a cost of less than one dollar a thousand dozen, and be kept in an ordinary room six months or more, thoroughly preserved; the yolk held in its normal condition, and the eggs as fresh and perfect as on the day they were treated, and will sell strictly "choice." The advantage in preserving eggs is readily seen when the season is long they can be bought for 8 or 10 cents a dozen, and by holding them, can be sold for an advance of two or three hundred to three hundred per ct. One man, with this method, can preserve 3,000 dozen a day.

FRUITS may be permitted to ripen in their native climate, and can be transported to any part of the world. The juice expressed from fruits can be held for an indefinite period without fermentation—hence the great value of these as flavorings for producing a temperance beverage. Cider can be held perfectly sweet for any length of time.

Jams filled with different articles, such as eggs, meat, fish, etc., can be treated at one time, without additional trouble or expense.

In fact, there is nothing that Ozone will not preserve. Think of everything that you can that is liable to decay, or spoil, and then remember that we guarantee that Ozone will preserve it in exactly the condition you want it for any length of time. If you will remember this it will save asking questions as to whether Ozone will preserve this or that article—It will preserve anything and everything you can think of.

There is not a town-ship in the United States in which a live man can not make any amount of money, from $500 to $5,000 a year, that he pleases. We desire to get a live man interested in each county in the United States to whom hands we can place this Preservative, and through him secure the business which every county ought to realize.

To convince parties of the absolute truth of everything we have said in this paper, **we propose to place in your hands the means of proving for yourself that we have not claimed half enough.** To any one who doubts any of these statements, and who is interested sufficiently to make the trip, we will pay all traveling and hotel expenses for a visit to this city, if we fail to prove any statement that we have made.

How to Secure a Fortune with Ozone.

A test package of Ozone, containing a sufficient quantity to preserve one thousand dozen eggs, or other articles in proportion, will be sent to any applicant on receipt of $2. This package will enable the applicant to pursue any kind of tests and experiments he desires, and thus satisfy himself as to the extraordinary merits of Ozone as a preservative. After having thus satisfied himself, and had time to look the field over to determine what he wishes to do in the future—whether to sell the article to others or to confine it to his own use, or any other line of policy which is best suited to him and to his township or county—we will enter into an arrangement with him that will make a fortune for him and give us good profits. We will give exclusive township or county privileges to the first responsible applicant who orders a test package and desires to control the business in his locality. The man who secures control of Ozone for any special territory, will enjoy a monopoly which will surely enrich him.

Don't let it pass until you have ordered a Test Package, and if you desire to secure an exclusive privilege we assure you that delay may deprive you of it, for the applications come in to us by scores every mail—many by telegraph. "First come first served" is our rule.

Cost do not care to send money in advance for the test package we will send it C. O. D, but this will put you to the expense of charges for return money. Our correspondence is very large we have all we can do to attend to the shipping of orders and giving attention to our working agents. Therefore we can not give any attention to letters which do not order Ozone. If you think of any article that you are doubtful about Ozone preserving remember we say **it will preserve it, no matter what it is.**

REFERENCES.

We do here call your attention to a class of references which no enterprise or firm based on any thing but the soundest business sense and highest commercial merit could secure.

We refer by permission, as to our integrity and to the value of the Prentiss Preservative, to the following gentlemen: Edward C. Royce, Member Board of Public Works, E. O. Eshelby, City Comptroller Amor Smith, Jr., Collector Internal Revenue, Wilkin & Worthington, Attorneys, Martin H. Harrell and B. F. Hopkins, County Commissioners, W. S. Cappeller County Auditor; all of Cincinnati, Hamilton County, Ohio. These gentlemen are familiar with the merits of our Preservative, and know from actual observation that we have without question

The Most Valuable Article in the World.

If you invest in a test package, will surely lead you to secure a township or county, and then your way is absolutely clear to make from $2,000 to $10,000 a year.

Give our full address on every letter, and send your letter to

PRENTISS PRESERVING COMPANY, (Limited,)
S. E. Cor. Ninth & Race Sts., Cincinnati, O.

THE
LANCASTER EXAMINER

OFFICE

No. 9 North Queen Street,

LANCASTER, PA.

THE OLDEST AND BEST.

THE WEEKLY
LANCASTER EXAMINER

One of the largest Weekly Papers in the State.

Published Every Wednesday Morning,

Is an old, well-established newspaper, and contains just as much desirable to make it an interesting and valuable Family Newspaper. The postage to subscribers residing outside of Lancaster county is paid by the proprietor. Send for a specimen copy.

SUBSCRIPTION:

Two Dollars per Annum.

THE DAILY
LANCASTER EXAMINER

The Largest Daily Paper in the county.

Published Daily Except Sunday.

The daily is published every evening during the week, and is delivered in the City and to surrounding Towns accessible by railroad and daily stage lines, for 10 cents a week.

Mail subscription, free of postage—1 month, 50 cents; 6 months, $3.00.

THE JOB ROOMS.

The job rooms of The Lancaster Examiner are filled with the latest styles of presses, material, etc., and we are prepared to do all kinds of Book and Job Printing at as low rates as any other establishment in the State.

SALE BILLS A SPECIALTY.

With a full assortment of new cuts that we have just purchased, we are prepared to do the most and the most attractive style of work for sales.

JOHN A. HIESTAND, Proprietor,

No. 9 North Queen St.,

LANCASTER, PA.

WHERE TO BUY GOODS IN LANCASTER.

BOOTS AND SHOES.

MARSHALL & SON, No. 12 Centre Square, Lancaster, Dealers in Boots, Shoes and Rubbers. Repairing promptly attended to.

M. LEVY, No. 3 East King street. For the best Dollar Shoes in Lancaster go to M. Levy, No. 3 East King street.

BOOKS AND STATIONERY.

JOHN BAER'S SON'S, Nos. 15 and 17 North Queen Street, have the largest and best assorted Book and Paper Store in the City.

FURNITURE.

HEINITSH'S, No. 15½ East King st., over China Hall; is the cheapest place in Lancaster to buy Furniture. Picture Frames a specialty.

CHINA AND GLASSWARE.

HIGH & MARTIN, No. 15 East King st., dealers in China, Glass and Queensware, Fancy Goods, Lamps, Burners, Chimneys, etc.

CLOTHING.

MYERS & RATHFON, Centre Hall, No. 12 East King st. Largest Clothing House in Pennsylvania outside of Philadelphia

DRUGS AND MEDICINES.

G. W. HULL, Dealer in Pure Drugs and Medicines Chemicals, Patent Medicines, Trusses, Shoulder Braces, Supporters, &c., 15 West King St., Lancaster, Pa.

JOHN F. LONG & SON, Druggists, No. 12 North Queen St. Drugs, Medicines, Perfumery, Spices, Dye Stuffs, Etc. Prescriptions carefully compounded.

DRY GOODS.

GIVLER, BOWERS & HURST, No. 25 E. King St., Lancaster, Pa., Dealers in Dry Goods, Carpets and Merchant Tailoring. Prices as low as the lowest.

HATS AND CAPS.

C. H. AMER, No. 39 West King Street, Dealer in Hats, Caps, Furs, Robes, etc. Assortment Large. Prices Low.

JEWELRY AND WATCHES.

H. Z. RHOADS & BRO., No. 4 West King St. Watches, Clocks and Musical Boxes. Watches and Jewelry Manufactured to order.

PRINTING.

JOHN A. HIESTAND, 9 North Queen st., Sole Bills, Circulars, Posters, Cards, Invitations, Letter and Bill Heads and Envelopes neatly printed. Prices low.

Thirty-six Varieties of Cabbages; 26 of Corn; 2 of Cucumber; 31 of Melon; 31 of Peas; 28 of Beans; 17 of Squash; 25 of Beets and 11 of Tomato, with other varieties in proportion, a large portion of which were grown on my five seed farms, will be found in my **Vegetable and Flower Seed Catalogue for 1882**. Sent post paid to all who apply. Customers of last Season need not write for it. All seed sold from my establishment warranted to be fresh and true to name, so far, that should it prove otherwise, I will refill the order gratis. The original introducer of Early Ohio and Hurbank Potatoes, Marblehead Early Corn, the Hubbard Squash, Marblehead Cabbage, Phinney's Melon, and a score of other New Vegetables, I invite the patronage of the public. New Vegetables a specialty.

JAMES J. H. GREGORY,
Marblehead, Mass.

Nov-6mo

EVAPORATE YOUR FRUIT.
ILLUSTRATED CATALOGUE
FREE TO ALL.

AMERICAN DRIER COMPANY,
Chambersburg, Pa.

A1-6tf

THE LANCASTER FARMER

FARMING FOR PROFIT.
It is conceded that this large and comprehensive book, advertised in another column by J. C. McCurdy & Co., of Philadelphia, the well-known publishers of Standard works, is not only the newest and handsomest, but altogether the BEST work of the kind which has ever been published. Thoroughly treating the best subjects of general Agriculture, Live-Stock, Fruit-Growing, Business Principles and Home Life; telling just what the farmer and the farmer's boys want to know, combining Science and Practice, stimulating thought, awakening inquiry, and interesting every member of the family, this book must exert a mighty influence for good. It is highly recommended by the best agricultural writers and the leading papers, and is destined to have an extensive sale. Agents are wanted everywhere. jan-5

CIDER MILLS!
Wine Presses!

Fruit Presses, Apple Slicers, Fodder and Ensilage Cutters,

GRAIN FANS,

Grain and Fertilizer Drills,

Broad-cast Seed Sowers,

Corn Shellers, Corn Mills,

Grain Mills, etc., etc.

FOR SALE BY
D. LANDRETH & SONS,
AGRICULTURAL AND HORTICULTURAL IMPLEMENT
AND
SEED WAREHOUSE,
Nos. 21 and 23 South Sixth Street,

BETWEEN MARKET AND CHESTNUT STS.,
— and —
No. 4 ARCH STREET,

apr-6m PHILADELPHIA.

MERCHANT TAILORING.
1848   The Oldest of All.   1881
RATHVON & FISHER,
MERCHANT TAILORS AND DRAPERS,

respectfully inform the public that having disposed of their entire stock of Ready-Made Clothing, they now do, and for the future shall, devote their whole attention to the CUSTOM TRADE.

All the desirable styles of CLOTHS, CASSIMERES, WORSTEDS, COATINGS, SUITINGS and VESTINGS constantly on hand, and made to order in plain or fashionable style promptly, and warranted satisfactory.

All-Wool Suit from $12.00 to $30.00.
All-Wool Pants from $5.00 to $10.00.
All-Wool Vests from $2.00 to $5.00.

Union and Cotton Goods proportionately less.
Cutting, Repairing, Trimming and Making, at reasonable prices.
Goods retailed by the yard to those who desire to have them made elsewhere.
A full supply of Spring and Summer Goods just opened and on hand.

Thankful to a generous public for past patronage they hope to merit its continued recognition in their "new departure."

RATHVON & FISHER,
PRACTICAL TAILORS,
No. 101 North Queen Street,
LANCASTER, PA.

1848   1881

GLOVES, SHIRTS, UNDERWEAR.
SHIRTS MADE TO ORDER,
AND WARRANTED TO FIT.
E. J. ERISMAN,
56 North Queen St., Lancaster, Pa.
7-1-12j

A HOME ORGAN FOR FARMERS.

THE LANCASTER FARMER,

A MONTHLY JOURNAL,

Devoted to Agriculture, Horticulture, Domestic Economy and Miscellany.

Founded Under the Auspices of the Lancaster County Agricultural and Horticultural Society.

EDITED BY DR. S. S. RATHVON.

TERMS OF SUBSCRIPTION:

ONE DOLLAR PER ANNUM,
POSTAGE PREPAID BY THE PROPRIETOR.

All subscriptions will commence with the January number, unless otherwise ordered.

Dr. S. S. Rathvon, who has so ably managed the editorial department in the past, will continue in the position of editor. His contributions on subjects connected with the science of farming, and particularly that specialty of which he is so thoroughly a master—entomological science—of a knowledge of which has become a necessity to the successful farmer, are alone worth much more than the price of this publication. He is determined to make "The Farmer" a necessity to all households.

A county that has so wide a reputation as Lancaster counts for its agricultural products should certainly be able to support an agricultural paper of its own, for the exchange of the opinions of farmers interested in this matter. We ask the co-operation of all farmers interested in this matter. Work among your friends. The "Farmer" is only one dollar per year. Show them your copy. Try and induce them to subscribe. It is not much for each one; send us to do but it will greatly assist us.

All communications in regard to the editorial management, should be addressed to Dr. S. S. Rathvon, Lancaster, Pa., and all business letters in regard to subscriptions and advertising should be addressed to the publisher. Rates of advertising sent on application at the office.

JOHN A. HIESTAND,
No. 9 North Queen St., Lancaster, Pa.

$5 TO $20 per day at home. Samples worth $5 free. Address STINSON & Co., Portland, Maine.
t26-1yr

ONE DOLLAR PER ANNUM.—SINGLE COPIES 10 CENTS.

THE LANCASTER FARMER

DEVOTED TO AGRICULTURE, HORTICULTURE, DOMESTIC ECONOMY AND MISCELLANY

THE FARMER IS THE FOUNDER OF CIVILIZATION.—WEBSTER

Dr. S. S. RATHVON, Editor.   LANCASTER, PA. MARCH, 1882.   JOHN A. HIESTAND, Publisher.

Entered at the Post Office at Lancaster as Second Class Matter.

CONTENTS OF THIS NUMBER.

EDITORIAL.
Our Apology, - - - - - - - - 33
"Our Winged Friends," - - - - - 33
Kitchen Garden for March, - - - - 33
Why not Write for the Farmer? - - - 33
The Bane and Antidote, - - - - - 34
  Death from Wild Animals in India—Snake Destroyers.
"Revised Fruit List," - - - - - - 34
Eating Before Sleeping, - - - - - 35
How Long are We to Live, - - - - 36
  The Extreme Limit of Human Life—Weak Lungs.
The Will and the Deed, - - - - - 37
Excerpts, - - - - - - - - 37

CONTRIBUTIONS.
Forestry, - - - - - - - - 38
Strawberries, - - - - - - - 39
Practical Poultry Notes, - - - - - 39
Domestic Hints, - - - - - - - 39
Practical Recipes, - - - - - - 39

ESSAYS.
The Growth and Consumption of Timber Trees
  In America, - - - - - - - 40
"Our Winged Friends, - - - - - 41
Seedling Fruits, - - - - - - - 44

OUR LOCAL ORGANIZATIONS.
Lancaster County Agricultural and Horticultural Society, - - - - - - - 45
  Crop Reports—Apples, Local vs. Foreign Remarks—Is Sub-Soiling Beneficial?—Can We Dispense with Division Fences on Farms?—Remarks—When is the best Time to Sow Clover Seed?—More About Apples.
Poultry Society, - - - - - - - 47
Fulton Farmers' Club, - - - - - 47
Linnaean Society, - - - - - - 47
  Twentieth Anniversary of the Founding of the Society Museum—Library—Historical—Anniversary—Science Gossip—History of the Society.
Literary and Personal, - - - - - 48

LIGHT BRAHMA EGGS
FOR HATCHING,
$1.50 FOR SETTING OF 13.
ALSO,
Three Barrels of Chicken Manure
FOR SALE.
L. RATHVON,
Examiner Office No. 9 N. Queen-st., Lancaster, Pa.

SEND IN YOUR SUBSCRIPTIONS
—FOR—
THE FARMER
FOR 1882.
The cheapest and one of the best Agricultural papers in the country.

Only $1.00 per year.
JOHN A. HIESTAND, Publisher,
No. 9 North Queen St., Lancaster, Pa.

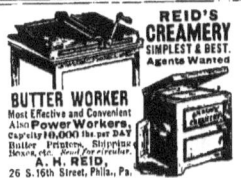

REID'S CREAMERY
SIMPLEST & BEST.
Agents Wanted

BUTTER WORKER
Most Effective and Convenient
Also Power Workers,
Capacity 10,000 lbs. per DAY
Butter Printers, Shipping Boxes, etc. Send for circular.
A. H. REID,
26 S. 16th Street, Phila., Pa.
feb-4m

Eggs! Eggs!

From all the leading varieties of pure bred Poultry Brahmas, Cochin, Hamburgs, Polish Game, Dorking and French Fowls, Plymouth Rocks and Bantams, Rouen and Pekin Ducks. Send for Illustrated Circular.
T. SMITH, P. M., Fresh Pond, N. Y.
feb-3m

SEEDS BULBS, PLANTS.
Beautiful Illustrated Catalogue Free.

J. LEWIS CHILDS, QUEENS, N. Y.
Jan-3m

WE WANT OLD BOOKS.
We Want German Books.
WE WANT BOOKS PRINTED IN LANCASTER CO.
We Want All Kinds of Old Books.
LIBRARIES, ENGLISH OR GERMAN BOUGHT.
Cash paid for Books in any quantity. Send your address and we will call.
REES WELSH & CO.,
23 South Ninth Street, Philadelphia.

 D. M. FERRY & CO.'s
ILLUSTRATED DESCRIPTIVE AND PRICED
SEED ANNUAL FOR 1882
Will be mailed free to all applicants, and to customers without ordering it. It contains five colored plates, 600 engravings, about 200 pages, and full descriptions, prices and directions for planting 1500 varieties of Vegetable and Flower Seeds, Plants, Fruit Trees, etc. Invaluable to all. Send for it. Address,
D. M. FERRY & CO., Detroit, Mich.
Jan-4m

$66 a week in your own town. Terms and $5 outfit free. Address H. HALLETT & Co., Portland, Maine.
jun-1yr§

PENSIONS For SOLDIERS,
children. Thousands yet entitled. Pensions given for loss of finger, toe, eye, or rupture, varicose veins or any disease. Thousands of pensioners and soldiers entitled to INCREASE and BOUNTY. PATENTS procured for inventors. Soldiers land warrants procured, bought and sold. Soldiers and heirs apply for your rights at once. Send 2 stamps for Pension and Bounty laws, blanks and instructions. Very liberal terms. We can refer to thousands of Pensioners and Clients. Address E. H. Gelston & Co.,
(C. St Claim Att'ys,) Lock Box 173, Washington, D.C.
Jan

LIGHT BRAHMA EGGS
For hatching, now ready—from the best strain in the county—at the moderate price of
$1.50 for a setting of 13 Eggs.
L. RATHVON,
No. 9 North Queen st., Examiner Office, Lancaster, Pa.

WANTED.—CANVASSERS for the
LANCASTER WEEKLY EXAMINER
In Every Township in the County. Good Wages can be made. Inquire at
THE EXAMINER OFFICE,
No. 9 North Queen Street, Lancaster, Pa.

$72 A WEEK. $12 a day at home easily made. Costly outfit free. Address TRUE & Co., Augusta, Maine.
jun-1yr§

SEND FOR
SPECIAL PRICES
On Concord Grapevines, Transplanted Evergreens, Tulip, Poplar, Linden Maple, etc. Tree Seedlings and Trees for timber plantations by the 100,000
J. JENKINS' NURSERY,
3-2-79   WINONA, COLUMBIANA CO., OHIO.

PENSIONS For SOLDIERS,
widows, fathers, mothers or children. Thousands yet entitled. Pensions given for loss of finger, toe, eye, or rupture, varicose veins or any disease. Thousands of pensioners and soldiers entitled to INCREASE and BOUNTY. PATENTS procured for inventors. Soldiers land warrants procured, bought and sold. Soldiers and heirs apply for your rights at once. Send 2 stamps for Pension and Bounty laws, blanks and instructions. Very liberal terms. We can refer to thousands of Pensioners and Clients. Address N. W. Fitzgerald & Co., Pension & Patent Att'ys, Lock Box 585, Washington, D.C.
dec-tf

WELL-AUGER.
Ours is guaranteed to be the cheapest and best in the world. Also nothing can beat our SAWING MACHINE. It saws off a 2 foot log in 2 minutes. Pictorial books free. W. GILES, Chicago, Ill.
-6m§

PENNSYLVANIA RAILROAD SCHEDULE.

Trains LEAVE the Depot in this city, as follows:

WESTWARD.	Leave Lancaster.	Arrive Harrisburg.
Pacific Express*	2:40 a. m.	4:05 a. m.
Way Passenger*	5:00 a. m.	7:30 a. m.
Niagara Express	11:00 a. m.	11:20 a. m.
Hanover Accommodation	11:05 p. m.	via. 10:05 a. m.
Mail train via Mt. Joy	10:20 a. m.	12:40 p. m.
No. 2 via Columbia	11:25 a. m.	12:55 p. m.
Sunday Mail	10:50 a. m.	12:40 p. m.
Fast Line*	2:30 p. m.	3:25 p. m.
Frederick Accommodation	2:45 p. m.	Col. 2:45 p. m.
Harrisburg Accom	5:15 p. m.	7:40 p. m.
Columbia Accommodation	7:20 p. m.	Col. 8:40 p. m.
Harrisburg Express	7:30 p. m.	8:40 p. m.
Pittsburg Express	8:50 p. m.	10:50 p. m.
Cincinnati Express*	11:30 p. m.	12:15 a. m.

EASTWARD.	Lancaster.	Philadelphia
Cincinnati Express*	2:55 a. m.	5:00 a. m.
Fast Line*	5:08 a. m.	7:40 a. m.
Harrisburg Express	8:05 a. m.	10:00 a. m.
Columbia Accommodation	9:10 a. m.	12:20 p. m.
Pacific Express*	1:40 p. m.	3:40 p. m.
Sunday Mail	2:00 p. m.	5:00 p. m.
Johnstown Express	3:05 p. m.	5:30 p. m.
Day Express*	5:25 p. m.	7:20 p. m.
Harrisburg Accom	6:25 p. m.	9:30 p. m.

The Hanover Accommodation, west, connects at Lancaster with Niagara Express, west, at 9:35 a. m., and will run through to Hanover.

The Frederick Accommodation, west, connects at Lancaster with Fast Line, west, at 2:10 p. m., and runs to Frederick.

The Pacific Express, east, on Sunday, when flagged, will stop at Middletown, Elizabethtown, Mount Joy and Landisville.

*The only trains which run daily.

†Runs daily, except Monday.

NORBECK & MILEY,

PRACTICAL
Carriage Builders,

COX & CO'S OLD STAND,

Corner of Duke and Vine Streets,
LANCASTER, PA.

THE LATEST IMPROVED

SIDE-BAR BUGGIES,

PHÆTONS,
Carriages, Etc.

THE LARGEST ASSORTMENT IN THE CITY.

Prices to Suit the Times.

REPAIRING promptly attended to. All work guaranteed.
79-2-

S. B. COX,
Manufacturer of

Carriages, Buggies, Phaetons, etc.

CHURCH ST., NEAR DUKE, LANCASTER, PA.

Large Stock of New and Second-hand Work on hand very cheap. Carriages Made to Order Work Warranted for one year.

EDW. J. ZAHM,

DEALER IN

AMERICAN AND FOREIGN
WATCHES,

SOLID SILVER & SILVER PLATED WARE,

CLOCKS,
JEWELRY & TABLE CUTLERY.

Sole Agent for the Arundel Tinted

SPECTACLES.

Repairing strictly attended to.

ZAHM'S CORNER,

North Queen-st. and Centre Square, Lancaster, Pa.
79-1-12

E. F. BOWMAN,

AT LOWEST POSSIBLE PRICES.
Fully guaranteed.

No. 106 EAST KING STREET,
79-1-12)   Opposite Leopard Hotel.

ESTABLISHED 1832.

G. SENER & SONS,
Manufacturers and dealers in all kinds of rough and finished

LUMBER,

The best Sawed SHINGLES in the country. Also Sash, Doors, Blinds, Mouldings, &c.

PATENT O. G. WEATHERBOARDING

and PATENT BLINDS, which are far superior to any other. Also best COAL constantly on hand.

OFFICE AND YARD:
Northeast Corner of Prince and Walnut-sts.,
LANCASTER, PA.
79-1-12]

PRACTICAL ESSAYS ON ENTOMOLOGY,
Embracing the history and habits of

NOXIOUS AND INNOXIOUS
INSECTS,

and the best remedies for their expulsion or extermination.

By S. S. RATHVON, Ph. D.
LANCASTER, PA.

This work will be Highly Illustrated, and will be put to press (as soon after a sufficient number of subscribers can be obtained to cover the cost) as the work can possibly be accomplished.
79-3-

$77 a month and expenses guaranteed to Agents. Outfit free. SHAW & CO., Augusta, Maine.
79-2-12

TREES

Fruit, Shade and Ornamental Trees.

Plant Trees raised in this county and suited to this climate. Write for prices to
LOUIS C. LYTE
Bird-in-Hand P. O., Lancaster co., Pa.
Nursery at Smoketown, six miles east of Lancaster.
79-1-12

WIDMYER & RICKSECKER,
UPHOLSTERERS,
And Manufacturers of

FURNITURE AND CHAIRS.

WAREROOMS:

102 East King St., Cor. of Duke St.
LANCASTER, PA.
79-1-12)

Special Inducements at the
NEW FURNITURE STORE
OF
W. A. HEINITSH,

No. 15 1-2 E. KING STREET
(over Bursk's Grocery Store), Lancaster, Pa.
A general assortment of furniture of all kinds constantly on hand. Don't forget the number,
15 1-2 East King Street,
Nov-1y)   (over Bursk's Grocery Store.)

For Good and Cheap Work go to
F. VOLLMER'S
FURNITURE WARE ROOMS,
No 309 NORTH QUEEN ST.,
(Opposite Northern Market),
Lancaster, Pa.
Also, all kinds of picture frames.
nov-1y

GREAT BARGAINS.
A large assortment of all kinds of Carpets are still sold at lower rates than ever in the

CARPET HALL OF H. S. SHIRK,
No. 202 West King St.

Call and examine our stock and satisfy yourself that we can show the largest assortment of their Brussels, three plys and Ingrains at all prices—at the lowest Philadelphia prices.

Also on hand a large and complete assortment of Rag Carpet.

Satisfaction guaranteed both as to price and quality. You are invited to call and see my goods. No trouble in showing them even if you do not want to purchase. Don't forget this notice. You can save money here if you want to buy.

Particular attention given to customer work.
Also on hand a full assortment of Counterpanes, Oil Cloths and Blankets of every variety   nov-1y

PHILIP SCHUM, SON & CO.,
36 and 40 West King Street.
We keep on hand of our own manufacture,
QUILTS, COVERLETS,
COUNTERPANES, CARPETS,
Bureau and Tidy Covers, Ladies' Furnishing Goods, Notions, etc.
Particular attention paid to cleaning Rag Carpet, and scouring and dyeing of all kinds.

Nov-1y   **PHILIP SCHUM, SON & CO.,**
Lancaster, Pa.

THE HOLMAN LIVER PAD.
Cures by absorption without medicine.
Now is the time to apply these remedies. They will do for you what nothing else on earth can. Hundreds of citizens of Lancaster say so. Get the genuine at

LANCASTER OFFICE AND SALESROOM,
22 East Orange Street.
Nov-1yr

C. R. KLINE,
ATTORNEY-AT-LAW,
OFFICE: 15 NORTH DUKE STREET,
LANCASTER, PA.
Nov-1y

The Lancaster Farmer.

Dr. S. S. RATHVON, Editor. LANCASTER, PA., MARCH 1882. Vol. XIV. No. 3.

EDITORIAL.

OUR APOLOGY.

An apology is due our patrons for the late appearance of THE FARMER in the month of February, and also the present number, and we can't tell how long this state of things may continue, but we can assure our readers that it is only temporary, and is not without a mitigating cause. Since the destruction of the *Inquirer* building by fire, a large portion of the printing executed by that establishment devolves upon our office; and, as the arrangement is only a temporary one, we can only make a temporary provision for it. When our patrons become aware of this fact we feel assured that they will sympathize with us in our efforts to accommodate those who in one fell swoop of the devouring flames have been deprived of the mechanical means to execute their business engagements. These contingencies cannot be foreseen, but when they do occur, charity dictates that we should help to bear each other's burdens.

"OUR WINGED FRIENDS."

We publish in this issue of THE FARMER the very interesting paper on insectivorous birds, read by S. P. Eby, Esq., of Lancaster, Pa., before the "State Horticultural Society," at its annual meeting, held at Harrisburg in January last; and we regret that we were not able to publish it sooner, for such papers are worthy of a permanent record; and especially when they come within the category of one of the leading specialties of this journal from its very origin, namely, to make it a record of the sayings and doings of the people of Lancaster county in relation to agriculture and its allies. The essay needs no commendation of ours, for, having it before them, our readers will be able to judge for themselves, and we can assure them that they will be both interested and instructed.

KITCHEN GARDEN FOR MARCH.

In the Middle States spring has arrived according to the calendar, but the experienced gardener is not caught by arbitrary terms; and though March and the almanac may indicate spring, frost and storm and biting winds caution him to care and patience. He will wait the progress of the month and bide his time.

Artichokes dress; plant. Asparagus sow; plant the *colossal* roots. Beets— Extra Early, Philadelphia Turnip, and Early Blood Turnip, sow. Cabbage sow in a sheltered place, it not already in a hot-bed. Test Landreth's new varieties—the Wakefield, Early Market, and Bloomsdale Brunswick. Carrots—Early Horn, sow. Cauliflowers—attend to those under glass. Celery sow. Cress sow. Compost prepare. Dung prepare for hot-beds, Horse-radish plant. Hot-beds make; also force. Lettuce sow; prick out. Mushroom beds attend to. Mustard sow. Onions put out in sets—those known as "Philadelphia Buttons" much the best. Parsnips sow—the Sugar is the best. Peas—Landreth's Extra Early and Invicta sow. Also, McLean's Advancer and McLean's Gem, which we commend with confidence. Potatoes—early, plant. The early Rose is admirable in every respect. Radish—the Long Scarlet, and Red and White Turnip sow. The "Strap-leaved Long Scarlet," an improvement on the Long Scarlet, we recommend. Rhubarb sow; plant roots. Sage sow; plant. Tomato sow in hot-bed. Turnip, Strap-leaved Early Dutch sow; but generally be it observed, so far north as Philadelphia, these directions will apply better to April than to March.—*Landreth's Rural Register*.

March has always been a fitful, capricious and uncertain month; and, under the most favorable circumstances, the spring season cannot be considered as fairly inaugurated before St. Patrick's day; and this, too, without regard to "Candlemas" or the "Ground Hog," traditional weathermarks very unsafe to give character to any practical enterprise. Of course, in matters relating to husbandry it is always well to be forewarned; but, as "a single swallow don't make a summer," even so the judicious farmer will not be deceived by a single "weather-breeder" in the month of March, but will defer his sheep-shearing and goose-plucking to a more reliable period.

If we cannot *plant* in March we can, at least, do something in the way of preparation which will advance and facilitate the work of the kitchen-garden when the proper time arrives. Those who manipulate a hot-bed will be not greatly hindered, whatever is likely to be the character of the weather in March. In any event, we would admonish our readers to plant and cultivate the best varieties of fruits, grains and vegetables, and especially the best adapted to their different soils. The *best*, although it may be the most expensive, is in the end the cheapest, both to the producer and the consumer. We confess that we are often surprised at the inferior quality of vegetables that are often found in our markets. Especially is this the case in the matter of green corn. That for table use should possess the *highest saccharine qualities*, and to obtain the seed of this, application should be made to the Landreths, of "Bloomingdale Seed Farm," and then "crop it," in order to have it for a more protracted period than is usually the case among our farmers in general.

WHY NOT WRITE FOR THE FARMER?

Mr. RATHVON: In your February number, I see a correspondent asks the question, "why don't the farmers of Lancaster county write for the FARMER," and you state that you cannot tell why.

The reason why, is plain and simple to me and a good many other people. We have many good, old, and experienced farmers in the county, but not so well educated—at least some of them—as other classes perhaps; but when they write what they *know by experience*, they are apt to be ridiculed by those of higher education, and so they keep it to themselves, and only communicate it in a private way, to a good friend or neighbor.

How was it when J. G. wrote about lunar influence, about two years ago?

Editors and publishers generally, make it a rule not to admit into their columns, or even to notice anonymous communications; but, on this occasion, we waive the rule, because we believe the reason assigned is honestly—although we think mistakenly—entertained, and is therefore insufficient in its conclusions. In the first place, the experiences of men differ; and hence, there always *have* been, and perhaps always *will* be different opinions among men on the same subjects. And, under our form of government all men have the privilege of expressing their opinions without restraint, so long as they are not contraband of law. But, because men may differ with us, or contest our opinions, it is no reason that we should be silent, especially when they are supported by our own practical experience. A thing that is *really true*, is none the less true, because some other man don't believe it is true. Does any Christian believe that he ought to cease to preach and pray, because Bob Ingersoll, and other infidels, ridicule christianity and the Bible? No more ought a farmer to cease to give his experimental views of farming, because others may honestly dissent from his views. When the Saviour advanced his views, he was "laughed to scorn;" but he continued to teach them, because he *knew them* to be true. When St. Paul proclaimed the truths of christianity he was declared to be "mad;" but he continued to preach and to *write*. He did not keep what he received "to himself," but he communicated it to others, whether they believed it or not.

It was the same with Fulton and the steamboat, Morse and the telegraph, and hundreds of others in the world, who have advanced their views on different subjects, only to meet with views in conflict with them, but time and experience ultimately demonstrated what was true and what was false. Moreover, a mere denial does not negative a proposition;", hence, if one man *disbelieves* us another may *believe* us; therefore, so long as *our* believes, there is as much reason to continue our writing as there is to discontinue it, because *one* disbelieves. When the "sower went forth to sow," his seed fell upon different kinds of ground, and the *effect* of his sowing was different in its results, and it perhaps will always be so.

We always give our correspondents and contributors a respectful hearing, when their views come within the scope of an agricultural journal. We cannot reject or suppress an article because it happens to be in conflict with opinions previously advanced by some other writer; nor, indeed, when it is in conflict with our own opinions on the same subject. When the views advanced are *true* ones, the *truth* will only become more apparent through free discussion. We, therefore, admonish our patrons that they should not feel discouraged because others—even those of a higher education—may differ from them in opinion, especially when their views are based upon *facts* elicited by actual experience and not mere theories. It is true that an educated man by fallacious arguments may be more successful in impressing *false* views than an uneducated man is in advancing *true* ones, but time and experience will eventually demonstrate who is right and who is wrong. As to the matter of education, it does not require a very high degree of it in order to be able to tell the truth when we *know* the truth. In conclusion, lunar influence upon the earth's surface involves questions that are in an unsettled condition among men—even among those who are educated—but there are plenty of other subjects of a more practical character, which are worthy the pen of the experienced farmer—subjects more tangible and nearer to his daily life, and these should become the objects of his occasional illustration.

THE BANE AND ANTIDOTE.

The two paragraphs adduced below illustrate a state of things in India that perhaps never enters the mind of the average individual who concerns himself but little as to how other people live and die in this world ; and, even those who *may* feel an interest in their fellow-men, and contribute to their alleviation, may be astonished at the aggregate of human exposure to the fatalities of animal ferocity. Venomous reptiles and carnivorous mammals seem to be the *bane* of India, however endowed she may be in other respects ; and, although we may suppose that she also, to some extent, possesses the *antidote*, yet it can not be sufficiently strong to overcome the preponderating bane. India possesses many snake destroying birds, notably the Secretary, the Cassowary and the Vulture, besides many others. If, therefore, with all these checks to the increase of venomous snakes, together with the hundreds of thousands destroyed annually, under the auspices of government, there is still such a fearful mortality from snake bites, what might it be if none of these counter-operations existed. The question involved in these two papers is one that brings before us on a large scale the relations that one class of animals bears to another, in maintaining the equilibrium of nature, in which it is plainly evident that if the one did not at all exist, even though its presence might be regarded by some people as an unmitigated nuisance—what a fearful redundancy there might be of the other more objectionable class. This rule may also be applied on a smaller scale to the noxious animals and their natural antidotes in our own country. We don't know what the mortality from snake bites is in the United States, but from all that gets into the prints we may infer that it is very trifling when compared with India. We have, however, a tolerable idea of the injuries sustained by agriculture from noxious insects ; but we cannot even guess what it might be if none of the natural antidotes existed; and yet because these often operate against the interests of certain individuals, they would have them all destroyed. Birds in our own country, as well as in India, are the natural enemies of insects, and to a greater extent, too, than we may be aware of ; and yet many people are restive and impatient under the presence of birds, because they also appropriate a little fruit, or other substance of human production.

Death from Wild Animals in India.

The total number of persons killed by snakes and wild beasts in the several Provinces of India during 1880 has gradually increased from 19,273 in 1877 to 21,990 in 1880. The largest number of deaths occurred in Bengal and the Northwestern Provinces and Oudh, in which Provinces the deaths during the year aggregated 11,350 and 5,281, respectively. In Bengal 10,064 deaths were caused by snake bites, 359 persons were killed by tigers, while in the Northwestern Provinces and Oudh, 4,723 persons died from snakebites and 205 were killed by wolves. The total number of persons killed by wild beasts and venomous snakes during the year 1880 was 21,990. The increase was common to all Provinces, except British Burmah. The number of cattle killed increased from 54,830 in 1876, to 55,914 in 1879, and 58,386 in 1880, (exclusive of the figures for Mysore, where the deaths in the previous year amounted to 5,899.) The increase compared with 1879 is common to all Provinces except the Northwestern Provinces and Oudh, the Punjab, and Ajmere-Merwara. In the Northwestern Provinces and Oudh the totals for the two years are nearly the same, and in the Punjab there was a decrease of about 1,200 in the number of cattle killed. The total number of wild animals destroyed has fallen year by year from 23,450 in 1876 to 18,641 in 1879, and 14,886 in 1880. As compared with the previous year the falling off was common to all Provinces, except the Central Provinces, Coorg. and Berar. The most remarkable decrease occurred under the heading "other animals" in the Madras Presidency, the figures for 1879 and 1880 having been 2,950 and 139, respectively. The number of snakes shown as destroyed was 211.775, as compared with 131,927 in the previous year, the increase being mainly due to the very large number (177,070) of snakes which were killed in the Bombay Presidency. The total amount of rewards paid for the destruction of snakes was 11,903 rupees, as compared with 6,663 rupees the previous year. It is chiefly in towns and villages that the destruction of snakes is desirable, and for this reason it is satisfactory to observe that so many municipalities are now beginning to offer rewards. These results are not regarded as satisfactory, because the falling off in the number of wild animals killed has been accompanied by an increase in the destruction of men and cattle. The government of India attributes this to the operation of the Arms act, although the reports assert that licenses are freely granted in tracts where wild animals abound.—*Science Gossip*.

Snake Destroyers.

Birds are, perhaps, the greatest snake destroyers, especially certain families of them. Even small insectivorous birds will devour a tiny serpent as readily as a worm when they find one, and storks, falcons, pelicans, cranes and some vultures are always on the lookout for this special delicacy. The secretary bird, *Serpentarius reptilicarous*, owes its scientific name to this habit ; the cassowary and sun-bittern are said to entertain a similar partiality ; while peacocks are so fond of snakes that they will actually desert the home where they are fed for a district where these reptiles are plentiful. A well-known London banker purchased a small island on the west coast of Scotland some time ago ; no attempt at cultivation had been made there, and it was uninhabited save by sea-birds and vipers. That the latter should have swarmed in such abundance in a situation so far north and isolated from the mainland is certainly remarkable; but there they were in force so strong that the banker found his newly acquired territory quite unavailable for the purpose he had intended it—a shooting and fishing station in summer. Acting under advice, he procured six pairs of pea-fowl and turned them loose on the island, which they very soon cleared of its unwelcome tenants, or at any rate reduced their numbers to such an extent that the remainder could be evicted without much danger or difficulty. Almost any bird will attack a snake of suitable size (of course it is not to be expected that a lark will swallow a boa-constrictor) ; and it is a curious thing that they eat venomous or non venomous species indiscriminately. They appear to first disable it by a sharp blow with the beak on the spine, then kill it by successive pecks and shakings which dislocate the vertebræ, and finally transfix the head ; then gobble it down. The presence of the venom in the bird's uninjured stomach would do it no harm, but one would have supposed that the sharp fangs or broken bones projecting through the mangled skin in its passage down must sometimes cause excoriations of the mucous membrane, and thus provide a means of inoculation, even if the aggressor did not get bitten in the combat. Neither accident, however, has been observed to occur by those who have repeatedly watched the operations. Pigs are tremendous fellows on snakes, too. They, as well as peacocks, have done good service in ridding entire islands of these dangerous pests ; and it is said that Maritius was cleared of poisonous reptiles by the wild hogs which were imported there in the first instance, and have now spread over the island.—*All the Year Round*.

"REVISED FRUIT LIST."

We insert the following from the *Germantown Telegraph*, not because we feel confident that it will accord with the preferences of *all* horticulturists—even in the same latitude—but because it has been compiled, and at various periods revised, by a veteran experienced in horticultural, as well as editorial lore ; and, also, because it may be more reliable and more general in its practical application than catalogues embracing the fruit stock of the entire Union, if not the entire world. Of course individual fruit growers will also have *their* preferences based upon their own experiences. Moreover, it has not only been demonstrated that one particular farm is better adapted to the thrift of some particular fruits than another one near it, but that even on the same farm, to a great extent, these diversities of adaptation exist. If a man possesses only a small farm, or desires to restrict his cultivation to only a few varieties, other things being equal, it would perhaps be most judicious in him to make his selections in the numerical order that they appear on the list, unless he is positive a variation from this would be better for him individually. Under any circumstances, that should be selected which is best adapted to the special locality. Every fruit season we still see a great quantity of "trash" in the way of fruit and vegetables exposed to sale in our markets, and we have often wondered whether people will ever discover that the *best* is the most profitable and finds a more ready sale, and can also be cultivated with as little labor as the inferior varieties ; and even if it cannot, it is quite certain that the gathering and bringing to market is the same, but when *there* the compensation is always favorable to the superior kinds. It would not do to say that farmers keep the best of their produce for their own use, and only sell the worst to others ; for, nine times in ten, the converse is the case. Doubtless it is altogether owing to indifference to the subject of fruit culture, or devotion to some other more absorbing interests or prejudice against "Book-farming." But it is never too late to learn, nor is there any rational source from which we cannot learn *something*, and those who think thus, we feel confident, may glean that "something" from the following list:

Since the last publication of our fruit list, we have for satisfactory reasons changed our opinion with respect to a few of the fruits which it contained. But in regard to the list as a whole we can see no just grounds for disturbing it. Indeed, we do not see how it can be improved for this section of the country, or as a general list for all the Middle States. Some of each of the separate selections may not do well upon our premises that will succeed admirably on another. Each grower must find out for himself the particular apples, pears, &c., especially adapted to his soil and location. This can be easily done by inquiries of those who are successful fruit-growers, whose soil is somewhat similar to their own.

According to our present preference, we should select the following for our own planting, and nearly all of which we are now growing more or less successfully: Standard Pears—1. Giffard; 2. Doyenne D'Ete; 3. Early Catharine; 4. Kirtland; 5. Bloodgood; 6. Summer Julienne; 7. Tyson; 8. Brandywine; 9. Bartlett; 10. Belle Lucrative; 11. Manning's Elizabeth; 12. Seckel; 13. Howell; 14. Anjou; 15. Shelden; 16. St. Ghislan; 17. Lawrence; 18. Reading; 19. Kieffer.

For those who may desire a smaller number we should select: 1. Giffard; 2. Early Catharine; 3. Bloodgood; 4. Tyson; 5. Bartlett; 6. Belle Lucrative; 7. Seckel; 8. Lawrence; 9. Reading; 10. Kieffer. They ripen in about the order they are arranged, except as to the three latter. The Lawrence, which begins to ripen, or can be made to ripen, early in November, will keep until March, it being the only pear of our entire stock still in our fruit vault.

In the above list, from No. 1 to 8, are summer varieties; from 9 to 16 autumn (early and late), and 17, 18 and 19 winter, thus affording a sufficient number for each of the periods of the best known sorts for this region.

Dwarf Pears.—1. St. Michael d'Archange; 2. Bartlett; 3. Coudec; 4. Rostiezer; 5. Diel; 6. Tyson; 7. Belle Lucrative; 8. Lawrence; 9. Ott; 10. Louise Bonne; 11. Base; 12. Bonsseck; 13. Giant Maxenu.

Apples.—1. Maiden's Blush; 2. Baldwin; 3. Smokehouse; 4. Northern Spy; 5. Smith's Cider; 6. Fallawater; 7. Cornell's Fancy; 8. Red Astrachan; 9. Wagoner; 10. Porter; 11. Gravenstein; 12. Tompkins King; 13. Roxbury Russel. We add to the foregoing list Tompkins King and Roxbury Russet, both most excellent varieties; indeed the King is regarded by some as unsurpassed. Northern Spy is also restored.

Peaches—1. Crawford's Early; 2. Hale's Early; 3. Troth's Early; 4. Old Mixon; 5. Crawford's Late; 6. Ward's Late; 7. Smock's Late; 8. Admirable, late.

We have substituted in the peach list Troth's Early for York' Early, and Admirable for Susquehanna. The former seems to have seen its best days, and the latter is too shy a bearer for profit.

Grapes—1. Telegraph; 2. Concord; 3. Hartford; 4. Clinton; 5. Salem; 6. Rogers' No. 32; 7. Brighton; 8. Prentiss.

We have added to the list Rogers' No. 32, which, should it maintain its present character will be the very best out-door variety cultivated. It is a beautiful pink, or rather maroon colored grape, and at times is transparent. It bears regular crops yearly with us. Clinton, in the foregoing list, is only for wine, and is probably the very best for that purpose. We add the *Brighton*, a maroon color, as promising well. It is, however, a small berry and rather straggling bunches, but almost pulpless, and of excellent quality. The *Prentiss* is also added. It is a new white grape, somewhat larger than the Delaware, of good quality and scarcely a perceptible pulp. It promises to take the lead of all the white varieties. The bunches are compact and of large size.

Cherries.—1. May Bigarreau; 2. Belle de Choisy; 3. Black Tartarian; 4. Black Eagle; 5. Black Hawk; 6. Elton; 7. Downer's Late; 8. Early Richmond; 9. Early Purple Guigne; 10. Delaware Breeding Heart.

The ripening of the list will range from the earliest to the latest, thus carrying one through the whole cherry season. No one can go amiss in adopting this list.

Raspberries.—1. Hornet; 2. Herstine; 3. Philadelphia; 4. Brandywine.

Strawberries.—1. Captain Jack; 2. Seth Boyden; 3. Sharpless; 4. Triomphe de Gand.

New kinds of strawberries are constantly appearing, but thus far we know of no improvements on the foregoing.

Currants.—1. Black Naples; 2. Red Dutch; 3. White Grape. These three varieties are the best among the different colors. The Red Dutch is a regular bearer and is of better quality than any other. There are others larger, but they are more acid. The white grape is transparent, of good quality, and ought to be more generally grown, but it is not a great bearer, and it is not profitable for market.

Gooseberries.—1. Houghton; 2. Downing.

These are two best gooseberries grown in this country. They bear every year heavy crops, are free from mildew, and are of excellent quality. They are large enough for all practical purposes. We cannot recommend the giants and their giant prices, and especially those of foreign origin.

Blackberries.—1. New Rochelle; 2. Missouri Cluster; 3. Wilson's Early; 4. Snyder.

The Snyder, a new Western Blackberry, is highly spoken of at distant points, and from the very respectable endorsers which it has we have no doubt of its value, at least in the West. We shall probably fruit it this year, having failed to do so last year.

It is better that those who intend to cultivate fruit and have to make purchases, to take this list with them to the nursery and adhere to it as far as possible.

In selecting fruit trees or any other, be careful to choose those with smooth, healthy-looking bark, have entirely shed their leaves, and have plenty of small fibrous roots. Trees on which the leaves remain after frost sets in, and stick to the branches in the spring, may be regarded as not healthy, and in some way lacking stamina.

EATING BEFORE SLEEPING.

Man is the only animal that can be taught to sleep on an empty stomach. The brute creation resent all efforts to coax them to such a violation of the laws of nature. The lion roars in the forest until he has found his prey, and when he has devoured it he sleeps until he needs another meal. The horse will paw all night in the stable,and the pigs will squeal in the pen, refusing to rest or sleep until they are fed. The animals which chew the cud have their own provisions for a late supper just before dropping off to their nightly slumbers. Man can train himself to the habit of sleeping without a preceding meal, but only after long years of practice. As he comes into the world nature is too strong for him, and he must be fed before he will sleep. A child's stomach is small, and when perfectly filled and when no sickness disturbs it, sleep follows naturally and inevitably. As digestion goes on the stomach begins to empty. A single fold in it will make the little sleeper restless; two will weaken it, and if hushed again to repose the nap will be short, and three folds put an end to the slumber. Paregoric or other narcotic may close its eye again, for without either food or some stupefying drug it will not sleep, no matter how healthy it may be. Not even an angel who learned the art of minstrelsy in a celestial choir can sing a babe to sleep on an empty stomach. We use an oft-quoted illustration, "sleeping as quietly as an infant," because this slumber of a child follows immediately after the stomach is completely filled with wholesome food. The sleep which comes to adults long

hours after partaking of food, and when the stomach is empty, is not after the type of infantile repose. There is all the difference in the world between the sleep of refreshment and the sleep of exhaustion. To sleep well the blood that swells the veins in our head during the busy hours must flow back, leaving a greatly diminished quantity behind the brow that lately throbbed with such vehemence. To digest well, the blood is needed at the stomach and nearer the fountains of life. It is a fact established beyond a possibility of contradiction that sleep aids digestion, and that the processes of digestion are conducive to refreshing sleep. It needs no argument to convince us of this mutual relation. The drowsiness which always follows the well-ordered meal is itself a testimony of nature to this twindependence.

The above paragraph has been "going the rounds" of the public press for some months, and those persons who are sufficiently intelligent to grasp the argument, will, of course, regard it from their own individual standpoint of experience, and will, perhaps, also reach different conclusions on the subject. Practically, we can endorse the whole of the foregoing, whatever the opinion of others may be; but, in doing so, we by means intend to encourage the *abuse* of eating at any time, whether noon, night or morning.

When we became addicted to "eating before sleeping" must have been very long ago, for we can well remember that when a mere boy, working on a farm, we on many occasions ate from three to half a dozen apples after 9 or 10 o'clock at night, while we were abed; then fell asleep and "slept like a top" until morning, and never even had a dream on them. Two physical conditions have ever been detrimental to our complete repose at night, and these are hunger and cold feet; hence, for the last quarter of a century, or more, we have never retired at night without eating *something*, whether much or little, unless we were unwell and had no appetite for food; nor have we gone to bed during all that period before 12, and often 1 o'clock at night. Reading and writing at night absorbs our vitality as much as do any of our labors during the day, and hence we naturally require food to sustain those labors, as much as a stove requires fuel to impart heat to the house it occupies. There is nothing irrational or physically hurtful in this when it is confined within rational bounds, and at regular periods. For instance, say we got our breakfast at 8, our dinner at 1 and our supper at 6 o'clock in the evening. Here we have three meals within ten hours, and six hours to work yet before 12 o'clock, and then eight more before we get our morning meal, which would be fourteen hours of fasting within the twenty-four. We don't know how we would feel if we were entirely idle, but this we know, that we would have to lie awake for hours if we attempted to retire and fall asleep on an empty stomach. True, our night meal is a light one, and never includes meats of any kind; but, in lying down from twenty to thirty minutes thereafter, we usually fall asleep within ten minutes, and wake up in the morning refreshed.

We never, or at least rarely ever, eat *anything* between meals—not even an apple, an orange, or a nut; and except the single cup of coffee which we drink every day at our regular meals—four times—we drink very little, if anything. We pass long summer days without even drinking water. But, "what is sauce for the goose," is not always "sauce for the gander;" or in other words, "what is one man's meat is another man's poison." Much depends on long continued habit, through which a sort of "second nature" is cultivated; we therefore do not pretend to set ourself up as a teacher as to how, when, and what men ought to eat, or abstain from eating. Much will depend upon their temperament, the texture of their physical constitutions, their secular occupations, their private habits, and freedom from extremes. There certainly have been diverse theories advanced on the subject of eating during the last half century; all of which perhaps, have some good in them, if honestly observed; and it is equally certain that men's minds have undergone a great change as to what is healthful and what injurious. Perhaps *what* people eat, and *when* they eat, is less essential than *how* they eat. The appropriation of nutriment, or eating, is the great moving force of the animal universe — and proximately also of the vegetable — and unless that all pervading want is supplied, everything animate would hopelessly perish; but the rationale of eating depends upon *assimilation*, in order to produce the most favorable result. *Hunger* is a great leveler, and has no respect to any condition in life. The rich and the poor, the intelligent and the ignorant, the high and the low, are all amenable to its absolute and universal demands. Perhaps the greatest mistakes, blunders, and willful perversions in eating, occur among the human family. All in the animal world, below the genus *Homo*, "eat to live," whilst it is very evident that many people "live to eat." Seeing that physical life is based upon this primary condition, too little regard is paid to rational eating, and also to healthful culinary preparation, to say nothing about social condition. The first thing that every living mortal craves- after fresh air- that comes into the world, is eating and sleeping, and if the first is not supplied, the second will not follow, and the subject is liable to perish. It *cannot* and *will not* sleep if *hungry*, whether man or beast.

HOW LONG ARE WE TO LIVE.

It is not every one who asks himself this question, because, strangely enough, it is the belief of many persons that their lives will be exceptionally lengthy. However, life assurance companies are aware of the credulous weaknesses of those whose lives they assure, and have therefore compiled numerous tables of expectancy of life for their own guidance, which are carefully referred to before a policy is granted. The following is one of the authenticated tables, in use among London assurance companies, showing the length of life at various ages. In the first column we leave the present ages of persons of average health, and in the second column we are enabled to peep, as it were, behind the scenes of an assurance office, and gather from their table the number of years they will give us to live. This table has been the result of careful calculation, and seldom proves misleading. Of course, sudden and premature deaths, as well as lives unusually extended, occasionally occur, but this is a table of average expectancy of life of an ordinary man or woman :

Age	Years to live.	Age	Years to live.
...	...	...	...

Our readers will easily gather from the above tabulated statement the number of years to which their lives, according to the law of averages, may reasonably be expected to extend.—*Harper's Bazar*.

Inasmuch as the above claims to be the basis upon which the London insurance companies operate, it may be regarded about as reliable as the subject could well be presented, although by no means absolute or infallible. If we understand the table rightly, if a child is fortunate enough to attain the age of *one year* there is a reasonable probability that it will attain the age of *forty years*. Of course there is a *possibility* that it may live longer ; but insurance companies would not be likely to deal in risks founded upon mere possibilities — probabilities are sufficiently dubious. But should the juvenile subject attain the age of *ten years*, then he may expect to reach life's prime, or *forty-one years* ; and if he should score *twenty*, then he is good for *three score and one*. If he is fortunate enough to number *thirty years* he may entertain a reasonable expectation to number *sixty-four*. But every decade after *thirty* the number of years still alloted to him are shortened, or supposed to be shortened. These calculations are presumably made upon the general life tenure of men in health, and take no account of epidemics and accidents.

The following article relating to the tenure of human life is rather argumentative than a statement made upon business experience. We do not think, however, that either Buffon, Heusler, Voltaire or Flourens could make any nearer approximation to the truth of the matter than persons less intellectually endowed than they were, when they attempted a *literal* explanation of the ages of the Bible patriarchs. They might as well have attempted to explain the speeches and actions of the animals in Æsop's fables, on a literal basis. The Bible on this subject has never yet been explained—never *will* be, and perhaps never *can* be—so as to be literally comprehended by the masses of mankind, nor is it morally, socially or philosophically essential that it should be, in order to be a text-book of morals to the human family. It may be regarded rather as a system of sacred symbols, tropes, figures and parables, having no special signification relating to the physical universe ; but, at the same time, a moral instructor to the human family through spiritual correspondence, illustrating mutual relations existing between the Creator and the created ; and the obligations of the latter to the former.

The Extreme Limit of Human Life.

Can man reach and pass the age of one hundred years ? is a question concerning which physiologists have different opinions. Buffon was the first one in France to raise the question of the extreme limit of human life. In his opinion, man, becoming adult at sixteen, ought to live six times that age, or to ninety-six years. Having been called upon to account for the phenomenal ages attributed by the Bible to the Patriarchs, he risked the following as an explanation : Before the flood the earth was less solid, less compact than it is now. The law of gravitation had acted for only a little time ; the productions of the globe had less consistency, and the body of man, being more supple, was more susceptible of extension. Being able to grow for a longer time, it should in consequence live for a longer time than now.

The German Heusler has suggested on the same point that the ancients did not divide the year, among some people of the East, was only three months, or a season ; so that they had a year of spring, one of summer, one of fall, and one of winter. The year was extended so as to consist of eight months after Abraham and of twelve months after Joseph. Voltaire rejected the longevity assigned to the patriarchs of the Bible, but accepted without question the stories of the great ages attained by some men in India, where, he says, "it is not rare to see old men of one hundred and twenty years." The eminent French physiologist, Flourens, fixing the complete development of man at twenty years, teaches that he should live five times as long as it takes him to become an adult. According to this author the moment of a complete development may be recognized by the fact of the junction of the bones with their apopyses. This junction takes place in horses at five years, and the horse does not live beyond twenty-five years ; with the ox at four years, and it does not live over twenty years, with the cat at eighteen months, and that animal rarely lives over ten years. With man it is effected at twenty years ; and he only exceptionally lives beyond one hundred years. The same physiologist admits, however, that human life may be exceptionally prolonged under certain conditions of comfort, sobriety, freedom from care, and observance of the rules of hygiene.

Weak Lungs.

Every one knows that physical exercise invigorates the muscular system ; that the constant action, within limits, of any muscle enlarges and strengthens that muscle. It is the working of the same law that gives fullness and vigor to the blacksmith's arm. This law is physiologically universal, and therefore applies to the lungs.

The one work of the lungs is to inhale and exhale air ; and this depends on the alternate expansion and contraction of the chest. Now, some persons are born with thin, narrow chests. The lungs of these persons are generally weak, and easily become diseased, because seldom brought into full, vigorous action.

The employments of other people—students, tailors, seamstresses, shoemakers, etc.—are such as do not call out the full actions of the lungs. In some cases, they interfere with it. If such persons are troubled with general weaknesses, have difficulty of breathing after exercise, and dull pains in the sides, the lungs should be looked after, although there may still be no organic disease. What is needed is to strengthen them—not by medicine—but by their own proper action. The *Medical and Surgical Reporter* gives an account of a young student whose pulmonary symptoms of weakness were wholly overcome. It was done by his simply breathing through a small tube the size of a quill, a dozen times every three or four hours each day. Every third respiration he withdrew the tube, when the lungs were thoroughly filled, and held his breath as long as he could without distress. Keeping this up during his student-life, he acquired the ability to enlarge his chest five inches by an inspiration, and to hold his breath without distress a full minute.

It is our belief that the same thing may be accomplished by breathing as above through a single nostril, closing the other with the finger.—*Youth's Companion*.

So far as the matter relates to our own personal experience, the above extract contains sound doctrine. When we were first bound an apprentice to the tailoring business (1827) we were jeeringly admonished that we would not live to serve out our term of five years, and we confess we sometimes felt some boyish anxiety about it; for our mother had died of consumption when we were just twelve years old, and since then our elder sister, our two brothers and one of our sister's daughters have died, and three of them unmistakably of the same fell disease, and at about the same age as our mother; moreover, quite a number of our mother's relations have died of the same disease. One day a physician well advanced in years, coming into the shop in which we were employed, and noticing our peculiar attitude on the "board," advised us to sit straight, expand our chest as much as possible, and to cultivate a habit of breathing through the nostrils. And furthermore to take as much active outdoor exercise, as we could possibly find opportunity to do, and when we sat at rest, to throw our arms backward over the back of the chair, or bench, if it had such a support. We followed this advise and with good results.

Notwithstanding, about forty-five years ago we were troubled with a pressure of the lungs and difficulty of breathing—perhaps a collapse of the cells of the lung—induced by a too incessant confinement to the shopboard. One day an agent for the sale of Dr. Fitch's "Inhaling Tube" called on us, and explained the nature and object of the instrument, and we purchased one, at a cost of $1.75. This instrument is hammer-shaped, and consists of a cylinder three-quarters of an inch long, and three-eighths of an inch in diameter; one end is closed and the other contains a small ball-valve. To this cylinder is attached, in the middle, a shaft or stem about four inches long and three-sixteenths of an inch in diameter.

This tube is taken in the mouth and a deep inspiration is taken, thoroughly filling the lungs. After a moment, or as long after as the breath can be conveniently held, when expiration commences, the ball-valve will be partially closed, and the breath will not escape as freely as it was inhaled. If the nostrils are then held shut, the effort to force the breath through the diminished aperture, will also force it into the collapsed cells of the lungs, and gradually open and expand them, and this effects their cure.

This practice was continued, at intervals, until 1848, when we abandoned the shopboard for more active employment, and the cause being removed, there was no necessity of continuing the remedy. The good effects do not follow immediately, but by continuing the process the respiration becomes gradually free and strong. We have on many occasions been able to take a full inspiration and hold our breath long enough to read a paragraph equivalent to forty or fifty lines in the columns of the LANCASTER FARMER. We by no means pretend to say that lungs in an absolute state of decay could be cured by the aid of this instrument—indeed, we have loaned it to persons so affected, and they have declared that they could not use it, or a continued use of it would kill them—but where there is only a weakness, or a compression of the cells of the lungs, it, or any substitute of it, cannot fail to be beneficial if judiciously and perseveringly used. About twenty-five years ago, through an inadvertent exposure, we contracted a stubborn and protracted cough, which, according to the opinion of our medical adviser, terminated in rupture of the lungs. It is only necessary to say that in this case our instrument was altogether useless, as we needed a different treatment; and we were finally relieved—we may say totally cured—by the use of medicated inhalations. These were progressively modified to the pending condition of the lungs.

The predisposition to pulmonary affections is said to be greater than is generally supposed by the thoughtless and unobservant; and, that the disease is not more frequently and fatally developed, may be owing more to favorable contingencies than to remedial agencies. Mental or emotional condition may also be a potent factor in the development of pulmonary diseases. It has been alleged that every violent paroxysm of anger, hate, envy, jealousy, fretfulness, anxiety, sorrow, chagrin, obstinacy or grief, adds so many nails to the coffin of the consumptive; and doubtless this may also be said of violent physical exposures, dissipations, debaucheries, or any irritating draft made upon the passions or the material energies. Inflamed lungs, no more than an inflamed cuticle, cannot heal as long as they are in a state of violent irritation. We have now attained our "three-score and ten;" and although we claim no special merit for such a fortuitous contingency, yet it is none the less a commentary upon the judgment of those who predicted our demise before we completed our apprenticeship, more than half a century ago.

THE WILL AND THE DEED.

There are circumstances under which charity compels us to regard the will as equivalent to the deed—in a moral sense at least. It is very true, that the will, or the wish, unaccompanied by the necessary food, would never save a needy man from starving; but, in a moral sense, it might exculpate a destitute person who had been appealed to for help; and in the same sense, it might even go farther than the real material gift of another. All would depend upon the motive which instigated the deed. "There, take it, and may it choke you," uttered in a snappish vein, when importuned for part of the loaf you are eating, would be a deed far inferior to a generous wish or will, in a moral sense, although the latter might not be so effective in alleviating one suffering from hunger. We are led to these reflections in considering the responses to our solicitations to "write for the FARMER." For instance, LEOLINE writes us: "Esteemed friend, I will now endeavor to write you a few lines. As my husband told you, I have my hands full just now. But, if I could have the opportunity to write as often as I wish, you would get a goodly number of contributions from this quarter." Now, we happen to know that LEOLINE is a self-educated farmer's wife, in medium circumstances, and has a large family of children, just at that age when they most need a mother's care, and that the general labors of the household devolve upon her. Under such circumstances the wish, or will, becomes equivalent to the deed, and yet from time to time she has contrived to do more than merely indulge in unsubstantial wishes; and we feel confident that she, nor any one in like circumstances, will sustain any moral injury in ultimating their wishes in corresponding actions.

The most effective way to educate the "million" is for the million to write for and become the instructors of the million. Domestic hints and recipes emanating from the "upper ten," are often inaccessible and impracticable to the million. They occupy a plane beyond the reach of the million. Culinary preparations that cost a dollar are of no practical use to the person that can't afford more than ten cents for the same. But the million is characteristically timid and diffident, and hence practical knowledge is often overawed by theoretical assurance; and because the million cannot write with the fluency of the upper ten it prefers to abandon the field and continue a "hewer of wood and a carrier of water." All, within their spheres are useful, no doubt, but the common people need the experiences of common people; instead of being spoken at they need to be spoken to.

By the common people, we by no means have reference to the "Tramps," sansculottes, the "Greasers," and the Lazaroni of the human family, but to those who wilfully labor for the benefit of mankind, whether from necessity, from love, or for its emoluments. Not that social exclusiveness or domestic antagonism necessarily should exist between the common people and the upper ten, for it must needs be that a diversity of classes will exist in the present constitution of society; but we should ever remember the "pots of earth and the pots of brass," as we float down the stream of time. If the lion were famishing with hunger, and the bull were to bring him a bundle of hay, the pig an ear of corn, the parrot a bunch of fruit, and the partridge a pint of seeds, it might be all very kind in them, but it would not meet the wants of the lion. His needs, under such circumstances, could best be supplied by an animal that had a clearer appreciation of those needs. Many books have been written on Domestic Economy by those who were characteristically neither domestic nor economical, and hence, so far as they concerned the common people, they were a dead letter. But if the common people will not place the results of their life experiences on record, then society will have to appropriate such domestic literature as it can find, whether adapted to its wants or otherwise.

EXCERPTS.

CHESTER white pigs have increased in price in the past two years.

LIKE the blackberry, the raspberry bears the fruit upon the cane of the previous year's growth, which, after fruitage, dies, the new cane coming forward for the next year's crop.

NEARLY, all kinds of fruits do well on a mixture of superphosphate and wood ashes. Lime is not suitable for strawberries, but excellent around apple, peach and pear trees.

GRAPEVINES should be pruned as early as the season will permit. If deferred too late they will allow an escape of sap (bleed), even if trimmed a little while before it begins to ascend.

PINE flowers require thoroughly rotted manure and wood mould mixed, and tomato or other early plants can be grown in boxes, and afterward transplanted with better results with such a mixture.

IN cold weather, eggs for hatching should be collected daily. They freeze easily when exposed, but will retain vitality for several weeks if gathered as soon as laid and then kept at a uniform temperature.

In Lancaster county, Pa., last season, one farmer sold his two acres yield of tobacco for $1,112; cost of labor, etc., $212; net profit, $900. Another farmer realized $430 an acre from three and a half acres, and another $550 an acre from one and a half acres. They think it pays.

The exportation of potatoes, cabbages and other vegetables from Germany to this country has constantly increased since it began last fall. One steamer in October took out 8,100 heads of cabbage; four others have since left with 6,000 bushels of potatoes, 11,000 head of cabbage and 30 bags of turnips. It is said in Germany, that additional shipments will be made during the winter.

A strip of land bordering on the Mediterranean, about 100 miles long and five or six wide, is the raisin-producing territory of Spain. The Muscatel grapes are carefully cut in August, laid on a sort of bed made of fine pebbles, and dried, being turned often until they are perfectly cured. Then they are taken to the wine presses, where, after being laid in trays, they are subjected to heavy pressure, when they are ready for market.

Our readers must remember that only recently has it been clearly demonstrated that a dead branch on a tree makes almost as great a strain on the main plant for moisture as does a living one. It is one of the most important discoveries of modern botanical science to the practical horticulturist, as by this knowledge he can save many a valuable tree. When one has been transplanted some roots get injured, and the supply of moisture in the best cases is more or less deficient. Any dead branch or any weak one should, therefore, be at once cut away.—*Gardener's Monthly.*

The Cincinnati *Tobacco Journal,* in order to answer the question of how much seed is necessary to plant an acre of tobacco, has pursued an investigation and found this: In one grain we found by actual count 1,494 seeds. This would make, by multiplying by 480, the number of grains in an ounce, 717,170 seeds to the ounce, and 8,605,440 seeds to the pound. Estimating 5,000 pounds to the acre, and supposing every seed will make a plant, every half ounce will plant nearly 72 acres, an ounce 144 acres, and one pound, 1,721 acres. As many farmers are contemplating planting largely this season, we recommend a careful study of these figures.

Hiram Warel, of Conestoga township, had eight hogs about a year old, which weighed as follows, dressed: Killed two on November 2, weighing 434 and 416 pounds, respectively; killed two on December 12, weighing 484 and 443 pounds, respectively; killed two on January 2, weighing 578 and 533 pounds, respectively; and killed two on January 31, weighing 520 and 628 pounds, respectively.

Queen Elizabeth granted the first royal patent conceded to players in 1576.

It takes 1920 silk worms to make a pound of cocoons.

Buffaloes are common in Ceylon, white ones being sometimes found.

Texas sells annually 400,000 head of cattle; at $20 per head it foots up a grand aggregate of $8,000,000.

The amount of fruit shipped from California during the present season will bring about $1,000,000 profit to the State.

A Wisconsin farmer, twenty-three years ago, planted a piece of waste land, unfit for cultivation, with black walnut trees. The trees are from sixteen to twenty inches in diameter and have been sold for $27,000.

It costs the people of Tennessee $1,000,000 annually to sneeze and use snuff. This is a Nashville merchant's estimate of the annual consumption of the article.

France produced last year 750,000,000 gallons of wine. Of these, 47,000,000 were made from sugar, 51,000,000 from raisins, while 154,000,000 gallons were imported from Spain and Italy, to "blend" with their home product. No wonder everybody wants to drink French wines; they are so pure.

In a small grove which adjoins the Schoenberger residence near Cincinnati, an army of crows take shelter every night. They assemble by thousands an hour before dark, and an old man living near the place says that to his personal knowledge the same grove has been their dormitory for sixty years.

During leisure hours this month make a simple hot-b d, even if it is no larger than a dry-goods box from which the bottom and top have been removed. This, if sawed in a diagonal direction, will make two frames one foot in height on the front side and twenty to twenty-four inches on the rear side when placed in position at the south side of a building or high plank fence. If no old sash are at hand, cotton cloth, saturated with boiled linseed oil, will answer a very good purpose. No manure will be needed within the frames, but fresh stable manure should extend one foot beyond the frames on each side.

Immense quantities of wheat straw are being shipped to this city from New York State—hundreds of car loads. It is used for bedding purposes, and afterwards for manure, and it is stated that much of it contains the Canada thistle, which is, by this process being spread broadcast over the country. Our farmers should be on their guard.

Adam Beam, of Carnarvon township, this county, was very successful in raising tobacco last year upon one acre of ground, which was carefully cultivated, and the crop has just been sold for $300. The prices paid were 22 cents per pound for leaves measuring over sixteen inches, and 7 cents per for the remainder, the average being 19½ cents per pound. This is considered the best sale made in the neighborhood.

The number of feet of merchantable pine left standing in this country May 31st, 1880, is given as follows :

	Standing pine. Feet.	Cut census yr.'80. Feet.
Texas,	67,508,500,000	174,410,000
Wisconsin,	41,000,000,000	2,097,289,000
Michigan,	35,000,000,000	4,497,299,000
Mississippi,	23,975,000,000	115,775,000
Alabama,	21,192,000,000	245,398,000
Florida,	6,615,000,000	208,036,000
Minnesota,	6,100,000,000	340,977,000

This does not include some of the most important timber regions—Oregon and Washington Territory, which will be given hereafter by the Census Bureau.

CONTRIBUTIONS.

For The Lancaster Farmer.
FORESTRY.

The timber question involves a subject that will not soon be exhausted: hence allow me to keep it afloat "all the time," for, from all we know to the contrary at present, the consumption of timber is likely to be perpetual. Therefore, its reproduction must necessarily become continuous. We live in an age of absolute necessity, and also in an age of great indifference and negligence, in regard to the reproduction of timber. Even the ancient Greeks and warlike Goths, were more careful, less profligate, and valued the forests more than we do at the present day. The orientals were like us Americans, they made no provisions for the replenishment of their exhausted forests. They became so impoverished in timber that they were compelled to abandon their country for the want of it, and migrate to Europe, where they learned to appreciate its value. Timber was held in classic veneration in Greece. The students of Athens habitually assembled under stately poplar trees to recite their lessons, and declaim before their fellows. Political gatherings, would assemble in timber groves, reserved for that purpose.

The ancient Druids recommended, and even enacted laws requiring states to make large reservations for all time to come, in order to supply the people with timber, and to avert if possible the timber panic of Asia. They especially professed great veneration for the oaks, under the wide expanding branches of which, they delivered their lectures, worshiped their Deity, and performed their mystic rites; believing *that* majestic tree of the forest to be the peculiar emblem of the residence of the Almighty. They would leave it unmutilated in some places in order to note its age, which has been known to exceed three or four hundred years.

It was through the example of the early settlers in Europe that these large forests were reserved, and have been preserved to the present day, and will continue to be kept up, for all time to come. These forests are generally owned by the different governments—whether large or small—who appoint officers, exercising a supervisory control over them. The "Wald-Herr," or "Forester" is quite an important personage, and exercises an indisputable authority within his domain. The matter of properly keeping a systematic forest is not a merely hap-hazard affair, and does not require all the trees to be left perpetually standing. When the cutting is finished at one end, then the other end is in a fit condition to begin afresh, and in this way they always have flourishing forests, and also always have timber. A judicious manipulation of a forest requires some science, more observation, and a great deal of experience, acquired through the exercise of common sense.

It appears that our generation, and especially "Young America," is more bent on imitating our Oriental than our European ancestry. There are a great many farmers in the *model* county of Lancaster without a single forest tree upon their premises, and very few trees of any other kind—even including fruit trees. Occasionally it happens that a corner or a few acres of the farm is preserved ; or a favorite oak, a walnut, or a chestnut is left standing as so many monuments to the memory of a grand old forest that has passed away ; and perchance if those farms pass into other hands under morbid ideas of improvement, the first thing done is to fell those venerable relics to the ground and utilize them, according to modern principles of economy. Under such circumstances we sensibly recall the sentiment of Morris' immortal lines—

"Woodman spare that tree,
Touch not a single bough,
In youth it sheltered me
And I'll protect it now."

There seems to be but one way to induce our people to commence the planting of forest trees, and that is for the Governments of the the States and the nation to offer premiums to those who plant a given number of trees, and assess an additional tax on those who refuse or neglect to perform that duty.

True, this might be construed into both a bribe and a threat, but it would not be the first instance in the world's history, where people have been bribed, or threatened, to do that which was their plain duty to do. Economically, as well as morally, men should do their duty as they understand it, or as it is made manifest to them in the *present*, and not be unnecessarily anxious about the *future*; but, it does not require an extraordinary amount of intelligence to perceive that the *present* may be so improved as to make it a pleasant and noble *past*, and at the same time amply provide for the *future*. Had our forefathers adhered to a similar rule in regard to our primitive forests there would be no necessity for their posterity to indulge in anxious discussions on the subject. But all *that* is now past ; they needed cleared land ; the forests were their *bane* and cutting down the trees the *antidote*. Things now are becoming reversed. Treeless, arid and sun-baked hills and valleys in time will be the *bane*, and tree replenishment the antidote. To those who are now advanced in life, it may make little difference what is done in this respect, but then we should never forget that the earth is "God's footstool."—*L S. R., Oregon*, 1882.

For The Lancaster Farmer.
STRAWBERRIES.

Among all the circle of fruits there is none that is so easily raised, or gives so much satisfaction to the amateur, as the strawberry. No other fruit gives as quick returns as it does. A bed planted in July or August, will, if well taken care of, make a full yield in less than a year. Sometimes, on account of drought, it is difficult to establish a good bed at that season, so upon the whole it is safer to plant in the spring as soon as the ground gets in good working order. The yield in good soil and proper conditions is simply enormous.

Last year on a plot 7 by 10 yards, there were raised over 100 quarts of the Sharpless variety, equal to 7,000 quarts to the acre. There are reports, apparently well authenticated, of twice that amount of berries to the acre. It can be seen from this that there should not be any difficulty in finding a plot of ground on almost every home in the land, large enough to raise a supply of this delicious fruit for family use. Two or three rods will be sufficient. Any soil that is rich enough for cabbage or corn will do for strawberries. It will be well to avoid a sod, or ground filled with a large amount of vegetable refuse, for in them the white grub is to be found, and where it is plentiful you may come to grief, as it (the grub) is a great eater, and will soon ruin a bed.

Dig deep, as you will thereby prevent the ground from drying out as soon as it will if shallow. The drought, by the by, is the greatest hindrance that the strawberry grower has to contend with. For some years past strawberry growers on a large scale have been quite unsuccessful on that account. But the amateur, with his two or three rows, need not suffer. Deep cultivation and mulching will, in a great measure, counteract drought. These small beds are easily irrigated. The soapsuds from the weekly wash will be excellent. These small beds will need no alley ways through them, for the work here can be done to the best advantage with the hoe. Set the plants eighteen inches apart every way. After the bed is planted give it a good raking once a week with a steel rake. Let no weed or runner grow. If any plants are missing train a runner in its place, and when well rooted, cut loose from its parent. Fifteen or twenty minutes' work every week will do all this.

In the fall, when the ground is frozen, cover with two or three inches of coarse manure, and then your work is done for the season. In the spring, when growing weather comes, rake off the coarse rubbish, but leave the fine stuff on for a mulch, and if you can add as much more as will keep the weeds from growing it will be all the better. Care must, however, be taken in putting on this mulch, that the crown of the plant is not covered. If mulching is scarce and your bed is inclined to be weedy, the weeds near the plant must be pulled by hand, as the strawberry roots are near the surface and are easily injured by the hoe. After the fruit is formed a little tanbark, leaves or chaff should be laid under it to keep the dirt off. If from a severe rain, however, the berries become dirty wash them. Some one has said you might as well try to wash sugar as strawberries, but that is all nix. Place a colander, or better, a square box with wire netting nailed on the bottom, in a tub, and pour water in until nearly even with the top of the box, then pour your berries in, a quart or two at a time, raise up your box two or three times, and they are clean. Set them in a shady place to drain. Persons not seeing you do this will never know that the fruit was washed.

After the fruiting season is over go over your beds the same as you did the first season. Beds thus treated may be kept productive for three or four years, after which it would be better to start a new bed. Now about varieties: Buist's Prize, Crimson Cone, Hovey, Longworth's Prolific, and hosts of others that were popular twenty years ago, are all superseded by others ; and even the Wilson, that so long reigned supreme, is being pushed to a back seat. Now we have Charles Downing, Cumberland, Miner's Prolific, Sharpless, &c. These combine quality and productiveness in a greater degree than the older varieties. These, a year or two ago, were sold at $2.60 and upwards a dozen; now they can be bought for a dollar or less per hundred.—*Casper Hiller.*

For The Lancaster Farmer.
PRACTICAL POULTRY NOTES.

As the time will soon be here, when good Farmers' wives will be sitting hens for early chicks, I will tell them how I do, though others may do better. I never give a hen more than eleven eggs if the weather is very cold, and 13 if the weather is warm.

Last spring I had a flock of 22 Leghorns. As soon as I cooped them, I greased the old hen under the wings, at the legs and breast, and every chick about the head; then I put them in the coop, to the old hen ; and this I did every two weeks. I fed them the first two weeks on stale bread, dry cheese, and onions chopped in the cheese once a week; after that I feed cracked corn, wheat, rye, thick milk, pure water to drink, till they are fit to eat or take to market, and only one died with the pip out of the whole flock. I keep the mother pen'd up, but let the chicks run at large, after they are 4 or 5 days old.

Out of another flock of 32 Cochins, treated the same way, I raised 30, but do not like them so well, they lay too little. We have plenty of eggs when we keep the Leghorns, summer and winter; never keep old ones over two years, and never keep over 30 through the winter; it does not pay to keep too many. A few well fed pay better than many ill fed.—*Leoline.*

For The Lancaster Farmer.
DOMESTIC HINTS.

To prevent small-pox pitting the face, keep a damp cloth on the face, with holes cut for the eyes, nose and mouth. This has been tried, and can be relied on ; no one likes to have such marks on the face.

To prevent sore eyes, wash in warm water, never cold, as the cold water will inflame weak eyes.

To increase the cream on milk, strain your milk in hot crocks and set in a cool room. The butter will come sooner, too, if milk is treated this way.—*Leoline.*

For The Lancaster Farmer.
PRACTICAL RECIPES.

FASTNACHT CAKES.—Set a sponge, as for bread, with 1½ quarts of good yeast. When it is raised, add 3 eggs, beaten, 1½ pounds of sugar, ¼ pint of butter and lard, mixed; knead it well for about twenty minutes; let it raise again, then roll on a board, and cut in cakes with a penny roller, with notches in; also cut through the cake three or four times, and bake in hot lard, having the pan about half full. Begin baking when you have about the half rolled.

SWISS CAKE.—Make a batter as you would for flannel cake, only so thick that it does not run; it should be pretty stiff, but not too stiff. Take 3 quarts of flour, 3 eggs, 2 teaspoonsful of saleratus, a large tablespoonful of salt; fry in hot lard, same as the others. By leaving out the salt and adding a small teacup of sugar, you can have them sweet. They should be very light when done, and should be draped in the lard in small spoonsful, otherwise the outside will burn before the inside is done.

To PENIFY DRIPPING.—Make it hot in a pan, and then pour it in clear water, when it is cold gather it and fry out the water, and it will not taste much any more like dripping.

PEOPLE who cannot bite radishes should grate them, and season them as they eat them; they are very nice in that way.—*Leoline.*

ESSAYS.

THE GROWTH AND CONSUMPTION OF TIMBER TREES IN AMERICA.*

When in a state of nature, and before Europeans penetrated far into the interior, this country was in all probability covered by a dense forest, for we find that Wm. Penn held a conference with the Indians under the spreading elm tree at Kensington; and all other information that has been handed down since 1681 confirms this belief. At the time of the settlement, says a distinguished writer, in 1682 the site of Philadelphia was a *dense forest*, a broad expanse of magnificent and illimitable wilderness, almost untrodden by civilized man. About the year 1720, thirty-eight years afterwards, John Bartram laid out on the banks of the Schuylkill below Philadelphia, a garden containing a large proportion of the various forest trees of North America. But even so early as the reign of Queen Anne, who occupied the English throne, from 1702 to 1714, an act of parliament was passed 'for the protection of forest trees in the English American Colonies; and by an act passed in 1750 prohibited the felling of white pine trees in the Colonies, unless within private enclosures. About the same time also some of the colonists petitioned the mother country for compulsory legislation regarding the planting of trees by the farmers. Between 1730 and 1750, furnaces for the smelting of iron had been erected in Virginia, Maryland, Pennsylvania, New York and New Jersey, and great fears were entertained that the fuel would give out. In later times, and during the early part of the present century, these apprehensions were renewed that the charcoal furnaces would surely cause a scarcity of fuel, but the forests held out until the introduction of coal into common use dispelled the popular delusion. But in our times, notwithstanding that the domestic consumption of wood for fuel has to a large extent been superseded by coal, other dangers confront us, that the railroads need immense quantities of white oak saplings for ties, and how to meet the demand has vexed some minds greatly. When the necessity arrives no doubt a substitute will be found. We well remember similar fears were entertained twenty-five years ago that the locomotives were eating up all the pine wood, but here coal again came to our relief.

Before the discovery of coal mines and inventions of cheap means of working them, wood was the general fuel of the earth, and in many counties where the arts have not much flourished, it is still the chief fuel. In our country as in all other civilized countries the consumption of timber is immense. Its aptitude to be shaped into a thousand various

*Read before the Lancaster County Agricultural and Horticultural Society, by C. L. Hunsecker.

purposes for the comforts, ornaments, and conveniences of society, enhances its value so that we could not well conceive how we could do without it.

It appears by a late official report that Minnesota, Mississippi, Alabama, Florida and Texas have an aggregate of 125,000,000,000 feet of standing timber, and that during the year 1880 there was cut nearly 1,500,000,000 feet, showing in these five States an enormous amount of growing timber trees.

In other portions of the States and Territories there is more or less forest, and in some of the Western Territories there is no calculation or numbers big enough to measure the amount of the magnificent trees that span the horizon of Washington, Oregon, Alaska, and the Indian Territory; also of Washington Territory. Governor Newell says there are on the borders of Puget Sound 15,000,000 acres of the finest timber land in the world. Thousands of trees are upwards of 300 feet in height and 10 feet in diameter at the base. The New Orleans *Democrat* estimates that Louisiana contains more than 17,000,000 acres of wooded land, and the saw mills have made very little impression upon this vast supply of timber, which comprises a large variety of valuable wood, although by the late census it appears there are 30,000 saw mills in the United States, doing a flourishing business. There is an immense amount of pine forest in California, in the State of Maine, in Michigan, Pennsylvania, Wisconsin, Virgina, Kentucky and the Carolinas.

There are in this country 760,000 square miles of timber, of which the South owns 400,000, or nearly two-thirds of the most valuable timber; whilst there are States in the American Union that were forestless a quarter of a century ago that are becoming wooded by the planting of trees, Iowa, Kansas, Utah, etc. When the Mormons settled at Salt Lake, in 1847, the country was destitute of trees, except what grew on the Wahsatch Mountains, which are covered with pine trees. The Utah valley is highly productive, but few farm houses are found beyond the limits of the towns, which to a distant observer present the appearance of immense orchards, with but here and there a chimney or steeple rising above the trees, indicating the presence of houses. And all this wooded appearance of the towns has been brought about by the policy of tree planting in thirty-three years.

The broad and rich prairies afford advantages to the settlers, which the settlers in the wooded districts of other States do not appreciate. But it seldom happens that any spot of land combines all the gifts of Providence. It is there that we find the richest lands, charged with the elements of agricultural success. There is an absence of trees, which has been considered a serious drawback. Experience, however, has shown the contrary. Those pioneers who weathered the storm and settled the timber lands of Pennsylvania, Ohio and Indiana, can testify to the weary life time of labor required to clear the breadth of a farm fit for cultivation. On the prairie it is entirely different, the farmer can go to work at once with his ox team and plow down the sod on which the tall grass has been growing uninterruptedly for years.

Chicago, in Illinois, and Toledo, in Ohio, commenced their career at the same time; the first becoming the mart of an extensive prairie country, was easily brought under cultivation, got ten years the start of the latter. Toledo, seated in the midst of the grandest old forest of the plain, had to cut the trees away to get room. The products of a soil of great fertility, which were an incumbrance to the first settlers and checked the early growth of the town, but have in later times become a source of great profit. It is even possible for a people to prosper greatly although they should inhabit a country destitute of forest trees. Holland, in Europe, during the seventeenth century its foreign commerce and navigation was greater than that of all Europe besides, and yet the country which was the seat of this vast commerce had no native product to export, nor even *a piece of timber fit for shipbuilding.*

Iowa was formerly a treeless country, but owing to favorable legislation and the efforts of its enterprising citizens, has by planting forests and orchards became a wooded country.

The head of the famous Mississippi river is a dense forest of magnificent pines.

Thousands of acres of valuable timber are annually destroyed in our country by the forest fires and large quantities of wood left to rot upon the ground, for want of a market near enough to pay the expense of moving it. Professor Buck, of Ontario, Canada, asserted lately that more timber had been destroyed in Canada by forest fires than had been exported, and one of the largest lumber operators of Ontario asserted that there will be no pine left in Canada at the end of twenty years.

The dwellings of the early settlers of New England, as well as Pennsylvania and other States, found the forests an incumbrance, and used them almost exclusively for building material. The houses and other buildings were mostly constructed with hewn logs, some of which are still standing and occupied, though brick, stone and mortar are fast replacing them.

Out West they have a yearly holiday called "Arbor Day," on which the people plant trees. Minnesota has already millions of saplings on her stretches and nooks. Iowa everywhere shows that her once bare prairies are to have their horizon broken into picturesqueness and color by the maple and the elm. Men plant trees, which is an emblem of civilization—Napoleon's willow, Shakspeare's mulberry; and Bryant's beautiful poem, "The Planting of the Apple Tree," sheds its variegated blossoms to the memory of the poet.

In consequence of the great consumption by the furnaces in England of timber, they were restrained by act of Parliament in 1581. Soon after this Lord Dudley invented the process of smelting iron ore with pit coal instead of wood fuel. Although of immense value to the country, the works were destroyed by an ignorant rabble, and the inventor was well nigh ruined. But in the early part of the eighteenth century the consumption of timber was so great and the complaint so well founded that the wood fuel would give out, that in 1740 Dudley's process for using pit coal instead of wood was generally adopted, and the iron business greatly increased up to the present time. Similar were the fears in England, about the same time, that the timber for shipbuilding would fail, but Sir Robert Seppings contrived the means of substituting straight timber for that of different forms and dimensions, before considered indispensable in shipbuilding. Although this want of timber for ship-building is not felt in the United States, the business is not flourishing very much. It may be said with truth that the cultivation of forests has been greatly neglected in most countries, and in many a very sensible want of wood is felt.¶ Trees' should be planted around country residences. Houses without being sheltered by trees against the wind and sun have a monotonous and lonely appearance.

In regard to the longevity of trees, Linnæus gives an account of an oak tree 260 years old, but we have traditions of some that have arrived to more than double that age. An English writer makes the Fortworth chestnut 1,100 years old, and the cedars of Lebanon are as old as the Bible. The great tree, Washington Elm, at Cambridge, has a surface of 200,000 square feet. Dr. Trimble, of New York, stated, some years ago, that he once saw a tree in the Dismal Swamp of Virginia that was 1,100 years old by the annual rings.

It is very difficult and almost impossible to get at the amount of the timber consumed, and the supply of growing timber remaining in so large a country as the United States and Territories. But at any rate it is a favorable sign, that year by year more stone, iron, steel, tin, slate, &c., is used in the construction of dwellings, bridges, ships, mills, and less wood, so that if building timber should much enhance in price, more substantial material would be used to construct buildings, and there would be fewer disastrous conflagrations.

Who shall plant trees. In the old world governments are paternal, and can decree and set apart land for the growth and protection of forests. In our country it is entirely different, the government has no such right; the duty of planting of trees depends upon the will of the citizen.

Mr. Faber, the manufacturer of the red cedar lead pencils, bought a large tract of land and planted it with cedars. In Virginia the Landreths, of Philadelphia, have been planting a large area of its worn out and abandoned fields, with catalpa, ailanthus, white oak, hickory, tulip, &c. A Scotch Emigration Company has purchased 110,000 acres of land in Barry co., Mo. To these lands the Company propose to draw families and communities of their countrymen. Some have already arrived near Purdy, a station five miles beyond the Waldensian settlement. These families have in a few months made a great change in the lands about Purdy. They have cleaned out the underbrush, and left standing the larger trees. This clearing off of the underbrush and the leaves annually would materially lessen the risk of forest fires.

The most valuable suggestions that I have met with are those of Mr. Williams, of Monongahela City, Pa., who proposes to plant 10,000 walnuts, sow them in rows, after two years' growth thin them out, leaving the thrifty trees; in five years cut or thin them for table legs; in eight years cut again, alternate trees for newel posts; in ten years begin to harvest nuts by the thousand bushels. This system of raising a forest is somewhat similar to that practiced in the cultivation of the pine forests in the Hartz Mountains of Germany. There the seed is sown; after growing two years the young shoots are transplanted into portions of the mountains. Five years later they are called a thicket, because the branches are then so closely interlaced that it is difficult to get through them. Ten years later the forester thins them out, leaving the best stems only for future growth. The growth of the tree is slow, the average age of the full-grown tree being 120 years.

"OUR WINGED FRIENDS."*

Since the time when man began to till the soil, he has called around him many assistants to lighten his labors and help him earn the bread he was to "eat in the sweat of his brow."

All our domestic animals have at some time been reclaimed from their wild state by man and trained to a higher condition of intelligence and usefulness.

Of the manner in which this is done we have no particular account. We can read that in those days there were "mighty hunters," and it is to be presumed that when they hunted and slew, they likewise captured and tamed some of the animals, and that the work of domestication was gradually brought about in that way.

The latest accession to the list, we believe to have been the wild turkey of the American forest. Why the work should have stopped with him, when other birds of equal merit, if not equal weight, are left to roam at large, is a question that remains unanswered.

Besides our domestic group, there is another and more numerous class that, on account of the kind of food on which they live, and their consequent migratory habits, cannot be domesticated, but are in their wild state equally the friends of man, doing him in many ways incalculable benefits, which by a little more protection and encouragement might be greatly increased.

It is in behalf of some of these neglected and oftentimes persecuted friends of the farmer and fruit-grower, that we desire to enlist your kind attention and sympathy for the brief space of time allotted to us.

When our remote ancestors, emerging from barbarism, began to build permanent homes, and settle down from a roving to a more civilized life, they observed that many of the tenants of the primitive forests began to approach their habitations and take part in the protection of the newly-planted fields and orchards, or busied themselves in clearing the surrounding atmosphere.

For instance, one kind of swallow left the hollow forest trees, to build in the newly-erected chimneys. The martin and rock-pewee forsook the savage cliffs, to rear their young under the friendly thatch. Robin and oriole came to assist, where the hand of the husbandman proved unequal to the work of keeping the growing buds and blossoms free from destructive insects.

---
*An essay read before the Pennsylvania Fruit Growers' Society, at Harrisburg, January 18, 1882, by Simon P. Eby.

Likewise came others, to prey upon the weaker ones, or to feed on the fruits of man's labor. These latter had to be driven off, and thus, between the two classes and man there sprung into existence a mutual feeling of friend and foe. He gave protection to the one, and waged war against the other.

The one coming in ethereal shapes with pleasant voices, to assume their labors at the time their services were needed, and again leaving for unknown lands when the season of usefulness was over. The other, issuing from their hiding places, to commit depredation at uncertain and unexpected hours, and again retreating to the depths of the adjacent forest, themselves unseen; they left behind them unmistakable evidence of their work—either friendly or hostile.

We can readily perceive how under such circumstances the untutored but imaginative minds of our remote forefathers, actuated by their love or hatred, invested some of these creatures with shapes and attributes half-human, and in that way peopled the streams and groves with strange beings "visible only by the uncertain glimpse of the moon."

"It is to be regretted," saith a writer, "that the light of modern science has frightened away all our elves and fairies."

This we believe to be a mistake. They are still with us; perhaps less numerous than formerly; but they are still here; as in the "olden time" the whims of the good require to be humored, and the tricks of the bad ones to be guarded against.

In the days of Æsop the beasts were made to speak and the birds to reason. The ancients accepted the fables not according to the letter any more than we moderns do; but for the lessons they conveyed. Even so with the creatures with which they were brought in contact. They represented the good or evil genii that haunted the ancient streams and groves, or hid within the dim recesses of the German forests.

Clothing them in shapes half human only served to bring them closer to man himself, and intensified the feelings already existing.

The transformation must have contributed to the welfare of such as were considered friendly and to the destruction of those looked upon as hostile.

Learning from the ancients let us interest ourselves a little more in the creatures by which we are surrounded. In the birds, for instance—as our good fairies, if we choose, or in the light of modern science. They will stand the test either way. Let us get our children and neighbors interested also. Teach them bird history, teach them to observe their habits, the manner of procuring their food and escaping their enemies; the skill with which they build their nests, the tender affection they show towards their mates, and the untiring industry with which they labor to rear their young. Direct their attention to the fine vocal powers some of them possess and the sweet and varied songs with which they help to swell the grand hymn of Nature. In short, let us learn that bird life has its labors, duties, difficulties, joys and sorrows, calling for sympathy, very much like human life; and the chances are we will love and protect our "winged friends" more, and in return reap the benefits of their multiplied labors in orchard and field.

We will not attempt to treat the subject scientifically, but in our own way—confining ourselves to birds wholly or in part insectivorous, and begin with those that take their food upon the wing. These constitute the scavengers of the air, and are provided with long and powerful wings, that sustain continuous flight with ease and great rapidity of motion.

Prominent among them are the different kinds of swallows, the night hawks and purple martin. They are old acquaintances in Pennsylvania; some of them great favorites with our people, and all of them deserving our fullest hospitality and protection.

They live altogether on dying insects, which they pursue with great diligence and dexterity from morning to night.

The number of insects a single pair of these birds destroy in a season, if it could be accurately computed, would be astonishing. Some faint idea may be formed by watching a nest of the young while the parent birds are feeding them. From our own observation we are satisfied that the visits of a pair of barn swallows at such a time was no less than once in every ten minutes each time, with their bills well filled with insects.

Mr. Palmer, of Massachusetts, states that he saw a parent bird visit a young purple martin on a church spire opposite his window five times in as many minutes, each time with an insect.

The barn swallow and purple martin, by no means homely in dress, are sociable in habits, and exceedingly graceful on the wing.

The first named, building its nest in or about the barn, follows the farmer to the field, and keeps him company while at work; skimming around and past him and his team —now close to the ground -now over the loaded hay-wagon—then away into the adjoining field, circling among the grazing cattle —it snaps up such insects as may be put to flight by the workmen or animals.

We remember on one occasion seeing a nest of winged ants issuing out of an old fence post. It was not long until a swallow discovered them, and must have communicated the fact to the others; for in a short time quite a flock of swallows and martins were swooping back and forth over the spot, snapping up the insects as fast as they took wing, and few, if any of them, escaped.

The purple martin is equally a favorite. Wherever these birds have once established themselves, which is usually in small colonies, among the habitations of man—they will, if not disturbed, return annually to the same boxes, and become, as it were, a part of the household during their stay. Their coming is anxiously looked for in the spring, their arrival is hailed with delight, and their departure, in the latter part of summer, more or less tinged with feelings of sadness, such as we experience in parting with a friend.

The presence of these birds, like the presence of the swallow, is by many persons considered as an assurance of continued prosperity; while their failure to return would be looked upon as an omen of impending misfortune to the house they have deserted.

They are a lively, garrulous and spirited bird. Not gifted with the power of song, they seem to make up for this deficiency by an increased love for gossiping. Their early morning salutations in front of their boxes are, however, very pleasant to listen to.

The male bird makes a model husband. During the time his mate is sitting, he becomes quite domestic, and spends part of his time in front of the box dressing and arranging his plumage, occasionally passing to the door of the apartment as if to inquire how she does. His notes, at this time, have assumed a peculiar softness, expressive of much tenderness. And yet he is a courageous bird, and will unhesitatingly attack with great spirit and audacity hawks, crows and other large birds, and even cats, if they show themselves in the vicinity of his home. Thus recalling to mind the closing lines of Bayard Taylor's "Song of the Camp:"

"The bravest are the tenderest,
The loving are the daring."

"Conjugal fidelity, even where there is a number together," says Mr. Wilson, "seems to be faithfully preserved by these birds."

The martin feeds upon the larger kind of insects; wasps and beetles forming his principal food. We are aware that he has been accused of a failure to discriminate between such legitimate prey and the honey bee, and that neighboring swarms have sometimes suffered in consequence. Be this as it may, for our part we shall find no fault with him on that account. Since its cross with the Italian our honey bee has become such a pest to fruit-growers that we might well be rid of it altogether.

There are two other well-known members of this group with which we could not well dispense.

Our summer sky could hardly be considered perfect without at least a pair of long-winged night-hawks sporting lazily through it, and descending occasionally with a sounding swoop.

Neither would our summer evenings be properly rounded off without a flock of twittering chimney swallows circling over our heads and dropping successively out of sight, as parting daylight is fading into darkness.

In the second group we will speak of those that watch for their prey from the perch, but take it while flying. These are called the fly catchers.

Prominent among them are the king bird, rock pewee and wood pewee.

"It seems a provision of nature," writes Mr. Samuels, "that all fly catchers shall only take those insects that have taken flight from the foliage of trees and shrubs, at the same time making the warblers and other birds capture those which remain concealed in such places."

"The king bird, in seizing a flying insect, flies in a sort of half-sitting hover and seizes it with a snap of the bill. Sometimes he descends from his perch and captures a grasshopper that has just taken a short flight and occasionally seizes one that is crawling up some tall stalk of grass."

"Those farmers who keep bees dislike this bird because of his bad habit of eating as many of those insects as show themselves in the neighborhood of his nest, but they should remember that the general interests of agriculture are greater than those of a hive of bees."

He is possessed of great courage and is more than a match for hawk or crow, which he attacks and drives off whenever they venture into his neighborhood.

The rock pewee, or house pewee, comes to us in the early days of spring, and announces his arrival by uttering the notes from which he derives his name.

Like the swallow, he generally seeks his last year's nest and makes such repairs as he fancies necessary; perhaps a small addition to strengthen the outside or a new lining.

The foundation of the nest is composed of pellets of mud mixed with fine roots and grasses, plastered to the wall or other object against which it is built, and lined with soft grasses, wool or feathers.

His favorite haunts is under arch of a bridge, or under the eave of a mill or dwelling. Here he can be seen during the breeding season, perched on the branch of some over-hanging tree, or upon the rail of the bridge, or neighboring fence post, flirting his tail, uttering his plaintive notes and darting about in all directions snapping up the insects which generally swarm plentifully in the locality he has chosen for his home.

The wood pewee is generally found foraging along the edge of the woods that hides his nest, or among the lower branches of the fruit trees near the gardens, and even among the trees growing on the city lots.

Here, like his less shy cousin, he can be seen perched on some projecting twig always on the alert, darting quickly forward and back, catching the flying insects that come within sight of his ever watchful eye. His notes, uttered while thus employed, are similar to the rock pewee only more plaintive and longer drawn out.

The next group, embraces those birds, that seek and capture their food among the foliage, buds and blossoms of the trees and shrubs.

Prominent among them rank the Baltimore oriole, orchard oriole, wood or song thrush, the vireoes and some of the warblers.

"The food of the oriole is almost entirely insectivorous, young peas and stamens of cherry and plum flowers forming the only exceptions. These small robberies are but a slight compensation for the invaluable services he renders the gardener in the destruction of hosts of noxious insects. At first beetles and hymenopterous insects form his diet and he seeks them with restless agility among the opening buds. As the season progresses, and the caterpillars begin to appear, he forsakes the tough beetle and rejoices in their juicy bodies. Even the hairy kind he does not refuse, and is almost the only bird that will eat the disgusting tent caterpillar of the apple trees."

To its usefulness it adds a plumage of rare beauty and brilliancy, a song of great cheerfulness and a nest wonderfully constructed.

"There is in his song," says Mr. Wilson, a certain wild plaintiveness, extremely interesting; that is uttered with the pleasing tranquillity of a careless plow boy, whistling for his own amusement."

"It is a joyous, contented song," says a writer in Harper's Magazine, "standing out from the chorus that greets our half awakened ears at daylight, as brightly as its author shines against the dewy foliage."

T. W. Higginson exclaims, "Yonder oriole fills with light and melody the thousand branches of a neighborhood."

He is a social bird—a bird of sunlight. His hammock-like nest is never found in the deep woods. His haunts are those grand old trees which the farmer leaves here and there in his fields as shade for his cattle, that lean over the brier-tangled fence of the lane, or droop toward the dancing waters of some rural river.

We are now among a host of feathered choristers, to which the song of the oriole is like the bugle notes for the opening of the grand winged orchestra.

Where all possess so much merit it is difficult to assign precedence. Out of the deep woods, however, comes a beautiful melancholy strain, which is not very common, but when heard cannot fail to arrest the attention.

"The prelude to this song," says Nuttall, "resembles almost the double-tonguing of the flute, blended with a tinkling, shrill and solemn warble, which re-echoes from his solitary retreat like the dirge of some sad recluse, who shuns the busy haunts of life."

"The whole air consists usually of four parts, or bars, which succeed in deliberate time and finally blend together in impressive and soothing harmony, becoming more mellow and sweet at every repetition. It is nearly impossible by words to convey any idea of the peculiar warble of the vocal hermit; but among his phrases the sound of "a'rioee," peculiarly liquid and followed by a trill, repeated in two separate bars, is readily recognizable."

We have followed this song, which seemed to recede before us deeper into the woods as we advanced, without getting a sight of the bird, until brought to a sudden halt by a sharp "chuck ;" when for the first time we saw the object of our search perched upon a twig of a neighboring tree and eying us sharply. It was the "song thrush" or "wood thrush ;" a bird in size between the blue bird and robin ; cinnamon brown on the back and whitish breast marked with well-defined dark triangular spots.

Its notes are uttered while engaged in hunting for insects among the foliage.

Next we have the "Vireoes," of which there are four reported as visiting this part of our country. The red-eyed warbling, white-eyed and blue-headed—all useful birds, that feed on insects, which, like the two preceding, they hunt among the foliage. They are in size about like the canary, of a grayish olive green, and variously marked as their names indicate. Their nests are pensile—or hanging—generally fastened to the fork of a horizontal twig, shaped not unlike a shallow, open-mouthed purse.

Mr. Samuels writes of the Red-eyed Vireo in the following commendatory manner : " I feel that no description of mine can do justice to the genial, happy, industrious disposition of this, one of our most common, and, perhaps, best-loved birds. From the time of its arrival, about the first week in May, until its departure, about the first week in October, it is seen in the foliage of elms and other shade trees, in the midst of our villages and cities, in the apple trees near the farm-houses, and in the tall oaks and chestnuts, in the deep forests—everywhere, at all hours of the day, from early dawn until evening twilight, his sweet, half plaintive, half meditative carol is heard ; and whenever we see him, we notice that he is busily searching in the foliage of trees for caterpillars and noxious larvæ, or pursuing winged insects that have taken flight from the trees.

"Of this beautiful and favorite family I feel that it is impossible to say too much in their favor ; their neat and delicate plumage and sweet song, their engaging and interesting habits, and their well-known insect-destroying proclivities, have justly rendered them great favorites ; and the farmer in protecting them and encouraging them to take up homes near his orchard and gardens, but extends a care and welcome to his best friends."

The wren and blue-bird may be considered together ; both being insectivorous, capturing their food alike upon the trees and on the ground, and building in crevices and boxes.

These birds seem to be getting more scarce in late years. In our school-boy days there was no season that we did not know of a wren or blue-bird's nest. We recollect instances when the wren contended for quarters with the martin and out-witted him by narrowing the entrance of the box with sticks, strongly and skillfully placed, so as to admit himself, but keep his larger antagonist out.

The wood-pecker family have been voted great scamps—fruit-stealing, sap-sucking rascals—a proper target for every idle boy, who could handle an old rusty gun, to blaze away at. Of late years their usefulness has become better understood, and a law enacted to save them from total extermination. They are the police of the trunk and woody part of our timber, fruit and shade trees. In fact, to us, the red-headed wood-pecker does not seem unlike a liveried policeman patroling his beat, up and down and around the trunk, and out along the limbs of some old tr e, tapping and rattling for concealed marauding insects, and dragging them from their hiding-places without mercy when discovered.

We have frequently noticed the trunks of old apple trees punctured in a regular succession of circles ; or have seen spots as large as a hand where the bark seemed dead, riddled like the bottom of a colander, all the work of this or a smaller speckled wood-pecker, known as the sap-sucker, in their efforts to dislodge the insects under the bark.

Shall these faithful servants be denied a few of the fruit of the trees they help to save ?

A few years ago we observed several Scotch pines in one of our cemeteries treated in this manner, and the resinous sap exuding and filling the punctured circles. Surely, we thought, this time the bird could have been after no honest purpose, and deserved the bad name it bore. Behold, in the following spring one of the pines was dead ; and taking a friend with us, we examined into the cause and found the inner bark of the upper part of the trunk and of some of the larger limbs reduced to the condition of fine sawdust, having been entirely eaten by worms. Here the borers had been too numerous or the trees too far gone. The other pines were no doubt saved by the timely interference of this much-slandered bird.

Closely allied in habits of life to the wood-pecker are the titmice and chickadees, of which ornithologists report three as visitants to this country. They feed on insects and the eggs of moths deposited on and in the crevices of the bark and in the buds of trees and shrubs.

During breeding season they are busy through the whole day in capturing vast quantities of caterpillars, flies and grubs. "It has been calculated," says Mr. Samuels, " that a pair of these birds destroy on the average not less than five hundred of these pests daily."

" The chickadee trips along the branches, trips under every leaf, swings round upon his perch, spies out every insect and secures it with a peck so rapid that it is hardly perceptible."

Last but not least in our list come some of our best known and most reliable friends. Prominent among which are the brown thrush, or mocking bird, robin, cat-bird, black-bird, meadow-lark, chipping-sparrow, song-sparrow and indigo-bird.

These feed on small fruit, seeds and berries, as well as on insects, grubs and worms. They help themselves to some of our early fruit, and in that way sometimes annoy us. Still if an account could be made up of what they take, and the good they do, the balance would show largely in their favor.

They compensate in still another way ; they cheer us with their presence and songs ; for amongst them are some of the most talented musicians. Unlike the oriole and vireo, which carol while they labor, this class lay aside other duties when they addressed themselves to song. Ascending some elevated perch and concentrating all their vocal powers, they pour forth their strains of melody, as if it were to a listening audience.

Mounted on the topmost spray of a neighboring tree or bush, the brown thrush welcomes the farmer planting seed at early morning with cries of " drop it, drop it, cover it up, cover it up. Pull it up, pull it up ; see, see, see ; there you have it ; work away, work away ; cover it up."

This bird, although often seen in the orchard and pasture field, generally builds his nest in the neighboring thicket and seems partial to sprout land, or woods having undergrowth.

A few years ago we considered ourselves highly favored when a pair selected a small evergreen upon the lawn for their nesting place, and we gave strict orders for no one to go near while the work was progressing, but unfortunately some unknown enemy must have discovered them, for one day we found the eggs broken and the nest deserted.

Of the robin a writer in the Atlantic Monthly says : " I shall not ask pardon for assigning to him the highest rank as a singing bird, while others may surpass him in some particular qualities ; the notes of the robin are ah melodious, all delightful—loud without vociferation, mellow without monotony, fervent without ecstasy, and combining more mellowness of tone, plaintiveness, cheerfulness and propriety of execution than those of any other bird. Without his sweet notes the mornings would be like a vernal landscape without flowers, or a summer evening sky without tints."

After the noon-day heat has silenced the early performers, the song sparrow, chipping sparrow and indigo bird continue to sing at intervals during the greater part of the day.

The song sparrow has been assigned a high place among singing birds. His song is certainly very soft and sweet, without a harsh note in it. We hear it mostly from the hedge rows, and along the edges of the grain or pasture fields.

The sprightly little indigo bird selects the highest twig of some tall tree in the vicinity of his nest to pour out his noon-day song.

Last and least is the chipping sparrow, greeting us from the fence posts, along the highways and country lanes, with its peculiar but pleasant little song not unlike that of a summer locust.

Having thus spoken in behalf of some of our "winged friends" as time would allow, leaving, however, many of them unmentioned, and many of the good things which might be said in their favor unsaid, the next question naturally suggests itself: How can we best preserve these winged institutions, which have become interwoven with some of our earliest and happiest recollections of rural life, and hand them down to posterity unimpaired?

The woods, of course, have ever been the great nursery for birds. We do not mean the endless forests, which at one time covered this country, but belts of timber with plenty of undergrowth lying between farms, adjoining the cultivated land, and along the streams. These gave plenty of room and material for nests, were within convenient reach of the sunlight, of the fields and the food there found: at the same time there was some protection from man against birds of prey.

As our woods are cleared away we should endeavor to provide other shelter, by saving the trees, wherever possible, upon the farm; by planting thickets of young timber in such places where the land cannot be profitably cultivated. Hedge-rows become good nesting places for the smaller kinds of birds, and afford protection when pursued by hawks. Evergreens planted for ornament or protection oftentimes attract birds. The summer-house or open building on the lawn or in the orchard is generally selected by the robin for a nesting-place. So the shrubbery and climbing vines around the house should be at the service of the chipping sparrow and warblers. The orchard, of course, we expect to have its full share of nests, and the elm, or weeping willow, or the old pear tree, to have one of its drooping limbs graced with an oriole's hammock. A row of boxes should be put up against the south or east side of the house for the marlins; shelvings under the forebay for the swallow, and an opening in the upper part of the barn for them to pass in and out freely, should they fancy that part of the building. Such chimneys as are not used in the summer should be left uncovered for the chimney swallows. Boxes should be put up against the outbuildings, and on the sheltered side of trees, for the blue-bird and wren; so that the whims of these our good fairies may be properly humored.

When these accommodations are provided and the birds happily do come to occupy them, or some of them, do not interfere with their housekeeping nor suffer anyone else to molest them, whether it be thoughtless man or sneaking cat. Do not approach their nests unnecessarily nor allow anyone else to do so; remember this is a tender point with all birds, and will cause them to change residence very soon. Do not allow the English sparrow to take possession of the boxes and drive the others out.

One more suggestion and I am done. If there is no running water on your farm or in the vicinity, provide a place for the birds to drink, and where they can get soft material to build nests. Swallows and martins love to skim near the surface of the water and take an occasional dip. Robins and cat-birds will help themselves at the water trough in the barn-yard, but the more shy birds, like the brown thrush, will not venture that far. Water should be kept for their use in a more secluded place.

A cheap bird fountain can be made with an old demijohn or carboy, which can be had at a drug or liquor store for a trifle. Select a suitable shady spot frequented by the birds and where they will not be disturbed. Place a trough or other shallow vessel on the ground; drive stakes for the demijohn to rest upon in an inverted position so that it s mouth will nearly touch the bottom of the trough and hold it in that position, then fill the demijohn and turn it upside down upon the stakes. The water will run out and keep the trough partly filled until the supply in the demijohn is all used; on the some principle as a small bird fountain.

We had a fountain made in this manner with a five gallon demijohn, which answered the purpose admirably, and required refilling about once a week. The depth of the water can be regulated by raising or lowering the mouth of the demijohn.

And now with your "winged friends" properly cared for, yourselves cheered and comforted by their presence and grateful song, your orchards saved from the ravages of insects and their golden fruit safely stored away for winter use, you may live as contented and happy as it is possible for mortals to be.

SEEDLING FRUITS.*

This question may be answered in a general way in a few words, viz.: Sow seeds and raise plants, shrubs, vines and trees. And further, does not nature attend to this matter without the aid of man? Are not a large proportion, if not the largest, of the most valuable fruits accidental seedlings.

I will not for a moment dispute the said assertion, but at the same time I hold tha, many valuable varieties of fruits are the result of seeds planted by the hand of man, for which he has received no credit.

Many trees have been planted throughout our country since its settlement, by missionaries, travelers and others, the result of which we can form no accurate estimate.

Within the last quarter of a century, however, many new fruits have been produced by more intelligent methods, viz.: by hybridization and by cross fertilization by design, which have produced the most gratifying results, we must be surprised at the result.

With the grape greater success has been attained than with any other kind of fruit. Of the value of grapes produced by design it is now impossible to estimate, when we compare the time when the Isabella and Catawba were the only popular grapes, with the present day, when scores of improved and superior varieties are being disseminated through the length and breadth of the land.

Thanks to Messrs. Rogers, Ricketts, Wiley and others for the choice we may now make in our selection for planting; and from present indications we are just on the threshold of what we may expect, and, unless the future shall belie the past, this country will, in the near future, be enabled to claim as great a variety of fine native grapes as any other.

The number of new and improved varieties of strawberries, raspberries, blackberries, cherries, pears and other fruits have of late years been multiplied to such an extent as to almost confuse the planter with limited room or means. This, however, should not deter any from raising new seedling fruits, as time and testing will eventually decide so as to lead to the "survival of the fittest." The venerable President of the American Pomological Society has in almost every annual address urged the production of new fruits by hybridization and cross fertilization, and is in his advanced years lending a helping hand in this laudable cause.

The only serious drawback in the multiplication of new fruits is the disposition to make too great a speculation of new things, and too many make extravagant claims for their pet products and cling tenaciously to those claims whether worthy or not, simply because it is their own production. May this, as well as other horticultural and pomological societies, be slow to recommend any new fruit or vegetable for general cultivation unless thoroughly and extensively tested; no new fruit should be added to our catalogues unless it has special merit not possessed by any already on the list.

But to the question. Prior to the formation of the flower and period of inflorescence, nature seems to work in the dark, but during and from this time until the fruit is perfected, her operations are intensely interesting to the close observer. Let us follow her progress in the development of the bud, the expansion of the corolla, the spreading of the petals, exhibiting the stamens and pistils—a perfect flower in all its beauty and fragrance. The most important parts, however, are the reproductive organs—the stamens and pistils; the former are termed male and the latter female organs. When the anther of the stamen is ripe it casts off its pollen in very minute particles which falls upon the stigma of the pistil, which when in a condition to receive the pollen is of a glutinous nature, to which the same adheres; thence it passes down through the style of the pistil into the ovary, when the fertilization is complete.

Some plants, shrubs and trees have flowers purely staminate and others purely pistillate on the same plant or tree, while others have staminate flowers on one plant or tree and pistillate on the other. The latter are termed dioecious, the former monoecious; but whatsoever the nature of the plant may be, unless

---
*Essay read before the Pennsylvania State Horticultural Association at Harrisburg, by Henry M. Engle, of Marietta.

the pollen reaches the ovary of the pistil' there can be no fertilization, and consequently no seed or fruit.

It is well known that seeds do not always produce the same fruit as their parent, showing that pollen is carried from other trees, by insects or by the wind, or both.

It is not at all strange, therefore, that by planting seeds that have been fertilized by nature the chances will be few and far between of the seedling being superior to its parent, although such cases have occurred. Stone fruits reproduce their kind truly, more commonly than pip fruits.

The object in producing new varieties is to combine desirable qualities of both parents in the progeny, on the same principle on which stock-breeders operate, and we must admit that they have more nearly attained their ideal in that particular than horticulturists have theirs.

The former, however, have followed their object practically for a longer period than the latter, who will, in my opinion, eventually, by judicious selection, breed out objectionable and breed in desirable qualities in fruit, as breeders of animals do in live stock. Let us not be surprised some day to hear of thoroughbred apples, pears, peaches, grapes, and other fruits; and that books of fruits, with their pedigrees, will be kept as well as herd books. The new and improved varieties of fruit produced by design by hybridization and cross-fertilization are too recent to prove the above assertions, for very few, if any, have been recrossed to test the theory of transmission.

Is it not reasonable that laws which govern the vegetable kingdom are as immutable as those that govern the animal kingdom, however limited our present knowledge of the subject may be?

In crossing a sweet fruit with an acid one, we would reasonably expect the new seedling to be sub-acid, but such will not certainly follow. It is therefore of the highest importance that those who propose to follow, or who are now following, this very interesting business, should search diligently the laws which govern its processes in all its details.

For instance, the question may arise whether the more vigorous plant or tree will transmit more of its nature than the weaker one; or, what will be the effect of applying the pollen to the pistil as early as it can be made effective, or as late as the nature of the case will admit, or by applying the pollen in its earliest available condition to the stigma as late as it will admit, and *vice versa*; the results of applying the pollen by sunshine or under a cloud; the effect of wet or dry weather following fertilization; also, whether the application of fertilizers to the plant or tree while the fruits, or the young seedling is growing, will produce different results. Whether it will ever fall to the lot of man to fully understand the laws which govern this delicate process or not, one thing is reasonably certain: that by crossing two varieties of fruit of great excellence, the resulting fruit will be superior to the product of two inferior varieties. But how to obtain the qualities we may desire, by crossing, is yet a hidden mystery.

If, however, stockmen could breed off horns, and almost reach their ideal in breeding beef, milk, and butter strains into fixed types, may not fruit-growers attain similar results in the vegetable kingdom by taking a thorough course in Nature's school of experience?

But whatever we may achieve, our calling is a noble one; and, with what has been done in the past, and the progress being made at present, our future looks bright. Let us thank a Divine Providence that we were born in the Nineteenth Century!

OUR LOCAL ORGANIZATIONS.

LANCASTER COUNTY AGRICULTURAL AND HORTICULTURAL SOCIETY.

The Lancaster County Agricultural Society met statedly on Monday afternoon, March 6th, in their rooms in the City Hall.

The following members were in attendance: Jos. F. Witmer, Paradise; M. D. Kendig, Creswell; H. M. Engle, Marietta; Calvin Cooper, Bird-in-Hand; S. P. Eby, Esq., J. M. Johnston, city; Casper Hiller, Conestoga; C. L. Hunsecker, Manheim; F. R. Diffenderffer, city; Ephraim Hoover, Manheim; J. C. Linville, Salisbury; W. W. Griest, city; Enos H. Weaver, Strasburg; John H. Landis, Manor; John G. Resh, West Willow.

On motion, the reading of the minutes of the previous meeting was dispensed with.

Crop Reports.

H. M. Engle said winter wheat and grass look well. The prospect for fruit is good.

E. H. Weaver reported old clover as frozen out in some places, but the young clover looks well.

M. D. Kendig reported a good many sales of tobacco in his township at fair prices.

Mr. Witmer thought the *young* clover was lifted considerably; whether it will take hold again was the question.

H. M. Engle said this was generally the case when young clover fields are pastured late and the following winter is an open one.

Apples—Local vs. Foreign.

Calvin Cooper read the following essay on the above subject:

By the term foreign, I do not intend to convey the idea that I allude to fruits brought from "foreign countries," but varieties from other sections of our *own* country. It is a well known fact that every country has its *native* fruits, adapted to its own *particular* climates, and when removed elsewhere, are often so materially changed in appearance, flavor and habit, as to be almost unrecognizable, and indeed often quite worthless. So changed, that many persons would assert they were entirely different. Although the change in location may not be very great, yet there is a certain something in soil and climatic influence so unsuited to its natural element that *man* cannot supply, and which we are unable to account for.

It will doubtless be asserted that continual changes are taking place in all newly settled neighborhoods. The removal of forests will, in itself, bring about changes not perceptible at the time. But as years of time intervene we are enabled in our comparisons of the seasons of former years with those of the later to perceive such a material difference as to lead us to pause and query what has been the cause. This may affect, to some extent, the local fruit of each section. Nevertheless, I believe it is so gradual as to be of little importance, as the power of the tree to adapt itself to the surrounding circumstances of its native place.

The idea that I more particularly wish to impress is the common error in bringing varieties of apples from distant parts of the same country, or even the same State, and, I might assert, of the same county, for, indeed, what might be considered first quality in the higher altitudes of Northern Lancaster county would be of little value in the southern section, and *vice versa*. Although they might be upon the same degree of latitude, the natural elements of soil would not supply the requisites of its native locality. Then, too, what would flourish in the eastern section, along the Mine Hill and Welsh Mountain ridges, might not be worthy of cultivation in the fertile valleys of the western end. It does not necessarily follow that a single variety will produce well and retain all its good qualities in every part of its own locality. But we have good reason to believe there is more certainty of receiving a good reward for the labor in removing sorts *native to each particular* section, provided, the same altitude, degree of latitude and natural aspects be maintained, and even then exceptions will occur.

The grave error of our own vicinity might be attributed to several causes. The high price of land and a want of interest by our agriculturists in horticulture has caused many to neglect their apple orchards, as unprofitable and let the trees perish for the want of proper nourishment and care; some, too, have become victims of the woodman's axe, and what was once the pride and comfort of the husbandman, supplying health giving luxuries to the household, has been supplanted to the growth of a noxious and poisonous weed. In conversation with a neighbor who had cut down a flourishing orchard, he said he could "buy his apples much cheaper than he could raise them;" and this was doubtless true in his case, and why? Because in the selection of varieties he did as thousands of others have done, was enthusiastic in his estimation and value of the fine fruits that were then put in our markets, from the northern part of this State and central and western New York, where Baldwins, Spitzenbergs, Twenty Ounce, Tompkins King, Gillyflower, Northern Spy, R. I. Greenings and dozens of others flourished to perfection, and planted mostly of such sorts, doubtless thinking that they would produce here equally well, never for a moment pausing to query whether they might be unsuited to this locality, and I am not surprised from the results that you perhaps have all observed that a lukewarmness has been created in the interest of the orchardist, and since the transportation facilities are such that apples of superior excellence can be brought cheaply to our markets from sections less valuable in agricultural wealth, and find a ready market at reasonable prices in the cities, while the poorer classes of the rural districts are not supplied and often suffer for the health-giving juices of a well-ripened Smokehouse or Rambo.

I might enter an apology here for aiding in the dissemination of varieties unsuited to our locality, but the nurseryman, like all business men, is not exempt from occasionally dealing in humbugs, especially when the demand was for the varieties brought to our markets from the northern districts above referred to. So great was the inquiry for the then newer sorts that our usual supplies of some of the older reliable stand-bys were left in the nursery rows to be dug and burned by the thousand to make room for other stock.

This experimental mania has brought its accompanying evil, and created a demand for new varieties, generally following in the wake of their predecessors, after years of care and expectation, to be cast aside for other novelties that in all probability would meet with the same fate. Thus from lists of from twenty to thirty at most, they are counted now by the scores and hundreds to suit the varied fancy of customers. Hence failures have become so numerous that some look with distrust upon all, and abandon the enterprise as discouraging and unprofitable.

Is there no remedy? Can we not grow apples as heretofore? I believe we can. Trees grow and flourish as of yore, and we have many instances of success in all sections where the proper care has been taken in the selection of varieties, and due attention to cultivation, pruning and the application of necessary fertilizers. I am fully convinced that every planter in selecting an apple orchard should first consult some fruit grower in his immediate neighborhood and ascertain what varieties are doing best in that section, and then plant nine-tenths of his orchard with those known to do well there. *Failures* would be the exception; the balance might be an experimental plot of those promising well and not fully tested. I do not, however, wish to be understood as disapproving of the introduction of new sorts.

But I do protest against them being planted to supersede old reliable kinds, until thoroughly tested. The prevailing habit of planting a long list for the sake of variety is not only vexatious to the nurseryman, but brings disappointments after years of waiting. I have frequently supplied orders of fifteen or twenty trees with as many varieties, while a lesser number would doubtless have been much more satisfactory after the trees had begun to bear, and the planter learned to his sorrow that the one-half were almost worthless. Not that there could *not* be that many varieties selected as reliable. But the lists are often taken from some distant nurseryman's catalogue whose glowing descriptions please the fancy of the prospective fruit grower.

To conclude, I would most emphatically discourage the experimental mania for varieties grown in remote districts, except to a very limited extent, and plant of those native to each particular section in connection with a few doing well generally; and I would also *caution* all from bringing *northern* apples into *southern* districts, expecting to get a *late winter* keeper, notwithstanding the fruit can be brought and kept through the winter. When grown where the season is longer, and the latter part of it often quite warm, the fruit ripens too early and begins to decay before the cool weather sets in.

Remarks

Casper Hiller said it was a fact that we can no longer keep apples as we once could. What the cause is he was not prepared to say, but the fact is indisputable. He was inclined to attribute it to a change in the seasons. But for all this we must not do away with the apple orchard. An acre of orchard is worth more than any other on the farm. Not in the money value, perhaps, but in other ways. He believed the warm dry summers are the reason we have not large crops of apples for winter use.

Ephraim H. Hoover thought the increase of insects had much to do with our poor apple crops and their non-keeping qualities. When we put the crop away it is imperfect, and therefore will not keep. He has tried turning in hogs when the apples begin to fall and keeping them there all the season. This keeps down the insects and preserves the crop. Insects are one of the causes that make our apples poor keepers. We must dispose of the insects before we can hope to increase the quality of our apples.

M. D. Kendig believed with the essayist in his theory of planting native varieties. He thought it was better and safer to buy either fruit or ornamental trees of home grown origin than to get them elsewhere.

C. L. Hunsecker spoke of the fine apple crop of 1855; we have had none like it since. How long will apple trees continue to bear; he gave some information on this subject. He mentioned a tree in Maine that lived 150 years and bore good crops yearly. He also believed that certain climatic changes had something to do with the failure of our apple crops. If trees could be protected against high winds it might be beneficial. He believed the insects had something to do with this, but we do not take enough care of our orchards.

Mr. Cooper said hot weather makes apples drop prematurely. If the temperature in September and October was not so high we would be able to grow as much fruit as ever and as good.

H. M. Engle concurred with nearly all the essayist had said. The question is, Are there any remedies to overcome the evils that are upon us? If we select better varieties our chances will improve. An important matter is at the will of the orchardist. The curculio and codling moth can be controled. But this can best be done by co-operation. There are good apple crops even when there are many insects, but the following year is generally a failure, because a large apple crop is the cause of a large crop of insects. The codling moth is our greatest enemy; it does more damage than all the rest combined. But we can control this by using the banding system, using straw, canvas, paper or any other article. By selecting varieties adapted to the locality and attending to the moths we can do much to help along our orchards.

S. P. Eby thought there were some means at our command we have not yet used. Plant orchards where they can be irrigated; plant shelter trees; leave the tree a large tap root when it is removed for transplanting. Trees that spring up of themselves and are grafted where they grow without removal are less liable, perhaps, to diseases and failures.

H. M. Engle also directed attention to the fact that orchards must have as much manure as other fields. We expect full crops, but make no effort to secure them by putting as much manure as we do when we grow wheat or corn. The orchard must be fed. Altitude has much to do with full crops. He lately saw a fine crop of apples grown in Virginia, at an elevation of 1,100 feet. They were northern varieties.

Is Sub-Soiling Beneficial?

John C. Linville said:

When I was quite a small boy my father made himself a sub soil plow. The late Jesse Buell was at that time editor of the Albany *Cultivator*, and in that excellent journal proved by irrefutable and convincing argument that the sure road to successful farming lay through sub-soil ploughing. The first trial of the new method was made in corn ground. The surface plow turned the sod to the depth of about six inches, and the sub soil plow followed in the bottom of the furrow and loosened up the clay from four to six inches deep. This plow did not throw the sub-soil on top, but merely lifted it up two or three inches and let it fall back to its former position. Two teams were used—a span of horses to each plow. It made it very laborious for the lead horse of the surface plow to walk in the loose furrow. The field of ten acres was sub soiled in alternate strips, the other strips left in the usual way. The sub soil in this field is rather stiff, red clay, and the land is limestone.

I do not know whether the season was wet or dry, but there was no perceptible difference in the corn nor in the succeeding crop of oats, wheat or grass. It was observed, however, that the sub soil in the strips that had been double ploughed retained its mellowness the following season when broken for oats and wheat. Of course, sub soiling doubles the cost of ploughing. This and its signal failure to increase the crops condemned the sub-soil plow, and it lay for years in the lumber loft.

At length there came another "boom" in sub-soiling. The lamented Prof. Mapes was at that time editor of the *Working Farmer*, and showed by invincible logic that a loose sub soil would let the surplus water down in a wet season and be equally beneficial to retain moisture in a dry one. The old long-legged sub-soil plow was brought out out from its long hiding place, the dust and cobwebs swept off, and the share sharpened for action. The corn field was subsoiled in alternate strips, as on the former occasion. The season was rather favorable for corn and there was no perceptible difference in favor of the subsoil ploughing either on the corn or succeeding crops. The subsoil plough was again consigned to the lumber loft and oblivion, where it remained until the sale of my father's personal effects, when it was bought by an enterprising farmer on the border of Chester county. I have no knowledge of its subsequent history.

These two experiments do not prove anything. If the seasons had been different or the soil different, the results might have been other than they were. Had the crops been roots, or vegetables, or orchard, or nursery, the sub-soiling might have been beneficial. There is, however, one convincing argument against sub-soiling. The system has been advocated time and again for a great many years and yet nobody uses the sub-soil plow now. If it has all, or even a few of the advantages claimed for it, farmers surely would not be so slow to discover its merits.

H. M. Engle has tried sub soiling and has not derived any benefit from it. Our soil does not seem to require this method.

Can We Dispense with Division Fences on Farms?

Ephraim S. Hoover gave his views of this question as follows:

This is a question which at this time, when lumber is getting scarce and valuable, is well worth the consideration of all who are owners of arable land. How may we avoid the expenses of division fences profitably? This may, we think, be done by the soiling system, which does away with inside or division fences except a large cattle yard in front of the barn surrounded with shade trees, and well supplied with an abundance of water for the use of stock. An average of the whole farm land of the State shows us that the fences cost us at the rate of $1,124.25 per one hundred acres, and in some localities where timber is scarce, the cost may be more. At this rate, the interest on the amount invested per farm of one hundred acres at six per cent. is $67.45 annually, to which add, as it is estimated that the repairs cost $6.23 per 100 rods or $60.35 annually per one hundred acres, making a total of $127.80 for fences alone, not including the value of space occupied by the fences, which would be valuable in adding to the land under cultivation. Besides the loss of this space of land occupied by fences, it creates a harbor for weeds, which would not exist if properly cultivated.

From five to six months of the year our cattle in this latitude are fed in stalls on products of the farm and are not allowed to roam over the fields of the prudent farmer. Hence, we claim that in view of the above facts it does not pay, all things being considered, to fence a farm for grazing cattle the remaining six or seven months of the year, when it is an admitted fact that cattle will subsist on less acreage under the soiling system than under the present system of grazing, where much pasture is trodden down and becomes unfit as food for cattle.

Another way of doing away with division fences is to have portable fences sufficient to enclose a few acres of pasture at one time, which could be done at a small expense and at such a time as not to interfere much with the other operations of the farm. The cost of such fences would be comparatively small and feed more stock than when allowed to run over many acres at once. We were convinced of this some years ago in our own experience, when a field lying in permanent pasture, well supplied with water, was divided into two parts by a temporary fence, changing the stock from one to the other; we found the same piece of ground fed more stock than when the whole field was pastured at once.

Another way by which we can in part do away with division fences is to remove division fences between certain fields so situated that they may be formed in pairs, or we may very much decrease the amount of fences by altering the shape of a field or fields. If our farm is of such a nature that we can turn a portion of it into permanent pasture land, we believe this system to work well.

The object of every farmer should be to have as few fences as practicable, and of this every owner of a farm should be the best judge—in other words, he should adapt himself to natural advantages, such as water, soil, locality, etc.; whether best adapted to grain growing or grazing—all of which must be taken into consideration, for while some particular system would do very well for some localities, it would not be practicable in some others. But, in short, to have the least capital invested in fences possible, and yet to be so arranged as to bring the best results, should be the object of every wideawake and progressive farmer.

Remarks.

J. C. Linville thought there was no doubt we could dispose with most of our division fences, and will have to do so before long. He has found nothing so good for temporary fencing as barbed wire fencing. He used only one wire, and it was sufficient, although the cattle were very tame. Such a fence is cheap and lasts a long time. The wire used was about three feet from the ground. Unless the barbs are close together they will not turn sheep, no matter how close the wires are.

Mr. Eby thought the fence question was a very important one. The law as it is at present is imperfect. We must make fences to keep out our neighbors' cattle. If we could have a cattle law passed for this county it would be advantageous, and he suggested that the members should make an effort to this effect.

Casper Hiller had no doubt it was possible to dispense with division fences. The expense of keeping them up was greater than to hire a man to look after the cattle. We are, however, so wedded to old customs that it would be difficult to bring about a change. He did not see how we could get around this question.

H. M. Engle thought by showing the farmers that it was putting money into their pockets, they could be induced to make the desired change. The law should make every man take care of his own cattle, instead of compelling him to protect himself against those of his neighbors. He was in favor of a law that would change the present cumbersome law.

H. M. Engle made a motion that a committee be appointed to examine into this question and report at the next meeting.

The Chair appointed S. P. Eby, Esq., Calvin Cooper and F. R. Diffenderffer as the committee.

When is the Best Time to Sow Clover Seed?

E. H. Weaver responded in answer that no particular day or week can be specified as the best time to sow clover seed, for that depends upon the weather and condition of the soil. From the middle of March to the first of April might be set down as the proper period, the farmer exercising judgment in selecting the best time in this interval. Earlier sowing is useless and attended with risk. When sown on frozen ground, as some do in February, a sudden thaw or heavy rain flood may wash the seeds from the slopes into the low lands, or a warm spell may sprout the seed and a following cold snap may freeze the germs, which has been the experience of some farmers. If the wheat ground is not previously harrowed, comparatively early sowing is best, as alternate freezing and thawing will more effectually cover the seed. The farmer cannot afford to run any known risk in sowing clover seed. A failure of the the crop is a serious loss to him and his land, as it is the great restorer of exhausted soil. The Hon. Geo. Geddes reports a field on his farm upon which had been grown crops of wheat, corn, oats, barley and grass, which has had no other manure but clover for seventy-four years, and the land shows no diminution of fertility.

S. P. Eby believed late sowing of cloverseed was, all things considered, the best.

H. M. Engle said there was a difference of opinion on this question. Many believe it cannot be sowed too early. There is much in having a good start.

More About Apples.

An answer to the question, "Should we encourage new varieties of apples?" was sent in by Levi S. Reist, who was not present. It was read. Whether we should encourage new varieties depended on circumstances to a great extent; but the referee would at least recommend the cultivation of Smith's Cider, Imperial, Dominie, Wine Sap, Seek no Farther, Sheepnose, Baldwin and others.

Questions to be Answered at Next Meeting.

What is the best method to raise a good crop of corn? Referred to John J. Rush.

Should wheat be harrowed in the spring? Referred to John H. Landis.

Can commercial fertilizers be profitably used on the potato crop, and how can they be applied? Referred to H. M. Engle.

How should lime be applied; on the surface or plowed under? Referred to J. C. Linville.

There being no further business before the society, a motion to adjourn was made and carried.

THE POULTRY SOCIETY.

At the monthly meeting of the Lancaster County Poultry Association on Monday morning, March 6th the following were present: Secretary, J. B. Lichty, city; Charles Lippold, city; John Schum, city; F. R. Diffenderffer, city; W. W. Griest, city; A. S. Flowers, Spring Garden; J. M. Johnston, city; J. B. Long, city; Henry M. Engle, Marietta; H. S. Garber, Mount Joy.

In the absence of President Geyer, Charles Lippold presided.

Amendments to the constitution, authorizing the annual election to be held in February instead of January, and requiring the treasurer to report in February and August the money in his hands and at the February meeting make a detailed written report of receipts and expenditures, were adopted.

A communication from T. Frank Evans tendering his resignation as treasurer was read. The matter was postponed until the next meeting.

Fifty-nine members were reported in good standing.

FULTON FARMERS' CLUB.

The Fulton Farmers' Club met on Saturday, March 11th, at the residence of Joseph P. Griest, in Fulton township.

Mr. S. L. Gregg asked the question, "Which is the best paying crop for farmers in this section, wheat or corn?"

Joseph P. Griest said that considering that wheat was not so hard on the land as corn, and as it required less labor and brought more per bushel, he thought it was the better paying crop of the two.

Day Wood: We generally put on more fertilizers for wheat than for corn, but if we would manure them alike, corn is the more certain crop, and while it is seventy cents per bushel and wheat one dollar and twenty-five, corn pays much the best. Montillion Brown and some others coincided with Day Wood.

Day Wood asked: "Is wheat going to advance in price soon, or would it be as well to sell now?"

S. L. Gregg said he could not see what would make it advance, unless there should come a foreign demand. There is now a prospect of a large crop. The winter wheat looks well and they are already sowing spring wheat in the West, where they are likely to put in an unusual amount in consequence of the early spring.

Several others spoke of the reports being favorable to a large crop, and could see no reason for any permanent advance in price.

Montillion Brown asked: "What kind of fertilizers are those present going to apply for corn?" Nearly all answered, South Carolina rock. Thomas Stubbs said he had good reports of the result of using Orchilla guano in York county, and he would try it.

Melissa Gregg inquired: "Is a soap or meat vessel built of brick or stone and cemented, satisfactory?" Joseph P. Griest and Mary A. Stubbs, both reported having them in use for soap and they answered very well. C. S. Gatchell said he had seen meat salted in such a vessel and it answered well.

Rebecca D Kegg: "How many tomatoes can be raised on an acre?" None of those present had had any experience in field culture of this plant and therefore could only guess at the amount, and the guesses ranged all the way from 100 to 1,000 bushels.

E. H. Haines: "Do seedling peach trees live longer than grafted ones?" Wm. P. Haines had not found any difference. S. L. Gregg said he had not noticed much difference, but a neighbor of his had found the natural fruit to live longer and bear better than the grafted.

Joseph C. Stubbs plants his peach trees in the fence corners and allows the cattle to keep the tops eaten off for two or three years. He thinks that by keeping the tops back until the trees are well rooted they do better. He is opposed to cultivating peach trees, and cited an instance where the trees in an orchard had been cultivated, and they did not live as long nor do as well as some that were planted along the fence at the same time.

Thomas Stubbs said he had not noticed any difference between seedling and grafted trees under similar treatment, but trees that come up along the fences do better than cultivated ones. C. S. Gatchell had found seedlings to do much better for him than the grafted ones.

Priscilla Coates said her husband had planted ten acres in grafted peach trees and they bore five good crops and one inferior crop, and then died. He took the worms out of the roots twice each season, the first time about the first of June and then in the fall again and scattered some salt around the trees once a year.

After dinner the male portion of the meeting took a look at things in and around the barn, where they found quite a difference made in the stock since the meeting here a year ago. The host was then feeding cattle and his stables contained some very fine fat steers, now he is dealing in horses and mules, and this kind of stock has taken the place of the former. We were shown a pair of gray mules, well matched, and weighing 2,400 pounds, and several smaller pairs, besides several horses, the good qualities of which I leave the owner to tell to his customers when they call to see him.

After reassembling at the house some criticisms were made, generally favorable to the host. The shed over the barnyard had been improved and a field of wheat sown about the first of October had made an extraordinary growth.

Joseph P. Griest read from the Century Magazine a description of a machine which had been on exhibition at the Atlanta Exposition, and which is intended to destroy potato bugs and other insects by sprinkling poisons mixed with water on the plants infested. It consists of a barrel, mounted on a cart and having several elastic tubes attached, to the ends of which are fastened muzzles of peculiar construction for delivering the poisoned water to the under side of the leaves of the plants.

Montillion Brown read an article on protecting grapes from insects and also from the sun, in which the writer recommends placing small paper bags, such as are used by grocerymen, over the bunches as soon as they are of the size of peas and tying them around the stems. This led to some discussion on the question of shading grapes from the sun.

Joseph C. Stubbs said he knew of a vine that did no good until it was allowed to grow under the eaves of the house and then it yielded perfect fruit; and also of a Catawba vine that did much better after being taken from a trellis and allowed to grow on a tree.

Priscilla Coates recited "Some Day."

The following list of officers were selected to serve the club for one year: President, Wm. King; Secretary, E. H. Haines; Treasurer, Lindley King; Librarian, Day Wood.

The next meeting will be held at the residence of Lindley King, on the second Saturday in April.

THE LINNÆAN SOCIETY.

Twentieth Anniversary of the Founding of the Society.

The society met in the room of the Mechanics' Library, on Friday evening, February 24, 1882. In the absence of the President and both Vice Presidents, Dr. Knight was called to the chair; Dr. Davis in place as Secretary. After formal opening and collection of monthly dues, the following donations were recorded:

Museum.

A very superior specimen of Sulphuret of Iron, as it occurs in coal beds. This specimen is a transverse section of an odious mass five inches in diameter, and exhibited a brilliant fracture. Obtained and donated by the curators.

Library.

"Statutes of the United States," in three volumes, imperial octavo, 600 pages, exclusive of copious indices to each volume.

"Messages and Documents," for 1880 and 1881, pp. 1059. Royal octavo, from Department of Interior. Also from the same, "Circulars 4 and 5," of the Bureau of Education. Royal 8vo. of 250 pages.

Report of Silver Commission, vol. 2, pp. 511.

Proceedings of American Philosophical Society, from June to December, 1881, from the society.

Report of the Department of Agriculture for 1880 673 8vo. pages, copiously illustrated.

Report of the Silver Commission, vol. 2, pp 511, octavo. Hon. A. Herr Smith.

Bulletin of St. Louis Public School Library, and Sunday Book Catalogue and Circulars.

Historical.

Two envelopes containing 24 biographical, historical and scientific scraps.

Anniversary.

This was the twentieth anniversary meeting of the society, and it is lamented that it was so poorly attended, especially since evening meetings were adopted in order to suit the convenience of those who alleged that they could not attend a meeting held during the day. Dr. Rathvon read a paper on the origin and history of the society, which, on vote was requested to be published.

Science Gossip.

After half an hour's pleasant intercourse under this rule of order, the society adjourned to meet on the last Saturday in March, of which notice will be given by the Secretary.

History of the Society.

Dr. Rathvon's paper was as follows:

Mr. President: The first stated meeting of the Linnæan Society was held in February, 1862, just twenty years ago. Preliminary meetings had been held in January, but on the 8th of February its organic laws were adopted, its first board of officers elected, and the days and hours of its meetings fixed. It possessed nothing save the unmanifested intents and purposes of its members, and these consisted exclusively of the Committee on Natural Science, of the "Athenæum and Historical Society." Of that nucleus but three now reside in Lancaster—namely, Prof. Wickersham, J. B. Kevinski, and the narrator. Prof. T. C. Porter was the first president, and continued in office until his removal from Lancaster in 18—. Jacob Stauffer was the first record ing secretary, and continued in office until his death, in March, 1880, and I have been its first and only treasurer. Those who seem to be the most earnest in its organization, have either removed to other localities, have died, or have become lukewarm. Among those who have died, were some of its most active members and correspondents. Although I was a member of the original committee which finally culminated in the organization of the Linnæan, I must confess that I was little more than passive in it, for I had a foretaste that it meant labor, and would infer fere with the complete unity of my specialty in natural science. I had been a member of the "Marietta Lyceum," as early as 1837, and of the "Lancaster Conservatory of Arts and Sciences" in 1840, and I had seen both of these institutions disbanded for the want of working members.

Institutions of this kind need a goodly number of wealthy patrons, who are liberal men of leisure, as a sustaining element, as it is in England, France and Germany, and to some extent in our larger towns and cities, especially in Massachusetts—notably, men like Peabody, Thayer and Dr. Morton.

As it is, in this country, they are generally composed of men who are compelled to earn their bread by the sweat of their faces, and when leisure they are obliged to abate their energies, to meet their secular obligations.

When the Linnæan Society was first organized, its object was the development of the natural history of Lancaster county and adjacent territory. This seemed to be the object at least of the few original members who participated in its organization. I hardly think they fully comprehend the magnitude of the undertaking; for, from the very beginning it seemed more intent on a species of scientific pic-nicking or recreation. Possessing no peculiar endowment, it was compelled to sustain itself by monthly contributions, and even these, merely nominal as they were, came forth in a too feeble streams to irrigate the ground it proposed to cultivate to a prolific fruitage. This was not owing to the absence of material, for this has continued to increase gradually from its infancy down to the present moment. The great drawback *was*, and still *is*, the proper digestion of the material it possesses, a matter that requires time for its development more than money. The early summer seasons of the society were mainly devoted to limited scientific excursions, mostly within the county of Lancaster, on which occasion large amounts of material were collected, but much of it was neither scientifically nor systematically utilized. Out of these field meetings grew the "Tucquan" excursions, and these were composed mainly of men who had little sympathy with the original objects of the society. They were too large, unwieldy and expensive, and much of the material collected on those occasions was almost a "dead litter" in the museum, for the want of the necessary time to classify, arrange and label them. At first an attempt was made to catalogue and number them; but this work nominally devolved on the secretary, and he finally became discouraged, on account of the rapid accumulation. Had each specimen been numbered, labeled and catalogued, as it was presented, and by the individual or member who donated or presented it, the result would have been more satisfactory to all concerned, and the general usefulness of the society as a scientific object educator would have been far in advance of its present condition in that respect.

From the records of the society kept by the secretary, it appears that over one hundred and thirty active members, and over one hundred and ten correspondents of the society have been elected, from its organization in February, 1862, down to the present time. Had the one hundred and thirty odd who were elected active members paid their initiation fees of one dollar each, and the small dues of ten cents per month; and become merely contributing members, from the organization of the society down to the present date, their contributions would have amounted to over *twenty five hundred dollars*. From the treasurer's report at the last annual meeting, it appeared that the total income of the society from all sources, during the twenty years of its existence, only amounted to about *twelve hundred dollars*, and this included the amounts received from sale of stock, extra contributions of a few members, a few outside donations, and the monthly 25 cent dues, which prevailed for about three years, during the rebellion. Therefore, regarding these dues as legal obligations there are fully *fifteen hundred dollars* due the society from those who have been, from time to time, elected members of it. It is not generally expected that all the members of any association will become *active* to the same extent, or in the same sense; and they are no special hindrance to its progress if they are not so. But all who have voluntarily become such members, should at least contribute to its pecuniary support. Otherwise, it must languish, become ineffective, or fail.

Notwithstanding these hindrances to the progress of the society, whether they may be regarded as *real* or only *apparent*, it has for twenty years continued to accumulate a large amount of material at least; indeed, a larger and a more valuable amount than the members themselves have a correct knowledge of. When in the winter of 1837 a natural history society, under the auspices of Josiah Holbrook, was organized at Marietta, in this county, the said Holbrook stated, in an introductory, that the most essential element in developing a practical knowledge of natural science was a well ordered *museum* of natural objects; and that, however essential it was to possess a library of scientific books, a *museum* was of primary importance, for it brings the subject practically down to the comprehension even of the illiterate; for, said he, in effect, I hold in my hand a rhombic crystal of *calcareous spar*, which, after seeing and handling it, with ten minutes' instruction as to its chemical composition, its lustre, form and action under acids, the amateur may recognize as soon as he sees it again, whether he can read a description of it or not; and even if he *can*, there is no description, however scientific it may be, that will convey as correct an idea of what calcareous spar is, externally, as the object itself. Not any of the members knew anything about mineralogy, and little more about any other branch of natural science. Mr. H. furnished the society with a suit of minerals and metals, each specimen being about the size of a chestnut, for which, I think, he charged $20, and we thought them cheap. I could go into our storeroom and carry away a larger and better collection in my pantaloons pockets, and yet you would not know that any were missing. At the same rates our collection of minerals alone would be worth $20,000. It is true, that the prices of minerals have depreciated, but fine specimens, especially if rare, are as expensive now as they were forty years ago, simply because there is a greater demand for them. If the Linnæan Society could command the leisure and the pecuniary means to select from its duplicates, suits of minerals properly classified and labeled, and present them to every village or school district in the county of Lancaster, it would approximate its legitimate function as a central scientific organization and *medium* of development.

We hardly comprehend the real value of any department of our *museum*. We have the life-labors of two working botanists. A large collection of reptilia, and much more in undetermined paleontology and archæology than appears to the superficial observer—the largest collection in entomology in the State of Pennsylvania, outside of Philadelphia. Indeed, this collection is remarkable, in that it includes the collections (or what remains of them) of Professors Hentz and Haldeman, the former of which was cotemporary with the Melsheimers, Say, and Harris, the fathers of American entomology. In this collection may be recognized specimens collected by Prof. Hentz, nearly seventy years ago. Perhaps no larger and more diversified collection of local mineralogy than ours exists in Pennsylvania. In ichthyology and ornithology, of a local character, it is no mean representation of Lancaster county. Its historical collection could be very much augmented, if it possessed the facilities for a permanent and secure preservation. The collection in oology could be very much increased through exchange of duplicates, if we had any one to take an active hold of that department. Perhaps it lacks more in mammology than in any other of the conspicuous branches, but even in that department, it possesses as much as it has any room for illustration.

On the whole, so far as the matter relates to the rough material, the accumulations of the Linnæan Society during the twenty years of its existence, and under the peculiar circumstances in which the society has almost single-handed labored, has been progressing. What it needs most is more space for the proper classification and arrangement of its collections, a publishing fund, and a few more earnest workers, in order to make its collection useful and accessible to the public. If I had nothing else to do it would be the delight of my life to devote the remainder of my days on earth to making our *museum* a credit to the county of Lancaster and an auxiliary to our educational institutions.

Allow me to congratulate the society on the twentieth return of its anniversary, as being endowed with a longevity that was hardly anticipated when it was first organized, and at a period, too, that seemed inauspicious to the perpetuity of the Federal Union.

LITERARY AND PERSONAL.

TWO THINGS WORTH READING.—We have received from Ehrich Bros., of New York, a brace of interesting pamphlets, which are sent out as the *avant couriers* of the spring number of that well-known magazine of fashion, *Ehrichs' Fashion Quarterly*.

The first is the "Premium List" of the *Quarterly*, and tells its readers what good things they may secure by subscribing to the *Fashion Quarterly* themselves, and inducing others to do the like. Among other novelties in the way of premiums, we notice a choice selection of vocal and instrumental music, issued at forty cents per piece, one piece of which (as selected) is sent to every subscriber without exception. Among the premiums for clubs are some really beautiful sets of jewelry, which will be sent in return for four, six, or eight subscribers. Of course, the jewelry is not pure gold, but the Ehrichs say it looks just as pretty as if it were, and they ought to know.

The second pamphlet is entitled "Shopping in New York," and is intended to convey an idea of the plan and scope of the *Fashion Quarterly*, and we must say that if the spring number of the *Quarterly* only carries out the promises made for it by its forerunner, it will be a very complete magazine indeed. In this pamphlet of sixteen quarto pages, almost every department of a large New York retail store is represented to a limited extent. The careful mother finds in it a few standard styles of underwear; of children's clothing for both boys and girls; of wonderfully cheap embroideries; of hosieries; of window curtains, and a dozen other necessities of housekeeping: while the less thoughtful daughter will enjoy the jewelry, the laces, and the knick-knacks for room adornment which are spread before her. A blank form for sending orders occupies the last page, and fittingly completes the little book.

The *Fashion Quarterly* ought to have a large circulation, and we think it will. For only fifty cents a year the publishers offer four beautiful fashion books, issued at the beginning of the successive seasons, and give to each subscriber, as well, the privilege of making a selection from a list of choice music, every piece of which is retailed by the music dealers all over the country, at forty cents.

The *Fashion Quarterly* is published by Ehrich Bros., of Eighth Avenue, New York, who will send the pamphlets referred to, free, on application.

THE SOUTHERN CULTIVATOR AND DIXIE FARMER.—The February number of this splendid and deservedly popular farm, plantation and family journal is before us. For many years the leading agricultural journal in the South, it not only maintains its former high reputation under the new management, but augments it with every successive number. A glance at the broad, beautifully printed and illustrated pages, its numerous and harmoniously arranged departments, and its choice original and selected reading, will convince any one that the South has at last a truly representative agricultural periodical of which our people may well feel proud. Dr. W. L. Jones, the veteran editor and writer, continues to occupy the editorial chair, ably assisted by Dr. J. S. Lawton and a host of prominent writers in every department, among whom we notice, in this number, Prof. Wm. Browne, of the University of Georgia; Prof. Allen Curr, of Scotland; Col. D. T. T. Moore, founder of the *Rural New Yorker*, and the inimitable "Bill Arp," who is a regular contributor. Really no intelligent and progressive farmer or planter in the South can do without *The Southern Cultivator* without serious harm to his own interests. Subscribe for it at once. The price is only $1.50 a year. Jas. P. Harrison & Co., Atlanta, Ga., are the publishers. Club with exchanges, $1.25.

THE SUGAR BEET. Devoted to the cultivation and utilization of the Sugar Beet. Philadelphia, February, 1882, No. 1 of the third volume (or third year) of this royal quarterly has appeared, and fully sustains the reputation acquired by its predecessors. Published, as it is, at the low price of 50 cents a year, it ought to be accorded an extraordinary support, in order to sustain it in the noble work to which it is devoted. We not only commend the superior quality of the paper, the finely executed illustrations, and the literary ability of its letter press, but the "grit" and indomitable perseverance of the editor, in so ably advocating an interest so nearly allied to the health, the domestic comfort, and the pecuniary prosperity of the nation. The universal use of sugar, and the feasability of its production, all over our country, must ultimately result in the success of the enterprise in the near future, whatever may occur seemingly adverse to such a contingency in the present. There is no great interest of which our country is now reaping the advantages, that has not been "pooh poohed" and otherwise discouraged, at its initiation—notably, the steamboat, the locomotive, and the telegraph. So long as 1,803,000 acres of land in Europe are devoted to the cultivation of the sugar beet there ought to be no apprehensions in this country that it will not ultimately pay; and, if cold Canada and Russia are able to make it a success how much greater the prospect on the generous soil of Lancaster county.

THE MILLER'S REVIEW.—"Devoted to milling, millwrighting, and mill-furnishing;" a royal quarto of 16 pages. Published monthly by Henry L. Everett, 703 Walnut street, Philadelphia, Pa., at $1.00 per year. The first number of the first volume of this able journal has reached our *sanctum*, and we feel complimented that we have been deemed worthy of it, for it certainly "fills the measure of its department's glory," and not its glory only, but its substantial use. In his salutatory the editor says: "It has become a custom or rather a necessity to have each of the trades represented by its journal in that part of the country where such trades are carried on," and this is the case in nearly all occupations, professions, institutions, &c., whether civil, social, scientific or religious; and those who are in a condition to compensate such journals are usually in advance of all others. As farming and the productions of the farm constitute the foundation stones upon which all other occupations are erected, it necessarily ought to have the largest number, and the most liberally compensated journals. From the contents of this first number of the *Miller's Review*, we feel assured that it will be an able and faithful representative of its industrial interest, and ought to elicit a corresponding reward. If there is any trade that is second to agriculture, that position may be legitimately accorded to milling, for its object is to reduce the staple productions of agriculture to practical use. On the first page of this journal is a description and a fine illustration of Malvern Mills, located on the Pennsylvania Railroad about twenty miles west of Philadelphia. These are not the largest mills in Pennsylvania, only having the capacity to turn out 250 barrels of flour per day, but they are well ordered, and produce flour of the highest quality. But still, how insignificant, when compared with the mill that went into operation on the 2d of January last at Winona, Minn., which has the capacity of 1500 barrels per day.

Important to Grocers, Packers, Hucksters, and the General Public.

THE KING FORTUNE-MAKER.

OZONE
A New Process for Preserving all Perishable Articles, Animal and Vegetable from Fermentation and Putrefaction, Retaining their Odor and Flavor.

"OZONE—Purified air, active state of Oxygen."—*Webster.*

This preservative is not a liquid pickle, or any of the old and exploded processes, but is simply and purely OZONE, as produced and applied by an entirely new process. Ozone is the antiseptic principle of every substance, and possesses the power to preserve animal and vegetable structures from decay.

There is nothing on the face of the earth liable to decay or spoil which Ozone, the new Preservative, will not preserve for all time in a perfectly fresh and palatable condition.

The value of Ozone as a natural preserver has been known to our older chemists for years, but, until now, no means of producing it in a practical, inexpensive, and simple manner have been discovered.

Microscopic observations prove that decay is due to septic matter or minute germs, that develop and feed upon animal and vegetable structures. Ozone, applied by the Prentiss method, seizes and destroys these germs at once, and thus preserves. At our office in Cincinnati can be seen almost every article that can be thought of, preserved by this process, and every visitor is welcomed to come in, taste, smell, take away with him, and test in every way the merits of Ozone as a preservative. We will also preserve, free of charge, any article that is brought or sent prepaid to us, and return it in the sender, for him to keep and test.

FRESH MEATS, such as beef, mutton, veal, pork, poultry, game, fish, &c., preserved by this method, can be shipped to Europe, subjected to atmospheric changes and return to this country in a state of perfect preservation.

EGGS can be treated at a cost of less than one dollar a thousand dozen, and be kept in an ordinary room six months or more, thoroughly preserved; the yolk held in its normal condition, and the eggs as fresh and perfect as on the day they were treated, and will sell as strictly "choice." The advantage in preserving eggs is readily seen; there are seasons when they can be bought for 8 or 10 cents a dozen, and by holding them, can be sold for an advance of from one hundred to three hundred per ct. One man, with this method, can preserve 5,000 dozen a day.

FRUITS may be permitted to ripen in their native climate, and can be transported to any part of the globe. The juice expressed from fruits can be held for an indefinite period without fermentation—hence the great value of this process for producing a temperance beverage. Cider can be held perfectly sweet for any length of time.

VEGETABLES can be kept for an indefinite period in their natural condition, retaining their odor and flavor, treated in their original packages at a small expense. All grains, flour, meal, etc., are held in their normal condition.

BUTTER, after being treated by this process, will not become rancid.

Dead human bodies, treated before decomposition sets in, can be held in a mineral condition for weeks, without puncturing the skin or mutilating the body in any way. Hence the great value of Ozone to undertakers.

There is no change in the slightest particular in the appearance of any article thus preserved, and no trace of any foreign or unnatural odor or taste.

This process is so simple that a child can operate as well and as successfully as a man. There is no expensive apparatus or machinery required.

A room filled with different articles, such as eggs, meat, fish, etc., can be treated at one time, without additional trouble or expense.

In fact, there is nothing that Ozone will not preserve. Think of everything you can that is liable to sour, decay, or spoil, and then remember that we guarantee that Ozone will preserve it as perfectly in condition you want it for any length of time. If you will remember this it will save us asking questions as to whether will preserve this or that article—it will preserve anything and every thing you can think of.

There is not a township in the United States in which a live man can not make any amount of money, from $1,000 to $10,000 a year, that he pleases. We desire to get a live man interested in each county in the United States, in whose hands we can place this Preservative, and through him secure the business which every county ought to produce.

A FORTUNE
Awaits any Man who Secures Control of OZONE in any Township or County.

A. C. Powen, Marion, Ohio, has cleared $2,000 in two months. $2 for a test package was his first investment.
Woods Brothers, Lebanon, Warren County, Ohio, made $6.00 on a test package purchased in August and sold November 1st. $2 for a test package was their first investment.
F. K. Raymond, Morristown, Belmont Co., Ohio, is clearing $2,000 a month in handling and selling Ozone. $2 for a test package was his first investment.
D. F. Webster, Charlotte, Eaton Co., Mich., has cleared $1,000 a month since August. $2 for a test package was his first investment.
J. B. Gaylord, 81 La Salle St., Chicago, is preserving eggs, fruit, etc., for the commission men of Chicago, charging 11/2c. per dozen for eggs, and other articles in proportion. He is preserving 5,000 dozen eggs per day on his business is making $5.00 a month clear. $2 for a test package was his first investment.
The Cincinnati Feed Co., West 6th Seventh Street, is making $3.00 a month in handling brewers' malt, preserving and shipping it as feed to all parts of the country. Malt unpreserved sours in 24 hours. Preserved by Ozone it keeps perfectly sweet for months.

There are instances which we have asked in the privilege of publishing. There are scores of others. Write to any of the above parties and get the evidence direct.

Now, to prove the absolute truth of every thing we have said in this paper, **we propose to place in your hands the means of proving for yourself that we have not claimed half enough.** To any person who doubts any of these statements, and who is interested sufficiently to make the trip, we will pay all travelling and hotel expenses for a visit to this city, if we fail to prove any statement that we have made.

How to Secure a Fortune with Ozone.

A test package of Ozone, containing a sufficient quantity to preserve one thousand dozen eggs, or other articles in proportion, will be sent to any applicant on receipt of $2. This package will enable the applicant to pursue any line of tests and experiments he desires, and thus satisfy himself as to the extraordinary merits of Ozone as a Preservative. After having thus satisfied himself, and had time to look the field over to determine what he wishes to do in the future—whether to sell the article to others or to confine it to his own use, or any other line of policy which is best suited to him and to his township or county—we will enter into an arrangement with him that will make a fortune for him and give us good profits. We will give exclusive township or county privilege to the first responsible applicant who orders a test package and desires to control the business in his locality. **The man who secures control of Ozone for any special territory, will enjoy a monopoly which will surely enrich him.**

Don't let a day pass until you have ordered a Test Package, and if you desire to secure an exclusive privilege we assure you that delay may deprive you of it, for the applications come in to us by scores every mail—many by telegraph. "First come first served" is our rule.

If you do not care to send money in advance for the test package we will send it C. O. D., but this will put you to the expense of charges for return money. Our correspondence is very large; we have all we can do to attend to the shipping of orders and giving attention to our working agents. Therefore we can not give any attention to letters which do not order Ozone. If you think of any article that you are doubtful about Ozone preserving remember we guarantee that it will preserve it, no matter what it is.

REFERENCES.

We desire to call your attention to a class of references which no enterprise or firm based on any thing but the soundest business success and highest commercial merit could secure.

We refer, by permission, as to our integrity and to the value of the Prentiss Preservative, to the following gentlemen: Edward C. Royce, Member Board of Public Works, E. O. Eshelby, City Comptroller; Amor Smith, Jr., Collector Internal Revenue; Wulsin & Worthington, Attorneys; Martin H. Harrell and M. J. Hopkins, County Commissioners; W. S. Cappeller, County Auditor; all of Cincinnati, Hamilton County, Ohio. These gentlemen are each familiar with the merits of our Preservative, and know from actual observation that we have, without question,

The Most Valuable Article in the World.

If you invest in a test package, it will surely lead you to secure a township or county, and then your way is absolutely clear to make from $1,000 to $10,000 a year.

Give your full address in every letter, and send your letter to

PRENTISS PRESERVING COMPANY, (Limited,)
No. 8 (Cor. Ninth & Race Sts., Cincinnati, O.

THE LANCASTER EXAMINER

OFFICE

No. 9 North Queen Street,

LANCASTER, PA.

THE OLDEST AND BEST.

THE WEEKLY

LANCASTER EXAMINER

One of the largest Weekly Papers in the State.

Published Every Wednesday Morning,

Is an old, well-established newspaper, and contains just the news desirable to make it an interesting and valuable Family Newspaper. The postage to subscribers residing outside of Lancaster county is paid by the publisher.

Send for a specimen copy.

SUBSCRIPTION:

Two Dollars per Annum.

THE DAILY

LANCASTER EXAMINER

The Largest Daily Paper in the county.

Published Daily Except Sunday.

The daily is published every evening during the week. It is delivered in the City and to surrounding Towns accessible by railroad and daily stage lines, for 10 cents a week.

Mail Subscription, free of postage—One month, 50 cents; one year, $5.00.

THE JOB ROOMS.

The job rooms of THE LANCASTER EXAMINER are filled with the latest styles of presses, material, etc., and we are prepared to do all kinds of Book and Job Printing at low rates and short notice as any established in the State.

SALE BILLS A SPECIALTY.

With a full assortment of new cuts that we have just purchased, we are prepared to print the finest and most attractive sale bills in the State.

JOHN A. HIESTAND, Proprietor,

No. 9 North Queen St.,
LANCASTER, PA.

WHERE TO BUY GOODS IN LANCASTER.

BOOTS AND SHOES.

MARSHALL & SON, No. 12 Centre Square, Lancaster, Dealers in Boots, Shoes and Rubbers. Repairing promptly attended to.

M. LEVY, No. 3 East King street. For the best Dollar Shoes in Lancaster go to M. Levy, No. 3 East King street.

BOOKS AND STATIONERY.

JOHN BAER'S SONS, Nos. 15 and 17 North Queen Street, have the largest and best assorted Book and Paper Store in the City.

FURNITURE.

HEINITSH'S, No. 15½ East King st., (over China Hall) is the cheapest place in Lancaster to buy Furniture. Picture Frames a specialty.

CHINA AND GLASSWARE.

HIGH & MARTIN, No. 15 East King st., dealers in China, Glass and Queensware, Fancy Goods, Lamps, Burners, Chimneys, etc.

CLOTHING.

MYERS & RATHFON, Centre Hall, No. 12 East King St. Largest Clothing House in Pennsylvania outside of Philadelphia.

DRUGS AND MEDICINES.

G. W. HULL, Dealer in Pure Drugs and Medicines Chemicals, Patent Medicines, Trusses, Shoulder Braces, Supporters, &c., 15 West King St., Lancaster, Pa

JOHN F. LONG & SON, Druggists, No. 12 North Queen St. Drugs, Medicines, Perfumery, Spices, Dye Stuffs, Etc. Prescriptions carefully compounded.

DRY GOODS.

GIVLER, BOWERS & HURST, No. 25 E. King St., Lancaster, Pa., Dealers in Dry Goods, Carpets and Merchant Tailoring. Prices as low as the lowest.

HATS AND CAPS.

C. H. AMER, No. 39 West King Street, Dealer in Hats, Caps, Furs, Robes, etc. Assortment Large. Prices Low.

JEWELRY AND WATCHES.

H. Z. RHOADS & BRO, No. 4 West King St., Watches, Clock and Musical Boxes. Watches and Jewelry Manufactured to order.

PRINTING.

JOHN A. HIESTAND, 9 North Queen st., Sale Bills, Circulars, Posters, Cards, Invitations, Letter and Bill Heads and Envelopes neatly printed. Prices low.

Thirty-Six Varieties of Cabbage; 26 of Corn; 28 of Cucumber; 11 of Melon; 37 of Peas; 28 of Beans; 17 of Squash; 25 of Beet and 8 of Tomato, with other varieties in proportion, a large portion of which were grown on my few seed farms, will be found in my Vegetable and Flower Seed Catalogue for 1882. Sent free to all who apply. Customers of last season need not write for it. All Seed sold from my establishment warranted to be fresh and true to name, so far, that should it prove otherwise, I will refill the order gratis. The original introducer of Early Ohio and Burbank Potatoes, Marblehead, Early Corn, the Oakland Squash, Marblehead Cabbage, Phinney's Melon, and a score of other New Vegetables, I invite the patronage of the public. New Vegetables a specialty.

JAMES J. H. GREGORY,
Marblehead, Mass.

Nov-6mo]

EVAPORATE YOUR FRUIT.
ILLUSTRATED CATALOGUE
FREE TO ALL.
AMERICAN DRIER COMPANY,
Chambersburg, Pa.
Apl-tf

FARMING FOR PROFIT.

It is conceded that this large and comprehensive book, (advertised in another column by J. C. McCurdy & Co., of Philadelphia, the well-known publishers of Standard works,) is not only the newest and handsomest, but altogether the BEST work of the kind which has ever been published. Thoroughly treating the great subjects of general Agriculture, Live-Stock, Fruit-Growing, Business Principles, and Home Life; telling just what the farmer and the farmer's boys want to know, combining Science and Practice, stimulating thought, awakening inquiry, and interesting every member of the family, this book must exert a mighty influence for good. It is highly recommended by the best agricultural writers and the leading papers, and is destined to have an extensive sale. Agents are wanted everywhere. jan-lt

CIDER MILLS!
Wine Presses!

Fruit Presses, Apple Slicers,
Fodder and Ensilage Cutters,
GRAIN FANS,
Grain and Fertilizer Drills,
Broad-cast Seed Sowers,
Corn Shellers, Corn Mills,
Grain Mills, etc., etc.

FOR SALE BY

D. LANDRETH & SONS,
AGRICULTURAL AND HORTICULTURAL IMPLEMENT

AND

SEED WAREHOUSE,

Nos. 21 and 23 South Sixth Street,

BETWEEN MARKET AND CHESTNUT STS.,
— and —
No. 4 ARCH STREET,

apr-6m   PHILADELPHIA.

MERCHANT TAILORING.

1848 (The Oldest of All.) 1881

RATHVON & FISHER,

MERCHANT TAILORS AND DRAPERS,

respectfully inform the public that having disposed of their entire stock of Ready-Made Clothing, they now do, and for the future shall, devote their whole influence to the CUSTOM TRADE.

All the desirable styles of CLOTHS, CASSIMERES, WORSTEDS, COATINGS, SUITINGS and VESTINGS constantly on hand, and made to order in plain or fashionable style; prompt y, and warranted satisfactory.

All-Wool Suit from $18.00 to $30.00.
All-Wool Pants from 3.00 to 10.00.
All-Wool Vests from 2.00 to 6.00.

Union and Cotton Goods proportionately less. Cutting, Repairing, Trimming and Making, at reasonable prices.

Goods required by the year to those who desire to have them made elsewhere.

A full supply of Spring and Summer Goods just opened and on hand.

Thankful for a generous public for past patronage they hope to merit its continued recognition in their "new departure."

RATHVON & FISHER,
PRACTICAL TAILORS,
No. 101 North Queen Street,
LANCASTER, PA.
1848   1881

GLOVES, SHIRTS, UNDERWEAR.
SHIRTS MADE TO ORDER,
AND WARRANTED TO FIT.
E. J. ERISMAN,
56 North Queen St., Lancaster, Pa.
-79-4-12]

A HOME ORGAN FOR FARMERS.

THE LANCASTER FARMER,

A MONTHLY JOURNAL,

Devoted to Agriculture, Horticulture, Domestic Economy and Miscellany.

Founded Under the Auspices of the Lancaster County Agricultural and Horticultural Society.

EDITED BY DR. S. S. RATHVON,

TERMS OF SUBSCRIPTION:

ONE DOLLAR PER ANNUM,

POSTAGE PREPAID BY THE PROPRIETOR.

All subscriptions will commence with the January number, unless otherwise ordered.

Dr. S. S. Rathvon, who has so ably managed the editorial department in the past, will continue in the position of editor. His contributions on subjects connected with the science of farming, and particularly that specialty of which he is so thoroughly a master—entomological science—some knowledge of which has become a necessity to the successful farmer, are alone worth much more than the price of this publication. He is determined to make "The Farmer" a necessity to all households.

A county that has so wide a reputation as Lancaster county for its agricultural products should certainly be able to support an agricultural paper of its own, for the exchange of the opinions of farmers interested in this matter. We ask the co-operation of all farmers interested in this matter. Work among your friends. The "Farmer" only one dollar a year. Show them your copy. Try and induce them to subscribe. It is not much for each subscriber to do but it will greatly assist us.

All communications in regard to the editorial management should be addressed to Dr. S. S. Rathvon, Lancaster, Pa., and all business letters in regard to subscriptions and advertising should be addressed to the publisher. Rates of advertising can be had on application at the office.

JOHN A. HIESTAND,
No. 9 North Queen St., Lancaster, Pa.

$5 TO $20 per day at home. Samples worth $5 free.
Address STINSON & Co., Portland, Maine.
jun-1yr*

ONE DOLLAR PER ANNUM.—SINGLE COPIES 10 CENTS.

THE LANCASTER FARMER

DEVOTED TO AGRICULTURE, HORTICULTURE, DOMESTIC ECONOMY AND MISCELLANY

THE FARMER IS THE FOUNDER OF CIVILIZATION.—WEBSTER.

Dr. S. S. RATHVON, Editor.   LANCASTER, PA. APRIL, 1882.   JOHN A. HIESTAND, Publisher.

Entered at the Post Office at Lancaster as Second Class Matter.

CONTENTS OF THIS NUMBER.

EDITORIAL.
Ensilage, ... 49
April Meeting, ... 49
Souls in Gardens, ... 50
Kitchen Garden for April, ... 50
Phenomenal, ... 51
Eating Between Meals, ... 51
Excerpts, ... 52

ESSAYS.
Fruit and Vegetables—their Culture, ... 53
The Bright Side of Horticulture, ... 54
Horticultural Fertilizers, ... 55
What are Best and Cheapest and How Applied.

SELECTIONS.
The New Wheat Region, ... 55
How to Deodorize Stables, ... 56
Utilizing Rough Ground, ... 56
The Building of Homes, ... 56
When to Cut Grass, ... 57
Feeding Poultry and Raising Chicks, ... 57
Vegetable condiments, ... 57
Trichinosis, ... 57
Testing Cream, ... 58
Application of Liquid Manure, ... 58
Early Price of Pennsylvania Lands, ... 58
A Home Fruit Canning Factory, ... 59

OUR LOCAL ORGANIZATIONS.
Fulton Farmers' Club, ... 59
    March Meeting—April Meeting.
Linnæan Society, ... 60
    Museum—Library—New Business.

ENTOMOLOGICAL.
Swarming Ants and Allied Phenomena, ... 60
Curculio in Plum Culture, ... 61
Birds and Canker Worms, ... 61

AGRICULTURE.
Sowing the Seed, ... 61
Clover and Grass, ... 61
Clover, ... 61
Ploughing, ... 61
Potatoes, ... 61
Onions, ... 61

HORTICULTURE.
The Rhubarb Plant, ... 61
The Mulberry Trees, ... 61
An Excellent Old Apple, ... 62
An Experiment in Potato Planting, ... 62

HOUSEHOLD RECIPES.
To make a Cheap Wash or Paint, ... 62
Rice, Milanaise Style, ... 62
Macaroni and Ham, ... 62
Poor Man's Plum Pudding, ... 62
Fig Pudding, ... 62
Yorkshire Pudding, ... 62
Warm Slaw, ... 62
Cold Slaw, ... 62
Lincoln Cake, ... 62
Pantry, ... 62
To Clean Marble, ... 62
Valuable Hints, ... 63
Cocoanut Cookies, ... 62
To Renovate Black Grenadine, ... 62
To Wash Silk Stockings, ... 62
Corn-Starch Cake, ... 63
Black Bean Soup, ... 63
To Clean Musty Barrels, ... 62
Cottage Gingerbread, ... 62
Household Weights and Measures, ... 62
Scotch Butter Candy, ... 62

LIVE STOCK.
Sawdust for Bedding, ... 62
Salting Stock, ... 62
Floors for Horse Stables, ... 63
Charcoal for Sick Animals, ... 63
The Hog Crop, ... 63
Tying Up Calves, ... 63
Man's Treatment of the Horse, ... 63
Advantages of Small Flocks, ... 63
"Loss of Cud," ... 63
Training Heifers to Milk, ... 63
Bedding for Cows, ... 63
Inoculation of Animals, ... 63

POULTRY.
Sunflower Seed for Poultry, ... 64
Grain and Vegetables, ... 64
Poultry Upon the Farm, ... 64
Dressing and Keeping Poultry, ... 64
Common Sense in the Poultry Yard, ... 64
The Roup in Fowls, ... 64
Poultry, ... 64
Literary and Personal, ... 64

SEND IN YOUR SUBSCRIPTIONS
—FOR—
THE FARMER
FOR 1882.

The cheapest and one of the best Agricultural papers in the country.

Only $1.00 per year.

JOHN A. HIESTAND, Publisher,
No. 9 North Queen st., Lancaster, Pa.

Eggs! Eggs!

From all the leading varieties of pure bred Poultry Brahmas, Cochin, Hamburgs, Polish Game, Dorking and French Fowls, Plymouth Rocks and Bantams, Rouen and Pekins Ducks. Send for Illustrated Circular.

T. SMITH, P. M., Fresh Pond, N. Y.

feb-3m

WE WANT OLD BOOKS.
WE WANT GERMAN BOOKS.

WE WANT BOOKS PRINTED IN LANCASTER CO.
We Want All Kinds of Old Books.
LIBRARIES, ENGLISH OR GERMAN BOUGHT.
Cash paid for Books in any quantity. Send your address and we will call.

REES WELSH & CO.,
27 South Ninth Street, Philadelphia.

LIGHT BRAHMA EGGS
FOR HATCHING,
$1.50 FOR SETTING OF 13.
ALSO,
Three Barrels of Chicken Manure
FOR SALE.
L. RATHVON,
Examiner Office, No. 9 N. Queen st., Lancaster, Pa.

$66 a week in your own town. Terms and $5 outfit free. Address H. HALLETT & Co., Portland, Maine.

Will be mailed FREE to all applicants, and to customers without ordering it. It contains five colored plates, 600 engravings, about 200 pages, and full descriptions, prices and directions for planting 1500 varieties of Vegetable and Flower Seeds, Plants, Fruit Trees, etc. Invaluable to all. Send for it.
D. M. FERRY & CO., Detroit, Mich.

Jan-4m

$66 a week in your own town. Terms and $5 outfit free. Address H. HALLETT & Co., Portland, Maine.
jun-1yr*

PENSIONS For SOLDIERS, widows, fathers, mothers or children. Thousands yet entitled. Pensions given for loss of finger, toe, eye or rupture, varicose veins or any disease. Thousands of pensioners and soldiers entitled to INCREASE and BOUNTY. PATENTS procured for inventors. Soldiers land warrants procured, bought and sold. Soldiers and heirs apply for your rights at once. Send 2 stamps for "The Citizen-Soldier," and blanks and instructions. Free to all. We can refer to thousands of Pensioners and Clients. Address N. W. Fitzgerald & Co., U. S. Claim Att'ys, Lock Box 585, Washington, D. C.

Jan

LIGHT BRAHMA EGGS

For hatching, now ready—from the best strain in the county—at the moderate price of
$1.50 for a setting of 13 Eggs.
L. RATHVON,
No. 9 North Queen st., Examiner Office, Lancaster, Pa.

WANTED.—CANVASSERS for the
LANCASTER WEEKLY EXAMINER
In Every Township in the County. Good Wages can be made. Inquire at
THE EXAMINER OFFICE,
No. 9 North Queen Street, Lancaster, Pa.

$72 A WEEK. $12 a day at home easily made. Costly outfit free. Address TRUE & Co., Augusta, Maine.
jun-1yr*

SEND FOR
SPECIAL PRICES

On Concord Grapevines, Transplanted Evergreens, Tulip, Poplar, Linden Maple, etc. Tree Seedlings and Trees for timber plantations by the 100,000.
J. JENKINS' NURSERY,
3-2-79   WINONA, COLUMBIANA CO., OHIO.

MARBLEHEAD
Early Sweet Corn

Is the most profitable of all, because it matures before any other kind, giving farmers complete control of the early market. I warrant it to be at least a week earlier than Minnesota, Narragansett or Crosby, and decidedly earlier than Dolly Dutton, Tom Thumb or Early Beginner. Of size of Minnesota, and very sweet. The original introducer, I send pure stock, postpaid, per package 15 cents; per quart, 50 cents; per peck, by express, $3.00. In my catalogue, (free to all) are emphatic recommendations from farmers and gardeners.

JAMES J. H. GREGORY,
apr-3t   Marblehead, Mass.

THE LANCASTER FARMER.

PENNSYLVANIA RAILROAD SCHEDULE.
Trains leave the Depot in this city, as follows:

WESTWARD.	Leave Lancaster.	Arrive Harrisburg.
Pacific Express*	2:40 a. m.	4:05 a. m.
Way Passenger†	5:00 a. m.	7:30 a. m.
Niagara Express	11:00 a. m.	11:20 a. m.
Hanover Accommodation	11:05 p. m.	Col. 10:40 a. m.
Mail train via Mt. Joy	10:20 a. m.	12:40 p. m.
No. 2 via Columbia	11:25 a. m.	12:55 p. m.
Sunday Mail	10:50 a. m.	12:40 p. m.
Fast Line*	2:30 p. m.	3:25 p. m.
Frederick Accommodation	2:15 p. m.	Col. 2:45 p. m.
Harrisburg Accom.	5:45 p. m.	7:40 p. m.
Columbia Accommodation	7:00 p. m.	Col. 8:20 p. m.
Harrisburg Express	7:30 p. m.	9:40 p. m.
Pittsburg Express	9:50 p. m.	11:10 p. m.
Cincinnati Express*	11:50 p. m.	12:45 a. m.

EASTWARD.	Lancaster.	Philadelphia
Cincinnati Express*	2:55 a. m.	3:00 a. m.
Fast Line*	5:08 a. m.	7:45 a. m.
Harrisburg Express	8:05 a. m.	10:00 a. m.
Columbia Accommodation	9:10 p. m.	12:00 p. m.
Pacific Express*	10 p. m.	3:40 p. m.
Sunday Mail	2:00 p. m.	5:00 p. m.
Johnstown Express	3:05 p. m.	5:30 p. m.
Day Express*	5:15 p. m.	7:20 p. m.
Harrisburg Accom.	6:25 p. m.	9:30 p. m.

The Hanover Accommodation, west, connects at Lancaster with Niagara Express, west, at 9:35 a. m., and will run through to Hanover.
The Frederick Accommodation, west, connects at Lancaster with Fast Line, west, at 2:10 p. m., and runs to Frederick.
The Pacific Express, east, on Sunday, when flagged, will stop at Middletown, Elizabethtown, Mount Joy and Landisville.
*The only trains which run daily.
†Runs daily, except Monday.

EDW. J. ZAHM,
DEALER IN
AMERICAN AND FOREIGN
WATCHES,
SOLID SILVER & SILVER PLATED WARE,
CLOCKS.
JEWELRY & TABLE CUTLERY.

Sole Agent for the Arundel Tinted

SPECTACLES.

Repairing strictly attended to.

ZAHM'S CORNER,
North Queen-st. and Centre Square, Lancaster, Pa.
79-1-12

E. F. BOWMAN,
Watches & Clocks
AT LOWEST POSSIBLE PRICES,
Fully guaranteed.
No. 106 EAST KING STREET,
79-1-12 Opposite Leopard Hotel.

ESTABLISHED 1832.

G. SENER & SONS,
Manufacturers and dealers in all kinds of rough and finished
LUMBER,
The best Sawed SHINGLES in the country. Also Sash, Doors, Blinds, Mouldings, &c.
PATENT O. G. WEATHERBOARDING
and PATENT BLINDS, which are far superior to any other. Also best COAL constantly on hand.
OFFICE AND YARD:
Northeast Corner of Prince and Walnut-sts,
LANCASTER, PA.
79-1-12]

PRACTICAL ESSAYS ON ENTOMOLOGY,
Embracing the history and habits of
NOXIOUS AND INNOXIOUS
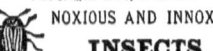
INSECTS,
and the best remedies for their expulsion or extermination.
By S. S. RATHVON, Ph. D.
LANCASTER, PA.
This work will be Highly Illustrated, and will be put in press (as soon after a sufficient number of subscribers can be obtained to cover the cost) as the work can possibly be accomplished.
79-2-

$5 to $20 per day at home. Samples worth $5 free. Address STINSON & Co., Portland, Maine

---

TREES
Fruit, Shade and Ornamental Trees.
Plant Trees raised in this county and suited to this climate. Write for prices to
LOUIS C. LYTE,
Bird-in-Hand P. O., Lancaster co., Pa.
Nursery at Smoketown, six miles east of Lancaster.
79 1-12

WIDMYER & RICKSECKER,
UPHOLSTERERS,
And Manufacturers of
FURNITURE AND CHAIRS.
WAREROOMS:
102 East King St., Cor. of Duke St.
LANCASTER, PA.
79-1-12]

Special Inducements at the
NEW FURNITURE STORE
OF
W. A. HEINITSH,
No. 15 1-2 E. KING STREET
(over Burak's Grocery Store), Lancaster, Pa.
A general assortment of furniture of all kinds constantly on hand. Don't forget the number.
15 1-2 East King Street,
Nov-1y] (over Burak's Grocery Store.)

For Good and Cheap Work go to
F. VOLLMER'S
FURNITURE WARE ROOMS,
No 309 NORTH QUEEN ST.,
(Opposite Northern Market),
Lancaster, Pa.
Also, all kinds of picture frames. nov-1y

GREAT BARGAINS.
A large assortment of all kinds of Carpets are still sold at lower rates than ever at the
CARPET HALL OF H. S. SHIRK,
No. 202 West King St.
Call and examine our stock and satisfy yourself that we can show the largest assortment of these Brussels, three plies and ingrain at all prices—at the lowest Philadelphia prices.
Also on hand a large and complete assortment of Rag Carpet.
Satisfaction guaranteed both as to price and quality. You are invited to call and see my goods. No trouble in showing them even if you do not wish to purchase. Don't forget this notice. You can save money here if you want to buy.
Particular attention given to customer work.
Also on hand a full assortment of Counterpanes, Oil Cloths and Blankets of every variety. [nov-,yr.

PHILIP SCHUM, SON & CO.,
38 and 40 West King Street.
We keep on hand of our own manufacture,
QUILTS, COVERLETS,
COUNTERPANES, CARPETS,
Bureau and Tidy Covers, Ladies' Furnishing Goods, Notions, etc.
Particular attention paid to customer flag Carpet, and scouring and dyeing of all kinds.
PHILIP SCHUM, SON & CO.,
Nov-1y Lancaster, Pa.

THE HOLMAN LIVER PAD,
Cures by absorption without medicine.
Now is the time to apply these remedies. They will do for you what nothing else on earth can. Hundreds of citizens of Lancaster say so. Get the genuine at
LANCASTER OFFICE AND SALESROOM,
22 East Orange Street.
Nov-1yr

C. R. KLINE
ATTORNEY-AT-LAW,
OFFICE: 15 NORTH DUKE STREET,
LANCASTER, PA.
Nov-1y

NORBECK & MILEY,
PRACTICAL
Carriage Builders,
COX & CO'S OLD STAND,
Corner of Duke and Vine Streets,
LANCASTER, PA.

THE LATEST IMPROVED
SIDE-BAR BUGGIES,
PHÆTONS,
Carriages, Etc.
THE LARGEST ASSORTMENT IN THE CITY.
Prices to Suit the Times.
REPAIRING promptly attended to. All work guaranteed.
79-9-

S. B. COX,
Manufacturer of
Carriages, Buggies, Phaetons, etc.
CHURCH ST., NEAR DUKE, LANCASTER, PA.
Large Stock of New and Second-hand Work on hand very cheap. Carriages Made to Order. Work Warranted or one year. [1--9-1y

The Lancaster Farmer.

Dr. S. S. RATHVON, Editor.   LANCASTER, PA., APRIL, 1882.   Vol. XIV. No. 4.

EDITORIAL.

ENSILAGE.

"Whatever is worth doing at all, is worth doing well," and this rule applies in a very special sense to the silo and ensilage. It is true, the subject may have been overestimated, or excessively lauded; and it may also have been underestimated, and needlessly disparaged; but these circumstances ought not to militate against the subject as a fundamental principle that is capable of practical illustration. There is a vast difference between *sauerkraut* and *rotten cabbage*, practically; although theoretically, they may be synonymous. Well-made and well-preserved saur-kraut, is healthful and nutritious, and millions in the world subsist upon it during a great part of the year, and would experience a great deprivation, if not a great distress, if they could not obtain it. But rotten cabbage is unhealthful, enervating, and the source of stomach cramps, inflammation of the bowels, diarrhœa, and possibly death. Doubtless, for lack of skill, there is a vast deal of this spurious saur-kraut made and consumed. Now, practically, ensilage is nothing more nor less than a sort of saur-kraut, made on a large scale, as food for cattle. If skillfully manipulated, and systematically preserved, cattle will eagerly appropriate it, and thrive upon it. The maker of good and nutritious saur-kraut, is careful to exclude it from the corroding and corrupting action of the air, and in proportion as he succeeds in this, in that degree will he have good and healthful saur-kraut. The same rule obtains in ensilage. The mere opinions of men, either *pro* or *con* on this subject, must go for just what they are worth, whatever may be their standard of intelligence; and even where ensilage has been tried and succeeded, or failed, in its results, it may not be sufficient to establish the system as a fixed *fact* in the domain of domestic economy; much less when such failure is the result of an insufficient or illy constructed silo.

It has been alleged that cattle—or at least some cattle—will not voluntarily appropriate the contents of the silo, and thrive upon it, unless it is mixed with other kinds of food, as a sort of condiment; but, admitting this to be the case, it does not "settle the question." No one eats his saur-kraut "naked," but on the contrary, he always accompanies it with pork, or bacon; but this does not prove that saur-kraut is not nutritious. We presume that the gastronomical faculty in all animated nature is subject to cultivation—indeed we see it in many animals, (including even insects) as well as in man. And especially in ensilage has it been demonstrated during the past year, that cattle, which at first rejected it, ultimately became the most fond of it. This is nothing new.

Mr. S. S. Spencer, whose model farm and dairy is in near proximity to the western suburbs of the City of Lancaster, thoroughly tested the silo last year, and in every respect the results were satisfactory—indeed, more than realizing all he anticipated. But his silo was mechanically constructed, and on philosophical principles. His cattle consumed every shred of its contents, and looked, and longed for more; and this year his silo will be increased, at least ten-fold. He considers that there is no risk whatever in it, if the silo is properly constructed and intelligently manipulated. But this is not all; his cattle thrived upon it, and produced more and better milk, cream, and butter, than on any other kind of food. Had we heard or read such testimony coming from an unknown source, in view of the conflicting experiences and opinions published on the subject, possibly we might have received it with a large margin of doubt; but we have it from Mr. Spencer himself, and we *know* him to be an intelligent witness, and one not likely to continue long in self-deception. True, it might not just suit every man to have a silo, nor might it in every case be profitable. Perhaps to a man possessing only a single cow it might not be as profitable as it would be to the man who owns a dozen cows; and yet if a quarter or a half barrel of saur-kraut can be preserved in a sound and healthy condition, there seems to be no good reason why the same quantity of ensilaged fodder could not be equally as easily kept. When canned fruits and vegetables were first introduced as an adjunct to domestic housekeeping, immense quantities were utterly spoiled and had to be thrown away every year, because of the inexperience of those who manipulated it. But this is not the case now; and hence, we find the shelves of every grocery filled with a great variety of that which in the beginning was deemed impracticable to preserve. Experience will doubtless work the same results in regard to ensilage and the silo.

APRIL MEETING.

For aught that we know, and for aught that the Daily and Weekly papers knew—at least, for aught t at they uttered in their columns, there was no meeting of the *Agricultural and Horticultural Society*, on the first Monday (3d day) in April. True, there is a law of the Society, that when New Year's day, First of April, Fourth of July, and any other public holiday, occurs on the *first* Monday in the month, then the stated meeting of that month shall be held on the *second* Monday in that month, and it shall be so stated, or proclaimed, at the preceding meeting; and, the object of said proclamation was for the purpose of making a record of it in the proceedings of the Society, and published with the other proceedings, in order to enable the members to act intelligently on the subject. But, no such statement or proclamation was made at the meeting held in March last, unless it was "suppressed" by the Reporters, or omitted in the proceedings.

This may have been neglect, forgetfulness, or inadvertence—we do not think it was the result of indifference—but, under any circumstances, it was contrary to that good order which should distinguish such a society, or any living society.

When it is necessary to omit, or dispense with a meeting in any society, and especially in one which is a "body politic in law," a record should be made of such omission or dispensation, to complete the chain of its existence; otherwise, it will only add to the apathy, indifference, or lukewarmness, into which societies are liable to lapse, when no attention is paid to their organic laws.

Attending the stated meetings of the society, and an active participation in its proceedings, are twin relatives to contributing to the literary columns of its representative journal; neither of which should be neglected or ignored. There are societies and journals in this country which have been in existence for more than half a century, and it is doubtful whether any one who had ever been connected with them, had ever been socially, morally, or intellectually injured by such connections. They may have had onerous duties imposed upon them, or may even have sustained pecuniary loss, but they have been compensated by a conciousness of having done something—or having *endeavored* to do something—for the benefit of mankind.

Man, in his normal condition, is essentially a gregarious—or, perhaps more properly—a social creature, and when any number of men are banded together under organic laws, for the advancement of legitimate objects by legitimate means, they will only be ultimating on earth those qualities which will fit them for a more useful hereafter; for, it seems totally at variance with all we know of the word and the works of God, that that *rest* which all seem to covet in the hereafter, is a state of inactivity, or slothful indolence. When a man *prefers* habitual solitude—all other things being equal—depend upon it, there is something deranged in his mental or moral organization. He can work out his moral salvation more effectually through social intercourse, than he can through solitary seclusion. Those who are incapacitated for social intercourse, through physical or mental infirmaties, are more to be commiserated, than enters into the imagination of the general run of men. Therefore, all institutions founded upon principles of equity and usefulness, should keep their organization intact.

Another great hindrance to the progress and the thrift of societies is, that so few enter their inner temples; the larger number are content to hang on the "ragged edges" of their outer borders. They take a merely temporary and irresponsible view of them instead of regarding them as permanent fixtures. A half-willed membership can only make a half-hearted organization, whatever the object of the society may be. "Whatever is worth doing at all, is worth doing well," applies as forcibly to social organization, as it does to any other human vocation. Take away all our social and organic institutions and we shall have a poor, miserable and selfish state of society.

It is true, there are many social evils in the world, but it is doubtful whether these will bear the least comparison with solitary evils. Moral influence is as likely to obtain through social proximity, as immoral. It is certain that no great interest can be as effectually advanced by individual effort, as it can by a unity of action.

"United we stand—divided we fall."

SNAILS IN GARDENS.

Dr. Rathvon:—Please tell us in your able monthly, The Lancaster Farmer, what can be done to prevent snails from destroying vegetation in gardens. I have a garden, and in some places they destroy the vegetation. Our cellar is also infested with them, where they destroy articles of food, flowers put there for wintering, the labels on fruit jars, and now they are making their slimy trails upon the kitchen floor. We had heard that salt strewed over the ground would dissolve them, but it has been tried by my wife, and did not remedy the evil. She is every few days asking, "What shall I do to prevent the destruction of my plants, and the cellar contents, by the nasty, slimy snails?" I hope you will be able to tell her in your next number of The Farmer what to do. They are beginning to eat off the flower stalks in the garden now already, as fast as they peep out of the ground.—J. F. W., Lancaster, March 24, 1882.

Salt is a very ancient, and generally considered a sovereign remedy for the extinction of snails; but its power is no more effectual in snails than as a means to catch birds, unless it is actually "dropped on their tails." We never knew an instance in which a snail survived a copious application of salt. Of course, one grain, or a dozen grains of salt might not hurt them much, but under a generous application of the saline mineral they invariably yield; but the salt, to be effectual at all, must come in actual and free contract with them. They will not cross a trail of salt, but they will "flank it," if they can.

Snails are not very rapid in their locomotion, and hence are easily surprised, and when surprised, they make no attempt to escape. They move cautiously, protruding their tentaculæ, and feeling their way.

Snails are great night prowlers; "they love darkness rather than light;" that, is their great advantage, and they freely avail themselves of it; but if, in their peregrinations, their tentaculæ come in contact with salt, they quickly draw them in, and that is the only quick motion they are capable of. Make a solid circle of salt around any plant you wish to preserve, and no snail will approach it. That is, however, only a preventive; if you wish to kill a snail outright, you must put the salt on its body in sufficient quantity. But salt does not dissolve them. They are very sensitive, and secrete a great deal of mucous or slime; that is their life-function. The salt contracts their pores, or organs of secretion, and hence they die, but the contracted body of the animal is still there, although greatly diminished in size.

"Salt-peter and burnt brandy;" ammonia; common lye; a strong infusion of tobacco, or a solution of whale-oil soap, will answer as well as salt, in destroying snails. Pulverized tobacco, (snuff) coal oil; spirits of turpentine; pulverized lime, and many other substances, including London-purple, Paris-green and Pyrethrum—we doubt not, would be equally efficacious, but they must be brought into actual contact with the body of the offending animal.

Some species of snails are very prolific; on one occasion we counted one hundred and fifteen under a flat stone, not more than six inches square. These were from a grain of powder to a buckshot in size—one of the many species belonging to the genus Helix, and we have found the common shelless species almost as numerous.

Our premises were at one time seriously infested with snails, and, we think, we finally extinguished them by starvation and salt. Snails must eat "for a living," and if they can get nothing to eat they die. In the house we closed every aperture from which a snail could possibly emerge, and those that could not conveniently be closed we salted, and if they issued from their cover at night at all, it must have been outside and not inside of the house.

Snails in gardens, may be taken in various kinds of traps, and then destroyed with hot water.

If pieces of board, say a foot square, are distributed through the garden, not pressed down too closely into the soil, the snails will gather under them before the sun begins to shine on them in the morning. They do not like sunshine. Any other object that will afford a hiding place may be as good as a board. If these are carefully examined during the day and the snails killed there will eventually be an end of them. This should be vigilantly continued as long as any snails are captured.

Cellars, with rough walls and numerous recesses, afford many hiding places for snails. Cellars should receive at least one good coat of plaster, to prevent the harboring of rats, mice, and snails in their walls. Old neglected cellars are prolific breeding places for snails, and other nocturnal species of vermin. One person may be greatly annoyed by snails, and his next door neighbor may have none of them, because he may have no harboring places for them.

Snails have many natural enemies that feed on them; chickens, birds, ducks, skunks and pigs, are very fond of them. They are also destroyed by night-roving beetles, and their larvæ. Both the imago and the larvæ of the Lampyridæ or "Fire-fly" family.—(Lightning-bugs) prey upon them. We have known a species of Telephorus to attack the common Helix, and "clean out" the whole shell. Mr. Hensel informs me that he saw nearly one hundred of these insects attack a large species of Helix in his garden, and they did not abandon it until they had eaten out the whole animal, and left nothing but the empty shell.

In mentioning the word shell, be it understood, that there are some species of snails that never have a shell, or if they have, it is too small to be readily seen. Others have a small scale-like shell, that does not seem to be of any use, as a protection to the animal; but others again have an ample and well developed shell, and into which they retire, whenever they are molested. Finally, In France and Germany, they breed, feed and rear snails for the table, and they esteem them as much as we do oysters. Why not take a hint from this? one is as much of a mollusk as the other.

KITCHEN-GARDEN FOR APRIL.

In the Middle States, now is the time to plant and sow, if we would hope to reap. Those of us who do not avail of the present, need not expect to profit in the future.

The exact time, however, in which certain seeds should be sown must depend not only on location in respect to latitude, but also on the nature of the soil; if it be heavy, a little delay will rather promote than retard our object. It is impracticable, under any circumstances, to always give undeviating directions—the common sense of each one must be brought in requisition.

Asparagus sow ; or plant roots, if not attended to last month. This vegetable is now coming into season. Wherever practicable, a bed of sufficient size should be permitted to afford an ample supply without cutting every feeble root which peeps above the surface ; indeed, wherever space and means admit, two beds should be maintained, and cut alternate seasons. The colossal asparagus appears to sustain its reputation. Beans, bush or bunch, sow. Broccoli, "Large Early White," is very fine. Beets, early and long, sow. Cabbage, Drumhead and Flat Dutch, sow freely, that there be enough for the fly and to plant ; also other sorts described in catalogues, which will afford an uninterrupted succession, so desirable in every country family. Carrots, Early Horn, and Long Orange, sow. Cauliflower, late, sow. Celery, sow, if not sown last month. Cress, sow. Cucumber, Early Frame, sow, in warm spot. Horse-radish, plant, if not done. Hot-beds attend to. Leek, sow. Lettuce sow in drills, also plant from beds of last autumn's sowing. Marjoram, sweet, sow. Mustard, for salad, sow. Nasturtiums, sow. Onions, Buttons, for table use, plant, and sow thickly for sets. Parsley, sow. Parsnips, sugar, sow. Peas, early and late, for a succession, sow. Potatoes, plant plenty of the Early Rose for the main supply during summer and autumn. Radish, Long Scarlet, and white and red turnip, sow, if not already sown ; also the Golden Globe and White Summer, for succession. Salsify, sow. Sage, sow or plant. Tomato, sow, to succeed those sown in hot-beds. Spinach, Bloomsdale, sow at short intervals. Thyme, sow or plant.

Turnips sow, if not sown last month—they may succeed. In short, this is the season for the main sowing and planting in the middle States. A small expenditure of time will yield large results.—Landreth's Rural Register.

The next thing in importance to timely sowing and planting, is good seeds, if it is not an absolute pre-requisite: for, we cannot reasonably expect to reap aught except that we sow. It is also a matter of some importance to sow such seeds as are adapted to the soil and the latitude of the locality where they are sown. No prudent man would plant lemons, oranges and bananas in the Arctic regions.

Landreth's "original sealed packages" of seeds, are perhaps the most reliable and convenient form in which they could possibly be presented to the public, and these can be obtained not only of Landreths themselves, but at almost every seed store in the Union. Of course, there are also other good seedsmen and good seeds in the country, but it would appear natural that Pennsylvanians, and especially Lancaster countians, should obtain seeds grown within their own belt of latitude. Those living in a district where certain seeds, fruits, roots and tubers are cultivated and matured, would be more likely to succeed with such, than with those grown and matured in a different latitude ; and this relates not only to latitude, but also in a greater or lesser degree to longitude. There are still some people who have an aversion to—if not a prejudice against "Book-farming," and "Book-gardening ;" and those people will condescend to follow the directions on a package of seeds (just as they follow the directions on a package or bottle of patent medicine) who would not go to the trouble of looking into a book ; and these "sealed packages" may furnish about all the garden literature that such people will patronize.

But, that is not all there is of it. There are many people who absolutely have not the time to pore over a book, even if they possessed one: or, if in the form of a periodical, it may not have been bound, and the particular number wanted may not just be at hand; hence when the season is at hand, and they possess a sealed package of seeds, with special directions for its use, they will know exactly what to do with it.

PHENOMENAL.

White-Spotted Tobacco.

Mr. Morris Gerschel recently left at the Lancaster *Intelligencer* office a leaf of tobacco that was very peculiarly marked with white spots and tracings, some of the lines being as beautifully curved and zigzagged as if they had been worked by some deft embroiderer. We handed the leaf to Dr. S. S. Rathvon, of the LANCASTER FARMER, with a request that he would examine and report upon it. He kindly furnishes the following paper :—*Ed. Intelligencer.*

White-Spotted Tobacco.

White spotted, like white-veined tobacco, is a phenomenon that comes distinctly within the scope of vegetable physiology ; and is the effect of a subtle cause or causes, about which there are various theories and opinions, even among intelligent and experienced tobacco growers. I am sure I noticed white-spotted cigars more than fifty years ago, and they were generally esteemed the best of cigars. I supposed then, that the spots had been produced by artificial means, because there were peculiar brands and boxes of them, all of which were more or less spotted—if, indeed, they were not fashionable. Perhaps, before the tobacco grower deplores the presence of white spots, he should submit his leaf to competent manufacturers, in order to learn to what extent the weed is injured by the presence of white spots. Perhaps, after all, the spots may be as conventional as those which sometimes occur on Berkshire pigs; which, although depreciating the extrinsic value of the animal, cannot possibly injure the quality of the pork.

Neither white-spotted, nor white-veined tobacco is therefore anything new, and may be present every year in some part of the country where tobacco is grown, although there may be certain years in which it may be more abundant than in other years.

It would be fortunate for the tobacco grower if spotted tobacco and spotted cigars could be raised to the dignity of fashion, provided the spotted crop and the fashion were coincidental events ; it would afford the dealers less opportunity to get the goods at prices below their real value, merely on account of the spots.

Being a physiological question then, the subject can only be elucidated through physiological laws and principles, and this being the case, I confess the subject is "too high for me ;" because, having no practical experience in vegetable physiology, I could, at best, only advance the experiences and theoretical deductions of others, with very limited corroborations of my own. I may be permitted to say, however, that the difference between white spots and white veins may not be so great as appears from a superficial view of the subject.

For instance, we cannot prick our bodies anywhere with a fine needle, but what a small globule of blood will exude from the wound; and this is also the case with succulent vegetation. If we closely examine a skeletonized leaf or plant, we will find that the whole disk is penetrated by innumerable nerves, nervures and nervelets, all of which have their absorbing and secreting functions ; so that we could hardly puncture it anywhere without rupturing one or more of these delicate organs ; hence, if the phenomenon is the effect of enervated circulation, which has been brought about by drouth or other weather contingencies, or by soil conditions, as is alleged, it is as likely to manifest itself among the smaller nervures as among the larger ones.

I have now before me a tobacco leaf from Mr. Morris Gerschel, of the firm of Gerschel & Bro., tobacco packers, also three leaves from Mr. William Roeting, of Elizabethtown, all of which are singularly marked with white, leathery spots, dots, rings and zigzag lines ; some of them like miniature streaks of "chain lightning," or Egyptian hieroglyphics ; and there does not seem to be any visible connection between these markings and the "mid rib," the lateral ribs, or any of the prominent veins or nervures ; hence it cannot be classed with white-veined tobacco.

Whilst manipulating these leaves in a moist condition, in order to expand them, I was particularly impressed with their peculiar fragrance, especially those from Mr. Roeting, which are also smaller in size and darker in color than that from Mr. Gerschel. And I cannot conceive that they are really injured by these peculiar markings, although they may affect their market value. On one of the leaves from Mr. Roeting the markings are much bolder, broader and more emphatic than the markings of any of the other leaves --indeed, no two of them are alike, the whole presenting an almost kaleidoscopic variety, and, if such effects could be produced artificially, I don't see why tobacco might not be cultivated as an ornamental plant, the same as the colias, calladiums and begonias, or the variegated grasses, and such like.

By the introduction of certain chemical substances into the soil, botanical experimenters have been able to produce visible effects upon the leaves and flowers of plants ; and, it is very probable that these markings have been produced through a diversion, or wrong direction of the sap-flow. A similar diversion or misdirection of the fluid circulation of plants, including trees and shrubbery, often develops discoloration of the leaves, protuberances, concavities, curled leaf, wrinkles, excrescences and numerous other outward manifestations. Analogous effects are produced by minute insects, in the form of an endless variety of galls, and also by bacterium fungi. As already intimated, I cannot hazard a theory of even sufficient outline to embrace this subject in a specific sense, and probably it will never be understood until some progressive tobacco culturist or vegetable physiologist discovers how to produce these markings by artificial cultivation, or chemical applications.

In conclusion allow me to refer the reader to page 26 (January number) of the LANCASTER FARMER for 1882, where may be found a paper by E. K. Hershey, of Creswell, Lancaster county, on the causes of "white veins in tobacco," which Mr. H. very plausibly regards as a *disease* engendered by meteorological conditions, operating upon the soil, or a sort of starvation of the plants through the prevailing droughts. In the same number of the FARMER, page 29, is a paper on the same subject, read by Mr. Hebron Herr before the Lancaster County Agricultural and Horticultural Society at its January meeting, 1882, in which Mr. Herr takes ground similar to that of Mr. Hershey, and nearly all who participated in the discussion coincided in sentiment with the essay. Of course both papers present mainly theories with such shadows of fact as their experiences have been able to reflect upon the subject, but I think both papers point in the right direction to the cause the phenomenon, and I am pretty sure that can point no nearer specifically. Some of them has come under my observation on such a complicated subject. The three city dailies of that date all contain Mr. Herr's paper.

EATING BETWEEN MEALS.

If your children are disposed to be greedy and desire food between meals, reason with them on the subject. A woman who has even a very superficial knowledge of the working of the stomach, can explain it to her child in such a way that it will make a strong impression upon his mind. To represent to an imaginative child that the stomach is like a man who, when he has eaten his breakfast, goes to work upon that with all his might, and who does not rest till he has ground the food up, and given the good part to the blood, so feeding each portion of the body, not forgetting the fingers and toes even, and who rejects all the bad, keeping you from sickness and pain will awaken intense interest in the child's mind and be a great aid to obedience. Put it before him, and ask him if it is not unkind and even cruel to give out another task before the first is finished and a little time for rest been given. It will help you greatly in enforcing it upon his mind that he must not eat at irregular intervals. A child's digestive organs may be weak ; he may need to eat more frequently than a grown person, but it should invariably be at some stated time. Cake or pastry should be given him but seldom, if at all ; there is nothing which is more ruinous to the digestive organs.

We have often been "poo pooed" for entertaining, and acting, on principles akin to those expressed in the above paragraph, by persons of acknowledged *wisdom*, on all other subjects—persons of no mean moral and intellectual standing, either. We were in harmony with them on pretty much all subjects save eating. No man, no matter how healthy, robust or physically strong he is, can afford to impose upon his organization by forming a habit of indiscriminate eating. Nor can he do so with impunity, for he will be drawing drafts upon his physical constitution that must some day be paid, whether that day matures "sooner or later." True, he may claim that "he *knows* himself"—that he knows exactly what to do, and what to leave undone—in which he is liable to be mistaken —and that with *him* it makes no difference at all when, where, what and how he eats. As a nation, it is said that Americans eat too rapidly; do not sufficiently masticate their food; eat too irregular, and too large a quantity at a time. Be this as it may, it is not the *quantity*, nor yet always the *quality* of the food we eat, that furnishes the greatest nourishment to the human system; but the amount that is assimilated; and assimilation is a process that is subordinate to physiological laws; and, whatever hinders, retards, or subverts the proper execution of those laws, in a degree defeats assimilation; and, instead of food being a physical support, it becomes a physical burden. Man may accustom himself to eat but one meal a day (although we hardly think it advisable) or two, three, four, or even five meals in a day; provided he eats them at regular periods, in reasonable quantity, and properly masticated, and also eats nothing between meals, the different number of daily meals may not materially affect the general health of the man. In other words, if it requires ten pounds of food, daily, to support a man, it perhaps would be little matter whether he ate it in *one* meal, or *five* meals, provided he ate it at regular periods, and did not exceed the daily amount of ten pounds. But if he ate ten pounds twice, thrice, four times, or five times a day, when he really only required ten pounds in all, it seems clear that he could not possibly assimilate it, and hence all over the ten pounds would be a "dead weight," and hence an injury.

Judging from outward appearances alone, it would seem that there are many people in the world who do not recognize, or are perfectly oblivious to the *fact*, that the human body is governed by fixed laws—that it is, as it were, a *machine*, which will endure for a longer period, and more efficiently, when *properly used*, than it will when *improperly abused*; and perhaps the habit of eating *between* meals is more injurious than "gormandizing" at meals. If the human stomach was endowed with speech, it doubtless would make many "awful complaints" against tyrannical masters, for the heavy burdens imposed upon it. If it were a copper-lined tub, into which all manner of food-substances were thrown—Dutch cheese, bologna sausages, salted herring, mince pies, orange peal, fat pork, cocoa crumbs, raw onions, saur-kraut, and a hundred other things—almost every moment during the day, expecting to form healthy food out of it, there would be as little *good* to hope from it, as there is from the heterogeneous mass of "all sorts" that is forced into the human stomach every hour of the day, by many untimely and indiscriminate eaters. Perhaps the health, happiness and prosperity of all animated nature, including the human family, depends more upon eating than upon any one other thing in life's operations, and perhaps, also, man is the only being that does not eat habitually, "in decency and in order"; and yet, it is a self-evident *fact*, that he cannot long exist usefully without eating. We are admonished that we should "eat and drink to the glory of God," and it is, perhaps, the only thing in which we are the least subordinate to the laws of God. We are not willing to allow the stomach rest, or time for digestion and assimilation, but continue to impose upon it one burden after another in rapid succession, before it has had time to dispose of its first burden.

Of course, something depends upon human exercise and occupation. The man who has much rugged, active and wearing exercise, is likely to have a better appetite, better digestion and assimilation than the man of sedentary or indiscriminate habits, and what might affect the latter injuriously, would have a different effect upon the former; but we cannot always know the evil effects of abusive eating, or any other debauch, immediately after indulgence in it. These evil effects sometimes only manifest themselves long after the wanton indulgence, when the physical energies become weakened by age. Then, when we have learned through stern experience, we wonder we could not have seen the folly while we were yet in the prime of life.

EXCERPTS.

Dr. M. S. Leslie, of Lexington, Ky., says that the best remedy in ordinary hiccoughs is about twenty-five grains of common table salt placed in the mouth and swallowed with a sip of water.

The pine forests discovered lately in the Province of Ontario are now estimated to yield 24,000,000,000 feet of timber, which can easily be brought to market.

A few drops of ammonia added to a gallon of water and applied once a week to all pots of flowers will do much good and keep the pots and earth from souring.

Never wear a good woolen dress into the kitchen without the protection of a large apron.

No flannel that has not been carefully washed, and is not perfectly soft and fleecy, should ever touch the skin of an infant.

Your fat must be boiling when you put your meat in it to fry.

"Middlings" flour contain the best elements of wheat.

Slow and long cooking will make tough meat tender.

There were zoological gardens in China more than two thousand years ago.

The use of green or damp fuel of any sort is very unprofitable. A large amount of the heat which it would yield, if dry, is absorbed and lost in the evaporation of the sap or moisture.

Dr. Luton, of Rheims, reports that he has discovered that the ergot of rye associated with phosphate of soda produces on those to whom it is administered a hilarious excitement similar to that which is brought on by laughing gas.

When oil is heated sufficiently in a closed vessel, from which the air is excluded, it turns to gas, which occupies many times the bulk that the oil did. This is the philosophy of pop-corn and explains its tendency to burst into curious forms on being roasted.

At a meeting of the American Association of Window Glass Manufacturers lately, in Washington, the product of the past year was reported to have been nearly 2,250,000 boxes, valued at about $6,000,000. The demand for consumption has taken the entire product.

The Prefetti of Italy have been instructed to use all their influence to prevent laborers from going out to Panama. There are more men on the isthmus already than can be employed and it is the intention of those in charge of the canal excavation to depend mainly on Chinamen.

Chloral hydrate has proved an efficient antidote in several recently reported cases of severe poisoning by belladona in England.

Charred corn is one of the best things which can be fed to hens to make them lay, not as a regular diet, but in limited quantities each day.

It costs but very little per head to raise sheep. Not only will they pay as farm stock, but as is commonly known, sheep restore fertility to land.

Anybody can have grape vines by cutting them properly. Trim off a portion of the old vine and leave a bud at each end. Stick one end in the ground and it will take root.

Strawberry plants should be set out as early in the season as possible in order to avoid a dry spell. Nothing is so fatal to young plants as lack of moisture after being transplanted.

In breeding fowls for eggs, use the Leghorns, Houdans, Black Spanish and Hamburgs—non-setters. For market chicks, the Plymouth Rocks, Brahmas and Cochins should be tried.

Those feeding bran should not lose sight of the fact that wheat ground by the "new process" converts the best part of the bran into flour, and what is left is of but little value as cattle food.

One cow well fed and comfortably cared for will produce quite as much milk and butter as two that are allowed to run at large, lie on the wet ground and be subject to the exposure of the weather.

An application of 100 pounds of nitrate of soda to an acre of wheat, where the crop looks weak, will show its benefit in a few days, not only improving it in growth but largely increasing the yield.

Ensilage is nothing new, as far as preservation is concerned. It is simply keeping green food through the winter by the exclusion of air. It is on the same principle as canning, only on a larger scale.

Onions are the first vegetables that get in the ground. The land should be very rich. They can be grown in the same place every year, as they are very nearly equally proportioned in the constituent elements derived from the soil.

Alum water is recommended for preventing bugs and worms from infesting flour mills. Dissolve two pounds of alum in three quarts of warm water and supply with a brush to crevices where insects may be concealed.

The best disinfecting agents, according to Mr. W. M. Hamlet, are in general those capable of exerting an immediate and powerful oxidizing action, and that it is active oxygen, whether from the action of chlorine, nitric oxide, or hydrogen peroxide, which must be regarded as the greatest known enemy to bacterial life.

Gas-fitters have recently made a most useful application of photography. They photograph the gas flames given by different burners or jets so that a customer can see if the shape and form of a light will suit him before he gives his order. As the flames are moreover depicted "life-size" the purchaser can always tell whether his jet is up to the standard.

It is suggested, with a view of avoiding the bursting of water-pipes by freezing, to make them elliptical in section. As the water expands to form ice, it will alter the shape of the pipe, causing it to become circular in section, and therefore giving more room for the ice. It is proposed to squeeze the pipes into their original shape, when, by a succession of frosts, they have been rounded.

Les Mondes reports that M. Bufourcet has in the exposed court of his house two bars of iron planted in the earth, to each of which is fixed a conductor of coated wire terminating in a telephonic receiver. He consults the apparatus twice or thrice every day, and it never fails through its indications of earth currents to give notice of the approach of a storm twelve or fifteen hours before it actually arrives.

In Reimann's process for rendering cloth water proof the fabric is passed slowly by machinery through a tank divided into three compartments, the first containing a warm solution of alum, the second a warm solution of lead acetate, and the third pure water, which is constantly renewed. The cloth on passing from the latter is brushed and beaten to remove the salt adhering to the surface, and finally hot-pressed and brushed. In this case lead sulphate is deposited on the fibres.

Steel tools should never be heated either for forging or tempering, in a fresh fire, unless it be charcoal. If coke is not at hand the fire should be allowed to burn until all the gas is burned out of the coal before the steel is introduced.

Some farmers think that a cow must eat all the time when confined in the stall. It is a poor economy which puts fresh hay into a manger on the top of older hay. A little tossing of the hay left in the manger will dry it and make it seem of renewed niceness to cows or horses.

Occasional sowing of little patches of ground with mustard, green peas, oats, etc., will do much to assist in keeping a cow on a small farm. They grow quickly and the same land will give several crops. A little discretion in this way will save much expense as to keep as well as furnishing early green feed.

Green peas are early crops. Most persons prefer the dwarfs, but the tall varieties yield better. A fault with the dwarfs is that they furnish families growing them with such few pickings. This is because they ripen nearly all at once. The better plan is to put them in the ground at intervals for a succession of crops.

What a Railroad Car will Hold.

Taking 26,000 pounds as a fair average load the ordinary railroad freight car will hold: Corn, 450 bushels; barley, 400 bushels; oats, 800 bushels; rye, 400 bushels; wheat, 423 bushels; bran, 1,000 bushels; flaxseed, 960 bushels; apples, 360 bushels; potatoes, 480 whisky, 60 barrels; salt, 70 barrels; flour, 90 barrels; flour, 200 sacks; cattle, 13 head; hogs, 50 head; sheep, 30 head; hard wood, 6 cords; soft wood, 7 cords; solid boards, 6,000 feet; shingles, 40,000; hard lumber, 20,000 feet; green lumber, 15,000 feet; joist, scattering and large lumber, 4,000 feet.

ESSAYS.

FRUIT AND VEGETABLES.—THEIR CULTURE.*

MR. PRESIDENT:—As we have met once more to have a friendly talk on fruit, its qualities and cultivation, I will give some of my experience. With persons of experience as well as with beginners, it is very difficult to know what to select, as there are so many kinds in the market and each person thinks he he has the best varieties. There is also such a quantity of fine fruit brought in for sale as to surprise one into wondering where it all comes from; but, considering the thousands of persons that are in the business the quantity of each kind is not so great after all. There is too much that is only passable which spoils the sale of the finest quality, for in twenty years experience I found about only ten per cent of my customers were willing to pay a fair price for fine fruit; they all preferred it, at the same time a cheaper fruit sold best. It is the common and poor fruit that is the most expensive, but most persons will not believe it. It is the quantity they want. Now if we did not have to pay so dear, for our experience, we could afford to sell fine fruit cheaper with a profit, but when one must wait years for the trees to bear and then be disappointed with the fruit, it is poor encouragement. Many persons will sell their fruit for just what they can get while others top and graft their trees, which has been my way with most of the trees I have bought. It was often discouraged and thought I was the only one cheated by tree agents, but find others were in the same boat as myself. Some years ago a nurseryman in our town received several hundred trees from a New York firm with most of the labels lost, he sold them as they were. I bought a dozen and grafted all but one (the finest in the lot) and that turned out to be nothing but the poorest kind of a seedling. The worst of it was, there were others heard that he was selling them cheap, who bought them all just as he sold them, and when persons asked them for certain trees "O yes, they had them in the nursery," they would label them as they wanted them for their customers. I have seen at least fifty of them come into bearing and nearly all worthless. The trouble is we have too many tree agents; all anxious to sell and not at all conscientious about what they sell; for, as one told me, "that before the trees came into bearing, the labels would be lost, or the property change owners and no one would know where they came from or what they were bought for. I have bought more poor trees than good, but as I said before when the fruit is not good, I either plant other trees or graft them. When I have small fruit that does not suit my soil I drop them out of cultivation and keep up with those that do. Out of fifteen or twenty kinds of strawberries I have laid aside all but three, namely: "Charles Downing," "Wilson's Seedling" and the "Sharpless,"

*Essay read before the Pennsylvania State Horticultural Association, at Harrisburg, by John C. Hepler, Reading, Pa.

which beats all I ever had for size, but not for quantity. I have also discarded all raspberries but the "Philadelphia," "Clarke" and the old "Antwerp." The two last named went back on me last summer, whether on account of the drought or not I cannot say. In regard to currants. I have very few beside the Cherry currant which bear exceedingly well and sell for two cents a quart more than the smaller ones, to those of my customers who appreciate good fruit. One of them (a gentleman) was persuaded to buy some grape vines which were to be something extra. I planted with great care and watched them until last summer, when the one that should have been the Lady grape, was one of Rogers' light grapes, and the others were only common dark grapes, not to be compared with the Concord or Union Village. The latter in my estimation is a No. 1 grape as it will remain on the vines longer than the Concord without dropping. Among apples, the Baldwin, Spitzenberg, Smokehouse Krouser, Northern Spy, Greening and Golden Pippin are the best for winter marketing, as they always sell and give entire satisfaction. I would not advise the keeping of too many varieties as they need considerable attention. As a general rule tenant farmers pay too little attention to their fruit. Instead of studying its nature and trying to improve it, they allow it to take its own course, whether from carelessness or ignorance I know not. I frequently see apples for sale that are quite out of season and would pay the owners much better if they kept them a while longer, even at the risk of some decaying. In our section last year the apple crop was a failure. Judging from the display at our County Fair, what few trees did bear, produced fine fruit. I have observed that when we have an extra heavy crop of apples and pears one year, the next is very light, for then the trees do not produce the bearing spurs but require all their strength for the fruit. The "year off" as we call it, is their rest; and then the bearing spurs are made for the following year. This you will notice holds good in all cases, unless we have too much heavy rain in the blossom season, which will destroy the crop.

We had no peaches in our neighborhood. Instead of the blossom being frozen, I think the trees were exhausted from overbearing the year before. Having two or three trees exposed to the northern winds and storms, which bore a few peaches, proves that the blossoms were not frozen. Although we are in the coldest place in the neighborhood, I never saw the peach trees make such a vigorous growth, and I think simply because they had rested from fruit-bearing last year. It is my belief that we will have a plentiful crop this year. If not out of place I would like to recommend our farmer friends to plant trees along the road and at intervals along their line and partition fences. Fruit trees have been suggested, but I would not recommend them, as our boys, and even men, do not respect personal rights to such property and a man would be worried more than the fruit would be worth. Walnut trees, if planted at a distance of fifty feet and a little attention given them at first to start them straight, in six or eight years they will bear, and the nuts will pay for the gathering. In fifty years the trees will be worth as much as the farm, providing they have been trained to grow straight. I am sure, he that can imagine a fine row of trees around his farm, will start to planting this winter; as the nut must be planted in the fall, so that the shell will burst by the action of the frost to give the germ a chance to grow. To him who has low or bottom land I would advise the planting of shellbark hickory, as it is becoming very scarce, and is in good demand. I know of a man, when in the spoke manufacturing business, he used on an average two hundred acres of hickory wood a year, and that he bought trees worth ten dollars each; which proves they are worth the little attention they need in the outset. There is one more tree I would like to bring to the notice of those who do not fancy the walnut. The tulip poplar, that fine, majestic tree which naturally towers to the skies and is the admiration of all, especially when in blossom. They are fast growers and will repay the planter. Most of our nurserymen have them for sale, from ten to twenty-five dollars per hundred. The larger size are the best to plant, as it is so much time gained. I have bought some very fine trees of Hoopes Bros. and Thomas, whom I know will be pleased to serve any who may call on them. If the walnut alone is not preferred, it may be used alternately with the poplar, which will make a novel appearance, and provide a chance for observing the difference in the growth and habits of both. There is so much to say about forest trees; but as that is not my subject, and I have already trespassed on time, I will hasten on.

In the vegetable kingdom, we do not have so many improvements as in fruit, and therefore there is not so much to say on this subject. We have many varieties of vegetables of the same kind ; still, a cabbage is a cabbage, and so a radish is a radish ; but there are the early and the late varieties, the strong and the mild, the large and the small, the sweet and the sharp. To take each one separately would be tedious, so I will only say that during my experience I have found more in the name than in the reality ; also, that good, fresh vegetables are a luxury. Most of them when brought to our markets are too stale for use, especially those brought from abroad. As a general thing our home gardeners (who do bring us fresh vegetables) are not sufficiently paid for their labor, as it costs too much to raise them ; besides being some weeks later than those brought from a distance, and prices by that time are low. Every Berks county farmer has a garden, some a very large one, and all extra vegetables they sell, is so much gained ; but whether the soil is adapted to gardening is seldom considered, if only their vegetables *grow*, it matters not *how*. I do think two-thirds of our market gardens are out of place. In my observation I find that most persons will take any place they can get, regardless of soil, location or water, and embark in raising vegetables; consequently, after several years' trial they fail and become discouraged in their work and seek some other occupation. Were we dependent on the market gardeners alone, we would often fall short. But thanks to those farmers who always have what they call "luck," we do not fall short. I think a person wishing to engage in the truck business should select low ground near running water, for then the summer drought would not affect them so much, as the night air is damp and the mist from the water all help to dispel the dry atmosphere and assist vegetation. After location comes the selection of seeds, which is a very important part. If you want early vegetables, it would not do to buy and plant late kinds that take nearly all summer to grow. I have planted to try the difference of seed in red beets, and have found that the early beet could be taken up at the end of eight or ten weeks, while the "Long Late" require five months, and when the late variety were in their prime, the early variety were tough and stringy. So with cabbage, what is intended for early must be planted early or it is a failure. Many persons think a bean is a bean, as far as planting is concerned, but that is a mistake ; all have their time, quality and hardiness. Some will stand a frost, while others by their side will be killed by it. As above mentioned, I think the selection of the seed quite as important as the ground. To be *successful* it is necessary to be well posted on the time of planting and what kind to plant first. I do not wish to advertise, but would say that Mr. P. Henderson's book on "Gardening for Profit," is the best I have ever seen and read ; even then one must take one's own soil and location into consideration, for his location does not suit all ; at the same time I would advise any person wishing to raise vegetables for profit to get the book, as it is a good guide. This having been a poor season and vegetables at a high price, many may be tempted into the business, but unless they are well acquainted with their soil, seed and location, they will fail. At the present time I can count at least twenty-five persons who started the farm and truck business when I did, and all but one or two have left it for want of success, simply because their soil did not suit and their experience was not sufficient to see them through. I would just advise any one going into the business to select the proper soil, location and seed, making frequent use of the fertilizer, working the ground to its best condition before sowing the seed, and I do not think he will fail. Many are afraid if they use too much manure the strength all goes in the tops and not in the vegetable, but they are mistaken. If the ground is too poor to make tops it is too poor to produce vegetables. The better for the better will be the profit for its owner, and the earlier they are taken to market the better they will pay. I have sold tomatoes for five cents a piece in March and April, when I could not get that for a quarter peck in July and August ; the same with lettuce, radishes and all other vegetables. He who wishes to profit by raising vegetables must be wide awake to his interests, and without hot-beds he will be left far in the rear of his neighbor who has them. They must be made to produce from two to three crops a season or they will not pay. I hope the advice here given will be of benefit to some who are interested in "Fruit and Vegetables and their Culture."

THE BRIGHT SIDE OF HORTICULTURE.*

MR. PRESIDENT :—You have had "Horticulture for Profit," and discussed the subject in almost every form ; as well the bright, and also the "Dark side of Fruit Culture." But I have not observed that any of your contributors have spoken of "Horticulture for Pleasure." Notwithstanding the trials, disappointments, vexations and discouragements that every season brings, there is, nevertheless, a charm connected with the occupation, that we all embrace for the love of it, to a greater or lesser extent.

How do plants grow, is a query that meets us in the beginning, and is a problem that cannot be solved. The spontaneous action of the plant, the self-determined shapes it assumes, its vitality, are subjects for thought beyond the conception of the most learned botanist. Is it not a direct emanation from the Supreme Will, the fountain of all life.

Vegetation is doubtless the lower order of life. It springs directly from inorganic matter, and is the first step in the formation of plant life. To see the tender germ as it peeps through the earth, a real thing of life, and watch its development from day to day, has a fascination for all who have an eye for the beautiful. How many of us pause to think of the transformations taking place, when we drop the little seeds in the bed, (some of them so diminutive as to be almost invisible) and in a few days find a living plant ready to battle with the enemies of vegetable life. Its progress day after day is a food for thought that the learned and the most ignorant may embrace, and yet be unable to conceive from whence the matter is brought for its development.

I propose, in the few remarks I shall make, to take the "Bright Side of Horticulture." Did we always weigh the cost, and take into consideration the trials of every undertaking, few would launch out into the sea of experiments and labors, and new ideas, would then be the exception. Our life, our joys, are in part derived from horticulture.

I believe a large proportion of those who plant, do it as much for amusement, comfort, and home adornment, as they do for profit. One of the first requisites of a home are trees

*Essay read before the Pennsylvania State Horticultural Association, at Harrisburg, Pa., by Calvin Cooper, of Lancaster county, Pa.

and plants to diversify the appearance, and add comfort and attraction acquired in no other way. How welcome is the shade of a well-formed cherry tree on a hot summer's day. Then, too, its fruit (if the robins have not been there first) will add food as well as drink. while the husbandman tarries beneath its branches.

Well do I remember the favorite tree selected by the harvest hands at the old homestead, under which they took their noonday rest. This was a fine large "Roberts Red Heart" cherry, planted in front of the house, with shapely top and spreading branches, adding a charm to the spot during the whole summer. The planting, training and care of which was all the work of a fond mother, now peacefully resting in a home that knows no waking.

Then, too, the quick-growing peach, with its bright, luxurious foliage, soon fills a vacant spot that well repays the trifling cost, labor and time required to make a tree.

We will not pass the plum and apricot, for here is a little nook and there a recess that needs filling. Then, too, they might furnish a few "Curculio" for the fowls of the yard while they are in search for the early worm. The apple. too, has its claims, with broad spreading branches inviting comfort to its cooling shade as the yeoman (as well as his stock) pass in their routine to and from the toils of the day. All this usually is a work of love, and their training to recreate during the hours of leisure, affords an amusement that diverts rather than tires. The cutting out of a branch here and there, and shortening another that is likely to outgrow the rest, and the observation of the habits and growth, will afford food for thought, always pleasing and of a tendency to quiet the nerves and strengthen the body for the duties of the hour.

And should an occasional crop of cherries, peaches, plums, apricots, apples, etc., be the fruit of our labor we are doubly paid. First by the attractive appearance given to the home, and the pleasure derived by the cooling influence of the shade during the hot summer months ; and also the deliciously flavored fruit with its health-giving properties, contribute wonderfully towards supplying the household with the daily requisites to sustain life.

Mark the busy merchant with his modern suburban home, how he delights in the attractions of his handsome, well-kept grounds, as he meanders from bush to plant, here and there training a vine or a tree, to suit his cultivated taste, and varying the systematically laid-out grounds, adding a charm to the spot, to divert his thoughts from the perplexing trials of his business hours ; this surely is not a work of necessity nor profit, for those who know will say it is quite a drain on the purse, according to the magnitude of the grounds. Such is the fascination connected with the planting and training, that the labor and care is rather a work of amusement, because of its pleasing attractions, and its power to divert rather than tire an overworked brain.

The industrious mechanic, too, has taken the infatuation, and often found during his leisure hours, in beautifying his home with beauteous gifts, the plants of the earth. Who but will halt and admire in passing a handsomely laid-out yard, with its gorgeous beds of bloom, interspersed with choice shrubs and trees, and the well-kept walks with gentle curves leading now to a shady nook, thence by a beautiful border of variegated foliage plants, and not give a praise of gratitude to the Giver of all good works ?

In conclusion I cannot but affirm there is a charm, an attraction, a fascination in the work of the horticulturist, that will far overbalance the labor connected therewith, and supply a heartfelt recreation to all who will embrace it. The comparison of the wonderful works of nature are sufficient of themselves to awaken thoughts of praise. Instance if you will the stately oak with its spreading branches, and the erect form of the Lombardy poplar; the weeping willow, with its ever pendant branches, and the pyramidal form of a sugar maple ; the uncouth weeping ash, and the symmetrical form of the European larch ; the graceful birch, and the straggling catalpa ; the giant form of the common poplar, and the trailing kilmarnock ; and yet all spring from a tender germ, and assume the form common to its kind. The power of the tree to reproduce itself is even more forcibly shown in fruits. Top-graft an apple, and though there may be as many varieties as there are limbs on the tree, each will produce its color; red, yellow and striped, and in flavor its peculiar taste : acid, subacid or sweet, and yet all is supplied from the same root. The floral kingdom has its endless attractions in the various forms of its plants ; the beautiful and harmonious blending of foliage and flowers, ever has its charms for old and young; the nobleman and his servant ; the princely aristocrat and the lowest menial; all pause to admire the wondrous productions of the vegetable kingdom.

HORTICULTURAL FERTILIZERS.*

What are Best and Cheapest and How Applied.

When your worthy Secretary referred the above subject to me, I presume it was expected that I should be able to suggest with some degree of authority, some special fertilizers especially adapted to our wants as horticulturists. If such was the case, I fear I shall disappoint him ; for though the theory of a special fertilizer for each particular crop, is a very plausible and comfortable one—yet in my experience, it is not at all reliable.

Past experience with our teachers of Agricultural Chemistry, somewhat verifies the old adage "That a little learning is a dangerous thing," as shown by the amount of money wasted by farmers in following their specious theory of soil-analysis, as a reliable guide for the selection of fertilizers; and their still later teachings that a special fertilizer can be formulated upon the chemical analysis of a growing crop or plant ! As tillers of the soil, we are greatly indebted to scientists for help in many ways—but there are yet many unrevealed mysteries in the bosom of mother earth, that defy the wisest of our scientists, and, that often mar the practical proof of their fine spun theories. For instance, we may determine very accurately, the amount of nitrogen, potash and phosphate, in the grain and straw of a forty bushel crop of wheat, and with some degree of certainty the apparent amounts of these ingredients in the soil. But we have seen a difference of 15 or 20 bushels of grain and a ton or so of straw, made by an application of two hundred and fifty pounds of guano per acre—the guano containing only two pounds each of ammonia and phosphate, and distributed through six inches depth of soil—making a quantity so infinitesimally small, compared with the bulk of soil with which it is mixed, that the nicest of tests would fail to detect it at all. Hence we must naturally conclude that there are other agencies at work in the soil, that have not entered into the calculation, and that with our present light cannot be accounted for. Again we might ascertain with reasonable certainty, that an acre of apple trees would take from the soil in its growth of wood and apples, a certain number of pounds of potash; yet nothing but a practical trial, after all, will determine whether it is necessary to apply potash to each individual orchard. There may be plenty of it in the soil. It may be active or latent. Individual experiments only can solve these questions. While I had charge of the Experimental Farm in Chester county, I made several thousand experiments with fertilizers upon all the crops usually grown upon a farm, including horticultural crops—and in a large majority of cases, the fertilizer richest in soluble phosphate of lime, gave the best results. And it was also shown that the fertilizer that did the best for corn, was likely to do best for all other crops, if applied in a manner to suit the especial cases.

The soil, then, evidently stood in most need of phosphoric acid –and hence the application of that manurial element really enriched the soil, by establishing a proper balance among the plant-foods—and thus aided all kinds of plant-growth. Different crops require different modes of application.

Growing trees, or small fruits, should have the mineral fertilizer applied in the spring and plowed down, where plowing is possible. It is thus prevented from being washed away; has a good opportunity to be thoroughly dissolved by the action of the moisture and solvents in the soil, and is ready to start a vigorous spring growth that will be healthy ; and mature even if it is luxuriant, because it has the whole season to perfect itself. What we call immature wood-growth, is more frequently the result of a *late or unseasonable growth*, than because it was too luxuriant ! But heavy applications of nitrogenous manures are sometimes dangerous, not so much because of the rapid or strong growth made, so much as the temporary nature of this kind of manuring; the stimulating supply is exhausted before the plant is perfected. The safest general recommendation of a fertilizer for orchards and small fruits, would be a well dissolved raw bone, or a good acid phosphate, those being likely to meet the wants of a majority of cases.

In the bone we usually have from 3 to 4 per cent. of ammonia, which might be useful in some cases, as an excitant to early growth. But the acid phosphate, or dissolved South Carolina rock, furnishes the phosphate in a cheaper form than we can find it in a pure bone fertilizer.

I feel safe in making the above recommendation because, to a certain extent, it involves the use of the plow, which in itself, in most cases, is a judicious thing to use. A well cultivated orchard or fruit garden, where the soil is made loose and fine, freely admitting the rays of light and heat, absorbing moisture and ammonia from the atmosphere, the whole strength of the soil going to the trees and shrubs, and not to weeds and other crops—is certainly standing a better chance than one standing in grass, waiting for the very uncertain prospect for the few loads of barn-yard manure the farmer may have left from his regular crops. We do not mean to discourage the use of barn-yard manure by any means. It can seldom come amiss, particularly if applied in the late fall or early winter. Any covering of vegetable matter, sods or composts, that will smother the grass, or shade the ground, will mellow it, and by increasing its absorbing power, be benefited by the air and light.

Potash would naturally suggest itself as a special tree food, and it would certainly be wise to give it a thorough trial—especially as potash is now not only cheap but easily obtained. The muriate of potash showing a percentage of 90 per cent. of the pure salt, can be bought for about $60, or Kainit, the German dung salt, showing 30 to 40 per cent. potash, can be bought for $16. Either of these goods are furnished by the trade, and can readily be got. I think, sometimes, that the value of ashes is often over estimated, particularly when we judge by the effects of burned brush heaps: the good results are as often due to heating the earth, as from the deposit of ashes. In summing up all I have to say on this subject, I would give this advice: "Stand not on the order of your manuring, but *manure*." You are more likely to err "In the breach, than in the compliance." Orchards and fruit gardens, though highly important adjuncts to the farm, are too often made entirely secondary, when the annual distribution of the manure-pile takes place. We seem to expect from our fruit trees what we could not from any other crop—a persistent yield without adequate manure. An attempt to raise successive crops of grain upon the same field without manure, would be deemed the height of foolishness. Yet we seem to expect our apple trees to yield ten times the weight of fruit that our fields do of grain—and do it too, with very infrequent manuring. I regret my inability to treat this subject in the manner expected, and failing to tell you *how* to manure, or with *what*, must content myself with commending you to put on plenty of *something*.

SELECTIONS.

THE NEW WHEAT REGION.

The rapid settlement of the wheat lands of Dakota is perhaps the most striking feature of recent Western development. It is estimated that the farming population of the Territory has been increased fully 30,000 since the census of 1880 was taken, and the immigration of 1882 promises to be much greater than that of any previous year. A majority of the new comers are no doubt attracted by the profits of wheat-raising. Making due allowance for the enthusiasm of new settlers and of the local journals, it appears from the census returns and from the published statements of farmers that a yield of not less than twenty bushels to the acre can be depended on year after year, and that twenty-five bushels is not an extraordinary crop. The cost of breaking new land is said to average $1,75 an acre; of "backsetting," as the second or spring plowing is called, $1,50; of seeding $2,50; of harvesting $2, and of threshing $1; making $8,75 per acre. After the first year only one plowing is necessary. Some statements place the cost of the grain thrashed and ready for market at $9 for each acre's yield; others give it as low as $6—the difference being mainly due, no doubt, to variations in the price of labor. Taking the highest estimate as a basis for calculation, with wheat at $1 a bushel, the profit of the farmer on each acre of ground cultivated, after all expenses of raising and marketing his crop are paid, cannot well fall below $10 an acre.

The enterprise of railroad companies eager to occupy a field of future profitable business makes transportation facilities in Dakota keep pace with settlement. In the southern part of the Territory the Chicago and Northwestern and the Chicago, Milwaukee and St. Paul Companies have built trunk lines to the Missouri river, which are being pushed on to the Black Hills, and have constructed numerous branches as feeders. The Red River Valley is traversed for nearly its entire length by two roads, one on each bank of the stream. Across the northern part of the Territory, from east to west, runs the Northern Pacific Railroad, which is building branches north and south to widen the area of wheat culture. Thus in most of the new wheat region the farmer finds a station within a few miles of his fields and a freight train ready to haul his grain to market.

The work on a wheat farm occupies only a few weeks of the year, and the business is attractive on that account apart from its profits. After the plowing and sowing are finished the farmer can look on and see Nature grow and ripen his crop until the harvest time comes. By the end of August the year'

---

*Essay read before the Pennsylvania State Horticultural Association at Harrisburg, by John I. Carter, Chester county, Pa.

work is done. Expensive farm buildings are not required, for the grain is threshed in the fields and hauled immediately to the nearest railway station. Very little fencing is needed on a wheat farm. Frequently the cultivated portion is left uninclosed, and a barbed wire fence put around the pasture lot to confine the cattle. Thus the outlay for improvements is comparatively light, and as the country is open and ready for the plow, the new settler makes a crop for the first season, and is tolerably independent from the start. A village, with school, postoffice, stores and churches, springs up as if by magic in the neighborhood of his home, and he suffers few of the privations which used to attend frontier life.

The extent of the new Northwestern wheat region cannot now be estimated, nor its future production foreseen. It appears to include nearly the whole of Dakota east of the Missouri River, and a considerable portion of the western half of the Territory. How far north and west in Manitoba it reaches, will only be determined as settlement advances in that little-known Province. One thing is to be borne in mind about this region—it is the ultimate and permanent wheat-field of the continent. The wheat-growing industry has been steadily moving west for more than half a century. Western New York and Eastern Pennsylvania were once the great wheat sections. Then Ohio had its turn. Now the counties of that State which forty years ago shipped large quantities of wheat to the East do not raise wheat enough to supply their own population with bread. Afterward Indiana, Southern Michigan, Northern Illinois and Wisconsin made wheat the chief crop, to be succeeded by Iowa, and now by Minnesota, Nebraska and Dakota. The wheat belt cannot move much farther west. It will soon reach the great grassy plains where there is not sufficient rain fall for successful agriculture. Beyond the Rocky Mountains, in Oregon and Washington Territory, a new wheat country of immense extent is being slowly developed, but on the Atlantic side of the continent the area where wheat-farming is the chief industry will not be pushed much beyond the present limits of Dakota. The rich lands in the valley of the Red River of the North and the vast rolling plains of Dakota and Manitoba are evidently destined to be a permanent granary, like Hungary and Southern Russia.

Their product, it is interesting to note, seeks the markets of the world by way of the harbor of New York. A glance at a map will show that the water route from the head of Lake Superior to Buffalo and thence to this city by the Erie canal and the Hudson river is almost a direct line from the new wheat country of Northern Dakota. A short stretch of rail transportation brings the wheat produced on the vast Northwestern plains to cheap water transportation extending to the seaboard. The commerce of New York cannot fail to profit directly and largely by the development of the new wheat region of the Far Northwest.—*N. Y. Tribune.*

HOW TO DEODORIZE STABLES

We often wonder why the occupants of large costly dwellings permit stables for horses and the pits adjoining holding the excreta so close to the house, and have hostlers and coachmen to come there, to kitchen and dining-rooms, with rank-smelling person and clothing. When yellow corn, mixed with mill feed is fed to horses generally, or hard-husked old oats given to old horses that cannot duly masticate and consequently fully digest them, the droppings and urine are unusually acrid, and will badly scent whatever absorbents are about. All this injurious unsavoriness may be avoided by simple and cheap means. Very dry, *waste* plaster of Paris, or fine powdered land plaster dusted on stable floors where said voidings generally accumulate, will cover or coat them and preclude the escape of ammonia.

When the bottom and sides of the vaults are dusted, and the ordure nicely leveled therein, then firmed by treading them down with the feet of stablemen standing on a thick board; finally, having a moderate coat of plaster scattered over, as painters sand-coat oiled walls, no effluvia will issue, because the ammonia is bound. On emptying these vaults the contents may be properly called manurial matter unless too much salt hay or long straw, not fully soaked, or badly carbonized litter be there. The wagon loads might likewise profitably be dusted, top and flanks, ere starting, and so further obviate the ungrateful sight and odor of offal openly passing through the streets. We have read of a prominent liveryman in Manchester, England, disinfecting his stables with Douglas' powder, made for that purpose. This did not only prove beneficial to man and beast, precluding sore eyes and coughs, etc., but the voidings were eagerly bought by truckers, for these got the full value of their money.

The *rubbish* so generally bought for manuring is almost worthless—hardly worth hauling —for the substance has largely evaporated, either before or during transit, and more yet, ere said stuff is covered with soil enough to prevent still more exposure. It might be well for the horse-car companies to try this process on a small scale.

UTILIZING ROUGH GROUND.

On many farms there are portions of land that cannot be plowed without great difficulty on account of ravines or stones. They may be seeded to grass and used for pasturage, but it is hard to cut the grass that grows on them. This broken land may generally be utilized to excellent advantage by planting it to crops that require considerable room. Grapes do well on rocky and broken land, if sufficient pains be taken to prepare the places where the vines are to stand. Quite a large hole should be excavated and partially filled with manure and loose earth. A rocky soil is ordinarily warm and well drained by the spaces between the stones. Many of the best vineyards in Europe are located on land so broken and rocky that it cannot be made to produce paying crops of grain, grass, or potatoes. Tomatoes can also be profitably raised on broken land. The vines require considerable space in which to spread their branches. There is some trouble in preparing the hills, but the warm location and good drainage will generally insure large crops that ripen early in the season. Pumpkins, melons and squashes may be planted on broken and rocky land to most excellent advantage. As the hills should be about ten feet apart, but little difficulty will be found in making them. Excavations can be made with the spade or pick if necessary, and filled with suitable manure and fine earth. The large space between the hills will require little attention except to remove the weeds, which will not be very troublesome in a poor soil. If a farmer has large tracts of broken and rocky land he can scarcely do better than to plant it to forest trees, giving a preference to those that will produce nuts.

THE BUILDING OF HOMES.

Double doors—folding or sliding—are a great social "institution." By them two rooms may be thrown into one. A good broad hall becomes in summer an extra room. The air circulates. There is a freedom, an openness about the house, which gives an air of superiority to even very humble dwellings. The superiority is real, too. If we invite a few friends for the evening, it is not necessary to confine them to the "parlor," but the doors are thrown wide open, our guests will fill parlor, and hall, and sitting-room and kitchen, perhaps, and yet all are one company, for the broad doors being open the whole house is thrown together. Music sounds through such a house delightfully, and people have a good time and love to come, because it is so cheerful and social. Another point in our home building which we too often overlook is the exposure of the principal living and sleeping-rooms to the direct influence of the sun. The effect of the sunlight is best gained when the house stands with its corners toward the cardinal points, for thus the sun shines with considerable power on all sides of the house every clear day in the summer, and yet its power is broken, because at noonday the rays strike two sides obliquely, and very soon leaves the southeastern side in the shade. We should not forget that the sunshine is healthgiving; dampness and shade, if slightly in excess, injure the health of both men and animals.

One thing more is the importance of having some provision for fire in the chambers. We build for health and not for sickness, and I do not hesitate to say that many a family mourns the loss of a member simply because the sleeping-room could not be easily heated.

The best mode of heating no doubt is by an open fire of some kind. It is very easy in building to make open fire-places in at least three chambers through which the chimney passes.

Of course, open fire-places are not economical of fuel, but in the chambers fire is seldom wanted, and stoves may be used, if preferred. As to economy of fuel, builders, as well as architects and proprietors, either frequently overlook one important fact, or they do not look at it, that is, that the warmest part of any room is farthest from the floor; so if we make our rooms ten or eleven feet high we must heat the air in all that upper part before a person sitting at a table begins to feel at all warm, unless he is where he gets radiation from the stove or open fire. Low ceilings effect the greatest economy of fuel, and even make open fires economical as compared with stoves and high ceilings. Nine feet is, I think, an extreme height for the ceiling of an ordinary country house, say one in which the largest room is not more than twenty feet square, or of equivalent area.

Besides, there are other numerous considerations which tend to the saving of fuel and at the same time increase the healthfulness and comfort of a home. Some of these are the material of the walls, their impenetrability to air and moisture, "deafening" of the floors, which adds greatly to their warmth, good joiner work about windows and doors, etc.—*A Farmer, in American Agriculturist.*

WHEN TO CUT GRASS.

The method of curing grass among farmers varies, some drying it more than others. Too much drying impairs the feeding quality of the hay. In curing some put hay into the mow while green in color, but not so green in condition as to heat. This method was deemed the best. One day of curing of grass that had been cut free from dew was ordinarily enough to cure the grass. When or at what stage of growth should grass be cut for hay was a question often discussed. It was generally conceded that early-cut grass made hay of a better quality than that cut late. Early cut fodder was more digestible than late cut, the digestible nutriment being the measure of value. Young plants were richer in proteine than later cut, and therefore more nutritious, but no only the quality but the quantity from a given area had to be considered, which complicated the problem. The proteine after the grass blossoms was transferred to forming seeds, the stem, or stalk. As the woody fibre was forming, the proteine decreased in both leaves and stalks. The older the plant the less digestible it was. The increase of quantity was at the expense of quality. Seeds were not masticable, and for practical purposes hay that was fully ripe was little, if any, better than straw. If but one crop had to be cut the cutting ought to be done when the plants begin to bloom. The lecturer then went on to give the results of experiments calculated to show that it was more profitable to cut two or three crops of young grass than one crop of ripe grass; in practice, however, it had to be remembered that the fertility of the soil, the length of the season and the cost of labor were all elements that must enter into the calculation. No general and inflexible rules could be laid down in these matters. Early cutting favored quality, while late cutting favored quanty. The quality of rowen on second crop varied in quality according to the richness of the soil and the time of cutting the first crop. If cut at a comparatively early date of its growth, and properly cured, it is a valuable fodder for milch cows and sheep. It requires more skill and care in curing than the first crop, or it suffers loss in quality.

FEEDING POULTRY AND RAISING CHICKS.

One of the secrets of successful poultry raising is the art of feeding properly, not merely at regular intervals, but on the most suitable food, and keeping the chicks growing as rapidly as possible from the very start. It is very poor economy to stint the poultry, especially young growing stock, for when once stunted, it takes a long while to recover, if it does occur at all. For the first twenty-four hours after the chicks emerge from the shell, they should remain under the hen unmolested, both to dry and gain strength and hardiness. They do not require any food, as the store nature provides will last over this time. As the chicks hatch sometimes irregularly, the older ones can be cared for in the house until the others are ready to be taken away, when the hen and her brood can be removed to a roomy coop, with a tight-board bottom and a rain-proof roof. They should be fed five times daily, but only just what they will eat up clean. The first food should consist of stale bread *moistened* in water or in fresh milk —the milk is decidedly preferable. Do not wet the food, as very moist or sloppy food will cause sickness and a high rate of mortality among young, tender birds. Keep the water (for drinking) away from them until they are six to eight weeks old, but if milk can be spared, give them occasional drinks of it. The too lavish use of corn meal has caused more death among young chicks than has cholera among grown fowls. Until the chicks are half-grown, corn meal should be but sparingly fed, but after that time, when judiciously used, is one of the very best and cheapest foods for fowls and chicks. Nine-tenths of the young turkey and guinea-fowls, which die when in the "down" state, get their death-blow from corn meal, as it is a very common practice (because it is so "handy," and snits lazy people so well) to merely moisten, with cold water, some raw corn meal and then feed it in that way. Young chicks relish occasional seeds of cracked wheat and wheat screenings, while rice, well boiled, is not only greedily eaten by the chicks, but is one of the very best things that can be given. It frequently happens that damaged lots of rice, or low grades of it, can be bought, at low figures, in the cities. As it increases so much bulk in cooking, it is not an expensive food for young chicks, even at the regular retail price, though it would not, ordinarily, pay to feed it to full-grown fowls very liberally or very frequently. In the absence of worms, bugs, etc., during early spring, cheap parts of fresh beef can be well boiled and shredded up for the little chicks, but care must be taken not to feed more frequently than once in two days, and only then in moderation. This feeding on meat shreds is very beneficial to young turkey and guinea chicks when they are "shooting" their first quill feathers, as then they require extra nourishment to repair the drain on immature bodies.—*American Agriculturist.*

VEGETABLE CONDIMENTS.

The best of all is watercress, and considering how easy it can be grown it seems astonishing how few people raise it, and how few have it on their tables. It can be produced o the best advantage on the banks of a spring, brook, as a supply may then be obtained at almost any season in the year. Grown on the banks of a stream of this kind it will be crisp and firm and require no care. It may also be grown on the banks of lakes or of streams of tolerably clear water. Experiments recently made in England show that it can be produced in any garden if pains be taken to apply considerable water to the plants whenever the ground becomes dry from lack of rain. It may be propagated by plants, which are easily transplanted if kept moist, or by planting the seed, which is kept by nearly all large dealers. The flavor of the leaves and stalk is pungent and very agreeable. The plant is promotive of health, and is recommended for curing impurities of the blood. It requires no preparation for the table, and is eaten with the addition of a little salt. The common garden cress, or "pepper grass" which resembles watercress in flavor, but is more pungent, is a desirable condiment early in the spring, when the leaves and stalks are quite tender. Celery is in most respects the favorite of all the vegetable condiments. It requires, however, more skill and care to raise, blanch and store it than most people are willing to bestow. Lately great improvements have been made in its cultivation. It is no longer found necessary to set the plants in deep trenches, which are gradually filled up as the leaves extend. Level culture is now generally adopted and dwarf varieties grown, which require very little banking up. Onions may be classed with condiments, although they are generally ranked among food plants. Radishes are very desirable condiments and are very easily produced by any one who has only a very small amount of land to cultivate. The quicker they are grown the more brittle they are. It is desirable to have a succession of them from early spring till winter. The first crop can be raised in hot-beds. Subsequently a few seeds should be sown every week. They may be dropped a few inches apart in rows when flower and vegetable seeds are planted. The seeds germinate quickly, and the roots are large enough to pull before the other plants attain much size.

TRICHINOSIS.

Few diseases have created more alarm both in this country and in Europe than the one caused by that insinuating parasite, the *Trichina Spiralis*. Although its ravages were little known, or at all events attracted little attention until a comparatively recent period, the people of both hemispheres have taken the alarm, and half the nations of Europe have legislated against the importation of that food which is generally supposed to be the medium of its introduction. How long the "pork worm," as it has been called by way of distinction, has infected the swine of this and other countries will probably never be known. It may have existed for many years, unsuspected and undetected; but a dread of its effects has been aroused which it is to be hoped will not abate until men shall cease to subject themselves to its insidious attacks, or discover an effectual remedy to overcome its deadly effects.

Up to the present time the trichina spiralis, we believe, has been found exclusively among the animals used for food—in the flesh of swine. Inasmuch as this meat is more or less freely eaten by a large majority of the people this country, it would seem to follow that most persons are liable to be attacked by trichinosis. Fortunately there is an undoubted remedy in the hands of every one who choses to avail himself of it. The parasite can exist only in the living animal or in pork while in its raw condition. The flesh of of an infected animal will, of course, convey the disease to all who eat it unless the parasite is destroyed. This can be easily and effectually done by first thoroughly boiling it. If this precaution was observed, the disease among the human family would be unknown. It is only when ignorant or thoughtless people indulge in eating raw pork that the insidious enemy finds a lodgment in the human system.

Numerous undoubted cases of this terrible disease have occurred in this country. The investigation of Dr. Belfield and Mr. Atwood show that eight per cent. of all the hogs slaughtered in Chicago were infected with this parasite. This fact shows how important it is that every one should know how to avoid infection. The parasite is exceedingly tenacious of life, often resisting the effects of powerful chemical agents, and the influence of putrefaction. Ordinary smoking or salting of infected meat does not destroy them. Thorough boiling is the only remedy that has been found at once easy and effectual. Pork may be boiled sufficient to kill those nearest the surface, while those in the interior may escape unharmed. If ordinary precautions are adopted there is no danger, but without them the danger is constant and great. It would be well if some feasible remedy could be discovered for its prevention in the animals themselves, but as this can only be done by a series of precautions which are impossible under ordinary circumstances, our main reliance against contagion must be by thoroughly disinfecting the meat. Most persons are well aware that thorough boiling will prevent any serious results, and cases of trichinosis ought to be excessively rare instead of numerous. There is really no excuse for people eating raw pork. It is a barbarous habit, and we are almost tempted to say the person who is guilty of it hardly deserves a better fate.—*New Era.*

TESTING CREAM.

The test adopted by creamery men to find the butter value of cream collected from the dairy farms, consists in churning each batch of cream separately, and if it does not produce one pound of butter from two inches, from an eight inch can, it is said to "fall short" and the patron is liable to the imputation of dishonesty in skimming. That there are dishonest practices resorted to on some farms, is undoubtedly true, but it is also true that the test adopted by the creamery is not always a sure indication that cream has been tampered with to the detriment of the creamery. Where the patron does a share of the skimming himself, he may stir in a quantity of milk to increase the measure, or stir in soda to cause an effervesence and thus increase the bulk of the cream. Besides these there are other devices resorted to for the purpose of defrauding the creamery. But in justice to those who do not resort to such practices, it should appear to the candid mind that there are certainly other causes than dishonesty from which a shortage may occur. On many farms it is entirely necessary that some of the milk shall be skimmed daily.

There are young calves to be fed, and young pigs that ought to have milk, from which only a portion of the cream has been removed. This necessity implies the skimming at a stage when the cream is as yet unripened, or has not had time to concentrate itself to the thick mass of butter globules. Cream thus prematurely skimmed is watery and will not yield the amount of butter to the inch that can be obtained from thoroughly ripened cream. This is also measurably true of cream raised during the Spring months when the young grass is yet watery; it will not produce the rich cream that milk will raise at a later season, when the grass has perfected its growth. Much of the "falling short" is also due to carelessness in skimming by the person who collects the cream, or carelessness in handling. It should also be noted that in taking a can out of the bank to be skimmed and setting it down on the floor (especially if it has stood a long time) that a layer of gas has formed between the cream and the milk, which by the jar of setting the can on the floor, starts this gas upward through the cream, puffing it up so that it will measure more than if allowed to rest a few moments to allow the gas to escape and the cream to gain its normal condition. We might detail many other conditions that interfere with a true measurement did space permit. What is needed is that all milk should be allowed to stand till it acquires a ripeness. Then it should be measured by a careful hand who has that rare faculty to do even handed justice between the patron and the party who employs him to collect the cream.

APPLICATION OF LIQUID MANURE.

The comparative advantages of applying fertilizers to land, in liquid form, or after the liquid has been taken up and made solid by absorbents, have not been fully settled by intelligent farmers generally. Liquids have the advantage of immediate action when applied; while, on the contrary the soluble portions of solids must be first dissolved or washed out, requiring a considerable length of time. On the other hand, liquid manure can only be secured by more perfect and expensive buildings, and the facilities for conveying it to the fields include pumps, tanks and sprinklers. In addition to these, care must be taken to prevent the inconvenience of freezing in winter.

On a large scale, and with complete tanks and appliances, the use of liquids may be attended with less labor in applying than if they were all first absorbed and then drawn out in solid form. With a good pump, and with the tank or tub not much higher than the cistern, a laborer will load a liquid ton, ready for drawing, with less labor than he can throw a solid ton on a wagon with a fork. Through the sprinkler he spreads it with no other labor than driving the wagon across the field, and it is more evenly distributed and finely diffused than by any spreading with the fork and breaking with the harrow—in which it is scarcely equaled by Kemp's spreading cart, which pulverizes and scatters the manure with no labor to the driver. This manure spreader is the most perfect contrivance yet brought into use for making manure available by fine pulverization; and next to this is the

fall spreading of manure and breaking it fine by suitable harrowing—the autumn and winter rains washing out the soluble parts into the soil.

Very few farmers have barns, stables, drains, cisterns, pumps and tubs for drawing, to enable them to use and spread liquid manure advantageously. Most of the stable floors are not water tight, and the liquids leak through and are absorbed by the earth beneath, or are lost. On grain farms, where there is an abundance of straw, enough may be used as litter to hold all or nearly all the liquid, and large quantities of this coarse or straw manure, thrown into a heap and exposed to weather and rains, will rot down and may be drawn out in the following autumn. A question here occurs, which we suggest for investigation—namely, how much gain in the labor of drawing out solid manure is obtained by the spontaneous evaporation of the water of the liquid manure as it lies in such a heap?

When absorbents cannot be had the liquid may be saved by excavating a shallow cistern beneath the stable, making the sides so sloping that the water-lime mortar may be spread on the smooth face of the earth. In this way the expense will be moderate. The contents of this cistern are pumped out as needed and drawn to the fields in the watercart. There are two difficulties with this method. If the cement is not made with the sharpest and purest sand, and the best water-lime, the frost of cold winters will crumble it. The fetid odor from the liquid may interfere with the maintenance of the pure air which should always exist about every good farmer's buildings. To prevent these two difficulties is a subject for careful investigation, which will be differently answered according to circumstances. A large use of absorbents in the cistern would defeat the attempt to carry off its contents by pumping.

The object of these remarks is to open the question for examination, and to invite the statements of those who have tried the different modes for securing and applying liquid manure under the most favorable circumstances. The three points to be taken specially into consideration are the comparative advantages of—1. Using the clear liquid with good and suitable appliances; 2. Employing an abundance of straw and other absorbents in the stable; 3. Conveying the liquids by means of tight floors and gutters to compost heaps of earth, peat or turf, placed under or near the barn.—*Country Gentleman.*

EARLY PRICE OF PENNSYLVANIA LANDS.

It is interesting to note the prices at which lands are selling in Pennsylvania to-day, compared with the prices that prevailed at the time of its settlement, and for a century thereafter. From the transfer of the colony to Penn down to 1762, the price was $41.33 per hundred acres, except in the lower counties, where it was only $9.33 per hundred. From 1762 until 1763 it was $24 per hundred. From 1763 to 1765 it was put up to the earlier figure, $41.33. From 1765 until 1784 it stood at $22.22. Under the Commonwealth the changes were as follows: From 1784 to 1792 it was $26.66⅔ per hundred acres. In the new purchase, made in 1784, including the northwestern portion of the State, and about one-third of its present area, land was sold from 1785 to until 1789 at $80 per hundred acres. From 1789 until 1792 the price was $53.33⅓; from 1792 until 1817 it was $13.33⅓ for all east of the Allegheny river, bought under the 1784 purchase. Lands secured by the Penn heirs under the 1768 purchase, which included the northeastern and southwestern portions of the colony, sold from 1792 until 1814 at $6.66⅔ per hundred acres; these latter lands have been selling since that time at $26.66⅔. Lands improved agreeably to the act of 1792 are sold respectively at $6.66⅔ and $13.33⅓ per hundred acres. Of course the lands still at the disposal of the State are neither very choice nor of great extent. All the vacant and unimproved lands of the State are now sold at $26.66⅔ per hundred acres, except lands lying north and west of the Ohio and Allegheny rivers and the Conewago creek, which are held at $20. It will be seen the extreme range of prices under the proprietary and the commonwealth for 500 acres ran all the way from $33.33 to $266.67, which included the choice lands of the State. We have progressed somewhat since the early days, when Lancaster county lands were to be had at a little more than six cents an acre. Three thousand acres could have been purchased then at the cost of a single acre today. What is more, the surveyors didn't look after the odd perches quite as closely as they do now. William Penn and his heirs not only gave full measure, but threw in six per cent. additional "for roads and highways." The purchasers of real estate in those days certainly got the worth of their money if ever men did."—*New Era.*

A HOME FRUIT CANNING FACTORY.

Among the little-known industries in this county is the fruit-canning factory of Messrs. C. Fell & Son, located at Kirk's Mills, in Little Britain township. We had a visit from the senior member of the firm recently who gave us some interesting particulars concerning this industry in which he embarked so extensively last year.

The Messrs. Fell were induced to go into the canning business because of the large amount of different kinds of fruits and vegetables that went to waste on their own place and also among their neighbors. The idea of utilizing these, and thus saving what would otherwise be lost, led them, in 1881, to begin putting up tomatoes, apples, plums and sweet corn. The result was when the season's work was over many thousands of cans of these fruits and vegetables had been prepared for market. They are put up in one and two-pound cans and find a market in this city and in Philadelphia.

Nearly all the articles canned are grown by the Messrs. Fell themselves, the rest being purchased from the farmers in the neighborhood. The tomatoes used were principally the "Beefsteak" and "Acme" varieties. The corn grown was Stoyle's evergreen. Some difficulty was had in procuring a sufficiency of tin cans for their purpose, and most of those used were made in Cecil county, Maryland. During the height of the season eighteen hands were employed in the various processes of canning. The quality of the articles put up at this establishment is first-class, and is more in demand as their excellent qualities become better known.—*New Era.*

OUR LOCAL ORGANIZATIONS.

FULTON FARMERS' CLUB.

March Meeting.

The second March meeting of the Club was held at the residence of Joseph P. Greist, in Fulton township. There was a very large attendance of members and their families, besides several visitors.

After the minutes of the last meeting had been read and approved, S. L. Gregg asked: "Which is the more profitable crop—wheat or corn?"

Joseph P. Griest: Wheat takes the longer season, but does not require the care and attention that corn does. It does not produce as many bushels per acre as corn, but brings a better price. On the whole, I would think it more profitable than corn.

Day Wood: Fifty bushels per acre might be considered a fair average yield of corn and twenty of wheat. At present prices—70 cents per bushel for corn and $1.30 for wheat—the difference would more than pay for the greater attention required by corn. He would say that if the same amount of fertilizers was used corn is the most profitable crop.

Thomas Stubbs: If fertilized alike, corn is more profitable than wheat. Several others concurred in this opinion.

Day Wood: Is it a good time to sell wheat now?

S. L. Gregg did not see anything to make it advance. There is enough wheat in the country to supply the demand and a fine prospect for a good crop next harvest. Reports also show that there has been a great deal of spring wheat sowed in the West.

Thomas Stubbs: Better sell anything when you have it ready.

E. H. Haines did not think it a good time to sell. At present prices wheat is not a paying crop and there was not much risk in holding it when prices were low.

Montillion Brown: What kind of fertilizers will be used this spring for corn by those present?

S. L. Gregg: South Carolina rock and Q. and L. (quick and lasting) bone seem to pay best at present prices. For either wheat or corn would plow in.

Jos. T. Greist will use barnyard manure; E. H. Haines and Day Wood will use rock; Thomas Stubbs, Orchilla guano; C. S. Gatchell rock and Q. and L. bone.

Melissa Gregg: Does any one know anything about meat or soap vessels built of brick and cemented?

Joseph P. Greist has one in his cellar. He uses it for a soap vessel, and it answers the purpose well.

Thomas Stubbs has one built partly of the cellar wall and partly of brick. After drying it was washed, cemented with a brush, to fill up the cracks. It makes a good soap vessel, but he would be afraid to use it for meat, as cement is said to taint.

C. S. Gatchell said that one of his neighbors had one which he used to salt meat in. It answers the purpose well.

R. D. King: How many bushels of tomatoes can be raised per acre?

William P. Harris: It depends very much on the land. One vine in his garden would produce as many bushels as eight would in his field.

E. Wilson: If they grow as they do in some places a great many bushels can be grown on an acre. He had known six bushels to be raised on a piece of ground about ten by twelve feet.

E. H. Haines: In Harford county 200 bushels are considered a good crop.

Day Wood thought that 200 bushels would be a large yield for an average. One hundred bushels per acre for 10 acres would be a fair average.

E. H. Haines: Planted 4 feet apart, 2,700 plants will stand on an acre, and at that rate it would take 27 plants to produce a bushel.

E. H. Haines: Does the seedling peach tree live longer than the grafted?

Wm. P. Harris did not find much difference; neither of them bears more than two crops.

S. L. Gregg: Not much difference; if anything, the seedling has it.

Joseph C. Stubbs had better luck when he planted in fence corners and gave them no care. He knew an old nurseryman that planted some peach trees in fields and some in fence corners, and the ones in the fence corners did the best.

C. S. Gatchel said that seedlings did best for him. He had not had a grafted tree to do any good for the last fifteen years.

Thomas Stubbs: A tree that has no care seems to do better and bear longer than those that are better cared for. He had known some trees to be brought to that were quite yellow by putting walnut hulls around them.

E. H. Haines said he made a garden several years ago. He planted some peach trees, both grafted and seedling, thinking that they would not live long. The grafted trees soon died. The seedlings are alive yet.

Priscilla Coates said that they had, some years ago, an orchard of ten acres. It bore five full crops. The ground was cultivated in potatoes. Every spring and fall the borers were taken out and salt was put around the trees—not against them. The trees were all grafted.

After dinner the usual inspection of the farm and live stock was made by the Club, and after it had again convened in the house criticisms were called for.

S. L. Gregg thought there was no improvement in the stock. The farm was looking fine, and the shed over the barnyard was much improved.

Montillion Brown was not here last year. The farm looks well now.

Thomas Stubbs: He has a rather remarkable field of wheat. It looks healthier than any in the neighborhood. It always has had a healthy appearance.

E. H. Haines: His stock has materially changed in kind. It is good. He has a new breed of hogs (small Yorkshire). There is not much wasted in none.

Joseph P. Greist read from *Scribner* an article describing an automatic machine for destroying insects that was exhibited at the Atlanta Exposition.

Montillion Brown read an article from the New York *Tribune* on covering grapes while ripening with paper bags, and asked if any one present had any experience with the method.

There was one who had tried tying paper bags around the clusters, but there were several that had noticed that grapes would ripen much nicer in the shade than in the sunshine.

A paper was found lying on the floor which proved to be another communication from the "Old Woman." She had been waiting impatiently for the new Scripture to come out. She wanted to know if it would make Adam to be so mean as the old one did, to eat the apple himself and then lay all the blame on Eve. Now she had got it and found that what was published did not go back as far as Adam by four thousand years, so she would have to wait a good while longer before she was satisfied.

But she had been looking over it and she did not find any more comfort for those who were in the habit of practicing sharp tricks in their dealings than she did in the old.

Priscilla H Coates recited "Some Day."

The Club next proceeded to elect officers for the ensuing year, with the following result: President, William King; Secretary, E. H. Haines; Treasurer, Lindley King; Librarian, Day Wood.

Mabel H. Greist, Sadie Brown, Joel King and Montillion Brown were appointed to furnish literary matter for the next meeting, which will be held at the residence of Lindley King on the second Saturday in April.

April Meeting.

The Fulton Farmers' Club met at the residence of Lindley King, near Wakefield, April 8th, 1882.

Grace A. King exhibited some apples to be named and they were pronounced to be Tewkesbury Winter Blush, a variety of remarkable keeping qualities.

Jesse Yocum exhibited a sample of some Russian oats which he had purchased for seed. Also a few potatoes of an unknown variety which he estimated had yielded for him in a patch in the garden at the rate of from 350 to 400 bushels per acre and he considered the quality good. They were raised under very favorable circumstances, the land being rich and the cultivation thorough. He believes in planting potatoes in ground previously made rich and then not to use any manure in the row. He also exhibited a few ears of hominy corn, this is a white, flinty, shallow-grained variety well adapted for making good hominy but not very productive, seldom yielding more than 40 bushels per acre.

Josiah Brown asked "which way can we raise the most corn, by drilling or checkering and planting with the hoe?"

S. L. Gregg said he thought that about as much corn could be raised in one way as the other; he drills in his, but not because he thinks it better but it is more convenient.

Montillion Brown prefers to checker his except when the field is hilly, as it is much easier to keep the corn clean.

Joseph P. Griest advocated drilling in rows, four feet apart, with two stalks in a hill and the hills thirty-nine inches apart in the rows. In this way he had raised seventy five bushels per acre.

Several others spoke in favor of drilling. They believed they could raise more corn, it stood dry weather better than when hilled and they could work close to it.

James Smedley said he once planted part of a field in hills 3½ feet apart each way and from two to three stalks in the hill; the rest of the field was planted in hills six feet apart one way and two feet the other with two stalks in the hill, and by the latter plan he raised about one third more corn than by the former. In each case the corn was worked both ways.

The following questions had been handed to M. Brown to be asked at the club: 1st. Would it be advisable in planting a field of corn to run the rows so as to render the field liable to wash for the sake of having the rows run north and south.

S. L. Gregg said he had no faith in the benefits to be had from making the rows to run from north to south. The air would circulate either way, and if the noonday sun did not shine well along the rows the morning and evening sun, would which answered just as well. He always runs the rows across and not up and down the hill. These views seemed to meet the approbation of all present.

2nd. How far apart should Lima beans be planted for a field crop? Nearly all were in favor of planting them about the same distance apart as corn. Jos. P. Griest plants his in rows; four feet apart, in hills one foot apart in the row, in this way he makes one pole answer for two hills. He plants two beans in a hill.

Jesse Yocum asked, "Where did the bean weevil come from and how can it be prevented from injuring the beans?" No one could tell where the weevils came from, but James Smedley said he had found that by heating the beans they could be saved from injury.

Emma King read a selected article encouraging all those who could to write something for the papers and maintaining that it is the duty of every person to try to say or write something for the instruction and encouragement of others so that the world may be the better for their having lived in it.

Mabel A. Haines recited "The Poetry of Housework."

The club having now been in existence twelve years, Montillion Brown read a history he had prepered of it, noting briefly its rise, progress and achievements. The first meeting was held at the residence of William Brown in March, 1870, and two of the six or seven persons who were present and effected its organization are still regular attenders of its meetings. During the twelve years of its existence it has lost two of its members by death, whose memories have been commemorated by appropriate resolutions. It held two fairs, at which the display and attendance exceeded the expectations of all who witnessed them. It has also held a public sale of stock, etc., at which the sale amounted to many hundreds of dollars, and last but not least it held a public meeting in the grove of the Hon. James Black at the now becoming famous Black Barren Spring.

After the reading of this several others spoke of the good effects of the club, dwelling more particularly on its social advantages and of the harmony that had always characterized its meetings.

Jesse Yocum, who was attending one of its meetings for the first time, said he was much pleased with the exercises, but was afraid that the members would allow it to take the place of small social gatherings which he considered so essential to the good of society.

The question, "Should a farmer make a specialty of one particular kind of business or follow what is known as mixed husbandry?" was adopted for discussion several months ago, but as the days were short there never seemed to be time for any discussion, and now most of the members had forgotten what the question was, which made the arguments come in rather a crude shape.

S. L. Gregg said that in this section of the country it does not seem possible for a farmer to follow raising any one particular thing; he must raise grain, hay and keep some stock, but he does not believe it pays to raise a few beans, a little broomcorn, buckwheat or sorghum to sell, for if he does he will neglect his regular crops. Several others coincided with this view of the subject, yet thought that some certain thing might be made the leading feature of our farming operations, and only such other things raised as are necessary to the carrying out of the main object in view. Dairying, for instance, might be the principal object in view, and then only such grains raised as furnished straw for bedding, and this could easily be done as there is in this section a necessity for breaking up the land occasionally and resetting it with grass in order to keep up a supply of hay and pasture.

The next meeting will be held at the residence of Montillion Brown on the first Saturday in next month.

THE LINNÆAN SOCIETY.

The society met in the hall of the Y. M. C. A., on Saturday afternoon, March 25, 1882, when, in the absence of the President and Vice President, Chas. A. Heinitsh was called to the chair. After organization in due form the following donations were reported by the curators:

Museum.

Four fine specimens of minerals and metals, namely: mic. ox. iron from Rochester, N. Y., by W. L. Hershey; asbestos from Wilmington, Delaware, and mica slate and quartz impregnated with malachite, from localities unknown, by students of the Lancaster High school. (A number of the students of the High and Secondary schools are manifesting an active, working interest in natural science, notably in botany and mineralogy, and they deserve encouragement.) A specimen of *Triton Jeffersonii*, by the curators. This reptile had been kept alive in a globe aquarium for a period of nine months, and died on the 2nd of March, 1882. As the animal belongs to the order Batrachia, the theory of its death is based upon the natural change in its organization, rather than upon its physical surroundings. During the whole nine months it had been provided with a pair of external ciliated gills, and was a water breathing animal, like the fishes. But the period of its final transformation had arrived, when it cast off its gills and became air-breathing, and as the form of the aquarium prevented it from reaching the surface to inhale the air, and the oxygen in the water in the tank being exhausted, the result was natural. We *know* now, however, that in an aquarium these animals may survive, at least nine months, whatever the case may be in their normal condition.

Library.

Report of the Chief Signal Office for 1879, 700 pages octavo, with 73 charts and maps. Proceedings of the Academy of Natural Sciences, for 1881 Messages and documents, from the Department of State, for 1881-2. 1 vol. 8vo., pp. 1,000, from Hon. A. Herr Smith. Congressional Record, two vols., 4to pp. 4,041, including index and appendix, Hon. A. Herr Smith. Nos. 1 to 9, vol. 21, Official Gazette of the United States Patent Office. Ten catalogues and circulars of rare and valuable scientific and historical books. Three envelopes, and thirty historical and biographical scraps.

New Businsss.

F. K. Hershey, of Creswell, Lancaster county, was proposed for active membership, which, under the constitutional rule, will receive final action at the next stated meeting.

J. N. Sloan, of Millville, Clarion county, Pa., was unanimously elected a corresponding member.

The next stated meeting will be held at the residence of Dr. Knight, on Thursday evening, April 27, 1882, being the last Thursday in the month.

It was marked that the evening meetings had not yet been as prolific of favorable results as had been anticipated. It was thought, however, that before the end of the year it would be demonstrated whether it would be best to continue them or not.

The meeting had about the average attendance during the winter sessions, and although there were no papers read, and little special business brought before the society, yet there were a number of visitors present, and on the whole there was a pleasant time spent in the Museum.

ENTOMOLOGICAL.

Swarming Ants and Allied Phenomena.

March, is usually a rough, boisterous month, and therefore it is not usual to find *ants*, or *bees*, swarming within that month. But, the March of 1882, "went out like a lamb," and on one of those lamb-like days, near the end of the month, we were brought a bottled swarm of these busy little insects, which issued from a small aperture in a pavement near "Penn Square," in Lancaster city. Males, females and neuters came forth in great numbers; the first two, amply provided with wings, and the last, entirely wingless. The phenomenon seemed strange to those who were unaware that ants ever possessed wings, and hence they were supposed to be "something new under the sun." Ants, however, belong to the same natural order that the various honey-bees do, (*Hymenoptera*) and have many traits in their habits that approximate those of our common "hive-bees" (*Apis mellifica*) in their domestic economies. The colonies are usually founded by a single fertilized female, or queen, in the spring, who deposits three kinds of eggs, from which are hatched male, female, and neuter ants, in time. We have often found, during winter, in turning over an old log of wood, or large flat stone, or ripping the bark from an old decaying tree, nestled in small cavities, a large ant—sometimes two or more—and sometimes a large "hornet," a "wasp" or a "yellow-jacket," in a state of torpor, but which would revive on exposure to the sun, or to artificial heat. These were the surviving females, or queens, of the previous season, and in their organisms contained all the elements of successive colonies in and during the following season; exhibiting a wonderful adaptation of means to ends in the economy of nature. Under similar circumstances, specimens of the common "bumble-bees," or of "wood-borer-bees," are found; and, before we knew that the "white-head" was the male of the wood-borer, we wondered why those found in winter had black-heads. This bee (*Xylocopa Virginica*), is also found in the winter occupying their galleries of the previous season in wood. Now, these insects also belong to the *Hymenopterous* order, and bridge over from one summer to another, with, perhaps, only a very few of the large colonies of a former season, and those few fertile females, each of which, will become the queen, or mother, of a future colony; perpetuating a dynasty that was founded "in the beginning," and doubtless will continue "until the end."

But the ants, like the honey-bees and bumble-bees, are "social" or gregarious in their habits, whilst the wood-borers are "solitary," and all the labor is performed by the females. Unlike the bees, however, ants are usually wingless until the swarming, or nuptial season arrives, and the wings are rapidly developed in the males and females, whilst the neuters, or workers (whether soldier, servant or slave) remain *apterous*, or wingless, throughout their entire lives. A case was reported to the Linnæan Society in 1881, of ants swarming on a warm day in November, and we have noticed this phenomenon in August, September and October. (*Termites* swarm in May.) Those that swarm in August would, perhaps, have time enough to mature a colony before the cold weather sets in, but this could not be said of those that delay until the month of November. As soon as they have accomplished the purposes for which they swarm they cast off their wings—and they do it quickly too—when the males and neuters perish, and the fertile females are preserved to found new colonies in the following spring and summer; but why they should be swarming in the month of March, is somewhat *anomalous*. It was much too early in the season for a colony to have multiplied from a fertilized female of last year, and hence must have been in active operation all winter, and this suggestion is based upon the fact that under the pavement, where the swarming occurred, is located a furnace and boiler, which generates steam-power for a number of printing presses; which seems to imply that ants are more influenced in their activities by temperature than by signs and seasons. We know that a few warm days, even in mild winter will develop the foliage and flowers of plants; and, if long continued, will also develop the fruit; and this is also the case when a protracted warm season occurs in autumn, producing what is termed a "second crop." Now, this may have been a continuous or *second crop* of ants, which matured during the winter under the influence of artificial heat. The same phenomenon frequently occurs in conservatories or warm "green-houses," and illustrates that, primarily, insects are of the same habits in their procreative powers all over the world; and that all departures from the primary habit are the results of temperature: of course there may be other conditions more or less affecting their development, but the main one is heat or cold—the one accelerating, the other retarding their progress. We can hardly name the species referred to in these remarks, because, before we could find time to commit them to paper, our specimens were lost or destroyed, but they seemed to be the "common brown ant," (*Fomica fusca*).

Insects—and especially ants—are supposed to be endowed with a very high order of instinct—many claim for them *reasoning* powers; and yet *they*, with many other animals, make many mistakes under the misguiding influences of heat and cold, or light and darkness. For instance, chickens instinctively hie them to their roosts about sunset, whether it "rains or shines"; and yet it is on record that chickens have retired to their roosts at mid-day, when a total eclipse of the sun prevailed. Truthful as they are generally esteemed as the heralds of the "break of day," they made a grave mistake in this instance, not being able to distinguish between noon and night. The ants under consideration, seemed to have been in a similar "fix," not being able to distinguish between summer and artificial heat. We knew of a "white cabbage butterfly" (*Pieris rapæ*) to evolve from the pupa in the month of February, when there absolutely was snow on the ground. Notwithstanding a snow had fallen, the weather was mild and the sun came out strong, and perhaps the snow may have assisted in bringing the pupa within the focal rays of the sun. Of course it starved to death, for there was no green thing for it to feed on for months.

On one occasion, on the 3d of December, we found about twenty "striped cucumber beetles (*Diabrotica vittata*) vigorously feeding on a plant on the sunny side of a hill, in proximity to a log, the plant having remained verdant, under the protection afforded by said log and continued warm weather, without the instinctive forewarning, that on the morrow it would snow, and then freeze, which actually occurred. The *fact* is they were deceived by the prevailing heat: they became hungry, and came forth from their place of hibernation without knowing whether it was June or December, and *being* hungry, they naturally must have something to eat. Last winter snakes were found abroad in February, as active as in summer, notwithstanding the "Groundhog" had admonished them that there would be six weeks of winter weather after "candle-mas." Indeed, intelligent human beings themselves, often commit the blunder of doffing their heavy winter under-clothing on the first warm dawning of spring, and thereby contract many injurious cold. The re-animation of insects during a period when they could obtain no food, and letting them starve to death, would be a much surer means of their destruction than to depend on them being frozen to death during a severe winter. Although the tenure of animal life seems to hang on a delicate thread, yet the vital forces of nature are often manifested in a most extraordinary manner, and under the, seemingly, most unfavorable circumstances. Many years ago we noticed the single stem of a "wild aster" that had fallen across a much-used foot path, where the apex was trodden off. It then threw out laterals at each axil, and the ends, too, of these were soon trodden off, and these laterals threw out others; and this continued until late in autumn. Then it bloomed, but instead of a dozen large flowers like its cogeners, it produced hundreds of very small flowers; its main stem and root were twice as large as those near it; and instead of a slender straggling plant, it was a dense green bush, full of flowers. It seemed determined to grow, and to bloom, notwithstanding the rough usage it met with. Fertilized female insects manifest the same determination under similar unfavorable auspices.

Curculio in Plum Culture.

Mr. Isaac Kauffman, Mountville, Pa., is reported in the Lancaster *Examiner* as having two plum trees of the same variety and apparently equal vigor, one of which bore nothing this year while the other was abundantly loaded with fruit in consequence of having been "shaken every morning for ten days while in blossom." Experienced plum-growers well know that merely shaking the trees with no effort to destroy the curculio as they are thus felled to the ground will do no possible good. Even if the weevils are destroyed, either by chickens kept cooped under the trees, or by catching on a sheet and destroying them, the work must be continued, not merely ten days, but five or six weeks, so long, in fact, as the curculio are caught. Apropos to the above is a remark made to me a few days since by Judge Ramsdell, the most extensive and successful plum-grower in Michigan. He said that the curculio was an advantage to him, as in bearing years they would so thin out the plums as to make his crop far more valuable, and by carefully gathering all the wormy fruit in such seasons, as soon as it fell, and burning it, he so thinned the insects that the expense or fighting them during the off-year, and saving the small crop, was quite light. He placed great stress on the importance of gathering and destroying all the wormy fruit as fast as it fell from the tree.—*Professor A. J. Cook, Michigan Agricultural College.*

Birds and Canker Worms.

The most serious losses of the farmer and gardener, due to insects, are not consequent upon the ordinary and uniform depredations of those species whose numbers remain nearly constant, year after year, but upon excessive and extraordinary depredations of those, the numbers of which are subject to wide fluctuations. Vegetation has become so far adjusted to our crickets, ordinary grasshoppers, etc., that the foliage which they eat can be spared without injury to the plant, and the damage done by them is commonly imperceptible. It is far otherwise, however, with the vast hordes of the Rocky Mountain Locust, the Colorado Potato Beetle, Chinch Bug, Army Worm and many other species, which occasionally swarm prodigiously and then almost disappear. The injurious species are therefore, chiefly the oscillating ones, and the dangerous species are those which show a tendency to oscillate. Anything which tends to limit the fluctuations of an oscillating species, or to prevent the oscillation of a stable species, is therefore highly useful; while anything which tends to intensify an oscillation, or to convert a stable species into an oscillating one, is as highly pernicious.—*Prof. S. A. Forbes, in American Agriculturist.*

AGRICULTURE.

Sowing the Seed.

There is no more prolific source of disappointment and failure among amateur gardeners, says D. M. Ferry, in his seed annual, than hasty, careless, or improper sowing of the seed. A seed consists of a minute plant minus the roots, with a sufficient amount of food stowed in or around it to sustain it until it can expand its leaves, form roots and provide for itself, the whole inclosed in a hard and more or less impervious shell. To secure germination, moisture, heat, and a certain amount of air are necessary. The first steps, are the softening of the hard, outer shell, the developing of the leaves of the plant from the absorption of water, and the changing of the plant food from the form of starch to that of sugar. In the first condition the food was easily preserved unchanged, but the plant with its undeveloped leaves and no root was incapable of using it, while in its sugary condition is easily appropriated; but if not used it speedily decays itself and induces decay in the plant. A seed then may retain its vitality and remain unchanged for years, while after germination has commenced, a check of a day or two in the process may be fatal. There is no time from that when the seed falls from the parent plant until it in turn produces seed, ripens and dies when the plant is so susceptible of fatal injury from the over abundance or want of heat and moisture as that between the commencement of germination and the formation of the first true leaves, and it is just then that it needs the aid of the gardener to secure favorable conditions. These are:

First. A proper and constant degree of moisture, without being soaked with water. This is secured by making the surface of freshly dug soil so fine that the smallest seeds may come in immediate contact on all sides with the particles from which they are to absorb the required moisture, and the pressing of the soil over the seeds so firmly with the feet or the back of a hoe that the degree of moisture may remain as nearly as possible the same until the plants are up. Second. A proper degree of heat, secured by sowing each variety of seed when the average temperature of the locality is that most favorable for its germination. Third. Covering the seed to such a depth that, while it is preserved at a uniform degree of heat and moisture, the necessary air can readily reach it, and the tiny stem push the forming leaves into the light and air. This depth will vary with different seeds and conditions of the soil, and can be learned only from practical experience. In general seeds of the size of the turnip should be covered with half an inch of earth pressed down; while corn may be an inch, beans two or three, and peas two to six inches deep. Fourth. Such condition of soil that the ascending stem can easily penetrate it, and the young roots speedily find suitable food. We can usually secure this by thorough preparation of the ground, and taking care never to sow fine seeds when the ground is wet. Occasionally a heavy or long continued rain, followed by a bright sun, will so bake and crust the surface that it is impossible for the young plant to find its way through it; or a few days of stong wind will so dry the surface that no seed will germinate. In such cases our only remedy is to try again.

Clover and Grass.

In all cases where land was laid down to grass last fall either with winter rye or without, a careful examination should be made to see if there are not portions winter-killed, and re-sow with seed all such places; the seed will catch and grow without harrowing. Clover seed may be added early this month at the rate of ten pounds per acre with success, the crackling of the ground by the action of the frost will afford ample covering. It is feared that the value of clover is not fully appreciated; it is very nutritious to stock, and on the whole increases the fertility of the soil rather than diminishing it, and may well be considered an important branch of good husbandry.

Clover.

N. Griffin says there is no subtitute for clover, so far as I know—nothing to take its place. It is better in its effect on land than any other forage plant. It is said that a good crop of clover—say such a crop as will yield two tons of cured hay" from an acre—will leave an equal weight of roots for the soil. That is like a coat of manure. I am sorry to hear that clover is falling into disrepute, for its renovating power is greater than that of any other plant. Lately clover does better than in a few years past, so I hope we shall soon have all the old measure of success. Many years ago the farms in Dutchess county used to give large crops of timothy and they were taken away and sold. Those farms are now exhausted—ruined because the crops were taken off. But clover is never all taken off when the roots are left. Forty five years ago a great deal of timothy was raised in Tompkins county, N. Y., and the land that produced it ran down under its production and the occupants had to turn their attention to clover. At first it was difficult to get it established, but little by little under its influence the lands grew better. Farmers had to ditch their lands as the first condition, then they used plaster, and at last got full crops of clover and better crops of grain, for their lands improved through clover. Hungarian grass has been tried, but, like timothy, when the crop is taken off nothing is left, and the soil becomes poor. The best crop is that which leaves most to the soil, and that is what clover does. I hope it will not lose its place in our farming, for there is no other plant so beneficial in its effect.

Ploughing.

Do not plough land until it is dry enough to be turned without packing like mortar under the trowel, and it is important to harrow before the furrows have dried much, else there will be hard lumps that will be difficult to dispose of all summer. It is believed that great mischief has been perpetrated by theoretical agricultural writers heretofore by advocating deep ploughing. A shallow soil may be deepened very gradually as the quantity of manure is increased, but not faster. The process of increasing the depth of the soil should be principally at fall ploughing. It is believed that farmers generally do not plough enough; there is no labor labor lost by an extra ploughing or two. A thorough pulverization of the soil is necessary for the best results at farming; some of the new harrows do excellent work in this direction and at small cost.

Potatoes.

It is better to plant potatoes early, then a second crop may be grown on the same land after they are off; either fall turnips, Hungarian grass, or land may be sown down to grass with the best success. Better plant in drills, cover with a plough, then harrow and drag just before they come up. Thorough cultivation should be made without any hand-hoeing. Economy of labor should be carefully studied, aiming to secure the best results with the least labor.

Onions.

It is folly to expect a good crop without the best of seed, thorough preparation of land and manure. They should be sown by the 25th of this month, if the land is in suitable condition. Between three and four pounds of seed is about right per acre.

HORTICULTURE.

The Rhubarb Plant.

No garden is complete unless it contains a few rhubarb plants. They are often grown on the edges of walks and sometimes near the fences, when but a few are wanted. Rhubarb is excellent for pies, and when prepared the same as when ready for pies it can be put away in jars and preserved for future use. Roots set out this year will produce good-sized stalks next year. The plants will do best if supplied with a dressing of well-rotted manure and an occasional watering with soapsuds. Plaster also benefits them, as likewise will common salt. In dividing the roots for planting a portion of the crown should remain with each section. Rhubarb needs good cultivation, and the soil should be kept mellow. Mulching also assists the plants.

The Mulberry Trees.

As the subject of silk culture is receiving considerable attention in the United States, a few items from this section may be of interest. I live near a colony of Russian Mennonites, who immigrated from Russia about seven years ago. The mulberry tree was introduced into their colonies in Russia by the Czar, for the purpose of silk culture and to facilitate rain fall. He compelled his subjects to buy trees of the government, and each land holder had to plant a certain number. They soon learned their value, and that silk culture was not the only consideration in raising them. They found the timber very desirable for fuel. It also furnished the finest material for cabinet work, and fence posts made from it would outlast those made from any other timber. The tree soon became the most highly prized of any Russian timber tree. It also bore edible fruit which was marketable in Russia.

When the Mennonites came to this country, they brought the seed of this tree with them. The mulberry grows quite rapidly. Trees, the seed of which was planted six years ago, are now twenty feet high, and large enough for fence posts. The tree resemble, the apple tree in its habits of growth. The Russians say that they grow quite large after reaching the height of forty to fifty feet, and from three to five feet in diameter. It bears fruit very young, frequently commencing at two years old, and bears every year. Last year the trees were densely loaded with fruit, and farmers came several miles to purchase this fruit for desert. It varies in flavor from sub-acid to sweet; color jet black and reddish white, ninety per cent. black. As the tree is different from any mulberry we know in this country, we call it "Russian mulberry." The bark is grayish white and branches drooping. The Mennonites also use it as a hedge plant, and it makes a beautiful hedge, and stands shearing as well as any tree, Prof. Budd, of the Iowa Agricultural College, says it is propogated more readily from cuttings than any mulberry with which he is acquainted. The Mennonites have interested themselves in the silk business somewhat since they have been in this country, and have some cocoons for sale.

An Excellent Old Apple.

The Yellow Bellflower—the "Belle Flour" of Coxe—one of our oldest apples, is most valued where best known. But the reason given in *The Prairie Farmer* for its unproductiveness in some soils, namely, an imperfection in the blossoms, is at most only partially correct. The tree is one of the very hardiest, and in rather poor, dry soils it is one of the surest croppers. In strong soils it makes a luxuriant profusion of growth, but often fails to bear. The cause of this unfertility is in that very profusion of leaves and shoots. Nurserymen find it remarkable in their nursery rows for the great number of twigs thrown out. While other sorts devote themselves chiefly to make spurs on the two-year old wood, with only a few wood shoots near the tips, the Bellflower sends out a brush of wood-shoots, or twigs, all the way up. The tree, of course, retains this disposition when set in the orchard. It is encouraged to it by rich soil. In such a location the countless drooping twigs that cover the whole head of the tree like hair, shut off the light from the fruit spurs on the older wood. The leaves of these drop off, or grow thin for want of sunshine, and of course but very few blossom buds can be properly formed. I have a magnificent tree of this first-class apple at the foot of a hill of rich loam. It is like a mountain of spray which rises far aloft, hides entirely all the interior frame of branches, and sweeps to the ground all around. Yet there is not a surer or better bearer in the orchard. Every winter I take a ladder and pruning shears and thin the exterior shoots all over the vast head, so that those left can leaf out without either shading each other or obscuring the leaves that will issue from the fruit spurs on the older wood immediately below. This is using many words and much room in the interest of one apple, but the Bellflower has such exceptional excellencies as justify the outlay, in the interest of all who enjoy apples, whether at dessert or from the oven, in their highest perfection of apple flavor, aroma, size and beauty.—*Penn.*

An Experiment in Potato Planting.

Last spring when planting my Beauty of Hebron potatoes, says a correspondent of the *Rural New Yorker*, I planted one row through the piece as follows: I took potatoes below the medium size, out of the seed and seed ends, cut out all the eyes but two, planted them and gave them the same care as the rest of the piece. The "seed" for the rest was of the same size potatoes cut in two and planted one piece in a hill about eighteen inches apart in the row. Now for the result: The first row, containing seventy hills, gave 195 pounds; one row by the side of it, with eighty-six hills, gave 143 pounds—a difference in yield per hill of over fifty per cent. in favor of the whole potatos with two eyes. This row could be distinguished from the rest as far as the piece could be seen, until the dry weather dried up the vines. The whole piece yielded a splendid crop, as did a piece of Snowflake in another part of the field.

Household Recipes.

To Make a Cheap Wash or Paint.—Put half a bushel of good lime in a clean barrel, and add enough water to make a thin whitewash, stirring with a flat stick until every lump is dissolved; then add fifty pounds whiting, fifty pounds road dust. Then thin to the proper consistency for spreading with a brush, by adding sweet buttermilk fresh from the churn in small quantities at a time, to give a chance for the ingredients to assimilate.

Rice, Milanaise Style.—Fry one ounce of butter (cost two cents) light brown; put into it half pound of rice (cost five cents) well picked over, but not washed, and one ounce of onion, chopped fine; stir and brown for five minutes, then add a pint of gravy from meat, season with a level teaspoonful of salt, quarter that quantity of pepper, and as much cayenne as you can take on the point of a very small pen-knife blade; the onion and seasoning will cost less than two cents; stew gently for fifteen minutes, stirring occasionally to prevent burning, and serve as soon as the rice is tender. This makes a palatable dish for about ten cents.

Macaroni and Ham.—Take one-half pound of macaroni, put it in boiling salt and water, and cook for twenty minutes or until tender. Drain the macaroni, put it in boiling salt and water, and cook for twenty minutes or until tender. Drain the macaroni, and put in cold water until you are ready to use it. Take a quarter of a pound of the lean of cold boiled ham, chop fine; take half a can of tomatoes, seasoned with salt and pepper; stew with a small onion; pass the tomatoes, when thoroughly done through a colander; thicken with a tablespoonful of butter, rubbed smooth. Lay the macaroni in a layer, spread on the chopped ham, add some of the tomato sauce, a little pepper, a little salt, and so on, until the dish is filled. Bake in a hot oven for twenty minutes.

Poor Man's Plum Pudding.—Take three cups of flour, one cup chopped suet, one cup stoned raisins, one-third cup molasses, one cup milk, one teaspoon of saleratus dissolved in the milk, half teaspoon of salt, one teaspoon each of allspice, cinnamon and cloves. Boil three hours. Serve with sauce as follows: One cup of sugar, half cup of butter, one egg, one tablespoon of flour, beat all together. When ready for the table pour in two thirds cup boiling wine. Add nutmeg, grated, and you will have a dish fit for a king, though planned for a poor man.

Fig Pudding.—One pound of flour, six ounces fresh beef suet, with half tablespoonful of salt; and one pound figs, with one tablespoonful of baking powder. Chop the suet as fine as possible, remove all strings; mix well with the flour, salt and baking powder; make this into a paste with iced water, and roll out into a sheet; cut the figs into long slices, cover the paste with them, tie in a cloth, and boil in fast boiling water for two hours. Eat with a sauce.

Yorkshire Pudding.—A quarter of a pound of flour, with a quart of water or milk; three eggs, well beaten, to be mixed with it; pepper and salt; butter the pan; put it under the beef so as to catch the gravy; have it in a good big pan, so as to be thin. Cut in pieces when served with the beef, and arranged around the dish.

Warm Slaw.—Slice a head of cabbage fine; put it in a stew pan, with a little water, and scald well; sprinkle salt, pepper and sugar over it; then take two thirds of a teacup of vinegar, one-third of a teacup of water, one egg, one-half teaspoon of flour, well mixed together; pour it over the cabbage, and let it come to a boil, when it is ready for the table.

Cold Slaw.—Slice one head of cabbage very fine; sprinkle a little sugar and salt over it; then pound the cabbage. For the dressing, take half teacup of cream, whip it into a froth, add to it one teacup of vinegar; stir the dressing well through the cabbage.

Lincoln Cake.—One and one half pounds sugar, one pound butter, one and three-fourths pounds flour, two pounds fruit, one pint sweet milk, one tablespoonful soda, six eggs, one nutmeg, one teaspoonful cloves, two teaspoonfuls cinnamon, one gill of brandy.

Pastry.—Fruit and custard pies are almost invariably spoiled by having a soggy undercrust. This may be remedied by coating the top of the lower crust of pies with the white of an egg; it will absorb no moisture from the fruit or custard, will come out of the oven crisp, and will remain so.

To Clean Marble.—To clean smoky marble brush a paste of chloride of lime and water over the entire surface. Grease spots can be removed from marble by applying a paste of crude potash and whiting in this manner.

Valuable Hints.—When ice is required at night for a sick person, break it into small pieces, and if scarce care must be taken to prevent its melting, put into a soup plate, cover with another plate; and put between two feather pillows.

Cocoanut Cookies.—One cup of milk, one cup of sugar, one cup of grated cocoanut, and prepared flour enough to roll out. Make very thin and bake quickly. The dessicated cocoanut may be used, but it is not quite so nice

To renovate black grenadine take strong cold coffee, strain it, and wring the grenadine out of it quite tight, after which shake out and fold up. Then iron it with a moderately hot iron over a piece of any old black material.

Silk stockings must be washed in cold water with white soap, rinsed in cold water, laid flat on a fine towel, rolled tightly until dry, and rubbed with a piece of flannel to restore the gloss.

Corn-Starch Cake.—Two cups of sugar, one cup of milk, one cup of corn starch two cups of flour, four eggs, one teaspoonful of soda, and flavor with lemon or vanilla.

Black Bean Soup.—One quart of black beans; soak them over night in cold water; drain off the water in the morning and add three pints of fresh water; let them stew gently four and one half hours. Add salt, pepper and a little clove. While cooking, put in meat, cooked or uncooked, as preferred. When done strain the soup; cut the lemon in slices; place in your dish; also add a hard boiled egg cut in slices. Pieces of bread toasted brown are an addition. Salt pork may be used instead of meat.

To Clean Musty Barrels.—A German paper gives the following directions for cleaning rusty or mouldy casks and barrels: First rinse them out well with water in which a little soda has been dissolved; then fill up with water slightly acidulated with muriatic acid, and let this stand for two days; then pour out the water and rinse with clean water, and the casks will be found perfectly sweet.

Cottage Gingerbread.—Take one cup of butter and lard melted together, add one cup of New Orleans molasses; stir into this one cupful each of sugar and cold water, two large teaspoonfuls of ginger, two eggs beaten, and four cups of flour having in three large teaspoonfuls of baking powder. Bake in a moderately hot oven.

Household Weights and Measures.—Wheat flour, one pound is a quart. Loaf sugar, broken, one pound is one quart. White sugar, powdered, one pound one ounce is one quart. Best brown sugar, one pound two ounces is one quart. Eggs, average size, ten are one pound. Liquid measures, sixteen teaspoonfuls are one pint.

Scotch Butter Candy.—One pound of sugar, one half pint of water. Boil as hard as possible without graining. When done add half a cup of butter and lemon juice to flavor, if desired. Turn on a buttered dish and when partly cool cut with a knife into small squares. When cold a slight tap will break it off.

Live Stock.

Sawdust for Bedding.

Many farmers claim that sawdust is not only worthless as a manure, but positively injurious to the soil. A farmer sends the following words in its favor: I use it when I can get it, and value it very highly for bedding the cow stable, as it will keep cattle cleaner than any other bedding I know of. It also makes the manure fine and mellow, so that it spreads more evenly and mixes with the soil more like composted manure. I also use it in the hen-house for filling the nest boxes, and on the floor to mix with the manure, as it absorbs all the ammonia and prevents the manure from sticking to the floor. For summer use it is not as good as dry dirt or sand in the hen house, because it tends to breed vermin, unless cleaned and replaced by a fresh lot quite often. A small quantity of it thrown into the privy vault will absorb all bad odors arising therefrom in hot weather. It is also one of the best dryers to mix with superphosphate. It makes it fine so as to handle well. I do not think sawdust is very valuable in itself as a fertilizer, yet it must be worth something. If it has no other value it contains all the saline properties found in wood ashes as well as some nitrogen: but these elements are found in small quantities and in a form which is unavailable for immediate use. Sawdust contains more nitrogen than straw, but less potash and phosphoric acid, and is probably not as good as cut straw for bedding or manure, but it is a better absorbent of bad odors, and is usually cheaper than cut straw. I believe its mechanical effect on the soil is excellent, especially to lighten heavy clays. Professor Johnson has said that "fresh sawdust in light, thirsty soils tends to increase their water holding capacity. In sticky clay it lightens the texture, and soil that forms a hard crust after rain it prevents, like other mulch, such puddling and backing of the surface." I think a cord of sawdust, well saturated with liquid manure, is worth as much if not more than a cord of solid manure.

Salting Stock.

Prof. James Law writes to the *Farmers' Advocate*, of London, Ontario, on the subject as follows:

In addition to its use as a condiment, salt is one of the best laxatives. In cattle, and sheep especially, in which a dry winter feeding is liable to induce a partial impaction of food between the leaves of the third stomach, the stimulus given by the salt to the free secretion and the muscular movement of the stomachs, together with the engendered disposition to drink more freely, serve to dislodge such obstructions and to restore perfect digestion. Even if a full purgative action is wanted, few agents will serve better than one or two pounds of salt, according to the size of the cow. But it should never be forgotten that water must be allowed without stint after the administration of such a dose, as a concentrated solution of salt is highly irritating to the stomach and bowels. An abundant consumption of water serves at once to remove the irritant qualities of the salt, and to hasten the action of the bowels.

Finally, salt is destructive to almost all intestinal worms. In Brazil, where cattle are very subject to parasites, the stock owners have discovered this virtue, and therefore dose their animals twice a year with one pound each of common salt, after the action of which they always manifestly improve in condition. It is to the young worms especially that salt proves destructive, hence a daily allowance of one or two ounces for the larger quadrupeds, or two drachms for the sheep, will go far towards warding off fatal attacks by destroying the young parasites as they are taken in, in the food or water. Thus for the liver worms in sheep (Rot) salt marshes or the free use of salt proves almost a specific, and the stomach, and, to a less extent, the intestinal worms of all domestic animals, may be kept in check by a daily liberal yet moderate allowance.

Floors for Horse Stables.

The long debated question as to the best material for stable floors is being again revived. A clay floor was adhered to by some for years, and such was the earnestness of the advocates and the many arguments brought to bear upon it that we were induced some twenty years ago to try it. In three or four months we had the planks back again, being satisfied of the disadvantage of clay for this purpose. Our present floor of plank is simply inclined a little from front to rear, where the usual gutter is made to carry off the liquid voidings. We do not believe in sand, coal ashes, sawdust, asphaltum, flags, cobble stones or any of these modern devices to injure horses. Thus far we have never noticed that this little inclination was in any way injurious, and we doubt whether the wooden grating that we frequently see placed over the planking that some use would be advisable on the ground that the animal would be more comfortable, while this movable grating or second floor might lead to accidents. When a person can keep horses in a good, sound, healthy condition for five or seven years, as we have done, on a carefully constructed plank flooring inclining a little to the rear, it is just as well to be satisfied with it. Do what one will, holes will be dug by the stamping of the feet in the clay, and these will be filled with moisture, which will necessarily result in scratches, quarter crack, etc. If the clay is leveled off and beaten down daily it will make no difference. Some time ago we visited a number of stables where many horses were kept and we encountered only one which was composed of anything but wood. Of course there will be new things—inventions—springing up which are to meet and overcome every objection, and there will be some to adopt them, but we will be satisfied with what we have until there is something produced about which there will be no mistake.—*Germantown Telegraph.*

Charcoal for Sick Animals.

In nine cases out of ten when an animal is sick, the digestion is wrong. Charcoal is the most efficient and rapid corrective. The hired man came in with the intelligence that one of the finest cows was very sick, and a kind neighbor proposed the usual drugs and poisons. The owner being ill and unable to examine the cow, concluded that the trouble came from overeating, and ordered a teaspoonful of pulverized charcoal to be given in water. It was mixed, placed in a junk bottle, the head turned downward. In five minutes improvement was visible, and in a few hours the animal was in the pasture quietly grazing. Another instance of equal success occurred with a young heifer which had become badly bloated by eating green apples after a hard wind. The bloat was so severe that the sides were as hard as a barrel. The old remedy, saleratus, was tried for correcting the acidity But the attempts to put it down always raised coughing, and it did little good. Half a teaspoonful of fresh powdered charcoal was given. In six hours all appearance of the bloat had gone, and the heifer was well.

The Hog Crop.

It is the opinion of many that we will have a large hog crop this year. In these days, when half of the hogs raised are marketed at or under one year of age, it does not take long to make good a shortage. It is true that a shortage of corn will very soon make itself apparent in the weight and general development of stock, but it does not necessarily interfere with the increase in numbers, and though the spring of 1881 was generally very unfavorable for pigs, everything has since been in favor of the breeder. The feeder, however, has been compelled to work on a more economical basis than usual, and many a lot of hogs that should have been kept gaining by full feeding has been allowed, or rather compelled to "root hog or die" sure enough. At the present time we are getting liberal runs of good hogs, from 30,000 to 50,000 per day at this point, and the indications do not point very strongly to any very serious falling off in the crop of marketable hogs for the spring and summer. A fact to be borne in mind, however, is that prices have recently advanced to tempting figures—as high as $7 for extra heavy hogs—well calculated to draw strongly on the available hog crop.—*Chicago Paper.*

Tying Up Calves.

Will you please give me your opinion in regard to tying up calves? Some say tie them up, others let them go with the sheep, and others turn them into a pen by themselves, and litter them well, and they will do first-rate. I think of taking up with the last advice. Yours truly, T. C. P.

If one has a good sized pen, and can keep it well bedded so as to keep the calves clean, it is the most natural way, and undoubtedly a better way, than tying them up, but if one is pressed for room, or has but a limited quantity of bedding, they will do very well if tied up with a halter, if they have been accustomed to be led and tied; if not, they will at first be very uneasy. It is always best to accustom calves to the halter when very young, for if brought up to be led, it is much easier to lead than to drive them.

It is not good policy to turn them in with the sheep and lambs, for when the lambs are young they are liable to get injured by the calves; it may be said that by letting them run with the sheep, they will eat up a considerable portion of what the sheep leave; but on the other hand, they will be sure to get a large share of the best hay given the sheep if it is within their reach, and if it is not, they cannot eat the poor. As the calves need to be fed different from sheep, this, if nothing more, is a sufficient reason for not keeping the calves with them.—*Massachusetts Ploughman.*

Man's Treatment of the Horse.

The man has cut away the frog, because he thinks the horse will be injured if the frog touches the ground. He has then cut a deep groove at the base of the frog. This is to give a well-opened heel, as he is pleased to call it. He has scooped away the sole to "give it spring." He has scored a deep notch in the toe for the shoe. This is evidently a conservative relic of the time when nails were not used, and the shoe attached by three pointed clips hammered over the edge, one in front and one on either side. Then he has improved the whole of the outer surface of the hoof. As the Creator has furnished this part of the hoof with a thin, hard, polished plate, forming a sort of varnish which is impervious to wet, the farrier, as a matter of course, rasps it all away up to the crown. And as the Creator has placed round the crown a fringe of hair, which acts as a thatch to the line of junction and throws off the rain upon the water proof varnish, he cuts this away with his scissors. Lastly, the Creator having given to the horny hoof a mottling of soft and partially translucent brown, gray-blue, yellow, black and white—never exactly the same in two hoofs, much less in two horses—the farrier takes a blacking pot and brush, polishes up the hoofs until they look like patent leather boots, all four exactly alike, and then contemplates his work with satisfaction. In his own words, he has "turned out a finished job of it."

Advantages of Small Flocks.

The reason why large flocks of sheep—and the principle applies to all farm stock—are less thrifty than a small number together is answered very truly by an address before the Indian Sheep Growers' Association, in speaking of pasturing: There is one thing about pasturing sheep that has been overlooked, viz., the damage done to the grass by being run over by the flock. While I believe one acre of good grass would keep five or maybe eight sheep well, I do not believe 100 acres would keep 500 sheep. Five sheep would probably do but little damage to one acre, even though they were confined to it; they would put a few tracks over it in a day, and would easily find fresh grass each day. But suppose we put five hundred sheep in a hundred-acre lot; if each sheep would confine themselves to their own particular acre they would probably do well in summer. But they will not do this, and right here is where theory and practice part company. Our five sheep start out to graze, and the 495 go along with them. Now, a sheep is a dainty creature, and likes clean food. So the hindermost part of the flock keep pushing ahead, paying little or no atention to what has been already run over, and being in each other's way each would go over ten times as much ground before it is filled as it ought. And having so much more work to get its food, it does not do so well as one that can satisfy itself with little or no exertion. Going over the trail too frequently and picking about dung and urine for grass is doubtless what makes large flocks so liable to disease.

"Loss of Cud."

"This is an ambiguous term," says the Kansas *Farmer*. "It may mean dropping of the cud from the mouth during rumination, or a suspension of rumination. But may occur from the same cause—viz., indigestion, or eating injurious or poisonous plants. If this is the cause the stomach and bowels should be cleared by a strong purgative, such as twelve ounces of Epsom salts with one ounce of ground ginger along with it. But the latter may occur from the animal having swallowed something which has injured the paunch, such as a thorn, a small piece of glass, or other rough, sharp substance. The symptoms which might be looked for in this case would be a rapid wasting and weakness, a staring coat and a dull colored skin, with much loose scurf upon it, irregular appetite and bowels, with discharge of gas from the throat and accumulations of it in the paunch. Nothing can be done in this case but to trust to chances and a natural recovery, leaving the animal to rest, to facilitate a cure or the expulsion of the intruding substance, if that is possible."

Training Heifers to Milk.

A heifer should be trained as soon as weaned. She should then be haltered and made used to being tied up and handled, and led by the halter. She should be carded and brushed, and her udder and teats handled frequently until she becomes used to it. A month or two before she calves she should be tied up and brushed, and the udder rubbed and the teats pulled; taught to lift the leg and keep it out of the way of the milker, and generally disciplined. All this should be done gradually and gently, and the young animal made to understand that there is nothing to fear by always exercising kindness to her. When she drops her calf no stranger should attend her, but one she knows well, and she will come to her duties as easily as an old cow. A newly calved heifer should always be tied when she is milked, as she may be very nervous and not to be depended upon until her disposition is shown.

Bedding for Cows.

Here in New England cords upon cords of good dry sawdust and turning shavings are put into the streams at the several sawmills, cabinet shops, etc., and farmer A's cows only a mile away never have one mite of bedding the whole winter. Having talked with some of the folks that have used turning shavings or sawdust for years, it is their opinion that it pays more than double the expense incurred in getting the sawdust, for the amount of manure it makes; it saves all of the liquid manure, makes the manure pile fine and easy to work over; and it will be obliged to be worked over unless you have hogs upon it to keep it from heating with considerable sawdust in it; but that can be done on leisure days; then it will be fine, all ready worked over for the spring work, and in its best shape for the crops to get the benefit. Green manure from cows that receive no bedding and not worked over only as it goes in and out of the cart, is not worth one-half as much as fine, properly prepared manure. It not only pays that way, but how much more comfortable the cows are; keeps them clean, much better milking, etc. You receive a profit in more ways than one.—*Mirror.*

Inoculation of Animals.

In the June number of the *Medical Record*, James Law has an exceedingly interesting article on the mitigation of the malignity of disease germs. A portion of the article is devoted to a consideration of the lung plague in cattle, and while not containing anything that is new to any one who is perfectly familiar with the disease, it does contain some things that will be new to the general reader. The lung-plague, so called, is not necessarily a disease of the lungs. Prof. Law in his article truly says that it is possible to inoculate the disease in the tail. This can be done, too, with the effect of inoculation or vaccination, and it will protect the animal from future attacks as certainly as if the disease had been developed in the lungs. The professor says that some who are more witty than wise have ridiculed the idea of thus inoculating an animal, but that their nomenclature was at fault and not the inoculatus, that the specific disease, whatever it may be called; has been really produced in the tail, and that the subject of the inoculation was made proof against what is called the lung plague. The plague is a local disease which will develop in any vascular structure of a susceptible animal in which it may be implanted. The germs inhaled into the lungs prey upon the lungs alone, and if other germs are placed upon the raw surface of the tail they will develop in the tail only, but in both cases the disease affects the system in such a way that the animal will not again have the disease, however much it may be exposed.

If the tail is inoculated, the severity of the disease will depend greatly upon the depth to which the poison is planted. The exudation and swelling rarely exceeds the size of a hen's egg. But in the lungs the air passages are closed, preventing a free ingress of oxygen, and it is not uncommon for the mass of exudation to weigh as much as thirty pounds, besides an enormous liquid effusion in the pleurae. In Australia, the professor says, the inoculation is clumsily but successfully performed, by drawing a worsted thread, smeared in the exudate, through the connecting tissue beneath the skin of the tail. This is a deep insertion, but the loose texture of the worsted, serves to favor the admission of air, and to counteract any dangerous change in the virus.

POULTRY.

Sunflower Seed for Poultry.

A correspondent, Mrs. M. J. C., Otter, Iowa, gives her experience in raising mammoth Russian sunflower seed for poultry and for stock. It is eagerly eaten, makes the hens produce eggs plentifully, keeps the feathers glossy and elegant. Our correspondent has grown it successfully on a variety of soils and even in fence corners, and regards the stalks to be used for kindling wood as by no means unimportant. In conclusion, she adds: "It grows to double the size of the common South American variety, and far excels it in the large heads of nice black seeds, if cultivated like other grains and kept free from weeds. I raised heads larger over than a water pail, and very heavy. I plant a patch every year for my chickens. If you plant near your barn the poultry will live and grow fat, and one would be astonished at the amount of eggs produced. It takes three quarts for an acre and plant as far apart as corn."

Grain and Vegetables.

There are none of the cereals raised in this country, says the *Poultry World*, but have their uses as feed for domestic poultry. And all the root vegetables, such as turnips, potatoes, carrots, beets, rutabagas, etc., when cooked and mixed with meal or bran, half and half each, are esteemed economical and healthful for fowl feed.

The fattening properties of some grains, and the undue allowance daily of boiled potatoes and corn or wheat meal are excessive, however. And where the Asiatics only are kept, discretion must be exercised in dealing out these hearty kinds of aliment, inasmuch as it is neither healthful, economical, or useful to stuff these birds with such feed.

The Cochins, the Brahmas, the Dorkings, and the Plymouth Rocks will very quickly become next to useless as *layers* if indulged in overfeeding with these rich grains and succulent roots. Care must, therefore, be had to avoid this error, particularly along through the late fall and winter, when we are preparing them for early spring laying.

If they get *fat* in cold weather old hens will gather this objectionable kind of flesh internally often. And this directly interferes with their laying, while it harms the quality of the egg they do lay, for future hatching.

All our fowls should be well fed in cold weather. But the *breeding* stock must be cautiously managed in this particular, or their eggs will very frequently prove infertile, from excessive cramming with these hearty sorts of food.

Keep a ready supply of oats and barley at hand for *these* birds. Allow them but little corn, and this crushed, and to be given at night only. Plenty of green vegetables will help them, steadily allowed all winter, such as cabbages, chopped rutabagas, etc. And if we avoid whole wheat, buckwheat, and whole corn—for our breeders—they will do much better next spring, when we want their eggs for incubation.

Poultry Upon the Farm.

As a general rule, fowls run the farmer in debt during the winter months, when, with proper care they could be made to pay a handsome profit. The first especial thing is good, dry, warm quarters. Second, is plenty of egg-producing food fed regular, with a dish of warm water (or warm skim milk is much better) for drink with each feed. The best food in the morning is a mush, made of equal parts of corn, barley, oat-meal with a few shorts all mixed thoroughly with boiling hot water (or milk if you have it), add a teaspoonful of fine salt to every quart of grain; once a week add a little sulphur and cayenne pepper; give it to them while warm, just as soon as they come off of the roost or can see to eat it.

Ten o'clock or just before dinner, feed with boiled meat, cut fine, or boiled fish; the offal from the markets and slaughter-houses is a very cheap way to get fish and meat for your fowls. Four o'clock, give all they will eat of dry, whole grain, equal parts corn, buckwheat, barley, burnt wheat or wheat screenings and oats. More depends upon the food than the breed of fowls; but most farmers would think it too much trouble to follow the above, but would go off to the nearest store or hotel and let poor biddy scratch in the hay-chaff for her breakfast, dinner and supper.—*Mirror*.

Dressing and Keeping Poultry.

In reference to dressing and keeping poultry, "J," of the *Philadelphia Record*, gives this sound and good advice: After the fowl is killed and plucked, cut off the head and feet, and draw out the entire inside parts. Then carefully wash and place aside to cool. After cooling sprinkle good table salt into every part of the inside, and well rub salt on the outside of the body; then nicely clean, and salt the heart, liver and gizzard, and place them into the carcass, as many buyers prefer to have them. Or, if preferred, clean the fowls as directed, and instead of salting, place them in an air-tight box, and at the bottom of the box burn a small quantity of sulphur. As soon as the sulphur fumes begin to rise close the box, and at the end of half an hour take out the fowls and pack them for market. They will keep for weeks by this process, as the sulphur fumes effectually prevent fermentation in all substances; and instead of being injurious are positively beneficial, in completely destroying everything that could by any means be injurious in the carcass. It will give a peculiar glazed appearance to the fowl, and a slight odor of the sulphur may remain (often none), but the moment the carcass is heated for cooking the sulphur gas passes off, and the meat is purer and better than before.

Common Sense in the Poultry Yard.

The "poultry" that everybody keeps are technically designated "fowls," or "barn door fowls." As a rule they are kept in small flocks, fed chiefly upon what no farmer misses. On most farms a flock of twelve to twenty hens will pick up a living without receiving a particle of grain from May to October, including both months. Their food consists of insects, seeds and grass or weeds; they need fresh water besides. What wonder is it that fowls thus kept are demonstrably more profitable than any class of stock or any crop on the farm? This is the best way to keep fowls, provided they can be induced to lay where their eggs can be found while fresh. To accomplish this a house of some kind is needed where the fowls may be shut in occasionally for a few days at a time, so as to make them roost and lay in convenient places. If fowls can roost in the trees, lay all over the farm and "dust" themselves in the road, they will almost surely be healthy, lay a good many eggs and keep in good condition. Besides, every now and then a hen will unexpectedly appear with a brood of ten or a dozen chicks, hatched under one bush where she had "stolen" her nest and done her hatching. That is all very well, so far as the hen is concerned, but no one wants it to happen. We wish the hens to lay and sit where we can put what eggs we please under them for hatching—and what is still more important, we wish to be able to collect the eggs for use or for sale daily. A fresh egg is a joy, a delight, a good gift of heaven—a "perfectly good" egg is an abomination, An egg to be fit to eat or for sale, must be fresh beyond peradventure, and utterly untainted with suspicion of having been brooded or weathered. For this reason it is a most untidy thing to use natural nest eggs. The nest egg after a while is almost surely gathered and of course is not "right."—*American Agriculturist*.

The Roup in Fowls.

Poultry is beginning to receive more attention from farmers of late years, and for the amount usually invested in that class of stock a much greater profit is derived than from any other. Of all the ills to which they are subject the most common is the roup. There are several forms of it, the disease presenting symptoms similar to the "colds" incident to humans. The signs are depression of the wings, running at the nose, sore throat and an occasional sound like "pip" from those that have it in the early stages. It does more damage than cholera, for the reason that it can be engendered in every yard without the knowledge of the farmer.

Roup comes from exposure. A single crack in a fowl house will allow a slight draught that often is more serious than a large opening. It develops itself among the fowls in damp weather generally, and is contagious if not at once arrested. It is a disease that seems to seek the *hard feathered* fowls, such as Games, Leghorns, Black Spanish and Hamburgs. Fowls with fluffy or downy feathers are not liable to it, as they are thus better protected from cold. Of this class the Asiatics and Plymouth Rocks are examples.

To know what fowls are subject to roup in preference to others let any farmer select the Black Spanish, for instance. The outer feathers are hard, seem full, and they really appear well protected. Now lift up the wing, and the skin is naked or free from downy feathers. Try a Brahma or Plymouth Rock in the same way, and the body cannot be seen, so thick are the fluffs of feathers. It is this covering which enables the Brahma to withstand the severity of our winters without passing through the different stages of roup; and, being so well protected, less food is appropriated to heating the body, and thus they are better winter layers. This is an important fact for poultrymen to understand—the feathering of the fowls in winter, for the better they are covered with the small downy feathers the more eggs will be gathered.

To cure the roup keep the fowl in a warm location, and give a teaspoonful of solution of chlorate of potash three times daily, at night swabbing the throat with strong copperas water. Wash the beak with warm water. Let the food be varied and soft, and keep sick fowls away from those that are well. In nearly all yards where roup appears the cause can be traced to inattention on the part of the breeder. Farmers seem to think poultry of but little importance, claiming that hens are unprofitable, but those who make this statement seldom do more than gather the eggs, leaving the fowls to care for themselves. Poultry finds quick sale in the markets, and, if the quality is good, high prices are obtained; and as to eggs, they are now selling for more than many buyers wish to pay. Were it not for roup, which is a nuisance in nearly all yards, fowls would be kept in larger numbers than usual. It is a very fatal disease, and from its habit of appearing in many different forms often misleads as to its presence. Droopy fowls in winter, if a large proportion, may be examined for roup. In its worst stages it causes swellings on the side of the head, the throat is white and slimy and the fowl refuses food.

Do not be afraid to handle the fowls, for if they are worth keeping they should be examined very often. Cholera comes but seldom in some localities but the roup is a wolf waiting at the door always. When once it gets a hold on a flock it will seize all if not stopped. The best preventives are warmth, cleanliness, changeable food and good shelter.

Poultry.

If you want fowls for general purposes take the Leghorns, Hamburgs or Spanish, or some would prefer Dorkings, Polish, Houdans or Crevecœurs. These last named breeds are what we call constant layers; but for eggs alone there is no fowl in existence that can compete with the Leghorn. They lay more eggs, consume less food, and for early, fast growing spring friers they will out travel any breed.

Perhaps at this time it would be in better place to say a little towards the care of fowls. There is no other class of stock on the farm, as a general rule, that is so sadly neglected as the domestic fowl. Why neglect this great source of human sustenance in such a way? Perhaps some of my readers will hoot at the idea, but it is true there are more fowls and poultry consumed in the United States than there is beef or pork. This looks like a big thing but the statistics show that such is the case. Look at the consumption of eggs alone; it is almost as great as as that of pork. Now is the time to clean and whitewash your roosts, and be sure and get ahead of all vermin, for they make their start in spring, and are more easily gotten rid of at the start than after they have your hen houses all polluted. A good way to keep them from starting is to pour coal oil on your roosts and any other place about your hen houses where they are likely to make a start. Spring generally brings disease with it, and a good way to keep fowls healthy is to keep a lump of alum in their drinking water; the sour from the alum mixed with water, helps to tone up their systems and keep them in healthy condition. To make fowls healthy and lay well, a good way is to give change of diet, say soft food in the former part of the day and whole grain in the evening; and green food is very essential for the health of fowls and also necessary to insure good success in hatching. But every farmer ought to see to it and have good fowls on his farm, for the first reason it takes no more to feed good ones than it does scrubs, and if he wants to sell he won't have one-half the trouble to sell blooded stock that is usually the case with common scrub stock. And I say there is not any stock on the farm that will pay more interest on capital invested than will well-fed fowls.—*Alex. Bickett in Journal of Agriculture.*

LITERARY AND PERSONAL.

Circular in Reference to Pyrethrum, issued by the Department of Agriculture; a demi quarto of 4 pages, with a full page illustration of *Pyrethrum roseum*, with a history of the plant, and ample direction for its cultivation, preparation for use as an insecticide, and modes of application. If our tobacco growers could be induced to devote a little corner of their tobacco enclosures to its growth, and give it the same care they give to the cultivation of their favorite "weed," they would at the same time produce the *antidote* to the *bane* which often diminishes the value of their crops.

The Agricultural Epitomist, John A. Woodward, editor; J. A. Everitt, publisher; semi-monthly, Watsontown, Pa., at 50 cents a year; devoted to the interest of American farmers. This is a new candidate for public favor, and judging from the number before us (No. 4, April,) it is amply worthy of it. It is a five columned folio, about the size of the *Daily Examiner*, and is replete with able and practical original and selected articles, illustrating that "*He that tilleth the land understandingly shall have plenty of bread.*" The material and typographical execution are unexceptionable, and there is not a single article in it that is not worthy of repetition; and, hailing from our own Pennsylvania, they are well adapted to the region of Lancaster county. We hail it as a valuable adjunct to our exchange list, and have no hesitation in commending it to the favorable consideration of our patrons.

THE WORLD OF NATURE

The world of animated nature is more splendidly represented under the canvas of Forepaugh's Great Show than in any zoological collection extant. Not since the day Noah lifted his hawser off the snubbing post have so many distinct varieties of rare animals been collected under one charge. This important fact should not be lost sight of by schools and parents. Boys and girls can learn more in an afternoon of natural history, in the great Menagerie of Forepaugh's Show, than by months of book study. Recognizing this, Mr. Forepaugh makes reduced rates to schools, and admits all children in orphan asylums free of charge. This Great Show will exhibit in Lancaster, Monday, April 24.

THE
LANCASTER EXAMINER

OFFICE
No. 9 North Queen Street,
LANCASTER, PA.

THE OLDEST AND BEST.

THE WEEKLY
LANCASTER EXAMINER

One of the largest Weekly Papers in the State.

Published Every Wednesday Morning,

Is an old, well-established newspaper, and contains just the news desirable to make it an interesting and valuable Family Newspaper. The postage to subscribers residing outside of Lancaster county is paid by the publisher.
Send for a specimen copy.

SUBSCRIPTION:
Two Dollars per Annum.

THE DAILY
LANCASTER EXAMINER

The Largest Daily Paper in the county.

Published Daily Except Sunday.

The daily is published every evening during the week. It is delivered in the City and in surrounding Towns accessible by railroad and daily stage lines, for 10 cents a week.
Mail subscription, free of postage—One month, 50 cents; one year, $5.00.

JOHN A. HIESTAND, Proprietor,
No. 9 North Queen St.,
LANCASTER, PA.

Important to Grocers, Packers, Hucksters, and the General Public.

THE KING FORTUNE-MAKER.
OZONE

A New Process for Preserving all Perishable Articles, Animal and Vegetable from Fermentation and Putrefaction, Retaining their Odor and Flavor.

"OZONE—Purified air, active state of Oxygen."—*Webster.*

This preservative is not a liquid pickle, or any of the old and exploded processes, but is simply and purely OZONE, as produced and applied by an entirely new process. Ozone is the antiseptic principle of every substance, and possesses the power to preserve animal and vegetable structures from decay.

There is nothing on the face of the earth liable to decay or spoil which Ozone, the new Preservative, will not preserve for all time in a perfectly fresh and palatable condition.

The value of Ozone as a natural preserver has been known to our abler chemists for years, but, until now, no means of producing it in a practical, inexpensive, and simple manner have been discovered.
Microscopic observations prove that decay is due to septic matter or minute germs, that develop and feed upon animal and vegetable structures. Ozone, applied by the Prentiss method, seizes and destroys these germs at once. Our office in Cincinnati can be seen almost every article that can be thought of, preserved by this process, and every visitor is welcome to come in, taste, smell, take away with him, and test in every way the merits of Ozone as a preservative. We will also preserve, free of charge, any article that is brought or sent prepaid to our office. It to the sender, for him to keep and test.

FRESH MEATS, such as beef, mutton, veal, pork, poultry, game, fish, &c., preserved by this method, can be shipped to Europe, subjected to atmospheric changes and return to this country in a state of perfect preservation.

EGGS can be treated at a cost of less than one dollar a thousand dozen, and be kept in an ordinary room as fresh or more, thoroughly preserved; the yolk held in its normal condition, and the eggs as fresh and perfect as on the day they were treated, and will sell as strictly "choice." The advantage in preserving eggs is readily seen; there are seasons when they can be bought for a few cents a dozen, and by holding them, can be sold for an advance of from one hundred to three hundred per cent. One man, with this method, can preserve 5,500 dozen a day.

FRUITS may be permitted to ripen in their native climate, and then be transported to any part of the world. The pulp expressed from fruits can be held for an indefinite period without fermentation where the great value of this process for producing a temperance beverage. Cider can be held perfectly sweet for any length of time.

VEGETABLES can be kept for an indefinite period in their natural condition, retaining their odor and flavor, treated in their original packages at a small expense. All grains, flour, meal, etc., are held in their normal condition.

BUTTER, after being treated by this process, will not become rancid.

Dead human bodies, treated before decomposition sets in, can be held in a natural condition for weeks, without puncturing the skin or mutilating the body in any way. Hence the great value of Ozone to undertakers.

There is no change in the slightest particular in the appearance of any article thus preserved, and no trace of any foreign or unnatural odor or taste.

The process is so simple that a child can operate as well and as successfully as a man. There is no expensive apparatus or machinery required.

A room filled with different articles, such as eggs, meat, fish, etc., can be treated at one time, without additional trouble or expense.

In fact, there is nothing that Ozone will not preserve. Think of everything you can that is liable to sour, decay, or spoil, and then remember that we guarantee that Ozone will preserve it in exactly the condition you want it for any length of time. If you will remember this it will save asking questions as to whether it will preserve this or that article—it will preserve any thing and every thing you can think of.

There is not a township in the United States in which a live man can not make any amount of money, from $1,000 to $10,000 a year, that he pleases. We desire to get a live man interested in each county in the United States, in whose hands we can place this Preservative, and through him secure the business which every county ought to produce.

A FORTUNE

Awaits any Man who Secures Control of OZONE in any Township or County.

A. C. Bowen, Marion, Ohio, has cleared $2,000 in two months. $2 for a test package was his first investment.
Woods Brothers, Lebanon, Warren County, Ohio, made $6,000 on eggs purchased in August and sold November 1st. $2 for a test package was their first investment.
F. K. Raymond, Morristown, Belmont Co., Ohio, is clearing $2,000 a month in handling and selling Ozone. $2 for a test package was his first investment.
D. P. Webber, Charlotte, Eaton Co., Mich., has cleared $1,000 a month since August. $2 for a test package was his first investment.
J. H. Layford, La Salle St., Chicago, is preserving eggs, fruit, etc., for the commission men of Chicago, charging 1½c. per dozen for eggs, and other articles in proportion. He is preserving 300 dozen eggs per day, and on his business is making $4,000 a month clear. $2 for a test package was his first investment.
The Cincinnati Feed Co., West 8th Seventh Street, is making $5,000 a month in handling brewers' malt, preserving and shipping it as feed to all parts of the country. Malt unpreserved sours in 24 hours. Preserved by Ozone it keeps perfectly sweet for months.

These are instances which we have asked in the privilege of publishing. There are scores of others. Write many of the above parties and get the evidence direct.

Now, to prove the absolute truth of what we have said in this paper, **we propose to place in your hands the means of proving for yourself that we have not claimed half enough.** To any person who doubts any of these statements, and who is interested sufficiently to make the trip, we will pay all travelling and hotel expenses for a visit to this city, if we fail to prove any statement that we have made.

How to Secure a Fortune with Ozone.

A test package of Ozone, containing a sufficient quantity to preserve one thousand dozen eggs, or other articles in proportion, will be sent to any applicant on receipt of $2. This package will enable the applicant to pursue any line of tests and experiments he desires, and thus satisfy himself as to the extraordinary merits of Ozone as a Preservative. After having thus satisfied himself, and had time to look the field over, to determine what he wishes to do in the future—whether to sell the article to others, or to confine it to his own use, or any other line of policy which is best suited to him and to his township or county—we will enter into an arrangement with him that will make a fortune for him and give us good profits. We will give exclusive township or county privileges to the first responsible applicant who orders a test package and desires to control the business in his locality. The man who secures control of Ozone for any special territory, will enjoy a monopoly which will surely enrich him.

Don't let a day pass until you have ordered a Test Package, and if you desire to secure an exclusive privilege we assure you that delay may deprive you of it, for the applications come in to us by scores every mail—many by telegraph. "First come first served" is our rule.

If you do not care to send money in advance for the test package we will send it C. O. D., but this will put you to the expense of charges for return money. Our correspondence is very large; we have all we can do to attend to the shipping of orders and giving attention to our working agents. Therefore we can not give any attention to letters which do not order Ozone. If you think of any article that you are doubtful about Ozone preserving remember we *guarantee that it will preserve it, no matter what it is.*

REFERENCES.

We desire to call your attention to a class of references which we can reproduce or firm based on any thing but the soundest business success and highest commercial credit could secure.

We refer, by permission, as to our integrity and to the value of the Prentiss Preservative, to the following gentlemen: Edward C. Royce, Member Board of Public Works; E. O. Eshelby, City Comptroller; Amor Smith, Jr., Collector Internal Revenue; Wulsin & Worthington, Attorneys; Marlin H. Harrell and B. F. Hopkins, County Commissioners; W. S. Cappeller, County Auditor; all of Cincinnati, Hamilton County, Ohio. These gentlemen are each familiar with the merits of our Preservative, and know from actual observation that we have without question

The Most Valuable Article in the World.

The $2 you invest in a test package, will surely lead you to secure a township or county, and then your way is absolutely clear to make from $2,000 to $10,000 a year.
Give your full address in every letter, and send your letter to

PRENTISS PRESERVING COMPANY. (Limited,)
S. E. Cor. Ninth & Race Sts., Cincinnati, O.

Nov-3m

WHERE TO BUY GOODS IN LANCASTER.

BOOTS AND SHOES.

MARSHALL & SON, No. 12 Centre Square, Lancaster, Dealers in Boots, Shoes and Rubbers. Repairing promptly attended to.

M. LEVY, No. 3 East King street. For the best Dollar Shoes in Lancaster go to M. Levy, No. 3 East King street.

BOOKS AND STATIONERY.

JOHN BAER'S SONS, Nos. 15 and 17 North Queen Street, have the largest and best assorted Book and Paper Store in the City.

FURNITURE.

HEINITSH'S, No. 15½ East King st., (over China Hall) is the cheapest place in Lancaster to buy Furniture. Picture Frames a specialty.

CHINA AND GLASSWARE.

HIGH & MARTIN, No. 15 East King st., dealers in China, Glass and Queensware, Fancy Goods, Lamps, Burners, Chimneys, etc.

CLOTHING.

MYERS & RATHVON, Centre Hall, No. 12 East King St. Largest Clothing House in Pennsylvania outside of Philadelphia

DRUGS AND MEDICINES.

G. W. HULL, Dealer in Pure Drugs and Medicines, Chemicals, Patent Medicines, Trusses, Shoulder Braces, Supporters, &c., 15 West King St., Lancaster, Pa.

JOHN F. LONG & SON, Druggists, No. 12 North Queen St. Drugs, Medicines, Perfumery, Spices, Dye Stuffs, Etc. Prescriptions carefully compounded.

DRY GOODS.

GIVLER, BOWERS & HURST, No. 25 E. King St., Lancaster, Pa., Dealers in Dry Goods, Carpets and Merchant Tailoring. Prices as low as the lowest.

HATS AND CAPS.

C. H. AMER, No. 39 West King Street, Dealer in Hats, Caps, Furs, Robes, etc. Assortment Large. Prices Low.

JEWELRY AND WATCHES.

H. Z. RHOADS & BRO., No. 4 West King St. Watches, Clock and Musical Boxes. Watches and Jewelry Manufactured to order.

PRINTING.

JOHN A. HIESTAND, 9 North Queen St., Bill Heads, Circulars, Posters, Cards, Invitations, Letter and Bill Heads and Envelopes neatly printed. Prices low.

Thirty-Six Varieties of Cabbage; 26 of Corn; 3 of Cucumber; 41 of Melon; 33 of Peas; 2 of Beans; 17 of Squash; 24 of Beet and 10 of Tomato, with other varieties in proportion, a large portion of which were grown on my five acre farm, will be found in my **Vegetable and Flower Seed Catalogue for 1882**. Sent FREE to all who apply. Customers of last season need not write for it. All Seed sold from my establishment warranted to be fresh and true to name, so far, that should it prove otherwise, I will refill the order gratis. The original introducer of **Early Ohio and Burbank Potatoes, Marblehead, Early Corn, the Hubbard Squash, Marblehead Cabbage, Phinney's Melon,** and a score of other New Vegetables, I invite the patronage of the public. New Vegetables a specialty.

JAMES J. H. GREGORY, Marblehead, Mass.

Nov-6mo)

EVAPORATE YOUR FRUIT.
ILLUSTRATED CATALOGUE FREE TO ALL.

AMERICAN DRIER COMPANY, Chambersburg, Pa.

Ap1-tf

FARMING FOR PROFIT.

It is conceded that this large and comprehensive book, (advertised in another column by J. C. McCurdy & Co., of Philadelphia, the well-known publishers of Standard works,) is not only the newest and handsomest, but altogether the BEST work of the kind which has ever been published. Thoroughly treating the great subjects of general Agriculture, Live-Stock, Fruit-Growing, Business Principles, and Home Life; telling just what the farmer and the farmer's boys want to know, combining Science and Practice, stimulating thought, awakening inquiry, and interesting every member of the family, this book must exact a mighty influence for good. It is highly recommended by the best agricultural writers and the leading papers, and is destined to have an extensive sale. Agents are wanted everywhere. jan1tf

BLOOMSDALE
LARGE LATE FLAT DUTCH CABBAGE.

Large, Flat, Solid Heads, Short Stems.

For a long period of time we have had this stock of Cabbage in cultivation, originally obtained from the German and Swedish market gardeners. It has been a part of our business occupation to keep it unrivalled, to-day we offer it in its original purity, equal in quality with the very best in the country, even though the best should cost a hundred dollars per pound.

We have made this crop a study and give our customers the result of many years close observation, for which our opportunities may be judged by the fact this we have, each and every year, about one hundred and fifty acres of cabbage raised expressly to produce seed for the ensuing season, and from which selections are made with scrupulous care, guided by experience. Not a single grain of seed is raised from Stalks all from Selected Heads.

We will mail our Catalogue free of charge to all applicants.

D. LANDRETH & SONS,
Nos. 21 and 23 South Sixth Street,

BETWEEN MARKET AND CHESTNUT STS.,
BRANCH STORE—S. W. COR. DELAWARE AVE. AND ARCH STREET,
apr-6m PHILADELPHIA.

MERCHANT TAILORING.
1848 (The Oldest of All.) 1881

RATHVON & FISHER,
MERCHANT TAILORS AND DRAPERS,

respectfully inform the public that having disposed of their entire stock of Ready-Made Clothing, therefore do, and for the future shall, devote their whole attention to the CUSTOM TRADE.

All the desirable styles of CLOTHS, CASSIMERES, WORSTEDS, COATINGS, SUITINGS and VESTINGS constantly on hand, and made to order in plain or fashionable style promptly, and warranted satisfactory.

All-Wool Suit from $10.00 to $30.00.
All-Wool Pants from 3.00 to 10.00.
All-Wool Vests from 2.00 to 5.00.

Union and Cotton Goods proportionately lower. Cutting, Repairing, Trimming and Making, at reasonable prices.

Goods retailed by the yard to those who desire to have them made elsewhere.

A full supply of Spring and Summer Goods just opened and on hand.

Thankful to a generous public for past patronage they hope to merit its continued recognition in their "new departure."

RATHVON & FISHER,
PRACTICAL TAILORS,
No. 101 North Queen Street,
LANCASTER, PA.
1848   1881

ERISMAN | GLOVES, SHIRTS, UNDERWEAR. | **ERISMAN**

SHIRTS MADE TO ORDER,
AND WARRANTED TO FIT.

E. J. ERISMAN,
56 North Queen St., Lancaster, Pa.

A HOME ORGAN FOR FARMERS.

THE LANCASTER FARMER,

A MONTHLY JOURNAL,

Devoted to Agriculture, Horticulture, Domestic Economy and Miscellany.

Founded Under the Auspices of the Lancaster County Agricultural and Horticultural Society.

EDITED BY DR. S. S. RATHVON.

TERMS OF SUBSCRIPTION:

ONE DOLLAR PER ANNUM,

POSTAGE PREPAID BY THE PROPRIETOR.

All subscriptions will commence with the January number, unless otherwise ordered.

Dr. S. S. Rathvon, who has so ably managed the editorial department in the past, will continue in the position of editor. His contributions on subjects connected with the science of farming, and particularly that specialty of which he is so thoroughly a master—entomological science—some knowledge of which has become a necessity to the successful farmer, are alone worth much more than the price of this publication. He is determined to make "The Farmer" a necessity to all households.

A county that has so wide a reputation as Lancaster county for its agricultural products should certainly be able to support an agricultural paper of its own, for the exchange of the opinions of farmers interested in this matter. We ask the co-operation of all farmers interested in this matter. Work among your friends. The "Farmer" is only one dollar per year. Show them your copy. Try and induce them to subscribe. It is not much for each subscriber to do but it will greatly assist us.

All communications in regard to the editorial management should be addressed to Dr. S. S. Rathvon, Lancaster, Pa., and all business letters in regard to subscriptions and advertising should be addressed to the publisher. Rates of advertising can be had on application at the office.

JOHN A. HIESTAND,
No. 9 North Queen St., Lancaster, Pa.

$72 A WEEK. $12 a day at home easily made. Costly Outfit free. Address TRUE & CO., Augusta, Maine

ONE DOLLAR PER ANNUM.—SINGLE COPIES 10 CENTS.

THE LANCASTER FARMER

DEVOTED TO Agriculture, Horticulture, Domestic Economy and Miscellany

THE FARMER IS THE FOUNDER OF CIVILIZATION.—WEBSTER.

Dr. S. S. RATHVON, Editor.   LANCASTER, PA. MAY, 1882.   JOHN A. HIESTAND, Publisher.

Entered at the Post Office at Lancaster as Second Class Matter.

CONTENTS OF THIS NUMBER.

EDITORIAL.
Pyrethrum..65
The Kitchen Garden for May..............................65
Gapes vs Entomology...66
   Gapes and Chicken Cholera—Gapes—Gapes.
A New History of Lancaster County....................67
Lime in Soil..67
Excerpts..69
Queries and Answers...69

ESSAYS.
Some Practical Points in Peach Culture.............69
The Management of an Orchard........................71
   1st. Location—2d. Selection of Soil—3d. Its Preparation—4th. Selection of Varieties—5th. Proper Planting—6th. Judicious Pruning—7th. Good Judgment, Close Attention and a Great Deal of Work.

SELECTIONS.
History of Pyrethrum...72
   Cultivation of Pyrethrum—Preparation of the Plants for Use—The Use of Pyrethrum as an Insecticide—Modes of Application.
Quince Culture..74
Poultry Farming..75
Poultry Abundant, but Dear..............................75
Notes on French Agriculture..............................76

OUR LOCAL ORGANIZATIONS.
Lancaster County Agricultural and Horticultural Society...76
   Crop Reports—Growing Corn—Commercial Fertilizers—How Should Lime be Applied?—The Poetry of Agriculture—Questions for Discussion.
The Poultry Society...77
   Miscellaneous Business—Straws in Poultry Breeding.
Fulton Farmers' Club...77
   Asking and Answering Questions—Inspecting the Host's Premises—Grapes Rent.

AGRICULTURE.
French Farming...78
Sand Farming..78
Crop Prospects..78
Fence Posts...78

HORTICULTURE.
Apples for Medicine...78
Greenhouse and Window Plants........................78
Profit in Onions...78
Celery Culture...78
How the Chinese Make Dwarf Trees..................78

HOUSEHOLD RECIPES.
Tapioca Pudding..78
Bread Pudding..78
Chili Sauce..78
Clam Chowder..78
Saddle of Lamb...78
Tomato Soup..78
Oyster Soup..78
Chicken Salad...79
White Sauce for Game......................................79
Sugar Kisses...79
Queen of Puddings..79
Lemon Pudding Sauce......................................79
Bird's Nest Pudding..79
Orange Pudding..79
Green Corn Pattics..79
Boston Cream Cake...79
Flake Pie Crust..79
Superior Doughnuts..79
Cookies..79
Custard Pie...79
Graham Rolls..79

Rice Waffles..79
Steamed Indian Loaf...79
Muffins...79
Lemon Pie..79
Pumpkin Pie...79
Graham Muffins..79
Turkey Soup...79
Fish Sauce..79
Cabbage Salad..79
Cottage Pudding...79
Suet Pudding..79
Boiled Bread Pudding.......................................79
Lowell Pudding...79
Hominy Muffins..79
Potato Cakes..79
Oyster Fritters..79
Corn Oysters..79
Boiled Leg of Lamb...79
Tapioca Pudding...79
Snow Pudding..79

LIVE STOCK.
Care of Horses' Legs...79
Care of Sheep...79
Watering Horses...79
Save and Care for the Pigs................................79
How to Grow a Pig..79
A Nevada Stock Raiser......................................80

POULTRY.
A Writer in the Poultry Monthly........................80
A House for 200 Fowls......................................80
Questions About Eggs and Fowls......................80
Raising Sunflowers for Hens.............................80
Care of Young Turkeys......................................80
How Chickens are Born.....................................80
A Cheap Chicken Coop.....................................80
Hawaiian Geese...80
Literary and Personal.......................................80

SILK-WORM EGGS.

Amateur Silk-growers can be supplied with superior silk-worm eggs, on reasonable terms, by applying immediately to
   GEO. O. HENSEL,
   No. 238 East Orange Street, Lancaster, Pa.
may-3m]

SEND IN YOUR SUBSCRIPTIONS
FOR
THE FARMER
FOR 1882.

The cheapest and one of the best Agricultural papers in the country.

Only $1.00 per year.
   JOHN A. HIESTAND, Publisher,
   No. 9 North Queen st., Lancaster, Pa.

Eggs! Eggs!

From all the leading varieties of pure bred Poultry Bramahs, Cochin, Hamburgs, Polish Game, Dorking and French Fowls, Plymouth Rocks and Bantams, Rouen and Pekins Ducks. Send for Illustrated Circular.
   T. SMITH, P. M., Fresh Pond, N. Y.
feb-3m

$66 a week in your own town. Terms and $5 outfit free. Address H. HALLETT & Co., Portland, Maine.

D. M. FERRY & CO'S ILLUSTRATED, DESCRIPTIVE PRICED SEED ANNUAL FOR 1882.

Will be mailed FREE to all applicants, and to customers without ordering it. It contains five colored plates, 600 engravings, about 200 pages, and full descriptions, prices and directions for planting 1800 varieties of Vegetable and Flower Seeds, Plants, Fruit Trees, etc. Invaluable to all. Send for it. Address,
D. M. FERRY & CO., Detroit, Mich.

Jan-4m

$66 a week in your own town. Terms and $5 outfit free. Address H. HALLETT & Co., Portland, Maine.
jan-1yr†

WE WANT OLD BOOKS.
WE WANT GERMAN BOOKS.
WE WANT BOOKS PRINTED IN LANCASTER CO.
We Want All Kinds of Old Books.
LIBRARIES, ENGLISH OR GERMAN BOUGHT.
Cash paid for Books in any quantity. Send your address and we will call.
   REES WELSH & CO.,
   27 South Ninth Street, Philadelphia.

LIGHT BRAHMA EGGS
For hatching, now ready—from the best strain in the county—at the moderate price of
   $1.50 for a setting of 13 Eggs.
   L. RATHVON,
No. 9 North Queen st., Examiner Office, Lancaster, Pa.

WANTED.—CANVASSERS for the LANCASTER WEEKLY EXAMINER
In Every Township in the County. Good Wages can be made. Inquire at
   THE EXAMINER OFFICE,
   No. 9 North Queen Street, Lancaster, Pa

$72 A WEEK. $12 a day at home easily made. Costly Outfit free. Address TRUE & Co., Augusta, Maine.
jan-1yr†

SEND FOR SPECIAL PRICES

On Concord Grapevines, Transplanted Evergreens, Tulip, Poplar, Linden Maple, etc. Tree Seedlings and Trees for timber plantations by the 100,000.
   J. JENKINS' NURSERY,
3-2-79   WINONA, COLUMBIANA CO., OHIO.

MARBLEHEAD
Early Sweet Corn

Is the most profitable of all, because it matures before any other kind, giving farmers complete control of the early market. I warrant it to be at least a week earlier than Minnesota, Narragansett or Crosby, and decidedly earlier than Dolly Dutton, Tom Thumb or Early Beguine. If size of Minnesota, and very sweet. The original introducer, I send pure stock, postpaid, per package 15 cents; per quart, 70 cents; per peck, by express, $3.00. In my catalogue, (free to all), are emphatic recommendations from farmers and gardeners.
   JAMES J. H. GREGORY,
apr-3t   Marblehead, Mass.

THE LANCASTER FARMER.

PENNSYLVANIA RAILROAD SCHEDULE.
Trains LEAVE the Depot in this city, as follows:

WESTWARD.	Leave Lancaster.	Arrive Harrisburg.
Pacific Express*	2:40 a. m.	4:05 a. m.
Way Passenger*	5:00 a. m.	7:30 a. m.
Niagara Express	11:00 a. m.	11:20 a. m.
Hanover Accommodation	11:15 p. m.	Col. 10:40 a. m.
Mail train via Mt. Joy	10:20 a. m.	12:40 p. m.
No. 2 via Columbia	11:25 a. m.	12:55 p. m.
Sunday Mail	10:30 a. m.	12:50 p. m.
Fast Line*	2:30 p. m.	3:95 p. m.
Frederick Accommodation	2:35 p. m.	Col. 2:45 p. m.
Harrisburg Accom	3:45 p. m.	7:40 p. m.
Columbia Accommodation	7:30 p. m.	Col. 8:20 p. m.
Harrisburg Express	7:30 p. m.	8:50 p. m.
Pittsburg Express*	8:50 p. m.	10:10 p. m.
Cincinnati Express*	11:30 p. m.	12:45 a. m.

EASTWARD.	Lancaster.	Philadelphia
Cincinnati Express*	2:55 a. m.	5:00 a. m.
Fast Line*	5:00 a. m.	7:40 a. m.
Harrisburg Express	8:05 a. m.	10:00 a. m.
Columbia Accommodation	9:10 p. m.	12:01 p. m.
Pacific Express*	:40 p. m.	3:40 p. m.
Sunday Mail	2:00 p. m.	5:00 p. m.
Johnstown Express	3:05 p. m.	5:30 p. m.
Day Express*	5:35 p. m.	7:20 p. m.
Harrisburg Accom	6:25 p. m.	9:30 p. m.

The Hanover Accommodation, west, connects at Lancaster with Niagara Express, west, at 9:35 a. m., and will run through to Hanover.
The Frederick Accommodation, west, connects at Lancaster with Fast Line, west, at 2:10 p. m., and runs to Frederick.
The Pacific Express, east, on Sunday, when flagged, will stop at Middletown, Elizabethtown, Mount Joy and Landisville.
*The only trains which run daily.
†Runs daily, except Monday.

NORBECK & MILEY,

PRACTICAL

Carriage Builders,

COX & CO'S OLD STAND.

Corner of Duke and Vine Streets,

LANCASTER, PA.

THE LATEST IMPROVED

SIDE-BAR BUGGIES,

PHÆTONS,

Carriages, Etc.

THE LARGEST ASSORTMENT IN THE CITY.

Prices to Suit the Times.

REPAIRING promptly attended to. All work guaranteed.
79-9-

S. B. COX,
Manufacturer of
Carriages, Buggies, Phaetons, etc.
CHURCH ST., NEAR DUKE, LANCASTER, PA.
Large Stock of New and Second-hand Work on hand very cheap. Carriages Made to Order. Work Warranted for one year. [7—9-12

DW. J. ZAHM,
DEALER IN

AMERICAN AND FOREIGN

WATCHES,

SOLID SILVER & SILVER PLATED WARE,

CLOCKS,

JEWELRY & TABLE CUTLERY.

Sole Agent for the Arundel Tinted

SPECTACLES.

Repairing strictly attended to.

ZAHM'S CORNER,
North Queen-st. and Centre Square, Lancaster, Pa.
79-1-12

E. F. BOWMAN,

AT LOWEST POSSIBLE PRICES.
Fully guaranteed.
No. 106 EAST KING STREET,
79-1-12] *Opposite Leopard Hotel.*

ESTABLISHED 1822.

G. SENER & SONS,
Manufacturers and dealers in all kinds of rough and finished

LUMBER,

The best Sawed SHINGLES in the country. Also Sash, Doors, Blinds, Mouldings, &c.

PATENT O. G. WEATHERBOARDING
and PATENT BLINDS, which are far superior to any other. Also best COAL constantly on hand.

OFFICE AND YARD:
Northeast Corner of Prince and Walnut-sts.,
79-1-12] LANCASTER, PA.

PRACTICAL ESSAYS ON ENTOMOLOGY,
Embracing the history and habits of
NOXIOUS AND INNOXIOUS

INSECTS,
and the best remedies for their expulsion or extermination.

By S. S. RATHVON, Ph. D.
LANCASTER, PA.
This work will be Highly Illustrated, and will be put to press (as soon after a sufficient number of subscribers can be obtained to cover the cost) as the work can possibly be accomplished.
79-9-

$5 to $20 per day at home. Samples worth $5 free. Address STINSON & Co., Portland, Maine

TREES
Fruit, Shade and Ornamental Trees.
Plant Trees raised in this county and suited to this climate. Write for prices to
LOUIS C. LYTE,
Bird-in-Hand P. O., Lancaster co., Pa.
Nursery at Smoketown, six miles east of Lancaster.
79-1-12

WIDMYER & RICKSECKER,
UPHOLSTERERS,
And Manufacturers of

FURNITURE AND CHAIRS.
WAREROOMS:
102 East King St., Cor. of Duke St.
LANCASTER, PA.
79-1-12]

Special Inducements at the
NEW FURNITURE STORE
OF
W. A. HEINITSH,
No. 15 1-2 E. KING STREET
(over Burek's Grocery Store), Lancaster, Pa.
A general assortment of furniture of all kinds constantly on hand. Don't forget the number.
15 1-2 East King Street,
Nov-1y] (over Burek's Grocery Store.)

For Good and Cheap Work go to
F. VOLLMER'S
FURNITURE WARE ROOMS,
No 309 NORTH QUEEN ST.,
(Opposite Northern Market),
Lancaster, Pa.
Also, all kinds of picture frames. nov-1y

GREAT BARGAINS.
A large assortment of all kinds of Carpets are still sold at lower rates than ever at the
CARPET HALL OF H. S. SHIRK,
No. 202 West King St.
Call and examine our stock and satisfy yourself that we can show the largest assortment of three Brussels, three plies and ingrain at all prices—at the lowest Philadelphia prices.
Also on hand a large and complete assortment of Rag Carpet.
Satisfaction guaranteed both as to price and quality. You are invited to call and see my goods. No trouble in showing them. See if you do not want to purchase. Don't forget this notice. You can save money here if you want to buy.
Particular attention given to customer work.
Also on hand a full assortment of Counterpanes, Oil Cloths and Blankets of every variety. [nov-1y

PHILIP SCHUM, SON & CO.,
38 and 40 West King Street.
We keep on hand of our own manufacture,
QUILTS, COVERLETS,
COUNTERPANES, CARPETS,
Bureau and Tidy Covers, Ladies' Furnishing Goods, Notions, etc.
Particular attention paid to customer Rag Carpet, and scouring and dyeing of all kinds.
PHILIP SCHUM, SON & CO.,
Nov-1y Lancaster, Pa.

THE HOLMAN LIVER PAD!
Cures by absorption without medicine.
Now is the time to apply these remedies. They will do for you what nothing else on earth can. Hundreds of citizens of Lancaster say so. Get the genuine at
LANCASTER OFFICE AND SALESROOM,
22 East Orange Street.
Nov-1yr

C. R. KLINE
ATTORNEY-AT-LAW,
OFFICE: 15 NORTH DUKE STREET,
LANCASTER, PA.
Nov-1y

The Lancaster Farmer.

EDITORIAL.

PYRETHRUM.

The present number of the FARMER we devote largely to the reproduction of the circular issued by the Department of Agriculture, on the history, cultivation, preparation, use, and modes of application of *Pyrethrum*, as an *insecticide;* and we ask for it the respectful and thoughtful perusal of our patrons and readers; and not only a *perusal* of the paper, but also an intelligent and determined effort to *cultivate* it—the same intelligence and determination that is evoked in the cultivation of tobacco, or any other plant possessing intrinsic value. In view of the bare *possibility* of an efflux of noxious insects at any time, without any forewarning whatever, it behooves the cultivators of the soil to know how to produce, prepare, and apply a simple antidote against the invasion and destruction of their crops by these pests and other noxious animals. Nothing seems more certain than that the higher the state of vegetable cultivation, the more liable it is to the destructive attacks of noxious insects, and therefore the bane and *antidote* should occupy parallel lines in the routine of agricultural production. Any man or woman that can successfully cultivate the "common Aster," as an ornamental plant, may be equally successful in the cultivation of *Pyrethrum*, as a useful plant. Noxious insects are animals that we may expect to have to deal with as long as a single blade of grass is grown upon this earth, and it seems a lack of wisdom even to expect thir *total* extinction, or perhaps to ever desire it. They certainly must be of some *use* or their existence would never have been permitted; but there is no *use* that may not be perverted, or be transmuted into *abuse*. Hence, against a redundancy of noxious insects, the providential farmer should always be forearmed, or forewarning would be of very little avail. The Agricultural Department has distributed a limited amount of the seeds of *Pyrethrum*, but they can now also be obtained at many of the seed-stores—especially those in the larger towns and cities.

We thankfully acknowledge the receipt of three papers of *Pyrethrum* seeds from the Department, one of which, (*P. cinerariæfolium*, from Transcaucasia,) we have placed in the hands of Mr. John Zimmerman, and another (*Pyrethrum roseum*, grown in Austria,) in the hands of Mr. George Hensel, who propose to make a practical test of their cultivation in this locality. The third paper (*P. cinerariæfolium*, from California,) we propose to test on our own premises, unless we feel convinced that it would be better to place it in other hands. The celebrated "Persian Insect Powder," which has been on the market for a dozen years or more, and which is represented to be "sudden death" to "bedbugs, rats and roaches," is nothing more nor less than the pulverized flowers of a plant of the *composite* order, and is allied to *Pyrethrum*, if it does not belong to the same genus. Some years ago, a vegetable powder called "Buhach," or C. N. Milco's California Universal Insect *Exterminator*, was brought out and widely distributed, but we have seen or heard nothing of it, *pro* or *con*, since its first production, either in an agricultural or an entomological journal, and we somehow came to the conclusion that it proved valueless for the purpose proposed.

The provident and foreseeing farmer is perfectly cognizant of the fact that a routine of domestic obligations annually devolve upon him, which cannot possibly be evaded or ignored, and hence he habitually makes ample provision for them. He requires a sufficient quantity of food, of fuel, of clothing, of shelter, and the usual concomitants of civilization, not only for his individual self, but for all that is subordinate to his social and domestic rule. And these things he provides understandingly, methodically and continuously, because he *knows* that both *he* and *his* will stand in need of them as long as life remains. They are not regarded as mere incidentals, or probabilities, or guess-work, but as things inevitable, and that cannot be compromised. Let him in addition to these, make provision for the continued destruction of noxious insects, for, depend upon it, like "the poor," we shall "always have them with us."

THE KITCHEN GARDEN FOR MAY.

"In the Middle States, during the past month, some of the hardier vegetables will have been sown, but perhaps not as freely as in former years; April having been unusually, and continuously cold; but by the middle of the present month, all will probably have been put in; hence the labor will now mainly consist of the various operations of transplanting, thinning, weeding, hoeing, &c. The following alphabetical directions will serve as a reminder to the unpracticed gardener who is also referred to the directions for April.

Beans, Bush, plant for succession: Lima, Carolina and other "pole-beans" may now be planted. *Beets*, long sow; *Cabbage* plant, sow seed, if not done last month. *Carrot*, long orange, sow. *Cauliflower*, in frames, remove glasses. *Celery*, weed. Crops which have failed when first sown, repeat sowings. *Cucumbers*, Early Frame, plant. *Lettuce*, large cabbage and Indian Dutch Butter, sow in drills to stand; thin out if too thick. *Melons* plant; the best is Landreth's Boss—see note below. *Parsnips* thin out, if ready. *Weeds* destroy as they appear, and hoe and otherwise cultivate the advancing crops; it is needless to particularize each duty. Where the interest and taste lead to gardening, directions for every operation are necessary to but few. It is not, however, discreditable to the character of many farmers who till their own land, and should reap the reward of well cultivated gardens, that none but the simplest vegetables may be found upon their tables, and in too many instances that scanty supply is the result of woman's labor?

We have in former issues of the RURAL REGISTER recommended a '*Farmer's Kitchen Garden*,' where nearly all the preparation of the land may be done by horsepower, and thus most ample supplies of vegetables be obtained at all seasons, without hand labor or occupation of time, which may not be readily spared from farm duties, and the women of the houshold relieved from toiling to supply household wants."—*Landreth's Rural Reg. for 1882.*

In this connection it may not be inappropriate to mention a new Water-melon of rare quality which has been originated by the Landreths and named the "Boss," which possesses qualities calculated to make it more popular than that term has become in the political world. When "Bossism" is founded upon real merit, there certainly can be no valid objection to its universal prevalence.

The special merits of this melon are the following: "Early, large in size, long in shape, and very heavy. Rind thin but very tough, dark green in color, slightly ribbed, showy in appearance. Flesh more highly colored than any other melon in existence, crystalline or granulated, melting, of unusually fine flavor, and extending within an inch of the skin. A variety certainly valuable either for shipping or home consumption." It is confidently recommended as the best melon in the market, by those who *know* all the sorts of this luscious and refreshing *gourd*.

Of course, it might be deemed more appropriate to discuss the subject of Water-melons in the months of July and August; but, as they are not a spontaneous production we must "begin in the beginning," and that *beginning* would be too late in those two months: for, from seed to matured fruit there is a pretty long "slip between cup and lip" in the development of the melon, as well as in other subjects of the vegetable kingdom.

GAPES vs. ENTOMOLOGY.

With all their knowledge of insect life, the entomologists have not yet solved the problem of gapes in chickens. A worm in the windpipe is the cause, but how it gets there, and where it lives during the season before and after it attacks the chicken, is unknown. In some localities it never appears, and elsewhere it is an annual pest, or nearly.

It is very easy to write an item like the foregoing, which we find in a column of the *Weekly Press*; and it would have been quite as appropriate to the subject to have said, "with all the knowledge of insect life, entomologists have not yet solved the problems of"—" What's blacker than a crow?"

"*Gapes*" in chickens," is not an entomological question, any more than *tapeworms* in human beings is, or than *measels* in pork is; although, an entomologist might happen to know about it as much as anybody else, or less about it than anybody else, without adding *to*, or detracting *from* his standing as an entomologist. An entomologist is such, not because he makes a special claim to that title *himself* 'so much, as because it is accorded to him through the courtesy of by others, on account of his specialty in natural history. He may, in this sense, be legitimately entitled to the designation of *entomologist*, without knowing anything about any other branch of natural science. Entomology, as a whole, or as a unit, embraces more subjects, and a greater variety in detail, than all other branches of natural history put together; hence, those most thorough in it—those who have made the most valuable contributions to its literature, are *specialists*, and never aspire to anything more; nevertheless, they are still *Entomologists*, just as much as those are *Botanists*, who make the study of trees, or shrubbery, flowering plants, lichens, mosses, or fungi their specialties. Scientific specialism is not as common in the United States as it is in Europe, where, amongst her entomologists are to be found many who are, or who *have been* Coleopterists, Orthopterists, Hemipterists, Lepidopterists, Neuropteris s, Hymenopterists, Dipterists, &c., and who aspire to nothing beyond these specialties; although, in the pursuit of any of these branches, it would be next to impossible not to know *something* about collateral branches;—indeed, even in the United States, we have many who devote themselves almost exclusively to special branches in entomology, and have distinguished themselves therein.

But considered from a *practical* standpoint, and as it stands related to the agricultural and domestic productions of the human family, as well as to the animal world in general; entomology and entomologists have had an immense responsibility thrust upon them, much of which they cannot know anything more about—and it is not their *business* to know anything more about—than any other people of equal intelligence; and through this promiscuous demand upon their scientific energies, their special studies are invaded or dissipated, and hence they are liable to become "Jacks of all trades, and masters of none." Even a *specialist* may know absolutely more about what many things *are not* in his specialty, than what they *really are*; and his humble confession to that effect may indicate an infinitely greater advance in scientific lore than an empty pretension to know all about things of which he may be profoundly ignorant. Any man, no matter how ignorant or stupid he is, may be able to propound a problem or a question that the most intelligent or profound scientist cannot satisfactorily answer—at least not to the satisfaction of the ignorant propounder—but that does not prove the former a philosopher, nor the latter a knave. A mechanic may be able to construct the most complicated philosophical instrument, and yet be totally unable to make a shoe or a coat, and yet, he may be eminently entitled to the name of a mechanic; but how absurd it would appear for any one to write—"With all their knowledge of mechanism, philosophical instrument makers are unable to construct a shoe or a coat."

How long has it been since the sciences of medicine, of anatomy, of surgery and of physiology have been introduced to the study of professional specialists? How many paid professors have been dispensing scientific lore? How many magnificent temples for their accommodation have been erected in different parts of the world? And how many pecuniary endowments have been bestowed upon them in order to facilitate their progress and their usefulness; and yet, how many cases occur in this line of science about which its students and its professors appear to know

absolutely *nothing*; seemingly just to illustrate how little is known about the branches they profess to study and to teach, and that men must be *ever learning* "a knowledge of the truth." The pursuit of any branch of natural science is something like exploring a seemingly endless stream, that ever and anon sinks into the earth, and bubbles up again at a more or less remote distance from where its traces have been lost.

The explorer may learn much, or all, of that part of it which comes under his immediate observation, but of that part of it which has sunken into the bowels of the earth—except theoretically—he may be profoundly ignorant. In like manner, the transformations and developmental progress of some animals are involved in conjecture, and amongst these are included the "gapes," the "hair-worms," and their congeners which are only known so far as their development has come under human observation. Observation, cannot draw an exact focus upon that which is under ground—which must be left mainly to theory, analogy or conjecture, for solution. The case is similar in the history of the "gapes" and its congeners. There is "here and there " an out-croping—as it were—in the development of these animals, and the *unseen* is "analogized" from that which is *seen*. True, it is of paramount importance that it should be known how the *strangula* or gapes, get into the *trachea*, or windpipes, of the fowls they infest, and also where and how they live "during the season before and after they attack the chickens, although it does not seem essential that the entomologist should know this as a qualification necessary to the successful pursuit of his specialty in natural science; and yet, he may occasionally have illustrations or something analogous to it in insect physiology. For instance, it has been demonstrated by those who are reputed to be competent authorities, that, like the spores of *fungi*, or the sporific germs of *epizooty*, the embryo of gapes may be in the soil, in the food, or in the water to which fowls have access; and not only this, but they retain their vitality for an indefinite period, even after they are perfectly dried; and also that they are perpetuated by carelessly throwing them aside, without first killing them, after they have been dislodged from the windpipes of the infested fowls.

But, the following, which we clip from the columns of a cotemporary, seems to deny that the gapes are *animal* organisms at all, which would remove them still farther from the category of entomology.

Gapes and Chicken Cholera.

The season is at hand when young chickens require attention, and a word on the subject may be read with some interest. It is an old saying that an ounce of prevention is better than a pound of cure, and the rule is eminently a good one with young chickens. One of the most necessary things to prevent gapes is to keep them dry and well protected from the chilly rains of spring, as this disease is a species of croup, similar to the chronic croup in children, when a false membrane forms in the windpipe and proves fatal in nearly all cases. This is usually caused by a neglected cold, and it is so with the young chickens; hence the necessity of keeping them dry and warm during the wet days common in spring. The membrane formed in the chicken and usually supposed to be a red worm, can be re-

moved by folding a horsehair and forcing the loop down the windpipe, and a sudden pull will bring out the membrane. Others use a feather, and I have seen a strong pinch of the windpipe loosen it, and the chicken cough it up; but all often fail to save the life of the chicken.

Formerly I lost many chickens in the spring, but for years, since learning the preventive measure of keeping them dry and warm during the cold, damp weather, I have not seen a chicken with the gapes.

The following remedy and preventive of chicken cholera is highly recommended as a sure thing: Permanganate of potash and chlorate of potash, of each 10 grains. Mix in one powder and dissolve in water enough to mix a quart of feed. This will be enough for twenty to thirty chickens, to be given several times during the spring.

And this:

Gapes.

Gapes in chicks are caused by the presence of minute worms in the windpipe, and when these worms are present in great numbers the chicks die of suffocation. I don't know how the worms get there, and it don't matter much; the main idea is to prevent them from getting there. In the whole list of chicken ailments there is no disease more easily prevented or cured than gapes. To prevent them feed cayenne pepper and sulphur with the soft food two or three times a week, and use the " Douglass mixture " in the drinking water three times a week.

Gapes may be cured by giving a piece of camphor gum the size of a small pea every day until the chick seems well. Sometimes two or three liberal doses of pepper will effect a cure. If the chicks are very bad fumigate with sulphur and give two or three drops of solution of carbolic acid and water; sixty drops of water to one drop of acid forms the solution. Don't hold the chicks directly over the fumes of burning sulphur, and don't fumigate too long, or the remedy may prove worse than the disease. Let the chicks inhale the fumes for two or three minutes, and in most cases that will be sufficient to effect a cure.—*Prairie Farmer*.

And this:

Gapes.

Chicks most subject to gapes are those that run on damp, low places. It is generally understood now that gapes are caused by small worms in the windpipe. These can be removed by the use of a fine horse hair twisted and run down the windpipe; a quick jerk after turning around will remove the worms or kill them. But one must be dexterous and practiced to do this. A small feather is perhaps better. Leave only the tip, which wet with one ounce of glycerine and twenty drops carbolic acid. Twist it quickly in the windpipe, withdraw and repeat. You will see the worms or a little blood come out.

Here are two good gape remedies. Give the chick a piece of camphor the size of a pea. The fumes will kill the little worms. Camphor in the drinking water will prevent gapes. Another good remedy is spirits of turpentine; dose five to ten drops at a time. Either of these two remedies will do. If not, increase the dose. If that fails use the feather or twisted loop in the windpipe. Change the chicks to high, dry ground and put camphor in the water, and it will save the rest of the flock.—*Journal of Agriculture and Farmer.*

And, if more is desired on the subject, we would respectfully refer the reader to volume 13, No. 6 (June 1881), of the LANCASTER FARMER, where he may find *eight columns* on the subject of the "gapes" in fowls (*Strongylus syngamus*), and its corelatives, discussed at large. To those of our readers who do not subscribe for the FARMER, we would respectfully suggest that they make immediate application to the publisher, perhaps they may be able to procure that number; and if not, they most probably could procure the whole volume.

From the advanced condition of *Gallinaculture*, and the intelligent minds now engaged in its development, the gapes is a subject that is clearly within that specialty; and from the wonderful progress made therein during the last fifteen or twenty years, one would naturally suppose that something should have been elicited in solution of this knotty problem. There is where the light must come from, and not from entomology, necessarily, which, scientifically restricted, operates entirely within a different sphere.

A NEW HISTORY OF LANCASTER COUNTY.

It has often been said—and with some show of truth—that "*the* history of Lancaster county has not yet been written." It may not be generally known to our readers that H. L. Everts, an experienced publisher of Philadelphia, has engaged the services of several competent citizens within our county as assistants in collecting and elaborating authenticated material towards the production of a *new history*; and, from the following synopsis of the portion alloted to Simon P. Eby, Esq., so far as the matter relates to the farm, the farmer, and farming, it has fallen into competent hands, and the reading public may reasonably expect to realize their most sanguine anticipations. If any one is in possession of important *facts* bearing upon the subjects embraced in the outline suggested by Mr. Eby, they ought to submit them, whether he may have occasion to use them or not.

1. The condition of the county when the first settlers arrived.—Its agricultural resources, soil, climate, timber, stone and water supply. A brief notice of some of the principal native trees, plants and fruits. Extent of Indian farming, Wm. Penn's opinion respecting our native fruits.

2. The first settlers and their early farming.—Who and what they were and whence they came, the different nationalities and their characteristics, combatant and noncombatant elements, a brief notice of some of the manners and customs, virtues and prejudices they brought with them from their mother country and fatherland.

3. How they began the work of establishing new homes, in a new country.—Some of the trials and difficulties they encountered and the encouragements and successes they met with, brief notice of their primitive log houses and thatched barns, how they had to depend for hay on watered meadows, their early implements of husbandry.

4. Secondary stage of farming.—Introduction of new seeds, clover and timothy, new fertilizers, rotation of crops and improvements in farming implements. How log dwellings gave place to more substantial stone mansions, with massive chimneys and wide open fireplaces, that welcomed newly arrived kinsfolk from across the Atlantic to the warmth of its hospitable blaze. How the newcomers lent helping hands. Saw-mills at work along the streams, converting the forest trees into more convenient building materials. Swisser barns (built after models brought over by the Swiss palatinates) now receive liberal additions; or, new square-timbered structures of increased capacity go up in more suitable places; and the flails of the threshers make lively music upon the newly laid barn floors. How the axe continued to extend the fields each year further into the timber lands. How flax and hemp were grown and dressed by the farmer and his assistants, and the fleece of the flocks prepared for the loom. How, during the winter, spinning wheels held high carnival in concert with the blazing logs upon the hearth within, and the roar of the tempest without. How early and late the deft fingers of matron and maidens plied the busy spindles, and chests and presses were filled with homemade linens and woolens. And how, some fine morning in spring, a joyous procession with a newly-married pair riding at its head, and loaded wagons and lowing cattle following after, issued from the parent farm and disappeared in the woods to settle down beside some pleasant fountain and begin the carving out of a new home and fortune.

5. Glimpses into the home life of the good old people.—Their time-honored customs, their thrift and industry. Their struggle against pride, extravagance and ungodliness. How they raised and educated their children. How they lived and labored and died, their dress, courtships, marriages and amusements. Old-time company of young men and maidens on horseback; apples and cider; markets and marketing, Conestoga teams. Concerning the peculiar non-combatant doctrine many of the people held and practiced. Their steadfastness of faith and reliance on the divine commandment not to draw the sword. How they were allowed to live in peace while three wars swept over other parts of the land. Their religious services, manner of preaching and holding of funerals; their dialect. Is "Pennsylvania German" a distinct language?

6. Tertiary stage of farms and farming.—The advancing wave of modern improvement and invention, introduction of new cereals and vegetables, labor-saving implements and farming machinery, sub-division of the old farms and a more thorough system of tillage, application of lime to the soil, waste land brought under cultivation, introduction of coal for fuel, railroads built, different views, and stable-tailings sometimes consequent increase of many of our native birds, consequent increase of destructive insects, partial failure of the apple crop, praiseworthy efforts of fruit growers to supply the deficiency by experiments in the propagation and introduction of native varieties and improved small fruits, the theory of the founder of Pennsylvania concerning the cultivation of native fruits, adopted after a lapse of nearly two centuries, change of climate and gradual diminution of the water supply, how springs and streams have been affected, public school system at work, should the intellect be educated at the expense and neglect of the morals? What education should do for the farmer. Concerning agricultural exhibitions—Improved live stock, tobacco farming, sewing machine taking the place of the spinning wheel, present day marketing, modern farm houses modeled after city houses with inadequate roofing and deficient ventilation, pernicious effects upon the health of their inmates. Effect of the accumulation of wealth, growing dislike of the young for manual labor, farms passing into the care of renters and the owners drifting to towns and villages; luxury, refinement and extravagance, inroads of fashion and expensive habits.

7. Are we getting better or worse?—Shall we disregard the experience of the past, or gather instructions from its lessons?" "The coming farmer"—who and what shall he be?

LIME IN SOIL.

Every farmer, in using lime on his fields, should first ascertain whether the soil *needs* lime. Until he knows this, his liming is done at random, and may be a positive injury instead of benefit. Doctor San Grado, in his "consultations" with his pupil (Gil Blas) always prescribed "*more blood-letting and additional draughts of warm water*"—without regard to previous treatment or condition—through which the undertakers and marble masons flourished, but the poor patients died. Analogous are the results of liming where the soil already contains a sufficient quantity. Such applications may be beneficial to the limeburner, but they are detrimental to the health of the plants. True, a farmer may not have a chemical laboratory of his own, nor access to one, but still, to a limited extent, he may be able to test his soil as to the *presence* of lime in it, although he may not be always able to determine the quantity or quality. A simple analysis can be made by mixing a small quantity of soil in a cup with water, and pouring over it some *muriatic acid*, which he can obtain at any drug store. If a free effervescence, like fermenting cider, or frothing beer takes place, it indicates the presence of lime. But if it remains perfectly still and dead, it contains no lime, or at least not sufficient to produce any beneficial effect on vegetation, and in such a case lime should be applied in some form or other. Take a piece of common limestone and drop on it a little muriatic acid, and you can immediately see the boiling or fermenting effect of effervescence. The freer it effervesces the purer the limestone is. If this result does not follow, it would be useless to waste fuel and labor in attempting to "burn" such limestone.

Sulphate of lime or plaster-of-paris, exercises additional beneficial action on soil, by its sulphuric acid stimulating vegetation and assisting in the decomposition of mineral and organic substances in the soil.

Plants require lime in the following proportion to one thousand pounds :

Barley	12 7-10 lbs.
Barley, straw	3 8-10 "
Spring wheat, straw	14 10 "
Winter wheat	2 1-16 "
Rye straw (winter)	3 1-10 "
Cornstalk and fodder	5 "
Peas	1 2-16 "
Pea straw	18 6-10 "
Beans	1 5-10 "
Bean, straw	13 5-10 "
String bean, straw	14 1-10 "
Potato vine	5 8-10 "
Hemp	12 2-10 "
Linseed	5 "
Tobacco	73 1-10 "
Clover Hay	10 4-10 "
Meadow hay	7 7-10 "

(*From "What of fertilizers?"*)

Many farms contain more than a sufficiency of lime, while in others it is wanting; hence an occasional application of lime alone will act as efficiently as if artificial or ordinary manure had been applied. But in the application of this mineral to the soil, the more intelligence is brought to its application the more prolific and profitable will be the result. The foregoing may be of some assistance to the farmer in making the proper discrimination in the use of lime.

EXCERPTS.

HEAVY work or driving soon after eating is bad treatment for a horse. Let him rest on a full meal, or use very moderately when use cannot be avoided.

IF a dull, backward, sleepy neighborhood desires to improve its agricultural resources, let the farmers start a creamery. More live stock is what the country wants.

A HORSE with no change of diet in a long time is apt to tire of it, and indigestion will soon result. Horses, like men, like a change now and then, and it does them good.

THE principal mule raising States are Illinois, Indiana, Kentucky, Tennessee, Missouri, Ohio and Texas. The beauty and musical qualities of the mule are only a small part of his attractions.

THE mutton consumption of this country is increasing, and also the quality of the meat. First-class, well cooked mutton is worth more than most other meats, and when dogs are scarce it will be cheaper, as well as more profitable to farmers.

AFTER the sudden disappearance of a cow at Florence, Ala., she was found two weeks afterwards alive in a cotton shed, wedged between two bales of cotton, but was thin and nearly blind. Perhaps with a refreshment of water when wanted, she might have lasted forty days as well as Tanner did.

ILLINOIS has an average of twenty horses to each square mile of territory. In the number of horses, cattle and hogs, Illinois leads all the States. The hog-population is fifty-three per square mile, and one county (Stark) has 100, while Cook County, in which Chicago is located, only has seventeen.

THE farmer of small means who desires to improve his live stock should start with care, after he has observed and read enough to make a moderate investment safe. It is of little use for an ignorant, bull-headed man to attempt this kind of work. It is brain work that tells in stock-breeding as well as in professional life.

INTELLIGENT foreigners often express surprise that with so much land as there is in the United States well adapted to sheep breeding, that we have so few sheep. This again raises the dog question. Outlaw the dogs and put a reward of $5 on every one killed, not licensed, registered and collared, and there will soon be a "boom" in sheep breeding.

THE improvement of farm live stock does not come simply from a mere investment of money, but from care, thought, observation, comparison and study of animal physiology, and the laws of breeding. Money might purchase a dozen first-class cows and a bull, but it takes something more than money to keep the stock up to the standard at starting, or to improve it. The benefit of improved stock comes largely from the fact that it is an educating force in farm life.

WILLIAM C. BLACKFAN, of Solesbury, Pa., kept twelve steers, averaging about 800 pounds each, through the past winter on cut and soaked cornstalks, along with one ton of wheat bran, 350 pounds cornmeal, and 150 pounds oil meal cake, all well mixed, with no hay whatever, and the animals are in better condition than usual. The bran, meal and oil cake cost less than $100, and he was enabled to sell ten tons of hay for $22 per ton. Hence Mr. B. don't feel that a little study of the nutritive qualities of cattle food did him harm. His milk cows got four quarts of this mixture twice a day, and never did better. We should not suppose that such results would make him hanker for ensilage.

WEALTHY stock breeders, who desire to see improved stock become general in order to do the country good as a whole, should not aim to keep prices at fancy figures. We notice that one writer in a stock journal advises to castrate all good bulls that cannot be sold for $100 at least. A man who does that for such a reason might as well acknowledge that his only purpose in breeding is to draw exorbitant profits from a class which ought to be benefited and yet cannot invest at fancy prices. Ordinary farmers sell their products for what they can get, and do not destroy it, and the example is a good one for rich stock breeders. If a $100 bull will bring only $80 or $75, to "get mad" and castrate him and then sell him to a butcher for $50 is rather mean sort of enterprise.

AN Iowa farmer put up twenty-one year-old hogs for fattening, and for the first twenty days fed them on shelled corn, of which they ate eighty-three bushels. During this period they gained 837 pounds, or upward of ten pounds to the bushel of corn. He then fed the same hogs for fourteen days on dry corn meal, during which time they consumed forty-seven bushels, and gained 535 pounds, or 11½ pounds to the bushel. The same hogs, next fed 14 days on corn meal and water mixed, consumed 55½ bushels of corn, and gained 731 pounds, or 13½ pounds of pork to the bushel. He then fed them fourteen days on corn meal cooked, and after consuming 45 bushels of the cooked meal the hogs gained 700 pounds, or very nearly fifteen pounds of pork to the bushel of meal.

CONSIDERABLE attention has been recently given to the differences between the rain of the city and the country. The country rain is neutral and is considered the best adapted for human consumption of any found above the earth, on the earth, or under the earth. The rain that falls in cities, on the other hand, is acid, corroding metals, stones and bricks and mortar crumble before it. Its evil effects are visible on every side—in paint, in all decorations, and, in fact, almost everything created by man. The purest rain is that collected at the sea coast, more especially at considerable heights; while organic matter in the air usually corresponds with the density of population.

THE best way with all grapes, and especially with those not quite hardy, is to prune in the autumn as soon as practicable after the fall of the leaves. If the vines are pruned and trained upon the renewal system it will be a very small matter to lay them upon the ground and give a covering of two or three inches of earth upon the shortened canes, which covering should be left on until all danger of severe freezing is passed in the spring.

AN inventor proposes to make machine gear wheels of raw buffalo hide by cementing and pressing together, as many layers as are required for the breadth of the wheel. The blanks thus prepared are cut to form the teeth in the usual manner with suitable tools. The advantages claimed are smooth and noiseless action at very high speed, and greater durability without lubrication.

THE most simple and best stain for mahoganizing cherry is ground burnt sienna, mixed in benzine or turpentine. Apply with a brush or sponge, let it stand for a short time and clean off with a cloth. It will be better to let it remain in this condition until the following day before commencing to finish.

HICKORY-NUT COOKIES.—Mix together two cups of sugar, two-thirds of a cup of butter, two eggs, six tablespoonfuls of sweet milk with half a teaspoonful soda dissolved in it and flour enough for a soft dough with a teaspoonful of cream of tartar sifted through it. Add a cupful of the chopped meats; drop, in spoonfuls on buttered tins, put into shape and bake to a light brown.

CORN cakes that are nice for breakfast are made of one quart of flour, one pint of meal, three teaspoonfuls of baking powder, one teaspoonful of sugar, three tablespoonfuls of melted lard, sweet milk enough to make a thin batter; add salt enough to suit your taste.

HOW TO TELL GOOD BUTTER.—When butter is properly churned, both as to time and temperature, it becomes firm with very little working, and it is tenacious; but its most desirable state is waxy, when it is moulded into any shape, and may be drawn out a considerable length without breaking. It is then styled gilt-edged. It is only in this state that butter possesses that rich, nutty flavor and smell, and shows up a rich golden-yellow color, which imparts so high a degree of pleasure in eating it, and which increases its value manifold. It is not always necessary, when it smells sweet, to taste butter in judging it. The smooth unctuous feeling in rubbing a little between the finger and thumb express at once its rich quality; the nutty smell and rich aroma indicates a similar taste, and the bright golden, glistening, cream-colored surface shows its height of cleanliness. It may be necessary at times to use the trier, or even use it until you become an expert in testing by taste, smell and rubbing.

WINTER PROTECTION OF GRAPEVINES.— The grape is a tender plant in almost every sense, and must be treated accordingly. We know how it is affected by great changes o temperature, extreme heat and humidity, severe pinching back and overbearing in summer. In winter it is still worse ; millions of vines are annually lost and more hurt, for the want of a little attention in protecting them. It is only necessary to lay them on the ground at the beginning of the winter, and weight them with something to keep them down. The object is to avoid, not so much the cold, as the draft of the wind, which, when the vines are frozen, dries them, and thus perish the smaller vines first, as they are soonest dried. Near the ground this is avoided, though where there is no obstruction at all to the wind, and the winter is an open one, leaving the vines exposed, harm will sometimes result.

THE Kansas Farmer says: The practice of forcing a horse to stand on his legs, or walk about, while laboring under an attack of colic, is most inhuman. The same remark is also applicable to the plan of exercising a horse during the time he is under the purgative action of a dose of physic. He should be moved gently about before the medicine commences to operate, but never after. Do those barbarians who knock the animal about while enduring the pains of colic or when suffering the purgative action of medicine, ever think of what they are doing? If they were treated themselves on the same plan under similar circumstances, they would soon come to their senses regarding the management of the unfortunate animal which is placed under their charge.

THE American to-rist passing through Germany is surprised at the number of fruit trees along the sides of the public roads. These trees are pruned and looked after by the "road makers," and three or four weeks before the fruit ripens are watched day and night by these guardians. In the province of Wurtemberg the sale of the fruit thus raised is said to have realized as much as $2,000,000 in a single year.

WASHING the leaves of the wax plant occasionally is the very best treatment for it. When washing, brushing with a soft brush about the axils of the leaves will tend to keep the plant free from mealy bugs, one of its insect enemies. When the plant commences its growth we would supply it once a week with weak manure water.

THE most profitable way to raise beef cattle is to keep them constantly in a thrifty and improving condition. It is not necessary to keep very young stock rolling in fat, but there should always be an abundance of nutricious food to help nature in its development. To allow stock to run down in flesh and become ill-conditioned, simply because it is not designed for market for some time, is the height of folly.

IN killing poultry, the French open the beak of the fowl, and with a sharp-pointed, narrow-bladed knife, make an incision at the back of the roof of the mouth, which divides the vertebra and causes instant death, after which the fowl is hung up by the legs to bleed. This is a neat and merciful way of doing it.

SEASONED posts treated over the lower third to two or three washes or soakings of cheap petroleum will make them last longer than by almost any other process. This is easier than to coat with boiled tar, and far more sensible than to set top end down. Parker Earle, Cobden. Ill., earnestly commends this treatment after experiments.

THE capacity of the glucose factories of the United States is said to be sufficient to use up about 11,000,000 bushels of corn per annum. While this aids a little to keep up the price of corn, it is all extracted back from the farmer's pocket in the shape of adulterated sugars and syrups. The glucose manufacturer is about as much a public benefactor as one who should adulterate our coin with an inferior metal.

A CORRESPONDENT in an exchange wants to know how to purify bad-smelling cistern water "by throwing something into it." The question does not indicate a surplus amount of "gumption" or taste. He might as well ask what will purify bad old cheese, or an egg six months past its prime. He should clean out his cistern and purify that, not the water, and see that only pure water goes into it. Let him apply the bad water to his garden.

AN old apple tree past its usefulness had better be cut down or dug out. It is a useless cumberer of the ground.

MULCHING always retards the ripening of fruit, but that is often advantageous. It also makes the fruit larger and better.

IT is not entirely creditable to men who have long been identified with fruit growing not to be able to tell what is the best system. It ought to be the aim of every specialist to find out.

CLAPP'S FAVORITE is a good market pear if picked early enough, so that it does not rot at the core. It ripens in advance of the Bartlett, and is of better quality for those who do not like the Bartlett's spicy flavor.

FRUIT TREES late in bearing can be hastened in this matter and permanently benefitted by root pruning. Cut a trench about them and fill up with vegetable or animal matter, including some rubbish, and see how they will boom.

THERE are many varieties of fruit on nearly every fruit farm which are unprofitable to grow in spite of excellent and popular qualities. Except a few for home use these had better be grafted to more prolific and profitable sorts. In many places the Sheldon and Seckel pears stand in this category.

A PROMINENT Illinois fruit grower (Parker Earle), states that the Wilson strawberry is still the popular sort for the Chicago market. The Wilson is a hard berry to root out, and in going to market it bears rough handling better than any other sort. The Wilson sometimes is shipped 600 miles successfully.

ONLY think of it! When a man eats strawberries grown on a patch fertilized with 300 pounds of rectified Peruvian guano, 250 pounds dissolved boneblack and 200 pounds muriatic of potash per acre, he eats 29.24 per cent of potassium oxide, 3.22 sodium oxide, 13.47 calcium oxide, 8.12 magnesium oxide, 1.74 ferric oxide, 18.50 phosphoric acid and 5.66 per cent of siticic acid. That is what ails them exactly.

PROFESSOR GOESMANN finds that an application of from three to four pounds of muriate of potash per tree to peach trees slightly affected with yellows, restores them to health. It would be a good plan undoubtedly to keep trees in health well supplied with this fertilizer, and then they might not get out of health as regards the yellows. It is quite certain that sick peach orchards are generally neglected peach orchards. Muriate of potash is also spoken of sometimes by chemists as chloride of potassium.

IN Professor Goesmann's application of muriate of potash to yellow-sick peach trees he recommends distributing it over a radius of eight feet or so on mulch. But no demonstration has yet been made as to the cause of the yellows.

QUERIES AND ANSWERS.

WEST CHESTER, PA., April 17, 1882.
Mr. S. S. Rathvon—Dear Sir: The enclosed curiously formed cocoons, I took this morning from a branch of the "*Stewartia Pentagynia*," a large flowering shrub, growing on my lawn in West Chester; the whole is so different from anything in the product of insect life, and so curiously striking in uniformity of shape in the silk and receptacles and their silken connection, that I have ever seen, I inclose it for your inspection. I crushed the smallest of the three, and from the exuding matter, I judge that it contained small egg, or perhaps larva or pupa, of some insect. Very Truly, *J. Rutter*.

The cocoons were duly received, and in tolerable good condition; but, I am unable to name the "insect," specifically, that constructed them; although, without much doubt, they belong to the genus *Theridion*—most likely *T. trigonum*—a small species of spider. The little globular cocoons are filled with eggs, and the species could only be positively determined after these eggs are hatched, and the animal can be compared with existing descriptions; because, so far as the mere cocoon is concerned, there are other species of *Theridion* that construct a similar receptacle—notably, *T. globosum*. But, as *trigonum* is a northern species, and *globosum*, southern, if it is not entirely new, it is more probably the former. Those specific names are derived from the abdominal forms of the spiders alluded to—triangular or globular—and hence it will require the presence of the arachnid itself to determine the question, unless with those who have made this class of animals a special study, and have become "experts" therein. There are several species of spiders that construct different shaped cocoons, and of different sizes, some of which are two inches, or more, in diameter. The fiber is stronger and more silky than the common spider's web, and efforts have been made to utilize it, and also to rear it, but the success, so far, I believe, has not been very promising. There is no knowing, however, what a hundred years hence may bring forth.— *R.*

ESSAYS.

SOME PRACTICAL POINTS IN PEACH CULTURE.*

Having reason to think that my experience with peaches the past season was something remarkable, inasmuch as I had a fair crop amidst almost universal failure, and as this is a subject of growing importance to the fruit growers of Pennsylvania, I propose to lay before the society a few points that may be of interest from some notes that I have made principally in regard to the relative hardiness of different varieties.

I see that Secretary Edge, of the State Agricultural Department, in his report of the crops for 1881, reports on fruit as follows; Comparative yield compared with average crop, apples 105, pears 100, cherries 100, plums 110, grapes 100, berries 100, peaches none. So far as I am able to judge this is about correct. And it thus appears, that while all other fruit crops were as good or better than common, in this part of the State, peaches were a total failure. And yet I had, as I have stated, a satisfactory crop, for though many varieties failed almost entirely, others were wonderfully fine, both in quantity and quality. And this in spite of the very severe drought which greatly injured some. I do not pretend to be able to give any explanation of this, but the question is certainly an interesting one, why, amidst otherwise universal failure, one should be even partially successful. The cause of the general failure of the peach crop last year, as every one knows, was the extreme cold of last winter. The fruit bud of the peach not being able to withstand a very low temperature. And yet it would seem that there must be some other causes or conditions not yet

*Essay read before the Pennsylvania State Horticultural Association, at Harrisburg.

understood, because it is not supposed that any particular locality could escape, when the severity of the cold was so great and long continued, as was the case last winter. It is not my purpose in this paper to advance any theory in explanation of the facts above stated, for though I have some crude notions on the subject, I prefer, in a paper of this kind, to confine myself entirely to the statement of facts.

Having noticed early in the season that while the fruit buds of some varieties were almost entirely killed and others only partially, and some had almost entirely escaped injury, and having about seventy-five varieties under cultivation, I determined to observe, particularly, which were most hurt, and which withstood best the trying ordeal. For this purpose, I carefully noted in a book kept for the purpose, how each variety came out, and I am thus able to lay before the society some facts in regard to the relative hardiness of most of the leading varieties under cultivation in this region, which may be of great value, particularly in localities where peaches are very liable to be injured by cold.

In the following list, which for convenience, I arranged in alphabetical order, the relative hardiness of the different varieties is given in a scale ranging from 1 to 100, the highest number representing a very full crop:

Amelia	00	Late Admirable	80
Alexander	50	Late Rareripe	15
Atlanta	40	Mary's Choice	5
Beer's Late	1	Mountain Rose	15
Beer's Smock	60	Nauticoke	100
Bernard's Early Yellow		Newington Free	15
Alberge	100	Old Mixon	20
Bilyen's Late October	90	Orange Cling	6
Bilyen's Late Comet	100	Piquet's Late	10
Brandywine	5	Price's Late	100
Brigg's Red May	50	President	100
Chinese Cling	1	Prince of Wales	100
Coolidge's Favorite	13	Princess of Wales	20
Crawford's Early	2	Reeves' Favorite	15
Crawford's Late	20	Richmond	2
Crockett's Late	80	Ruding's Late Red	40
David Hill	80	Salway	20
Early Beatrice	60	Smock	80
Early Louise	75	Snow	10
Early Rivers	75	Shipley's Late Red	10
Early Filloteon	50	Stump the World	40
Early York	40	Steadley	20
Foster	15	Susquehanna	1
Freeman	50	Temple's Wilie	25
Geary's Holder	10	Thurber	90
Golden Eagle	100	Troth's Early	90
Hence's Golden Rare.		Transom's Free	1
ripe	2	Tuckahoe Late	50
Hale's Early	90	Van Buren's Golden	
Harker's Seedling	75	Dwarf	2
Jarrett's Late White	15	Ward's Late	50
Keyport Witie	50	Wilkin's	
Kilrell's Favorites	25	Yellow Alberge	50
La Grange	15	Yellow St. John	25
Leatherbury's Late	100		

I should state here that there was no difference worth noticing as to the exposure. Some trees were, of course, more exposed to the cold winds, and some were on higher ground than others, but where the same varieties were in different exposures and different altitudes, there was no noticeable difference, except that old trees did much the best. All had about the same treatment as to cultivation, manuring, &c.

I do not propose to say more at present on the question of varieties, except that of all the early kinds I have tried, none are worth having. They all thrive, bear abundantly and look very promising up to the time of ripening, when every one rots, just before they are fit to pick. If there is one variety earlier than Mountain Rose worth anything, I have not been fortunate enough to get it, though I have tried many. My object in this paper more than anything else, is, to bring before this Society some questions connected with peach growing that appear to me to be very important, and on which light is very much needed, and one of these is, the subject of the universal rotting of all the early varieties. I wish to ask, first, is there a good variety earlier than Mountain Rose or Early York that don't all rot before ripening? Second, does anyone know the reason why, or can anybody throw any light on the question, why all the early varieties rot, when later ones in the same ground and same kind of weather, do not?

I have also a few questions concerning yellows that I consider very important. It is claimed now by experts, that the cause of yellows is a specific fungoid affection, which is in some way communicated from one tree to another and when once the poison finds its way into any part of the tree it spreads itself by the circulation of the sap or otherwise until the whole tree is affected, and in time destroyed. And when this infection has once entered a tree there is no remedy but its removal to prevent further spread of the disease. Admitting this much to be settled, there yet remains much more to be learned in connection with this all important question. Mr. Rutter says, and I believe all other practical writers on the subject say, remove at once, root and branch, and some say the diseased tree should be immediately burned. Now, what I want to get at is, is this really necessary? even if it were practicable. To remove a tree immediately root and branch, is a very easy thing to recommend, but to put in practice next to impossible. The roots of a peach tree 10 or 12 years old, in rich and well cultivated soil, will be found to have run 50 and perhaps 100 feet or more, and their total eradication would require besides an immense amount of labor, the destruction of all the other crops and trees within a circle of 50 to 100 feet in diameter. I have peach trees 8 inches in diameter, some of the roots of which I have no doubt run for 100 feet, and most of them must be more than a foot beneath the surface, as the ground is constantly ploughed about that depth. The total eradication of one of these trees would involve the destruction of perhaps a quarter of an acre of strawberries or some other crop, besides a dozen or more pear, or other peach trees. Some time these trees will get the yellows. For according to my experience none escape, it being only a question of time. It is needless to say that I do not "totally eradicate" such trees. I do not attempt to remove the roots, because to do this thoroughly, would cost in labor and destruction of other crops, where at least five per cent. of the trees have to be removed every year, more than the whole peach crop would be worth. I am aware that it is the practice, perhaps in most of the peach growing regions and particularly in very light soils, to remove old and diseased trees by drawing them out with a strong team, perhaps a yoke or two of oxen. But in that case I imagine only a small portion of the roots are removed. I am sure it would take at least a dozen yoke of oxen to pull out some of my trees and then the greater part of the roots would be left in the ground. To cut down and remove even a large peach tree is a very simple thing to do and costs but little time and labor. But taking out by the roots, or, as Mr. Rutter says, totally eradicating root, body and branch, and that perhaps at the busiest season of the year, is what I imagine no one ever gets done. Now the question I want to get at is, is this really necessary? And this brings me to the other great question in which all other questions connected with peach yellows are involved. How is the disease communicated from tree to tree? Is it by actual contact alone, by being conveyed by the knife or saw in pruning; by contact of the roots, or as some have supposed by bees flying from tree to tree and carrying the pollen from diseased trees to the flowers on a healthy one and impregnating that with the poison. Or does it spread by sporadic infection or any way other than by contact? And then again if the disease can be spread otherwise than by actual contact, it is very important to know at what season of the year are trees liable to be infected. Is it at all times, or only when the trees are growing or in leaf. This is important, because if a tree that is not in leaf cannot receive or impart the disease by actual contact, it is difficult to understand what harm a dead stump of a tree could do by remaining a few years, until it rots and gets out of the way of itself, which it soon does. If, as Mr. Rutter says, the disease is communicated by contact of the roots, the mere grubbing out, or even drawing out with oxen, only removes a small portion of the roots of an old tree. And then again another question of great importance presents itself. How long after a tree has been cut down, or grubbed out, or drawn out with oxen, if you please, will the roots that remain in the ground, retain the disease so as to communicate it to another tree. I profess to know very little about Fungi, but it seems contrary to all that I have heard on the subject, to suppose that a Fungus which would thrive in living wood or bark would also live and thrive in dead or decayed wood.

I have dwelt perhaps longer than I ought, upon what may seem to most persons a matter of little importance, but it is because I have found it to be a matter of great practical import, so much so, that if I really believed it necessary to do what Mr. Rutter says must be done, I would at once abandon peach culture. Though I have never pretended to understand the yellows, and certainly have no theory about it, I cannot forbear saying here, that I have some doubts about the roots being affected to any great extent, because while all peach trees get the yellows, sooner or later a tree with peach roots, the body and limbs of which are plum, will never take it. I have often thought it would be an interesting experiment to try how far the peach and plum might be grown together in one tree, without being liable to this disease; suppose the plum is worked on the peach several feet above the ground, or suppose a peach tree, say three years old, and free from disease, has all its branches worked with plum and no peach buds allowed to grow, so as to have, as near as possible, a peach tree with plum leaves. Would such a tree be liable to the yellows? If not, it would go to show that the infection is only received through the medium of the foliage or blossoms. Or else that the development of the disease is something incidental to the growth of the peach, and the Fungi that are found in trees affected with the yellows, are only the effect and not the cause of the disease.

Of all the theories that have been advanced in regard to the spread of the yellows, the bee theory seems to me to accord best with the facts. How else can we account for the fact, that if a peach orchard of perfectly healthy peach trees is planted a mile or more from any other peach trees, as soon as the trees begin to bear, the disease will show itself, not, however, in regular rotation from tree to tree, as might be supposed would be the case, if the disease was spread by sporadic contagion, by contact of the roots, or by means of pruning implements, but jumping about; sometimes one, and sometimes two or three trees in a spot affected; just as bees are observed to fly, skipping about from one part of an orchard to another.

But, as I before said, it is not my purpose to go into a theoretic discussion, my only object being to bring up practical questions with which we are brought face to face in our daily practice, a proper solution of which is of the utmost importance.

For fear that the points that I have endeavored to bring out may not be clear, I will recapitulate. 1st.—What varieties will best withstand severe cold? On this I hope I have thrown some light. 2nd.—Why do all early varieties invariably rot prematurely, or are there any that do not? In regard to the yellows. 1st—How is it communicated from tree to tree? 2nd.—Will the disease spread otherwise than by actual contact, except during the period of growth? These questions are very important in their bearing on the all-important practical questions—Must a tree be immediately removed on showing the first symptoms of the yellows? and must it be totally eradicated root and branch? To these may be added the following: If, as is supposed, the disease is a fungus that pervades the bark of the living tree, will this same fungus live in dead or decayed wood or bark, so as to communicate the disease to another tree, the roots of which may come in contact with them. There are of course a great many other practical points in peach culture, which might have been introduced that might be interesting to many, but it was not my purpose to bring up questions that have been often discussed here, and which are, or should be, now considered settled. I think we ought to make some progress, and I desire to take a step in advance, and if I only succeed in awakening an inquiry that may throw some light on what are, as yet, some dark places in the path of the pomologist, my object will be accomplished.

MANAGEMENT OF AN ORCHARD.*

The subject upon which I have been requested to write is one of such vital importance that it is with diffidence I present my views before this Society. I have no new discoveries to present. My success has been mainly due to paying strict attention to the following:

1st. Location.

*Essay read before the Pennsylvania State Horticultural Association, at Harrisburg, by J. H. Funk, Boyleston, Berks county, Pa.

2d. Selection of soil.
3d. Its preparation.
4th. Selection of varieties.
5th. Proper planting.
6th. Judicious pruning.
7th. Good judgment, close attention and a great deal of labor.

I will treat of these in the order in which they stand, and I prefer giving them just as I have treated my orchard which I planted a few years since, the same for which the Committee appointed by the Berks County Agricultural Society awarded the premium as the best regulated orchard in the county of Berks.

1st. Location.

Under this head comes the exposure. In my orchard I have every exposure; protected on all sides by low mountain ranges, except the northwest, which opens into a narrow valley, and through which the northwest wind has a clear sweep over a portion of the orchard. My choice would be a northern exposure, protected on the north and west by mountain or forest. Trees thus located are less endangered by late frost.

2d. Selection of Soil.

This is one of the most important considerations in planting an orchard. Be not governed by price. Better pay $300 per acre for good, suitable land, than take indifferent soil as a gift. In the first it will be pleasure and profit; in the other, disappointment and loss.

In describing my soil, I have sand loam, loam and clay loam. The loams are to a depth of two to four feet. We then come to a stratum of micaceous deposit, averaging from four to twenty feet in depth; beneath this a soft rock strongly impregnated with iron and small veins of plumbago. The clay loam is underlayed with a light clay sub-soil which never breaks. Were I again to make a selection, I would take the same, if obtainable. All these different soils are not equally good for any one variety of fruit, but each superior for such varieties adapted to it. I have my peach and cherry on my lightest soil. Apple on the loam. Pear, plum and quince on the clay loam.

3d. Its Preparation.

When I purchased this tract, the soil was completely exhausted, being farmed continually, and for twenty years receiving no manure. I reversed the usual mode of farming. Instead of plowing a furrow 5 inches deep and 15 inches wide, I put in a strong team and plowed a furrow but 8 inches in width and 15 inches deep, thus throwing up and intermixing 10 inches of subsoil with the exhausted surface. I then spread 75 bushels of good lime to the acre, and let it lay thus over winter. The following spring I applied 500 pounds best dissolved bones to the acre, plowed shallow, harrowed well, and considered it in good condition to plant.

4th. Selection of Varieties.

On this depends to a great extent, whether your orchard will be a profit or a loss. The best guide is to select such varieties as do well in your own immediate neighborhood. Do not be tempted by fine, showy plates of fruit you know nothing about. If you are not acquainted, ask some one on whom you can rely, who has had experience. Do not plant too many varieties. Select as near as you can, trees that are good growers and annual bearers of showy, good flavored and good keeping fruit. My selection would be, in the order named, Grimes' golden, Hubbertston nonsuch, Krauser, Cole, Hays' winter, Smith's cider, Ben Davis, Falawater, Westfield seek-no-farther, for winter; Maiden's blush and Duchess of Oldenburg, for fall; Red Astrachan and Early Harvest, for summer. Many may ask why I have omitted the Baldwin, Rhode Island Greening, Roxberry, Russet, &c. My reply is, I can buy them cheaper than I can raise them, and if I have plenty of such apples as Grimes' Golden pippin, others are welcome to Baldwins, &c. Of pears, Duchess, De Angouleme, (dwarf) Bartlett, Beurre de Angeau, Seckel and Louise Bonne de Jersey (standard). Plums, German, prune, imperial, gage and Lombard. Quinces, Orange and Rea's Mammoth. Cherries, Early Richmond, Mayduke, Black Tartarian and Gov. Wood. I never plant large trees, preferring 2 years old of stocky growth; peaches 1 year old.

5th. Proper Planting.

Have the holes dug large, not less than 2½ feet square, and 18 to 20 inches deep. When ready for planting, throw the surface soil in below, filling the hole to such a height, keeping the centre slightly convex, that when the tree is set in, the roots take their natural position, and leave the tree when planted nearly the same depth as it stood in the nursery row. Trim off all mutilated roots, and set the tree in place, spreading all the roots out evenly; throw on some loose, mellow ground, filling up all vacant places around the roots. When sufficiently covered, press the earth gently but firmly to the roots with the foot, then finish filling the hole. When young, thrifty trees are thus planted, they need no stakes. I have planted 3,000 trees in my orchard and have never staked one, and I have scarcely a-half dozen crooked trees on the place. After planting, it is very beneficial to mulch with any loose material, long manure, straw, weeds, leaves, tan bark or even coal ashes; anything that will retain moisture and keep the ground loose. This is more necessary in sod than in cultivated ground, where the loose surface soil acts as a mulch.

6th. Judicious Pruning.

Here considerable good judgment is necessary. First know for what you are pruning; do not lop limbs indiscriminately; there are several objects to be obtained by pruning. First we prune a young tree when planting, to assist nature and relieve her from overdue exhaustion; for when a tree is dug up the greater part of the fine rootlets are cut off, thus diminishing the supply of nourishment by cutting off from one-half to two-thirds of the last year's growth. You relieve the tree of that much material to supply with food, and you do more. By careful pruning at the proper, buds you start the foundation for a low round symmetrical top. By proper pruning you can keep up a good, thrifty growth. For this, always prune in early spring and if the work be properly done the tree will need but little pruning in after years.

7th. Good Judgment, Close Attention and a Great Deal of Work.

We now come to a question that has caused

more contention than any other in the management of an orchard. Cultivation or non-cultivation. I adopted the following plan and have been so well pleased with the result that I should follow the same course were I to plant another orchard. I planted my young orchard the first year with potatoes and corn. The trees all made a good growth except 14 cherry and 13 peach. Apple, pear, &c., all grew fine. The second year I applied good, strong manure, 20 tons to the acre, and again planted corn and potatoes, adding superphosphate of lime to the hill. The result was a good crop of potatoes and corn, and an enormous growth in the trees, excepting 3 acres which I put in sod. These trees made a very menger growth, although they received the same application of fertilizers. The third year I plowed the fallow ground again, applied 500 lbs. phosphate of lime per acre, planted corn and potatoes. The result this year was very marked. The trees occupying the cultivated ground made a strong healthy growth, while those standing in sod made a very small growth. The contrast between the two was so marked as to be seen at a distance. This was the year the committee visited the orchard. They all noticed the mark contrast between them. Fourth year I plowed under a heavy cont of manure; also turned under the sod around two tree rows; planted potatoes. Result, a fine crop of cherries, a heavy crop of peaches and a great many apples, and a strong growth from the trees under cultivation, none from those in sod, but the two rows which had the sod turned under made an enormous growth, leaving those continuing in sod far behind. Fifth year, ran the cherry block in sod, cultivated the peach, apple, pear, &c., without crop. Result, less growth in cherry, but a heavy crop, the apple and pear continued their strong growth under cultivation, and bore heavily, some as much as half a bushel to the tree. The peach also grew enormous, but bore no fruit, owing to the buds being winter killed. This year the trees in sod made a better growth than any year since cultivation was stopped.— I have measured several of the apple trees, under cultivation now five years, and they average 4 to 5 inches in diameter, are 15 to 16 feet high, and have a spread of from 10 to 12 feet, with heads nearly to the ground. My plan would be to cultivate a young orchard for eight to ten years, then run into sod, and top dress, mowing the grass and leaving it decay on the ground. But this is not all. Eternal vigilance must be the watchword of the successful orchardist. I wash my trees every spring with a wash, 1 lb. of caustic potash to 5 gallons of water, washing with a stiff hand scrub. This keeps the bark clean, smooth and healthy, destroying thousands of insects. At the same time potash is a very necessary ingredient to all vegetable matter. It is also very necessary to examine for the borer at least twice a year. During winter is a good time to destroy thousands of eggs of such insects as infest the tops. They can be readily seen, gathered and destroyed.

It is useless for a lazy or indifferent man to endeavor to be a successful fruit raiser. He must take pride in his work. He must apply himself to work, not entrusting it to others. Nor is the exercise of muscle alone necessary.

The mind must be brought into action; he must devote a portion of his time to reading the various works relating to his business, as well as good horticultural papers, thereby becoming familiar with the causes of failure or success of others, learning how to avoid the first, and benefit by the latter. With these remarks I will close. If what little I have said will help any one, in any particular, I am amply repaid.

SELECTIONS.

HISTORY OF PYRETHRUM.

There are very few data at hand concerning the discovery of the insecticide properties of Pyrethrum. The powder has been in use for many years in Asiatic countries south of the Caucasus mountains. It was sold at a high price by the inhabitants, who successfully kept its nature a secret until the beginning of this century, when an American merchant, Mr. Jumtikoff, learned that the powder was obtained from the dried and pulverized flowerheads of certain species of Pyrethrum growing abundantly in the mountain region of what is now known as the Russian province of Trans-caucasia. The son of Mr. Jumtikoff began the manufacturing of the article on a large scale in 1828, after which year the Pyrethrum industry steadily grew until to-day the export of the dried flower-heads represents an important item in the revenue of those countries.

Still less seems to be known of the discovery and history of the Dalmatian species of Pyrethrum (*Pyrethrum cinerariafolium*), but it is probable that its history is very similar to that of the Asiatic species. At the present time the Pyrethrum flowers are considered by far the most valuable product of the soil of Dalmatia.

There is also very little information published regarding either the mode of growth or the cultivation of Pyrethrum plants in their native home. As to the Caucasian species we have reason to believe that they are not cultivated, at least not at the present time, statements to the contrary notwithstanding.† The well-known Dr. Gustav Radde, director of the Imperial Museum of Natural History at Tiflis, Transcaucasia, who is the highest living authority on everything pertaining to the natural history of that region, wrote us recently as follows: "The only species of its genus, *Pyrethrum roseum*, which gives a good, effective insect powder, is nowhere cultivated, but grows wild in the basal-alpine zone of our mountains at an altitude of from 6,000 to 8,000 feet." From this it appears that this species, at least, is not cultivated in its native home, and Dr. Radde's statement is corroborated by a communication of Mr. S. M. Hutton, Vice-Consul General of the United States at Moscow, Russia, to whom we applied for seed of this species. He writes that his agents were not able to get more than about half a pound of the seed from any one person. From this statement it may be inferred that the seeds have to be gathered from the wild and not from the cultivated plants.

As to the Dalmatian plant it is also said to be cultivated in its native home, but we can

*"From recent communications by him in the "Americana Naturalist."
†Report Comm. of Patents, 1857, Agriculture, p. 130.

get no definite information on this score, owing to the fact that the inhabitants are very unwilling to give any information regarding a plant the product of which they wish to monopolize. For similar reasons we have found great difficulty in obtaining even small quantities of the seed of *P. cinerariafolium* that was not baked or in other ways tampered with to prevent germination. Indeed, the people are so jealous of their plant that to send the seed out of the country becomes a serious matter, in which life is risked

Cultivation of Pyrethrum.

The seed of *Pyrethrum roseum* is obtained with less difficulty, at least in small quantities, and it has even become an article of commerce, several nurserymen here, as well as in Europe, advertising it in their catalogues. The species has been successfully grown as a garden plant for its pale rose or bright pink flower-rays. Mr. Thomas Meehan, of Germantown, Pa., writes : "I have had a plant of *Pyrethrum roseum* in my herbaceous garden for many years past, and it it holds its own without any care much better than many other things. I should say from this experience that it was a plant which will very easily accommodate itself to culture anywhere in the United States." Peter Henderson, of New York, another well-known and experienced nurseryman, writes : "I have grown the plant and its varieties for ten years. It is of the easiest cultivation, either by seeds or divisions. It now ramifies into a great variety of all shades, from white to deep crimson, double and single, perfectly hardy here, and I think likely to be nearly everywhere on this continent." Dr. James C. Neal, of Archer, Fla., has also successfully grown *Pyrethrum roseum* and many varieties thereof, and other correspondents report similar favorable experience. None of them have found a special mode of cultivation necessary. In 1856 Mr. C. Willemot made a serious attempt to introduce and cultivate the plant* on a large scale in France. As his account of the cultivation of Pyrethrum is the best we know of, we quote here his experience with but few slight omissions: "The soil best adapted to its culture should be composed of a pure ground, somewhat siliceous and dry. Moisture and the presence of clay is injurious, the plant being extremely sensitive to an excess of water, and would in such cases immediately perish. A southern exposure is the most favorable. The best time for putting the seeds in the ground is from March to April. It can be done even in the month of February if the weather will permit it. After the soil has been prepared and the seeds are sown they are covered by a stratum of ground mixed with some vegetable mold, when the roller is slightly applied to it. Every five or six days the watering is to be renewed in order to facilitate the germination. At the end of about thirty or forty days the young plants make their appearance, and as soon as they have gained strength enough they are transplanted at a distance of about six inches from each other. Three months after this

*"Mr. Willemot calls his plant *Pyrethre du cancase* (*Py etreum Willemoti* a nevaeter), but it is more than probable that this is the * * * species of *Pyrethrum roseum*. We have drawn liberally from Willemot's paper on the subject, a translation of which may be found in the Report of the Commissioner of Patents for the year 1861, Agriculture, pp. 323-331.

operation they are transplanted again at a distance of from fourteen to twenty inches, according to their strength. Each transplantation requires, of course, a new watering, which, however, should only be moderately applied. The blossoming of the Pyrethrum commences the second year, toward the end of May, and continues to the end of September." Mr. Willemot also states that the plant is very little sensitive to cold, and needs no shelter, even during severe winters.

The above-quoted directions have reference to the climate of France, and as the cultivation of the plant in many parts of North America is yet an experiment, a great deal of independent judgment must be used. The plants should be treated in the same manner as the ordinary Asters of the garden or other perennial Compositæ.

As to the Dalmatian plant, it is well known that Mr. G. N. Nileo, a native of Dalmatia, has of late years successfully cultivated *Pyrethrum cinerariæfolium* near Stockton, Cal., and the powder from the California-grown plants, to which Mr. Mileo has given the name of "Buhach," retains all the insecticide qualities and is far superior to most of the imported powder, as we know from experience. Mr. Mileo gives the following advice about planting—advice which applies more particularly to the Pacific coast: "Prepare a small bed of fine, loose, sandy, loamy soil, slightly mixed with fine manure. Mix the seed with dry sand and sow carefully on top of the bed. Then with a common rake disturb the surface of the ground half an inch in depth. Sprinkle the bed every evening until sprouted; to much water will cause injury. After it is well sprouted, watering twice a week is sufficient. When about a month old, weed carefully. They should be transplanted to loamy soil during the rainy reason of winter or spring."

Our own experience with *Pyrethrum roseum* as well as *Pyrethrum cinerariæfolium* in Washington, D. C., has been so far quite satisfactory. Some that we planted in the fall of 1880 came up quite well in the spring, and a few plants bloomed in November of 1881, though such blooming was doubtless abnormal. The plants from sound seed which we planted this spring are also doing finely, and as the soil is a rather stiff clay and the rains were in early summer many and heavy, we conclude that Mr. Willemot has over stated the delicacy of the plants. We have observed further that the seed often lays a long time in the ground before germinating, and that it germinates best when not wet red to heavily. We think that the too rapid absorption of moisture often causes the seed to burst prematurely and rot, where slower absorption in a soil only tolerably moist affords the best conditions for germination.

Preparation of the Plants for Use.

In regard to manufacturing the powder, the flower-heads should be gathered during fine weather, when they are about to open, or at the time when fertilization takes place, as the essential oil that gives the insecticide qualities reaches, at this time, its greatest development. When the blossoming has ceased the stalks may be cut within about four inches from the ground and utilized, being ground and mixed with the flowers in the proportion of one-third of their weight. Great care must be taken not to expose the flowers to moisture, or the rays of the sun, or still less to artificial heat. They should be dried under cover, and hermetically closed up in sacks or other vessels to prevent untimely pulverization. The finer the flower-heads are pulverized the more effectually the powder acts and the more economical is its use. Proper pulverization in large quantities is best done by those who make a business of it and have special mill facilities. Lehn & Fink, of New York, have furnished us with the most satisfactory powder. For his own use the farmer can pulverize smaller quantities by the simple method of pounding the flowers in a mortar. It is necessary that the mortar be closed, and a piece of leather through which the pestle moves, such as is generally used in pulverizing pharmaceutic substances in a laboratory, will answer. The quantity to be pulverized should not exceed one pound at a time, thus avoiding too high a degree of heat, which would be injurious to the quality of the powder. The pulverization being deemed sufficient, the substance is sifted through a silk sieve, and then the remainder, with a new addition of flowers, is put in the mortar and pulverized again.

The best vessels for keeping the powder are fruit jars with patent covers, or any other perfectly tight glass vessel or tin box.

The Use of Pyrethrum as an Insecticide

Up to a comparatively recent period the powder was applied to the destruction of those insects only which are troublesome in dwellings, and Mr. C. Willemot seems to have been the first in the year 1857, (?) to point out its value against insects injurious to agriculture and horticulture. He goes, however, too far in his praise of it, and some of his statements as to its efficacy are evidently not based upon actual experiment. Among others he proposes the following remedy: "In order to prevent the ravages of the weevil on wheat fields, the powder is mixed with the grain to be sown, in proportion of about ten ounces to about three bushels, which will save a year's crop." This is simply ridiculous, as every one who is familiar with the properties of Pyrethrum will understand. We have during the past three years largely experimented with it on many species of injurious insects, and fully appreciate its value as a general insecticide, which value has been greatly enhanced by the discovery that it can be most economically used in liquid solution; but we are far from considering it a universal remedy for all insects. No such universal remedy exists, and Pyrethrum has its disadvantages as has any other insecticide now in use. The following are its more serious disadvantages: 1, the action of the powder, in whatever form it may be applied, is not a permanent one in the open air. If *e. g.*, it is applied to a plant, it immediately effects the insects on that plant with which it comes in contact, but it will prove perfectly harmless to all insects which come on to the plant half an hour (or even less) after the application; 2, the powder acts in the open air—unless, perhaps, applied in very large quantities—only upon actual contact with the insect; if *e. g.*, it is applied to the upper side of a cotton leaf the worms that may be on the underside are not affected by it; 3, it has no effect on insect eggs, nor on pupæ that are in any way protected or hardened.

These disadvantages render Pyrethrum in some respects inferior to arsenical poisons, but on the other hand, it has the one overshadowing advantage that it is perfectly harmless to plants or to higher animals; and if the cultivation of the plants in this country should prove a success, and the price of the powder become low enough, the above-mentioned disadvantages can be overcome, to a certain degree, by repeated applications.

In a closed room the effect of Pyrethrum on insects is more powerful than outdoors. Different species of insects are differently affected by the powder. Some resist its action most effectually, *e. g.*, very hairy caterpillars and especially spiders of all kinds; while others, especially all Hymenoptera, succumb most readily. In no case are the insects killed instantaneously by Pyrethrum. They are rendered perfectly helpless a few minutes after application, but do not die till some time afterward, the period varying from several hours to two or even three days, according to the species. Many insects that have been treated with Pyrethrum show signs of intense pain, while in others the outward symptoms are much less marked. Differences in temperature and other meteorological changes do not appear to have any influence on the effect of Pyrethrum.

Modes of Application.

Pyrethrum can be applied—1, as dry powder; 2, as a fume; 3, as an alcoholic extract diluted; 4, by simple stirring of the powder in water; 5, as a tea or decoction.

The following recommendations are based on repeated experiments in the field:

1. *Applications of Pyrethrum as a dry Powder.*—This method is familiar to most housekeepers, the powder being used by means of a small pair of bellows. It is then generally used without diluent, but if it is unadulterated and fresh (which cannot be said, in many instances, of the powder sold at retail by our druggists) it may be considerably diluted with other pulverized material without losing its deadly effect, the use of the powder thus becoming much cheaper. Of the materials which can be used as diluents, common flour seems to be the best, but finely-sifted wood-ashes, saw-dust from hard wood, etc.—in short, any light and finely-pulverized material which mixes well with the Pyrethrum powder will answer the purpose. If the mixture is applied immediately after preparation, it is always less efficacious than when left in a perfectly tight vessel for about twenty-four hours, or longer, before use. This has been proven so far only with the mixture of Pyrethrum with flour, but holds doubtless true also for other diluents. Mr. E. A. Schwatz experimented largely under our direction with the mixture of Pyrethrum and flour for the cotton worm, and he found that one part of the powder to 11 parts of flour is sufficient to kill the worms (only a portion of the full-grown worms recovering from the effects of the powder), if the mixture is applied immediately after preparation; but if kept in a tight glass jar for about two

days, one part of the powder to 22 parts of flour is sufficient to kill all average-sized worms with which the mixture comes in contact. For very young cotton worms a mixture of one part of Pyrethrum to 30 parts of flour, and applied one day after preparation, proved most effective, hardly any of the worms recovering.

An ordinary powder bellows will answer for insects infesting dwellings or for plants kept in pots in rooms, or single plants in the garden but it hardly answers on a large scale outdoors, because it works too slowly, the amount of powder discharged cannot be regulated, and there is difficulty in covering all parts of a large plant. Another method of applying the dry powders is to sieve it on to the plants by means of sieves, and this method is no doubt excellent for insects that live on the upper side of the leaves. For large, more shrub-like plants with many branches, and for insects that hide on the underside of the leaves, this method will be found less serviceable. A very satisfactory way of applying the powder on large plants, in the absence of any suitable machine or contrivance, is to throw it with the hand after the manner of seed-sowing. This method is more economical and rapid than those mentioned above, and it has, moreover, the advantage that, if the plants are high enough, the powder can be applied to the underside of the leaves.

2. *Application of Pyrethrum in Fumes.*— The powder burns freely, giving off considerable smoke as an odor which is not unpleasant. It will burn more slowly when made into cones by wetting and molding. In a closed room the fumes from a small quantity will soon kill or render inactive ordinary flies and mosquitoes, and will be found a most convenient protection against these last where no bars are available. A series of experiments made under our direction indicates that the fumes affect all insects, but most quickly those of soft and delicate structure.

This method is impracticable on a large scale in the field, but will be found very effective against insects infesting furs, feathers, herbaria, books, etc. Such can easily be got rid of by inclosing the infested objects in a tight box or case and then fumigating them. This method will also prove useful in greenhouses, and, with suitable instruments, we see no reason why it should not be applied to underground pests that attack the roots of plants.

3. *Alcoholic Extract of Pyrethrum Powder.*— The extract is easily obtained by taking a flask fitted with a cork and a long vertical glass tube. Into this flask the alcohol and Pyrethrum is introduced and heated over a steam tank or other moderate heat. The distillate, condensing in the vertical tube, runs back, and at the end of an hour or two the alcohol may be drained off and the extract is ready for use. Another method of obtaining the extract is by re-percolation after the manner prescribed in the American Pharmacopœia. The former method seems to more thoroughly extract the oil than the latter; at least we found that the residuum of a quantity of Pyrethrum from which the extract was obtained by re-percolation had not lost a great deal of its power. The first method is apparently more expensive than the other, but the extract is in either case more expensive than the other preparations, though very conveniently preserved and handled.

The extract may be greatly diluted with water and then applied by means of any atomizer. Professor E. A. Smith, of Tuscaloosa, Ala., found that, diluted with water, at the rate of one part of the extract to 15 of water, and sprayed on the leaves, it kills cotton worms that have come in contact with the solution in a few minutes. The mixture in the proportion of one part of the extract to 20 parts of water was equally efficacious, and even at the rate of 1 to 40 it killed two-thirds of the worms upon which it was sprayed in 15 or 20 minutes, and the remainder were subsequently disabled. In still weaker solution, or at the rate of 1 to 50, it loses in efficacy, but still kills some of the worms and disables others. Professor Smith experimented with the extract obtained by distillation, and another series of experiments with the same method was carried on last year by Professor R. W. Jones, of Oxford, Miss.* He diluted his extract with twenty times its volume of water and applied it by means of an atomizer on the cotton worm and the boll worm with perfect success. Mr. E. A. Schwarz tried, last summer, the extract obtained by re-percolation,† and found that 10 drachms of the extract stirred up in two gallons of water and applied by means of Whitman's fountain-pump was sufficient to kill all cotton worms on the plants. Four drachms of the extract to the same amount of water was sufficient to kill the very young worms.

4. *Pyrethrum in Simple Water Solution.*— So far as our experiments go, this method is by far the simplest, most economical, and efficient. The bulk of the powder is most easily dissolved in water, to which it at once imparts the insecticide powder. No constant stirring is necessary and the liquid is to be applied in the same manner as the diluted extract. The finer the spray in which the fluid is applied the more economical is its use and the greater the chance of reaching every insect on the plant. Experiments with Pyrethrum in this form show that 200 grains of the powder stirred up in two gallons of water is amply sufficient to kill the cotton worms, except a very few full-grown ones, but that the same mixture is not sufficiently strong for many other insects, as the boll worm, the larva of the *Tortrix nicippe*, and such species as are protected by dense long hairs. Young cotton worms can be killed by 25 grains of the powder stirred up in two quarts of water.

The Pyrethrum water is most efficacious when first made and loses power the longer it is kept. The powder gives the water a light greenish color, which after several hours changes to a light brown. On the third day a luxuriant growth of fungus generally develops in the vessel containing the liquid, and its efficacy is then considerably lessened.

5. *The Tea or Decoction.*—Professor E. W. Hilgard, of Berkeley, Cal., is the only one who has experimented with Pyrethrum in

*Vide "American Entomologist, Vol. III, pp. 232-3.
†From one pound of the powder one pint of extract is made, each drop of the extract representing one grain of the powder. The actual cost of making the extract was 50 cents.

this form, and expresses himself most favorably as to the result. He says:

"I think, from my experiments, that the tea or infusion *prepared from the flowers* (which need not be ground up for the purpose) is the most convenient and efficacious form of using this insecticide in the open air; provided that it is *used at times when the water will not evaporate too rapidly*, and that it is applied, not by pouring over in a stream, or even in drops, but *in the form of a spray from a syringe* with *fine* holes in its rose. In this case the fluid will reach the insect despite of its water-shedding surfaces, hairs, etc., and stay long enough to kill. Thus applied, I have found it to be efficient even against the armored scale-bug of the orange and lemon, which falls off in the course of two or three days after the application, while the young brood is almost instantly destroyed. As the flower tea, unlike whale soap and other washes, leaves the leaves perfectly clean and does not injure even the most tender growth, it is preferable on that score alone; and in the future it can hardly fail also to be the cheaper of the two. This is the more likely, as the tea made of the leaves and stems has similar although considerably weaker effects; and if the farmer or fruit grower were to grow the plants, he would save all the expense of harvesting and grinding the flower-heads by simply using the header, curing the upper stems, leaves, and flower-heads all together, as he would hops, making the tea of this material by the hogshead, and distributing it from a cart through a syringe. It should be diligently kept in mind that the least amount of *boiling* will seriously injure the strength of this tea, which should be *made* with briskly boiling water, but then simply covered over closely, so as to allow of as little evaporation as possible. The details of its most economical and effectual use on the large scale remains, of course, to be worked out by practice."

The method of applying Pyrethrum in either of the three last-mentioned forms is evidently far more economical in the open field and on a large scale than the application of the dry powder, and, moreover, gives us more chance of reaching every insect living upon the plant to which the fluid is applied. The relative merits of the three methods can be established only by future experience, but so far we have found the simple water solution most convenient and satisfactory.

QUINCE CULTURE.

There is some difference of opinion as to the best length for a cutting. Eight or ten inches are recommended. My experience in Vineland gives the preference to a cutting of about fifteen inches, planted a foot in the ground. The advantages of so deep a setting are, that it guards against drouth, and furnishes a greater length for the formation of roots, which comes out through the bark all the way from the lower end as high as the soil is moist.

Cuttings can be made to grow if taken at any stage of their development. If green and soft they depend on conditions of heat and moisture in the soil and air, requiring the skill of a professional gardener with the appliances of the hot-house. For out-door cultivation the wood must all be well ripened and taken in its dormant state, after the trees have shed their leaves in autumn. I have found February and March favorable to success. Any time before the buds start in the spring may succeed. A few grew one year taken in May. The sprouts often growing on the part of roots near the surface, suggested *root cuttings* as an additional means of multiplying trees. Any large root cut off near the collar of the tree is almost sure to develope buds of seed and send up sprouts. The best time to make root cuttings is just before the usual season for the buds of the tree to swell in the spring.

Propagation by grafting on the stocks and roots of other trees as well as its own, is a successful method of quince culture. The thorn and apple have been used for the quince, as that has been for the pear; and here all the different methods of grafting are available. The thorn is much valued by some on account of its strength, and freedom from borers. Roots of apple trees, as well as other quince trees, are available. Trees without apple roots were exhibited at the late meeting of the New Jersey Horticultural Society in Vineland. The scions of a fruitful tree grafted into one that was barren have borne the second year. Crafting on older stocks in this way will enable us to test new varieties; and also to gain time in proving the value or worthlessness of all our seedlings. It is also a convenient way of comparing the relative merits of different varieties, by securing the perfect equality of all conditions.

A good grafting wax, to be applied warm, is made by melting together six parts rosin, with one part each of bees-wax and tallow or linseed. It can be applied with a brush, or spread on strips of muslin for wrappings. For a wax to be applied by hand, cold, Downing recommends bees-wax and rosin each three parts, to two parts of tallow.

Scions for grafting are best if cut after the leaves have fallen and before the stimulation of the spring. They can be wintered in sand or sawdust. If not cut till spring they may as well be set at once. By the aid of an ice-house the season of setting them can be greatly extended. Vigorous stocks often produce a profusion of sprouts. It may be best to leave some of these to direct the circulation of the sap and thus secure a supply to the scion, but all should at length be removed leaving the graft to enjoy every advantage.

Propagation by budding, or inoculating, is a favorable method for some trees, as the peach and apricot; but is only recommended for the quince where grafts have missed, or where we want to increase the sorts for which the other methods are not available. It differs from grafting mainly in being confined to the season when the cell circulation is most active, and the union of parts much quicker than with grafts. Budding is most successfully performed in that part of the growing season when the *cambium* or gelatinous matter between the bark and wood is in greatest activity organizing new cells. The "pulp," as gardners call this cambium, must be present between the bark and wood of the stock, so that the bark can be easily separated for the insertion of the bud. It buds of the previous year are to be worked, the scions should be kept dormant till the young leaves of spring indicate that the bark will slip. If buds of the current year are used they should be well developed; and this perfection may be accelerated by cutting off the tips of the shoots from which they are to be taken. As soon as the scion is separated from the tree the blade of every leaf should be removed, so that its evaporation may not injure the vitality of the bud. If dormant buds have been used in spring the stocks should be cut away above them as soon as they begin to swell, and the shoots from the stock below rubbed off. If buds of the current year have been successful, then the removal of the stocks should be deferred to the next spring. With a vigorous stock, a bud, like a graft, should make a handsome tree the first season.—*The Weekly Press.*

POULTRY FARMING.

The cost of adequate fencing still strikes me as one of the main difficulties of the poultry business. The easiest put-up fence is wire netting fixed to posts or stakes at proper intervals, and, all things considered, it is, perhaps, as cheap as any unless exception be made in favor of tarred twine netting, but that is not so durable. Employing 2¼ inch meshed wire netting, size eighteen for the bottom width, and 3½ inch mesh and number nineteen gauge for the top width, a yard wide and buying in quantities, the cost of netting, six feet high, with a wire to run through and stiffen the top, will be about fifteen to eighteen cents per linear yard without the stakes and building. After allowing for this, it will bring the cost of each run, allowing one side for each plot, to $5, and the fencing in the hedges for the permanent lines will cost an additional $3 for each plot.

But in my opinion the netting alone will not best answer the purpose. There would be some fighting through it; it affords no shelter; and bad habits would be communicated. Birds kept as they must be kept in such yards are observant and ready to adopt vices very quickly. If the hens in one yard take to egg-eating or feather plucking, the vice will be learned by hens in adjoining yards when there is no obstruction to sight. To separate different flocks from sight of each other, even partially, is of some importance, and this entails increased expense. Stakes or posts of sufficient size and height may be set in the ground at proper intervals and then nine-inch thin boards attached for the bottom of the fence, and a width of netting stretched on top will make a fence that will overcome all the difficulty, but it increases the expense to double, or nearly double that of the netting alone. But it is a great deal better fence for the partition. It is to be made in sections so as to be easily removed from one side of the house to the other, as needed. We must therefore incur an expense of $13 for the fencing of each yard, or at the rate of fifty cents a head. This will cover the whole expense for fencing, if a man falls in with a streak of more than ordinary good luck. But it would perhaps be safer to make estimates on $15 per fenced plot, and the whole fencing and building per acre will reach a cost of at least $75.

It will be objected that the expense and time required are too great before the farm could be put into successful operation, but lapse of time in arrangements and investment of capital is a part of all successful business. I cannot see why people should demand a system of poultry farming that is capable of springing into existence all at once. No other kind of farming or of fencing ever did spring into existence in that way. The ordinary farmer finds to his land buildings and fences and arrangements which have been the growth of years. I might almost say of centuries. It is unreasonable to expect that land is to be made equally adapted for an entirely new pursuit without time and expense and labor, and if I point out, therefore, the kind of fences and other arrangements adapted to the end in view it is no kind of answer to say these things take too much time and money. These are only some of the difficulties of the undertaking, and are a very good reason why no one should embark lightly in such an enterprise, but go gradually to work, feeling his way as he goes.

Thus the planning and the fencing of a poultry farm absorbs relatively the largest amount of capital. If any man can present a system that will work permanently without this preliminary labor and investment of capital he will have a large audience ready to hear him. My own experience of fowls is against all and any seductive theories of cheapest instantaneous arrangements in poultry farming, and I believe any one who tries to force success on that basis will come to grief. Granite can be boiled as easily as water if you will take the necessary steps in organizing your boiler—but not otherwise.—*Dr. A. M. Dickie, Doylestown, Pa.*

POULTRY ABUNDANT, BUT DEAR.

Since the year 1, of the days of Peter, the raising of poultry has been a certain gauge of civilization. The wild Indian keeps no fowls; but, as man advances from the savage state he gathers the feathered tribe around him, because they make some of the richest delicacies of his table, as well as the softest down for his couch. Although poultry is raised in large quantities in this region, it is very dear. In price it stands alongside of beef, mutton and pork, all of which are about double American prices. A pair of chickens costs about $2, and they are not very large at that. Of course the common day-laborer earning not more than thirty sous per day, cannot often indulge in such luxury, or the mechanic either, who earns but five francs a day. But as poultry is always abundant in market, somebody eats a great deal of it. Of the various kinds, chickens are kept in the greatest numbers. Like the farm stock they are a great deal mixed; but mostly dark-colored or black. The Black Spanish are the most common, and like cats and dogs, are kept more or less all over the city, hence you may hear the crow of Chanticleer in almost any direction in the morning, but it is not loud like his Shanghai relations in America. It is almost as different as the little ear whistle here and the big one at home. Chickens are outlawed in town and country—they must keep in bounds, or their heads will come off before their time. In the country—there is plenty of good pasture for them so that they cannot help but thrive; as we have before remarked, every spot of ground which has not been recently harvested or plowed is green as grass can make it. We hear of no serious diseases. —*Philadelphia Weekly Press.*

NOTES ON FRENCH AGRICULTURE.

Nitrogen, the most valuable and costly element of bone manure, occurs largely in an insoluble form, and may remain. All cultivated soils contain large quantities of it, so that in soluble form there would be sufficient for large crops for hundreds of years.

Many farmers, however, are not aware of the existing facts. Experiments by Messrs. Laws and Gilbert will enlighten them. In raising barley on the same ground during nineteen years in succession they found that as much barley was obtained by applying chemical fertilizers containing forty-one pounds of nitrogen in ammonical salts, which were readily soluble, as from applying bone manure containing 200 pounds of nitrogen. In other words, the nitrogen in soluble salts, which were available for plants, proved nearly five times as effectual as nitrogen in bone manure. If the latter could be as readily soluble as the nitrogen in ammonical salts four times the immediate effect usual would be obtained from it. Here is a chance for the young farmer with a "large intellect."

What can be done to render the fertilizing elements of barn manure more soluble and available for the use of plants? Dr. Lawes, who has given the subject much attention, after what he calls a "scientific prelude," says: "I am bound to confess that I am just as helpless in regard to the management or improvement of dung as the most old fashioned farmer." This is certainly not very encouraging. Prof. S.C. Caldwell, of Cornell University, in commenting upon some of the results of experiments by Messrs. Lawes and Gilbert, in which the crops obtained contained only a part of the nitrogen contained in the barn manure, says: "These considerations teach us to convert the nitrogen of stable manure, as far as possible, into more assimilable forms by judicious rotting before putting it in the soil; since the proportion immediately recovered is so much larger, the more soluble the nitrogen with which the plant is fed." Much may be done by allowing it to ferment and decompose. The process, however, must be conducted with care or the ammonia formed will escape. If horse manure is allowed to ferment in a heap loosely thrown together, as is usually the case around stables, it becomes dry and ammonia escapes freely. By making the heap more compact, as may easily be done by allowing pigs access to it, and keeping it moist, very little ammonia will be lost. Water has so strong an affinity for it that a gallon of ice-cold water, it is said, will absorb 1,150 gallons of ammonia gas. By keeping the heap well moistened very little ammonia will escape. According to Dr. Voelcker ulmic, humic, cramic and approcramic acids are produced during fermentation, and these uniting with the ammonia form salts which are retained in the heap, and preserve the ammonia in a form easily available for the use of plants. Pigs aid in this work, and the occasional addition of soil or muck will serve as another precaution against loss. Earth and muck readily absorb ammonia and tenaciously retain it. If by decomposition half the nitrogen in barn manure could be made immediately as available for plants as the nitrogen in ammonical salts twice the effect would be obtained from it that Lawes and Gilbert obtained in their experiments in barley raising.—*Henry Reynolds, M. D., Auburn, Maine.*

OUR LOCAL ORGANIZATIONS.

LANCASTER COUNTY AGRICULTURAL AND HORTICULTURAL SOCIETY.

The Lancaster County Agricultural Society met statedly Monday afternoon, May 1st.

The following members were present: J. F. Witmer, Paradise; W. L. Hershey, Chickies; Daniel Smeych, city; John C Linville, Salisbury; Peter H. Hershey; F. R. Diffenderffer, city; J. M. Johnston, city; John Monk, Chickies; Levi S. Reist, Manheim; S. P. Eby, city; Harry G. Rush, West Willow; Mr. Haws, New England.

Mr. John Monk, of West Hempfield, was nominated and elected to membership.

S. P. Eby, as chairman of a special committee on the laws relating to fencing lands, reported progress and asked to be continued.

Crop Reports.

H. M. Engle said the wheat crop prospects are good. Along the river some fields are exceedingly fine. Clover has suffered. The lookout is not favorable to a heavy hay crop. The fruit prospect very good. The peach bloom is profuse; so is that of pears and plums. Potatoes are just coming out of the ground. Rain fall for February was 3 14-16 inches; for March, 3 2-16, and for April 2 14-16 inches.

L. S. Reist said the wheat and fruit crops are good, but clover was never poorer. Some fields have almost none.

P. H. Hershey remarked the singular fact that the best lands seem to have the poorest clover. Why this was so he could not understand. He wished to know why this was so.

J. C. Linville also remarked the fact mentioned by the former speaker. The best clover on his farm to day is on finty and stiff clay ground, which is contrary to the usual experience. He has noticed some wheat is far better than the rest.

S. P. Eby has also observed that clover is very poor. A promising field of his own is frozen out completely.

John G. Rush noticed that the poorest lands this year have the best clover, something that is unaccountable.

H. M. Engle has been accustomed to sow rye for green food for his cattle. It makes rough hay, but it comes early as green food, and he has been feeding it for several weeks already this season. The clover sown this spring got an excellent start, and if the season is favorable we ought to have a good hay crop next spring; but we have had poor hay crops for a number of years, as all know. He believed during dry seasons much hurt is done to the young clover. It is pastured closely when the ground is dry and hard, and the life is tramped out of it by the cattle which are kept on it long after they should be taken off.

Jos. S. Witmer reported a good wheat crop, but the grass was rather poor. A little corn has been planted. There is still a good deal of tobacco on hand. Young plants are coming along rapidly.

Growing Corn.

H. M. Engle thought growing a good crop of corn depended on many things: good land, properly prepared; good seed, and careful after cultivation. A two-year old sod he thought best for corn. Don't plant too early nor too deep. The longer corn requires to come up the weaker the plant. When it comes up rapidly it grows much faster; from half an inch to an inch is deep enough to plant. The largest average crops are grown in hills while the largest yields have been taken from drilled fields. When checkered it receives more attention. Cultivate shallow. Remove the suckers early; when left they draw the vitality of the ear plants.

S. P. Eby gave an instance of a farm on which at one time no corn could be grown. Gradually the land was brought to a good condition and fine crops were grown. Much trouble was experienced from crows. The seed was then soaked in tar water, which put an end to this trouble, and also brought the plants along faster.

John G. Rush did not think our farmers should make the corn crop a specialty. The West can grow it cheaper than we can. We cannot afford, therefore, to give so much time and attention to corn.

P. H. Hershey thought our corn crop a very important one. Thorough culture is an important point. Don't plant too deep, and begin working the moment the corn is up. You can't work your corn land too much. The best crop of corn he ever grew was in a rather dry season; he worked it eight times; it never stopped growing and gave 75 bushels to the acre.

Mr. Monk asked whether any one had experience with Chester county Mammoth corn.

Jos. F. Witmer had had some experience. He did not like it and will not plant it any more.

H. M. Engle never plants corn dry. He soaks it until it shows signs of sprouting. If planted dry in dry weather, it lies there weeks in a dry season without coming up. He never plants corn without putting on coal tar. A very small quantity is enough. Put a little plaster over it; after it has been thus treated it can be easily handled. Birds will never touch corn treated in this way. He has no fear of crows. The lands along the river are very well adapted to corn, and when a two year sod is plowed under no manure is required. A good crop is nearly always to be relied on. Failures are uncommon in that vicinity. Large corn is not so good for fodder. Small corn is better and cures better, being easier to handle besides. It has also subdued it to a disease, like the yellows in peaches, that sometimes come upon the corn. It is caused by a minute aphis that operates on the roots, and he did not know of any remedy against its ravages.

Commercial Fertilizers.

H. M. Engle believed not only commercial fertilizers, but fertilizers of almost every kind, are valuable when applied to the potato crop. Anything that contains potash will benefit potatoes. Even ashes from anthracite coal are serviceable. Potash, nitrogen and phosphoric acid are the three great fertilizers. Nitrogen is not so useful as the other two for potatoes. It will pay, however, to apply some of the high-priced manures to the potato crop. Nothing but experience will tell the farmer what kind of manure his fields need. Every farmer must find out what is best adapted to his lands.

J. C. Linville said some writers hold commercial fertilizers are somewhat uncertain when applied to the potato crop. Fertilizers that combine the three articles above mentioned are the best with which to grow potatoes. Nothing is better for all purposes than well-rotted barnyard manure, which contains all three of them.

Mr. Engle asked why we should buy potash when our soil may already have plenty of it—more than will be used in a generation.

How Should Lime be Applied?

John C. Linville thought we should keep it on the surface, but only under certain conditions. His practice is to apply it to stubble. Lime works down into the soil. Surface application of lime is desirable because it at once absorbs carbonic acid, which is valuable to the soil. If plowed under this process does not take place. If lime is to be applied he believed it should be done during warm fall weather and on the surface only.

H. M. Engle believed the best results are when applied in a fine mealy condition to dry soil, and left on the surface.

J. G. Rush thought there was as much in the mode of liming as in the lime itself. He did not believe in placing it in large heaps. He slacks his lime at the very time he places it on the land. Lime must come into contact with the vegetable matter, and the finer it is the better this result is accomplished.

The Poetry of Agriculture.

Mr. Haws, from New England, was introduced, who delivered a very flowery address on the constitution of the earth, the beauties of vegetation and the part the atmosphere plays in producing crops He covered a great deal of ground in the course of his remarks, and was listened to with attention from the beginning to the close.

Questions for Discussion.

At what period of growth should grass be cut to make the best hay? Referred to Casper Hiller.

How can the best results be obtained from barn yard manure? Referred to M. D. Kendig.

At what stage of ripeness is it best to cut wheat? Referred to H. M. Engle.

J. C. Linville was appointed essayists for the next meeting.

There being no further business, the society adjourned.

THE POULTRY SOCIETY.

The Lancaster County Poultry Association met in their rooms Monday morning, May 1.

The following members were present: J. B. Lichty, Charles Lippold, W. W. Griest, Charles E. Long, John A. Schum, J. M. Johnston, Dr. Witmer, Neffsville; F. R. Diffenderffer, city; Isaac H. Brooks, Marticville.

The minutes of the previous meeting were read and approved.

Miscellaneous Business.

The resignation of T. Frank Evans, offered at the last meeting, was called up, and, on motion, accepted.

John A. Schum was nominated to the position occupied by Mr. Evans. Under the rules, action on the nomination was deferred until next meeting.

Samuel Brubaker, of Neffsville, and Lawrence Knapp, of Lancaster, were nominated to membership and elected.

On motion the Secretary was instructed to inform all the members of the time of meeting hereafter by postal card.

Strains in Poultry Breeding.

The Secretary read the following essay, written by T. F. McGrew, Jr., of Ohio, for the society, by special request. It was as follows:

To write an essay on fowls to be read by a stranger to a number of men of whom I have no knowledge, is to me a very hard task. Should I be able to please you for the time being, and also furnish a few points of such worth that a few will gain just a little benefit from them, my reward will be gained.

You may all be better informed than myself that about 1850, or a little later, what is called the hen fever had its start in this country. About that time the Shanghai fowls were imported into this country, and from them, or others like them, our Cochins and Brahmas have been bred. Twenty-five years ago but little interest was taken in the breeding of fine fowls in this country, but to-day thousands all over the land are paying close attention to their culture. American breeders after some twenty-five years of close attention have to their credit the production one of the very best fowls now known, viz., the highly prized Plymouth Rock.

But all thoughtful breeders are satisfied with this venture at a cross breed fowl, and have no desire to try any more such experiments, as long as there is so much room to improve what we now have.

The furor for new breeds has taken a very strong hold on fanciers, and it is to be feared will not do us any good in the long run; pardon me, if I am intruding on any of your pet themes, but to me, it looks very much as if we, as breeders, should try to improve what we now have on hand, for, after twenty five years of hard labor and close attention, the Light Brahmas, the so-called "kings of the poultry fancy," are very far from perfection. Why should we turn aside for new breeds while there is so much to be done yet for those we have had with us so long.

The great desire to make a few paltry dollars from the fancy is doing the interest more harm than any other one feature we have to contend with.

The term "strain," as used by us, is very much abused, and to this point let us turn our attention.

Because a breeder has for three or four years bred a certain kind of fowl does not give him the right to claim it as his strain; but to put forth a strain, the certain prominent features of the birds he breeds must be so established in them that their fine qualities will vindicate themselves on any stock with which they may be crossed.

To illustrate this let me give you the groundwork of the gold dust strain of Buff Cochins, not for aggrandizement, but because of the knowledge of their origin.

Sixteen years ago next October I was first taken with Buff Cochin fowls, and kept the best I could get at that time. These birds were loaned to an inveterate exhibitor at State and county fairs, who won with them for years.

Up to 1870 they were considered by me about perfect; at this time a trip was made to the East, and the yards about New York and Philadelphia visited, and stock much better procured; the cross with these fowls proved the point above mentioned. The birds spoken of as purchased were of a true strain and stamped their good qualities so plainly on their progeny that I was convinced a strain must be established, and to that end set to work.

After some seven years close attention it now appears that the start of such a strain has been accomplished.

The aim has been to eclipse the standard and the only reward hoped for is the accomplishment of the feat.

The start was made with the best birds to be had from my old stock. To these have been added, from time to time, the best birds to be had, always using new male blood on the best females in hand. By so doing the points gained were retained in the offspring as strongly as possible. The only kind of blood allowed in the yards has been that which was known to be of English pure bred strains.

At times wrong crosses have been made and all the young have come out a head shorter, until now not a bird is in the yards except those which are from $\frac{3}{4}$ to $\frac{7}{8}$ full English blood.

To these have been added the whole of Mr. Doolittle's stock, and from them only the very best have been retained; these crossed with my own should in a few years establish what can be called the groundwork of a true strain. This is my notion of what can be called a "true strain," not a new breed of a few years' breeding, that cannot be counted on to even breed like themselves.

Long years of close attention is the only way to establish a strain, and it is to be hoped the breeders of this country will soon drop chance work and settle down to establishing true breeding strains of the many fine varieties we now have.

The establishing of a society like yours must be a great benefit to you as breeders, and I will venture to mention for your consideration a plan that, in my opinion will be both a pleasure and profit to you all. Let each meeting day be set apart for some breed, and have the specimens of this variety brought to your rooms, and let each man present take a score card and score birds as per his judgment; after which, compare the cards, and let the different scores be your subject of discussion.

This will be the most profitable way you can, in my opinion, spend an hour or two each month. In a short time the very best judges among you will place themselves prominently before you, and without doubt some will spring up who will make themselves the equals of Pierce or Ball.

The aim all breeders should be, first, to breed the very best stock he can, and not to be content with medium, but to try and be the very first in his class. Should this be his aim, and full force and determination put to the work, he who wins over him will have so close a shave that the honors will be about even, and his efforts should be doubled the next year or until he does reach the point; second, do all you can for each other, without fear of doing yourself an injury, for it should be the public we wish to benefit, and not ourselves entirely.

Trusting these disconnected lines may have proved of momentary interest to some of you, I will wish you all a prosperous season, and say "Good-day."

On motion, the thanks of the society were tendered to Mr. McGrew for his essay.

On motion, the society adjourned.

FULTON FARMERS' CLUB.

The April meeting of the club was held at the residence of Lindley King, in Fulton township. Members present: Montillion Brown, E. H. Haines. Joseph R. Blackburn, Joseph P. Griest, Josiah Brown, Grace A. King and Solomon L. Gregg. There were also quite a number of visitors in attendance.

In the absence of the President, Joseph R. Blackburn was elected President pro tem.

Grace A. King exhibited some very fine Tewksbury Winter Blush apples.

Jesse Yocum, a visitor, exhibited White Russian Oats and Hominy Corn.

Asking and Answering Questions.

Josiah Brown: Which method will raise the most corn to the acre, planting in the hill or drilling?

This question created quite an animated discussion. Nearly all present, both members and visitors, were of the opinion that drilling was preferable to planting in the hill, not only because more corn could be raised in that way, but also because it would allow working nearer to it, and it was not so liable to be taken up by birds.

James Smedley, a visitor, asked if it made any difference which way the corn rows were run—north and south, or east and west?

Alvan King had noticed on his way to Lancaster rows that had been taken straight up and down a hill in order to have them run to north and south, and the consequence was the corn had been washed out. The water in time of rain had followed the rows. Some others present had seen the same effect when the rows were up and down hill. The general opinion was that it would be better to run the rows to suit the grade of the ground than to run to the points of the compass.

Montillion Brown asked how lima beans should be planted and cared for.

The only answer to this question was to plant in rows four feet apart, the hills about three feet apart in the row, three beans in the hill. The poles should be put up when the beans were planted. Work with horse like corn.

Jesse Yocum asked if any one present had been troubled with the bean weevil.

Quite a number had had their beans destroyed with it.

James Smedley said that his wife last fall had heated a part of their beans, but not hot enough to destroy the germ. The beans so treated had not been disturbed. Those that had not been heated were destroyed.

Inspecting the Host's Premises.

After dinner the club made the usual inspection of the farm and buildings, and again convened in the house, when some very complimentary criticisms were given in regard to their management. One member remarked that the bachelors of the neighborhood could show the neatest farms. (The host is a bachelor.)

Papers Read.

An excellent article on "Agriculture" was read by Emma King. Mabel A. Haines gave a recitation. Montillion Brown read an original essay on the "Origin and Progress of the Fulton Farmers' Club."

Twelve years ago a few farmers met in the parlor of William Brown, in Fulton township, for the purpose of forming a farmers' club, and, although the prospect looked rather gloomy, a few of the number agreed to try the experiment. Since then the club has kept up regular meetings, and during the whole time there had never been manifested any desire for leadership among its members, but perfect harmony has prevailed throughout. There appeared to be a kindly feeling existing between all the members, uniting them together as a band of brothers and sisters. During that time they had held one public sale, amounting to nearly one thousand dollars, and two fairs or exhibitions of farm products, which were little, if any, inferior to fairs held under the auspices of the Agricultural Society of this or adjoining counties.

The pic-nic held last summer at the Barren Springs was quite an enjoyable affair.

In all these the public had been invited to participate on perfect equality with club members, and the proceeds of the fairs, after paying the expenses, had been divided among the exhibitors as premiums. He paid a deserved tribute to the memory of William Brown, the father of the club.

How, lacking the advantages of a good education, he often spent the hours that his neighbors were sleeping over his old Pike's arithmetic and other books, in order to keep up with the times. In his death the club had lost a valued member, and the community at large a live and progressive man.

(We are sorry that the essayist neglected to speak of the virtues of his estimable wife, also deceased, who was a help meet for him in the true sense of the word.—Ed.)

John Gregg, another member, had also been removed by death, and of the original number, but two were now members of the club, viz., William King and himself.

The question, Is it better for the farmer to pursue a mixed husbandry or make a specialty of some one of its branches? was next taken up and discussed at some length, the majority being of the opinion that it would be better to give particular attention to some particular branch, as dairying, cattle feeding, &c., and if they lost money at times trust to the future to make it up, than to try a little of everything.

The May meeting will be held at the residence of Montillion Brown, in Fulton township, first Saturday in the month.

AGRICULTURE.

French Farming

Every square foot of ground is put to use, has been in use for unnumbered generations. Here and there in the distance appear patches of wood, carefully preserved and guarded, but the rest of the land is almost bare of shade. There is no brush or tangle of weed and wild flower by the roadside, no thicket by the stream. The last of these trespassers were eradicated ages ago, along with the last stump. A gray stone wall borders the highway. The cross-roads are often sunk two or three feet below the general level. Narrow ridges of earth mark the boundaries of the fields, and the furrows are driven so close to them that it is a wonder how the plough is turned. Single rows of poplars stretch with exasperating regularity across the landscape. They are trimmed close, and sometimes every twig is removed except a bunch at the extreme top, then they look like liberty poles with bushes tied to them. There are willows by the brook, but they are pollard-willows, kept for their twigs which are scrupulously cut off, and they lift their scarred and knotted trunks like hands from which all the fingers have been amputated.

Sand Farming.

What is sand? Writers differ so much in their ideas of soils that it is puzzling often to define their meaning. Pure sand makes a poor soil, or no soil at all, on which to attempt to grow any kind of crop. Our common, sandy soil contains more or less clay, and this it is which gives them their capacity for being improved or made productive. The sand farming referred to by Mr. Loomis is on sandy loam. Of this soil there are grades: the lightest is a quick, warm soil, and crops grown on such mature early, while at the same time they partake of the nature of the soil; this gives to melons and similar products their rich, delicate flavor. These light and heavy sandy loam soils are the best of all our arable soils in New England for any crop we produce. There are considerable tracts where, a few years since, a person could buy any quantity for three to ten dollars per acre, which, rightly handled, are very productive of all kinds of crops; but they need constant manuring, to which they are very sensitive, showing its effects, in the crops, quicker than heavier soils. Some of these lands will now command a price from ten to fifty times as high as twenty years since.

Crop Prospects.

Taking the most recent returns from the great wheat-growing States for the basis of an estimate, we are justified in concluding that the wheat crop of the present year will exceed in quantity anything which the country has yet produced. The winter wheat is at all points in the best condition, and a largely increased acreage is reported. Spring wheat, too, will be sowed over a much larger number of acres than before, thus bringing to the market an abundance which persons reputed to be not over-sanguine on the subject estimate in the gross at five hundred million bushels. It is almost too enormous to be conceived by the mind. The decidedly good, or bad, prospect of a war in Europe adds wonderfully to the anticipated value of such a crop.

But allowing that no such war occurs, and that the production of Europe itself is as large as not to compel an extraordinary draft on this country, there is the cotton crop on which to rely for keeping foreign exchange in our favor and thus retaining our gold at home. There is already a large remaining surplus for foreign spinners from last year's crop, but there will be a demand for all that we shall ordinarily have to sell. But the calculation is, in any event, according to the experience of past years, that the cotton crop of the coming season will be the largest ever yet produced, owing to the prolonged overflow of the richest cotton producing region of the country. So that, as a nation, we have everything to encourage and very little to cloud our hopes for the immediate future.

Fence Posts.

An experimental writer on this subject very rationally remarks:

"To have a fence that will last we must have good posts, for that is the part that gives out first by rotting off at the surface of the soil. Then the fence has to come down, new posts set, and the boards replaced. Sixteen years ago I experimented with fences, and find seasoned oak posts oiled and then tarred with boiling coal tar makes them last the longest. I took green posts that were sawed five inches square at one end and two by five inches at the other, and seven feet long. I tarred half as many as would build my fence, and the other half I put in the ground green with nothing done to them. In five years after the tarred posts were nothing but a shell under the ground, all the inside being decayed. Some of the other posts were rotted off, and some were about half rotten.

"Two years after I built another fence with seasoned oak posts, same size as the first, giving them all a good coat of oil, and in a few days after tarred them as I did before with coal tar, heated in a can made for the purpose, four feet deep and large enough to hold four posts set on end; left them in the boiling tar about ten minutes, then took them out and sanded them. And now, after fourteen years, not one in ten need replacing. I shall never build a fence for myself requiring posts without first thoroughly seasoning, then oiling and then tarring them. If they are tarred when green the tar does not penetrate the wood, and in a short time will all scale off. When the wood is seasoned the oil penetrates the wood, and the coating of coal tar keeps out the moisture, thereby preserving the wood from decay."

HORTICULTURE.

Apples for Medicine.

Apples, in addition to being a delicious fruit, make a pleasant medicine. A raw, mellow apple is digested in an hour and a half, while boiled cabbage requires five hours. The most healthy desert that can be placed on the table is a baked apple. If eaten frequently at breakfast, with coarse bread and butter, without meat or flesh of any kind, it has an admirable effect on the general system, often removing constipation, correcting acidities, and cooling off febrile conditions more effectually than the most improved medicines. If families could be induced to substitute apples, ripe and sound, for pies, cakes and sweetmeats, with which their children are frequently stuffed, there would be a diminution in the total sum of doctor's bills in a single year sufficient to lay in a stock of this delicious fruit for the whole season's use.

Greenhouse and Window Plants.

The increasing sun will bring many plants into flower, and at the same time encourage the insect. Free use of tobacco smoke or tobacco water, where it is convenient to use smoke, will destroy many. A small collection of plants, tended by one really fond of them, may be kept free of insects by mere "thumb and finger work." Daily examination, the use of a stiffish brush, like an old tooth-brush, and a pointed stick to pick off mealy bugs and scale, will keep insects from doing harm. Neglect to examine in time, and nip the trouble in the bud, is the cause of much of the difficulty. More water will be needed by plants in bloom and making their growth. Bulbs, if any remain in the cellar, may be brought to the heat and light. When the flowers fade on the earlier ones, cut away the stalk and let the leaves grow on; when they begin to fade dry off the bulbs, which may be planted in the garden afterwards.

Profit in Onions.

More money can be realized from a given amount of land in onions, taken one season with another, than from any other crop that can be raised. A large amount of hand labor is required, however, to produce the crop, which must be put in very early. The labor of old persons and children can be utilized to good advantage in raising onions, as most of the work required is light. The best land for onions is black muck containing a good deal of loam. The manure should be the most thoroughly rotted part from the farmyard. Too much manure cannot be used. It should be well mixed with the soil, say by spreading, turning under and cross harrowing. This should be done in the fall to secure the best results. In the spring the ground should be cultivated and harrowed till it is fine as it can be made. Then the onion seed should be drilled in rows fourteen inches apart. It will take four or five pounds of seed to the acre. As good varieties as any are yellow Danvers, red Wethersfield, and silver skin. The latter are not good keepers, but sell well. As soon as the young onions appear they should be hoed or cultivated. The great secret in growing onions is to keep them free from weeds. Therefore, hoe and cultivate frequently though no weeds may at that moment be above the surface. When the onions are ripe they should be pulled and left on the ground till the tops are dry; then they are gathered up and bagged for market.

Celery Culture.

The demand for celery increases every year. Lately the demand has been greater than the supply in all parts of the country. The past season was a most unfavorable one for this crop. The spring was cold and wet, and the summer hot and dry. The leaves were generally small and were often ill-shapen. The culture of this plant has been greatly simplified during the past few years. The system of planting in trenches has been abandoned by nearly all market gardeners. This effects a great saving of labor. The plan of starting the plants in hot-beds has also been given up by most persons. The seed is sown in well prepared beds in the open ground, but great care is taken to prepare the soil for the growth of the tender young plants. Some burn the soil as they do when preparing the seed beds for tobacco, so as to have no trouble with weeds. The seed is sown in rows about eight or ten inches apart, and the soil between them frequently stirred to hasten the growth of the plants to prevent the springing up of weeds. The young plants are rendered stocky by shearing off the tops two or three times before they are put in the rows where they are to mature. In June and July they are placed in rows three feet apart and six inches in the row. As soon as they become established, the soil is kept well supplied with water. Unless there are seasonable rains, water is supplied by means of pipes or rubber hose. Some have located celery plantations on the banks of streams or the side of lakes, so that water may be easily supplied. Gardeners have been slow in finding out that celery is by nature an aquatic plant, and they are now treating it to all the water it wants.

How the Chinese Make Dwarf Trees.

We have all known from childhood how the Chinese cramp their women's feet, and so manage to make them keepers at home; but how they contrive to grow miniature pines and oaks in flower-pots for half a century has always been much of a secret. They aim first and last at the seat of vigorous growth, endeavoring to weaken it as much as may be consistent with the preservation of life. Take a young plant—say a seedling or cutting of a cedar—when only two or three inches high, cut off its tap-root as soon as it has other rootlets enough to live upon, and replant it in a shallow earthen pot or pan The end of the tap-root is generally made to rest on a stone within it. Alluvial clay is then put into the pot, much of it in bits the size of beans, and just enough in kind and quantity to furnish a scanty nourishment to the plant. Water enough is given to keep it in growth, but not enough to excite a vigorous habit. So likewise is the application of light and heat. As the Chinese pride themselves on the shape of their miniature trees, they use strings, wires and pegs, and various other mechanical contrivances to promote symmetry of habit or to fashion their pets into odd fancy figures. Thus, by the use of very shallow pots, the growth of the tap-root is out of the question; by the use of poor soil and little of it, and little water, any strong growth is prevented. Then, too, the top and side roots being within easy reach of the gardener, are shortened by his pruning knife or seared with his hot iron. So the little tree, finding itself headed on every side, gives up the idea of strong growth, asking only for life, and just life enough to look well. Accordingly each new set of leaves become more and more stunted, the buds and rootlets are diminished in proportion, and at length a balance is established between every part of the tree, making it a dwarf in all respects. In some kinds of trees this end is reached in three or four years; in others ten or fifteen years are necessary. Such is fancy horticulture among the Celestials.

HOUSEHOLD RECIPES.

TAPIOCA PUDDING.—Take one and one-half cups of tapioca and soak over night; three eggs beaten thoroughly, and reserving the white of one for frosting; one cup of white sugar; one teaspoonful of butter; one and one-half pints of milk; a little salt and nutmeg. Bake until well done. Frost same as directed for lemon pie, and return to oven until brown.

BREAD PUDDING.—Take one pint of bread crumbs soaked in one quart of sweet milk; one-half cup of white sugar; two eggs, beaten thoroughly; one cup of raisins if desired; heaping teaspoonful of butter, and salt to suit the taste; stir well together and bake.

CHILI SAUCE.—Forty-eight ripe tomatoes, ten peppers, two large onions, two quarts vinegar, four tablespoons salt, two teaspoons each of cloves, cinnamon, nutmeg and allspice; one cup sugar. Slice the tomatoes, chop peppers and onions together; add vinegar and spices, and boil until thick enough. Mustard and curry powder improves this.

CLAM CHOWDER.—Put in a pot a layer of sliced pork, chopped potatoes, chopped clams, salt, pepper and lumps of butter, and broken crackers soaked in milk, cover with the clam juice and water, stew slowly for three hours, thicken with a little flour, it may be seasoned with spices if preferred.

SADDLE OF LAMB.—Time, a quarter of an hour to the pound; one hour and a half to two hours. Cover the joint with buttered paper to prevent the fat catching, and roast it at a brisk fire, constantly basting it, at first with a very little butter, then with its own dripping. Mint-sauce.

TOMATO SOUP.—Three pounds of beef, one quart canned tomatoes, one gallon water. Let the meat and water boil for two hours, or until the liquid is reduced to a little more than two quarts. Then stir in the tomatoes, and stew all slowly for three quarters of an hour longer. Season to taste, strain and serve.

OYSTER SOUP.—Take one quart of water, one teacup of butter, one pint of milk, two teaspoonfuls of salt, four crackers, rolled fine, and a teaspoonful of pepper. Bring to full boiling heat as soon as possible, then add one quart of oysters. Let the whole come to a boiling heat quickly and remove from the fire.

CHICKEN SALAD.—For one good sized chicken take one bunch of celery chopped fine, a little pepper and salt. For dressing for the above quantity take the yolks of two eggs boiled hard, make them fine, and add mustard, vinegar, oil and a little Cayenne pepper and salt, to suit taste, and the liquor of the chickens boiled in is very nice to use, mixing it. Put in just enough to moisten it nicely. When it becomes cold it is just like a jelly, but it is a great improvement to the salad.

WHITE SAUCE FOR GAME.—Boil an onion in a pint of milk till it is like a jelly; then strain, and stir into the boiling milk sifted bread crumbs enough to make it like thick cream when well beaten. Beat while boiling, and season with salt, black and Cayenne pepper and a little nutmeg.

SUGAR KISSES.—Whites of two eggs, beaten as for frosting; one cup of sugar added to them. Mix well and drop in small cakes on a buttered tin. Bake in a moderate oven until lightly touched with brown.

QUEEN OF PUDDING.—One pint fine bread crumbs, one quart sweet milk, three ounces of loaf sugar, small piece of butter, yolks of four eggs, grated rind of one lemon; bake till done, then spread over a layer of preserves or jelly; whip the whites of the eggs stiff, add three ounces of pulverized sugar, in which has been stirred the juice of the lemon. Pour the whites over the pudding and replace in the oven. Let it brown lightly. To be eaten cold.

LEMON PUDDING SAUCE.—One large cup of sugar, nearly half a cup of butter, one egg, one lemon—all the juice and half the grated peel, one teaspoonful nutmeg, three tablespoonfuls boiling water. Serve with lemon sauce.

BIRD'S NEST PUDDING.—Pare and core apples sufficient to fill a pudding-dish. Make a batter of one quart of milk, three eggs, two cups of flour. Pour over the apples, and bake in a quick oven. Eaten with a sauce.

ORANGE PUDDING.—Take four good-sized oranges, peel, seed, and cut into small pieces. Add a cup of sugar, and let it stand. Into one quart of *nearly* boiling milk, stir two tablespoonsful of corn starch, mixed with a little water and the yolks of three eggs. When done let it cool, and then mix with the orange. Make a frosting of the whites of the eggs and a half cup of sugar. Spread it over the top of the pudding, and place for a few minutes in the oven to brown.

GREEN CORN PATTIES.—Grate as much corn as will make one pint, add one teacupful of flour and one teacupful of butter, one egg, pepper and salt to taste. If too thick add a little milk. Fry in butter.

BOSTON CREAM CAKE—*The Cake.*—One-half pint of milk, five ounces flour, four ounces butter, and five eggs. Boil milk and butter together, stir in flour while boiling, then add eggs.

FLAKE PIE CRUST.—Take one-half cup of lard to a pint of flour; rub well together; take water sufficient to make a dough (not too stiff); roll out and spread with butter; fold over evenly, and make a second fold in the opposite direction; roll out again, being careful not to squeeze the butter out.

SUPERIOR DOUGHNUTS.—Take two cups sugar; one and one-half cups sweet milk; five eggs; three spoonsful of butter; three teaspoonsful of baking powder; salt and flavor to suit the taste. Mix as soft as possible, roll out, cut in proper sizes and drop into hot lard; when removed from lard and partly cool, dip in powdered sugar.

COOKIES.—Take one and one-half cups of white sugar; one-half cup of lard; one-half cup of butter; sufficient caraway seeds or nutmeg to season to suit the taste; one cup of sour milk, with a teaspoonful of soda, and flour sufficient to make dough. Mix thoroughly, roll very thin, and bake quickly.

CUSTARD PIE.—Take three eggs, beaten thoroughly; two tablespoonfuls of white sugar, one pint of milk, nutmeg to suit the taste, a little salt, stir all together, adding the eggs last.

GRAHAM ROLLS.—Two cups of wheat meal, one and a half cups of flour, salt, three-quarters of a cup of sugar, two and one-half cups of sour milk, one teaspoonfull of soda.

RICE WAFFLES.—One cup boiled rice, one-half teaspoonful soda, one pint milk, one teaspoonful cream tartar, two eggs, one teaspoonful salt, lard size of a walnut, flour for a thin batter.

STEAMED INDIAN LOAF.—Four cups of corn meal, two cups of flour, two cups of sweet milk, two cups sour milk, one teaspoonful soda, a little salt, one cup of molasses. Steam three hours.

MUFFINS.—One quart milk, two eggs, quarter of a cup of butter, same of lard. Raised with yeast.

LEMON PIE.—Take juice and grated rind of one lemon; stir together with three fourths of a cup of white sugar and one cup of water; lastly, stir in four eggs, well beaten (reserving the whites of two for frosting). Fill into crust and bake. For frosting, beat the whites of two eggs reserved, to a stiff froth, with a tablespoonful of powdered sugar, spread over top evenly, and return to oven until slightly browned.

PUMPKIN PIE.—Take one quart of pumpkin, stewed and pressed through a sieve, two quarts of milk, two cups of sugar, seven eggs, beaten very light, a teaspoonful of butter, ginger and cinnamon to suit the taste, stir well together and bake with plain crust.

GRAHAM MUFFINS.—One quart of Graham flour; two tablespoonsful of sugar; two eggs, one-half tablespoonful of butter, one tablespoonful of baking powder, and a little salt; moisten and mix thoroughly with little milk. Bake in pattypan at once in a quick oven.

TURKEY SOUP.—Take the turkey bones and cook for one hour in water enough to cover them, then stir in a little of the dressing and a beaten egg. A little chopped celery improves it. Take from the fire, and when the water has ceased boiling add a little butter, with pepper and salt.

FISH SAUCE.—Yolks of two raw eggs. Add salad oil, drop by drop, until it is of the consistency of thick cream; add the juice of half a lemon.

CABBAGE SALAD.—One small head of cabbage, one half bunch of celery, one quarter cup of vinegar, one tablespoonful of mustard, one egg well beaten, one tablespoonful of sugar, pepper and salt. Take a little of the vinegar to wet the mustard, put the rest over the fire; when boiling, stir in the ingredients and cook until it becomes thick: pour it over the cabbage while hot, and mix it well. When cold it is ready for the table. The same sauce, when cold, will do for lettuce.

COTTAGE PUDDING.—One cup of sugar, one egg, two tablespoons of melted butter, one cup sweet milk, two cups of flour, two cups flour, one teaspoonful of cream tarter, half teaspoonful soda. Bake one-half hour. Eat with hot sauce.

SUET PUDDING.—One pint of milk, one pint of syrup, half pound of raisins, half pound of currants, half pound of suet; add prepared flour as stiff as pound cake. Spice to suit taste.

BOILED BREAD PUDDING.—To one quart of bread crumbs, soaked in water, add one cup of molasses, one tablespoonful of butter, one cup of fruit, one teaspoonful each of all kinds of spices, one teaspoonful of soda, about one cup of flour. Boil one hour.

LOWELL PUDDING.—One coffee cup of milk, one cup raisins, half cup molasses, half teacup of brown sugar, one teacup suet, one teaspoonful saleratus, half teaspoonful salt; flour to make a stiff batter. Boil three hours. Serve with sauce.

HOMINY MUFFINS.—Two cups of boiled hominy; beat it smooth, stir in three cups sour milk, half cup melted butter, two teaspoons of salt, two tablespoons of sugar; add three eggs well beaten; one teaspoon of soda, dissolved in hot water; two cups of flour. Bake quickly.

POTATO CAKES.—Roast some potatoes in the oven. When done, skin and pound in a mortar, with a small piece of butter, warmed in a little milk. Chop a shallot and a little parsely very finely, mix well with the potatoes, add pepper and salt; shape into cakes; egg and bread crumb them, and then fry a light brown.

OYSTER FRITTERS.—Time, five or six minutes. Some good-sized oysters, four whole eggs; a table spoonful of milk; salt and pepper; crumbs. Bread some good-sized oysters, make a thick omelet batter with four eggs and a tablespoonful of milk, dip each oyster into the batter, and then into the grated bread, fry them a nice color and use them to garnish fried fish.

CORN OYSTERS.—One pint grated green corn, one cup flour, one spoonful of salt, one teaspoonful of pepper, one egg. Drop by the spoonful in hot lard, and fry.

BOILED LEG OF LAMB.—Time, one hour and a quarter after the water simmers. Select a fine fresh leg of lamb, weighing about five pounds; soak it in warm water for rather more than two hours, then wrap it in a cloth and boil it slowly for an hour and a quarter. When done, dish it up and garnish with a border of carrots, turnips or cauliflower around it. Wind a cut paper around the shank bone, and serve it with plain parsley, and butter sauce poured over it.

TAPIOCA PUDDING.—Three-fourths of a cup of tapioca, three pints of milk. Boil the tapioca with a portion of the milk and the yolks of four eggs, until soft; pour into a pan, and add the whites of three eggs, with the rest of the milk, and two tablespoonfuls of sugar.

SNOW PUDDING.—Take a little more than the third of a package of Coke's Gelatine; pour a pint of cold water over it, and let it stand ten minutes; add the juice of one lemon and one cup of white sugar (sweeten and flavor to taste); add a pint of boiling water; stir and beat till worked up to a light froth, adding to it the well-beaten white of the eggs that are used for the soft custard. Do not commence to beat the gelatine till nearly cold; when well-frothed up, put it into a mould in a cold place. Have a nice soft custard to pour round it when taken from the mould. It is very nice and a pretty dessert.

LIVE STOCK.

Care of Horses' Legs

Few men, who handle horses, give proper attention to the feet and legs. Especially is this the case with the farmer. Much time is often spent in rubbing, brushing and smoothing the hair on the sides and hips, but the feet are not properly cared for. The feet of a horse require ten times as much, for in one respect they are almost the entire horse. All the grooming that can be done won't avail anything if the horse is forced to stand where his feet are filthy, for the feet will become disordered and then the legs will get badly out of fix, and with bad legs and feet there is not much hope for anything. In short, to those owning horses we would say attend to the feet and legs.

Care of Sheep.

There are some points in the care of sheep, which, if rightly heeded during the winter months, would add greatly to the profit of the shepherd and the comfort of his flock. First, the waste of fodder resulting from the slovenly practice of feeding on the ground is greater than the farmer can afford. Let him once adopt the practice of feeding from racks and he will soon see the economy of it. Sheep accustomed to pull the hay from the racks will be loth to take it from the ground. The thick fleece is thought by many to be ample protection from the cold; but the sheep is an animal of low vitality, and give them access to warm quarters and they will quickly avail themselves of the profered shelter. Regularity of feeding should be strictly observed, and no more given at a time than will be readily consumed. If the hay is coarse the shorts that are left in the feed racks will be readily eaten by colts or horses not steady at work, and thus all may be utilized.

Watering Horses.

One thing in the treatment of work horses in hot weather we are disposed to deprecate, viz: the custom of watering them three times a day and no more. It is simply cruelty on the part of man toward his beast, to compel the team to plow or mow from early morning until noon, or from noon until night without allowing it the privilege of a refreshing draught. It is inconvenient, many times, to water the team during the forenoon or afternoon, and we are apt to think the time thus taken lost, but when the farmers' millennium comes, there will probably be drinking troughs in every field, supplied from some elevated spring, or from a running stream. In the meanwhile time "lost" in doing good, even though it may be in behalf of the dumb animals, is well "lost"—it may be regained. Could they speak it might be to say that they would like to be treated, in the matter of *times* for food and drink, somewhat as we—their wise masters—are accustomed to treat ourselves.

Save and Care for the Pigs.

For many years past no spring season has found so few swine in the country in proportion to the coming wants. Owing to the scarcity and high price of corn, and the demand for hog products at figures far above average years, the last hog, grown and half grown, that could be got into anything near a fit condition to be slaughtered, is then sent to market. It is reported that a good many breeding sows have gone into the barrel, and also the lard can. This being the case, with the probability that the markets of this country and elsewhere will be cleared up and nearly bare of pork, bacon, hams and lard before next winter, makes it important to look well after the pigs; to see that not one is lost for want of care and protection until warm, settled weather arrives; also to give the young porkers a good start and continuous vigorous growth by liberal feeding; also to do all that can be done to multiply the number. The foreign demand for hog products is always large, and 10 lbs. of corn, when converted into 1 lb. of pork or lard, is transported at one-tenth cost.—*American Agriculturist for April.*

How to Grow a Pig.

Editor Mass. Ploughman: Will you please inform me through the *Ploughman* how I can grow a pig through the warm weather and not get him fat? I have milk, and would flaxseed meal be good for him, with milk, and how long would you give that kind of meal, if you gave it at all? Or would something else be better for him? It is a sucking pig, and I want to grow him for winter use as much as I can.—*Truly yours, C. P., New Salem, Mass., March 20th,* 1882.

Shorts mixed with the milk would, in our opinion, be better than flaxseed meal. At first the pig should be fed often, and principally on milk, but as he grows older more shorts may be added, and when it is desired to fat him there is nothing better than Indian meal to make good pork; but the health of a pig may be improved by giving plenty of fresh grass and weeds. When it is convenient to do so, it is best to let a pig run over a quarter of an acre of grass land. Many pigs are injured by keeping in close, dirty quarters. Pigs, like most other animals, want light, air and room to exercise in, if they are to be kept thrifty and healthy.—ED.

A Nevada Stock Raiser.

W. D. Todhunter branded last spring over 9,000 calves, and has sent to the market this season 6,000 beef cattle. These figures prove Mr. Todhunter to be the largest stock raiser in the country. There are others who send more cattle to market, but they buy them instead of raising them. He has over 20,000 head of stock cattle and over 100,000 acres of patented land. He got patents last month for 25,000 acres of swamp land in one bunch. He has about 1,000 bulls and 200 saddle horses. He employs fifty men, and puts up 2,500 tons of hay to guard against hard winters. He keeps 100 work horses, and raises grain enough to feed all his saddle and work stock. Besides his cattle, he has 700 or 900 stock horses, four jacks and fifty stallions.

His stock is divided among four ranches—one known as the White Horse Ranch, lying just inside the Oregon line, where 5,000 head are kept; one in Long Valley, in the northwest corner of Nevada, lying alongside of Surprise, supports 4,000 head; the Pyramid Ranch lying at the northeast corner of the lake, has 1,500 and a lot of horses; the Abbott Ranch at Steen's Mountain, feeds about 6,000, and Harney Valley 6,000 more. The home ranch is 25 miles from a neighbor.—*From the Reno Gazette.*

POULTRY.

A WRITER in the *Poultry Monthly*, whose neighbors have lately been paying forty eight cents per dozen for eggs, truthfully says that with proper attention to chickens there is no difficulty in obtaining an abundant supply of eggs at all seasons. The chicken house should have a southern exposure; fowls should be given moderately warm water two or three times daily during the inclement portion of the year up to about the tenth of April. The correspondent feeds his fowls four times daily, the first feed being always hot, consisting of scraps from the table well-seasoned with pepper; the second feeding consists of buckwheat, the third wheat screenings, and the fourth corn. In cold weather he never allows his chickens to roost out doors.

A House for 200 Fowls.

To accommodate 150 to 200 fowls, it would be best to have two houses, or rather one so divided in the middle as to make two, with a door at each end. A very cheap and good house may be made of boards; 4 feet high at the back, 10 feet in front, 16 feet wide, and 36 feet long. The roosts should be made at the rear, and in the form of a ladder, sloping back from the floor to the roof. In the middle there may be a room for nest boxes. If the front, which should face the South, is of glass, it will be much improved. As large a yard as possible may be provided, and fenced with lath, so that the fowls may be kept in when desirable. The materials for such a house need not cost more than $20, and $12 additional for sash for the front.

Questions About Eggs and Fowls.

Manitoba. We do not believe half the reports current about extraordinary production of eggs. Yet it may easily be true that a hen of the non sitting breeds may lay an egg every day for a long period. The Black Spanish hens often do this, and we have personally known one to lay two eggs in one day, but there was none the next day. Hens cannot well cover goose eggs and keep them warm; they are too large. If the bottom of the nest is made warm with down or feathers, a hen may probably keep six of these eggs warm enough. A young bird needs no help to get out of the shell; unless it is very weak, in which case it is just as well for it to remain there, as it would probably fail to thrive. If the chicks should need any help, this can be very easily given when an incubator is used. Light Brahma chicks, when newly hatched, are all white; Dark Brahmas are black and brown; Plymouth Rocks are black and yellow, and Black-red Games are black and yellow, or brownish.

Raising Sunflowers for Hens.

The necessity for a variety of feed for chickens is generally understood, but very few people are aware of the value of sunflowers as hen feed. They are very productive of oil, are eaten greedily, and give a peculiar luster to the feathers. I have one-eighth of an acre planted to this crop, and propose to bind them into bundles and stow them away in a dry place for winter use. The heads can be thrown into the hen-house, where the chickens will soon pick out the seeds, thus giving them exercise as well as variety. With plenty of other grain within reach they will eat no more sunflower seeds than are beneficial to them. The seed can be bought at our feed stores for one dollar per bushel, at which price it ought to be more generally used than it is. I think a plot of sunflowers, with their great yellow faces turned to the sun, an agreeable sight.—*Kansas Farmer*.

Care of Young Turkeys.

See that your turkeys come home every night. At first, if you raise them with a turkey mother, you will have to hunt them up and drive them home, but if you feed regularly every morning and always at night they will soon learn to come home as regular as the cows. After they have fully feathered, and have thrown out the red on their heads, which usually occurs at about three months, young turkeys are hardy, and may be allowed unlimited range at all times, and from that time on, as long as the supply of insects lasts, they will thrive on two meals a day. Keep your turkeys growing right straight from the shell, and you will find that it will pay when pay day comes. Some farmers, as soon as their young turkeys are feathered up, turn them out to get their own living the best way they can until a few weeks before Thanksgiving. Then they stuff them for a few weeks and wonder why they do not equal in weight those of their neighbor, who has kept his turkeys growing all the time from the day they were hatched.—*Cincinnati Grange Bulletin.*

How Chickens are Born.

Take an egg out of a nest on which a hen has had her full time, carefully holding it to the ear; turning it around, you will find the exact spot which the little fellow is picking on the inside of the shell; this he will do until the inside shell is perforated, and then the shell is forced outward as a small scale, leaving a hole. Now, if you take one of the eggs in this condition from under the hen, remove it to the house or other suitable place, put in a box or nest, keeping it warm and moist, as near the temperature of the hen as possible (which may be done by laying it between two bottles of warm water upon some cotton or wool,) and lay a glass over the box or nest, then you can sit or stand, as is most convenient, and witness the true *modus operandi*. Now watch the little fellow work his way into the world, and you will be amused and instructed as we have often been. After he has got his opening he commences a nibbling motion with the point of the upper bill on the outside of the shell, always working to the right (if you have the large end of the egg from you, and the hole upward,) until he has worked his way almost around, say with one-half of an inch in a perfect circle; he then forces the cap or butt end of the shell off, and then has a chance to straighten his neck, thereby loosening his legs somewhat, and so, by their help, forcing the body from the shell.—*N. E. Homestead.*

A Cheap Chicken Coop.

A "Jerseyman" describes in the *Tribune* his neighbor's cheap arrangement for raising chickens:

For coops he uses tight old barrels laid lengthwise on the ground, with the head taken out. On the bottom of each, for nests, he places some very dry earth and then a little straw or leaves from the woods, if early in the spring; if later, the earth alone; There is nothing better on which to set a hen than a dry sod laid with the grass side down, and just enough of the soil scraped off from the center of the top to make a hollow to hold the eggs. In these barrels the hens laid and sat. When the chickens were hatched the barrels were cleaned and enough narrow sticks driven in front to keep in the hen and allow the young to run out at pleasure, which they would only do in dry weather. To let out the hen for sun and for exercise it was only necessary to roll the barrel a little on one side or withdraw a stake or two from the front. When the chickens got to be a few weeks old the hen was allowed to come out at will. Each generally kept a remembrance of its barrel, and went back to it with her brood for food and water during the day and to hover in it at night. If likely to rain it was necessary to see that all got into their coops for shelter before it began to fall. As the staves were set tight the barrels shed the rain perfectly.

Hawaiian Geese.

The Hawaiian geese (*Bernicla sandvicensis*) which I brought over in the spring of 1879 have proved hardy, and I trust will prove reproductive. They were all sheltered and cared for last winter, and came through in good order. Both geese commenced laying in April; one laid three and the other four eggs, but only one showed a disposition to sit upon the eggs, and she, after attending to her business faithfully for ten days, tired of it and quitted the nest, so they produced no goslings. In the wild state they lay but two or three eggs, while in domestication they sometimes lay eight or ten. Mr. Brickwood, Postmaster-General of the kingdom, who had them in domestication for many years, sometimes raised as many as ten in a brood. In domestication they seem to have strong attachments, and are fond of human society; one gander in particular has become very fond of me and always greets me cordially, and will talk with me in a low, soft, plaintive tone so long as I will indulge the humor. They are less aquatic than the other geese. The foot is not more than half webbed. They take a bath scarcely once a day, and rarely remain in the water long. I once saw one with the tail under water, as we see a hen when forced to swim. Their native habitat are the high volcanic mountains in the island of Hawaii, where they breed among the lava beds, depending upon the pools which they find among the rocks for water, never going down to the sea. They are of strong flight in the wild state, though in domestication they show little disposition to fly. Altogether they are the most interesting water fowl I possess, and I hope another year to raise some of them from the only pair I have left. A few weeks ago I lost the other pair by a mink.—*Judge Caton in American Naturalist.*

LITERARY AND PERSONAL.

THE VERDICT OF THE JURY.—We have just received a copy of the most popular piece of music ever published in this country, called the "Verdict March," composed by Eugene L. Blake. It is written in an easy style, so that it can be played on either piano or organ. The title page is very handsome, containing correct portraits of Hon. Geo. B. Corkhill, Hon. J. K. Porter, and Judge W. S. Cox; also a correct picture of the twelve jurymen who convicted the assassin of our late beloved President. This piece of music should be found in every household throughout the entire country. Price, 40 cents per copy, or 3 copies for $1. Postage stamps taken as currency. Address all orders to F. W. Helmick, music publisher, 180 Elm street, Cincinnati, Ohio, United States of America.

THE RECORD.—A new bi-weekly educational journal devoted to general information, popular science, agricultural news and the work of the Young Men's Christian Association. Published every alternate Saturday at the Y. M. C. A. building, Lancaster, Pa., at $1 a year, including postage. This is a quarto of 16 pages, very creditably gotten up, and illustrated, and proposes to present to its patrons in each issue 16 pages of useful information, embracing chemistry, electricity, photography, agriculture, natural history, botany, astronomy, microscopy, optics, archæology, explorations, local history, &c. And, judging from the copy now before us, (Vol. 2, No. 8) it has faithfully kept its word. Moreover, published as it is, under the auspices and for the benefit of the Young Men's Christian Association, it ought to receive a more liberal support from the community than has been heretofore accorded to that worthy association. The illustrated article in this number on the sponge is especially interesting and instructive. Its motto—"*Liberty can only be safe where suffrage is illuminated by education*," breathes a truth that needs to be more fully apprehended, and widely extended than now appears on the surface of society, and we hope it may find a very large *vacuum* to fill in this community.

THE FREE-TRADE BULLETIN: A four collumned "half sheet" (Vol. 1, No. 6), devoted to the political doctrine of "Free-trade," has found its way to our *Sanctum*. It is a handsomely printed journal, and advocates its specialty with singular ability: and, whether truthful or fallacious, in perusing it, "almost thou compelest me to be a christian," is powerfully suggested to the mind not fettered by previous prejudices. Will we, as a nation, ever learn to know what is best for the interests of *all*, in this respect? Price 50 cents a year, monthly, New York.

THE SOUTHERN CULTIVATOR: The April number of this popular and well established Agricultural journal has been received. The issue was delayed a few days, owing to the fact that Messrs. J. P. Harrison & Co., the publishers, were removing their immense printing establishment to a much larger building.

It should be a matter of pride with our Southern farmers to sustain THE CULTIVATOR, because it is their representative, published alone in their interest, and is by far the neatest, most reliable and best filled Agricultural publication in the South. The publishers are certainly spending large sums of money, in making it the best of all journals of a like kind, judging from the fine paper used, the handsome engravings that adorn its columns and from the men of brains employed as contributors.

It is sold for the low price of $1.50 per annum.

In this issue it is announced that Mr. H. H. Cabaniss, recently of Forsyth, becomes the Business Manager. $2.00 in advance will secure the *Cultivator* and the LANCASTER FARMER for one year.

THE WORLD OF NATURE

The world of animated nature is more splendidly represented under the canvas of Forepaugh's Great Show than in any zoological collection existent. Not since the day Noah lifted his hawser off the snubbing post have so many distinct varieties of rare animals been collected under one charge. This important fact should not be lost sight of by schools and parents. Boys and girls can learn more in an afternoon of natural history, in the great Menagerie of Forepaugh's Show, than by months of book study. Recognizing this, Mr. Forepaugh makes reduced rates to schools, and admits all children in orphan asylums free of charge. This Great Show will exhibit in Lancaster, Monday, April 24.

THE
LANCASTER EXAMINER

OFFICE
No. 9 North Queen Street,
LANCASTER, PA.

THE OLDEST AND BEST.

THE WEEKLY
LANCASTER EXAMINER

One of the largest Weekly Papers in the State.

Published Every Wednesday Morning.

Is an old, well-established newspaper, and contains just the news desirable to make it an interesting and valuable Family Newspaper. The postage to subscribers residing outside of Lancaster county is paid by the publisher. Send for a specimen copy.

SUBSCRIPTION:
Two Dollars per Annum.

THE DAILY
LANCASTER EXAMINER

The Largest Daily Paper in the county.

Published Daily Except Sunday.

The daily is published every evening during the week. It is delivered in the City and to surrounding Towns accessible by railroad and daily stage lines, for 10 cents a week.
Mail Subscription, free of postage—One month, 50 cents; one year, $5.00.

JOHN A. HIESTAND, Proprietor,
No. 9 North Queen St.,
LANCASTER, PA.

Important to Grocers, Packers, Hucksters, and the General Public.

THE KING FORTUNE-MAKER.
OZONE

A New Process for Preserving all Perishable Articles, Animal and Vegetable from Fermentation and Putrefaction, Retaining their Odor and Flavor.

"**OZONE—Purified air, active state of Oxygen.**"—*Webster.*

This preservative is not a liquid pickle, or any of the old and exploded processes, but is simply and purely OZONE, as produced and applied by an entirely new process. Ozone is the antiseptic principle of every substance, and possesses the power to preserve animal and vegetable structures from decay

There is nothing on the face of the earth liable to decay or spoil which Ozone, the new Preservative, will not preserve for all time in a perfectly fresh and unpalatable condition.

The value of Ozone as a natural preserver has been known to our abler chemists for years, but, until now, no means of producing it in a practical, inexpensive, and simple manner have been discovered. Microscopic observations prove that decay is due to septic matter or minute germs, that develop and feed upon animal and vegetable structures. Ozone, applied by the Prentiss method, seizes and destroys these germs at once, and thus preserves. At our office in Cincinnati can be seen almost every article that can be thought of, preserved by this process, and every visitor is welcomed to come in, taste, smell, take away with him, and test in every way the merits of Ozone as a preservative. We will also preserve, free of charge, any article that is brought or sent prepaid to us, and return it to the sender, for him to keep and test.

FRESH MEATS, such as beef, mutton, veal, pork, poultry, game, fish, &c., preserved by this method, can be shipped to Europe, subjected to atmospheric changes and return to this country in a state of perfect preservation.

EGGS can be treated at a cost of less than one dollar a thousand dozen, and be kept in an ordinary room six months the day they were treated, and will sell as strictly "choice." The advantage in preserving eggs is readily seen; there are seasons when they can be bought for 4 or 5 cents a dozen, and by holding them, can be sold for an advance of from one hundred to three hundred per ct. One man, with this method, can preserve 5,000 dozen a day.

FRUITS may be permitted to ripen in their native climate, and can be transported to any part of the world. The place expressed from fruits can be held for an indefinite period without fermentation—hence the great value of this process for producing a temperance beverage. Cider can be held perfectly sweet for any length of time.

VEGETABLES can be kept for an indefinite period in their natural condition, retaining their odor and flavor, treated in their original package at a small expense. All grains, flour, meal, etc., are held in their normal condition.

BUTTER, after being treated by this process, will not become rancid.

Dead human bodies, treated before decomposition sets in, can be held in a natural condition for weeks, without puncturing the skin or mutilating the body in any way. Hence the great value of Ozone to undertakers.

There is no change in the slightest particular in the appearance of any article thus preserved, and no trace of any foreign or unnatural odor or taste.

The process is so simple that a child can operate as well and as successfully as a man. There is no expensive apparatus or machinery required.

A room filled with different articles, such as eggs, meat, fish, etc., can be treated at one time, without additional trouble or expense.

☞ In fact, there is nothing that Ozone will not preserve. Think of everything you can that is liable to sour, decay, or spoil, and then remember that we guarantee that Ozone will preserve it in exactly the condition it is in when you want it for any length of time. If you will remember this it will save asking questions as to whether Ozone will preserve this or that article—it **will preserve anything and every thing you can think of.**

There is not a township in the United States in which a live man can not make any amount of money, from $1,000 to $10,000 a year, that he pleases. We desire to get a live man interested in each county in the United States, in whose hands we can place this Preservative, and through him secure the business which every county ought to produce

A FORTUNE Awaits any Man who Secures Control of OZONE in any Township or County.

A. C. Bowen, Marion, Ohio, has cleared $2,000 in two months. $2 for a test package was his first investment.
Woods Brothers, Lebanon, Warren County, Ohio, made $6,000 on eggs purchased in August and sold November 1st. $2 for a test package was their first investment.
F. K. Raymond, Morristown, Belmont Co., Ohio, is clearing $2,000 a month in handling and selling Ozone. $2 for a test package was his first investment.
D. F. Webber, Charlotte, Eaton Co., Mich., has cleared $1,000 a month since August. $2 for a test package was his first investment.
J. B. Gaylord, 80 La Salle St., Chicago, is preserving eggs, fruit, etc., for the commission men of Chicago, charging 1½c. per dozen for eggs, and other articles in proportion. He is preserving 5,000 dozen eggs per day, and on his business is making $5,000 a month clear. $2 for a test package was his first investment.
The Cincinnati Feed Co., West 498 Seventh Street, is making $3,500 a month in handling for seven mall, preserving and shipping it as feed in all parts of the country. Malt improves 40 hours. Preserved by Ozone it keeps perfectly sweet for months.

These are instances which we have asked in the privilege of publishing. There are scores of others. Write many of the above parties and get the evidence direct.

Now, to prove the absolute truth of every thing we have said in this paper, **we propose to place in your hands the means of proving for yourself that we have not claimed half enough.** No any person who doubts any of these statements, and who is interested sufficiently to make the trip, we will pay all traveling and hotel expenses for a visit to this city, if we fail to prove any statement we have made.

How to Secure a Fortune with Ozone.

A Test package of Ozone, containing a sufficient quantity to preserve one thousand dozen eggs, or other articles in proportion, will be sent to any applicant on receipt of $2. This package will enable the applicant to pursue any line of tests and experiments he desires, and thus satisfy himself as to the extraordinary merits of Ozone as a Preservative. After having thus satisfied himself, and had time to look the field over to determine what he wishes to do in the future—whether to sell the article to others or to confine it to his own use, or any other line of policy which is best suited to him and to his township or county—we will enter into an arrangement with him that will be extremely lucrative for him and give us good profits. We will give exclusive township or county privileges to the first responsible applicant who orders a test package and desires to control the business in his locality. **The man who secures control of Ozone for any special territory, will enjoy a monopoly which will surely enrich him.**

Don't let it by. Pass until you have ordered a Test Package, and if you desire to secure an exclusive privilege we assure you that delay may deprive you of it, for the applications come in to us by scores every mail—hurry by telegraph. "First come first served" is our rule.

If you do not care to send money in advance for the test package we will send it C. O. D., but this will put you to the expense of charges for return money. Our correspondence is very large, we need you to be willing to the shipping of orders and giving attention to our working agents. Therefore we can not give any attention to letters which do not order Ozone. If you think of any article that you are doubtful about Ozone preserving, remember we guarantee that it will preserve it, no matter what it is.

REFERENCES.

We desire to call your attention to a class of references which no enterprise or firm based on any thing but the soundest business sense and highest commercial merit could secure.
We refer, by permission, as to our integrity and to the value of the Prentiss Preservative, to the following gentlemen: Edward C Royse, Member Board of Public Works, E. O. Eberly, City Comptroller; Amor Smith, Jr., Collector Internal Revenue; William E Worthington, Attorney; Martin H. Harrell and H. F. Hopkins, County Commissioners; W. S. Cappeller, County Auditor; all of Cincinnati, Hamilton County, Ohio. These gentlemen are each familiar with the merits of our Preservative, and know from actual observation that we have without question

The Most Valuable Article in the World.

The $2 you invest in a test package, will surely lead you to secure a township or county, and then your way is absolutely clear to make from $2,000 to $10,000 a year.
Give your full address in every letter, and send your letter to

PRENTISS PRESERVING COMPANY. (Limited),
S. E. Cor. Ninth & Race Sts., Cincinnati, O.

Nov-3m

WHERE TO BUY GOODS
IN
LANCASTER.

BOOTS AND SHOES.

MARSHALL & SON, No. 12 Centre Square, Lancaster, Dealers in Boots, Shoes and Rubbers. Repairing promptly attended to.

M. LEVY, No. 3 East King street. For the best Dollar Shoes in Lancaster go to M. Levy, No. 3 East King street.

BOOKS AND STATIONERY.

JOHN BAER'S SONS, Nos. 15 and 17 North Queen Street, have the largest and best assorted Book and Paper Store in the City.

FURNITURE.

HEINITSH'S, No. 15½ East King st., (over China Hall) is the cheapest place in Lancaster to buy Furniture. Picture Frames a specialty.

CHINA AND GLASSWARE.

HIGH & MARTIN, No. 15 East King st., dealers in China, Glass and Queensware, Fancy Goods, Lamps, Burners, Chimneys, etc.

CLOTHING.

MYERS & RATHFON, Centre Hall, No. 12 East King St. Largest Clothing House in Pennsylvania outside of Philadelphia.

DRUGS AND MEDICINES.

G. W. HULL, Dealer in Pure Drugs and Medicines, Chemicals, Patent Medicines, Trusses, Shoulder Braces, Supporters, &c., 15 West King St., Lancaster, Pa.

JOHN F. LONG & SON, Druggists, No. 12 North Queen St. Drugs, Medicines, Perfumery, Spices, Dye Stuffs, Etc. Prescriptions carefully compounded.

DRY GOODS.

GIVLER, BOWERS & HURST, No. 25 E. King St., Lancaster, Pa., Dealers in Dry Goods, Carpets and Merchant Tailoring. Prices as low as the lowest.

HATS AND CAPS.

C. H. AMER, No. 29 West King Street, Dealer in Hats, Caps, Furs, Robes, etc. Assortment Large. Prices Low.

JEWELRY AND WATCHES.

H. Z. RHOADS & BRO, No. 4 West King St, Watches, Clock and Musical Boxes. Watches and Jewelry Manufactured to order.

PRINTING.

JOHN A. HIESTAND, 9 North Queen st., Sale Bills, Circulars, Posters, Cards, Invitations, Letter and Bill Heads and Envelopes neatly printed. Prices low.

Thirty-Six Varieties of Cabbage; 26 of Corn; 28 of Cucumber; 41 of Melon; 33 of Peas; 28 of Beans; 17 of Squash; 21 of Beet and 41 of Tomato, with other varieties in proportion, a large portion of which were grown on my five seed farms, will be found in my **Vegetable and Flower Seed Catalogue** for 1882. Sent FREE to all who apply. Customers of last season need not write for it. All Seed sold from my establishment warranted to be fresh and true to name, so far, that should it prove otherwise, I will refill the order gratis. The original introducer of **Early Ohio** and **Burbank Potatoes**, **Marblehead Early Corn**, the **Hubbard Squash**, **Marblehead Cabbages**, **Phinney's Melon**, and a score of other New Vegetables, I invite the patronage of the public. New Vegetables a specialty.

JAMES J. H. GREGORY,
Marblehead, Mass.

Nov-5mo]

EVAPORATE YOUR FRUIT.
ILLUSTRATED CATALOGUE
FREE TO ALL.
AMERICAN DRIER COMPANY,
Chambersburg, Pa.

FARMING FOR PROFIT.

It is conceded that this large and comprehensive book, (advertised in another column by J. C. McCurdy & Co., of Philadelphia, the well-known publishers of Standard works,) is not only the newest and handsomest, but altogether the BEST work of the kind which has ever been published. Thoroughly treating the great subjects of General Agriculture, Live-Stock, Fruit-Growing, Business Principles, and Home Life; telling just what the farmer and the farmer's boys want to know, combining Science and Practice, stimulating Inquiry, awakening inquiry, and interesting every member of the family, this book must exert a mighty influence for good. It is highly recommended by the best agricultural writers and the leading papers, and is destined to have an extensive sale. Agents are wanted everywhere. jan-1t

BLOOMSDALE
LARGE LATE FLAT DUTCH CABBAGE.

Large, Flat, Solid Heads, Short Stems.

For a long period of time we have had this stock of Cabbage in cultivation, originally obtained from the German and Sweedish market gardeners. It has been a part of our business occupation to keep it unmixed, and to-day we offer it in its original purity, equal in quality with the very best in the country, even though the best should cost a hundred dollars per pound.

We have made this crop a study and give our customers the result of many years close observation, for which our opportunities may be judged by the fact that we have, each and every year, about one hundred and fifty acres of cabbage raised expressly to produce seed for the ensuing season, and from which selections are made with scrupulous care, guided by experience. Not a single grain of seed is raised from stalks all from Selected Heads.

We will mail our Catalogue free of charge to all applicants.

D. LANDRETH & SONS,
Nos. 21 and 23 South Sixth Street,
BETWEEN MARKET AND CHESTNUT STS.,
BRANCH STORE—S. W. COR. DELAWARE AVE. AND
ARCH STREET,
apr-6m PHILADELPHIA.

MERCHANT TAILORING.
1848 (The Oldest of All.) 1881
RATHVON & FISHER,
MERCHANT TAILORS AND DRAPERS,

respectfully inform the public that having disposed of their entire stock of Ready-Made Clothing, they now do, and for the future shall, devote their whole attention to the CUSTOM TRADE.

All the desirable styles of CLOTHS, CASSIMERES, WORSTEDS, COATINGS, SUITINGS and VESTINGS constantly on hand, and made to order in plain or fashionable style promptly, and warranted satisfactory.

All-Wool Suit from $10.00 to $30.00.
All-Wool Pants from 3.00 to 10.00.
All-Wool Vests from $3.00 to 6.00.

Union and Cotton Goods proportionately less. Cutting, Repairing, Trimming and Making, at reasonable prices.

Goods ruined by the yard to those who desire to have them made elsewhere.

A full supply of Spring and Summer Goods Just opened and on hand.

Thankful to a generous public for past patronage they hope to merit its continued recognition in their "new departure."

RATHVON & FISHER,
PRACTICAL TAILORS,
No. 101 North Queen Street,
LANCASTER, PA.
1848 1881

GLOVES, SHIRTS, UNDERWEAR.
SHIRTS MADE TO ORDER,
AND WARRANTED TO FIT.
E. J. ERISMAN,
56 North Queen St., Lancaster, Pa.
19-1-12]

A HOME ORGAN FOR FARMERS.

THE LANCASTER FARMER,

A MONTHLY JOURNAL,

Devoted to Agriculture, Horticulture, Domestic Economy and Miscellany.

Founded Under the Auspices of the Lancaster County Agricultural and Horticultural Society.

EDITED BY DR. S. S. RATHVON.

TERMS OF SUBSCRIPTION:

ONE DOLLAR PER ANNUM,

POSTAGE PREPAID BY THE PROPRIETOR.

All subscriptions will commence with the January number, unless otherwise ordered.

Dr. S. S. Rathvon, who has so ably managed the editorial department in the past, will continue in the position of editor. His contributions on subjects connected with the science of farming, and particularly that specialty of which he is so thoroughly a master—entomological science—some knowledge of which has become a necessity to the successful farmer, are alone worth much more than the price of this publication. He is determined to make "The Farmer" a necessity to all households.

A county that has so wide a reputation as Lancaster county for its agricultural products should certainly be able to support an agricultural paper of its own, for the exchange of the opinions of farmers interested in this matter. We ask the co-operation of all farmers interested in this matter. Work among your friends. The "Farmer" only one dollar per year. Show them your copy. Try and induce them to subscribe. It is not much for each subscriber to do but it will greatly assist us.

All communications in regard to editorial management should be addressed to Dr. S. S. Rathvon, Lancaster, Pa.; and all business letters in regard to subscriptions and advertising should be addressed to the publisher. Rates of advertising can be had on application at the office.

JOHN A. HIESTAND,
No. 9 North Queen St., Lancaster, Pa.

$72 A WEEK. $12 a day at home easily made. Costly Outfit free. Address TRUE & Co., Augusta, Maine

ONE DOLLAR PER ANNUM.—SINGLE COPIES 10 CENTS.

Dr. S. S. RATHVON, Editor.   LANCASTER, PA. JUNE, 1882.   JOHN A. HIESTAND, Publisher

Entered at the Post Office at Lancaster as Second Class Matter.

CONTENTS OF THIS NUMBER.

EDITORIAL.
The Proposed New Department of Agriculture....81
Increase of our Crops............................81
Potash in Plants.................................8
Kitchen Garden for June..........................81
Exports of Cheese................................82
  Exports of Butter—Exports of Oleo-margarine
  —Marketing Farm Products.
The Consologa Flying Fish........................82
Pyrethrum Roseum.................................82
Venor Predicts a Bad Summer......................83
Caddice Flies....................................84
  An Insect that walls itself up.
Eggs...84
Our Crops..84
Excerpts...84

CONTRIBUTIONS.
Comparative value of farms between now and
  Fifty years ago................................85
  Fictitious Value—Good Crops—Good Governments
  —Tariffs, etc.
On Wheat Crops...................................86

ESSAYS.
Insects and some of their relations to the vegetable Kingdom.

SELECTIONS.
The Benevolent Sunflower.........................89
Our Timber Lands.................................90
Roots and how to grow them.......................91
  Composition and Value of Root Crops.
Green Manures....................................29

OUR LOCAL ORGANIZATIONS.
Lancaster County Agricultural and Horticultural
  Society..92
  New Members Elected—Crop Reports—Pruning
  Apple Trees—When to Cut Grass—White-marked
  Tobacco—Yellow Locust—Double Peaches—
  Books for the Library.
The Poultry Society..............................93
Linnæan Society..................................93
  April Meeting—May Meeting.

AGRICULTURE.
Rotation of Crops................................94
Manure Made Under Cover..........................94
Exports of Breadstuffs...........................94
Corn Culture in Gardens..........................94

HORTICULTURE.
An Abundant Apple Crop...........................94
What Kills Fruit Trees...........................94
Early Turnips....................................94

HOUSEHOLD RECIPES.
Beefsteak Rools..................................95
Devilled Ham.....................................95
Yankee Plum Pudding..............................95
French Beefsteak.................................95
Squash Pie.......................................95
Delightful Pudding...............................95
To Make Tough Meat Tender........................95

Cabbage Salad....................................95
Scallopped Oysters...............................95
Roast Shoulder of Veal...........................95
Western Cookies..................................95
Fairy Apple......................................95

LIVE STOCK.
Improving the stock on the farm..................95
Keep up the Flow of Milk.........................95
Care of Dairy Vessels............................95
Raise the Good Cows Heifer Calf..................95

POULTRY.
One Variety......................................95
Treatment of Young Ducks.........................96
A Profitable Hennery.............................96
Literary and Personal............................96

SILK-WORM EGGS.

Amateur Silk-growers can be supplied with superior silk-worm eggs, on reasonable terms, by applying immediately to

GEO. O. HENSEL,

may-3m]   No. 236 East Orange Street, Lancaster, Pa.

SEND IN YOUR SUBSCRIPTIONS
—FOR—

THE FARMER

FOR 1882.

The cheapest and one of the best Agricultural papers
in the country.

Only $1.00 per year.
JOHN A. HIESTAND, Publisher,
No. 9 North Queen st., Lancaster, Pa

Eggs! Eggs!

From all the leading varieties of pure bred Poultry
Brahmas, Cochin, Hamburgs, Polish Game, Dorking
and French Fowls, Plymouth Rocks and Bantams,
Rouen and Pekius Ducks. Send for Illustrated Circular.

T. SMITH, P. M., Fresh Pond, N. Y.

feb-3m

JOB PRINTING.

The EXAMINER PRINTING ESTABLISHMENT is one of the most complete in the State of Pennsylvania, and is prepared to do all kinds of

BOOK, JOB AND NEWSPAPER WORK,

PROMPTLY,

and at as low prices as can be obtained at any place.
Write for prices for any Printing you have to do.

JOHN A. HIESTAND,

PROPRIETOR.

$66 a week in your own town. Terms and $5 outfit
free. Address H. HALLETT & Co., Portland, Maine.

Will be mailed FREE to all applicants, and to customers without ordering it. It contains five colored plates, 600 engravings, about 200 pages, and full descriptions, prices and directions for planting 1500 varieties of Vegetable and Flower Seeds, Plants, Fruit Trees, etc. Invaluable to all. Send for it. Address,
D. M. FERRY & CO., Detroit, Mich.

Jan-4m

$66 a week in your own town. Terms and $5 outfit free
Address H. HALLETT & Co., Portland, Maine.
jun-1yr

WE WANT OLD BOOKS.

We Want German Books.

WE WANT BOOKS PRINTED IN LANCASTER CO.
We Want All Kinds of Old Books.
LIBRARIES, ENGLISH OR GERMAN BOUGHT.
Cash paid for Books in any quantity. Send your address and we will call.

REES WELSH & CO.,
23 South Ninth Street, Philadelphia.

LIGHT BRAHMA EGGS

For hatching, now ready—from the best strain in the county—at the moderate price of

$1.50 for a setting of **13 Eggs.**

L. RATHVON,

No. 9 North Queen st., Examiner Office, Lancaster, Pa.

WANTED.—CANVASSERS for the
LANCASTER WEEKLY EXAMINER
In Every Township in the County. Good Wages can be made. Inquire at
THE EXAMINER OFFICE,
No. 9 North Queen Street, Lancaster, Pa

$72 A WEEK. $12 a day at home easily made. Costly
Outfit free. Address TRUE & CO., Augusta, Maine.
jun-1yr

SEND FOR
SPECIAL PRICES

On Concord Grapevines, Transplanted Evergreens, Tulip,
Poplar, Linden Maple, etc. Tree Seedlings and Trees for
timber plantations by the 100,000
J. JENKINS' NURSERY,
3-2-79   WINONA, COLUMBIANA CO., OHIO.

MARBLEHEAD
Early Sweet Corn

Is the most profitable of all, because it matures before any other kind, giving farmers complete control of the early market. I warrant it to be at least a week earlier than Minnesota, Narragansett or Crosby, and decidedly earlier than Dolly Dutton, Tom Thumb or Early Boynton. Of size of Minnesota, and very sweet. The original introducer, I send pure stock, postpaid, per package 15 cents; per quart, 70 cents; per peck, by express, $3.00. In my catalogue, (free to all,) are emphatic recommendations from farmers and gardeners.

JAMES J. H. GREGORY,
apr-3t   Marblehead, Mass.

THE LANCASTER FARMER.

PENNSYLVANIA RAILROAD SCHEDULE.
Trains LEAVE the Depot in this city, as follows:

WE TWARD.	Leave Lancaster.	Arrive Harrisburg.
Pacific Express*	2:40 a. m.	4:05 a. m.
Way Passenger	5:00 a. m.	7:50 a. m.
Niagara Express	11:00 a. m.	11:20 a. m.
Hanover Accommodation	11.05 p. m.	Col. 10:40 a. m.
Mail train via Mt. Joy	10.20 a. m.	12:40 p. m.
No. 2 via Columbia	11.25 a. m.	12:55 p. m.
Sunday Mail	10.50 a. m.	12:40 p. m.
Fast Line*	2:30 p. m.	3:25 p. m.
Frederick Accommodation	2:35 p. m.	Col. 2:45 p. m.
Harrisburg Accom	5:15 p. m.	7:40 p. m.
Columbia Accommodation	7:20 p. m.	Col. 8:20 p. m.
Harrisburg Express	7:30 p. m.	8:40 p. m.
Pittsburg Express*	8:50 p. m.	10:10 p. m.
Cincinnati Express*	11:50 p. m.	12:45 a. m.

EASTWARD.	Lancaster.	Philadelphia
Cincinnati Express*	2.55 a. m.	5:00 a. m.
Fast Line*	5:08 a. m.	7:40 a. m.
Harrisburg Express	8:05 a. m.	10:00 a. m.
Columbia Accommodation	9:10 p. m.	12:01 p. m.
Pacific Express*	1:30 p. m.	3:40 p. m.
Sunday Mail	2:00 p. m.	5:00 p. m.
Johnstown Express	3:05 p. m.	5:50 p. m.
Day Express*	5:15 p. m.	7:20 p. m.
Harrisburg Accom	6:25 p. m.	9:30 p. m.

The Hanover Accommodation, west, connects at Lancaster with Niagara Express, west, at 9:35 a. m. and will run through to Hanover.
The Frederick Accommodation, west, connects at Lancaster with Fast Line, west, at 2:10 p. m., and runs to Frederick.
The Pacific Express, east, on Sunday, when flagged, will stop at Middletown, Elizabethtown, Mount Joy and Landisville.
*The only trains which run daily.
†Runs daily, except Monday.

NORBECK & MILEY,

PRACTICAL
Carriage Builders,
COX & CO'S OLD STAND.
Corner of Duke and Vine Streets,
LANCASTER, PA.

THE LATEST IMPROVED

SIDE-BAR BUGGIES,
PHÆTONS,
Carriages, Etc.

THE LARGEST ASSORTMENT IN THE CITY.

Prices to Suit the Times.

REPAIRING promptly attended to. All work guaranteed.
79-2-

S. B. COX,
Manufacturer of
Carriages, Buggies, Phaetons, etc.
CHURCH ST., NEAR DUKE, LANCASTER, PA.
Large Stock of New and Second-hand Work on hand very cheap. Carriages Made to Order. Work Warranted for one year. [78-9-12

EDW. J. ZAHM,
DEALER IN
AMERICAN AND FOREIGN
WATCHES,
SOLID SILVER & SILVER PLATED WARE,
CLOCKS.
JEWELRY & TABLE CUTLERY.
Sole Agent for the Arundel Tinted
SPECTACLES.
Repairing strictly attended to.
ZAHM'S CORNER,
North Queen-st. and Centre Square, Lancaster, Pa.
79-1-12

E. F. BOWMAN,
Watches & Clocks
AT LOWEST POSSIBLE PRICES.
Fully guaranteed.
No. 106 EAST KING STREET,
79-1-12] *Opposite Leopard Hotel.*

ESTABLISHED 1832.

G. SENER & SONS,
Manufacturers and dealers in all kinds of rough and finished
LUMBER,
The best Sawed SHINGLES in the country. Also Sash, Doors, Blinds, Mouldings, &c.
PATENT O. G. WEATHERBOARDING
and PATENT BLINDS, which are far superior to any other. Also best COAL constantly on hand.
OFFICE AND YARD:
Northeast Corner of Prince and Walnut-sts.,
LANCASTER, PA.
79-1-12]

PRACTICAL ESSAYS ON ENTOMOLOGY,
Embracing the history and habits of
NOXIOUS AND INNOXIOUS
INSECTS,
and the best remedies for their expulsion or extermination.
By S. S. RATHVON, Ph. D.
LANCASTER, PA.
This work will be Highly Illustrated, and will be put in press (as soon as a sufficient number of subscribers can be obtained to cover the cost) as the work can possibly be accomplished.
79-2-

$5 to $20 per day at home. Samples worth $5 free)
Address STINSON & Co., Portland, Maine

TREES
Fruit, Shade and Ornamental Trees.
Plant Trees raised in this county and suited to this climate. Write for prices to
LOUIS C. LYTE,
Bird-in-Hand P. O., Lancaster co., Pa.
Nursery at Smoketown, six miles east of Lancaster.
79-1-12

WIDMYER & RICKSECKER,
UPHOLSTERERS,
And Manufacturers of
FURNITURE AND CHAIRS.
WAREROOMS:
102 East King St., Cor. of Duke St.
LANCASTER, PA.
79-1-12]

Special Inducements at the
NEW FURNITURE STORE
OF
W. A. HEINITSH,
No. 15 1-2 E. KING STREET
(over Bursk's Grocery Store), Lancaster, Pa.
A general assortment of furniture of all kinds constantly on hand. Don't forget the number.
15 1-2 East King Street,
Nov-1y] (over Bursk's Grocery Store.)

For Good and Cheap Work go to
F. VOLLMER'S
FURNITURE WARE ROOMS,
No 309 NORTH QUEEN ST.,
(Opposite Northern Market),
Lancaster, Pa.
Also, all kinds of picture frames.
nov-1y

GREAT BARGAINS.
A large assortment of all kinds of Carpets are still sold at lower rates than ever at the
CARPET HALL OF H. S. SHIRK,
No. 202 West King St.
Call and examine our stock and satisfy yourself that we can show the largest assortment of these Brussels, three plys and ingrains at all prices—at the lowest Philadelphia prices.
Also on hand a large and complete assortment of Rag Carpet.
Satisfaction guaranteed both as to price and quality.
You are invited to call and see my goods. No trouble in showing them even if you do not want to purchase.
Don't forget this notice. You can save money here if you want to buy.
Particular attention given to customer work.
Also on hand a full assortment of Counterpanes, Oil Cloths and Blankets of every variety [nov-1y-

PHILIP SCHUM, SON & CO.,
38 and 40 West King Street.
We keep on hand of our own manufacture,
QUILTS, COVERLETS,
COUNTERPANES, CARPETS,
Bureau and Tidy Covers, Ladies' Furnishing Goods, Notions, etc.
Particular attention paid to customer Rag Carpet, and scouring and dyeing of all kinds.
PHILIP SCHUM, SON & CO.,
Nov-1y Lancaster, Pa.

THE HOLMAN LIVER PAD!
Cures by absorption without medicine.
Now is the time to apply these remedies. They will do for you what nothing else on earth can. Hundreds of citizens of Lancaster say so. Get the genuine at
LANCASTER OFFICE AND SALESROOM,
22 East Orange Street.
Nov-1yr

C. R. KLINE
ATTORNEY-AT-LAW,
OFFICE: 15 NORTH DUKE STREET,
LANCASTER, PA.
Nov-1y

The Lancaster Farmer.

Dr. S. S. RATHVON, Editor. LANCASTER, PA., JUNE, 1882. Vol. XIV. No. 6.

EDITORIAL.

THE PROPOSED NEW DEPARTMENT OF AGRICULTURE.

Its Importance, Its Necessity and Its Rights in the Category of Progressive Civilization.

"The total money value of all the farms in the United States foot up the immense sum of $10,196,890,645; the value of farm implements, $406,516,902; the live stock, $1,500,482,187. The exports of agricultural products for 1881 amounted to $729,650,016, being an average of 78½ per cent. of all our exports."

These figures may illustrate the magnitude, the importance and the value of our agricultural interests, and are sufficient to afford light to indifferent Congressmen in regard to the claims of agriculture to a distinct departmental recognition in the Presidential Cabinet, endowed with all the powers, influences and means within its legitimate sphere that distinguishes any other department of the Government. Indeed, in view of the pregnant fact that all we eat, all we wear, all that shelters us is either an agricultural product, or in some way connected with it; and that even commerce and manufactures could not exist independent of agriculture—we repeat, in view of all this, it seems like an unaccountable omission that the founders of the government did not establish a co-equal department of agriculture from the very beginning, especially since agriculture, at that period, embraced so large a part of the industrial interests of the New Government. Nothing but the inherent modesty of its representatives and the remnants of a veneration for the class role introduced from the mother country, could have withheld the farmers of those days from asserting their *right* to a departmental position in the constitution of the Executive Cabinet. But, instead of such a wise and generous recognition of an industry involving the physical vitality of the government itself, the subject of agriculture has been practically regarded as a sort of tail-piece (something like Nast's caricatures of Gratz Brown on the Greeley Presidential ticket) to the U. S. Patent Office, almost entirely eclipsed by a department that under any circumstances could only have been secondary to it. It is hoped *now*, however, that Congress will see what it has failed to see for many years and make amends in the near future for its habitual delinquencies of the past.

"INCREASE OF OUR CROPS."

Hon. William Fullerton, in a very excellent paper on this subject, published in *Southern Industries*, says: The doctrine I would enforce may be thus briefly stated:

1st. Constant attention should be given to the home manufacture of manure, and made the farmer's chief dependence, not forgetting to protect it from waste until applied to the land.

2nd. Practice green manuring as a *system*, using commercial fertilizers, if necessary to promote a vigorous growth for that pupose.

3rd. Keep the surface of the land mulched by letting something remain on it, to protect the roots of the grasses and imprison the fruits of decomposition.

4th. Feed on the farm the most of its products, and make beef, pork, mutton, wool, &c., &c., rather than depend upon raising and selling grain for a livelihood.

He also states with emphasis that the importance of this last injunction cannot be over estimated. Money could be profitably expended to raise food to be fed on the farm, whilst the same amount expended on the same land for raising grain to sell, would result in loss.

Mr. Fullerton further alleges that he advances no theory, the value of which he has not practically tested and proved. According to his reasoning, the great want of the farmer is manure. This, in some form, he must have, to cultivate profitably. Barn-yard manure must be the chief reliance, and, when made, it must be better cared for than is usually the habit.

A story is told of a farmer whose lands failed to produce a crop, upon which he finally applied to his minister, to pray over his fields. The good man consented, on condition that he would accompany him and point the fields and crops he desired to be prayed for. In going along, they arrived at a particularly unpromising field, and here the farmer thought a very special prayer should be offered; but the minister only shook his head, and very sensibly replied in his own vernacular, "*Es ist gor net der eardt das mer do baidet dool, do kar'dt misht.*" The minister doubtless had had sufficient experience to know that the Lord does not work arbitrarily in man's behalf, but through *media* best adapted to ends, and, that the media best adapted to poor lands, is manure, *manure*, MANURE.

POTASH IN PLANTS.

Potash is one of the absolute necessities of all plants, and the time was when in order to obtain this substance for other purposes it was extracted largely from plants by mechanical means. While the phosphoric acid directs itself mostly to the development of the seed, potash applies in the greater part to the perfecting of roots, leaves and stems, as exhibited in the following table:

Plants require potash in the following proportion to one thousand pounds:

	LBS.		LBS.
Wheat,	$5\frac{1}{2}$	Peas,	$9\frac{8}{10}$
Wheat Straw,	$4\frac{2}{10}$	Pea Straw,	$10\frac{7}{10}$
Spring Wheat,	7	Beans,	12
Barley,	$4\frac{9}{10}$	Bean Straw,	$25\frac{9}{10}$
Barley Straw,	$9\frac{3}{10}$	Potatoes,	$5\frac{6}{10}$
Oats,	$4\frac{5}{10}$	Green potato vines	$7\frac{1}{10}$
Oat Straw,	$9\frac{7}{10}$	Beet-root (sugar)	4
Rye,	$5\frac{8}{10}$	Beet-tops,	4
Rye Straw, (winter)	$7\frac{6}{10}$	Hemp, whole plant,	$5\frac{8}{10}$
Rye Straw, (sum'r)	$11\frac{1}{10}$	Linseed,	$11\frac{3}{10}$
Corn,	$3\frac{3}{10}$	Clover-Hay,	$19\frac{5}{10}$
Corn fodder and stalk,	$16\frac{6}{10}$	White-clover hay,	$10\frac{6}{10}$
Meadow Hay,	$17\frac{1}{10}$	(From "What of Fertilizers.")	

"The readiest and most acceptable method of furnishing potash to the field is by the application of wood-ashes, even those of bituminous and anthracite coal are very useful, when seperated from the grosser particles. In default of these, recourse must be had to the various low priced potash salts, now so abundantly supplied by the recently opened enormous deposits in Germany."

Potass, potash, or pearlash, is an oxide of *potassium*, the two latter names being applied to the article as found in commerce. At one period it was entirely produced or obtained by burning various plants, hence its name *potash*. But it is also obtained native in various parts of the world, notably in Germany, but not of as pure a quality as that produced through chemical manipulation. It is widely diffused, and of course that which exists in vegetation must have been absorbed from the soil, and also in a gaseous state from the atmosphere. It readily combines with other chemical substances, and forms various compounds. The ordinary potash is a *carbonate* in an impure form.

KITCHEN GARDEN FOR JUNE.

The labor of the gardener in this month, will mainly consist in the tillage of the growing crops in this latitude. The rapid growth of weeds at this season will admonish him of the necessity of timely exertion.

The aid of appropriate tools in the culture of crops, and extermination of weeds will be commended. Good implements are indispensable to success, and he who has provided them will not only have greater pleasure in his labors, but the profit which attends the judicious application of them in both time and labor.

Asparagus-beds keep clean. *Beans*, bush or bunch, plant for succession, and cultivate those in growth. *Beets*, thin the later planting. *Broccoli*, plant out those sown in April. *Cabbage*, ditto, especially the sorts which it is desired shall come into use in September and October, in advance of winter varieties. *Celery*, plant out a portion for early use. *Cucumbers*, sow successive crops. *Corn, Sugar*, plant for succession. *Endive* sow. *Leeks*, thin or transplant. *Peas*, a few may be planted for succession.—*Landreth's Rur. Reg.*

As the foreging directions are intended to be effective from the very begining of the month, and as our Journal is never issued before the middle of it, yet, as the season is fully half a month later than the average, they are not inappropriate to the period at which our patrons will receive them. Indeed, at any season, there are few so far beforehanded in their work, as not to be benefitted by such advice if they heed it, and avail themselves of its practical benefits. Especially are those items which relate to *succession*, matters of interest, not only to the gardener, but also to consumers of garden crops. "Succession," or "cropping" of garden vegetation, is very little more of a specialty now than it was a quarter of a century ago, except in the vicinities of large cities. In Lancaster county we are beginning to find green corn in market late in October, but that is pretty much all of the early summer vegetables we find at that season. During the "Crystal Palace" exhibition in New York—nearly a quarter of a century ago—we were surprised to find green peas and green corn on the table near the close of the month of October, a State that is meteorologically two weeks later than Pennsylvania.

EXPORTS OF CHEESE.

The following are the exports of cheese from New York to the under-mentioned ports since May 1, 1881 (beginning of the trade year), and for the same time last year:

	1881-1882	1881-1880
Liverpool	76,025,512	81,060,783
London	17,610,512	13,075,326
Glasgow	40,359,607	17,640,760
Bristol	8,417,517	11,714,528
Cardiff	395,000	1,377,721
Hull	1,336,013	373,841
Newcastle	914,530	592,600
Havre	148,670	121,145
Hamburg	975,524	119,373
Bremen	1,130,225	375,480
Other ports	7,489,076	2,693,831
Total	131,442,726	120,303,008

Exports of Butter.

The following are the exports of butter from New York to the under-mentioned ports since May 1, 1881 (beginning of the trade year), and for the same time last year:

	1881-1882	1881-1880
Liverpool	6,195,025	10,775,296
London	475,017	936,144
Glasgow	3,123,986	6,481,720
Bristol	864,000	1,781,002
Cardiff	357,000	672,080
Hull	68,000	105,150
Newcastle	84,300	171,076
Hamburg	537,050	340,898
Havre	674,510	1,252,301
Bremen	684,209	1,048,347
Other ports	5,050,449	3,913,442
Total	17,927,093	27,782,566

Exports of Oleo-margarine.

The following are the exports of oleo-margarine from New York to the under-mentioned ports since May, 1, 1881, and for the same time last year:

	1881-1882	1881-1880
Liverpool	451,174	790,250
London	13,165	64,180
Glasgow	1,641,654	1,980,500
Bristol	47,000	179,744
Rotterdam	5,886,607	5,200,180
Antwerp	1,447,863	490,875
Hamburg	25,430	78,007
Bremen	45,830	81,712
Other ports	1,136,985	561,250
Total	10,464,709	8,983,968

The above, which we clip from the columns of *The American Dairyman*, exhibits an appreciation in our exportations of Cheese during last year, of 2,130,628 pounds, which, at only *ten cents* per pound, would amount to the handsome sum of $213,062.80. That is certainly soon advance, so far as the exportation of cheese is concerned.

Our exports of Oleo-margarine during the same period shows an increase of 1,481,101 pounds. (Whether Oleo-butter or cheese, the tables don't state,) but, at the same rate per pound, it would amount to $148,110.10, also a very respectable advance as a domestic exportation.

These two items of increase aggregate 3,620,738 pounds, amounting to $362,072.90. Does this indicate that oleo-margarine—whether in the form of butter or cheese—is becoming more popular than it formerly was in the foreign market? If oleo-margarine is healthful and can be furnished at a lower price than genuine butter or cheese there certainly will grow up a market for it, because the masses of the people cannot afford to pay the prices that are now demanded for the genuine article, especially *butter*. And, as to *quality*, nine out of ten would prefer good oleo-margarine to rancid, oily butter.

The exhibit of the butter exportation does not look so favorable. From the same tables we discover that there was a depreciation in the item of butter, during the same period, of 9,854,873 pounds, which, at the nominal price of *twenty cents* per pound, would amount to $1,970,974.60, which absorbs the increase in cheese and oleo-margarine, and exhibits a reduction in last year's operations amounting to 6,234,135 pounds, and $1,608,901.70.

The absence of rains and the short grass crop of last year may have been the cause of the short butter crop, although we might naturally suppose it would have had the same effect upon the production of cheese, unless, indeed, the oleo-margarine had been " smuggled in " as genuine cheese.

We ought to do better in the butter business the present year; and yet, just *now*, (May 13th,) it looks more likely that we may be " drowned out," or " rotted out," than be " dried out."

As pertinent to the subject we append the following from the source above named:

Marketing Farm Products.

Whatever may be said against oleo-margarine, truthfully or otherwise, it is an undeniable fact that since it has been put upon the market butter has presented itself in better garb, sweeter, sounder, cleaner, and in every way more worthy of being recognized as a prime product of the American dairy.

Mr. Starr, of Echo Farm, was one of the first to get a dollar a pound for the delicious butter sent to New York, Boston and other cities. The task to market in neat half-pound packages wrapped in snow-white linen, and was as fragrant and sweet as the June grasses upon which the cows fed. If there is a *paradise* for cows on earth Echo Farm is one, and a worthy model, creditable to the heart of a humane farmer.

Now, we have many dairies sending sweet, waxy, golden and aromatic butter to the market, perfectly gratifying the most fastidious tastes of our citizens. These dairies and these products honor such names as Havemeyer, Coe, Crozier, Holly, Dinsmore, Park, Valentine, and scores of others.

The great Western States are worthy competitors in gilt-edged butter.

Cheese, eggs, poultry and fruits, put up in a neat manner, are always acceptable to the purchaser, and bring remunerative prices to the producer.

In Baltimore and Philadelphia, for many years, poultry came to market nicely drawn, fresh, sweet and ready for the cook; and now, in New York and Boston, the hotel-keepers demand drawn poultry. They are posted in such matters, for they cater to the most extravagant tastes; and a man who knows how to keep a first-class hotel knows what human provender should be.

Compare our first-class retail groceries now with what they were twenty years ago. The demands of consumers require goods neatly put up, the stores to be kept clean, and the clerks aproned in immaculate white. In fact, some of the spruce clerks now wax their mustaches, à la Napoleon III., to please the ladies.

The neat and tasty marketing of farm products pays a handsome profit on all the extra taste and labor bestowed upon them.

Our best merchants understand the art of displaying their goods and the profit it brings. A visit to Thurber's will convince the most skeptical. In this house, where twenty millions are annually sold, the goods are put up in the best possible style. Even the canned goods are radiant with colors and rich in gilt.

The packages of coffee, tea and spices are clothed with beautiful pictures of the Oriental shrubs that produced them. Thurber's labels are exquisite specimens of taste and art. " Straws tell which way the wind blows."

Let farmers' wives and daughters tastefully decorate the packages of farm products and they can afford to dress in silk.

THE CONESTOGA FLYING FISH.

In regard to the rumor of a flying fish having been caught in the Conestoga, at Wabank, some time ago, if not a canard, it has, at least, turned out to be a " gurnard." Before the week was out I handled three specimens, and that " settled it." Now, it is not impossible that the fish in question should have been caught in the Conestoga, but it is altogether improbable. About forty years ago a genuine sturgeon was caught in the Susquehanna, above Marietta, in a " fish basket," and, I think, is still extant, in possession of Judge Libhart, of Marietta; and just here I would suggest that that specimen ought to be in the museum of the Linnæan Society. This supposed Conestoga fish is a species of the " Flying Gurnard," *Prionotus carolinus* of Dekay, also called " Sea-Robin" and " Grunter," from a grunting noise it makes when taken out of the water. It belongs to Cuvier's first order, and second family of Bony-Fishes, the first family (*Percidæ*) being typified by the common perch. The family to which this subject belongs includes the " hard-cheeked " fishes, and the attempt to " palm it off " as a Conestoga fish smacks very much of a hard-cheeked adventure. I have now two specimens of it, obtained from second persons, who could not tell whence they originally came, and representing them to have been caught in the Conestoga and the Susquehanna, may have been more to enhance the value of the fishes than to " sell " the naturalists. They are an Atlantic coast fish, and abound from the Carolinas as far northeast as Nantucket, feeding, according to Dekay, on small mollusks and crustaceans. They have the power of making a short flight by the aid of their large pectoral fins, when pursued by their enemies, but they are not the true flying fish (*Exocetus volitans*) and do not belong either to the same family or the same order. The pectoral fins of the true flying-fish are longer than the body of the fish, but in this subject they are only about one-third the length of the body. These fishes attain a length of from twelve to eighteen inches, and their edible qualities are not of a very high order—too dry and insipid.

PYRETHRUM ROSEUM.

The illustration of this comparatively new *Insecticide*, which we publish in this number of the FARMER, was originally intended to have accompanied the history, etc., of the plant, which appeared in our *May* number, but which we did not receive in time. To those of our patrons who possess regular files of our journal, it will be little or no inconvenience to have the history and the illustrations in two consecutive numbers, if it does not facilitate reference thereto. We believe that both amateur and professional flower gardners might do many worse things than to cultivate this plant, both for utility and ornamentation. By carefully gathering the flowers after they had accomplished their ornamental functions and preserving them for future use, they would have ready access to an antidote against those insect and other pestiferous vermin which so often damage or destroy the fruits of their labor in the house, the garden and the field.

PYRETHRUM ROSEUM.

a, Flowers and upper leaves from plants grown at Washington city, D. C.; *b*, flower; *c*, leaf from flower-stalk; *d*, involucre from the *Botanical Register*, Vol. XII, Fig. 1024.

VENNOR PREDICTS A BAD SUMMER.

Vennor, the Canadian weather prophet, was written to concerning the significance of the recent aurora. In his answer he says:

"The approaching summer will be cold and wet over a very considerable portion of the continent, south and west. He should not be surprised should each month for the remainder of the year bring frosts. In past years brilliant auroras at this time in April at Toronto, New York and more southern points, have most invariably been succeeded by cold and wet summers."

It is becoming a serious question whether, after all, Vennor may not be something of a weather prophet. His prophecies up to the present date (June 6th) have come much nearer the true state of the weather "than a goose is to a turkey," hard-shelled literalists to the contrary notwithstanding. No prophecies, perhaps, have ever come literally true, from the beginning of history down to the present time, and perhaps never *will*, nor is it necessary that they ever *should*. Of course, if the above predictions of *Vennor* come literally true an abundant harvest of hay, corn and pignuts will follow them just as a shadow follows its substance. To Vennor himself, it may be no prophecy nor intended as such, but simply a mathematical conclusion, as *twice 2 are 4*.

CADDICE FLIES.
An Insect that Walls Itself Up.

In answer to an inquiry made of him, our learned scientific friend, Dr. Rathvon, writes as follows:

W. U. HENSEL, ESQ.—*Dear Sir:* The curious little worms, enclosed in a gravel-covered follicle, which you and your editorial companions recently observed in a spring in Franklin county, are commonly called "caddice worms" or "case worms," and are the larvæ of a Neuropterous (nerve-winged) insect, commonly called "caddice flies," of which there are various species, the most common of which, perhaps in this latitude is the *Phryganea Cinerea* of Mr. Walker. They are tolerably abundant in nearly all the springs throughout Lancaster county, especially the southern portion of it, as Gen. Steinman some years ago sent me a large number from a spring on his farm in Martic township. Most likely, however, it may be the *Phryganea Senificiata* of Mr. Say. Species cannot be determined without having the specimens of the mature insect. These are four-winged flies, from 23 to 28 millimeters in length, and from 4 to 5d to nine-expansion, which means from tip to tip of the expanded wings. The general characters are long antennæ, compressed body, wings longer than the body, nerved longitudinally with a few transverse veins, and generally of a grayish, brownish and blackish color, but not very brilliant. Their flight is sluggish and they are usually found near ponds, streams and springs of water, in which the larvæ or worms are found. Their development is very interesting. The female fly deposits her eggs either on the water or on some plant or other object in the water, and as soon as the young worm issues from the egg it seeks the bottom of the pool and begins to construct a sort of oblong cocoon out of finely spun silk or webbing, and as the body of the insect increases in size it increases the length and diameter of its case, incorporating with it, on its outer surface, small particles of whatever it may find on the bottom of the pool or spring; if sandy it will be covered with the larger grains of sand or gravel. But if such material is not at hand it will use small portions of leaves, leaf stems, wood, or anything it can conveniently appropriate.

The fly makes its appearance annually in June, July and August, according to species or other circumstances, but their lives are short; during the larger part of the year (ten months or more) they are found in the water, in the form of case or caddice worms. As a fly they eat nothing, but the worm feeds on vegetation.—*Algæ, &c.—Intelligencer.*

EGGS.

"Hens in France produced $300,000,000 worth of eggs last year."

This is, perhaps, the most *eggs*-traordinary *egg*-sample of *egg*-culture in *eggs*-istence, and ought to be *eggs*-tended with literal *eggs*-actness to all places *eggs*-posed to contingent *eggs*-igencies, without *egg*-ception.

Punning aside, the above-quoted paragraph is only another practical illustration of "La Belle France" in the role of oviculture, and is intensely *French*; manifesting economical resources in a domestic production which *we*, and other nations, are only beginning to see, in a commercial sense. There is more profit and fewer vicissitudes in the production of small things than there is in greater things.

An egg may be regarded as a condensed chicken, containing all the nutritive elements that are to be found in a fully-developed fowl, differing only in a quantity, the quality of the former being decidedly superior.

Estimating those $300,000,000 worth of eggs at twenty cents per dozen (which is a pretty high figure for France, where, 'tis said, a good dinner can be obtained for ten cents) shows the product to have been 1,500,000,000 dozens, or 18,000,000,000 of eggs, in a country not four times larger than Pennsylvania, without special reference to domestic consumption, a matter that is seldom taken an account of by producers.

'Tis also said that that prolific nationality has *one hundred and thirty-two* ways of cooking an egg, and perhaps the object in producing so many eggs is to further duplicate their modes of culinary preparation. In any event, let our oviculturists keep the foregoing before them as a text in their efforts in that direction. Perhaps it would be well to import French chickens, for surely they must be of the "Old Grimes" stock of "Bunties," that "every day laid me an egg and Sunday she laid three." EGGS-IT.

OUR CROPS.

The readers of the FARMER are respectfully referred to the proceedings of our local society for an epitome of the crop reports from the different parts of Lancaster county; from which it will be seen that there is a tolerable, although not an entire uniformity. We prefer to let our farmers speak for themselves upon a subject about which they ought to know more than those who do not possess their experimental knowledge. There seems, however, to be a general opinion prevailing that just now (June 12) there never was a better prospect before of an "A, No. 1" crop of wheat—not a bad field of wheat in the county. We hope their most sanguine anticipations may be fully realized.

EXCERPTS.

IN choosing a cow the crumpily horn is a good indication; a full eye another. Her head should be small and short. Avoid the Roman nose; this indicates thin milk, and but little of it. See that she is dished in the face, sunk between the eyes. Notice that she is what stockmen call a hauuler—soft skin and loose like the skin of a hog. Deep from the loin to the udder, and very small tail. A cow with these marks never fails to be a good milker. There is more difference in cows than is usually supposed, and but few really good cows are offered in our markets.

PROF. COOK says: "After several years' experience I have only one point on which to discount the Light Brahma; there is not quite enough white meat. Brahmas should be hatched in March and April; then we shall have abundant eggs during the succeeding winter. Let no one who keeps light Brahmas forego the important suggestion to devote all their fowls to table use before they pass their second birthday.

DOGS are at present the chief obstacle to sheep raising in Georgia. There are something like 120,000 worthless curs in the State, and their fondness for illicit mutton leads to an annual slaughter of from 30,000 to 40,000 sheep. Sometimes entire flocks are killed. A sweeping dog law would no doubt interfere with 'possum hunting, but it would be worth many thousands of dollars a year to the State and sheep-raisers. A Georgia newspaper estimates the profits of sheep-raising at 53 per cent., notwithstanding the loss by dogs; hang the dogs, and the profits would rise to 73 per cent.

D. G. ROBERTS says: "Now sowed corn is like a good many other forage plants. There is a right and wrong way to raise it. Planted and grown as it should be it makes valuable feed. In traveling about the country I notice but few places properly planted. A great deal is fed before it is matured sufficiently. At certain stages of its growth it is very valuable as a butter food. At just the right stage of growth it is very valuable for that purpose. I have experimented in feeding this plant for butter many times and it has always proved best just at the time that the ears are in best condition for the table.

THE agricultural editor of the New York *Times* publishes an elaborate article on the comparative value of manures, in which he attempts to show that the manure obtained by feeding a ton of cotton-seed meal contains phosphoric acid, nitrogen and potash which it would cost $27.86 to buy in any of the commercial fertilizers, as against $6.05 worth from a ton of corn meal, $6.48 worth from a ton of good hay, and $14.50 worth from a ton of bran. This is the result of chemical analysis and good figuring. Experience probably would not show as much difference but it is certain that farmers as a rule fail to take into account the comparative value of manures which come from feeding different foods. With them manure is manure, and they do not stop to ask what it was made from or what it contains.

RYE FOR WINTER AND EARLY SPRING PASTURE.—Rye sown among standing corn will do almost equally as well for winter and early spring pasture as if it had been done at the last working of the corn, as the first rain will cause it to sprout and take root just as well as if it had been put in with cultivators. Sow not less than a bushel to the acre. Ewes and lambs and yearlings may be then turned on it after Christmas, and kept on until the 1st of April, when it may be set apart either for turning under as a manure or saved for a crop.

INCREASING FARM MANURE.—A very good plan for increasing the supply of home-made manure, may be adopted by farmers generally with equal success. It is merely by placing in alternate layers rich stable manure and turf and sods until the heap is some six feet high and as long as you please, and then, after a time, beginning at one end of the pile to turn the whole over. As the turf and sods rot they will absorb the rich gases generated by the manure, and which might otherwise escape thus forming a most excellent compost for all kinds of crops.

AN Illinois farmer who keeps twenty horses, some of them worth $1,500 each, writes that he pastures them at all times in fields fenced with barbed wire, has done it for years, and had no harm result from it. Before turning them out he first leads them to the fence and lets them rub their noses against the barbs, and the hint is sufficient. They know enough after that to keep away from the fences.

To prevent falling off of hair from a horse's mane, or to restore the growth, rub the skin of the part with the following mixture, viz: One pint of alcohol and one drachm of tincture of cantharides. Give the horse a dose of salts (12 oz.) and feed them wheat bran, which will allay the irritation of the skin, to which the loss of hair is due.

WHEAT is more valuable cut at a stage which would be commonly considered a little early than when left to become over-ripe. The cellulose or woody fibre rapidly increases in the days of over-ripening, giving more bran and less flour, thus materially reducing the milling value of the wheat.

A cubit is two feet.
A pace is three feet.
A fathom is six feet.
A palm is three inches.
A league is three miles.
There are 2,759 languages.
A great cubit is eleven feet.
Two persons die every second.
Bran twenty pounds per bushel.
Sound moves 743 miles per hour.
A square mile contains 640 acres.
A barrel of ice weighs 300 pounds.
Slow rivers flow five miles per hour.
A barrel of pork weighs 200 pounds.
A barrel of flour weighs 196 pounds.
An acre contains 4,840 square yards.
Oats, thirty-two pounds per bushel.
Barley, forty-eight pounds per bushel.
A hand (horse-measure) is four inches.
A span is ten and seven-eighths inches.
A rifle ball moves 1,000 miles per hour.
A storm blows thirty-six miles per hour.
A rapid river flows seven miles per hour.
Buckwheat, fifty-two pounds per bushel.
Electricity moves 228,000 miles per hour.
A hurricane moves eighty miles per hour.
The first lucifer match was made in 1829.
A firkin of butter weighs fifty-six pounds.
Coarse salt, eighty-five pounds per bushel.
A tub of butter weighs eighty-four pounds.
The average human life is thirty-three years.
Timothy seed forty-five pounds per bushel.

AGRICULTURAL.

SORGHUM seed is readily eaten by poultry, and is better for small chickens than corn.

MILK should stand at least thirty-six hours before skimming to get good results. Farmers take notice.

KILL the dog first and hunt for his owner afterwards, is the maxim of certain Georgia farmers who mean to make sheep-raising profitable.

COCKLE seed will remain in the ground many years if untouched by the plow. As soon as brought to the surface they begin to sprout.

IF sulphur is well dusted around the sheds and hog-pens it will effectually drive off lice. Dust it on the hogs also, and leave a little in the trough for them to eat.

HORSERADISH is a profitable crop to grow as it finds sale at from five to six cents per pound unprepared. It is bought readily by manufacturers of the prepared article.

NEARLY all kinds of fruit do well on a mixture of superphosphate and wood ashes. Lime is not suitable for strawberries, but excellent around apple, peach and pear trees.

EVERY farmer should select a portion of rich soil, clear from weeds, which should be devoted to roots, such as beets, turnips, rutabagas or carrots for feeding cattle and hogs. They are good starters for fall feeding.

CONTRIBUTIONS.

COMPARATIVE VALUE OF FARMS BETWEEN NOW AND FIFTY YEARS AGO

Fictitious Value—Good Crops—Good Governments—Tariffs, Etc.

Are our Lancaster county farms worth what they now bring at a sale? Or, are these prices fictitious, like they have been at different periods in our history for several generations back? These are questions of some importance, raised among many classes of people. Even politicians make it a point to discourse on the subject of Lancaster county farms, apparently with a view to lessen the bulk of taxes. I say *apparently*, for really it seems when politicians get into office they are so much absorbed with ideas of personal emolument, in the form of salaries, fees and perquisites that one must doubt all protestations made in *obtaining* office. Many politicians, who are such for the sake of pelf and gain—for self and spoils—are prone to "set up" and manipulate political tickets that cannot but extort high taxes (from those fine Lancaster county farms) in order to fill their own pockets, whilst the farmers themselves are doomed to rustic toil, earning their bread by the sweat of their brows. * * * Many years ago we heard that in the east—Connecticut for instance—farmers and tobacco growers, were buying their manure in New York, and other large towns, at a cost of $10.00 per cord, five cords to the acre, making $50.00 per acre for the manure alone, which we thought simply enormous. But we are fast following in their footsteps, and in addition, apply 100 bushels of lime to the acre; in all footing up $60.00; and, after repeating this operation for a number of years, our lands, together with the outlay for fences, buildings and other improvements—such as houses, barns, sheds, &c.—these coveted farms, the eyes of which some of the avericious officeholders are fixed upon, will cost from $100 to $200 per acre, not including the original cost of the land. But a farm in itself alone, even with fences and farm buildings, is like a well without water, unless it is intelligently and practically operated, and for this purpose we must add the usual live-stock and implements—for instance—four good horses $800, four cows $200, besides implements such as wagons, plows and harrows, running up to $2,000 on a one hundred acre farm. A good practical farmer will also need $1,000 worth of cattle, and $1,000 worth of sheep in the fall, to make manure, finally running up a bill of $4,000, for all of which he is compelled to pay a heavy burden of taxes.

Now comes the great problem, "does it pay," or, in other words, can a man realize any amount of interest at these figures, or are the prices fictitious? The old sayings are "experience is the best teacher," and "practice makes perfect;" but, an inexperienced man may say, "Why, do not farmers realize so many bushels of wheat, corn, &c., from an acre, and lastly so many pounds of tobacco?" which must surely cover all the outlay. Well I will only say, "try it." Buy a farm at the present prices of land—fictitious or otherwise—if you ever get hold of so much money, and farm for yourself, or rent a farm, and at the end of the year, or a number of years, tell us all about your experience and success; but, the best proof you can adduce, will be the ability to buy another farm in eight or ten years thereafter. This question is one open for consideration, but to me it looks as though lands and other properties, at the present day, have a more real value than at any previous period, notwithstanding the small profit realized out of them.

Taking into consideration the interests of the railroad as a standard of values, in addition, we are still better off with the $200 per acre farm, than we were with the $50 or $100 per acre farm; moreover, in Pennsylvania there is yet a great deal of room for improvement. Farmers and growers of produce are not men up to England and other countries. We are even gathering up the bones of dead cattle to be shipped or exported to other countries, which we could so necessarily use here at home. Thousands of dollars worth of manure, and material which would make manure, or fertilizers, are lost here annually. I contend that we should not stop short of raising—as an average—*thirty* bushels of wheat or one *hundred* bushels of corn, or *sixty* bushels of oats, or *twenty-five hundred* pounds of tobacco to the acre, and make our farmers pay a compensating interest at these figures. I hope the time will come when all, without exception, shall become so educated as to be really practical farmers and mechanics, and also up to the requirements of the times, and not so merely in pretention, or name. That we may become able to discriminate, select, and vote intelligently for such lawmakers as will make laws for the people; to subserve *their* interest and not their own, except so far as they are integral parts of the people; and not to legislate in the service of "treason stratagems, and spoil," merely. Right here I would respectfully ask my friend—who so politely criticised me sometime ago, when in an essay I stated, in allusion to the tariff, that the "balance of trade," was in our favor, and that it was a better sign of prosperous times than when the balance was against us—what he has to say *now*, since the balance of trade is going strongly against us? Is it better to sell our surplus to Europe than to sell it here at home, unless we cannot possibly use it here? We are importing about $3,000,000 worth of goods per week, and our exports during last year are far below the year 1880. It seems to me that this must ultimately result to our disadvantage.—*P. S. R., Lititz, June 8, 1882.*

A FARMER should so arrange his kitchen garden so that he can use both plow and cultivator in its management.

ON WHEAT CROPS.

Recollections of Over Fifty Years Ago.

The winter of 1827-8 was so mild that the oats that fell to the ground in the oat fields of the previous summer remained all winter in the soil uninjured, and grew up the following spring almost as rank as if it had been specially sowed. There has been no such a mixture of oats and wheat since that time. The wheat crop was very good; the grains of wheat were very plump and full, weighing from 60 to 65 pounds to the bushel, so that the wheat and the oats was easily separated in winnowing.

We generally have but one good wheat year in six; the intervening years averaging from three-quarters down to one-half, or even one-quarter of a crop. The price of wheat in 1828 was seventy-five cents per bushel. In 1836 and in 1837 the wheat crops were very poor—cause, the Hessian-fly. Many fields were pastured. The kinds of wheat then generally cultivated were the "blue-stem" and the "red-beardy." The latter variety was usually sown on oat-stubble. "Supple" on wheat-stubble, or second crop blue stem did very well in good soil, but soon commenced to get smutty. The price of wheat in 1835 was $1.25 per bushel; in *June 1837, $2.31, and in 1836, $3.00 per bushel. That was the great† "grass-hopper" year, and a drought also prevailed. That year we had the poorest crop of corn ever known in Lancaster county. About this time the Mediterranean wheat was imported. It was at first thought a little rough, but was supposed to be fly-proof; it was at least not affected for some time, and took the place of the three old varieties and improved in quality. In 1844 or 5, we had again a very good wheat crop, on high grounds. On low and level grounds the wheat was entirely destroyed by a black frost in June. The Mediterranean was generally sown up to 1850, when a man in Paradise township, Lancaster county, noticed a bunch of Red-beardy wheat growing in a wheat of a different variety. He secured it and propagated a new variety. It was an improvement on the old, or "white Mediterranean," as it was named. Sometimes the new was named the red, and by 1865 it had almost entirely taken the place of the old. About that time "something new" appeared, several years in succession. The wheat looked remarkably well all spring; made straw enough apparently to yield from twenty-five to thirty bushels of grain to the acre, but many fields only yielded from five to ten bushels to the acre. The cause was never definitely discovered. Some said it was the "weevil;" others believed it was caused by atmospheric poisoning when the wheat was in bloom. About 1870 we had again a severe wheat failure in the larger portion of Lancaster county in a new way. The season was a little dry

*In the spring of 1867 we paid $14.00 for a barrel of flour which we believe was the only time we ever paid that price.

†Either our contributor or we are in error as to the "Grass-hopper year," unless there were two such years in succession. On the 1st of September 1829 we became a resident of Lancaster, and remained here until April, 1841, boarding at the Cooper House, which was largely patronized by the farmers of Lancaster county, and we distinctly recall the advent of the destructive "hoppers" and the groups of people in North Queen and East and West King streets, looking skyward where millions of these insects seemed to be hovering over the city of Lancaster.—Ed.

the previous fall; the winter was without snow, dry and very cold, the thermometer registering from *fifteen* to *twenty-five* degrees below zero; cold bleak winds prevailed, with occasionally a few inches of snow, which was drifted from the fields mixed with surface soil and the bare ground could be seen in many fields in the middle of May. The yield was only from five to fifteen bushels per acre, in the greater part of the county. About the year 1875 the "Foltz" wheat was introduced into Lancaster county, and at first had the appearance of supplanting the Mediterranean again, but the millers did not like it in the manufacture of superfine flour and it is a wheat also that will not make much straw, unless sown on very rich, low lands. A great many of the farmers who raised it, went back again to the Red Mediterranean, and at the present time we have about three-foh it in a growing condition. Unless the Foltz wheat improves soon, it will likely be abandoned altogether. As for the present crop of wheat, it so far promises to be the very best ever grown in Lancaster county, and a poor field of wheat cannot be seen in a whole day's travel through the county. If nothing unforeseen happens the yield might be from twenty-five to forty-five bushels to the acre. It will depend entirely on the state of the weather in maturing and harvesting, but the earlier it matures, the less it will be in danger of mildew. We have had the different varieties of wheat in our county. Some far preferable to others. I have not the least doubt that new and improved varieties could be raised, if farmers or the men who operate the reapers, would keep a vigilant watch for stray heads of wheat that are superior to those around them; and in that way we might produce a hearty and prolific variety of wheat that would be better adapted to our climate, and make a more sure crop, and less liable to degeneration than any of our past and present varieties have been, excepting only the variety that originated in Paradise township about twenty years ago. It would be a very good thing if our local Agricultural Society would offer a liberal premium for the best new variety of wheat, selected from among our own wheat fields. It might originate a variety that would regularly average a yield of from twenty-five to thirty bushels per acre.—*L. S. R., Oregon June, 1882.*

ESSAYS.

INSECTS AND SOME OF THEIR RELATIONS TO THE VEGETABLE KINGDOM.*

"Creative wisdom never works in vain nor merely to sport."

Sir John Lubbock estimates that there are *seven hundred thousand species* of animals inhabiting this world of ours, the smaller moiety of which have been recorded and described, and perhaps the larger number of those described belong to the class INSECTA. It may assist you in fully comprehending the true import of *species*, when I state that it means different *kinds* of animals, and that a single species may comprise millions of individuals, and almost an endless number of varieties; and that by a consecutive series of procrea-

*Read before the Lancaster Plant-club, June 5, 1882, by the Editor.

tions each species is capable of reproducing its specific kind to an almost limitless extent, and doubtless *would* do so, if it were not that in the economy of nature itself there are counter-influences, through which a sort of *equilibrium* is preserved.

It has also been carefully estimated by intelligent and experienced authorities, that the losses annually sustained, by the United States alone, through the depredations of noxious insects, amounts to the enormous sum of *four hundred millions of dollars;* and *this*, from the standpoint of experience, do not I consider an exaggerated statement. It is not difficult to make an estimate of this kind when we reflect that in the single State of Kansas, only a few years ago, the almost entire crop of the vegetable productions of certain districts was *totally* destroyed by the influx of the "Rocky Mountain Locusts," commonly called *grasshoppers*.

These preliminary statements bring me immediately to the door of the leading topic of this essay, namely: "The relations existing between insects and plants in the economy of *nature;*" and, not only in nature's economy, but also, correlatively, in social, commercial and domestic economy. When I say *plants*, I mean the entire vegetable kingdom, although my remarks must necessarily be confined to a few incidental references to either insects or plants, and those mainly of a general or popular character.

Notwithstanding the admitted destructive character of noxious insects, it cannot be *positively* demonstrated that the immediate extinction of insect life throughout the world, is a thing to be seriously desired. If the universe, and all the living beings therein, are the outbirths of a *Divine Economy*, overruled by an *Infinite Intelligence*, then we may rationally conclude that the existence of noxious insects would not have been *permitted* except as a lesser evil, and hence that they are of some *use*. All other things, therefore, remaining just as they at present *are*, the sudden removal of the insect world from the category of animal life, might leave this globe of ours an "unwholesome" place for the human family to dwell on. Even the most annoying, repulsive, or disgusting among them, may be the scavengers, the fertilizers and purifiers of the physical world; and in *that* degree, useful. It is mainly through their occasional and destructive redundancy, and our ignorance of their useful functions, as well as their general utilization, that they become a pest, or a scourge.

I pass by the little domestic house-fly, and his immediate congeners, as being carnivorous, or *carrionarious*, in their developmental habits, and therefore not specially germain to the present subject. But, if we were entirely ignorant of the uses of the common "silk-worm," it might be legitimately regarded—where it is native—as one of the most destructive insects to plant foliage that is known to its *class*. Compared with its size and weight it can consume a greater amount of vegetation, in a given time, than any other insect extant. Instead, however, of being an injury to the human family, it has become one of the greatest factors in commercial and domestic economy. The product of this, to many, repulsive worm, amounts annually to many millions of dollars, and the various fabrics wrought out of its silken fibre are amongst the most elegant and brilliant that adorn the human habitation and the human body. The traffic in silk-worm eggs alone amounts annually to many hundreds of thousands of dollars, and whole trains of cars freighted with them pass through our territory in transit to European ports every year.

Again ; if we were entirely ignorant of the value of the little "cochineal insect"—so small, indeed, that when they are dried it requires 70,000 to make a pound—their presence would be as destructive to the vegetation they feed on, as the famous "chinch-bugs" of the Western States are to the wheat and corn crops. Where the cochineal insect is indigenous, the chief anxiety of the people is to *increase* its numbers, because "there is money in it." Therefore it becomes a prolific object of human husbandry, and its failure to multiply would be as much of a disaster as the failure of our wheat or fruit crops. You will perceive, then, that the increase of the silk-worm and the cochineal insect are the results of human manipulation, for it is not probable that either of them would multiply in states of unprotected and unassisted nature, as they do under human intervention. The cactus, the mulberry and other silk-producing trees are carefully cultivated, and an abnormal increase of the insects is thus facilitated.

We have a striking instance of the effect of cultivation upon the increase of noxious insects in the notorious "Colorado potato-beetle." Far up in the Rocky Mountains this insect was discovered about sixty years ago, in moderate numbers, feeding on a wild species of *solanum*, and it probably would have remained there, content with its rustic fare, but for human intervention. As soon as the domestic potato was cultivated in its vicinity, it abandoned the harsh and comparatively sapless native plant and betook itself to the more succulent vines of the domestic potato ; and perhaps the whole secret of insect prolification depends more or less upon the conditions produced by cultivation.

The palatable quality of meats and drinks, even among human beings, tends to their increased consumption, and it appears that insects are no exception to the rule. The common house-moth will eat a fine, soft, woolen fabric when it is sandwiched between two coarser and harsher pieces and leave the latter untouched. Progressive culture seems to engender progressive taste even among such insignificant creatures as insects. When we discover some mode of utilizing the Colorado potato-beetle, its presence and its increase may become objects of as much solicitude as the increase of the cochineal and silk-worm.

These three examples—although the utility of the potato-beetle has not yet been discovered—out of scores of others that might be named, amply illustrate *one* phase of the intimate relations existing between *insects and plants.* The *silk-worm,* by the process of mastication, digestion, absorption and secretion, elaborates the tissue known as silk ; and no matter how complicated the mechanical machinery is, or how endless the varieties of the silken fabrics produced ; nor how brilliant and gaudy their external sheen, they are all subordinate to the humble *worm* and the leafy *plant.* And although the cochineal does not consume the plant by mastication, yet it absorbs its fluid circulation by suction ; and, by a peculiar chemical sublimation, or distillation, appropriates the fine particles of coloring matter, which become latent in its body and rivals the famous "*Tyrian Dye.*"

Passing to another phase of the subject I would briefly remark, that even among the most destructive insects to vegetation, there is some compensation in specific cases. Dr. Asa Fitch, of New York, records an instance of an oak tree on his premises which was infested by the common "oak-pruning beetle," and no human manipulator of the pruning-knife could have produced the healthful and symmetrical effect upon the tree that this little industrious pruner did—cutting off and heading in the struggling branches—making it "a thing of beauty."

A further illustration of the peculiar relations under consideration may be found in the growth of various species of *cryptogamic* plants in the bodies and the tissues of insects and other animal substances. Waiving all the microscopic species, I would only remark that it has long been known to entomologists, and doubtless also to botanists, that frequently a large fungoid plant has been found growing out of the body of a subterranean larva of the "May-beetle" or "June bug," commonly called the "white grub-worm." This grub when mature is about two inches in length, and one inch in circumference, and the largest plant, as recorded in the *American Entomologist,* sometimes attains a length of five inches, and grows out of the under side of the first segment of the body.

Nor is this an isolated case. Mrs. Mary Treat, a distinguished lady entomologist, of Vineland, N. J., states that in the spring of 1865, whilst botanizing in Benton county, Iowa, she saw great numbers of them. "There were literally thousands of them scattered over quite a district." So far as the matter has been explored the fungus seems to be *unique,* although it probably would grow out of different species of the *white grub.*

Another link in the chain of connection, or relation between insects and plants, finds illustration in what may be termed *Carnivorous plants.* The carnivorous character of some mammals, birds, reptiles, fishes and insects is well known, but it may not be so well known that there are certain species of plants that capture and appropriate the juicy substance of insects. They doubtless have other sources of subsistance, but it is quite certain that they also capture and absorb the fluid parts of insects. And this is not all, for they also possess, apparently, in *form,* in *structure,* in *fragrance* and in *beauty,* the facilities for attracting and decoying unwary insects into their fatal embraces. But, as this subject is to be, or *has* been, exemplified in the series of papers read before this society, it is not necessary to make any more than this general reference on the present occasion.

But the most important function of insects in the economy of nature, so far as relates to the vegetable kingdom, is their *mediumship* in the fertilization of plants. Although they may assist, or render more complete the fertilization of nearly all *plants,* yet there are quite a number of plants that could not become fruitful—so far as human observation extends—without the aid of insects ; and especially those known as *Diœcious plants,* in which the male and female flowers occur on two *different individuals.* In plants bearing *hermaphrodite* flowers, where the *pistillate* and *staminate* organs are in the same flower, the usefulness of insects may not be so apparent ; and yet it is very probable that insects assist even in energizing *these,* and especially pollen gathering insects.

As to *monœcious* plants those in which the staminate and pistillate flowers are entirely distinct, although both occur on the same plant. I feel confident that the function of insects in their fertilization, is much greater than may be generally supposed.

Take for instance the "Gourd family"—the *cucurbitaceous* plants of botanists—in which the male flower grows on a stem, or *peduncle,* from the axils of the leaves, and the female flower is *sessile* and attached to the apex of the embryo fruit. The first-named flowers are usually erect and open upwards, or nearly so, whilst the last-named are usually horizontal or nearly so. Although it is not impossible that these flowers should become fertilized without the aid of insects, yet it is very probable that many of them would become abortive without such aid. In any event, those who pay any attention at all to the habits of insects will find these flowers often visited during the day by a rollicking family of polleniferous *hymenoptera* that seem to be perfectly intoxicated with their foraging manipulations. These insects are provided—on the broad *tibiæ* of their posterior limbs, or on the under side of the body, near the apex—with bristling brushes, and with these they gather the pollen which they bear off and convert, by the admixture of nectar, into a sort of "bee bread," or *propolis,* which is stored in their cells for the sustenance of their young after they are "excluded" from their eggs.

Of course, the bees are influenced by no *motive,* either rational or instinctive, that relates to the fertilization of the plant. They rush into the flower cup impulsively, and seem to be in a state of buzzing agitation, gyrating hither and thither, detaching the pollen and scattering it about, but all the while gathering a portion of it with their brushes and then departing as hastily as they entered ; visiting half a dozen or more of flowers before they make their final departure for their *homes.* It is easy to perceive that these manipulations of the bees facilitate the fertilization of the plants they visit—indeed, Professor Riley has demonstrated very conclusively that a certain species of *yucca* could not be fertilized at all if it were not for the intervention of the "yucca moth." The carpenter-bee, commonly called the "wood-borer," and the various species of bumble or "bumble-bees," are especially imbued with this habit. The little "leafcutters" are remarkable as pollen gatherers, and they perform a similar use—not only in *monœcious* and *diœcious* plants, but also in those bearing *hermaphrodite* flowers. Darwin has written a clever sized book, in which he conclusively demonstrates that the *orchids,* or "air plants," are fertilized largely if not exclusively by insects; and the most effective workers in that direction belong to the *Hymenopterous* order; and this order, too, happens to be one that contains the fewest enemies to vegetation and to man in proportion to the number of species it contains. The destructions of the "saw flies" and "wood borers" of this order are compensated by the saccharine fluid and the wax of the domestic "honey-bee," to say nothing about the great number among them which are parasitic upon the bodies of the noxious members of their own and correlative orders.

The relations existing between the whole animal world and plants are too obvious to need special illustration. It may not be inappropriate, however, here to introduce a quotation from Darwin, exhibiting in a peculiar manner a short section of the great chain of interrelation. Whatever we may think of Darwin's ultimate philosophical deductions, we need not ignore his facts.

"Many of our orchidaceous plants absolutely require the visits of moths to remove their pollen-masses and thus to fertilize them. I have also reason to believe that humble-bees are indispensable to the fertilization of the heartsease (Viola tricolor), for other bees do not visit this flower. From experiments which I have lately tried, I have found that the visits of bees are necessary for the fertilization of some kinds of clover; but humble-bees alone visit the red clover. (*Trifolium pratense*), as other bees cannot reach the nectar. Hence I have very little doubt that if the whole genus of humble-bees became extinct or very rare in England the heartsease and red clover would become very rare, or wholly disappear. The number of humble-bees in any district depends in a great degree on the number of field-mice, which destroy their combs and nests; and Mr. H. Newman, who has long attended to the habits of humble-bees, believes that 'more than two-thirds of them are thus destroyed all over England.' Now the number of mice is largely dependent, as every one knows, on the number of cats, and Mr. Newman says: 'Near villages and small towns I have found the nests of humble-bees more numerous than elsewhere, which I attribute to the number of cats that destroy the mice.' Hence it is quite credible that the presence of a feline animal in large numbers in a district might determine, through the intervention first of mice and then of bees, the frequency of certain flowers in that district."

There is a wonderful parallel which characterizes the animal and the vegetable worlds, and this parallel is as manifest in relation to insects as to any other class of animals. The organic world consists of plants and animals, and the line of demarkation between them is not as obvious as appears from a superficial view; nevertheless the revelations of geology and rational inference suggest that plants were *first* created in the category of organic life; and, by a divine adaptation of means to ends, it is manifest that insects were not necessary to the fertilization of the primitive plants, for these were procreated by countless millions of *spores*, and it "must needs be" that a sufficient number of these would always germinate to afford sufficient aliment to the different *herbiverous* animals as they successively appeared on the face of the earth. In process of time, however, insects were created, and abundant fossil remains of these have been brought to light; and, as among these are butterflies and bees, we may naturally infer that prior to their advent the pollenaceous and honey-bearing plants appeared.

From that remote period down through all the intervening ages to the present time, these two grand divisions of the organic world have progressed side by side in parallel lines of interdependence. Without the advent of animal life, the vegetable kingdom would have become either superabundant, rank, and putrescent, or sterile, abortive, and unfruitful. Without the prior development of the vegetable world, the existence and subsistence of animals would have been a physical impossibility.

The normal balance of these two organic kingdoms was mainly instrumental in rendering the physical world a fit place for the habitation of the human family. Man's progress has not been uniformly and uninteruptedly *onward*; it has also been digressive, and sometimes retrogressive, and through repeated vicissitudes the *equilibrium* of nature has been disturbed, it not destroyed. Instead of *diminishing*, as man advanced in civilization and intelligence, his mental and physical wants *increased*, and hence he struck out into new channels of *improvement*, especially in agricultural, mechanical and domestic productions.

But, in destroying the normal haunts of insects and insectivorous animals, and devoting them to the cultivation of improved species of plants, he unconsciously improved, and rendered more palatable, the aliment upon which most of the noxious insects feed; and this illustrates one of the disastrous relations between insects and plants. Fifty years ago, or before the general cultivation of the tobacco plant, as a crop, in Lancaster county, the "Sphinxes" or "Horn-worms," were mainly confined to the common potato vines, and subsequently to the tomato vines, as foodplants. The "Tree-crickets" and "Spectres" fed upon the foliage of trees and shrubbery; the "Cut-worms" and "Boll-worms" fed upon garden-vegetables and the young ears of corn; the "Grass-hoppers" on the various species of grass; the "Flea-beetles" on the various cucurbitaceous plants, and the "Wireworms" on cereals.

But *now*, all these insects and many more, and also in greatly increased numbers, feed on the tobacco plant; not only because of its greater succulency, but because of its greater abundance and accessibility.

And just here it may be suggested that these *facts* may possess a significance that can only be realized in the undeveloped future. For instance; more than a score of *species* belonging to the insect world have been recognized and described as depredators upon the tobacco plant; and the day may come when the cultivation of this plant will be as precarious as that of the plum, which has been so largely damaged by the notorious *Curculio*. This little "Turk" enhances the price of plums and prevents the possibility of a glut in the market, illustrating one of the many commercial and domestic relations between insects and plants.

If these things exist in relation to a fruit so luscious, so healthful and so popular as the plum, what may not come to pass within this century in regard to the culture of tobacco. I am not arraigning tobacco, I am merely alluding to *possibilities* that may be realized in the future in reference to the cultivation of and traffic in tobacco. Is there any other industry in Lancaster county involving more massive buildings for its accommodation, more capital to carry it forward, more anxiety in its development and more solicitude in its results, and yet more barren in *real use*, than tobacco; or one, in case of every other crop failure, that would furnish less support to physical life (than tobacco). But conceding its universal *utility*, the relations between this plant and the insect world are such that it is doubtful whether its production will ever so far exceed its consumption as to work that ruin which has been so often anticipated. The insect world will more and more furnish that check upon its redundancy, which will culminate in commercial and domestic *equilibria*. When its production is involved in those uncertainties which now distinguish the plum crop, the farmer will turn his attention to something else that will pay him better. If tobacco *is*, or ever *becomes a bane*, it will find its *antidote* in devouring insects. Whatever may transpire within the next hundred years, we may feel assured that only the least evil will predominate, and that we shall have one of the most striking illustrations of the relations between insects and plants in maintaining nature's *equilibrium*.

The mutual relations existing between the vegetable and the insect worlds are likely to continue as long as plants exist and insects subsist upon them. Where one is found, there also will be found the other; and improved cultivation of the former is likely to result in the increased multiplication of the latter.

All that advancing civilization and human progress can accomplish, is perhaps the subordination of the insect tribes to human dominion, by discovering and applying antidotes against the possible redundance of the destructive species. And to bear out these relations with additional emphasis, it is now becoming manifest that one of the most effective *insecticides*, and at the same time the most harmless to human beings, comes from the vegetable kingdom. Prominent among the plants that are destructive to insect life are the different species of *Pyrethrum*, but especially the *roseum* and the *cinerariæfolium*. These plants belong to the composite order, and are as simple in their cultivation as common *asters*. When the flowers of these plants are dried and pulverized, they yield a powder that is fatal to insect life.

Some insects are indiscriminate visitants or feeders on different species or varieties of plants, but others manifest a decided partiality for a particular species or genus of plants, and are seldom or never found on any other, and this is the case too, where they are not known to *feed* on the plant. The Scarlet *Tetraopes* is uniformly found on the *Asclepius*, or wild cotton. The beautiful gold and green *Chrysochus* is always found on the "Dogbane," the pretty little *Languria* always on the *cacalius*, and the repulsive *Chrcus* or "squash-bug," always on the squash or pumpkin vines. Some predaceous insects and also some spiders hide themselves in the flowers of certain plants, for the purpose of capturing the visiting insects, upon which they prey. What special benefit these insects may be to the plants where they are usually found, is not particularly manifest, but it seems very clear that the plants subserve a useful purpose to the insects, either as food, or in attracting the harmless species upon which the rapacious species feed.

There is a mucilaginous exudation from the flowers of some plants which captures or disables certain species of insects that visit them for the purpose of extracting their nectar or gathering their pollen; whilst other species habitually visit them, and ramble all over them, with perfect impunity. I might instance the *Asclepias cornuti*, or common "milkweed," if indeed, this habit is not shared by all the members of the genus. It is not unusual to find bees, wasps, butterflies, moths, and various species of flies, with their feet entangled in this treacherous gum, where they finally perish; but the scarlet *Sctruopes*, the *Harlequinized Lygœus*, and the larva of the beautiful *Danais* butterfly, are perfectly at home upon it. What the significance of this peculiar relation may be in all its details, is not in every instance apparent, but it is quite certain that the females of butterflies or moths so captured, are entirely defeated in the deposition of their eggs where the development of their progeny would be assured—if they had not performed that important function before they had been so fatally captured. There are also many species of subterranean insects which are *carnivorous* in their habits; and these, both in their larva, and in their adult states, feed upon the bodies of the noxious larva, which feed upon the roots of vegetation; so that a very direct and intimate relation between insects and plants may be here recognized. The French gardeners have for many years colonized and protected these predaceous species. Not only because they destroy subterranean intruders, but also because they come forth at night and ascend plants, shrubbery, and even trees, in quest of those which feed upon the foliage of vegetation. I might also mention the Carrionarrous species—notably the Burying-beetles—which bury the carcases of dead animals upon which they feed, and which adds something to the fertility of the soil. And this is also the case with *stercorarious* insects, which bore holes into the soil and deposit manurial pellets therein, as food for their young after they issue from the eggs, all of which benefit the soil and the plants that grow therein.

Finally—the Book of nature is a sealed volume to those who have not taken the trouble to learn its alphabet, and, after acquiring this much, to patiently persevere in "spelling out" and constructing its prolific sentences, until the import of its language is measurably learned, and its various phenomena are fully understood.

There is little danger that the subject will become exhausted in a single human life-time: It seems like a never-failing spring of pure water; and, after imbibing it until our locks become silvered with age, we shall still find new aqueous globules bubbling up from its subterranean depths. Entomology and Botany have been subjects of systematic culture from the days of ARISTOTLE and PLINY down to the present time, and may continue so for centuries to come, and still leave a margin for the future novitiate to work upon. Reasoning from these analogous, and often anomalous, phenomena, which manifest themselves at every progressive step we take, in exploring natures vast and varied domain, we cannot but be impressed with the reflection—if not with the absolute conviction—that no created object is isolated, or stands entirely alone; but that all bear a near or more remote relation to each other; and that when we have contemplated the length, and breadth, and depth of each, we may discover the elements of a homogeneous and harmonious whole, culminating in *material* if not in *moral use*. We have no right, therefore to conclude, that any object of the physical world has been permitted to exist in vain. If we cannot comprehend its use, or understand its relations to other things, the fault may be in our own want of intelligent perception; or in our failure to grasp the normal tenor of natures operations; and rightly interpret the significance of her symbolic language.

"Nature hath nothing made so base, but can
Read some instruction to the wisest man."

SELECTIONS.

THE BENEVOLENT SUNFLOWER.
(*Helianthus globosus flatuosus*.)*

It is not the æsthetical nor sentimental view of the sunflower that at present commands our attention, but rather its sanitary powers in warding off disease.

Agriculture is always lavish of its gifts. It feeds the hungry, clothes the naked and shields mankind from disease, sickness and death. The grass, the tree, the flower, all add to man's pleasure, comfort and health. Trees drain the wet places, and slowly but surely fill up disease-breeding swamps. But, in proportion to size, no plant is so beneficent in warding off malaria as the sunflower. Sections of the once malarious West have became salubrious from the growth of sunflowers, accidentally dropped by some enterprising citizen seeking a new home on the generous acres of the West. These uncared for seeds took root, grew, and the plants ripened their seeds. Then, the birds, or the winds, or both, scattered broadcast until an annual crop is furnished for whomsoever will partake of it. These plants have furnished for the emigrants' horses, oxen and other stock on his road to a new home a grateful shade in midday; and the old stalks convenient fuel to cook the breakfast dinner and supper for the weary traveler. But the greater blessing conferred by the sunflower is the protection from malaria of the settlers on the rich lands of the praries.

Whether the leaves inhale or absorb the malarial elements of disease ; or whether, by exhaling a superabundance of oxygen, sunflowers protect man and beast from sickness, physiologists haven ot yet determined ; but that they protect from malaria, experience and experiment have abundantly and convincingly proven.

All plants absorb carbonic acid gas, and exhale oxygen; while living animals exhale carbonic acid gas and inhale oxygen. Plants are largely composed of the carbon obtained from the air, while oxygen is the vitalizing element in animal organisms.

Homes, districts, army stations, hamlets,

*Paper read at the regular weekly meeting of the American Institute Farmer's Club by Dr. A. S. Heath.

villages and cities have been protected from malaria by trees and plants; but of all the plants, none exert so benign an influence against malaria as does the sunflower.

Recent experiments have shown that persons may be inoculated with the malaria contained in the water of swamps, and in the algæ growing and decaying in them. Whether the large exhalations of oxygen from great numbers of sunflowers or the excessive transpirations of water through the broad excreting leaves of these plants exert the sanitary influences attributed to them, or whether some unknown agency operates or co-operates to produce this desirable result is not material, so long as the result is obtained by liberally planting sunflowers around, or on the swampy side of habitable places; so that there may be interspersed between the human domiciles and the malaria-producing regions this efficient preventive agency.

Efficient engineering doubtless is the most effective means of overcoming malaria—by thorough drainage. Arboriculture ranks next. But for the quick and efficient aids to both of these, the planting of sunflowers in a proper manner is the most prompt and reliable means.

The necessary excavations of the engineer at first intensifies evil, by liberating the pent-up miasm. So indeed does tree planting, but in a less degree. The sunflower cultivation, however, produces immediate good results while these more permanent measures are being perfected.

Another plant, the Jerusalem artichoke—*Helianthus tuberosus*—near akin to the sunflower in its anti-malarial influence, and having the advantage in not requiring to be planted annually, and of also yielding a valuable preventive.

Washington is a veritable hot-bed of malaria. That this state of things should have been so long permitted to have existed is not creditable to Congress, the governing power. Many of our most valuable representatives have been sacrificed by exposure to Washington malaria; and vastly more have suffered in health in consequence of the unsanitary conditions surrounding the capital of a great, intelligent and rich nation.

While engineering and arboriculture are laying great sanitary plans, let the simple, efficient and immediate offices of the sunflower be brought to bear to protect the President, the Cabinet, Senators, Congressmen and the citizens of Washington from a pestilence that constantly hovers over the capital.

This valuable protecting power of the sunflower may be utilized in any locality where miasma is rife.

To protect that part of the city near the Potomac flats there should be planted a broad belt of sunflowers between that part of the flats upon which the engineers will operate and the unoccupied land; as broad and long a belt as practicable should be well plowed and planted with the Russian mammoth sunflower, four feet apart in rows at right angles, so that a single horse-plow may cultivate both ways. One plant in the square thus laid out will be best, as the growth is rapid and vigorous.

Similar management will protect other localities. The occupants of farm houses and country residences can be thus secured against the baneful influences of malaria.

A few sunflowers planted about the farmhouse might be sufficient to satisfy the æsthetic taste of Oscar Wilde, but they would not be numerous enough to ward off malaria. A belt of sunflowers and Jerusalem artichokes is required. Though there would be but little variety in these plants alone, there might be interspersed a few plants of pearl millet, golden millet, or some others to please the fancy and relieve the homely monotony of the sunflowers and artichokes. Judging from the display of artificial sunflowers in the shop windows in New York City, one might imagine that the sentimental malaria of æsthetical society has been utterly banished, yet the sunflower æsthetical malaria has spread far and near. The subjects most susceptible are those of a peculiar organization—those who are more sensitive than sensible.

It is to be hoped that artificial sentiment and artificial sunflowers will not in any way impede the rational employment of natural sunflowers to protect mankind from real ills.

Even a considerable belt of sunflowers planted on hard ground without cultivation, will make a poor show and prove ineffectual as a prophylactic. In this, as in everything else, a corresponding effort must be made to secure an object of great importance. The means must be commensurate with the magnitude of the object sought. To depend upon a few sickly and neglected sunflowers for protection against malaria is a sad, sorry and chilly prospect, enough to bring down the vengeance of an ague chill upon such a cultivator. Plant, cultivate and harvest a large crop of sunflowers, and a large crop of health at the same time. And at your harvest home festivities, bestow a thank-offering upon the Dispenser of all gracious gifts.

Thousands of valuable lives have been extinguished by the remorseless venom of malaria and if its full powers can be overcome by the simple act of planting trees and sunflowers, God bless the generous hearts that plan, and the benevolent hands that plant these life-preserving gifts for man.

OUR TIMBER LANDS.

"Our National Legislature," tritely observes Bryant, is almost wholly indifferent to the fate of our forests, and betrays a destitution of statesmen like forecast that is painful. If this was all it would not be so bad ; but, aside from their indifference, the Congress is constantly squandering large bodies of our forest lands on public corporations, who are obtaining them only for profit, and who will destroy them with more rapacity even than private individuals. Candidly, I believe that very many of our Congressmen do not credit the statements and theories that, by denuding a country of its forests, you can injure its productiveness. Some of them have lived a great many years and as yet have seen no evil effects from the cutting down of forests, nor have they experienced any scarcity of firewood at home. Wise men ; to them there is no other land than Spain, and no other age than that in which they live. It is now nearly fifty years since Dr. Drake, of Cincinnati, proposed to Congress the importance of saving our forests. Failing in this he begged the government to at least reserve tracts of woodland around the headwaters of the principal streams as a means of preventing their diminution. The wise doctor was poohed at and thought a little cracked. Well, some of the streams he proposed to save are almost valueless, and in a half century more will be entirely useless for purposes of navigation. Probably the doctor did not anticipate the time would come when these reserves would become important as a source of timber supply ; and if he had proposed such a thing he would have been laughed at outright. It is needless to say that Congress disregarded Dr. Drake's advice, and to-day the children of the very men who poohed at the doctor are suffering for the follies of their fathers. Maine, New York and Pennsylvania are practically ruined as timber States, and their streams are gradually drying up. In twenty-five years more the Northwestern States will be as bad, or even worse off, for timber than the Eastern States are, and in twenty-five years more the timber famine in the United States will begin. Good, say the Congressmen and timber vandals of to-day, we shall be dead by that time, and why should we care what happens then? Americans owe more than any other people on earth to the toils, sacrifices and forethought of their forefathers, and it is their duty—every man's duty—to transmit the inheritance they received from them to their descendants unimpaired by waste or neglect. Says Bryant, "the length of time required for the growth of timber from the seed to maturity shows conclusively that it was never destined in the order of nature for the exclusive use of a single generation." Nor is this all. The man who wantonly destroys that which he cannot reproduce in his lifetime, is not only a coward and a fool, but he commits a flagrant crime against nature and nature's God. I never see a man cutting down a fine tree but I feel like crying out " stop thief!" What is his life as compared to the life of the tree ? If he were to immediately plant another, not in his lifetime, in that of his children or his children's children would the tree attain to maturity. All this he knows, yet he fells it to the earth and does not even plant another to replace it for future generations. Is not this man a vandal? Surely ; and worse, for he is a criminal and his seed shall suffer for his sins. If the trees could talk what a pitiful tale they would tell. How they had for ages drawn moisture from the earth and distributed it through ten thousand leaves into the air to descend again in showers refreshing the earth and watering the gentle flowers. Even the tiny blades of green grass would cry out :

"Oh woodman spare the tree,
Touch not a single bough."

But they must perish from the earth ; the fiat has gone forth and we shall soon be able to say no more

"Thank God! for noble trees!
How stately, strong and grand
These bannered giants lift their crests
O'er all this beauteous land."

They will be cut down and gone and the shifting sands alone will mark where they once stood. The bleakness and barrenness of death will cover the earth, the sun pour down his vertical rays and the scorching winds unchecked howl over the sterrile plains.

I fear you will think I am becoming excited over this subject, and I do warm up a little when speaking or writing of the murder of the beautiful trees which in atrocity is little short of human murder itself. But it is not line phrases or grand, eloquent expressions we want in this case, but facts, cold arguments to convince the unreasoning and the ignorant. The voracious monster who threatens to devour all our young timber in his insatiable maw is the railroad interest of the United States. Last year there were 101,000 miles of railway in this country, and this year we are building 16,000 miles of new railway. All these roads have to be tied with comparatively young timber. I have not at hand an estimate of the number of ties used per mile, but the annual consumption is very large. Some years ago to build 71,000 miles of railway required 184,000,000 ties. Ties have to be replaced every seven years, and it is fair to set down the number of ties required annually for future consumption at 100,000,000. As every one knows, railroad ties are cut from young timber, the trees being from eight to twenty inches in diameter, and this demand strikes at the very source of our timber supply.

It is a fact that the fences of the United States have cost more than the land, and they are to day the most valuable class of property in the United States, except buildings, railroads and real estate in cities. To keep up the fences requires annually an enormous consumption of timber. The 125,000 farms in Kentucky require 180,000,000 panels of fence to enclose them. The number of rails required is set down at 2,000,000,000 costing, $75,000,000. To repair and keep in good order the fences in this one state alone, costs annually $10,000,000. Illinois, a comparatively new state, has $250,000,000 invested in fences, but it costs her only about $300,000 annually for repairs, many of her fences being constructed of wire. The whole value of the fences in the United States, may be set down at $2,000,000,000, and it costs $100,-000,000 annually to keep them in repair.

The City of Chicago alone last year employed 17,800 men in handling lumber. There were 500 clerks, 4,000 wood-workers, 2,000 sailors, 1,000 men to load and unload the vessels, and 10,000 men to handle and prepare the lumber for market, besides 300 proprietors. The lumber brought to Chicago in 188. exceeded 2,000,000,000 feet and would have loaded one train of cars 2,000 miles long. No less than 300 square miles of land were stripped of trees last year to supply the Chicago market with lumber. These figures are indeed appalling and may well alarm any one as to the future source of our timber supply. There is no hope of any diminuition in the future, for Chicago will require more lumber this year than she did last. The demand is ever increasing and the supply ever diminishing. Between the two the end must come soon and the grand old forests disappear. After the Saginaw, Muskegon, Menomonee, Manistee and Ludington sources are exhausted the Rocky mountain slope and Washington territory will be stripped of their forests, and then we will have all that is worth taking. Every year we demand 8,000,000 acres of trees and plant less than 1,000,000 acres to replace them. The end is so plain even a fool may read it as he runs.—*Gen. Jas. T. Brisbin in N. Y. World.*

ROOTS AND HOW TO GROW THEM.

The root-grower is the competitor of the ensiloer, if, indeed, he is not the original ensiloer himself. For in preserving roots in pits the process is nothing more nor less than ensilage, the fresh roots being preserved in pits covered with earth to protect them from decay and from drying, so that they may be fed during a long season when no crops are grown. With abundance of roots no farmer needs ensilage, excepting a small supply to carry him through the summer months until roots come around again. The advantages and economy of feeding roots are in no way lesser or fewer than those appertaining to ensilage, and, in fact, the balance is even in favor of roots, because they can be preserved much more cheaply than ensiloed fodder, and do not lose any portion of their nutritive elements by chemical changes during the period of their storage. The average yield of roots, too, is considerably in excess of that of corn-fodder or those other crops which are used in ensilage, and no expensive silo and troublesome process of pressure under heavy weights are required for their keeping. With a good stock of roots a farmer or dairyman, or a feeder of beef, mutton or pork, can succeed perfectly well with pasture and a few acres of soiling crops to help him out until the winter comes around, when the necessary succulent and digestible food is in readiness for the animals. In short, if every farmer should have a silo, as he is advised by some persons, he should also have a root cellar as well, both to give his stock a change of food and to reduce the cost of the construction of silos large enough to furnish fodder for the whole season. But it is hardly necessary to try to prove the enormous value of a good crop of roots to the farmer; every one admits that the trouble is that few farmers know how to grow them or will take the trouble to learn. They fear the cost, the labor and the manure required, forgetting that labor and manure are the first essentials to profitable crops, and that without these the soil has no inducement to be generous, and refuses to grant any favors whatever. Nothing comes out of nothing, and it is in vain to expect large and valuable crops without furnishing the elements out of which they are produced.

But before we proceed further it may be well to enumerate and describe the various root crops that are known in our ordinary agriculture. These are—to begin with the best known—turnips, rutabagas, carrots, mangels and sugar-beets; of these the first is the least, and the last the most valuable. Every one can grow turnips, that is, to some extent, but it is not easy to grow a maximum crop of 30 or 40 tons to the acre, and it is, in fact, no easier to grow this than to produce the same quantity of mangels or sugar-beets. The white turnip is, however, a very poor root, as may be seen by comparing the figures of the following table:

Composition and Value of Root Crops.

	Water.	Albuminoids.	Carbohydrates.	Fat.	Food units per 100 lbs.
Turnips	92.0	1.1	5.1	0.1	16
Carrots	87.0	1.2	9.8	0.2	24
Rutabagas	87.0	1.6	9.3	0.1	23
Mangels	88.0	1.1	9.2	0.1	22
Sugar-beets	81.5	1.0	15.4	0.1	30
Green corn fodder	83.7	0.9	7.8	0.1	11

The composition of corn fodder is given for comparison. Turnips and rutabagas, which are, in fact, turnips, are open to the serious objection that they are not suitable food for milch cows, giving a strong odor and flavor to the milk and the butter, which cannot be altogether avoided by any device or method of feeding. Turnips have the advantage of very quick growth so that a crop sown in August or early in September may yield a very considerable amount of feed. Rutabagas should be sown early in July, but white turnips not before August. Two pounds of seed per acre are used for either. Phosphate of lime, either in the form of bone dust or superphosphate, is the dominant fertilizer for turnips, and always helps to produce a good crop. One good use for the turnip crop is for seeding down with grass; the broad leaves shade and shelter the young grass and the small turnips left after pulling the larger ones afford shelter in the winter and manure in the spring. The best grass and clover seedings we have had have been with turnips in August. The best farm crop for cattle feeding, no variety but the large orange Belgian should be grown, for the roots of all but this kind penetrate so deeply and the crowns grow so near the surface as to make it very troublesome to harvest them. The first growth of carrots is very small and slow, and unless the ground is very free from weeds the plants are smothered before they can be seen. But while we have mangels and sugar beets, carrots may be confined to the garden, where hand-weeding may be tolerated. Mangels are a species of beet, sometimes called the mangold wurzel-beet, and are familiarly spoken of by the English farmers as "wuzzles." Here we have mangled the name in our own fashion, taking the left handle of it, while the English have taken the right. But they are a magnificent root, call them by whatever name we may. Reaching a weight of 24 to 40 pounds, the single specimen, growing half out of the ground and holding but very loosely to the soil, they are harvested with the greatest ease. We have ourselves loaded a two-horse wagon with the huge roots, of which 120 filled a 40-bushel box level with the edge and made a full ton, taking them in the row as they came; but it was, to tell the truth, in a spot where the compost heap which manured the field had stood for three months and where the soil was, of course, unusually rich. But of that crop fully 10 per cent. would weigh over 20 pounds and one root out of 20 would reach a weight of 24 pounds. One of these roots made a good meal for a cow, and the tender, crisp flesh literally melted under, or we should, strictly speaking, say over, her teeth. But to our text again. Mangels are of several kinds—the long top rooted; the ovoid or egg-shaped, and the globe—they are of several colors—red, yellow, and orange. The best, to our mind, are the long red, of which kind was the crop above referred to; the yellow globe is said to be the best suited for light soils, although our long reds were grown on a sand that sometimes blows a lively fashion on a breezy, dry day; the yellow ovoid is said to be the largest cropper, although our long reds yielded at the rate of 1,260 bushels per acre, or an equivalent of 36 tons. Then there is the Norbiton Giant long red mangel, one whose name certainly justifies a large crop if length of name could do this; and it is credited with being enormously productive, single roots weighing 100 pounds, and the whole crop reaching 72 tons per acre, or nearly half a ton to a square rod. But this enormous yield is by no means incredible, for roots growing 14 inches apart in 3 foot rows and weighing 16½ pounds each only would make a ton to two square rods, or about 80 tons to the acre, and what could be done on one two square rods might surely be done on 80 of them, if it would pay to do it. But, as a rule, enormous crops cost more than they come to, and it is the medium-sized crops that are the most profitable, and any farmer may be well satisfied with 36 tons of good sound roots to the acre, which is equivalent to the feeding of six cows for a period of six months of three cows for a year.

Another excellent root, and even more excellent than all the rest, is the sugar-beet, with 18½ per cent of solid dry matter, of which 15 per cent. is carbo-hydrates and 1 per cent is albuminoids; and which thus makes an exactly complementary food for clover hay and wheat bran, or cotton-seed meal, all together forming a perfectly nutritious, complete and well-balanced food. Of this root there are two kinds, the small French sugar-beet, extremely rich in sugar, so as to be a tempting morsel to the village boys on their way to or from school, and the larger improved sugar-beet produced by the Hon. Henry Lane, of Cornwall, Vt., after many years of cultivation and inbreeding, so to speak, and which has yielded 35 and 40 tons to the acre.

All the best tribe require the same sort of cultivation. A light, warm, sandy loam made rich with well-rotted compost, and reinforced by 500 pounds per acre of the special beet fertilizer and 350 pounds per acre of salt, these huge roots revel in; they grow so fast that they cannot make their way into the soil, and so make their way out of it, standing in all sorts of grotesque and comical ways—upright, leaning, and nodding to each other, twin roots trying to divorce themselves, and roots separated trying to embrace each other, but all stout and robust and ruddy, doing their best to make the farmer look as comical as themselves. To reach this result we must plow the ground early in April, and harrow; mark it out with a furrow-marker made of 4 strips of 2 by 8, 12 feet long plank set on edge 30 inches apart, and connected by 3 cross-pieces gained in and firmly spiked on to the upper edge, and attached to a draught-pole well braced, that the machine may not wabble, but go steadily and evenly, and mark out straight lines across the field; then we open a furrow right on each of these lines, deep and broad, and returning on it double it and make a double open furrow. Then we should drop the manure in this furrow and cover it with the plow, returning the soil previously thrown out, and thus we form a sort of ridge. After this has settled a few days, we must go with seed planter—the handy little "monitor" seed drill is the one we use—and drop the seed immediately over the manure and cover it and roll the surface at one operation. Then the fertilizer is scattered along the ridges and all between, and left for the rain to carry it down to the roots, which will spread from row to row and meet each other. When the slender little twin leaves appear we must run on each side of them with the Planet Junior hand wheel-hoe, which scrapes the soil on each side of the plants within an inch of them, and kills the weeds and loosens up the ground. By and by we run through the middle with the Planet horse-hoe, a sort of universal tool, which stirs and plows the middle, and either scrapes the soil from the rows or throws it to them, just as we may wish to do. But the four pounds of seed we have sown to the acre is four times too much, but necessary to secure a close and even stand, and the excess of plants is to be cut out with a hoe, leaving spaces of 14 inches between the plants, or if one wants a few extra roots to win a premium with at the fair, let him leave them 30 inches apart and add a little extra fertilizer, or give them a little liquid manure once a week. Last of all, let the cultivator be kept going, not to keep the weeds down so much, but to stir the soil and let in the warm rays of the sun and the rain, which will carry the fertilizer down to the feeding roots and fill the great main tap root of the plant, not one-tenth so large and heavy as its root, with rich sap, and force a rapid and healthful growth. By this manner of growing roots, the crop will bring no disappointment with it; but if three acres are grown one may be sure of finding the bulk of the feeding for a dozen or fifteen cows for at least 200 days; that is from the first of November to the end of May.—*H. Stewart in N. Y. Tribune.*

GREEN MANURES.

Dr. Alfred L. Kennedy, the chemist and geologist of the Pennsylvania State Agricultural Society, has issued an address "To the Farmers of Pennsylvania" on the subject of the use of green manuring. We think that we may be doing a service to the agricultural interests of the State by giving this circular a place in this department, though it occupies more room than we have just now to spare, and at the same time to call the attention of farmers to the propositions it embraces:

"In many parts of the State the fertility of the soil is economically increased, by sowing it down with red clover and plowing under the crops. The crop which follows next, frequently finds in the decaying green manure the fertilizing materials it needs, and finds them, too, in the form most readily assimilable. Large tracts of land, both in Pennsylvania and Maryland, have, at a comparatively small cost, thus had their fertility so far restored, as to be made productive. In many respects the red clover is admirably adapted to the purpose. Two seasons are, however, often required before it is sufficiently well-rooted and grown, to be plowed under with the greatest advantage.

"On the continent of Europe the yellow lupin is preferred in green manuring. It is a vigorous grower, and it matures in one season. Here, as well as there, it attains a height of over two feet, sending down its strong tap-root to an equal distance, penetrating the subsoil, and bring to the surface fertilizing agents lying below the the reach of the plow. To these qualities it adds the yet more valuable one of producing a foliage more than *eleven and three-quarters per cent.* (11.79) *richer in nitrogen* than the red clover.

"Nitrogen in the soil is indispensable to our crops. Applied to them, as it is in the form of nitrate of soda and Peruvian guano, it is the most costly of chemical fertilizers. The plants which, like the yellow lupin, gather it and store it up, must, *under certain conditions*, be the most valuable of green manures.

"To determine what these conditions are, is so important to our agriculture, that to do so would be one of the first duties of American agricultural experiment stations, were they multiplied and organized. At present they are too few and too isolated to render the results of their 'soil tests,' etc., truly valuable to the mass of our farmers, whose locations, soils, subsoils, and atmospheric and other conditions differ so widely. Fortunately every county in the State contains farmers who are perfectly competent to determine by experiment the *comparative value of green manures*, and they are cordially invited to aid in settling the interesting question of the relative advantages of the red clover and the yellow lupin.

"These advantages are to be ascertained through the effects which the green manures have upon the crop of grain which immediately follow them. A portion of a field which was last year in corn, and which this year is to be put in oats or potatoes, will be found convenient, and that portion will not be thrown out of the regular order of crop rotation.

"Measure off for experimental purposes one-fourth of an acre, uniform in quality and exposure, and plow and work it as one 'land.' Forty-five by two hundred and forty-two feet will be a good proportion, divided into two plats twenty-two and a half by two hundred and forty-two feet, being one-eighth of an acre each. In April, sow or drill one of them marked No. 1, with the quantity of red clover usual in the neighborhood, noting the quantity; the other plat, marked No. 2, with one-eighth of a bushel of yellow lupin. Promptly after each crop comes into full flower, plow it under to a uniform depth, and when the field is being made ready for the fall grain, whatever harrowing and rolling other portions receive should be given also to the plats, and the whole field be similarly seeded. If a potassic or phosphatic fertilizer be used on the plants, the fertilizer must be absolutely free from nitrogenous matter and must be carefully applied in equal quantity to each, time and quantity to be entered under 'Additional Remarks.'

"Next spring (1883) carefully stake off the middle *eighteen feet* of each plat, making each central plat eighteen by two hundred and forty-two feet, or exactly one-tenth of an acre. At harvest, begin by cradling, binding and cleaning up the space between the two central plats, and also that outside of them, and then reap the plats. Thresh and clean the product of each separately, and note the weight and measure of the grain, and the weight of the straw in each.

"A blank form accompanies the foregoing to fill up as the experiment progresses, which can be obtained by application to Dr. Kennedy, at the Polytechnic College, Philadelphia. By noting the details accurately a vast fund of information can be obtained, which may prove of great value to this important branch of industry.—*Germantown Telegraph.*

OUR LOCAL ORGANIZATIONS.

LANCASTER COUNTY AGRICULTURAL AND HORTICULTURAL SOCIETY.

A stated meeting of the Lancaster County Agricultural and Horticultural Society was held in their room in City Hall, on Monday afternoon, June 5th.

The following named members and visitors were present:

Messrs. Henry M. Engle, Marietta; John C. Linville, Gap; Casper Hiller, Conestoga; James Wood, Kirk's Mills; Simon P. Eby, city; C. L. Hunsecker, Manheim township; P. S. Reist, Lititz; John H. Landis, Manor; William H. Brosius, Drumore; W. W. Griest, city; J. M. Johnston, city; Levi S. Reist, Manheim; Peter Hiller, Conestoga; Frank Griest, city; Eph. Hoover, Manheim.

The president and secretary being absent, Vice President Henry M. Engle took the chair, and John C. Linville was appointed secretary pro tem.

New Members Elected.

John H. Landis, of Manor, proposed for membership Washington B. Paxson and Francis N. Scott, of Colerain township, and both were elected.

Crop Reports.

Casper Hiller, of Conestoga, reported the cherry crop almost a failure, the pear crop not much better, the apples very thinly set, the peaches more promising, grass rather thinly set but healthy looking, giving promise of a good crop of hay.

Peter S. Reist reported the wheat in Warwick and Manheim as very promising, possibly a little too rank in growth; grass in general looks well; new clover not so well; cherries, except in low-lying places, good; currants greatly damaged by worms, both on the leaf and at the root; other fruits promise a fair average.

John H. Landis said that in Manor township, the wheat never looked better than it did a week or ten days ago, but now it is growing too rank and beginning to lodge, and the straw, near the ground, is getting black; the grass looks well; there are indications of a full fruit crop; apples and peaches are plentiful, though the cherries are not very full.

John C. Linville, of Salisbury, reported wheat as growing very rank, with straw full of sap, and therefore liable to rust if the weather should become hot; oats looks better than he has seen it for twelve years past; the grass is good as far as it goes, but is short and in some places thin; peaches and cherries are nearly all killed by unfavorable weather; potatoes plenty and so are the potato bugs.

James Wood, of Little Britain, said that in his neighborhood the fruit trees did not blossom well and there would not be much fruit; wheat is strong and healthy; oats don't look so well; potatoes coming up nicely; corn healthy but backward in growth for this time of the year; clover well set.

Wm. H. Brosius, of Drumore, said that Mr. Wood's report will answer for Drumore—wheat fine; grass fair, but fruit unpromising.

H. M. Engle, of Marietta, said the wheat looked remarkably well, but some of it is beginning to lodge. Its outcome will depend on the condition of the weather for the next four weeks; if it should remain cool all will be well; but if it should become hot and dry the crop will suffer. The grass is thin in many places; the corn well set, owing largely to the superiority of the seed; the heavy rains, however, have packed the ground and the cold weather retard its growth. Potatoes look well, but there are a great many bugs. In regard to the controversy had some months ago as to whether exclusively cold weather would destroy the eggs and larvæ of insects, Dr. Rathvon had taken the position that it would not affect them. Mr. Engle was inclined to think that it would destroy some species, but not the iron clad potato bug. He noticed that the cabbage worm and circulio are very scarce this year and he attributes their scarcity to the cold and wet weather. Peaches and apples are fairly promising, but many of them are dropping from the trees, and there will not be a full crop; cherries are very poor, there not being one-tenth of a crop; all kinds of small fruits look exceedingly well. The rainfall for the month of May was nearly five inches.

Pruning Apple Trees.

John C. Linville read the following essay:

Now is the right time to prune the apple trees if we want the wounds to heal over quickly. If the wounds are large it is well to cover with grafting wax or varnish, to exclude the sun and drying winds. It is seldom necessary to remove large limbs if the trees have been properly cared for at the start. I think there is more harm done by too much than too little pruning. Vigorous growing trees are very impatient at having part of their branches cut away. Dormant buds along the upper side of the limbs push out into "suckers." These have to be removed again and again, until finally the bark along the top of the limbs dies and the tree is ruined. I have always met with this trouble in attempting to thin out dense headed trees. Great care should be taken in summer pruning not to jar the bark loose.

In old orchards there will every year be found some dead limbs that must be cut away. This had better be done in winter, but may be done now. In many orchards may be seen unsightly stumps of limbs left from six inches to a foot in length. I have never heard any reason given for this mode of pruning. It is pitiful to see nature making desperate efforts to cover over these deformities, the "sap wood" creeping year after year further and further out on the useless member, until finally the dead stump decays away and leaves a hole in the trunk of the tree. This makes a capital place in which the flickers may build their nests, but it is death to the apple tree. Dead limbs should be sawed off at the shoulder, and they will heal over even if four or five inches in diameter.

In order to cut off large limbs neatly the saw should be put in good order. After the saw should be laid on the work bench and a flat file run from heel to point along each side of the teeth. This prevents the points of the teeth from scratching the wound and leaves it smooth.

Casper Hiller said there can be no particular rule laid down for pruning; one kind of tree will require one method and another kind another. In pruning apple trees, his plan is to commence when they are young and cut off the tops so as to keep the tree low. Ordinarily the large limbs if cut off close to the shoulder and painted to prevent the escape of sap; but if the tree is old the limbs may be cut off a few inches from the shoulder; the stump will then decay gradually for several years and not affect the tree until it has become too old to be useful. He thought an orchard ought not to be allowed to stand more than thirty years, and that a young orchard might be safely planted on the same ground by placing the young trees in rows between the old ones, removing the latter when the former comes into bearing.

John H. Landis took exception to Mr. Linville's statement that too much pruning was worse than too little, especially as applied to peaches. If the trees are allowed to go unpruned the fruit is sure to be small and scrubby.

The question was further discussed by S. P. Eby, Esq., Peter S. Reist, Levi S. Reist, Henry M. Engle, and John C. Linville.

When to Cut Grass.

In answer to a question referred to him at last meeting, Casper Hiller answered as follows:

I am not prepared to answer the question: "At what period of growth should grass be cut to make the best hay?" from actual test of the feeding quality of hay made at different periods of cutting. If my opinion would be of any value, I would say that the proper time to cut is when the plant is past full bloom, and up to the time when the seed is half formed. When grass and clover are cut too young they are too watery, make no weight, and are difficult to cure. If left until the seed is ripe the plant becomes woody and will lose its best feeding quality. I believe it has been satisfactorily shown that sorghum contains the most saccharine matter if cut when nearly ripe, and I think the same principle holds good with the grasses.

Henry M. Engle said that owing to a misapprehension he thought the above question had been referred to him for answer, and had accordingly prepared a paper on the subject, which if there was no objection he would read.

He read it as follows:

In order to answer this question satisfactorily I present it in an essay rather than in a few verbal remarks. According to statistics (which we have no reason to question) the value of the grass crop of this county exceeds that of any other crop. It requires but a small amount of either loss or gain to each farmer in utilizing it to best advantage or otherwise to swell the aggregate to millions. The period to cut grass for hay in order to realize the greatest value therefrom is the subject of a wide difference of opinion, much wider indeed than should exist in the use of so important an article during thousands of years. It is, however, interesting to know that for some years past practical tests have been made in Europe and also in this country by chemical analysis as well as feeding tests, which seem to agree that grass cut in bloom will make the most valuable hay; that in proportion beyond that period until ripe it approaches woody fibre and consequently loss of important nutritive qualities. My own experience would dictate to err (if error it be) by cutting a little earlier than later, i. e. before full bloom, especially clover.

The difficulty of curing is claimed as a strong objection to cutting early, and is perhaps the leading cause of error on the other side; the result is, a very large proportion of hay not much better than straw. As proper curing is as important as the proper time of cutting, I would add that possibly one half of the hay made (in this section at least) loses much of its value by being gathered too dry. The old adage "make hay while the sun shines" I believe to be applied to excess. Were it always practicable hay as well as herbs and seeds would be better if dried in the shade and housed or stacked as damp as will allow, only so as not to mould or mow burn, in fact a little of the latter is preferable to housing it in a very dry and brittle condition.

Although grass is more difficult to cure when cut young than when more nearly ripe, in the former condition it will bear much more rain without injuring it, than when cut nearer ripe. The difference in bulk is in favor of late cut grass, but the difference in weight is trifling, besides stock will relish the early cut much better than late cut, and fed to milk cows, the butter will have a richer color from early than from late cut grass. Another important consideration applies to grasses which make second growth, is that early cutting does not stunt the plant, so much, consequently the second growth will be of so much more value as to outweigh any amount of weight gained by delay in cutting the finest crop. My experience and observation of forty years has confirmed me so strongly in favor of early cutting of grass that I do not hesitate to recommend a trial to all who have any doubts of its advantages, and would suggest that as many members of this society as can to test the matter fairly by cutting some early and other a week or ten days later, and the coming winter feed to milk cows each kind alternate two or three weeks at a time and report results to this society, after which we may talk more intelligently on this important subject.

Peter S. Reist, S. P. Eby, esq., John C. Linville, James Wood and C. L. Hunsecker, discussed the matter further, reviving some pleasant recollections of the old mode of curing hay, and comparing its advantages with the present mode.

White-Marked Tobacco.

Prof. Rathvon sent to the meeting several leaves of tobacco upon which there were very pretty and curiously wrought lace-like tracings. [The leaves were presented by Wm. Roeting, of Elizabethtown, and Morris Gershel, of this city, and a description of them formed the subject of a paper by Dr. Rathvon which has already appeared in the INTELLIGENCER—REPORTER.]

Yellow Lucan.

A printed circular from Alfred M. Kennedy, of the State Agricultural Society, in which he recommends farmers to try the experiment of substituting yellow lucan for red clover, was read; but as the season is too far advanced to make the experiment, no action was taken.

Double Peaches.

Casper Hiller brought to the meeting several twigs pulled from peach trees, on which there were a great many remarkable growths of double peaches. Other members present stated that they had noticed the same phenomenon in their several neighborhoods. This abnormal growth was attributed to the exceptionally dry and hot weather of last autumn.

Books for the Library.

Mr. Engle presented to the library bound copies of the last Pennsylvania agricultural reports; annual report of the Michigan Pomological Society, and the sixth annual report of the Agricultural Society of Kansas.

Ajourned.

POULTRY ASSOCIATION.

The regular monthly meeting of the Lancaster County Poultry Society was held on Monday morning June 5th.

The following members were present: J. B. Lichty, F. R. Diffenderffer, city; George A. Geyer, Florin; M. L. Grider, Mount Joy; J. M. Johnston, city; Dr. E. H. Witmer, Neffsville; Charles Lippold, W. W. Griest, city; T. Frank Evans, Lititz; John A. Schum, city; E. C. Brackbill, Strasburg; L. H. Brooks, Martieville.

The meeting was called to order by President Geyer.

The minutes of the last meeting were read and approved.

On motion, John E. Schum was elected treasurer by acclamation.

Charles E. Long was elected to fill the vacancy created in the Executive Committee by the transferance of Mr. Schum to the treasurership.

The time for holding the next society show was taken up and discussed.

On motion, Thursday, the 11th of January, was chosen as the opening day of the exhibition, to continue during a period of six days, namely the 11th, 12th, 13th, 15th, 16th and 17th of the month.

William Powden was nominated and elected to membership.

On motion of J. M. Johnston, F. R. Diffenderffer was requested to prepare a paper on gapes in chickens for the next meeting.

There being no further business, the society adjourned.

LINNÆAN SOCIETY.

April Meeting.

The Linnæan Society held a stated meeting at the residence of Dr. H. D. Knight, North Queen street, on Thursday evening, April 26, 1882, Prof. Stahr occupying the chair.

The Curators reported no donations to the Museum during the month.

The librarian reported the following donations to the library: Annotated list of the birds of Nevada, by W. I. Hoffman, M. D., 256 pp. octavo, with a map. Donated by the author. Bulletin of the United States National Museum, 264 pp. octavo, with a large folded map—a guide to the flora of Washington—from the Department of the Interior. Report of the Commissioner of Education for 1879. Same. Internal Commerce of the United States for 1880. Same. Production of gold and silver in the United States, 1880. Same. Transportation routes of the seaboard, Nos. 10 to 15, *Official Gazette* of the United States Patent Office, volume 21. From the Department of the Interior. Volume 12, Congressional Record, from Hon. A. Herr Smith, M. C. Lancaster *Farmer* for April, 1882. Two envelopes containing twelve biographical, historical and miscellaneous "scraps." Sundry catalogues and circulars.

E. K. Hershey, of Creswell, was balloted for and unanimously elected an active member.

S. M. Sener was proposed for associate member, which, under the rules, is laid over to the next meeting for definite action.

The treasurer was authorized to subscribe for "Ward's Quarterly Bulletin of Natural History."

Dr. Knight kindly tendered his office for the use of the society in holding its evening meetings, which was thankfully accepted.

Mrs. Zell read some notes on technical terms, which elicited quite a spirited discussion, participated in by Prof. Stahr, Dr. Davis and others.

After an examination of Dr. Knight's objects of *vertu*, and continued scientific gossip, the society adjourned to meet in the ante-room of the Museum on Saturday afternoon, May 27, 1882.

May Meeting.

The society met on Saturday afternoon, May 27, 1882, in the ante-room of the Museum, the president, Prof. J. S. Stahr, occupying the chair.

After the usual preliminaries the Curators reported the following donations to the Museum : Two specimens of the "Sea Robin" or "Flying Gurnard," represented to have been caught in the Susquehanna river and Conestoga creek, but which were identified as *Prionotus carolinus*, of Dekay—marine fishes, which inhabit the Atlantic Coast from Nantucket as far south as the Carolinas. As a specimen of the "common sturgeon," *Acipenser sturio*, was caught in the Susquehanna river some years ago, near Bainbridge, in this county, it is not *impossible* the "Gurnards" may pass from the ocean to the bays, and from the bays into the rivers, although it is not very probable. One fine specimen of well-defined graulis, presumably from the "Granite State." Sundry botanical specimens for determination.

Donations to the Library : Volume 3 of *American Entomologist*; No. 1, volume 1, of the *Pennsylvania Farm Journal*, dated April, 1881, which originated in Lancaster city thirty-one years ago, under the auspices of A. M. Spangler; Nos. 17 to 20, volume 20, of the *Official Gazette* of the U. S. Patent Office ; circulars 1–6, from the Bureau of Education, Department of the Interior ; two book catalogues, and sundry circulars ; the Lancaster FARMER for May, 1882 ; five envelopes containing forty-five historical, biographical and scientific scraps.

S. M. Sener, Esq., was unanimously elected an active member of the society.

After indorsing a small bill reported by the Curators, the meeting passed under the rule of "Science Gossip" for a brief period, and then adjourned to meet at the rooms of Dr. H. S. Knight, North Queen street, Lancaster, Pa., on the last Thursday evening in June (29th), 1882.

AGRICULTURE.

Rotation of Crops.

In a well planned system of farming, the subject of crop rotations should be carefully considered, as one of the essential elements of success in its highest and best sense. It seems to be the prevailing opinion that the alternation of crops, in systematic order, is a modern invention that was gradually developed as a direct result of the application of science to the art of agriculture. The early writers on agriculture, even from the time of the Romans, have, however, quite uniformly urged the advantages of a succession of crops from the teachings of experience. They were satisfied that a variety of crops grown in succession, all other conditions being equal, would give a greater aggregate yield than could otherwise be obtained. The reasons for the success of the system could not, it is true, be given, but practical men were fully agreed in urging its importance, and many systems of rotation, more or less perfect, were planned, some of which became the prevailing rule of farm practice in particular localities. That these practical rules of alternating crops of different habits and modes of growth are based on correct, but not explained, principles, has been shown by direct experiment.

Manure Made Under Cover.

Of course all the advantages of making manure in covered yards may be secured by box feeding, with less outlay for roofing, since no more space must be allowed for a given number of animals turned loose together than when confined in stalls. It is the protection from rain and sun, the abundant use of litter, and its thorough incorporation with the excrements and the exclusion of air by compact treading, which go to make the superior manure; all these features of the method work against the loss of valuable plant food. Nor does box feeding and constant accumulation of the manure under the feet of the animals necessarily imply offensive stalls. Mr. Lawrence said that everybody noticed the general sweetness of his stalls. It is only essential that enough litter be used to absorb all liquid, and this absorption is more effectual if the straw is cut up. One method or the other, box-feeding or covered yards, should be adopted by every farmer who lives where manure is worth saving, and who finds himself compelled to supplement his stable manure with commercial fertilizers. Stable manure must not be lost sight of in this increasing interest to those concentrated fertilizers, for we cannot produce our crops and have enough for ourselves and others without its aid ; and there is nothing in all the list of commercial mixtures which gives so good an average return for the money invested in it as well made stable manure.—*American Agriculturist.*

Exports of Breadstuffs.

The following shows the decrease in our exports of breadstuffs for the eight (8) months ending February 28, 1882, as compared with the same time last year :

	1882.	1881.	Decrease.
Flour, bbls.	3,863,474	5,307,432	1,443,958
Wheat, bush.	85,913,154	107,659,416	21,746,262
Corn, "	37,048,841	58,770,782	21,921,942
Rye, "	502,098	1,587,578	1,085,480
Barley, "	172,525	883,376	661,050
Oats, "	440,473	268,564	*171,817
Corn Meal, bbls.	214,194	265,147	50,953

Of the above we reduce wheat and corn to bushels which give us a total decrease in our exports of over forty-eight million bushels (*excepting oats) and a total of over forty-seven million dollars.

Corn Culture in Gardens.

In field culture corn is planted in hills. Some have tried growing it in lines or drills, and have obtained more corn. There is not the same chance for three or four plants feeding together in one hill, that there is for a single plant alone, and with nothing nearer to it than a foot or so. Three plants a foot apart will give more corn than three plants in one hill three feet from another hill. This is not only reasonable but has been verified by actual facts. But the increased crop does not pay. The horse-hoe cannot work but one way when the corn is in drills, and then the horse has to be idle in the stable while the driver takes the slow hand-hoe to clean out the weeds in the row. But the field-practice, proper enough in the field, has been carried to the garden, and sweet corn for the table is treated just as if it were a field crop. In gardens where hand labor is exclusively used, there is no reason whatever for growing corn in hills. One can have better sweet corn by sowing in rows than in hills, while the labor is in no wise any more.

To insure a constant supply of sweet corn for the table, there should not be less than *seven* different plantings through the season. The first-planting (of dry seed) should be made not earlier than the 20th of April ; the next planting ten days after, and then follow the five other plantings from ten to twelve days apart, the last being toward the end of June.

HORTICULTURE.

An Abundant Apple Crop.

During a recent ride through a large portion of the State, we could not fail to be struck with the enormous masses of apple blossoms wherever there were trees. The promise is of a great crop and vast surplus. Many thousand bushels will be wasted if the fruit ripens as well as in other seasons, unless efficient efforts are made to secure a foreign market, and to manufacture large quantities at home into evaporated fruit, or into apple jelly or vinegar. [This report of an abundant crop refers to New York State. In Pennsylvania, or at least in this part of the State, the promise for apples is far less encouraging].

It is to be hoped that the supplies which are sent to Europe will be selected and put up in a manner creditable to the fruit growers of this country, and that some means may be devised to distinguish such growers and shippers as do the work in honorable and skillful manner, from those who by carelessness and fraud do a lasting injury to themselves and others, as was too frequently the case in 1880.

A question of importance is asked many times in this connection, "How can we prevent this uneven bearing in alternate years, so as to have a fair supply every season ?" In answer, there are three remedies. One is to cultivate the ground well, so as to keep up the vigor of the trees to such a degree that the abundant crop the even year will not exhaust the trees and prevent bearing the odd year. This remedy, although operating more or less in all cases, is uncertain or incomplete. The second, manuring the trees at the right time, is more efficient. The best time to apply the manure—which must, of course, be broadcast—is on the surface in autumn or during winter, giving the trees when they start in spring so much vigor that the abundant crop the same year will not check the force of the trees and prevent bearing the following season. If the manure is applied in spring or early in summer and worked into the surface soil, it would have a similar tendency in less degree. Liquid manure, applied now to the whole surface through a sprinkler, would probably answer nearly as well as winter-spread manure, and is well worth trying by those who have facilities for this purpose. The third, and most certain way of changing the bearing year, is to prevent a crop this season, by which all the strength of the growth will be thrown into the young shoots for a crop next year. The best time is when the trees are in blossom, because they are easily seen ; and the best tool for the purpose is a pair of common sheep shears. The work should be done when the trees are young, for two reasons—the labor is much less, and the change is more likely to be permanent. We find that it requires a man three hours to shear off all the blossoms from a tree fifteen years old and large enough to bear twelve bushels and only one hour for a young bearing tree seven or eight years old. The value of the crop on either, during a scarce year, is much more than the cost of the labor. The work may be done when the young apples are as large as cherries, but they are not so easily seen as the blossoms.—*Country Gentleman.*

What Kills Fruit Trees.

Deep planting is one error. To plant a tree rather shallower than if formerly stood is really the right way, whilst many plant a tree as they would a post. Roots are of two kinds—the young and tender rootlets, composed entirely of cells, the feeders of the tree, always found near the surface getting air and moisture, and roots of over one year old, which serve only as supporters of the trees and as conductors of its food. Hence the injury that ensues when the delicate rootlets are so deeply buried in earth. Placing fresh or green manure in contact with the young roots is another great error. The place to put manure is on the surface, where the elements disintegrate, dissolve and carry it downward. Numerous forms of fungi are generated and reproduced by the application of such manure directly to the roots, and they immediately attack the tree. It is very well to enrich the soil at transplanting the tree, but the manure, if to be in contact with or very near the roots, should be thoroughly decomposed.—*Massachusetts Ploughman.*

Early Turnips.

The earliest and perhaps the best variety of turnips for table use is the Early Flat Dutch. It is universally popular, and it takes only a small plot to furnish a supply for a medium sized family. One reason why they so frequently fail in gardens is the richness of the soil and their frequent growing in the same bed. In preparing a plot for turnips dig down full spade deep, for the purpose of getting some of the virgin earth, and especially a little clay. As a fertilizer there is none equal to bone dust, and nothing else. The turnip should grow slowly, with as little top as possible. It will not bear pushing or forcing.

HOUSEHOLD RECIPES.

BEEFSTEAK ROLLS.—Cut a beefsteak quite thick, then split it open lengthwise, and cut in strips four or five inches wide; rub over the inside with an onion, and in each strip roll up a thin slice of bread, buttered on both sides; stick two cloves in the bread, and sprinkle some salt, pepper, celery seed (cut or thin slices of nice celery stalk if in season), and put into the gravy. Tie each roll with a thread; dredge it with flour, and fry in hot butter. Then put these, when a delicate brown, into a stewpan, with only water enough to stew them. Make a nice thickened gravy from the liquor in which the steaks were stewed, and serve with the rolls, very hot. The rolls should stew slowly two hours. Veal or mutton is good prepared in this way.

DEVILLED HAM.—One pint of boiled ham chopped fine with a good proportion of fat, one tablespoonful of flour, one half cup of boiling water. Press in a mould and cut in slices.

YANKEE PLUM PUDDING.—Take a tin pudding boiler that shuts all over tight with a cover. Butter it well. Put at the bottom some stoned-raisins, and then a layer of baker's bread cut in slices, with a little butter or suet, alternately, until you nearly fill the tin. Take milk enough to fill your boiler (as vary in size), and to every quart add three or four eggs, some nutmeg and salt, and sweeten with half sugar and half molasses. Drop it into boiling water, and let it boil three or four hours, and it can be eaten with a comparatively clear conscience.

FRENCH BEEFSTEAK.—Cut the steak two thirds of an inch thick from a fillet of beef; dip into melted fresh butter, lay them on a heated gridiron and broil over hot coals. When nearly done, sprinkle pepper and salt. Have ready some parsley chopped fine and mixed with softened butter. Beat them together to a cream, and pour into the middle of the dish. Dip each steak into the butter, turning them over, and lay them round on the platter. If liked, squeeze a few drops of lemon over and serve very hot.

SQUASH PIE.—Make the same as pumpkin pie with the addition of one egg to each pie.

DELIGHTFUL PUDDING.—Butter a dish, sprinkle the bottom with finely minced candied peel, and a very little shred suet, then a thin layer of light bread, and so on until the dish is full. For a pint dish make a liquid custard of one egg and one-half pint of milk, sweeten, pour over the pudding, and bake as slowly as possible for two hours.

TO MAKE TOUGH MEAT TENDER.—Soak it in vinegar and water; If a very large piece, for about twelve hours. For ten pounds of beef use three quarts of water to three-quarters of a pint of vinegar, and soak it for six or seven hours.

CABBAGE SALAD.—Shave a hard white cabbage into small white strips; take the yolks of three well-beaten eggs, a cup and a half of good cider vinegar, two teaspoonfuls of white sugar, three tablespoonfuls of thick cream, one teaspoonful of mustard mixed in a little boiling water; salt and pepper to suit the taste. Mix all but the eggs together and let it boil; then stir in the eggs rapidly; stir the cabbage into the mixture, and stir well. Make enough for two days, as it keeps perfectly and is an excellent relish to all kinds of meats.

The regulation French salad dressing is composed of three parts of salad oil to one of vinegar, with a palatable seasoning of pepper and salt.

SCALLOPED OYSTERS.—Crush and roll several handfuls of Boston or other friable crackers. Put a layer in the bottom of a buttered pudding dish. Wet this with a mixture of oyster liquor and milk slightly warmed. Next have a layer of oysters; sprinkle with salt and pepper, and lay small bits of butter upon them. Then another layer of moistened crumbs and so on until the dish is full. Let the top layer be of crumbs, thicker than the rest, and beat an egg into the milk you pour over them. Stick bits of butter thickly over it, cover the dish, set in the oven, bake half an hour. If the dish is large, remove the cover, and brown by setting it upon the upper grating of the oven or by holding a hot shovel over it.

ROAST SHOULDER OF VEAL.—Time, twenty minutes for each pound. A shoulder of veal some oysters, or mushroom sauce. Remove the knuckle from a shoulder of veal for boiling and roast what remains as the fillet, either stuffed or not with veal stuffing. If not stuffed, serve it with oysters or mushroom sauce, and garnish with sliced lemon.

WESTERN COOKIES.—One cup of sour milk, one cup of powdered sugar, a little salt, one teaspoonful of soda, mix as soft as possible; roll thin, sprinkle with sugar; slightly roll out, and bake in a quick oven.

FAIRY APPLE.—Bake ten nice tart apples. When soft, remove skins and cores, and mash fine with a silver or wooden spoon. While hot, add the white of one egg beaten to a stiff froth, and beat one minute. Place in a glass dish and pour over it a soft custard made of the yolk of the egg, one teaspoonful of corn starch, three tablespoonfuls of white sugar and one pint of milk. Pour this over the apple, flavor or not, as you like. Serve cold. This is nice for tea.

LIVE STOCK.

Improving the Stock on the Farm.

The season for calves and lambs is about over, and on a majority of farms there will be a surplus to be fitted for slaughter, and sold to the itinerant butcher or huckster. In making the selection of this surplus be sure to choose the inferior ones and keep the best on the farm for breeding purposes. Remember that the young stock grown this year are destined to produce the calves and lambs which you will want to sell a few years hence, and in order to have them average better than this year's product you must retain and breed only from the best. The dollar or so extra offered by the purchaser for the best animals of your lot will prove a temptation to part with them, but it will be money in your pocket to resist it and regard the money as an evidence of the greater worth of the animals to you as breeders. Look upon the differences in price simply as an investment in improved stock. Pursue this course constantly, from year to year, and you will effect a gradual improvement in your stock which will in time result in your having none but "the best" to offer. If, on the contrary, you suffer the young animals to be carried away and slaughtered year after year, you will just as certainly be depreciating the quality of your stock, and approaching the time when you will have none of "the best" to offer.—*Agricultural Epitomist.*

Keep Up the Flow of Milk.

The month of August, is perhaps, the most trying of the year to the dairyman who has determined to keep up an even flow of milk during the whole season. And all dairymen who have studied the profits of the business know that any large falling off to the flow of milk in midsummer is seldom or never recovered. This reduction of milk will largely affect the whole yield, and thus the profits of the whole season. Special green crops of corn, millet, etc., are seldom ready early enough to reach an August shrinkage in pasture. Those who make a specialty of clover will be best able to meet the short pasture with green food. A second crop of clover will be just in its glory, and when that is fed, corn, millet, etc., may be ready. As cows can only produce a large yield of milk on full feeding, let the dairyman be as liberal to his cows as he desires to be rewarded by them in return. When the second cutting of clover and green corn are both ready at once, they should be both fed together; for the clover is rich in the nitrogenous element, which is deficient in corn; they compliment each other.

Care of Dairy Vessels.

Prof. Arnold has the following to say about the absolute cleanliness required in dairy utensils and the influence of such neatness upon the quality of the butter :

"It is hardly necessary to say that wherever the finest butter is made the milking is done in the most cleanly manner. It is so neatly done that straining is of very little use; it might even be dispensed with but for the occasional dropping of a stray hair. Whoever places much dependence on the strainer for securing clean milk will never make gilt-edge butter. Allowing dirt to get into the milk, and then depending on the strainer to get it out, is a poor apology for cleanliness. More or less of the dirt, especially everything of a soluble nature, and some that is not, is sure to find its way through the meshes of the strainer with the crowded current of milk. The practice of using one cow's milk to wash the filth collected from another cow's milk, as is frequently done by continuing to strain mess after mess through the same strainer without cleaning, does not contribute anything toward gilt-edge, and is not allowed where the best butter is made. Then the tin pails for I notice wooden pails are not used where I find the best butter) and all the vessels used for holding or setting milk are kept scrupulously clean. When used, they are not left for the milk, and particularly the milk sugar, to dry and form a gummy coating to serve as a reservoir for infection, and which it is difficult to get off. They are attended to promptly, rinsed in cold water, washed in warm and scalded in water actually boiling hot, to avoid contamination from a sour dish cloth, are left to drain and dry without wiping. They are kept bright by scouring with salt, and as a protection against greasy and infectious matter sal soda is employed instead of soft soap, which, though it may possibly be clean, is generally too filthy to be used about milk vessels, to say nothing of the injury it does to tinware from the potash it contains.

Raise the Good Cow's Heifer Calf.

A large majority of dairymen have cows in their herds that do not pay their keeping; and, as they do not apply a test to individual cows, they continue not only to keep them, but to breed from them. This is a most suicidal policy. Although we strongly recommend dairymen to raise their own cows, we are far from advising them to perpetuate their poor cows. It would be even better policy to give them away to some favorite brother-in-law. The heifer calves from only the best cows should be raised, and the weeding out should go on still further. When there heifers come into milk, those that do not come up to the proper standard should be discarded. A careful test should always be made of each cow in the herd and of each heifer during her first period of milking. If the heifer has the appearance of a well-formed milker and of having had a good dam, it may not be judicious to pass upon her during her first milking season if her quality is below the standard, for the next season may develop her satisfactorily.—*Live Stock Journal.*

POULTRY.

One Variety.

As a rule, one variety of fowls is enough for almost any person to manage successfully and profitably, and this is especially true with beginners, who have to gain their experience in all the varied details of poultry management. If a breeder has been successful with one variety, has not merely made good sales, but has produced birds of such a high order of merit that the stock makes a good advertisement, and a permanent one, for the breeder, it can be taken for granted that it will pay to take up one or more breeds, provided the same care is bestowed upon each variety as was formerly accorded to the single breed, and provided there are ample conveniences, room and quarters for them. It seldom pays to attempt raising pure bred poultry, and several varieties, unless there is ample room, both in yards and houses, for they must have this to insure their healthfulness and consequent profit.

Treatment of Young Ducks.

I haven't much faith in the maternal instinct of ducks. They have a way of taking to the water with their offspring that is not at all to my liking; and for that reason I generally set my duck eggs under hens, who do not seem at all anxious to go in swimming with their wed footed charges. I never believe in giving a hen all the eggs she can possibly cover. Duck eggs are very large, and five of them are enough to give a small hen; a Brahma or Cochin will cover seven or eight. Duke eggs, like turkey eggs, should either be set on the ground, or on several inches of fresh earth in a nest-box, and should be sprinkled often with tepid water during the last two weeks of incubation. As soon as the ducklings are all hatched, remove the hen to a coop previously prepared for her. If the coop was used last season, it should be thoroughly cleaned with an old broom and hot soapsuds, and then whitewashed inside and out several days before it is wanted. When there is a running stream or pond on the premises, the coop must be placed at a respectable distance from the water, for ducklings are liable to cramps, and must be kept away from cold water until settled warm weather comes.

Novices in duck raising should always remember that turtles, minks, muskrats, stray cats and rats are remarkably fond of young ducks, and take suitable precautions to guard against their depredations. Ducklings should be fed cooked food until they are six weeks old. For the first two or three days after they are hatched they should be fed on boiled eggs and stale bread soaked in milk; afterward almost any kind of cooked food will do. Do not feed too soon after they are hatched; twelve hours is soon enough, and do not feed too much. Young ducks are not overburdened with sense, and if permitted, will eat until they kill themselves. Many promising broods of ducklings are killed by over feeding. If hatched before the young grass and insects make their appearance, ducklings will require an addition to their bill of fare in the shape of green food, and an occasional feed of boiled meat. For this reason I don't think it pays to hatch ducks very early. If hatched in May and the fore part of June, they will attain a good size for w nter market. Give plenty of water to drink, and after they are two weeks old give water that has the "chill" taken off to bathe in.—*Fanny Field.*

A Profitable Hennery.

James Wilson, who resides a short distance from Milford Square, Bucks county, has a two-story hennery, 18 by 20 feet, well lighted and ventilated, with all the appurtenances and conveniences of and in accordance with the most approved plans of buildings of this kind. Mr. Wilson has at present 270 chickens, only eight of which are males. During the severe weather of the last three months his hens have yielded him thirty-eight eggs per week, averaging over five dozen per day. The chickens are in the main of the Leghorn variety. In the morning the feed is a mixture consisting of six quarts of wheat bran and six quarts of chopped oats, or wheat screenings, at noon six quarts of oats and in the evening six quarts of corn. A handful of salt and pepper thrown into the morning feed. Mr. Wilson believes in giving his hens a warm breakfast during the winter season, and always uses warm water in mixing the feed, and frequently gives them a morning mess of boiled turnips and potatoes. The drinking water is also slightly heated. Pounded bones and oyster shells are scattered in the hennery once or twice a week. The chickens are allowed to be out several hours each day, when the weather is not extremely severe. About half a dozen barrels of manure are collected each month, worth several dollars, which with the sale of eggs at the present market price makes a total yield from the hennery of about $70 per month, the net profit being not far from $50.

LITERARY AND PERSONAL.

THE PLANTERS' JOURNAL.—The official organ of the National Cotton Planters' Association of America, which represents all the cotton States. Office of publication at Vicksburg, Miss., with branches at New Orleans, Memphis and Philadelphia, 115 South 5th street. Pronounced by a united press the most important movement of modern times for the South. The May number of this most excellent journal has reached the table of our *sanctum*, and is not surpassed by any publication that comes there. It is a Royal quarto of 36 pages, in illustrated colored covers, and contains a rich fund of knowledge, not only relating to the cotton industry of the country, but also to science, mechanics, agriculture and general literature, and with all endowed with more than ordinary ability. Its illustrations are very fine, and especially the full page picture in this number of "The convalescent," which is one of the prettiest and most characteristic "expressions" we have seen for a long time. The fashion illustration of "Cotton Dresses" is a reflection of living intelligence, and not merely stiff, simpering dollism. The determination of choice between two such beings as are here represented, could only be made by the "casting of the die;" and we are reminded by it of one of the songs of our early days—

"To lady's eyes around, Boys,
We cant refuse—we cant refuse,
Where bright eyes abound, Boys,
'Tis hard to choose—'tis hard to choose."

In the Literary department, we notice a poem entitled "*My Sleeping Pet,*" by Dr. I. E. Nagle; the sentiment of which is beautifully sad, from the standpoint on *this side*, but would be beautifully joyful, when viewed from the *other side* of the boundary line between time and eternity. But, it is the *name* of the author, just now, that elicits our attention more than the poem; for, it carries us back more than half a century. Dr. N., we presume, is the son of an old esteemed friend and "fellow-craftsman" of ours—in the art decorative of all artizans—and many years ago was a resident of Mount Joy, in the county of Lancaster. By a marginal imprint, we observe also that Dr. N. is the *editor* of the journal which is the subject of this notice. He need not be ashamed of the responsible fatherhood of such an offspring.

SOUTHERN INDUSTRIES, devoted to agriculture, horticulture, fish culture, live stock, mining and manufacturing, Rolfe S. Saunders, editor. Published by the Southern Industries Publishing Company, Nashville, Tennessee, at $2.00 a year. Office, 102½ Union street. This (No. 1, May, 1882,) is the novitiate of a demi-quarto of forty-eight pages of characteristic letter press, exclusive of ten pages of advertisements and embellished covers. The editor, in his introductory, says, among other equally important things, "We propose to make the *Southern Industries* rigidly authentic in its matter;" and a realization of such a proposition is just what is needed all over the country, and what we presume every editor and publisher aims to accomplish. It is a great promise, and in scanning the the pages of the journal before us, we think we discover a very encouraging effort in that direction. Nothing speaks, just now, more emphatic in behalf of the industrial progress of the "Sunny South" than the rapid increase of her industrial journals and the ability with which they are conducted.' In point of material, typographical execution, illustration and literary ability they compare favorably with the best journals of the North, if they do not, as a whole, excel them. This number contains ten first-class contributions, from as many able contributors, and if it can command such a continuation, it will not only succeed, but it will deserve success. The editorials are pointed and practical, but "too numerous to mention." A progressive spirit seems to pervade the columns of this juvenile in the ranks of industrial journalism that would do honor to older heads. "Mourn no more over the past, but rise to the times," which the editor suggests "should be the motto of every Southern man," is advice every one should heed, both North and South, who desires to march to the music of industrial progress. The pregnant events of the last twenty years were bound to eventually come, and happy is that man, or that people, in whom exist the elements of a harmonious acquiescence in the inevitable.

"In spite of pride, in erring reason's spite,
One truth is clear, whatever is, is right."

There is nother arbitrary or absolute about this, because it is in perfect harmony with that law of Divine *permission*, through which a lesser evil only transpires in order to prevent a greater evil. The sooner any country can comprehend this philosophy, the sooner it will work out its own political, social, industrial and domestic salvation.

BANNER OF CHOSEN FRIENDS.—"Fraternity, Aid, Protection." Indianapolis, Ind., April 29th, 1882. A supplementary issue of this little sheet, (4 p. 4 to.) No. 7, Vol. 2, has been placed upon our table, which seems to be the official organ of a secret beneficial or insurance organization, now extending over portions of our country—especially in the west—which affords terms more favorable than other societies and companies organized for a similar purpose. The order which this journal represents for the administration of its affairs is divided into *Supreme, Grand* and *Subordinate* Councils, and the following *expose* illustrates the objects of its establishment.

1. To unite in bonds of fraternity, aid and protection all acceptable white persons of good character, steady habits, sound bodily health and reputable calling.

2. To improve the condition of its membership morally, socially and materially, by timely counsel and instructive lessons, by encouragement in business, and by assistance to obtain employment when in need.

3. To establish a relief fund, from which members of this organization, who have complied with all its rules and regulations, may receive the benefit of a sum not exceeding $3,000, which shall be paid upon the following conditions, viz: First. When a member reaches the age of seventy-five years. Second. When by reason of disease or accident a member becomes permanently disabled from following his usual or some other occupation. Third. Upon satisfactory evidence of the death of a member.

Seven reasons are given why persons should join the CHOSEN FRIENDS for protection.

1. It furnishes its benefits at actual cost
2. Its plans are equitable and just.
3. Its fraternal obligations are binding and forcible.
4. No distinction made between ladies and gentlemen.
5. Benefits paid at the age of *seventy-five* years.
6. One half of benefits paid when a member becomes permanently disabled.
7. It offers all the special inducements that other orders do and these additional.

This much for the benefit of our patrons and readers who may desire to take stock in insurance associations on the score of economy. As for our self we are *ten years* beyond the bounds of reception, as no member is received who is over 61 years of age, and none who are less than 18.

No councils are organized within *yellow fever* districts, showing more than the ordinary degree of caution in such associations.

THE NATION, WEEKLY EDITION OF THE NEW YORK POST: Price 10 cents a number. A 3 column, 30 page Royal Quarto of choice current literature, embracing politics, science, domestic affairs, reviews, criticisms, commerce, manufactures, and general news, both foreign and domestic. (Volume 34, No. 880), and is teeming with able and interesting matter; and, as the above title indicates, contains the choice daily gleanings of New York condensed in a charming form. But the *Nation* is too well known to need any commendation of ours; for, anything *we* could say, would not "change one hair white or black." It is a journal of *rare* merit, especially valuable to men of letters.

ILLUSTRATED CIRCULAR of Bee-keeping Supplies and Bees, being the 11th annual edition, for 1882, by Alanson C. Hill, Kendalville, Ind. Invaluable to bee-keepers, both amateur and professional, for it tells them what they want, and how and where to get it.

REPORT upon the condition of winter grain, and upon the numbers and condition of farm animals of the United States, to April 1882, being special report No. 42 of the Department of Agriculture, 82 pages octavo, executed in the usual good style of the recent publications of the department.

HANEY'S JOURNAL. A workshop companion, published by Jesse Haney & Co., 119 Nassau street, N. Y. An 8 paged Demi-folio, being a repository of practical recipes and useful information. New series, vol. 1, No. 12, established in 1868. The subjects of mural decoration, historic reviews, texture, the *pleasures* of nearsightedness, how to prevent plugs from showing, silvering glass, the use of colors, painting tin roofs, sign writing and glass embossing, how to repair the watch, anatomy for artists, and many other items of a similar character, attest the usefulness of such a journal to the practical mechanic, or to amateurs in art.

PRICE LIST and manual of Prize Holly and Demas scroll saws, lathes and detachments for buzz sawing, dove-tailing, moulding, grinding and polishing, N. W. Ayre & Son, corner Eighth and Chestnut streets, Philadelphia. Illustrated.

THE WORLD OF NATURE

The world of animated nature is more splendidly represented under the canvas of Forepaugh's Great Show than in any zoological collection existent. Not since the day Noah lifted his hawser off the snubbing post have so many distinct varieties of rare animals been collected under one charge. This important fact should not be lost sight of by schools and parents. Boys and girls can learn more in an afternoon of natural history, in the great Menagerie of Forepaugh's Show, than by months of book study. Recognizing this, Mr. Forepaugh makes reduced rates to schools, and admits all children in orphan asylums free of charge. This Great Show will exhibit in Lancaster, Monday, April 24.

THE
LANCASTER EXAMINER

OFFICE
No. 9 North Queen Street,
LANCASTER, PA.

THE OLDEST AND BEST.

THE WEEKLY
LANCASTER EXAMINER
One of the largest Weekly Papers in the State.

Published Every Wednesday Morning,

Is an old, well-established newspaper, and contains just the news desirable to make it an interesting and valuable Family Newspaper. The postage to subscribers residing outside of Lancaster county is paid by the publisher.

Send for a specimen copy.

SUBSCRIPTION:
Two Dollars per Annum.

THE DAILY
LANCASTER EXAMINER
The Largest Daily Paper in the county.

Published Daily Except Sunday.

The daily is published every evening during the week. It is delivered in the City and to surrounding Towns accessible by railroad and daily stage lines, for 10 cents a week.
Mail Subscription, free of postage—One month, 50 cents; one year, $5.00.

JOHN A. HIESTAND, Proprietor,
No. 9 North Queen St.,
LANCASTER, PA.

Important to Grocers, Packers, Hucksters, and the General Public.

THE KING FORTUNE-MAKER.
OZONE
A New Process for Preserving all Perishable Articles, Animal and Vegetable from Fermentation and Putrefaction, Retaining their Odor and Flavor.

"**OZONE**—Purified air, active state of Oxygen."—*Webster*.

This preservative is not a liquid pickle, or any of the old and exploded processes, but is simply and purely OZONE, as produced and applied by an entirely new process. Ozone is the antiseptic principle of every substance, and possesses the power to preserve animal and vegetable structures from decay.

There is nothing on the face of the earth liable to decay or spoil which Ozone, the new Preservative, will not preserve for all time in a perfectly fresh and palatable condition.

The value of Ozone as a natural preserver has been known to our abler chemists for years, but, until now, no means of producing it in a practical, inexpensive, and simple manner have been discovered. Microscopic observations prove that decay is due to septic matter or animic germs, that develop and feed upon animal and vegetable structures. Ozone, applied by the Prentice method, seizes and destroys these germs at once, and thus preserves. At our office in Cincinnati can be seen animal every article that can be thought of, preserved by this process, and every visitor is welcomed to come in, taste, smell, take away with him, and test in every way the merits of Ozone as a preservative. We will also preserve, free of charge, any article that is brought at our expense to us, and return it to the sender, for him to keep and test.

FRESH MEATS, such as beef, mutton, veal, pork, poultry, game, fish, &c., preserved by this method, can be shipped to Europe, subjected to atmospheric changes and return to this country in a state of perfect preservation.

EGGS can be treated at a cost of less than one dollar a thousand dozen, and be kept in unordinary room six months or more, thoroughly preserved; the yolk held in its normal condition, and the eggs as fresh and perfect as on the day they were treated, and will sell as strictly "choice." The advantage in preserving eggs is readily seen; there are seasons when they cannot be bought for 5 or 6 cents a dozen, and by holding them, can be sold for an advance of from one hundred to three hundred per ct. One man, with this method, can preserve 5,000 dozen a day.

FRUITS may be permitted to ripen in their native climate, and can be transported to any part of the world. The price expressed from fruits can be held for an indefinite period without fermentation—hence the great value of this process for producing a temperance beverage. Cider can be held perfectly sweet for any length of time.

VEGETABLES can be kept for an indefinite period in their natural condition, retaining their odor and flavor, treated in their original packages at a small expense. All grains, flour, meal, etc., are held in their normal condition.

HIDES, after being treated by this process, will not become rancid.

Dead human bodies, treated before decomposition sets in, can be held in a natural condition for weeks, without puncturing the skin or mutilating the body in any way. Hence the great value of Ozone to undertakers.

There is no change in the slightest particular in the appearance of any article thus preserved, and no trace of any foreign or unnatural odor or taste.

The process is so simple that a child can operate as well and as successfully as a man. There is no expensive apparatus or machine ry required.

A room filled with different articles, such as eggs, meat, fish, etc., can be treated at one time, without additional cost or expense.

In fact, there is nothing that Ozone will not preserve. Think of everything you can that is to sour, decay, or spoil, and then remember that there is only one thing which will preserve it in exactly the condition you want it for any length of time. If you will then consider this it will save asking questions as to whether Ozone will preserve this or that article. **It will preserve anything and every thing you can think of.**

There is not a township in the United States in which a live man can not make any amount of money, from $1,200 to $10,000 a year, that he pleases. We desire to get a live man interested in each county in the United States, in whose hands we can place this Preservative, and through him secure the business which every county ought to produce.

A FORTUNE
Awaits any Man who Secures Control of OZONE in any Township or County.

A. C. Powers, Marion, Ohio, has cleared $2,000 in two months. $2 for a test package was his first investment. Woods Brothers, Lebanon, Warren County, Ohio, made $6,000 on eggs purchased in August and sold November 1st. $2 for a test package was their first investment.

F. K. Raymond, Morristown, Belmont Co., Ohio, is clearing $200 a month in handling and selling Ozone. $2 for a test package was his first investment.

H. P. Webber, Charlotte, Eaton Co., Mich., has cleared $1,000 a month since August. $2 for a test package was his first investment.

J. B. Gaylord, 80 LaSalle St., Chicago, is preserving eggs, fruit, etc., for the commission men of Chicago, charging 1½c. per dozen for eggs, and other articles in proportion. He is preserving 5,000 dozen eggs per day, and on this business is making $1,000 a month clear. $2 for a test package was his first investment.

The Cincinnati Test Co., West 7th Seventh Street, is making $375 a month in handling brewers' malt, preserving and shipping it as feed to all parts of the country. Malt unpreserved sours in 24 hours. Preserved by Ozone it keeps perfectly sweet for months.

These are instances which we have asked in the privilege of publishing. There are scores of others. What many of the above parties and get the evidence direct.

Now, to prove the absolute truth of every thing we have sold in this paper, **we propose to place in your hands the means of proving for yourself that we have not claimed half enough.** To any person who doubts any of these statements, and who is interested sufficiently to make the trip, we will pay all traveling and hotel expenses for a visit to this city, if we fail to prove any statement that we have made

How to Secure a Fortune with Ozone.

A test package of Ozone, containing a sufficient quantity to preserve one thousand dozen eggs, or other articles in proportion, will be sent to any applicant on receipt of $2. This package will enable the applicant to pursue any line of tests and experiments he desires, and thus satisfy himself as to the extraordinary use of Ozone as a Preservative. After having thus satisfied himself, and had time to look the field over to determine what he wishes to do in the future—whether to sell the article to others or to confine it to his own use, or any other line of policy which is best suited to him and to the township or county which we will enter into an arrangement with him that will make a fortune for him and give us good profits. We will give exclusive township or county privileges to the first responsible applicant who orders a test package and desires to control the business in his locality. **The man who secures control of Ozone for any special territory, will enjoy a monopoly which will surely enrich him.**

Don't let a day pass until you have ordered a Test Package, and if you desire to secure an exclusive privilege we assure you that delay may deprive you of it, for the applications come to us by scores every mail—many by telegraph. "First come first served" is our rule.

If you do not care to send money in advance for the test package we will send it C. O. D., but this will put you to the expense of charges for return money. Our correspondence is large; we have all we can do to attend to the shipping of orders and giving attention to our working agents. Therefore we can not give any attention to letters which do not order Ozone. If you think of any article that you are doubtful about Ozone preserving remember we guarantee that it if no matter what it is.

REFERENCES.

We desire to call your attention to a class of references which no enterprise or firm based on any thing but the soundest business success and highest commercial merit could secure.

We refer, by permission, as to our integrity and for the value of the Prentice Preservative, to the following gentlemen: Edward C. Royce, Member Board of Public Works; E. O. Eshelby, City Comptroller; Amor Smith, Jr., Collector Internal Revenue; Walsin & Worthington, Attorneys; Martin H. Harrell and B. P. Hopkins, County Commissioners; W. S. Cappeller, County Auditor; all of Cincinnati, Hamilton County, Ohio. These gentlemen are each familiar with the merits of our Preservative, and know from actual observation that we have without question

The Most Valuable Article in the World.

The $2 you invest in a test package, will surely lead you to secure a township or county, and then your way is absolutely clear to realize from $2,400 to $10,000 a year.

Give your full address in every letter, and send your letter to

PRENTISS PRESERVING COMPANY, (Limited,)
S. E. Cor. Ninth & Race Sts., Cincinnati, O.

Nov-3m

WHERE TO BUY GOODS IN LANCASTER.

BOOTS AND SHOES.

MARSHALL & SON, No. 12 Centre Square, Lancaster, Dealers in Boots, Shoes and Rubbers. Repairing promptly attended to.

M. LEVY, No. 7 East King street. For the best Dollar Shoes in Lancaster go to M. Levy, No. 7 East King street.

BOOKS AND STATIONERY.

JOHN BAER'S SON'S, Nos. 15 and 17 North Queen Street, have the largest and best assorted Book and Paper Store in the City.

FURNITURE.

HEINITSH'S, No. 15½ East King st., (over China Hall) is the cheapest place in Lancaster to buy Furniture. Picture Frames a specialty.

CHINA AND GLASSWARE.

HIGH & MARTIN, No. 15 East King st., dealers in China, Glass and Queensware, Fancy Goods, Lamps, Burners, Chimneys, etc.

CLOTHING.

MYERS & RATHFON, Centre Hall, No. 12 East King St. Largest Clothing House in Pennsylvania outside of Philadelphia

DRUGS AND MEDICINES.

G. W. HULL, Dealer in Pure Drugs and Medicinal Chemicals, Patent Medicines, Trusses, Shoulder Braces, Supporters, &c., 15 West King St., Lancaster, Pa

JOHN F. LONG & SON, Druggists, No. 12 North Queen St. Drugs, Medicines, Perfumery, Spices, Dye Stuffs, Etc. Prescriptions carefully compounded.

DRY GOODS.

GIVLER, BOWERS & HURST, No. 25 E. King St., Lancaster, Pa., Dealers in Dry Goods, Carpets and Merchant Tailoring. Prices as low as the lowest.

HATS AND CAPS.

C. H. AMER, No. 29 West King Street, Dealer in Hats, Caps, Furs, Robes, etc. Assortment Large. Prices Low.

JEWELRY AND WATCHES.

H. Z. RHOADS & BRO, No. 4 West King St. Watches, Clock and Musical Boxes. Watches and Jewelry Manufactured to order.

PRINTING.

JOHN A. HIESTAND, 9 North Queen st., Sale Bills, Circulars, Posters, Cards, Invitations, Letter and Bill Heads and Envelopes neatly printed. Prices low.

Thirty-Six Varieties of Cabbage; 26 of Corn; 28 of Cucumber; 41 of Melon; 33 of Peas; 28 of Beans; 17 of Squash; 23 of Beet and 40 of Tomato, with other varieties in proportion, a large portion of which were grown on my five seed farms, will be found in my **Vegetable and Flower Seed Catalogue for 1882.** Sent FREE to all who apply. Customers of last season need not write for it. All Seed sold from my establishment warranted to be fresh and true to name, so far, that should it prove otherwise, I will refill the order gratis. The **original introducer of Early Ohio** and **Burbank Potatoes, Marblehead, Early Corn, the Hubbard Squash, Marblehead Cabbage, Phinney's Melon,** and a score of other New Vegetables, I invite the patronage of the public. New Vegetables a specialty.

JAMES J. H. GREGORY,
Marblehead, Mass.
Nov-6mo]

EVAPORATE YOUR FRUIT.
ILLUSTRATED CATALOGUE
FREE TO ALL.

AMERICAN DRIER COMPANY,
Chambersburg, Pa.
Apl-tf

FARMING FOR PROFIT.

It is conceded that this large and comprehensive book, (advertised in another column by J. C. McCurdy & Co., of Philadelphia, the well-known publishers of Standard works,) is not only the newest and handsomest, but altogether the BEST work of the kind which has ever been published. Thoroughly treating the great subjects of General Agriculture, Live-Stock, Fruit-Growing, Business Principles, and Home Life; telling just what the farmer and the farmer's boys want to know, combining Science and Practice, stimulating thought, awakening inquiry, and interesting every member of the family, this book must exert a mighty influence for good. It is highly recommended by the best agricultural writers and the leading papers, and is destined to have an extensive sale. Agents are wanted everywhere. jan-lt

BLOOMSDALE
LARGE LATE FLAT DUTCH CABBAGE.

Large, Flat, Solid Heads, Short Stems.

For a long period of time we have had this stock of Cabbage in cultivation, originally obtained from the German and Sweedish market gardeners. It has been a part of our business occupation to keep it undefiled, and to-day we offer it in its original purity, equal in quality with the very best in the country, even though the best should cost a hundred dollars per pound.
We have made this crop a study and give our customers the result of many years close observation. For which our opportunities may be judged by the fact that we have, each and every year, about one hundred and fifty acres of cabbage raised expressly to produce seed for the coming season, old from which selections are made with scrupulous care, guided by experience. Not a single grain of seed is raised from Stalks all from Selected Heads.
We will mail our Catalogue free of charge to all applicants.

D. LANDRETH & SONS,
Nos. 21 and 23 South Sixth Street,
BETWEEN MARKET AND CHESTNUT STS.,
BRANCH STORE—S. W. COR. DELAWARE AVE. AND ARCH STREET,
apr-6m PHILADELPHIA.

MERCHANT TAILORING.
1848 (The Oldest of All.) 1881
RATHVON & FISHER,
MERCHANT TAILORS AND DRAPERS,

respectfully inform the public that having disposed of their entire stock of Ready-Made Clothing, they now do, and for the future shall, devote their whole attention to the CUSTOM TRADE.
All the desirable styles of CLOTHS, CASSIMERES, WORSTEDS, COATINGS, SUITINGS and VESTINGS constantly on hand, and made to order in plain or fashionable style promptly, and warranted satisfactory.
All-Wool Suit from $10.00 to $20.00.
All-Wool Pants from 3.00 to 10.00.
All-Wool Vests from 2.00 to 6.00.
Union and Cotton Goods proportionately low.
Cutting, Repairing, Trimming and Making, at reasonable prices.
Goods en tailed by the yard to those who desire to have them made elsewhere.
A full supply of Spring and Summer Goods just opened and on hand.
Thankful to a generous public for past patronage they hope to merit its continued recognition in their "new departure."

RATHVON & FISHER,
PRACTICAL TAILORS,
No. 101 North Queen Street,
LANCASTER, PA.
1848 1881

GLOVES, SHIRTS, UNDERWEAR.
SHIRTS MADE TO ORDER,
AND WARRANTED TO FIT.
E. J. ERISMAN,
56 North Queen St., Lancaster, Pa.
79-1-12]

A HOME ORGAN FOR FARMERS.

THE LANCASTER FARMER,

A MONTHLY JOURNAL,

Devoted to Agriculture, Horticulture, Domestic Economy and Miscellany.

Founded Under the Auspices of the Lancaster County Agricultural and Horticultural Society.

EDITED BY DR. S. S. RATHVON.

TERMS OF SUBSCRIPTION:

ONE DOLLAR PER ANNUM,

POSTAGE PREPAID BY THE PROPRIETOR.

All subscriptions will commence with the January number, unless otherwise ordered.

Dr. S. S. Rathvon, who has so ably managed the editorial department in the past, will continue in the position of editor. His contributions on subjects connected with the science of farming, and particularly that specialty of which he is so thoroughly a master—entomological science—some knowledge of which has become a necessity to the successful farmer, are alone worth much more than the price of this publication. He is determined to make "The Farmer" a necessity to all households.

A county that has so wide a reputation as Lancaster county for its agricultural products should certainly be able to support an agricultural paper of its own, for the exchange of the opinions of farmers interested in this matter. We ask the co-operation of all farmers interested in this matter. Work among your friends. The "Farmer" only one dollar per year. Show them your copy. Try and induce them to subscribe. It is not much for each subscriber to do but it will greatly assist us.

All communications in regard to the editorial management should be addressed to Dr. S. S. Rathvon, Lancaster, Pa., and all business letters in regard to subscriptions and advertising should be addressed to the publisher. Rates of advertising can be had on application at the office.

JOHN A. HIESTAND,
No. 9 North Queen St., Lancaster, Pa.

$72 A WEEK. $12 a day at home easily made. Costly Outfit free. Address TRUE & Co., Augusta, Maine

ONE DOLLAR PER ANNUM.—SINGLE COPIES 10 CENTS.

THE LANCASTER FARMER

DEVOTED TO Agriculture, Horticulture, Domestic Economy and Miscellany

"THE FARMER IS THE FOUNDER OF CIVILIZATION."—WEBSTER.

Dr. S. S. RATHVON, Editor.   LANCASTER, PA. JULY, 1882.   JOHN A. HIESTAND, Publisher

Entered at the Post Office at Lancaster as Second Class Matter.

CONTENTS OF THIS NUMBER.

EDITORIAL.
Egg Culture in France..97
Gapes in Chickens..97
Entomological Notes..97
  Directions for Sending Insects.
Kitchen Garden for July..98
  Quality and Vitality of Seeds.
How to Kill Wheat Moth..98
  Remedies for the Army Worm—Melons—Bugs—Coal-Tar—Insect Powder.
Our Local Crops..98
Destroying Weevil..99
Effects of Baking on Flour..99
Phosphoric Acid in Plants..99
A Mare's Nest...100
Excerpts..100
Three Wonders..100

CONTRIBUTIONS.
The Uses of Pruning...102
Balance of Trade..102

SELECTIONS.
U. S. Department of Agriculture.................................103
  To the Manufacturer of Sugar from Sorghum, Beets and other Sur-Producing Plants in the United States—Sugar Beets—Other Sugar-Producing Plants.
The Happy Granger..103
Underdraining...104
  Professor J. M. McBryde in Journal of American Agricultural—Silo and Ensilage.
Education for Farmers...104
Success in Farming..104
  Importance of Rotation and Clover and Grass Crops.
The Department of Agriculture...................................105
Fancy Butter..106
All about Poultry..106
Talks About Fruit..107

OUR LOCAL ORGANIZATIONS.
Lancaster County Agricultural and Horticultural Society........107
  Crop Reports—Shall we Have a Fair? No—The Immigration Question—Grain and Fruit Exhibited.
The Poultry Society...108
  Gapes in Chicks—How Do They Get There?
Linnæan Society..109

AGRICULTURE.
Green Crops..108
Loading Hay..108
Manure Under Cover..108
Plaster..108
The Largest Land Owner on the Continent......................108
Best Pasture Grass..109
Pacific Coast Wheat Items......................................109

HORTICULTURE.
Summer Grape Pruning..109
The Care and Pruning of Peach Trees...........................109
The Delaware Peach Crop......................................109
Strawberry Beds..109
Quince Culture..109

HOUSEHOLD RECIPES
Deep Apple Pie...110
Pan-dowdy..110
Fried Apples..110
Apple Toast..110
Apple and Bread Pudding.......................................110
Racket Club Pudding...110
Jelly Pudding...110
Cheese Crusts..110
Pumpkin Pie..110
Plain Mince Pie...110
Welsh Rare-Bit...110
Omelette..110
Chicken and Green Peas..110
Bean Soup..110
Codfish..
Broiled Birds..
Sage and Wine..
Beef Juice..
Wine Jelly..110
Toast..110
Barley Water..110
Egg and Wine...110
Milk Punch..110

LIVE STOCK.
Spoiling a Young Horse...110
The Pig in Agriculture..111
Sheep Raising in Dakota..111
Treatment of the Cow...111

POULTRY.
Floors for Poultry Houses......................................111
Fowl Fattening..111
Onions for Chicken Cholera....................................111
Cramming Poultry..111
Wild Chickens..111
Good Hatching...111
Literary and Personal..111

SILK-WORM EGGS.
Amateur Silk-growers can be supplied with superior silk-worm eggs, on reasonable terms, by applying immediately to
GEO. O. HENSEL,
No. 228 East Orange Street, Lancaster, Pa.
may-3m)

SEND IN YOUR SUBSCRIPTIONS
—FOR—
THE FARMER
FOR 1882.
The cheapest and one of the best Agricultural papers in the country.
Only $1.00 per year.
JOHN A. HIESTAND, Publisher,
No. 9 North Queen St., Lancaster, Pa.

$66 a week in your own town. Terms and $5 outfit free. Address H. HALLETT & Co., Portland, Maine.

D. M. FERRY & CO., Detroit, Mich.
Jan-6m

$66 a week in your own town. Terms and $5 outfit free. Address H. HALLETT & Co., Portland, Maine.
jun-1yr*

WE WANT OLD BOOKS.
WE WANT GERMAN BOOKS.
WE WANT BOOKS PRINTED IN LANCASTER CO.
We Want All Kinds of Old Books.
LIBRARIES, ENGLISH OR GERMAN BOUGHT.
Cash paid for Books in any quantity. Send your address and we will call.
REES WELSH & CO.,
23 South Ninth Street, Philadelphia.

LIGHT BRAHMA EGGS
For hatching, now ready—from the best strain in the county—at the moderate price of
$1.50 for a setting of **13 Eggs.**
S. RATHVON,
No. 9 North Queen st., Examiner Office, Lancaster, Pa.

WANTED.—CANVASSERS for the
LANCASTER WEEKLY EXAMINER
In Every Township in the County. Good Wages can be made. Inquire at
THE EXAMINER OFFICE,
No. 9 North Queen Street, Lancaster, Pa

$72 A WEEK. $12 a day at home easily made. Costly Outfit free. Address TRUE & Co., Augusta, Maine.
jun-1yr*

**SEND FOR
SPECIAL PRICES**
On Concord Grapevines, Transplanted Evergreens, Tulip, Poplar, Linden Maple, etc., Tree Seedlings and Trees for timber plantations by the 100,000.
J. JENKINS' NURSERY.
3-2-19   WINONA, COLUMBIANA CO., OHIO.

MARBLEHEAD
Early Sweet Corn
It is the most profitable of all, because it matures before any other kind, giving farmers complete control of the early market. I warrant it to be at least a week earlier than Minnesota, Narragansett or Crosby, and decidedly earlier than Dolly Dutton, Tom Thumb or Early Boynton. Of size of Minnesota, and very sweet. The original introducer, I send pure stock, postpaid, per package 15 cents; per peck, per express, $3.00. In my catalogue, (free to all,) are emphatic recommendations from farmers and gardeners.
JAMES J. H. GREGORY,
apr-3t   Marblehead, Mass.

THE LANCASTER FARMER.

PENNSYLVANIA RAILROAD SCHEDULE.
Trains leave the Depot in this city, as follows:

WESTWARD.	Leave Lancaster.	Arrive Harrisburg.
Pacific Express*	2:40 a. m.	4:05 a. m.
Way Passenger*	5:00 a. m.	7:30 a. m.
Niagara Express	11:00 a. m.	12:30 p. m.
Hanover Accommodation	11:45 p. m.	Col. 10:05 a. m.
Mail train via Mt. Joy	10:20 a. m.	12:50 p. m.
No. 2 via Columbia	11:25 a. m.	12:55 p. m.
Sunday Mail	10:50 a. m.	12:40 p. m.
Fast Line*	2:30 p. m.	3:25 p. m.
Frederick Accommodation	2:55 p. m.	Col., 2:45 p. m.
Harrisburg Accom	5:45 p. m.	7:30 p. m.
Columbia Accommodation	7:20 p. m.	Col. 8:30 p. m.
Harrisburg Express	7:30 p. m.	9:40 p. m.
Pittsburg Express	9:50 p. m.	12:10 a. m.
Cincinnati Express*	11:50 p. m.	1:245 a. m.

EASTWARD.	Lancaster.	Philadelphia
Cincinnati Express*	2:55 a. m.	5:00 a. m.
Fast Line*	5:08 a. m.	7:44 a. m.
Harrisburg Express	8:05 a. m.	10:00 a. m.
Columbia Accommodation	9:10 p. m.	12:10 p. m.
Pacific Express*	1:40 p. m.	3:40 p. m.
Sunday Mail	2:00 p. m.	5:00 p. m.
Johnstown Express	3:05 p. m.	5:50 p. m.
Day Express*	5:35 p. m.	7:20 p. m.
Harrisburg Accom	6:45 p. m.	9:50 p. m.

The Hanover Accommodation, west, connects at Lancaster with Niagara Express, west, at 9:45 a. m., and will run through to Hanover.
The Frederick Accommodation, west, connects at Lancaster with Fast Line, west, at 2:10 p. m., and runs to Frederick.
The Pacific Express, east and Sunday, when flagged, will stop at Middletown, Elizabethtown, Mount Joy and Landisville.
*The only trains which run daily.
Runs daily, except Monday.

NORBECK & MILEY,

PRACTICAL

Carriage Builders,

COX & CO'S OLD STAND.

Corner of Duke and Vine Streets,
LANCASTER, PA.

THE LATEST IMPROVED

SIDE-BAR BUGGIES,

PHÆTONS,

Carriages, Etc.

THE LARGEST ASSORTMENT IN THE CITY.

Prices to Suit the Times.

REPAIRING promptly attended to. All work guaranteed.

S. B. COX,
Manufacturer of
Carriages, Buggies, Phaetons, etc.
CHURCH ST., NEAR DUKE, LANCASTER, PA.

Large Stock of New and Second-hand Work on hand very cheap. Carriages Made to Order Work Warranted for one year.

EDW. J. ZAHM,
DEALER IN
AMERICAN AND FOREIGN

WATCHES,

SOLID SILVER & SILVER PLATED WARE,

CLOCKS.

JEWELRY & TABLE CUTLERY.

Sole Agent for the Arundel Fluted

SPECTACLES.

Repairing strictly attended to.

ZAHM'S CORNER,
WAREROOMS:
North Queen-st. and Centre Square, Lancaster, Pa.

E. F. BOWMAN,

AT LOWEST POSSIBLE PRICES,
Fully guaranteed.
No. 106 EAST KING STREET,
Opposite Leopard Hotel.

ESTABLISHED 1832.

G. SENER & SONS,
Manufacturers and dealers in all kinds of rough and finished

LUMBER.

The best Sawed SHINGLES in the country. Also Sash, Doors, Blinds, Mouldings, &c.

PATENT O. G. WEATHERBOARDING

and PATENT BLINDS, which are far superior to any other. Also best COAL constantly on hand.

OFFICE AND YARD:
Northeast Corner of Prince and Walnut-sts.,
LANCASTER, PA.

PRACTICAL ESSAYS ON ENTOMOLOGY,
Embracing the history and habits of
NOXIOUS AND INNOXIOUS
INSECTS,
and the best remedies for their expulsion or extermination.
By S. S. RATHVON, Ph. D.
LANCASTER, PA.

This work will be Highly Illustrated, and will be put in press (as soon after a sufficient number of subscribers can be obtained to cover the cost) as the work can possibly be accomplished.

$5 to $20 per day at home. Samples worth $5 free! Address STINSON & Co., Portland, Maine

TREES
Fruit, Shade and Ornamental Trees.
Plant Trees raised in this county and suited to this climate. Write for prices to
LOUIS C. LYTE.
Bird-in-Hand P. O., Lancaster co., Pa.
Nursery at Smoketown, six miles east of Lancaster.

WIDMYER & RICKSECKER,
UPHOLSTERERS,
And Manufacturers of

FURNITURE AND CHAIRS.
WAREROOMS:
102 East King St., Cor. of Duke St.
LANCASTER, PA.

Special Inducements at the
NEW FURNITURE STORE
OF
W. A. HEINITSH,
No. 15 1-2 E. KING STREET
(over Bursk's Grocery Store), Lancaster, Pa.
A general assortment of furniture of all kinds constantly on hand. Don't forget the number,
15 1-2 East King Street,
(over Bursk's Grocery Store.)

For Good and Cheap Work go to
F. VOLLMER'S
FURNITURE WARE ROOMS,
No 309 NORTH QUEEN ST.,
(Opposite Northern Market),
Lancaster, Pa.
Also, all kinds of picture frames.

GREAT BARGAINS.
A large assortment of all kinds of Carpets are still sold at lower rates than ever at the
CARPET HALL OF H. S. SHIRK,
No. 202 West King St.

Call and examine our stock and satisfy yourself that we can show the largest assortment of them Brussels, three ply and ingrains at all prices—at the lowest Philadelphia prices.
Also on hand a large and complete assortment of Rag Carpet.
Satisfaction guaranteed both as to price and quality.
You are invited to call and see my goods. No trouble in showing them even if you do not want to purchase.
Don't forget this notice. You can save money here if you want to buy.
Particular attention given to customer work.
Also on hand a full assortment of Counterpanes, Oil Cloths and Blankets of every variety.

PHILIP SCHUM, SON & CO.,
38 and 40 West King Street.
We keep on hand of our own manufacture,
QUILTS, COVERLETS,
COUNTERPANES, CARPETS,
Bureau and Tidy Covers, Ladies' Furnishing Goods, Notions, etc.
Particular attention paid to customer Rag Carpet, and scouring and dyeing of all kinds.
PHILIP SCHUM, SON & CO.,
Lancaster, Pa.

THE HOLMAN LIVER PAD!
Cures by absorption without medicine.
Now is the time to apply these remedies. They will do for you what nothing else on earth can. Hundreds of citizens of Lancaster say so. Get the genuine at
LANCASTER OFFICE AND SALESROOM,
22 East Orange Street.

C. R. KLINE
ATTORNEY-AT-LAW,
OFFICE: 15 NORTH DUKE STREET,
LANCASTER, PA.

The Lancaster Farmer.

Dr. S. S. RATHVON, Editor. LANCASTER, PA., JULY, 1882. Vol. XIV. No. 7.

EDITORIAL.

EGG CULTURE IN FRANCE.

Many small farmers in France pay their rents from their poultry yards. The fowls in Normandy, France, are almost exclusively of the Creve Cœur breed in its different varieties; and the number of poultry in Normandy is estimated at three million five hundred thousand, valued at £2,400,000, and the annual value of fowls' eggs alone, is £250,000 to the farmers; the average annual produce per hen is about one hundred eggs, and a hen will continue to lay for five years.

In 1875 England imported eight hundred million eggs, valued at $12,500,000, including charges, of which France furnished five-sixths; that is to say, more than two millions per day during the year. In France, hardly a meal is eaten at any table without eggs or poultry forming a part of it. Normandy furnishes nearly two million head of poultry of various kinds to the Paris markets annually, yet falls behind the supply from other provinces. Six millions of eggs are sold weekly in the Paris markets. Many are used in glazing ornamental cakes and sweetmeats. One pastry cook alone buys two millions of eggs a year for these purposes. Another large dealer uses five hundred thousand, of which he separates the whites from the yolks—the whites being sent to the manufacturing districts in the north, and the yolks being employed in dressing skins for gloves. Agricultural writers in France are continually urging that more attention should be paid to poultry raising by farmers, and they declare the production might be easily doubled.—*English Dairyman.*

Our punning remarks on eggs in the May number of the FARMER, were regarded by some readers as a sheer *eggs-aggeration*, but the above, from unquestionable authority, will illustrate that the real amount of the French egg-traffic is far in advance of estimates made in said remarks. The above relates to the traffic some years ago; hence, at the present date, it may be *progressively* larger; for in matters of domestic production France is not retrogressive, especially if it pays. The egg-statistics in our own country do not appear to have elicited sufficiently that detailed attention, through which alone the amount and value of the product could be accurately stated. The impulse given to Gallinculture of late years will, however, ultimately manifest this, for "Hen-Fruit" cannot be ignored any more than can the hen.

GAPES IN CHICKENS.

A correspondent of the London *Agricultural Gazette* says:

"I have frequently lost large numbers of chickens from gapes, and have never until this spring been successful in curing them. About six or seven weeks ago the old complaint made its appearance in about thirty chickens, some the size of pigeons and others less. As an experiment I tried sulphur, commonly called flour of brimstone, and salt, namely, two parts sulphur and one part salt mixed with water to the consistency of thick cream (it is best to use the finger in mixing, as sulphur will not readily mix with water.) I then applied it with a feather from a fowl's wing, dipping it in the mixture, and putting it down in the chicken's throat about three inches, worked the feather up and down a few times, and then applied some more in the same way again.

"I soon found they were improving very rapidly and so repeated the operation three or four times, two or three days between each application. They are now all cured and doing well. I have not lost one, although some of them were very bad indeed when the remedy was first applied. I may add that the feather requires to have about half the broad side clipped off, or it would be too large for the purpose required."

It is fully half a hundred years or more since we first knew of the "gapes in chickens," and it is questionable whether the average poultry breeders know anything more about it now than they did then. We think it was then called "pips," but it was all the same—little red worms in the windpipe—and the chickens would *gape* and *pip* and *die*, almost without remedy. Mechanical means were already employed fifty years ago for their removal, although perhaps not so skillfully as it can be done now. We think a thin wire was used, but about nine out of every ten died, if not under the operation, a short time after it. Mr. D., in his essay before the Lancaster Poultry Society, does not advance the subject one peg beyond where it was before, nor did he pretend to do so. We think, however, he is in error when he states that the subject is one that belongs to the domain of the scientist alone. We believe, the man who habitually contemplates the chicken in the egg, who rears it from its pristine condition to its full development, who has a natural and a peculiar interest in its physical existence, who sees it every day and provides its food and shelter, is the very man who is in a situation to get at the origin and cure of the disease. Even if he never should be able to discover its origin, if he discovers a certain and safe cure, he will be a benefactor.

ERCOLANI has found gapes living thirty days after they had been expelled from a fowl and exposed to the weather. From their peculiar organization they must necessarily be very local; hence, they may exist in one enclosure and not in another, although there may be only a fence between them, provided the chickens have been kept separate. They appear to be something like the California resurrection plant, becoming vitalized as soon as moisture is given them, although they may have remained dry for years. We think not sufficient importance may have been attached to the total annihilation of the gapes, after they have been expelled from the fowl. We must not, however, be too sanguine in any direction, with all the light we have on the subject, at the present time (of their origin). In the meantime, expertness in the mechanical removal of them should be carefully cultivated.

ENTOMOLOGICAL NOTES.

PROF. RATHVON—*Dear Sir:* I send you a worm by mail, and enclose a stamp—and wish you would write me and give the name of the worm, &c. We noticed them last year for the first time. Then, there were but few—now, they are much more plentiful. They destroy the tomato plants, night and day.

Very respectfully,
W. H. H. W.

DARLINGTON, MD., June 26, 1882.

Being in the midst of a multitude of secular engagements, we sent the box and worm to Prof. Riley, Entomologist of the Department of Agriculture, Washington, D. C. Not, however, because it was entirely new to us, but because we had not heard any complaints about it here; and, we supposed it *might* be something new. Prof. R. writes as follows:

Dear Sir: I have your note of the 30th ult., with accompanying box. The larva destructive to the tomato plants is that of *Prodenia lineatella*, which is known to feed upon a great variety of plants. The unusually moist weather we had this spring, greatly favored the development of this and other cut-worms, and complaints at their destruction have reached me from almost every State east of the Rocky Mountains.

Yours truly,
C. V. R.

WASHINGTON, D. C., July 6, 1882.

Since the rapidly increased cultivation of tobacco in Lancaster county, the tomato plant as well as its fruit, is tolerably free from insect infestation—only here and there and now and then—a Sphinx, a Potato-beetle, or a Cut-worm, are to be found feeding upon them. It will be observed that the "worm" under discussion belongs to the great family of "Cut-worms" (NOCTUADÆ), many of which are destructive to vegetation, "day and night," and we may infer that any remedy that would kill the one, would kill the other.

In conclusion, we commend the efficient manner in which this larva, in a living state, was sent to us through the U. S. mail: and also append the directions of the Department of Agriculture, for sending insects to it by mail, for the benefit of those who may be concerned:

Directions for Sending Insects.

All inquiries about insects, injurious or otherwise, should be accompanied by specimens, the more the better. Such specimens, if dead should be packed in some soft material, as cotton or wool, and inclosed in some stout tin or wooden box. They will come by mail for one cent per ounce. Insects should never be inclosed loose in the letter. Whenever possible, larvæ (i. e., grubs, caterpillars, maggots, etc.) should be packed alive in some tight tin box—the tighter the better, as air-holes are not needed—along with a supply of their appropriate food sufficient to last them on their journey; otherwise they generally die on the road and shrivel up. Send as full an account as possible of the habits of the insect respecting which you desire information; for example, what plant or plants it infests; whether it destroys the leaves, the buds, the twigs, or the stem; how long it has been known to you; what amount of damage it has done, etc. Such particulars are often not only of high scientific interest, but of great practical importance. In sending soft insects or larvæ that have been killed in alcohol, they should be packed in cotton saturated with alcohol. In sending pinned or mounted insects, always pin them securely in a box to be inclosed in a larger box, the space between the two boxes to be packed with some soft or elastic material, to prevent too violent jarring. *Packages should be marked with the name of the sender.*

KITCHEN GARDEN FOR JULY.

In the Middle States, this month, like June, is the month of labor in the garden. Weeds are in rapid growth, plants are to set out, seeds saved, and various matters require attention.

Beans, plant for succession. *Beets*, the long blood and sugar; also Mangold Wurzel may be planted for stock as late as July. June is, however, much better. *Beets*, for late winter and spring use, may now be sown. *Cabbage*, plant. The winter sorts of cabbage should now be planted out. Where many are to be transplanted it is proper to await a suitable time—a heavy rain, or showery weather—but in a small garden cabbages may be transplanted almost at any season, by careful watering, and, if need be, shading. *Celery*, plant. *Endive*, sow. *Peas*, a few may be sown; they seldom do well this season. *Turnips*, sow.—*Rural Register*.

Quality and Vitality of Seeds.

Seeds properly ripened are, with few exceptions, as good the second year as the first—indeed, many are so well protected by natural envelopes that they germinate freely after many years. The vitality or germinating power of seeds is *not*, however, the *most important* question of the gardener, for if seeds fail to sprout, the first cast is the principal loss. The *quality* of the vegetables seeds may produce is the *all important* question, and that can only be determined when, perhaps, it is too late in the season to remedy the imposition of bad seeds which we may have suffered. Absolute security against seed-frauds can only be found in patronizing seed-houses of acknowledged reputation. It no doubt, in the "long run," would be well to reject seeds peddled through the country and sold on commission by irresponsible, and often unknown, if not unprincipled, seed venders. Time is too precious, and the outcome too important, to hazard much, speculatively, in garden seeds.

HOW TO KILL WHEAT MOTH.

A gentleman of experience remarks as follows in regard to the wheat moth:

"I know of but one efficient remedy for this insect, and that applies as well to the weevil and to the Angoumois grain moth, which is said to do no little damage in the southern and southwestern part of the country. I have frequently seen every kernel of corn in samples from the Gulf States perforated by this moth larva. The remedy proposed is bisulphide of carbon. We have only to pour a quantity of this into the bin at the bottom of the grain to kill all of the insects. It is very penetrating and volatile and equally deadly to all the insect tribes. I think that half a pint of the liquid would destroy the insects in a bin of 50 to 100 bushels of grain. Not having experimented with grain in such quantities I cannot give the precise quantities of the liquid to be used in the different sized bins of grain, but this can be easily determined by trial. To try this remedy it is desirable to pour the liquid in at the bottom of the grain. To do this we can take a hollow iron cylinder—a gas pipe will do—and fit into it a wooden rod, which should be a little longer than the iron tube. One end of the rod is to be made sharp; now place the rod inside the tube, and with the sharp end down force then both to the bottom of the grain: then, having withdrawn the rod, turn in the liquid through the tube, which should then be pulled out. The insecticide, of course, is left at the bottom of the grain, and being very volatile, soon diffuses through the mass and converts the bin into an insect cemetery.—*American Miller.*

If such is absolutely a *fact*, established by experience, we confess we have more confidence in it, even for destroying the larvæ of the weevil, than we have in any "best remedy" involving the mere "stirring of the grain." Exactly what insect is meant by the "wheat moth" in the above, we do not clearly understand. Perhaps *Pyralis farinalis*—perhaps something else, for wheat or grain moths are many. Many years ago a small ear of corn was sent to us by mail, every grain of which contained the larva of a moth, which was determined for us as the "Angoumoise grain moth" (*Butali cerealis*). When the moths evolved they left a hole in the centre of the grains as round and sharp as if drilled in; and we do not think they could have been destroyed save by some remedy analogous to that described in the above paragraph.

Remedies for the Army Worm.

To meet a general demand that will probably soon be felt and made for the best means of coping with the army worm, I would here repeat in condensed form what I have in previous years recommended. Experience has established the fact that burning over a meadow, or prairie, or field of stubble, either in winter or spring, usually prevents the worms from originating in such meadow or field. Such burning destroys the previous year's stalks and blades, and, as a consequence of what I have already stated, the *nidi* which the female moth prefers. Burning as a preventive, however, loses much of its practical importance unless it is pursued annually, because of the irregularity in the appearance of the worm in injurious numbers. Judicious ditching, *i. e.*, a ditch with the side toward the field to be protected perpendicularly or sloping under, will protect a field from invasions from some other infested region when the worms are marching. When they are collected in the ditch they may be destroyed either by covering them up with earth that is pressed upon them, by burning straw over them, or by pouring a little coal-oil in the ditch. A single plow-furrow, six or eight inches deep and kept friable by dragging brush in it, has also been known to head them off.

From experiments which *I have made* I am satisfied that where fence-lumber can be easily obtained it may be used to advantage as a substitute for the ditch or trench, by being secured on edge and then smeared with kerosene or coal-tar (the latter being more particularly useful) along the upper edge. By means of laths and a few nails the boards may be so secured that they will slightly slope away from the field to be protected. Such a barrier will prove effectual where the worms are not too persistent or numerous. When they are excessively abundant they will need to be watched and occasionally closed with kerosene to prevent their piling up even with the top of the board and thus bridging the barrier. The lumber is not injured for other purposes subsequently.—*Prof. C. V. Riley.*

Melons—Bugs—Coal-Tar.

Among the most effective applications that I have ever known to keep bugs off of vines is tar-water. Stir coal-tar in a vessel of water, let it stand over night till the water is scented and colored with the coal-tar; then, morning, noon, and evening, or as often as convenient, go and sprinkle the vines and hill with the liquid; it will both keep the bugs away and make the plants grow more vigorously, being a good stimulant to such plants. Sprinkling the ground freely over the hill will almost wholly kill or keep away the cutworms and grubs. Very freely applied it does much to kill off the potato beetle, which is so destructive in some localities.

Insect Powder.

Wm. Saunders, of London, Ontario, well known for his horticultural experience, as well as distinguished as the editor of the *Canadian Entomologist*, finds the Dalmatian Insect powder, made from Pyrethrum cinerariæfolium, an excellent insecticide. He says: "House flies are very sensitive to the effects of these powders. A few puffs of the dust from an insect gun, blown into the air of a room with the doors closed, the discharges directed toward those parts where flies are congregated, will stupefy and kill them within a very short time. The powder is somewhat pungent, and to breathe an atmosphere charged with it will frequently cause a slight sneezing, but beyond this the operator need not anticipate any annoyance. Frequently during the past summer, when flies have been troublesome, we have pretty thoroughly charged the air in our dining-room and kitchen at night, closing the doors, and in the morning found all, or nearly all the flies lying dead on the floor. A few minutes after its use they begin to drop on their backs, and after a very short time die; if a room be closed for half an hour after using the powder, few, if any will escape."

He finds it as good against Aphides and other plant lice. Much superior in its results to tobacco smoke.

OUR LOCAL CROPS.

As we go to press our farmers have about finished gathering their hay, wheat and rye crops; and the present indications are that they have been unusually bountiful—considerably more than a fair average. Of course there will be some exceptions to the general result, influenced by local causes, both favorable and unfavorable.

The oats crop is also promising, and some very "tall oats" is reported in various localities. Perhaps no season has passed for a long time in which a more vigorous cereal growth has occurred.

The late rains have also had a stimulating effect upon the corn, potato, and tobacco crops, although in some localities great injuries from noxious insects have been reported, and especially by the notorious *cut worm*.

The term "cutworm" covers a large number of species, and many varieties, all of which are "maliciously" destructive to vegetation—cutting off much more than they can possibly devour. This season we have many complaints against them, as being severe upon the young tobacco plants, often necessitating two or three different plantings. The tobacco growers cannot go far wrong in concluding that this enemy to their cherished plant has "come to stay." It has in fact always been here, but nothing furnishes it such a luscious food as tobacco, and therefore it must be classed with "consumers." If this plant had no enemies at all, it would soon become a mere drug, and no sale could be found for all of it. The cutworm will be the great regulator of the quantity, of the quality, and also of the price. Like the "Colorado Potato Beetle," means must be found for its destruction, and this will involve a perpetual labor. It can never be said that "they are now extinguished," for perhaps when least expected, they will be most abundant. Fortunately for themselves, but unfortunately for the tobacco grower, they can and do thrive on other plants than the tobacco. When they attack this plant they are already well grown, and nearly mature, and hence must have been in the soil before the tobacco was planted, or in proximity to it. We raise as many and as good potatoes now as we did before the advent of the potato beetle, because we apply the remedies for their destruction when they become too abundant, and this must also be resorted to in regard to the tobacco. Of course, it will be more difficult to contend with the cut worm than with the potato beetle, inasmuch as the former is a "midnight marauder," whilst the latter is an "open enemy." With all these counter influences, there will be an immense crop of almost everything the present season and there may be some anxiety to know what to do with it—it would be unchristian to wish for war as an outlet.

DESTROYING WEEVIL.

The best remedy yet found for their extermination is frequently stirring the grain. It is more than probable that fully saturating the bins with the fumes of sulphur will kill the insects, and this would not be difficult to do by means of a suitable apparatus. Although Curtis says that turpentine and the fumes of sulphur did not seem to incommode the insects. Kiln-drying at a heat of 130 degrees will kill them without injuring the germinating powers of the grain. Placing the infected grain in close bins, without moving, is the best possible way to continue their ravages, since they delight in darkness, and in grain that is not handled. We do not suppose that the present scare in weevil in grain in the Chicago elevators will affect prices seriously, but if the weevil becomes generally disseminated in the west, it will become a most serious matter; not among farmers themselves, if the means we have indicated are closely followed (moving and fanning the grain often), but in elevators the means of destruction will not be so easily managed, though there is little doubt if the fumes of sulphur be driven into the bins, and there retained for ten hours, the destruction of the insects will be complete.—*Prairie Farmer.*

"Frequently stirring the grain," seems a very simple remedy for the extermination of the weevil, and if it is the "best yet found," it should be by all means universally adopted. It might be of some use in expelling the mature beetles from the bin, but we don't see how it would *exterminate* the larva, which is snugly ensconsed within the grain, and which *could* not leave it if it *would*, until its final evolution from the pupa state. This stirring the wheat is equivalent to jarring fruit trees for the expulsion of the *curculio*, but the jarring is only intended to disturb or expel the mature beetle, and not at all the larva, which is beyond the influence of the jarring, it being *inside* the fruit, and not inclined to come forth until it has fully matured as a larva. The grain weevil is the *Sitophilus granarius*, of Linnæus, and when it has fully matured it leaves the grain of its own accord, and hides itself in some convenient nook or crevice and there hibernates, and comes forth in due season to deposit its eggs on the grains of wheat of a subsequent crop.

EFFECTS OF BAKING ON FLOUR.

Good bread should be full of small pores, and *uniformly* light. Such bread is produced by strong flour; that is, such as will rise well, retain its bulk and bear the largest quantity of water. The largest proportion of gluten usually contained in the flour of wheat, gives the higher value it has over that of other grains. If the gluten be washed out, and put alone in the oven, it will swell and become full of pores, and the comparative baking qualities of different samples of flour can be tested by the height to which specimens, so treated, rise.

Dry starch, when heated, is generally changed into a species of gum, and of sugar completely soluble in water. According to Vogel 100 parts of flour, and of the bread made from the same wheat, respectively tested, shows a gain in the latter of 18 parts of gum at the expense mainly of the starch. The yeast added to the dough induces fermentation, by which the sugar of the flour is changed into carbonic acid and alcohol. The carbonic acid, liberated in the form of minute bubbles of gas, permeates the whole substance of the dough, causing it to rise. If too much water has been added—or if not sufficiently kneaded—or if the flour be too finely ground—or the paste not sufficiently tenacious in its *nature*—the bubbles will run together, forming large airholes, and that irregular appearance so disliked by the skillful baker. The quantity of water which bread retains, when baked, depends in some degree on the quality of the flour. The Acts of Parliament, England, assume that 280 pounds of flour will produce 320 pounds of bread—thus calculating the retention, when baked, of one-seventh of its weight of water. But the quantity of water retained by the flour now in use is much greater.

Johnston, in his lecture on Agricultural chemistry, states that home-made bread (white and brown) baked in his own house, whether of first or second quality, as well as that baked in two other private houses, lost by prolonged heating, at a temperature not exceeding 230° F., from 42.9 to 44.1 per cent. of water. So that wheaten bread, one day old, contains about 44, and two days old, 43 per cent. of water. This proportion is almost exactly the same as Dumas estimates the white bread of Paris.

Bread baked for public institutions, not generally being so well fixed, or baked with many loaves stuck together, contains more water. The barracks bread of England and Paris contains about 51 per cent of water. English wheaten flour contains naturally, on an average, 16 per cent. of water. If, therefore, the bread baked from it contains 44 per cent., 33⅓ per cent. will have been added to the natural amount, or the flour in baking takes up half its weight of water. A sack then of 280 pounds of flour ought to give 421 pounds of well baked bread. Deducting, say 5 per cent., for fermentation and dryness of the crusts, there would remain 400 pounds of bread of the best quality.

Chemical writers have assumed that the quantity of water absorbed depends mainly upon the proportion of gluten the flour contains. The following facts, says Prof. Johnston, do not accord with this supposition. (1) Household bread, made respectively from the flour of French wheat, and of wheat from Taganrog, Russia, retained nearly the same amount of water; tho' a sample of the latter contained more than twice as much gluten as the French. (2) The flour from Odessa wheat contains about one-fourth more gluten than French flour in general, yet it absorbs very little more water. (3) Rice is said to contain very little gluten—not estimated at more than 6 or 7 per cent.—and yet, as the result of numerous trials, it is said that an admixture of a seventh part of rice flour causes wheaten flour to absorb more water. (4) If hard wheats are ground too fine they lose a part of their apparent strength, the flour refuses to rise as it would do if sent to the baker in a more gritty and less impalpable state. (5.) Lastly, the admixture of very minute quantities of foreign matter, by way of adulteration, increases the water absorbing power of flour. In some parts of Belgium it is said to have been the practice to adulterate the bread with a small quantity of blue vitriol (sulphate of copper). A solution of the salt added to the dough, in proportion of about one grain to two pounds of flour, gives the bread a fawn color and thus permits the use of inferior flour, and causes the bread to retain about 6 per cent. more water, without appearing more moist. Alum improves the color of bread, raises it well and causes it to *keep* water, but requires to be added in larger quantities than the poisonous salt of copper. Common salt also strengthens the paste and causes it to retain more water, so its addition is a real gain to the baker.—*American Miller.*

PHOSPHORIC ACID IN PLANTS.

The substance especially important to the farmer is undoubtedly phosphoric acid, which is found in combination with lime, as plants assimilate the same in considerable quantity, while it is sparingly contained in the soil.

Plants require phosphoric acid in the following proportion to 1,000 pounds:

Wheat...............	8 1-5 lbs. equal to 17½, lbs. bone phos.	
Wheat Straw......	2½ "	5 "
Barley...............	7 1-5 "	15 "
Barley Straw.....	19-10 "	41-10 "
Oats..................	6½ "	12 1-10 "
Oats Straw........	18-10 "	39-10 "
Rye...................	8 1-5 "	17½ "
Rye Straw.........	19-10 "	41-10 "
Corn..................	5⅗ "	12 1-10 "
Corn Stalk & Leaf.	3 8-10 "	8 "
Pea Straw........	9 1-10 "	19-3 10 "
Beans...............	11 6-10 "	25½ "
Bean Straw......	41-10 "	81-5 "
Potatoes..........	1 8-10 "	4 "
Green Potato		
plant.............	6 4-10 "	13-10 "
Beet Roots sugar	11-10 "	2 4-10 "
Green Beet Tops.	13-10 "	2 8-10 "
Hemp (whole		
plant)............	33-10 "	7 1-5 "
Linseed............	7 4-10 "	16 1-5 "
Tobacco...........	7 1-10 "	15 6-10 "
Clover hay).....	5 6-10 "	12 "
Meadow (hay)..	41-10 "	81-5 "

If grain, potatoes, etc., are to nourish us and our cattle, they *must* contain phosphoric acid, as our growing bones require one-third of this substance in the form of phosphate of lime, in addition to considerable contained in blood and muscles.

Innumerable experiments have proven—

1st. That plants cannot perfectly develop unless the soil contains sufficient phosphoric acid.

2nd. That the application of phosphate increases the weight and quality, and frequently shows a difference of more than twenty per cent. in the particles of starch.—*From "What of Fertilizers."*

We cannot ignore the fact that all vegetable, as well as animal, growth, require for their normal development a sufficient quantity of inorganic and mineral substances as stimulants to that end, and that phosphoric acid is one of the most prominent among them.

COMMON hydraulic cement mixed with oil forms a good paint for roofs and out-buildings. It is waterproof and incombustible.

"A MARE'S NEST,

I have been unable to find any explanation of the origin of this oft used phrase, unless a German story, often heard in my childhood in Pennsylvania, may furnish it.

The Swabians (called "Schwopes" in Pennsylvania) are among Germans what the Irish are among English-speaking people, but less volatile, witty and frolic-loving—like them in their aptitude for blundering, and in confounding and intertangling subjects so as to form what are called "Irish Bulls." Of course every comic blunder and burlesque speech is, by the rest of the Germans, ascribed to the Schwopes. So much by way of preface. Now for the story.

A Schwope in passing through a cornfield saw a number of pumpkins, and inquired what they were. He was told they were mare's eggs. He bought one of the largest and carried it it on his journey until he reached the top of a long hill. Wearied with his walk and his burden, he laid down the pumpkin and sat down on it for the double purpose of resting himself, and also aiding in hatching out the mare's egg. As he rested, meditating on the advantages and pleasure of having a horse on which to ride, instead of trudging on foot, he fell asleep, lost his balance, and away rolled the pumpkin down the hill! Now there was a heap of dried brush at the foot of the hill, and in that brush heap a rabbit had made his home. On rolled the pumpkin with increasing speed, and, striking the brush heap, broke into pieces. The astounded "Bunny," thinking "the day of doom" had come, bounded away in affrighted haste. The poor Swabian, who awoke in time to see all this, verily thought that the rabbit was a colt released from his pumpkin, ran after "Bunny," whinneying like a mare after her foal, and crying out, in what he meant for the most endearing terms in horse language—"Ilee-haw! Ilee-haw! Hutchelie, da ist dein mutter !"—in English—"Ilee-haw! Ilee-haw ! Coltie, here is your mother!"

If any of your numerous readers can give us a better or truer explanation of the origin of the phrase, "finding a mare's nest," I cheerfully "yield him the floor," and will be obliged for his explanation. That the above has an unsatisfactory ending, but makes it in accord with the result of "finding a mare's nest."—G.

[Although the term "mare's nest" was known long before the childhood of the writer of the above, and also beyond the borders of Pennsylvania, it is doubtful whether any better account can be given of its origin than the one he alludes to; be the locality of its birth Pennsylvania or Swabia. It is a sort of paragram, which probably had its origin in some trivial circumstance that never was recorded, but which was sufficiently expressive to become popular among common people. To find a mare's nest is to make what you suppose to be a great discovery, but which turns out to be all moonshine. According to Dr. Brewer what we call a "nightmare" was by our forefathers supposed to be the Saxon demon mara or mare, a kind of vampire, sitting on the sleeper's chest. The vampires were said to be the guardians of hidden treasures, over which they brooded as a hen does over her eggs, and the place where they sat was termed their nidus or nest; hence the big-eyed, many-horned and long-tailed nightmares which so many see, may only be imaginary personations hatched out of a superabundance of soft-crab, buckwheat cakes and sausages, packed into an overwrought stomach just before going to bed.

When any one supposes he has made a great discovery we ask if he has discovered a mare's nest, or the place where the vampire keeps guard over hypothetical treasures. "Why dost thou laugh? What mare's nest hast thou found?"—Beau. and Fletcher.

Dr. Brewer says, farther, in some parts of Scotland the people use a skate's nest instead of mare's nest, and in Gloucestershire a long-winded tale is called a horse-nest. In Devonshire any kind of nonsense is called a blind mare's nest, and in Cornwall they say you have found a mare's nest and are laughing over the eggs. The word mare in England has various legendary phrases associated with it. For instance, the Cromlech at Gorwell—a large stone resting on two or more others, like a table—is called the white mare, and the Barrows, near Hambleton—tumuli or mounds —are called the grey mare. Away with the mare meant off with the blue-devils, or good-nye to care. This mare is the incubus called the nightmare.

To win the mare and lose the halter, was to play "double or quits," a reckless kind of speculation or gambling, which impoverishes nineteen where it enriches one.

In Herefordshire and Shopshire, to cry the mare was a singular harvest custom. When the ingathering was completed, a few blades of corn, left for the purpose, would have their tops tied together. The reapers then placed themselves at a certain distance and flung their sickles at the "mare." He who succeeded in cutting the knot would cry out "I have her." "What have you?" "A mare." "Where is she?" The name of some farmer whose fields had been reaped would here be mentioned. "Where will you send her?" The name of some farmer whose corn had not yet been harvested would then be given, and then all the reapers would give a final shout—"the mare."

The gray mare is the better horse: means that the woman is paramount. It is said that a man wished to buy a horse, but his wife took a fancy to a gray mare, and so pertinaciously insisted that "the gray mare was the better horse," that the man was obliged to yield the point. When a woman is paramount, the French say: "'Tis a hawk's marriage," because the female hawk is generally both larger and stronger than the male bird. Prior wrote :

"As long as we have eyes, or hands, or breath,
We'll look, or write, or talk you all to death,
Yield, or she—Pegasus will gain her course,
And the gray mare will prove the better horse."

In a work on Old Glees and Catches, the following is given as the origin of that popular maxim, "money makes the mare go."

"Will you lead me your mare to go a mile?"
"No, she is lame leaping over a stile."
"But if you will her to me spare,
You shall have money for your mare."
"Oh, ho! say you so?
Money will make the mare go."

It will be observed, however, that all this historical evidence is based upon the traditional—"It is said :" but who said it, where it was said, or when it was said, no deponent sayeth. We must therefore take it as we find it, and for what it is worth. If we limited our knowledge within the scope of our own practical experience, perhaps we should know but precious little, and that little would be circumscribed by our opportunities, and our habits of observation. If a long-winded story may be properly regarded as a mare's nest, then our readers may have found one in these cogitations. [Ed.

THREE WONDERS.

It is related of an aged Friend (or Quaker) that, "moved of the spirit" to rise and speak in meeting, she said there were three things in life which caused her to wonder greatly. The first was that boys worried themselves by throwing sticks and stones into the trees to knock down the apples; when, if they would but wait, the apples would fall of themselves. The second was, that men took so much pains and spent so much money in going to war to kill each other; when, if they would but wait a few years, their enemies would die of themselves. And the third was, that the boys took so much trouble and spent so much time in running after the girls; when if they would but mind their work and stay at home, the girls would run after them!

I, too, have a triad of wonders, but they are not exactly like those of the good old Friend preacher—as the reader may see.

The first is, that the makers and vendors of alcoholic drinks assert that all prohibitory laws only increase the sale and use of intoxicants, when they oppose all such laws as being injurious to their business: The second is, that the makers of Oleomargarine declare that their article is better than most kinds of butter—equal to any butter except the very finest—cheaper, more pure and wholesome—and that it is preferred to common butter by all who have tried it; and yet they do not advertise it as Oleomargarine, nor label it as Oleomargarine, but palm it off as butter, and oppose all legislation requiring it to be sold only for what it is. And the third wonder is, that farmers, who color their butter with unknown coloring matters, that they may palm it off as grass or June butter, and who declare that people prefer it colored, and pay more for it, do not label or advertise their butter as "colored," so as to induce buyers to take it, and to pay more for the adulterated article than they would pay for the "Simon-pure" and honestly genuine article!

Some may wonder that brewers and distillers, and Oleomargarine men, and all lard-cheese and colored butter makers do not combine, and urge our legislatures to pass prohibitory laws, and laws compelling all adulterators of cheese and butter to label their articles, and sell them for what they are!—G.

EXCERPTS.

GOOD feeding is the secret of success in sheep-husbandry.

THERE is no portion of our country where sheep husbandry can be more profitably carried on than in the Virginias, the Carolinas, Tennessee, and portions of Georgia and Alabama.

GARDENING is regularly and practically taught in more than 20,000 primary schools in France. Every schoolhouse has its garden, and teachers must be not only good gardeners, but qualified to teach horticulture, or they cannot pass examination.

The celebrated English farmer, Alderman J. J. Mechi, Tictree Hall, has but six acres of permanent pasture, and yet manages to keep as an average 200 sheep, and from fifteen to twenty head of cattle. All food is cut up, no roaming at large is allowed, and supplemental food is invariably given. The sheep are always within iron hurdled folds, removed morning and evening.

HEN manure should not be composted with unleached ashes unless it is to be used immediately. It is better to mix it after being thoroughly pulverized with dry earth, which is one of the very best absorbents. All of the valuable constituents of the fertilizer will then be saved.

WHAT is needed is that our American farmers should exhibit the same intelligent spirit of enterprise displayed by the inventor and manufacturer, that they should appreciate the new spirit of American civilization and rush forward to make agriculture not the mere follower and servant of manufactures, but the great leading and advancing interest.

Says the *Live Stock Journal:* The stock ranch and summer residence of ex-Governor Stanford, of California, contains about 300 highly-bred horses, and it requires a mile of stable to accommodate them. He is breeding his thoroughbred mares to trotting stallions; not especially with a view to the production of fast trotters, as some of our contemporaries would have us think, but as a means of laying the foundation of permanent improvement in the horse stock of this State, for general purposes; and in this he is not far out of the way.

THE longest line of fence in the world will be the wire fence extending from the Indian Territory west across the Texas Panhandle, and thirty-five miles into New Mexico. We are informed that eighty-five miles of this fence is already under contract. Its course will be in the line of the Canadian river, and its purpose is to stop the drift of the northern cattle. It is a bold and splendid enterprise and will pay a large percentage on the investment. The fence will be over 200 miles long.

THE oat crops of Georgia, South Carolina, and North Carolina, according to all accounts is the largest ever made in those States. The crop is now being harvested. It is estimated that Wilkes, Lincoln and Hancock counties, in Georgia, will produce one million bushels each. The Washington *Gazette* says the entire small grain crop of Wilkes county has been estimated at one and a quarter million bushels. One planter in that county has a thousand acres of oats and the yield will be fully forty thousand bushels. A planter near Augusta will make twenty thousand bushels of oats and wheat. With this immense crop there will be more than sufficient for home consumption and a large quantity can be sold, bringing a considerable amount of money into the State.

THE capital investment in railroads in this country has been divided as follows: Jay Gould and associates, $565,000,000; the Pennsylvania Central, $629,000,000; Vanderbilt combination, $564,000,000, Huntington combination, $321,000,000; Jewett and the Erie combination, $317,500,000; Garrett, of the Baltimore and Ohio combination, $194,000,-000; the Pennsylvania coal roads, $508,000,-000; Alexander Mitchell management, $129,-000,000; Garrison management, $62,000,000. —*Exchange.*

A HINT FOR COFFEE DRINKERS.—While "dining out" one day recently, the coffee, which, though the last, was by no means the least of the good things furnished, was so unusually excellent that it was the subject of general remark, and a word in the ear of the charming hostess after retiring to the drawing room called forth the following explanation of how the good result was obtained: The coffee furnished was a clear amber in color, rich in flavor and deliciously aromatic. To give the hostess' method a fair test it will be no more than just to don one's apron and adjourn to the kitchen. The coffee to be used is Maracaibo and Java, equal parts of each, finely ground. One large cup of coffee, one cup of cold water, one well beaten egg, mix thoroughly; add four cups of cold water and place over the fire. After it reaches the boiling point allow five minutes to finish the process; strain and serve immediately. This seems a very simple process, but in the hands of a servant, if allowed to boil too long, it would be easily spoiled.

THE STRENGTH OF HORSES.—Lieut. Roder of the German Army, has been riding to Granada from Strasburg in order to find out how far it is possible, under certain conditions, to draw upon the strength of horses. He left the latter place on September 29, and arrived in the former on November 20, a period of 53 days, including 8 days of rest and a distance of 2,500 kilometres. His animal was a Prussian mare, 9 years old, and when he arrived in Granada he found no difficulty in selling her to advantage. He wore no spurs, and his baggage comprised only a water-proof and a pair of capacious saddle pockets, in which were a guide-book, some maps and a few other objects. The pace at which he rode was a steady trot when the ground permitted, and a fast walk when he could not trot. Roder concludes from this experience, and in spite of the apparent good results of it, that so much work is too great for good horses and vigorous men.

HOW TO COOK RICE.—Rice is becoming a much more popular article of food than heretofore. It is frequently substituted for potatoes at the chief meal of the day, being more nutritious and much more readily digested. At its present cost, it is relatively cheaper than potatoes, oatmeal or grain-grits of any kind. In preparing it only just enough cold water should be poured on to prevent the rice from burning at the bottom of the pot, which should have a close-fitting cover, and with a moderate fire the rice is steamed rather than boiled until it is nearly done; then the cover is taken off, the surplus steam and moisture allowed to escape, and the rice turns out a mass of snow-white kernels, each separate from the other, and as much superior to the usual soggy-mass, as a fine mealy potato is superior to the water-soaked article.

HOW TO CATCH CROWS.—A gentleman writes us that he has succeeded in catching several crows from his corn-field in the following novel manner: "I arranged a number of large twine strings with a slip-noose in each, and placed them on stumps in the fields in such a manner that when pulled the stump would not interfere with the closing of the noose. I stood hidden at a convenient distance, and would almost invariably catch the crow when he alighted on the stump. I caught eleven in one morning in this manner."

THE FIRST BALLOON.—In June, 1783, Stephen and Joseph Montgolfier sent up the first balloon. To commemorate the centenary of the event, it is proposed that an International exhibition of "aerial arts" be held at Paris next year. The "aerial arts" are to include every industry, science or art, relating to gas or the atmosphere, which is supposed to have any connection directly or indirectly with aerostatic experiments.

FATTENING SHEEP IN WINTER.—In the first place a good way is to begin early in December by giving, in addition to straw, to each sheep, each day for a couple of months, a pound of meal, grain, or oil cake.

IF the roots of tulips and hyacinths are left in the bed where they have bloomed and the stalks cut after blooming and the bed sufficiently protected in the winter there will be annual blooming. The reason why hyacinths that are flowered in water-glasses are exhausted and make so poor a growth is that the flowers and stems are produced at the expense of the bulb, and this is not renewed in any way. When grown in rich soil this exhaustion does not occur and the bulbs are able to bloom repeatedly.

SAVE the middle grains of the fine ears of corn for seed.

HOGS should be allowed to have a heap of coal ashes. They will be all the healthier for it.

BEEF and mutton are not flavored by feeding turnips to the animals—at least this is the statement of some who have tried it.

THE amount of fruit shipped from California during the present season will bring about $1,000,000 profit to the State.

IT costs the people of Tennessee $1,000,000 annually to sneeze and use snuff. This is a Nashville merchant's estimate of the annual consumption of the article.

AN orchard should never be planted in a clay soil unless the latter is underdrained, after which it becomes one of the best soils for apples and pears.

LET every farmer keep all the stock he can possibly afford to—and generally he can afford to keep more than he does. The dependence of farming for all time must be mainly on stock.

A WISCONSIN farmer, twenty-three years ago, planted a piece of waste land, unfit for cultivation, with black walnut trees. The trees are from sixteen to twenty inches in diameter and have been sold for $27,000.

FRANCE produced last year 750,000,000 gallons of wine. Of these, 47,000,000 were made from sugar, $1,950,000 from raisins, while 154,000,000 gallons were imported from Spain and Italy, to "blend" with their home product. No wonder everybody wants to drink French wines; they are so pure.

IN a small grove which adjoins the Schoenberger residence near Cincinnati, an army of crows take shelter every night. They assemble by thousands an hour before dark, and an old man living near the place says that to his personal knowledge the same grove has been their dormitory for sixty years.

Don't Do It.

Don't sleep in a draught.

Don't go to bed with cold feet.

Don't stand over hot-air registers.

Don't eat what you do not need just to save it.

Don't try to get cool too quickly after exercising.

Don't sleep with insecure false teeth in your mouth.

Don't start the day's work without a good breakfast.

Don't sleep in a room without ventilation of some kind.

Don't stuff a cold lest you be next obliged to starve a fever.

Don't try to get along without flannel underclothing in winter.

Don't use your voice for loud speaking when hoarse.

Don't try to get along with less than eight or nine hours' sleep.

Don't sleep in the same undergarment you wear during the day.

Don't toast your feet by the fire but try sunlight friction instead.

Don't try to keep awake upon coffee and alcoholics when you ought to go to bed.

Don't drink ice water by the glass; take it in sips, a swallow at a time.

Don't strain your eyes by reading or working with insufficient or flickering light.

Don't use the eyes for reading or fine work in the twilight of evening or early morn.

Don't try to lengthen your days by cutting short your nights' rest; it is poor economy.

Don't wear close, heavy fur or rubber caps or hats if your hair is thin or falls out easily.

Don't eat anything between meals excepting fruits, or a glass of hot milk if you feel faint.

Don't take some other person's medicine because you are troubled somewhat as they were.

Don't blow out a gaslight as you would a lamp; many lives are lost every year by this mistake.

WHAT THE FARMERS MUST FEED.—The Census Bureau has issued a bulletin showing that the live stock of the United States on farms on June 1, 1880, was as follows: Horses, 10,357,981; mules and asses, 1,812,932; working oxen, 993,976; milch cows, 12,443,593; other cattle, 22,488,500; sheep, 35,191,656; swine, 47,683,951. The rate of increase from 1870 to 1880 was, in horses, 45 per cent; mules and asses, 61 per cent; working oxen, a decrease of 25 per cent; milch cows, an increase of 30 per cent; other cattle, 66 per cent; sheep, 24 per cent, and swine, 90 per cent.

SELECTIONS.

THE USES OF PRUNING.

Pruning is to the tree what education is to the mind, or the "polishment" of the marble after it is taken from the quarry. Pruning is absolutely beneficial to all kinds of fruit trees at least. Of course we mean pruning as a *use*, and not as an *abuse*.

As to the best *time* for pruning, in my view, it is to begin as soon as the trees are two feet in height. You then can use your pocket-knife, which ought to be used constantly whenever "suckers" appear. This gives the tree a good shape and takes all the surplus wood away. This work can be done any time during the year, with little exception. There are only about two or three weeks during which I generally avoid pruning—that is, from the time the sap begins to flow until the leaves are developed. Most fruit trees require continual pruning and shaping, to make them bear better and larger fruit, and with-.al, impart them to *beauty*. Yet, there is a great difference, especially in apple trees. Some trees need much more pruning than others, nearly all the time, or they would become like a hedge-fence. The Pennsylvania Red-streak, Munson-sweet, and the Wagner require very little pruning with me. Cherries also require little shaping, but plums, prunes and pears, are much improved by early and judicious pruning. The peach is also improved, and we all know that the grape needs a yearly thinning-out to bring it to perfection, unless it is mainly desired for shade, over an arbor, pump, shed, or a south-side exposure to the summer sun.

Nearly all kinds of trees need training and pruning, unless growing in a dense forest, and no man possessing an "arboricultured" eye and mind, can even pass through a forest, without speculating on improvements, here and there, that would have resulted from pruning, or the removal of obstruction. I have now a limited, but dense, forest of locusts, poplars, walnuts and chestnuts, in which the trees are growing straight up from forty to fifty feet in height, with the side branches dying and dropping off; but the same trees elsewhere, want trimming, or they would get too "forky", or spreading. Along a road or in a yard, such trees require constant shaping, in order to make good "butts," and beautiful and symmetrical tops. This is however controled very much by fashion, fancy or individual tastes.

When on a recent visit to the *Central Park* in New York, I was astonished at the luxuriant growth of the many varieties of trees in that magnificent enclosure. They have all kinds of ornamental and common forest trees, all over the park. These trees are almost invariably trained to grow with low tops, and long side or lateral branches, from two feet from the ground upward, many of them looking very strange, if not unslightly, for what purpose I could not understand. If I had had the control of the park, I would have trimmed every tree up from eight to ten feet from the ground, so that persons could easily promenade under their branches. But as it is now, it could not be done any more, as the trunks of some of them are a foot in diameter near the ground. The Elm is a favorite tree in the park, and in fact is also a leading tree all over the west, as well as many parts of the east.—L. S. R., *Oregon, July.* 1882.

[In relation to the Elm, we do not hazard much in saying that it is and always has been a favorite tree in Pennsylvania, and, the very first event which signalized the origin of the State, transpired under the spreading branches of an Elm, on the banks of the Shackamaxon, in the old "Northern Liberties" of Philadelphia. Boston common had at one time, and perhaps still has, some fine old elms. There were many of them in Lancaster, and some are still remaining. But the fact is, of late years they have been so much subjected to the ravages of the "Elmleaf Beetle," that many persons have been compelled to cut them down. This beetle is so exclusively destructive to the foliage of the Elm, and occurs in such immense numbers, producing two or three broods during the season—that when the trees are large, there seems to be no practical remedy but to remove them entirely. In regard to the low branched trees to which our contributor alludes, perhaps the authorities don't want people to promenade under them, lest they also trespass upon the grass. They provide special promonades, seats, canopies, pavilions and trees, sufficient for the shelter and the exercise of pedestrians, unless there should happen to be an unusual crowd in the park, and as to sightliness, or unsightliness, *that* depends altogether on the peculiar taste of the viewing individual.—*Ed.*]

BALANCE OF TRADE.

EDITOR FARMER.—Your correspondent P. S. R., in the last number of *The Farmer*, referring to a discussion between him and myself two or three years ago on the "Balance of Trade" question, credits me with having denied that it was a better sign of prosperous times when the balance of trade was in our favor than when the balance is against us, and asks what I have to say *now*, since the balance is going strongly against us and seeing that we are importing about $3,000,000 worth of goods per week, and our exports last year were far less than in 1880. This state of affairs, Mr. R. thinks, must result to our disadvantage, and I do not dispute it.

But according to my recollection I never denied that it is better to have the balance of trade "in our favor." I simply denied that the fact that we import more value than we export proves there is a balance against us; and the reasons I gave for that opinion have not been answered, or scarcely attempted to be, from that day to this, so far as I have seen. See my several articles in *The Farmer* of February, April and June, 1879.

It is of course disadvantageous to us that we find short crops last year, and that consequently we had less of agricultural products to export. And it is also disadvantageous to the country if the $3,000,000 of goods imported per week are not paid for, but are bought on credit, to be paid for out of our future earnings, just as is the case with an individual who runs in debt beyond his earnings or his means to pay. The reason there is so much confusion of ideas and wrong notions on this subject, it seems to me, is that people have been led to imagine that the exchange of productions between two nations is governed by different principles, and its advantages or disadvantages are gauged by an entirely different rule than the trade between individuals. That this is an erroneous notion is evident from the fact that the trade of one nation with another is not between the two nations *as* such, but merely between individuals of those nations; and the profits or losses, the advantages or disadvantages of any trading transaction in which a man engages, of course are not in the least affected by the nationality of the person with whom he deals.

Everybody knows and will acknowledge that if an individual sells property, the more value he gets for it the better he is off, but strange to say, there are thousands of people in this country who will seriously contend that in our trade with foreign countries the less value we receive in return for what we part with, the more prosperous we must become!

Now if a farmer in Lancaster county sends abroad—exports—to Europe or elsewhere, grain or tobacco worth at home $100, his aim and object of course is to get in return more than $100, either in money, or money's worth in some other property. If he did not expect that, he would not send it away. Well, suppose he gets for it, say $125 (after paying all expenses) either in cash or clothing or anything else that he may prefer, it is perfectly clear that he is $25 richer, and so is the country. But here the import has exceeded the export by $25, and if the balance of trade theory of Mr. R. is correct, it is a most unfortunate transaction for the country! On the other hand, if owing to a fall in the market at the place to which the produce was exported, or other cause, only $80 is realized for it and brought home, then, according to the same theory—our exports having exceeded the imports—it shows a highly prosperous condition of our foreign trade! It seems to me that a theory leading to such a conclusion ought to be explained or abandoned.

Is it not clear, in the light of common sense, that the only advantage to the country from its exports is, that we are thereby enabled to import in their place something more valuable or desirable than what was exported? Is not every dollar's worth exported for which we do not or cannot import something of equal or greater value, effort wasted and money thrown away?—*J. P., Lancaster, July 7, 1882.*

U. S. DEPARTMENT OF AGRICULTURE.

To the Manufacturers of Sugar from Sorghum, Beets and Other Sugar-Producing Plants in the United States.

Congress in the appropriation for this Department, for the fiscal year commencing July 1st, 1882, has provided for "experiments in the manufacture of sugar from sorghum, beets and other sugar-producing plants."

In view of the experiments which have already been made at this department, I have determined to institute the following plan for the coming season, in obedience to the act referred to.

Provision has been made for continuing the chemical analyses of sorghum at the laboratory of the department, should this be deemed necessary, in order to add to the information already obtained by investigations not only here but also in the Agricultural Colleges of this country.

On assuming the duties of my office in 1881, I found 135 acres of sorghum containing 52 varieties which had been planted in Washington for use of the department. On being informed that time had arrived for manufacturing sirup and sugar, I engaged the services of an expert in sugar making who had been highly recommended for the position of superintendent, and operations were commenced on September 26, at the mill erected by my predecessor, on the grounds. These operations were continued with slight interruptions until the latter part of October, at which time the supply of cane became exhausted. Forty-two acres of the crop were overtaken by frost before being sufficiently ripe for use, and this portion of the crop was so badly damaged as to be unfit for manufacture. The yield of cane per acre, on the 93 acres gathered was two-and-a-half tons ; the number of gallons of sirup obtained was 2,977 ; and the number of pounds of sugar was 165. The expense of raising the cane was $6,589.45; and the expense of converting the cane into sirup and sugar was $1,967.59—an aggregate of $8,557.04.

The manufacture of sorghum at the department therefore has been found to be so expensive and unsatisfactory that the work can evidently be better conducted elsewhere. To repeat the experiment of last year would be unwise under any circumstances, and it is made doubly so by the impossibility of procuring the sorghum cane at any reasonable price in this neighborhood, after the discouraging crops of last year, and by the additional fact that the appropriation is not available until too late in the season for planting to begin.

While therefore such scientific investigation as is deemed necessary at this department will be continued—the experiment of manufacturing can better be conducted by those who have thus far furnished us all the valuable information we have ; and this work I refer to the manufacturers themselves, to whom I submit the following proposition.

Each manufacturer is requested to submit an account of his work to this department, covering the following points, viz :

1. An accurate account of the number of acres of sorghum brought to his mill; the number of tons of cane manufactured; the yield of sorghum per acre; the mode of fertilizing; the time of planting; the time required for maturing the plant; and the value of the crop as food for cattle after the juice has been expressed.

2. The amount of sugar manufactured ; the amount yielded per ton of cane; the quality of the sugar; the amount of sirup manufactured; the process of manufacturing; the machinery used; the success of the evaporator, the vacuum-pan and the centrifugal in the work of manufacturing.

3. The number of hands employed in the mill; the cost of fuel; the cost of machinery; the wages paid for labor; and the price of sorghum raised at the mill if not raised by the manufacturer.

The returns when received will be submitted to a competent committee for examination, and in order to compensate the manufacturers for the work of making these returns I propose to pay for the ten best returns the sum of $1,200 each,—the decision to be made by the aforesaid committee. Each return must be sworn to before a competent officer.

Sugar Beets.

I have distributed to ninety persons a supply of the best sugar beet seed which I could obtain; and I would request each person having received this seed to send to this department a statement of the amount of land planted by him; the yield per acre; the fertilizers used; the value of the crop in the market. I also request each person making this experiment to forward to this department a sample of the crop for analysis. The directions for this will be issued hereafter. An accurate statement of the process of manufacturing beet sugar in this county is of great importance, and I propose to compensate the manufacturers for preparing such statement by the payment of the sum of $1,200 for each of the two best returns submitted to a committee as in the case of sorghum.

Other Sugar-Producing Plants.

The promise of 1000 pounds of corn-stalk sugar per acre, which was made in 1841, and has often been repeated with great confidence but at the expense of the corn crop and in addition to it, not yet having been fulfilled in manufacture, the experiments not having been satisfactory, and the business not having been followed up, it is not deemed necessary to institute sugar making experiments in this direction during the present year. The same may be said of many esculents which have been classed as sugar producers.

All proposals to enter upon this work for the department must be laid before the Commissioner on or before August 1st, 1882.—*Geo. B. Loring Commissioner of Agriculture, Washington, D. C., June 6th, 1882.*

THE HAPPY GRANGER.

Statistics show that so far this season the South has drawn on the North for wheat to the value of $35,000,000 ; corn, $50,000,000 ; provisions, $72,000,000—making an aggregate of $177,000,000. This sum indicated will make a very large hole in the net value of the South's cotton crop. The lesson of the past season, it is satisfactory to know, however, has not been without some excellent results in inducing the planting of an extended acreage in breadstuffs during the present year. As far as Texas is concerned at least, the prospect for crops is simply the best that has ever blessed the State. From all quarters and in all directions reports come in that the prospect for corn, wheat, oats, barley, millet, etc., was never better than at the present time, and that the State is fairly groaning under the abundance. The oat and wheat crops are the largest ever grown in the State, the question now being to find markets for the product. Both these cereals are assured, subject to the contingencies of harvesting. Fruit of all descriptions is plentiful and assured. Corn is in fine condition and well advanced, while fat cattle and splendid grass are the universal rule in every portion of the State. In fact, everything in the eatable line that grows in Texas has never been known to be in greater abundance or in better condition than at the present time, with the prospect that the State will not only have enough to supply all wants for those who are here and who are daily coming in, but large surplus to sell to the outside world. There has not been a great increase in the acreage planted in cotton in Texas, perhaps 5 per cent. covering the excess over last year, as applied to the whole State. The plant is backward throughout, from all that can be ascertained, with the exception of the Brazos bottom district and some few spots in Southern Texas. Yet the stand is generally very fine, the crop clean, and the plant is healthy. With favorable weather from now on as much cotton will be made in the State as can be well picked, but of course the contingencies based on a late crop have always to be considered in this connection. The bread and meat question has been attended to, however. Should the cotton crop turn out short the chances are that fair prices for the staple will more than make up the difference, and Texas this year will be a buyer of neither bread nor meat in Western or Northern markets. Altogether, the sturdy granger has a right to be happy over the prospect generally, in Texas, and with him the commercial interests of the State, so closely identified with agricultural prosperity.—*Galveston, Texas, News.*

UNDERDRAINING.

Professor J. M. M'Bryde in Journal of American Agriculture.

Modern writers on underdraining generally assume that the practice is of comparatively recent origin. Waring, in his work on draining, remarks:

The effort (probably an unconscious one) to make the theories of modern underdraining conform to those advanced by the early practitioners seems to have diverted attention from some more recently developed principles which are of much importance.

He then goes on to observe:

Joseph Elkington, of Warwickshire, England, about 100 years ago discovered that tapping underground springs where the land was wet would relieve and and improve the soil, and this, the Elkington system, may hence be considered the germ or beginning of the present practice of thorough drainage.

He admits, however, that catch-water drains, made so as to intercept a flow of water, have been in use from time immemorial, and are described by the earliest writers. Now, without dwelling upon the passage wherein Virgil speaks of "drawing off from the absorptive soil water there collected after the manner of a marsh," I would ask what is to be thought of the following passage, written by Columella nearly 17 centuries before Elkington was born. In his chapter on soils, while treating of wet land, he observes:

If it be wet, let the abundance of moisture be first dried up by ditches. Of these we are acquainted with two kinds, covered and open. In compact and calcareous soils they are left open; but where the ground is more porous, some of them are left open and some covered, so that the free vents of the latter may discharge into the former. It is necessary however, to make the open ones wider at the top and sloping and contracted at the bottom, like inclined pan-tiles, for those with perpendicular sides are soon damaged by water and filled up by the falling in of the sides. In addition to this the covered ones should be sunk three feet deep, and after being half filled with small stones and coarse gravel, should be made level with the surface by returning the earth thrown out in digging them. If neither stones nor gravel are convenient, then a bundle of twigs twisted together like a rope should be made of such thickness as to exactly fit and fill the bottom of the ditch. This should be stretched along the bottom and cypress or pine branches, or any other kind if these cannot be obtained, pressed down above it and the soil thrown back over all, first placing at the head and mouth of the drain two large stones, one against each of the sides, and a single stone across these after the manner of a little bridge, in order to support the sides and keep them from filling in and obstructing the ingress and egress of the water. (Lib. II, Cap. 4.)

Pliny, in Lib. XVIII, Cap. 6, evidently has this passage before him when he writes, a few years afterwards:

It is highly advisable to cut up and drain a wetter field with ditches—moreover, in clayey places, that the ditches should be left open; in looser soils, that they should be strengthened with supports or pantiles, or sunk with sloping sides in order that they may not fall in; that certain kinds should be covered and led into others larger and more open, and, if occasion required, filled in below with pebbles or gravel, also that the mouths of these should be strengthened on each side with two stone and covered on top with another.

Palladius also, nearly three centuries later discusses the same subject in almost similar language. Here we have assuredly something more than "the germ" of underdraining.

Silos and Ensilage.

Not a few of our farmers are prejudiced against the so-called new process of preserving green forage by reason of the novelty of the descriptive terms employed—"silo" and "ensilage." Who ever heard of these before? It will perhaps surprise them to learn that the French word *silo* is identical in form with the latin ablative *siro*, the substitution of *l* for *r* being a cometymological change. The word "siro" can in fact be traced back to the Persian. In all these languages its meaning is the same—an underground excavation or pit used for the storage of grain or perhaps forage. Columella speaks of grain being in pits as "in certain transmarine provinces, where the ground hollowed out into excavations resembling wells, which are called *siros*, receives back its own produce." (Lib. 1, Cap. 6; 15.)

Varro also mentions these siros and states that they were in use in Cappadocia and Thrace, and also formerly in Spain and around Carthage. Their bottoms, he says were covered with straw, and every precaution taken to prevent the access of moisture and air to the grain until it was brought out for use, for it was held that the weevil would not breed where the air was excluded. He adds that the wheat thus stored away kept 50 years and millet upward of 100. (Lib. 1, Cap. 57.)

Pliny, referring to different methods of preserving grain, and quoting from Varro, says:

They (corns) keep well-stored away in the ear, but they are best preserved in trenches which they call siros, as in Cappadocia and Thrace and Spain and part of Africa. They use every precaution to make these in a dry soil, next strew them with straw, and then store the grain away in them in the ear. If no air penetrates the cereals thus stored it is certain that they continue uninjured. Varro is authority for saying that wheat thus buried keeps 50 years, and millet even 100; that the bean and pulse smeared with ashes are preserved for a long time in olive oil casks, and that the bean continued uninjured in a certain cave of Ambracia from the reign of King Pyrrhus even down to the piratical war of Pompey the Great, a period of 220 years.—*Nat. Hist. Lib. XVIII., Cap.* 30.

Several months ago an article appeared in an agricultural paper warning farmers against descending incautiously into a partially filled silo in the morning. The writer stated that the carbonic acid produced during the process of filling collects over night, and that a laborer near Sing Sing, N. Y., very nearly lost his life by going down early in the morning into a half-filled silo. After this it seems strange to learn from Varro, in the days when Priestly and oxygen were not, that whenever they opened these siros they waited for some time before going down into them for fear of the noxious air collected therein.

It would appear, then, that the process of ensilage (or ensirage) has claims to a very respectable antiquity, and that it was used, only for preserving grain, but very probably green forage also, for the amount of carbonic acid given off by grain as long as it was perfectly preserved and germination prevented, would scarcely have been sufficient to attract the attention of the husbandman. This supposition is greatly strengthened by a passage in Curtius, a Latin historian of the first century. In his 6th Book he remarks:

The barbarians around Caucasus call these siros which they conceal so incongeniously that none save those who dig them are able to find them. In these their crops are stored away.

Now the word *fruges*, which occurs in this passage is a much broader term than the word *frumentum*, used by Varro or Pliny, or than *fructus*, the one employed by Columella. The classical writers carefully distinguish between these several terms. According to the best authorities, *frumentum* signifies grain, (balm-fruit), while *fructus* denotes more particularly tree fruits, and *frux* (*fruges*) "the fruits of the earth, or the produce of the fields, pod fruits," &c.

And finally, Ansonius Popma, an accurate grammarian and scholar of the 16th century, in his treatise on Farm Implements, ("De instrumento Fundi,") a work which concerns itself chiefly with the ancient instruments of husbandry, in referring to the subject of granaries, and citing authorities, appears to use the term *fruges* advisedly. In Chap. XV he writes.

Instead of these (granaries above ground) In some provinces, siros are used, dug out in the ground after the manner of caves or wells for receiving and preserving the crops.

It should be remarked in this connection that the term *silo* was in common use in French husbandry long before the days of Goffort. For example, the pits in which root crops, &c., are stored are called silos. See in Cassanova's *Pes Premiers Pas Dans' l'Agriculture* (edition of 1866,) page 112, under the head of *Silos*, the passage beginning "*Dan, ce cas il faudra faire des silos.* &c.

Palladius speaks of a modification of this process not altogether unworthy of the attention of the vine-dresser of to-day:

The Greeks [so he stated] assert that you can preserve the grapes on the vine even to the beginning of spring if you will dig near the plant, on the shady side. a ditch three feet deep and two feet wide, and fill in the bottom with gravel and straw reeds upon this. You must entwine the branches full of fruit among these reeds, binding together the uninjured branches so that the soil cannot touch them, and after filling up the trench with earth cover it over in order to keep out the rain.

EDUCATION FOR FARMERS.

To the average mind the word education is limited in its definition to what one learns at school, but that is altogether too narrow. Education means growth, culture, development, as well as the acquisition of knowledge and knowledge again is not monopolized by the schools; indeed, one who knows only what he learns at school is much more justly entitled to the epithet of ignoramus than he who, having no opportunity to attend school, has been a diligent student of nature and of men. There were wise men before letters were invented, or schools established. Schools, good schools, are excellent auxiliaries to education, but they are nothing more. It is admitted by all that no amount of book-learning will suffice to fit a young man for the duties of a physician, a lawyer, or a clergyman, and the idea that it would fit him for the profession of agriculture is absurd. Yet each profession has its literature, which can be reached only through the portal of the school or the aid of private instructors, and the literature of each profession is of prime importance to those who would pursue successfully a profession.

The literature of a profession, farming, for example, conserves the wisdom of the past and records the experiments of the present. But the wisdom of the past preserved in books is like wheat before it is winnowed, mixed with the chaff of ignorance and the cheat of prejudice. So, also, is much of the scientific knowledge of the present. They are both misleading and injurious to him who accepts them without question. But they are great helps to him whose mind has been trained to criticise all things, and who accept only that which stands this crucial test.

Colleges confer degrees, yet these are often misleading; the young man with A. M. or M. D, after his name is not necessarily a master of arts or of medicine. He is only prepared to enter upon a career of practical experiment, which, if he possesses the talent, the industry, and the perseverance necessary to the completion of his education, may ultimately make him worthy of the title conferred upon him prematurely by the school.

No amount of theoretical training will fit a man for the successful pursuit of agriculture; yet, without theoretical training, a man rarely rises to the dignity of an intelligent farmer.

Farming is a profession in the same sense that the practice of law or of medicine is a profession; hence the youth who is destined to become a farmer should be educated with reference to that profession. The public schools of this country furnish the facilities for all the literary training absolutely needed, and, in the larger cities, the scientific branches are taught as well as they are in our colleges, and these are important. While it were a waste of time to study the dead languages, the prospective farmer should become familiar with the elements of natural history, botany, chemistry, geology, and natural philosophy. These branches of science have a direct relationship to his future business, and the young farmer who enters the profession versed in them will find that he is not only prepared for a larger measure of success, but that his mind is fitted for communion with nature, whose secrets, hid from others are constantly revealed to him, affording an inexhaustible source of pleasure as well as profit. To him every expanding leaf or opening flower has a beautiful significance, and every phenomenon involved in the growth of plants has for him a meaning unknown to the ignorant plodder. All nature is to him one grand illustrated encyclopedia filled with lessons of wisdom, from the pen and pencil of the original author and artist of the universe.

To the educated farmer the rocks present their own history, written in unmistakable characters by the finger of God. The soil whispers to him of its fertility or complains of its poverty in language perfectly intelligible, and the treasures of Flora, Pamona and Ceres, are shown, in rich abundance at the feet of him who wields the magic wand of intelligent labor

SUCCESS IN FARMING.

Importance of Rotation and Clover and Grass Crops.

The necessary steps toward an improved husbandry are:

1. To cultivate less land.

2. To make that which is cultivated rich in plant food, so that it may produce large crops.

3. The practice of a rigid system of rotation of crops and mixed farming.

4. The cultivation of the grasses and less of the cereals, and the feeding upon the farm the most of its products.

5. Raising clover and enriching the land by turning under green crops.

I believe that the faithful practice of such a system of tilling would in ten years increase the value of real estate 100 per cent., and place the farming population in an independent position. All observation and experience go to show that those sections of the country are more prosperous where a mixed system of farming prevails. The farmer who finds in his own garners that which is needed to supply his daily wants is far removed from the vexation and losses attendant upon outside purchases, which so severely tax his means. It is not infrequently the case, when he produces but a single article for the market, that it commands a price which but poorly compensates him for his labor, while he has to pay exorbitant prices for that which he is compelled to purchase. This is "selling the hide for a penny and buying back the tail for a shilling," which surely is not a profitable transaction. Mixed agricultural necessarily leads to a system of rotation of crops, which is the key to successful farming. That there is a vast recuperative power in the land where a succession of different crops are grown, no one can deny in the light of universal experience. Thousands of those who have hitherto devoted themselves to a single production, such as cotton, tobacco or grain, now acknowledged this error.

Successive crops of the same character exhaust lands of the particular food they require with great rapidity. The aid which nature so freely renders, where crops rotate, is withheld in such a system of civilization, because the farmer is violating her laws. To fight against nature is to war at fearful odds, and it is not difficult to forcast the result. To work in harmony with her insures a comparatively easy victory. One of the most beautiful of her provisions is, that while one crop exhausts the soil of that element which enters most largely into its composition by the operation of some mysterious law, it prepares that some soil for some other crop of a different character. This is a very curious and interesting process of nature, which results immensely to our advantage if we accept her aid. As an illustration of this principle, we know that clover does not successfully follow itself, although it leaves the ground in the best possible condition for corn or wheat. One crop, therefore, restores in a measure what another has taken. By raising continuously the same plant you interfere with this beautiful contrivance of nature to rebuild her wasted strength. How this is done is imperfectly understood. We do know, however, that the deep rooted plants like clover, will pump from the depths below for the use of those that grow near the surface that food which has been carried beyond their reach. And not only that this element, when brought to the surface, acts chemically upon what it finds there, and renders soluble and available as plant food what before was inert and resisted assimilation.

Nature, therefore, will do much of our work for us if we only second her efforts and give full cope to her beneficial laws. It is, therefore a question for the farmer to determine whether he will, by a rotation of crops, have this soil enriched by drafts on nature's treasury or draw entirely upon his own. I do not mean to argue that there is nothing for the farmer to do but follow this rotation to make his lands productive. Far from it. But I do argue that he may make nature a co-worker with him in attending a desirable end. Change is a prominent feature in nature's economy. Cut down the forest of hard wood and the pines succeed. Again, remove the pine and the hard wood reappears. One kind of grass succeeds another, and nature supplies the seed. These changes give the soil rest, to the end that the process of re-invigoration may go on.—*Hon. Wm. Fullerton in Nashville, (Tenn.) Southern Industries.*

THE DEPARTMENT OF AGRICULTURE.

Nothing is more remarkable in our history than the fact that the most important of our national interests should be entirely unrepresented at the national capital. Agriculture, which at all periods of our progress has been the most prominent of our productive powers in the creation and development of our natural resources and positive wealth, is wholly unrecognized as an element of national power, or as an object of legislative concern.

The army of 25,000 has a department to manage its minutest movement. It expends $40,000,000 annually. It produces nothing.

The navy, limited to 11,000, almost destitute of ships, a mere burlesque on efficiency, as compared with any European power—made up of officers, navy stations and foreign squadrons to that favorite commanders in foreign climes, expends $20,000,000 annually.

The post-office is an institution by itself; it is worthy of the Government, the people, and the age.

The State Department is what it is venerable in precedent, dogmatic in practice; slow, aristocratic, it is the least American of our departments. If it were to drop out it would not be missed. It is the Rip Van Winkle element in our Government machinery.

The Interior Department is, after the Post-Office, the only real representative of the people. It is the source of titles for all our public lands : it issues all our patents ; it controls, manages, and provides for our Indians ; it distributes and settles our pension-rights ; it regulates our mines and controls our railroad grants. Its duties are immense ; they are performed with consummate ability, but red tape hangs from every window, garlands every alcove, and ties up in stupid uniformity of dullness every intellect not bold enough to say its soul is its own.

The Treasury is a marvel. More than $1,000,000 daily passes under its control. The care, precision, accuracy, and brilliancy of the management is equal to the grandest hopes of American supremacy. It is the treasure-home of the people. Its vaults to-day hold more coin than is treasured in any other government building in the world.

But agriculture, which creates the wealth managed by the Treasury, and without which neither the army nor navy could exist, has no department at Washington. But the voice has gone forth demanding the establishment of a department for agriculture. There is no government in the world whose progress in agriculture development has been equal to ours. All the European governments have special departments for agricultural protection, improvement and encouragement. Agriculture is the bed-rock on which we build; it is the foundation of wealth; it gives us subsistence, and subsistence is life.

Twenty-eight million of our people are directly or indirectly dependent on the products of the farms. The value of our farms, according to the last census, was $10,197,161,-905. The yearly product is now nearly if not quite $4,000,000,000. We have more than 5,000,000 farms, and out of the $883,925,947 of our foreign exports, $720,650,016 was agricultural. Last year we paid for $642,664,628 for foreign exports besides bringing $91,160,-000 of European gold to enrich our people with farm products.

We have 10,357,981 horses, 1,812,932 mules, 993,970 working oxen, 12,443,593 milch cows, 22,448,590 other cattle, 35,191,656 sheep, and 47,683,951 swine, making an aggregate of farm stock worth $1,500,503,807. Behold the means of production a single century has accumulated. And yet we are but in the dawn of our achievements. We have the broadest fields, the finest climates, the grandest resources, and the most limitless opportunities to become the most independent, the best supplied, and by all means the most thoroughly educated agriculturists of the globe. The last two weeks have developed the national interest in agricultural advancement in a manner worthy of Congress, worthy of the people, and worthy of the country. Le Fevre and Updegraff, of Ohio; Grant, of Vermont; Lacy, of Michigan; Mr. Morey, of Ohio; Mr. Dwight, of New York; Mr. Scales, of North Carolina; Mr. Williams, of Wisconsin, and others, have discussed the question of an agricultural department, with an earnestness and ability deserving of its importance.

The fact that during the year ending June 31, 1881, we imported into the United States $285,081,008 in agricultural products is sufficient evidence that we have yet much to learn in the way of adapting our infinite variety of soils and climates to the production of prime articles of necessity we are capable of producing, for which we are yet paying tribute to other lands. It has been well said that "the application of machinery, steam, and electricity to agriculture is but in its infancy." They are all to be applied to lessen toil and increase production. Every wheel, every lever, every physical appliance that relieves a human muscle wakes up the brain and gives it a chance. The farm-house of to-day is a palace in comparison to what it was in 1830, light has illumed it, machinery has elevated and refined it; the school-room and the newspaper have made it a home of intelligent comfort. The tiller of the soil is sovereign over nature, just in proportion as he is educated to comprehend it, and why should not the Government of the United States devote itself by all the appliances, concentrated ability and intensified means can bring together in departmental instruction to make the science of production equal to the opportunities our unequalled country affords. As Mr. Updegraff truly says, no country on earth has an agricultural interest comparable with ours. "It is confessedly the largest interest in the nation," and yet is without a department to enlarge, enlighten, protect, and increase its benificence. Our grain crop in 1880, was 2,697,362,465 bushels. The grain crop of California for ten years is shown to have been of the value of $313,231,046, or nearly double the gold and silver taken from its mines, which amounted to $186,406,248 for the same period. A single attested fact is enough to demonstrate the importance of Governmental aid in securing the best seeds and the best modes of cultivation. The seeds distributed by the Government in 1878, increased the yield nearly 50 per cent. wherever they were tested. In Prussia, Austria, Italy, Spain, Russia, France, and Brazil, the Agricultural Departments of the Government are regarded as of the first importance.

"The farmers are the tax-payers," and, as Jefferson says, "the revenue is the State." And, as Mr. Updegraff truly says, "when our great financial fabrics went down, burying fortunes and enterprise in their ruins when commerce was stagnant, when our manufactories were overwhelmed and pulseless, then the great agricutral productive forces of the country displayed its full measureless affluence to bring back prosperity and to fortify the nation's credit with the bounty of the nations surest wealth."

There is every reason why we should have an Agricultural Department worthy of the nation ; there is not one why we should not.

FANCY BUTTER.

For fancy butter, says Dr. Heath, the first requisite is the perfect cow. The Guernsey and Jersey cows are undoubtedly the first choice for making high-priced butter. But by this choice, the Aryshire, Holstein, grades or common cows are not excluded, for any and all of them, with the proper requisites, may be made to produce fine butter.

Pasture and food are also essentials elements in the production of fancy butter.—Weeds, sour grass, nor coarse swamp tufts, will fill the pasture requirements. Well kept old pastures, containing blue grass, meadow fescue, sweet-scented vernal, orchard grass, red and white clover, timothy, red-top and wire grass as the prevailing forage plants, together with the sweet grasses, which naturally carpet the mature and well-kept pastures, are the prime necessity for the stock of cows from which we would make good butter.

Next in order is an unfailing supply of good, cool running water, for every one hundred parts of milk contains eighty-seven parts of water, and unless the cows can have free access to good water, no matter how good the pasture, the milk must be defective.

Though we have good cows, good pasture and good water, yet there are many other considerations of indispensable necessity in the treatment and management of the butter diary. The cow must be treated with kindness—yes, even with affectionate care. She must not be driven far or fast to or from pasture. When stabled, she must be clean, comfortable and fed with good, sound and rich food. The cow is a quiet, easy-going, luxurious living animal, manufacturing her best products under the most favorable circumstances and only from the best materials. The milking must be regularly performed, and absolute cleanliness is a necessity with the cow; her food, her care, her milk, with the cream, the butter and the atmosphere of the cow, must be pure and sweet. The temperature must be proper, from the pasture to the butter package.

Cool, shady pastures are most desirable. No cow ever manufactured her best products at 90 degrees Fahrenheit. The milk should be 58 degrees as near as possible summer and winter, either by means of flowing water or the never varying temperature of the air vault.

The cream, when set for butter, should be frequently stirred to prevent irregular scouring, or more important, the formation of dried casein on the surface, which flecks and embitters the butter.—*Rural World.*

ALL ABOUT POULTRY.

In whitewashing a hennery put some kerosene into the mixture, for the benefit of the hen lice.

The time is coming when eggs will be sold by weight. It is the only fair way. Massachusetts has already a law to that effect.

Every nest box should be scalded after hatching, or painted with kerosene in order to kill lice. This sort of vermin is the worst pest of the heunery.

A hen that is very quiet for the first two or three days after hatching is better than a fussy or gadding one. She knows that chicks of that age can't travel much.

A flock of fowls that are frequently chased by dogs or often frightened by the owner, can not be expected to return heavy dividends in eggs. They want quiet, and constant anxiety for their lives does not conduce to natural development.

When a chicken picks a hole within the shell at hatching the access of air is apt to dry its down to the shell, and then it fails to turn over, and must be helped out. This is always a bad sign. A little warm water on the shell then may be of service.

Hens need to be in good order and sound health before they begin incubation, and given plenty of good food while continuing in it. A sitting hen's "sedentary habits" are poorly calculated to promote an increase of flesh. Always give her access to food, water and the dust bath.

If guinea hens will eat potato bugs, and make a business of it, get a few, or many, according to your needs. One guinea screecher to each half acre of potatoes is hinted at as the proper average.

Give fowls as much liberty as is compatible with a general good of the farm. Restraint is in opposition to nature, and tends to bad and dangerous habits. But when restraint is necessary, see to it that they have as many comforts as is possible in confinement, or you will suffer from it.

When hens do not sit on the ground their eggs should always be lightly sprinkled with tepid water every day or two after the first week or ten days. This is a matter real importance, and if attended to will prevent a good many disappointments, because many chicks will otherwise die in the shell.

Plant sunflowers now. The seeds are just as good for poultry as ever, but you can also wear the delicate flower in your buttonhole if you are a Wilde æsthete, or may, perhaps, sell them to some æsthete snob or snobess. Plant sunflowers, we say. The flour or meal is also good for feeding cows.

For ducks, if there is no good stream or pond at hand, a big extemporized basin, if not more than a mud hole, will do. But while they will not scratch much in the flower-beds or gardens, they are worse nuisances among flowers or vegetables than chickens, and their feet are anything but favorable to grass production.

When a chicken has to be assisted out of its shell, it is a nice point to do it at just the right time. The food for the first day is derived from the yolk absorbed, and some should be left in the shell. On the other hand, too long a delay is equally bad. Chickens that bleed when assisted out will generally die, but not always. It is a sign that they are not quite "ripe."

A separate room for setting hens where they can have food and water at will, and bathe in ashes or dust, is perhaps the most convenient way to manage them where quite a number are hatching at once. If an outdoor run can be provided in which they can get grass, it is still better. But as some hens don't know enough to always go back to their own nests, they need a good deal of superintendence when a dozen or two are quartered in one room.

Pure breeds are rather more satisfactory to most poultry raisers than all sorts of odds and ends; still, if well cared for, the difference is not so very important. When a farmer desires frequently to draw upon his poultry for a meal he don't want an entire flock of "everlasting layers," which are usually small and wild, and not much to boast of as to quality. The Brahmas, Plymouth Rocks or Cochins are better, and will furnish a good supply of eggs without running all over the farm or neighborhood.

H. P. Clarke, of Indiana, makes a recommendation in the Germantown *Telegraph* about breaking up a broody hen, that may have sense in it. It is to shut her up in a box with a raised bottom of narrow strips or laths, so that when she sits her breast is constantly exposed to cold air. If one side of the box is elevated so that it does not stand exactly level, it might add to her dissatisfaction. But after all, close confinement with plenty of food and water and the company of a social rooster, is probably as good way as any.

It is sad to see men, assuming to be teachers of farmers, to say that poultry "will preserve plum trees against the ravages of the curculio." The curculio is a winged insect, with no occasion to visit the ground, and fowls cannot catch it if they would, and will rarely eat them when offered—at least, not when dead, for we have seen it tried. How is a clumsy hen to catch a curculio that lights on a plum eight or ten feet above her head, and the plum at the end of a long limb, perhaps? The statement is as absurd as that bottles of sweetened water will keep the insect away.

Talks About Fruit.

Plenty of soil stirring will always be a partial substitute for manure in fruit growing. It is better than piles of manure with no culture. That means weeds, and weeds mean ruin.

The cold weather in April destroyed a good deal of the grape blossoms in the vicinity of Nashville, Tenn. Peaches there, as elsewhere, promise the best on high land.

Coal oil will kill any insect it touches, and hence, as it is easily applied to the trunks of young trees some fruit-growers are tempted to use it in this way. But they had better not. It will kill the tree also if heavily applied. Better experiment with it first on some tree of little value before applying to a good tree.

We see it stated that if half a pound of ammonia and the same of nitre be put into a hogshead of rain-water, it makes an excellent fertilizer for strawberries. Very likely; but the rain-water itself, applied to strawberries during a dry period, will be excellent, and no doubt did much of the good which has been referred to this experiment. There is little reason to believe that infinitesimal doses of costly fertilizers are to produce extraordinary results.

Judge Edmund H. Bennett, writing on the legal rights of farmers, says : "That when a fruit tree stands exactly in the line of two properties, it belongs jointly to both owners ; but if it merely stands near the line, but overhangs in part the next owner's land, the latter has no legal claim to the fruit, nor any right to destroy its limbs. The common impression that any man is entitled to the fruit which drops upon or overhangs his land, he says is incorrect.

Fruit trees are often scraped, not specially to make them look well, but to prevent insect enemies from having a hiding place under the scales of the bark which accumulate. Scrape the trunk in some way so as not to injure the living bark, then wash it with whale oil soap. Then you have a trunk tree from insects, and one that looks also "as nice as can be." Apple and pear trees are the ones most in need of this sort of care.

Some writers recommend fruit growing for women as "a light, pleasant and profitable occupation." Parts of the business are light and pleasant; but when a woman attempts to manage all departments of it—planting, manuring, hand and horse culture, picking, packing, loading, marketing, handling crates, etc., etc., it will be found that a good deal of the work is anything but "light." But women can greatly assist in fruit growing, and this is where their agency is most needed. Some women can also be managers, but men must aid in the heavy work.

OUR LOCAL ORGANIZATIONS.

LANCASTER COUNTY AGRICULTURAL AND HORTICULTURAL SOCIETY.

The July meeting of the Agricultural Society was held on Monday afternoon, July 3d, and was attended by the following named persons :

Levi S. Reist, Oregon ; Henry Kurtz, Mt. Joy ; F. R. Diffenderffer, city ; W. W. Griest, city ; Joseph F. Witmer, Paradise ; John H. Landis, Millersville ; M. D. Kendig, Creswell ; C. L. Hunsecker, Manheim twp.; J. M. Johnston, city ; J. L. Landis, city ; Peter S. Reist, Lititz ; W. B. Paxson, Colerain.

In the absence of the secretary and his minutes, M. Kendig was temporarily elected to the seat of the former, and the latter were not read. What the questions for discussion were only the minutes knew, and consequently the meeting was much shortened.

Crop Reports.

Henry Kurtz said that around Mount Joy the wheat is very promising, and he expects forty bushels per acre ; grass pretty good, and sells from $18 to $25 per load ; some tobacco is middling good and some not yet planted ; the cut worm is unusually plenty ; oats and corn look well.

In Levi Reist's section wheat is better than for years ; apples and dropping off, through the York Imperial and Baldwin are hanging well ; tobacco Mr. Reist never saw so indifferent. As a whole the crop is promising. In regard to cherries it is very curious that in some spots they hang plenty while in others, not a quarter of mile distant, the limbs are bare.

C. L. Hunsecker said that Lancaster county never had a better wheat promise ; oats has not looked better for years ; corn crop will probably be immense ; tobacco may yet equal former crops in this county ; potatoes, there will be enough of and some to spare. Prospects are very good all along the line.

John H. Landis reported that the rankness of the Manor wheat has disappeared and it now looks as fine as ever was seen there or elsewhere ; hay full crop ; oats fine ; tobacco is being cut in spots all over the township ; apples will not be so plenty as indicated four weeks ago ; the peach crop will be tolerably good in quality, though lacking in quantity ; no cherries.

Mr. Paxson, of Colerain, said that in his country there never were better prospects for a wheat crop ; hay long and well set ; corn healthy but backward ; oats rather poor ; no peaches, not so much as five bushels in the township ; of cherries there are none ; those who set tobacco out early did well, but cut worms and drought are doing damage.

President Witmer gave a promising report for Paradise, with the exception of tobacco. One gentleman says he has lost 9,000 out of 12,000 plants.

Shall We Have a Fair? No.

Henry Kurtz brought up the well worn fair question. Some of his neighbors were taking an interest in the matter, and they wanted to know whether there was any chance of having a fair. Though silence sometimes gives consent, yet the ominous quiet which followed Mr. Kurtz's remarks showed plainly that those present had no desire to undertake the getting up of a fair, and the president then suggested the inadvisability of action on this question when so few were present.

The Immigration Question.

C. L. Hunsecker proposed that as the society was doing nothing, it discuss the immigration question including the Chinese, Dutch, Irish, and everybody else. Mr. Hunsecker pictured the poverty-stricken condition of the over-crowded foreign lands, and thought it but right and humane that this country with all its unsettled lands should extend a helping hand to suffering humanity.

The business on the programme for this meeting was continued until the August meeting.

Grain and Fruit Exhibited.

Henry Kurtz displayed some stalks of Fultz wheat raised on the Kurtz farm near Mount Joy. They were five feet eight inches long and were a fair sample, Mr. Kurtz declared, of the whole of his forty-two acres of wheat.

Levi S. Reist had three varieties of cherries—the Little Britain, the large black juicy ones ; a seedling somewhat similar in appearance ; and Molkenkinsche, small, red and sweet—and some sharpless strawberries of fine flavor.

POULTRY ASSOCIATION.

The Lancaster County Poultry Association met in the agricultural room of city hall, on Monday, July 3rd, 1882.

The following named members were present:

Charles Lippold, Jacob B. Lichty, F. R. Diffenderffer, C. A. Gast, John E. Schum, Charles E. Long, J. M. Johnston.

Frank R. Diffenderffer read the following essay on

Gapes in Chicks.

"I may as well say at the beginning of these remarks that I am not posted in the literature of gapes. I do not know what has been written on the subject. I am only aware that they have been on my own place during the past four years, and what I shall say is confined entirely to my own experience with them. Most persons who raise poultry know what the gapes are, and are acquainted with the cause. A thread like worm, of a bright red color, and fully an inch long when full grown, is the cause of all the mischief. It is a member of the *Entozoa* family, a name given to parasites that live within other living bodies. There are many kinds of these, but the peculiar one under consideration is called by naturalists *Syngamus tracheolis*. Its a pretty bad name to begin with and the doings of the little worm are still worse.

This parasite has its its home in the windpipe of young chicks and turkeys. Even when quite small, while still only a quarter of an inch long, it begins to inconvenience the young peeps, and as they increase in size, which they do very rapidly, they gradually close the windpipe, making respiration, difficult, and unless removed in most cases bring about the death of the young chick. The earliest symptom of their presence is a frequent opening of the mouth, a gaping for air as it were, and the gravity of the case grows rapidly until it terminates in death.

How does this parasite find its way into the chick's windpipe? A good many theories have been advanced from time to time, but nobody knows. The general opinion is that they are taken up from the ground with the food, while in an undeveloped state, and finally find full development in their natural home. It may be so, but as somewhat similar parasites are found in the brains of certain birds, and in the eyes of horses, that theory, reasoning from analogy, falls to the ground. A current belief with which even our worthy secretary seems tainted, is that they are developed from that other troublesome parasite, the louse or its eggs. Now, the louse belongs to a family as widely separated from the gape worm as a monkey is from a whale, and the Darwinian theory, when stretched to its utmost, fails to bring them together. The nit of a louse will produce a louse, and it won't produce any thing else. Besides, the gape worm is amply provided with means of perpetuating its species. It is, like many other members of the genus *lumbricus* or worms, sexually perfect in itself. It is in fact a forked or double creature, the one part being the female and the other the male. The body is lengthened beyond the point of attachment, and the prolonged portion, we are informed by microscopists, contains numerous ova or eggs, each of which is a microscopic gape worm. This fact seems fatal to all these theories, and is in fact decisive as to the origin. But how these embryo ovules reach their development, and where, is a mystery no one has yet been able to fathom, and there we will leave them.

How Do They Get There?

How the gapes first get to a place I do not know. Where mine came from four years ago I can't imagine. None of my neighbors had them to my knowledge, and none have them now; but I think I can confidently say I have not raised a chicken in all that time that was not attacked by gapes. I have tried every plan to keep them away that I have ever seen suggested and all were equally worthless. Between the ages of two and six weeks the gape worm is invariably developed. I never tried the camphor cure until this spring. I had a promising brood of young turkeys and I put them in a patch of lawn about 40 feet wide and 100 feet long. Twice a day I put camphor in their drinking water and otherwise took extra care of them. On the day they were two weeks old one developed gape symptoms, and I promptly removed *nine* worms from its windpipe—the largest number I ever took from one bird. The rest got them, and I have relegated the camphor preventative to other innumerable "humbugs." The latest remedy comes from a Chester county farmer who feeds whole corn to his chicks when the gapes come on. If this was a remedy it would act by compression, thus killing the worm. But who ever had turkeys or chicks two weeks old capable of swallowing whole corn? Mine never could, and I believe you will agree with me when I say that they would not at that age even if they could.

Another thing I confess myself unable to understand. Why are chickens and turkeys afflicted with this parasite and not ducks? I believe there is no case on record of this kind. I have now running together 14 young ducks and 17 peeps nearly of the same age—three weeks; most of the chicks have had the gapes and the rest will have them, while the ducks running and feeding with them escape the disease altogether. If the origin of the disease lay in filth, or was taken up from the ground with the food, why would not ducks also be subject to it? Ducks also get lousy, I believe, so that is another heavy blow at the louse theory, which, however, was not necessary to kill it.

Most poultry raisers have their remedy for extirpating the gape worm. Most of these are mechanical, and, of course, effective. There is one which is often recommended, but I have never known of a solitary cure effected by it. I allude to shutting the bird in a closed box and subjecting it to the fumes of powdered lime. This, it is alleged, will set the chick to sneezing or coughing, if I may so call it, and in this way the worms are dislodged. I have no faith whatever in any remedy except a removal by actual force through mechanical means. Cat-gut, twisted wire and feathers—all have their advocates. The latter, I believe, does as much harm as good. The fluffy part is cut down on both sides nearly to the quill. This leaves a rough edge, which irritates and injures the delicate coating of the windpipe. I have tried most of these, but have given them all up. I now use several stands of horse hair doubled, samples of which I have present. This is smooth, flexible and d es not readily suffocate the chicks. My method of operating is this: An assistant holds the chick in its natural sitting position; with my left hand I open the bill and seize the tongue, which is gently drawn forward; this brings the oriface of the windpipe well forward, and into this the horse hairs are then thrust. I have several sizes to suit chicks of various ages; the hairs are pushed down until the end of the wind pipe is reached, when the end in the hand is rapidly twirled around in both directions; in this way the entire inner surface of the windpipe is brought into contact with the horse hair; the worms are dislodged from their place of attachment, become entangled among the hair, and are then slowly withdrawn. Generally all are not brought out at the first attempt; I make two and even three when I think it necessary. It is of the utmost importance that this operation should be performed at the first symptoms of the disease. If let run on the chick soon ceases to eat, becomes enfeebled and may die during the operation, as many did for me before I caught the trick of operating early. It is a most severe remedy and the strongest chick feels it severely, but if done in the early stages it recovers rapidly, will begin to eat in a few hours and never show any bad result. If performed in time, a chick ought rarely to be lost by gapes. It takes time and is a little troublesome, but not more so than the nurseries, raised platforms and other devices recommended to ward off the disease, and what is still better is effectual.

But it is not a remedy we need so much as a preventative. It is a little to the credit of the thousands of poultry fanciers that they have not been able to discover means to prevent or eradicate this fatal disease. Perhaps this can only be done when all the various metamorphoses of the gape worm ova are known. Strictly speaking this is the work of the scientists, and poultry raisers are seldom such. Once every stage of progression in the existence of these parasites is known, we shall have no difficulty in keeping them from our poultry yards."

The thanks of the society were tendered to Mr. Diffenderffer for the essay.

Mr. Chas. E. Long suggested that before the horse hair is inserted into the wind pipe of the chick the hair be immersed in a weak solution of carbolic acid. The worms not drawn out would by this means probably be killed.

The secretary reported that the executive committee were at work on the catalogue for the next annual exhibition and would have it ready for distribution by November.

LINEAN SOCIETY.

The society met on Thursday evening, June 29, in the office of Dr. H. D. Knight, Prof. Stahr in the chair. The meeting was "sparely" attended, the weather exceedingly warm and nothing of special interest was brought before it.

The following donations and additions were made to the library: Parts 23, 24 and 25, Vol. XXI, Official Gazette of the United States Patent Office. Proceedings of the Wyoming Historical and Geological Society to February, 1882. Proceedings of the Philadelphia Academy of Natural Science. The Lancaster FARMER for June, 1882. Sundry catalogues, circulars and book notices.

After a brief session under Science Gossip the society adjourned to meet in the anteroom of the museum on the last Saturday afternoon of July, 1882 (29th).

A MERINO ram crossed on a flock of common sheep, will double the yield of wool through the first cross alone, thus paying for the ram the first season.

AGRICULTURE.

Green Crops.

Green crops for manuring should not be plowed deeper than four inches; if they are turned under more than this they will not receive enough of solar heat and atmospheric air to insure rapid decay, and when covered too deep their beneficial effect cannot be realized till the next plowing, when they are brought nearer the surface.

Loading Hay.

To properly dispose of the hay as it is pitched upon the wagon requires considerable skill. Long, wide and low loads are much better than the opposite, for both the pitcher and the loader; besides, there is much less danger of the load slipping off, or the wagon being upset by an unequality in the surface of the fiel. If a horse-fork is used for unloading the person who manages the loading should bear this in mind, and so place the hay as it is pitched to him that the fork will work to the best advantage.

Manure Under Cover.

Of course all the advantange of making manure in covered yards may be secured by box feeding, with less outlay for roofing, since more space must be allowed for a given number of animals turned loose together than when confined in stalls. It is the protection from rain and sun, the abundant use of litter and its thorough incorporation with the excrements, and the exclusion of air by compact treading, which go to make the superior manure. All these features of the method work against the loss of valuable plant food. Nor does box feeding and constant accumulation of manure under the feet of the animals necessarily imply offensive stalls.

One method or the other, box-feeding or covered yards, should be adopted by every farmer who lives where manure is worth saving, and who finds himself compelled to supplement his stable manure wi h commercial fertilizers. Stable manure must not be lost sight of, in this increasing interest in these concentrated fertilizers, for we cann t produce our crops and have enough for ourselves and others, without its aid; and there is nothing in all the list of commercial mixture, which gives so good an average return for the money invested in it, as well-made stable manure.

Plaster.

Land plaster, or gypsum, is sulphate of lime. One hundred pounds of common gypsum consist of forty-six pounds of sulphuric acid, thirty-three pounds of lime and twenty-one pounds of water. It is ground fine and thus applied to land or crops. When it is heated to reduces the water is driven off and the residue is easily reduced to a very fine powder, and is known as the plaster of Paris used by masons. The theory of the beneficial action of land plaster upon crops has long been, and still is, a subject of dispute. That it supplies lime and sulphuric acid to plants to some extent is probably true, but it is now generally admitted, we believe, that gypsum is chiefly useful by its powder of solidifying and retaining the ammoniacal gases of the earth and air. For wheat and corn it has not proved satisfactory, but on clover, sanfoin and luguminous plants generally its useful effects are not questioned.—*Prairie Farmer.*

The Largest Land Owner on the Continent.

Col. Dan Murphy, of Halleck's Station, Elko county, came to California in 1844, and may be said to have made the country pay him well for his time. He is now probably the largest private land owner on this continent. He has 4,000,000,000 acres of land in one body in Mexico, 60,000 in Navada, and 23,000 in California. His Mexican grant he bought four years ago for $200,000 or five cents an acre. It is sixty miles long and covers a beautiful country of hill and valley, pine timber and meadow land. It comes within twenty miles of the city of Durango, which is to be a station on the Mexican Central. Mr. Murphy raises wheat on his California land, and and cattle on that in Navada. He got 55,000 sacks last year and ships 6,000 head of cattle a year right along.—*Reno (Nev.) Gazette.*

Best Pasture Grass.

The best pasture grasses have creeping or wholly fibrous roots, the creeping root running horizontally under ground and pushing up stems every few inches from this creeping part of the root or rhizome. This creeping root is not likely to be injured by close cropping, and retains its vitality better through severe droughts after close feeding, when a bulbous roots would be destroyed. The function of the bulb in bulbous grasses is evidently to store up materials for future growth, and if these bulbs are injured or eaten off the root is destroyed. The nutriment in all grasses is gathered by fibrous roots alone, and these fibrous roots are joined to the rhizome or the bulb in creeping or bulbous roots. The best specimens of creeping rooted pasture grasses are blue grass, June grass (*Poa pratensis*) and wire grass, also called blue grass (*Poa compressa*). Both of these grasses, when well established in the soil mentioned, will retain their foothold against many discouragements. Both of these grasses start quickly after cropping. Orchard grass (*Dactylis glomorata*) is one of the very best pasture grasses when once established. It starts, perhaps, more rapidly after cutting or cropping than any other grass. It will grow in the night almost as much as cropped off in the day. Red top (*Agrostis vulgaris*) should be included. White and red clover should always be mingled with the seeds for pasture. There are many other grasses that might be sown, but the seeds are difficult to be obtained. A good mixture of these seeds is the following: Timothy, six pounds; Kentucky blue grass, four pounds; wire grass, three pounds; orchard grass, four pounds; red top, three pounds; red clover, four pounds; white clover, three pounds, and sweet-scented vernal grass, two pounds. A pasture well stocked with these grasses and clovers will certainly produce the milk for "gilt-edge" butter. Too little attention has been as yet, paid to the stocking of pastures. The subject needs careful examination and discussion, and we shall be glad to have the views of some of our experienced readers upon it.—*National Live Stock Journal, Chicago.*

Pacific Coast Wheat Items.

Washington Territory promises to be as great a wheat-growing State as Oregon or California. A few items will interest farmers.

Walla Walla *Union*: At the depot in Walla Walla tons of wheat are being stacked out of doors, the warehouses being full. At Valley Grove, (Nelson's place on dry creek) a large platform is nearly covered with stacked wheat. At Hadley's another platform is full. At Prescott a platform is full and tons are piled upon the ground. At Waitsburg, Kinnear & Weller's warehouse is overflowing and great piles of grain being made outside. W. N. Smith's platform is nearly full, and wagons are being constantly unloaded at both places. A mile above Waitsburg another platform is full. At Huntsville there is wheat, at Long's there is more wheat, and at Dayton the wheat is piled up in warehouses and on platforms, "till you can't rest." Parties from Blue Mountain station and Milton report the warehouses and platforms there filled to overflowing with sacked wheat, and great stacks of sacks in adjacent fields. Buyers are asking producers to "let up" on delivery, while the railroad men are worked night and day trying to carry the wheat away. But is like trying to empty a barrel by the spigot while a big stream is flowing into the bung. Parties who have visited the farming region say "the farmers have not begun to haul in wheat yet. Just wait until they have got through threshing if you want to see wheat."

A farmer on Whidby Island harvested a field of wheat which harvested 60 bushels to the acre and a field of oats which cut 103 bushels per acre. Such a yield for any other than reclaimed tide land is remarkable.

HORTICULTURE.

Summer Grape Pruning.

About this, as in nearly every other horticultural subject, there is considerable difference of opinion. We have known vines to be "pruned to death" in following out some wild theory that some addle-headed fellow had started, while others would prune so sparingly as to be of no benefit at all. Many strip the vines of three-fourths their leaves to allow the sun and air to get in, as they say, while others allow the grapes to be smothered for want of a judicious removal of the leaves. Pinching the ends off the vines, or clipping off a portion of the sprouts where they are growing rampantly, so far as it appears to be necessary to any reasonable judgment, will greatly benefit the crop, just as the reverse will damage it. The thinning out of the surplus bunches, by removing from a third to a half of them as they usually show themselves, is of the greatest importance. In doing this be sure always to remove the weakest and most imperfect. The laterals of the fruit bearing branches, which have been pinched or clipped, will throw out more branches, and these also should be pinched, so as to leave only a single leaf. The laterals on the canes, remember, are to be the fruit bearing canes for next year, and should be allowed to grow unchecked. Care must be taken to tie up such of the branches containing bunches which are too heavy to bear its own weight. There should, also, be no more wood allowed to grow than is needed for the following year's fruiting. These simple general hints may be of service to those whose knowledge of grape-growing is limited. In a little while—a few years of experience, which may be greatly aided by examining the way that good grape-growers follow—will soon put one in the plain road to success.—*Germantown Telegraph.*

The Care and Pruning of Peach Trees.

It is a rare thing to find an orchard that has been kept properly pruned and cut back, and most of them are found with bean-pole stems or main branches bare of any fruit or foliage, except such as are crowded closely together at their extreme tips, resulting in overcrowded leaves and fruit, and poorly-colored, late ripening and small fruit, with a tendency to rot from overcrowding and shade.

The model peach tree, for the best results, we think, should have a clean stem about three feet. At this point a regular whorl of four or five branches should be started. When these are started, the tree should have vigor enough to give each a growth of at least three feet the first season. These, early the next spring, should be cut back to eighteen inches, being careful to leave on them any sub-branches near their base. The next spring the resulting or next crop of branches should be cut back in about the same way, and the sub-branches half of them cut clear away, leaving every other one, and those not cut away cut back one-third to one-half. The summer after this the trees should give a splendid crop of fine fruit that will need no thinning. The after-cutting back and pruning should be after the same general plan, thinning out and cutting back the upper and outer branches, but never thinning out the small branches, except as above. As the trees grow older it will be necessary to cut back and thin out more year by year, and eventually it will be necessary to cut back half of the main branches to near their base, at some point just above where a thrifty young twig is growing, so as to form a new, vigorous head, and to cut back the remaining branches the next year, and then follow again the same system of training given above. We think that this system, carefully followed, will give continuously crops of fine fruit, with but little or no thinning; or, in other words, that by this renewal system of training trees can be kept in a young, vigorous condition for a great many years. Who can find fault with it? Who will give us a better system? Our preference would be to have our trees with lower heads, rather than higher, were it not necessary to run the curculio-catcher in the orchard. On strong soils trees might do well with four feet of bare trunk.—*Prairie Farmer.*

The Delaware Peach Crop.

In view of the certainty that announcement will be duly made in the early spring of the melancholy fact that the Delaware peach crop was almost totally destroyed by the terribly cold weather in January, it is interesting to note the following paragraph in *The Wilmington Republican*: "Especially in the cold snap bright with promise to the fruit growers of the peninsula. Their great dread of short crops has always been open mild winters, which forced on the buds prematurely only to be killed by the more severe spring frosts. The late Samuel Townsend, considered good authority on peach growing, has maintained that healthy peach trees could stand a temperature of five degrees below zero, depending somewhat upon the forwardness of the buds when the frost occurs. Assuming this to be correct, the crop of peaches south of us is still safe, as the mercury only fell two and three degrees below zero in some of the most exposed points around Wilmington, which is the entering door of the great peach grounds of the peninsula south of us. So far, therefore, as the present season has developed, the indications are favorable to next spring's agricultural operations."

Strawberry Beds.

A writer in the New York *Tribune* says: "The time for seeing to the security of *next year's strawberry yield* is immediately on stopping picking this year. Dig, plough or scarify deeply between the rows or in lines through the mass, and clear the hills or rows left of every weed, however small. Some add to this severe-looking treatment that of mowing off the old leaves, and they declare that the plant gets its summer rest all the better and more completely for it, starting then with the August rains into a luxuriant September growth which is the making of the fruit beds for next year's expansion.

Quince Culture,

Almost every good housekeeper who has a garden wishes there were quinces in it. No fruit seems more desirable in the kitchen, but it is seldom that it is seen there They are planted in the garden time and again, but seldom seem to do much good. They just live, growing but little, and that little seldom of the vigorous, healthful kind. The whole plant is knotty and scrubby, and though they may flower freely, the young fruit drops prematurely, and a bush of a dozen years old will often not give a dozen sound fruit.

Now, some say that the trouble is in the soil, that it is very peculiar and particular in this respect, but but we think this is an error. Certainly we have now and then seen quince trees doing well in all sorts of soil and in all sorts of situations. It is more than probable that much of failure comes from injuries by the borer, which saps the strength of the whole tree. The borer enters the stem at or near the ground, and boring into it cuts off a large portion of its supplies. Some trees, like the apple and plum, when attacked by the borer soon die, but the quince roots out so readily from every part of its bark that, unless very badly attacked, it will manage to live on in a lingering sort of way for a good many years without any but a practical eye suspecting what the real matter is.

But sometimes the quince gets what gardners call hide bound. The bark has a hard scrubby look, and the growth is puny and not at all what we expect to see on a healthy tree. Whether this hide bound condition is the result of some disease, or is a disease, in itself, is not clear; but it is removed tolerably well by scraping and washing the stem with soapy water occasionally, and a trimming out of the weaker shoots. This course seems to lead to a vigorous growth, after which the bark seems to expand as naturally as any one can desire.

It is frequently recommended in the newspapers that salt should be given as a manure to the quince, and perhaps in some cases it may do good. The quince does not send its roots far away, but has an immense number in a small compass. It will therefore require good feeding to a greater extent than those trees which can send their roots long distances in search of feed. Salt is a great promoter of moist ure, and as these numerous roots will make the earth about them very dry it may be beneficial in this respect. But any good manure will benefit the quince, and it should have plenty of it.

HOUSEHOLD RECIPES.

DEEP APPLE PIE.—To make plain pastry mix together lightly quarter of a pound of lard or butter, a teaspoonful of salt, a pound of flour, and sufficient cold water to make a paste stiff enough to roll out. One way of mixing is to put these ingredients into a chopping tray, and chop them together with a large knife; another is to make a paste, stiff enough to roll, of the flour, salt, and water, roll it half an inch thick, spread quarter of the shortening over it, fold it and roll it out again, and use another quarter of the shortening, repeating this process until all is used; the pastry is then ready for the making of pies.

For a deep apple pie, pare and slice tart apples enough to fill a deep earthen baking dish heaping full; line the edges of the dish an inch down with a strip of pastry; put in the apples, sweeten them to taste, and flavor the pie with a little grated lemon rind or a little ground cinnamon; cover the top with pastry wet at the edges with cold water to make it adhere to the strips on the side of the dish; cut small holes in the top crust, brush it over with beaten egg or with a little sugar dissolved in water, and bake it until the apples are done in a moderate oven. For a test for the proper heat of the oven refer to the recipe for *Home made Bread.*

PAN-DOWDY.—Wash a quart of dried apples, soak them over night in cold water, stew them soft in the same water with sugar and spice to make them palatable; put the sauce thus made into an earthen baking dish with a teaspoonful of butter, and cover it with pastry made as directed in the recipe for *Deep Apple Pie;* bake the dowdy until the crust is done; then remove it from the oven, and break the crust down into the apple with a spoon; use it hot or cold. Apple sauce made from green or ripe apples can be used in the same way.

FRIED APPLES.—Pare sound apples, slice them half an inch thick, remove the cores without breaking the slices, fry them in hot butter until tender, lay them in little piles with sugar and spice dusted over them, and serve them on slices of toast.

APPLE TOAST.—Pare and core tart apples without breaking them; put them on slices of stale bread, fill them with sugar, put a little butter and spice on each one, and bake them tender in a moderate oven.

APPLE AND BREAD PUDDING.—Soak a quart of stale bread in cold water five minutes; pour off as much water as will escape without squeezing, and put the bread in a buttered baking dish; pare and slice a quart of apples, lay them on the bread, add sugar and spice to taste, and bake the pudding in a moderate oven.

RACKET CLUB PUDDING.—Buttered slices of stale bread, enough to cover the bottom of a two-quart baking dish; put a layer of raisins on the bread; add another layer of bread, pour over it a custard made of four eggs beaten with four tablespoonfuls of sugar and pint of milk; pare, quarter and core a quart of apples, lay them on the pudding, dust them with powdered sugar, and bake the pudding half an hour in a moderate oven. Serve it hot with powdered sugar or jelly sauce.

JELLY PUDDING.—Mix together one teaspoonful of corn starch or arrow root, one tablespoonful of jelly, four of sugar, and a pint of cold water; put the sauce over the fire and stir it until it boils one minute; then use it.

CHEESE CRUSTS.—Cut some slices of stale bread two inches square and half an inch thick, butter them, lay them on a baking-pan, put one tablespoonful of grated cheese on each, and brown them in a quick oven; serve them hot or cold.

PUMPKIN PIE.—Peel and slice a pumpkin, or part of one, boil it in boiling water until it is tender enough to rub through a sieve with a potato-masher; mix with each quart a custard made of six eggs beaten with eight tablespoonfuls of sugar and a quart of milk; flavor the mixture with spice and grated lemon rind, and bake it in deep earthen pie-plates lined with plain pastry. Squash pie is made in the same way.

PLAIN MINCE PIE.—Chop fine half a pound of cold boiled beef or cold boiled tongue; remove the fibre from half a pound of suet and chop that fine; stone half a pound of raisins, cutting them in halves; pick over and wash half a pound of currants; slice thin two ounces of citron; pare, core, and chop a pound of apples; grate the rind and squeeze the juice of an orange and a lemon, if they are available; mix all these ingredients in a glass or earthen jar with enough sweet cider to moisten them, sufficient sugar to sweeten them palatably, salt enough to be just perceptible, and plenty of mixed ground spices; last of all add quarter of a pint of good brandy for the purpose of preserving the mince-meat. Pack it down tight in the jar, and keep it closely covered two or three weeks before using it. When brandy is not used the mince-meat should not be kept long. In making pastry for mince pies use from half to three-quarters of a pound of shortening to a pound of flour. If mince-meat has become dry by long keeping, moisten it with cider before using it.

WELSH RARE-BIT.—Stir together in a saucepan over the fire one-quarter of a pound grated cheese, two tablespoonfuls of butter, a quarter of a teaspoonful each of salt, dry mustard, and pepper, with a dust of cayenne, pour these on a large slice of buttered toast and serve at once.

OMELETTE.—Break three eggs and beat for one minute with a half spoonful of salt and a fourth as much pepper; have your pan hot, with a tablespoonful of melted butter in it, pour in the eggs, scatter over them three crushed square crackers, and when cooked sufficiently roll the omelette toward one side of the pan by slipping a fork under one side and turning it over. Place the omelette on a hot dish and serve at once.

CHICKEN AND GREEN PEAS.—Cut cold roast or boiled chicken in small pieces, brown them in butter, stir in a tablespoonful of flour, and when it is brown add a pint of stewed peas with their liquor, (or one can if green peas are not in season,) add salt and pepper, heat five minutes, and serve on toast.

BEAN SOUP.—Pick over one pint of dried beans and wash them in cold water; peel and slice an onion, put in a saucepan and fry it brown, with a tablespoonful of drippings; ham or bacon fat preferable. When brown, put the beans in with the onion pour on three quarts of cold water, and boil slowly; every fifteen minutes add one cup of cold water until a quart has been used; mix one tablespoonful each of flour and butter to smooth paste, and fry some half-inch bits of stale bread with a little butter. As soon as the beans are soft put them through a sieve with a potato masher; put them again in a saucepan with their broth, stir in the paste, let the soup boil once, and serve with the fried bread in it.

CODFISH.—Parboil fish in successive waters until freshened, taking care to have skin upward—if below it will gather and hold the salt. Peel and slice a pint of onions, and when the water is changed on the fish the last time put the onions into another frying pan, with two tablespoonfuls of hot fat and fry slowly; when the fish is hot remove it, take off the skin and the bones which are on the surface, then put it in the pan with the onions, brown slightly on both sides, dust it with pepper, and serve with the onions over it.

BROILED BIRDS.—Carefully pluck and singe the birds; cut off the head and feet, or if the head remains be sure that no feathers are left on it; remove the crop and windpipe, and wipe the birds on a wet towel; split them down the back, take out the entrails without breaking them; lay the birds, without washing, between the bars of a buttered wire gird-iron, and brown the inside first over a quick fire; then turn the outside toward the fire and brown that, but be careful to avoid burning; the birds may be cooked rare or well done, as the physician permits, and slightly seasoned. Toast is usually served under them.

SAGO AND WINE.—Wash an ounce of sago in cold water; put it over the fire in a pint of cold water, let it slowly approach the boiling point, and boil it gently until tender; then stir into it two tablespoonfuls of sugar and a glass of Madeira or sherry wine, and serve it hot or cold.

BEEF JUICE.—Slice juicy lean beef, from the round, an inch thick; broil it quickly over a very hot fire, but without burning, until it is brown on both sides; lay it to a hot soup plate, cut it through in all parts with a very sharp knife, and set another hot plate on it, with the bottom against the meat; then grasp both plates firmly and press them together, squeezing the juice from the meat; let it run into another dish, or upon a slice of delicate toast, and serve it at once; the physician will indicate the seasoning.

TOAST.—To prepare toast suitable for invalids cut stale bread in slices half an inch thick, and trim off the crust; then hold it far enough away from the fire to dry it before browning it; it should be of a delicate brown color and quite dry in the middle of the slice; in this condition it is more easily digested than when made so quickly that the moisture of the bread remains in it.

WINE JELLY.—Dissolve one ounce of isinglass or gelatin in half a pint of hot water; add one ounce of sugar and one pint of wine, and cool the jelly in a mould.

BARLEY WATER.—Wash two ounces of pearl barley in plenty of cold water until the water is clear; put it over the fire with half a pint of water, let it slowly approach the boiling point, and boil five minutes; then strain it, put it again over the fire in two quarts of cold water, and boil it until the water is reduced to one-half; then strain and cool it; it may be sweetened and flavored, if desirable, according to the physician's direction.

EGG AND WINE.—Beat one egg to a froth with two teaspoonful of wine and use at once.

MILK PUNCH.—For hot punch mix together quarter of a glass of brandy, rum or whisky, three-quarters of a glass of hot milk; add sugar and nutmeg to make the punch palatable. For cold punch use the same proportion of liquor, but fill the glass with shaved or finely-cracked ice, with spice and sugar to taste.

LIVE STOCK.

Spoiling a Young Horse.

When a young horse acts badly in harness, it is because he has not been properly taught his business. To whip and misuse him is to spoil him. A horse is naturally willing and docile, if well used, and much may be done by kindness, patience and judgment in removing the ill effects of wrong treatment. A colt should be trained when young, and gradually taught his duties, the greatest care should be taken to avoid frightening or irritating the animal, and much patience should be exercised. If the animal refuses to do what is required, punishment will make matters worse, something should be done to distract his attention when it will generally become docile.—*American Agriculturist.*

The Pig in Agriculture.

The pig has been recently spoken of in contempt when compared with our other domestic animals. But if we examine his good qualities at all critically we must award him a high place in our agriculture. He is found to produce a pound of product from less food than either cattle or sheep, and is, therefore, the most economical machine to manufacture our great corn crop into marketable meat. Our people are becoming wiser every year, and exporting less, proportionately, of the raw material and more of condensed product. If it takes seven pounds of corn on the average to make a pound of pork, as is no doubt the case, the farmer begins to see the economy of exporting one pound of pork, bacon or ham, instead of seven pounds of corn. The difference in cost of freight makes a fine profit of itself; besides, the pound of meat is usually worth more than seven pounds of corn in the foreign market. The production of pork should be encouraged on the further consideration that it carries off less of the valuable constituents of the soil than beef. The fat pig contains only three-fourths as much mineral matter per cwt. as the fat steer, and only two fifths as much nitrogen per cwt.; and therefore the production of a ton of pork on the farm will carry off only a little more than half the fertility carried off by a ton of beef. Besides, a ton of beef will require nearly fifty per cent. more to produce it. This gives in round numbers the comparative effect of producing pork and beef. It is thus evident that the pig should have a high place in our agriculture; should be fostered in every way; his capabilities studied and pushed; his diseases carefully noted and prevented—for he is the most profitable meat producing animal on the farm. The pig is an excellent adjunct to the dairy, turning all refuse milk and even whey into cash. As he is king of our meat exports, so let us treat him with great consideration.—*Moor's Rural New Yorker.*

Sheep Raising in Dakota.

Sheep farming in Dakota has been demonstrated by practical men, who have had experience in and understand the business, to be a safe and profitable enterprise. The dryness of the atmosphere in winter time, and its purity and healthfulness at all seasons, the abundance and nutritiousness of the native grasses, and other favorable conditions, insure the health and good condition of the flocks at all seasons. These facts are becoming known to and are being taken advantage of by practical wool-growers, and a number of them have recently located in our territory and engaged in the business. Among the number is B. C. Bagley, who owns a range about fifteen miles from Yankton, in the northwestern part of Clay county. He owns a fine flock of Spanish Merino bucks and Cotswold sheep, in fine condition and health, and returning a handsome profit on the capital invested. Mr. Bagley called on us a few mornings since and showed us a fleece weighing 20½ pounds, and it is the finest and best quality, worth in the market at present prices from thirty-two to thirty-five cents per pound. Mr. Bagley was formerly engaged in sheep raising in Vermont—a State which produces the finest sheep in the world, and may be said to be the world's market for the purchase of the best and the purest blood—and is, therefore, thoroughly posted in the business. He is confident that Dakota possesses as good, if not superior, natural advantages as Vermont for successful and profitable sheep farming, and founds his faith upon his personal experience in our territory.—*Yankton Press.*

Treatment of the Cow.

There are conflicting opinions among good dairymen in regard to the treatment of cows after calving—some preferring a low or moderate diet, at most nothing more than good hay, with free access to the usual watering place and an avoidance of all warm drinks.

The arguments in favor of this course for the first few days after parturition are, that it is better calculated to allay fever and sooner brings the cow round to a healthy condition. On the other hand, it is urged that the animal during labor becomes more or less exhausted, and that if the weather is cold, the taking of considerable quantities of cold water to slake thirst, has a tendency to chill the animal and impede circulation, and hence, a gruel made of bran or oaten meal, and tepid water can be given soon after calving with the best results. The latter course has been our practice, and uniformly with success. After a lapse of several hours the cow is allowed to drink as usual. It is perhaps unnecessary to say that stock at this season of the year should be entrusted to careful hands—they demand almost constant oversight and attention. They should not be hurried in or out of the stables, or allowed to fight or worry each other at the water trough or in the yard. Accidents, of course, will occasionally occur, under the most careful treatment, but by the adoption of a uniform system of kindness to all neat stock, with a reasonable share of attention, there need be little, if any "bad luck" to be anticipated.

One thing we regard as imperatively demanded for success in the management of stock. Never allow a cow to be kicked or in any way abused by hired help. Whatever good qualities a man may have, better part with him at once if found disobeying orders in this respect. State the case plainly at the time of hiring, and make as a condition the forfeiture of a part or the whole of the man's wages who is found guilty of kicking or beating cows. The practice has become common and should be broken up. The animal losses from this source are immense. If every dairyman would make it a rule that his milch cow, must be treated kindly, and that no excuse can be taken for blows and kicks, and that no person would be employed who maltreats stock, the whole country would be greatly benefitted. We have known of valuable animals being lost by a kick, and others rendered valueless for the season by an apparently slight thump with a milking stool from bad tempered persons. Laborers of this kind are dangerous, and the sooner one is rid of them the better. Much can be effected in this matter by good example, for if the owner so far forgets himself as to abuse stock he cannot expect the men in his employ to do otherwise. The business of the year is about to commence, and the start should be made with sound, healthy and vigorous stock, and from such, reasonable results may be anticipated.—*Western Rural.*

POULTRY.

Floors for Poultry Houses

Experience has proved board floors for poultry houses to be injurious to the fowls. No amount of cleaning can keep them free from vermin and bad odors. Clean, dry earth is the proper flooring for hen houses. It should have an under strata of solid, packed earth. This should be scraped at least once a week and again sprinkled with road dust mixed with air-slaked lime.

Fowl Fattening.

The greatest curiosity in the Jardin d'Acclimation is the singular fowl fattening machine which has been in operation for but a short time, but which is a great success—remarks a lady, writing from Paris. Imagine the top of a round tea table divided off into sections, with a partition between each section and a board in front of with a half moon shaped aperture in it. In each of these sections an unhappy duck or chicken is confined by a chain to each leg, and under each is fitted a tray, which receives the dirt and is emptied daily. Through the centre of this structure goes a round post, and there is a series of such tea-table tops to the roof of the building, each with its divisions and imprisoned fowls. At stated intervals a man comes around with a somewhat complicated machine, filled with a kind of thin gruel, and fitted with a pipe at the end of a long India-rubber tube. He introduces this pipe down the throat of a duck, presses down a pedal with his foot, and a certain quantity of food is forced into the creature's craw, a disc above showing exactly what amount of force he is to use, and how much food passes. This process is gone through with each fowl till all are fed, and it is repeated four times a day for ducks and three for chickens. Two weeks suffice to fatten a duck, but three are necessary for a chicken. Apart from the necessary confinement of the birds, the process does not seem to be at all a cruel one, as the amount of food forced down their throat is not excessive. The ducks which I saw fed did not seem to suffer in the least; and, in fact, when they saw the man approach, most of them became clamorous for immediate attention and plucked at his clothes, as he passed, with eager beaks.—*Journal of Agriculture.*

Onions for Chicken Cholera.

A correspondent of the *Poultry Yard* thus describes his new remedy for chicken cholera: "While our neighbors for several miles around us, have lost nearly all their chickens from the so-called cholera, ours are in fine condition. They were attacked with the premonitory symptoms of the disease, which seemed to be endemic here, but we cured them and have no trouble with them since, having accidentally found a cure. Cut up onions with food, and administer once a day for several days, afterward once a week will answer. Also mix a little ground ginger with their meal, once every day or two. We also give them a little salt every two or three weeks, which we deem highly necessary, and, above all things, keep watermelons, muskmelons and cucumbers away from them. The tops of celery cut up with their food will be found beneficial, and they appear to like it very well. Do not get these statements mixed up. The onions and ginger only for cholera, the remainder constant attention. Too much whole corn we have found injurious; we give meal of this only once in three or four days. Raw onions and a very little ginger against the world for curing cholera, if the disease has not been allowed to run too far. We endorse heartily the raw onions and ginger, but have never found melons injurious. Last summer we raised, in an amateur way, nearly three hundred chickens and turkeys. Bushels of melon rinds and imperfect melons of both kinds were thrown to them daily and eaten eagerly. Over-ripe cucumbers and seeds of muskmelons were likewise devoured. We had no losses from any disease.

Cramming Poultry.

Poultry of all kinds can be well fattened, if in fair order previously, in three weeks. The method of cramming poultry to fat them is as follows: Oatmeal and cornmeal are boiled with milk and some sugar into a thick mush. When this is nearly cold it is rolled with dry meal into large pellets of the size of chestnuts, and that will be readily swallowed. The bird is taken between the knees on an apron, and its mouth held open while another person puts the pellets of food down the throat until no more can be put down. The bird is then put into a small coop, in which it cannot even turn, and shut up in darkness. It is fed four times a day and no water is given. The flesh of birds so fattened is very white and clear, and brings a good price in the market.

Wild Chickens.

Some years ago, several families settled in a frontier region in Commanche county, Texas, but becoming discouraged, they abandoned the enterprise and returned to the old settlement leaving their domestic fowls in possession of the clearings. These multiplied rapidly, and in a few years became as wild as any other birds of the forest. At the present time there are said to be thousands of these wild chickens in that region. They will probably become permanent inhabitants of the mountains of western Texas. In this manner the horse became a wild animal on the pampas of South America.—*Ex.*

Good Hatching.

We have often recommended Leghorn eggs for hatching, having found in our experience that they produce a large proportion of healthy chicks, than any other variety we ever tried. W. D. wrote us from Jackson county, Wisconsin, under date July 25th, that out of fifty two White Leghorn eggs received from us he got forty-five chicks. The day we received his letter, a game hen turned us out fourteen White Leghorn chicks, all smart and lively We find the chicks from this variety very smart and healthy from the start.

LITERARY AND PERSONAL.

THE AMERICAN MILLER.—A monthly journal devoted to the art and science of milling. If any outward manifestation were necessary to illustrate that this journal had "come to stay," and could afford to stay, it seems to us that the number now on our table (No 6, Vol. X.) would be ample evidence to that effect. Published by Mitchell Bro.'s Company, Chicago, Ill., at ONE DOLLAR PER ANNUM. We can best express its size by stating that it is a 12 by 15 demi folio, of 58 pages, including the covers, and has three columns of beautifully printed matter to the page. We think it can "afford to stay," from the fact that this number contains 210 advertisements, 200 of which are embellished with illustrations executed in the finest style of wood engraving, and that a number of these are full page advertisements, all relating to milling and machinery connected with milling and its corelatives. To any person, or any company making milling and grain-dealing a specialty, this journal would be a perfect *vade mecum*, at least within the realm of its circulation, if not beyond it. *Thirty-three* cards appear under the heading "Mills for Sale." *Twenty-two* "Flour and Commission Cards." *Eighteen* "Miscellaneous Notices;" *thirty-eight* under the head of "Wanted;" *ninety-four* "News Items" from mills, ranging from one line to fifty, in addition to which the "Minneapolis Budget" alone contains over *forty*. Special notices *seven*, business notices *eleven*, editorial notes *twenty-eight*, besides *ten* half column to whole column editorial papers. The "New Mills Items" number *one hundred and seventy-one*; "Foreign Milling News" *twenty*; "Canadian" do *twenty-two*; "Scientific and Practical," *seventeen*. Notes and Queries, *twenty-eight*, bringing the number up to 136 since the beginning of the present volume; "Changes," *fifty-seven*; list of "new patents," *thirty-five*, from April 25th to May 23rd, 1882. Besides communications, extracts from proceedings of miller's associations, notes on steam-power, descriptions of mill machinery, mill factories, analysis of grains; "trade gossip," observations on new patent milling aparatus, discussions of questions connected with milling, improvements, buildings, experiments, expressions of opinions, quotations from old inventors, corrections, &c., &c. We have been thus minute in scanning the contents of this journal (only limited for want of time and space) because milling is so intimately connected with *good bread*, which is literally the "staff" of physical life; and more of man's physical and moral health is involved in good bread than the world at large seems to apprehend, nor can it be too soon enlightened thereon.

AMERICAN SILK AND FRUIT CULTURIST. It may have been a dictate of wisdom in tying these two domestic interests together in one enterprise, for it seems clear that the first named is not yet able to stand alone, having never fully recovered from the attack of *Multicaulus* of five and forty years ago. No. 1, Vol. 1, of this lively little 32 page 8 vol., is before us, and we bid it "God Speed," for the interests it represents, ought to be crowned with success, especially in a country claiming to be "free and independent." Published by Campbell & Pepper, 1328 Chestnut street, Philadelphia, at $1.00 a year, monthly, devoting itself exclusively to the industrial interests included in its title. In the silk department it seems to be the organ of the "Women's Silk-Culture Association of the United States," and as the representative of *that* alone, it ought to receive the encouragement of all who wear silk in any form whatever. The quality of the material and typography are "A No. 1," and its silk and fruit literature are able and practical. When we reflect that for the fiscal year ending with June, 1880, we imported from abroad *raw silk* to the value of $11,688,822, Cocoon and waste silk, $1,206,806; and manufactured silk, $30,506,509, making an aggregate of $43,492,137, we cannot help concluding that we are missing one opportunity to participate in one of the greatest domestic trades that characterized human civilization. Silk culture to become a *permanent* success must be conducted on an economical and wide expanding basis, and on limited scales—it must become the occupation of the "common people," and the people must subscribe for a silk journal, pay for it and *read* it, before an intelligent beginning can be made, and a profitable result attained. We gather from the report of the Woman's Silk Culture Association, that cocoons are worth from $1.50 to $2.50 per pound, (pierced ones $1.10), and eggs from $4.00 to $5.00 per ounce, and as soon as a sufficient number of such "depots" are established the thrift of the business will begin. Having such a representative journal as the one which is the subject of this notice, the silk producers of our country will be provided with a solid staff that will bear them on to a successful issue. Although we do not expect to live long enough to see its crowning success, yet that *will eventually come*.

OUR HOME AND SCIENCE GOSSIP.—A sixteen-page royal quarto, published monthly, at Rockford, Ill., at $1.00 per annum, by Andruk Illingworth. No. 6 vol. 6 (June, 1882,) has been laid on our table and is a specially interesting and instructive number including in its scope, practical science, poetry and general literature, in their most latitudinal application. Printed in fair type, on tinted paper and worthy the patronage of an intelligent public.

THE IRON HALL.—"One thousand dollars safe in seven years." A demi-folio monthly, devoted to the interests of a secret beneficial organization, called "The Order of the Iron Hall," Indianapolis, Indiana, May 15, 1882, vol. 1, No. 11. The details of this publication is very similar to those given in our June number, noticing the "Banner of Chosen Friends," which need not be repeated here. If any of our readers desire to "take stock" in these associations they had better send for a representative number.

PROGRESS, published by the State Sunday School Committee, Boston, Massachusetts, at twenty-five cents per annum, monthly. This is a demi-folio of eight pages, and contains a large amount of practical matter on the organization and conduct of Sunday schools.

THE SIDEREAL MESSENGER.—A monthly review of astronomy, in ten numbers annually, at $2.00, and is the only periodical in the United States devoted exclusively to popular astronomy, conducted by Wm. W. Payne, Director of the Carleton College Observatory, Northfield, Minnesota .No. 4, vol. 1 of this splendidly printed octavo magazine of thirty-four pages in tinted covers, has been placed on our table, and we call the special attention of our "Star Club" to it, as in every way worthy of their liberal patronage, and in which they may realize that—

"The voice that rolls the stars along
Speaks all the promises."

The material and typographical execution are unexceptionally good, and the contributions and editorial notes all that a practical "star-gazer" could desire. It will be observed that this is a new claimant of public patronage, and there is a freshness and vigor about it which indicates that it has "come to stay." We hope the votaries of the beautiful science of astronomy may manifest an appreciative sense of its worth by a liberal patronage of the enterprise.

PROCEEDINGS of a convention of agriculturists held in the department of agriculture, January 10th to 18th, 1882, Washington, D. C. 204 pp. octavo. Report upon the condition of winter grain, and upon the condition of farm animals of the United States, April 1882, 82 pp. octavo. Florida—its climate, soil, productions, and agricultural capabilities, 1882. 98 pp. octavo. Report upon the acreage and condition of cotton, the condition of all cereals, and the area of spring grain. 15 pp. octavo; and report upon the condition of winter grain, the progress of cotton and corn planting, the rate of wages and labor, and results of the draining. 20 pp., octavo. All neatly and uniformly printed bulletins, issued by the department of agriculture, and all containing many items of solid information, and also much that has no value except in a local sense, and then only to those who *read and heed*.

ELEVENTH REPORT of the State Entomologist, on the Noxious and Beneficial Insects of the State of Illinois, being the Sixth Annual Report by Prof. Cyrus Thomas, Ph. D., State Entomologist. This is an octavo of 104 pages, with title page and index, and without the usual illustrations. Prof. Thomas has valuable aid in the State through the entomological labors of Mr. D. W. Coquillett, of Woodstock, Ill., who contributes largely to this report; also in Prof. G. H. French, who contributes the second part of the report.

The report is mainly confined to new phases of old insects—subjects heretofore described, but have since developed some new characteristic features in their histories, for instance, *Heliothis Armigera* as a boll-worm, a corn-worm, a tomato worm, etc., showing its flexibility of character, and its adaptation of means to ends. Two years ago we bread this insect from larva sent to us from Spring Garden in Lancaster county, where it was feeding on tobacco plant. We do not know that it was plentiful, but we have had its existence as a feeder on the tobacco plant and its species, identified by competent authorities outside of our own experience. We acknowledge the receipts of this report with thanks for this and many other favors extended to us by the State Entomologist of Illinois.

THE LADIES FLORAL CABINET, for July 1882. A large royal embellished quarto of 22 pages; a true "Pictorial Home Companion," devoted to the flower garden, and polite and domestic literature, New York, $1.25 a year. This journal is gotten up in the highest style of typographic, pictorial and literary art, and worthy the patronage of at least the women of our country.

THE SUGAR BEET, devoted to the cultivation and utilization of the Sugar Beet, 3rd year, No. 2, Philadelphia, May, 1882, price 50 cents per annum. This excellent quarterly quarto seems to have come to stay, and abates not in its faith in the ultimate success of the Sugar Beet industry in the United States; and from the fact that 38,666 pounds of Sugar Beet seed valued at $4,165 had been imported into the United States in 1881, we have reason to believe that the people are cultivating an abiding, although, perhaps, a somewhat tardy faith in it. *Slow but sure* is considered a normal progress.

WE have just received a "PAMPHLET OF POT-GROWN AND LATER STRAWBERRY PLANTS," with instructions for their cultivation, and for sale by J. T. Lovett, Monmouth Nursery, Little Silver, Monmouth county, New Jersey, for the Summer and Autumn of 1882. Six pages octavo, with a beautiful illustration of the *Manchester Strawberry*, natural size, in colors, including an announcement of the *Hansall Raspberry*, two varieties of small fruits that have received the endorsement of some of the most prominent fruit-growers of New Jersey, New York and Pennsylvania, after the most thorough and practical tests. We somehow never have too much of either of these fruits in our markets, nor yet of too good a quality or too low in price, and as it is to the interest of the producer as well as the consumer, to have good prolific and hardy varieties of these fruits, we believe it would be to the advantage of both if our fruit growers were to extend their inquiries in the direction above indicated. For further particulars we would suggest that they send for catalogues, especially as the proprietor offers pamphlets post free to all.

STATE, DISTRICT AND COUNTY FAIRS.—We have received a copy of the Premium List—50 pages octavo—of the Thirtieth Indiana State Fair, profusely embellished, and in tinted covers. We always receive similar documents from Indiana, Illinois, Kansas, Ohio, and elsewhere, long before we receive anything of the kind from Pennsylvania. The book also contains a diagram of the Indiana State fair grounds, and a map of the State, illustrating its entire railroad system. Perhaps there is no city in the Union that is so central in its State, as the City of Indianapolis, nor none that has more railroads ramifying its entire domain. The list is large and the premiums liberal, with a department especially for boys and girls under sixteen years of age, together with a large list of "special premiums" from outside enterprising business houses. Indiana certainly has some faith in the uses of fairs to stimulate progress in agriculture. She advertises 56 county fairs for 1882, between August 8th and October 11th. Also, 18 district fairs in the State, and one great general fair. The diagram of the fair-grounds looks like a clever town, having about thirty buildings for the accommodation of the various exhibits. About one-half of the enclosure is devoted to a race-course. That is a feature in agricultural exhibitions that seems indispensable almost everywhere, and the idea is to improve it, and not abolish it.

THE WORLD OF NATURE

The world of animated nature is more splendidly represented under the canvas of Forepaugh's Great Show than in any zoological collection existent. Not since the day Noah lifted his hawser off the ark blog post have we so many distinct varieties of rare animals been collected under one charge. This important fact should not be lost sight of by schools and parents. Boys and girls can learn more in an afternoon of natural history, in the great Menagerie of Forepaugh's Show, than by months of book study. Recognizing this, Mr. Forepaugh makes reduced rates to schools, and admits all children in orphan asylums free of charge. This Great Show will exhibit in Lancaster, Monday, April 24.

THE
LANCASTER EXAMINER

OFFICE

No. 9 North Queen Street,
LANCASTER, PA.

THE OLDEST AND BEST.

THE WEEKLY
LANCASTER EXAMINER

One of the largest Weekly Papers in the State.

Published Every Wednesday Morning,

Is an old, well-established newspaper, and contains just the news desirable to make it an interesting and valuable Family Newspaper. The postage to subscribers residing outside of Lancaster county is paid by the publisher.
Send for a specimen copy.

SUBSCRIPTION:

Two Dollars per Annum.

THE DAILY
LANCASTER EXAMINER

The Largest Daily Paper in the county.

Published Daily Except Sunday.

The Daily is published every evening during the week. It is delivered in the City and to surrounding Towns accessible by railroad and daily stage lines, for 10 cents a week.
Mail Subscription, free of postage—One month, 50 cents; one year, $5.00.

JOHN A. HIESTAND, Proprietor,

No. 9 North Queen St.,
LANCASTER, PA.

Important to Grocers, Packers, Hucksters, and the General Public.

THE KING FORTUNE-MAKER.
OZONE
A New Process for Preserving all Perishable Articles, Animal and Vegetable from Fermentation and Putrefaction, Retaining their Odor and Flavor.

"**OZONE—Purified air, active state of Oxygen.**"—*Webster.*

This preservative is not a liquid pickle, or any of the old and exploded processes, but is simply and purely OZONE, as produced and applied by an entirely new process. Ozone is the antiseptic principle of every substance, and possesses the power to preserve animal and vegetable structures from decay.

There is nothing on the face of the earth liable to decay or spoil which Ozone, the new Preservative, will not preserve for all time in a perfectly fresh and palatable condition.

The value of Ozone as a natural preserver has been known to our ablest chemists for years, but, until now, no means of producing it in a practical, inexpensive, and simple manner have been discovered. Microscopic observations prove that decay is due to septic matter or minute germs, that develop and feed upon animal and vegetable structures. Ozone, applied by the Prentiss method, seizes and destroys these germs at once, and thus preserves. At our office in Cincinnati can be seen almost every article that can be thought of, preserved by this process, and every visitor is welcomed to come in, taste, smell, take away with him, and test in every way the question of Ozone as a preservative. We will also preserve, free of charge, any article that is brought or sent prepaid to us, and return it to the sender, for him to keep and test.

FRESH MEATS, such as beef, mutton, veal, pork, poultry, game, fish, &c., preserved by this method, can be shipped to Europe, subjected to atmospheric changes and return to this country in a state of perfect preservation.

EGGS can be treated at a cost of less than one dollar a thousand dozen, and be kept in an ordinary room six months or more, thoroughly preserved; the yolk held in its normal condition, and the eggs as fresh and perfect as on the day they were treated, and will sell as strictly "choice." The advantage in preserving eggs is readily seen; there are seasons when they can be bought for 6 or 8 cents a dozen, and by holding them, can be sold for an advance of from one hundred to three hundred per ct. One man, with this method, can preserve 500 dozen a day.

FRUITS may be permitted to ripen to their native climate, and can be transported to any part of the world. The juice expressed from fruits can be held for an indefinite period without fermentation—hence the great value of this process for producing a temperance beverage. Cider can be held perfectly sweet for any length of time.

VEGETABLES can be kept for an indefinite period in their natural condition, retaining their odor and flavor, treated in their original packages at a small expense. All grains, flour, meal, etc., are held in their normal condition.

BUTTER, after being treated by this process, will not become rancid.

Dead human bodies, treated before decomposition sets in, can be held in a natural condition for weeks, without puncturing the skin or mutilating the body in any way. Hence the great value of Ozone to undertakers.

There is no change in the slightest particular in the appearance of any article thus preserved, and no trace of any foreign or unnatural odor or taste.

The process is so simple than a child can operate as well and as successfully as a man. There is no expensive apparatus or machinery required.

A room filled with different articles, such as eggs, meat, fish, etc., can be treated at one time, without additional trouble or expense.

☞ **In fact, there is nothing that ozone will not preserve.** Think of everything you can that is to come, decay, or spoil, and then remember that Ozone will preserve it in exactly the state you want it for any length of time. If you will remember this it will save asking questions as to whether Ozone will preserve this or that article—**it will preserve anything and everything you can think of**. Ozone will preserve this or that article—**it will preserve anything and everything you can think of**, from $4,000 to $10,000 a year, that he pleases. We desire to get a live man interested in each county in the United States, in whose hands we can place this Preservative, and through him secure the business which every county ought to produce.

A FORTUNE Awaits any Man who Secures Control of OZONE in any Township or County.

A. C. Lowen, Marion, Ohio, has cleared $2,400 in two months. $2 for a test package was his first investment.
Woods Brothers, Lebanon, Warren County, Ohio, made $6,000 on eggs purchased in August and sold 1st November 1st. $2 for a test package was their first investment.
F. K. Raymond, Morristown, Belmont Co. Ohio, is clearing $250 a month in handling and selling Ozone. $2 for a test package was his first investment.
D. F. Webber, Charlotte, Eaton Co., Mich., has cleared $1,000 a month since August. $2 for a test package was his first investment.
J. H. Gaylord, 81 La Salle St., Chicago, is preserving eggs, fruit, etc., for the commission men of Chicago, charging 1½c. per dozen for eggs, and other articles in proportion. He is preserving 5,000 dozen eggs per day, and on his business is making $50.00 a month clear. $2 for a test package was his first investment.
The Cincinnati Feed Co., West 6th Seventh street, is making $3,500 a month in handling brewer's malt, preserving and shipping it as feed to all parts of the country. Malt unpreserved sours in 24 hours. Preserved by Ozone it keeps perfectly sweet for months.

These are instances which we have asked in the privilege of publishing. There are scores of others. Write to any of the above parties and get the evidence direct.

Now, to prove the absolute truth of every thing we have said in this paper, **we propose to place in your hands the means of proving for yourself that we have not exhibited half enough.** To any person who doubts any of these statements, and who is interested sufficiently to make the trip, we will pay all traveling and hotel expenses for a visit to this city, if we fail to prove any statement that we have made.

How to Secure a Fortune with Ozone.

A test package of Ozone, containing a sufficient quantity to preserve one thousand dozen eggs, or other articles in proportion, will be sent to any applicant on receipt of $2. This package will enable the applicant to pursue any line of tests and experiments he desires, and thus satisfy himself as to the extraordinary merits of the Ozone as a Preservative. After having thus satisfied himself, and had time to look the field over to determine what he wishes to do in the future—whether to sell the article to others or to confine it to his own use, or any other line of policy which is best suited to him and to his township or county—we will enter into an agreement with him that will make a fortune for him and give us good profits. We will give exclusive township or county rights to the first responsible applicant who orders a test package and desires to control the business in his locality. **The man who secures control of Ozone for any special territory, will enjoy a monopoly which will surely enrich him.**

Don't let it lay thus until you have ordered a Test Package, and if you desire to secure an exclusive privilege we advise you that delay may deprive you of it, for the applications come in to us by scores every mail—many by telegraph. "First come first served" is our rule.

If you do not care to send money in advance for the test package we will send it C. O. D., but this will put you to the expense of return money. Your correspondence is very large, we have all we can do to attend to the shipping of orders and giving attention to our working agents. Therefore we can and give any attention to letters which do not order Ozone. If you think of any article that you are doubtful about Ozone preserving remember we guarantee that it will preserve it, no matter what it is.

REFERENCES.

We desire to call your attention to a class of references which no enterprise or firm based on any thing but the soundest business success and highest commercial merit could secure.

We refer, by permission, as to our integrity and to the merits of the Prentiss Preservative, to the following gentlemen: Edward C Royce, Member Board of Public Works; E. O. Eshelby, City Comptroller; Amor Smith, Jr., Collector Internal Revenue; Wulsin & Worthington, Attorneys; Martin H. Burrell and B. F. Hopkins, County Commissioners; W. S. Cappeller, County Auditor, all of Cincinnati, Hamilton County, Ohio. These gentlemen are each familiar with the merits of our Preservative, and know from actual observation that we have without question

The Most Valuable Article in the World.

The $2 you invest in a test package, will surely lend you to secure a township or county, and then your way is absolutely clear to make from $2,000 to $10,000 a year.

Give your full address on every letter, and send your letter to

PRENTISS PRESERVING COMPANY, (Limited,)

S. E. Cor. Ninth & Race Sts., Cincinnati, O.

Nov-3m

WHERE TO BUY GOODS IN LANCASTER.

BOOTS AND SHOES.

MARSHALL & SON, No. 12 Centre Square, Lancaster, Dealers in Boots, Shoes and Rubbers. Repairing promptly attended to.

LEVY, No. 3 East King street. For the best Dollar Shoes in Lancaster go to M. Levy, No. 3 East King street.

BOOKS AND STATIONERY.

JOHN BAER'S SONS, Nos. 15 and 17 North Queen Street, have the largest and best assorted Book and Paper Store in the City.

FURNITURE.

HEINITSH'S, No. 15½, East King st., (over China Hall) is the cheapest place in Lancaster to buy Furniture. Picture Frames a specialty.

CHINA AND GLASSWARE.

HIGH & MARTIN, No. 15 East King st., dealers in China, Glass and Queensware, Fancy Goods, Lamps, Burners, Chimneys, etc.

CLOTHING

MYERS & RATHVON, Centre Hall, No. 12 East King St. Largest Clothing House in Pennsylvania outside of Philadelphia.

DRUGS AND MEDICINES.

G. W. HULL, Dealer in Pure Drugs and Medicines, Chemicals, Patent Medicines, Trusses, Shoulder Braces, Supporters, &c. 15 West King St., Lancaster, Pa.

JOHN F. LONG & SON, Druggists, No. 12 North Queen St. Drugs, Medicines, Perfumery, Spices, Dye Stuffs, Etc. Prescriptions carefully compounded.

DRY GOODS.

GIVLER, BOWERS & HURST, No. 25 E. King St., Lancaster, Pa., Dealers in Dry Goods, Carpets and Merchant Tailoring. Prices as low as the lowest.

HATS AND CAPS.

C. H. AMER, No. 30 West King Street, Dealer in Hats, Caps, Furs, Robes, etc. Assortment Large. Prices Low.

JEWELRY AND WATCHES.

H. Z. RHOADS & BRO, No. 4 West King St. Watches, Clock and Musical Boxes. Watches and Jewelry Manufactured to order.

PRINTING.

JOHN A. HIESTAND, 9 North Queen St. Sale Bills, Circulars, Posters, Cards, Invitations, Letter and Bill Heads and Envelopes neatly printed. Prices low.

Thirty-Six Varieties of Cabbage: 26 of Corn: 24 of Cucumber; 41 of Melon; 37 of Peas; 28 of Beans; 17 of Squash; 27 of Beet and 40 of Tomato, with other varieties in proportion, a large portion of which were grown on my five seed farms, will be found in my **Vegetable and Flower Seed Catalogue** for 1882. Sent free to all who apply. Customers of last Season need not write for it. All Seed sold from my establishment warranted to be fresh and true to name, so far, that should it prove otherwise, I will refill the order gratis. The original introducer of **Early Ohio** and **Burbank Potatoes, Marblehead, Early Corn, the Hubbard Squash, Marblehead Cabbage, Phinney's Melon**, and a score of other New Vegetables, I invite the patronage of the public. New Vegetables a specialty.

JAMES J. H. GREGORY,
Marblehead, Mass.
Nov-6mo]

EVAPORATE YOUR FRUIT.
ILLUSTRATED CATALOGUE
FREE TO ALL.

AMERICAN DRIER COMPANY,
Chambersburg, Pa.
Ap1-tf

FARMING FOR PROFIT.

It is conceded that this large and comprehensive book, (advertised in another column by J. C. McCurdy & Co., of Philadelphia, the well known publishers of Standard works,) is not only the newest and handsomest, but altogether the BEST work of the kind which has ever been published. Thoroughly treating the great subjects of general Agriculture, Live-Stock, Fruit-Growing, Business Principles, and Home Life; telling just what the farmer and the farmer's boys want to know, combining Science and Practice, stimulating thought, awakening inquiry, and interesting every member of the family, this book must exert a mighty influence for good. It is highly recommended by the best agricultural writers and the leading papers, and is destined to have an extensive sale. Agents are wanted everywhere. Jan-tf

LANDRETH'S
BLOOMSDALE SWEDE, OR RUTA BAGA,

Is the result of critical selection, and has proved to be unquestionably the most desirable of all known strains of

PURPLE TOP YELLOW RUTA BAGA.

The foliage is not superabundant, the shape is nearly globular, the crown deep purple, and the flesh a deep yellow. The illustration conveys a good idea of the shape assumed by the strain.

Also, strap-leaved Garden Ruta Baga Turnip, white fleshed, Purple top Ruta Baga Turnip, Hanover Long French or Sweet German Turnip, Yellow Aberdeen, or Scotch Yellow Turnip, Pomeranean White Globe (strap leaved) Turnip, Amber Globe (strap leaved) Turnip, Yellow Stone Turnip, Early Flat Dutch (strap leaved) Turnip, the Flat Red, or Purple Top (strap leaved) Turnip, Cow Horn Turnip, Early White Egg Turnip, Large Early Red Top Globe Turnip, White Norfolk Globe Turnip, Seven Top Turnip.

Bloomsdale Swede or Ruta Baga.

Every farmer should sow Turnip Seeds. A good stock of turnips is the best and most economical food for cattle during the winter and early spring months. Also, turnips grown on the ground, and plowed in, make very valuable manure.

Descriptive and Illustrated Catalogue free on application.

D. LANDRETH & SONS.
AGRICULTURAL AND HORTICULTURAL IMPLEMENT AND SEED WAREHOUSE,

Nos. 21 and 23 South Sixth Street,
BETWEEN MARKET AND CHESTNUT STS.,
AND S. W. CORNER DELAWARE AVENUE AND ARCH ST.,
apr-6m   PHILADELPHIA.

MERCHANT TAILORING.

1848 (The Oldest of All.) 1881

RATHVON & FISHER,
MERCHANT TAILORS AND DRAPERS,

respectfully inform the public that having disposed of their entire stock of Ready-Made Clothing, they now do, and for the future shall, devote their whole attention to the **CUSTOM TRADE**.

All the desirable styles of CLOTHS, CASSIMERE, WORSTEDS, COATINGS, SUITINGS and VESTINGS constantly on hand, and made to order in plain or fashionable style from 15, and warranted satisfactory.

All-Wool Suit from $10.00 to $30.00.
All-Wool Pants from 3.00 to 10.00.
All-Wool Vests from 2.00 to 6.00.

Union and Cotton Goods proportionately less. Cutting, Repairing, Trimming and Making, at reasonable prices.

Goods retailed by the yard to those who desire to have them made elsewhere.

A full supply of Spring and Summer Goods just opened and on hand.

Thankful to a generous public for past patronage they hope to merit its continued recognition in their "new departure."

RATHVON & FISHER,
PRACTICAL TAILORS,
No. 101 North Queen Street,
LANCASTER, PA.
1848   1881

GLOVES, SHIRTS, UNDERWEAR.
SHIRTS MADE TO ORDER,
AND WARRANTED TO FIT.

E. J. ERISMAN,
56 North Queen St., Lancaster, Pa.
75-1-12]

A HOME ORGAN FOR FARMERS.

THE LANCASTER FARMER,
A MONTHLY JOURNAL,

Devoted to Agriculture, Horticulture, Domestic Economy and Miscellany.

Founded Under the Auspices of the Lancaster County Agricultural and Horticultural Society.

EDITED BY DR. S. S. RATHVON.

TERMS OF SUBSCRIPTION:

ONE DOLLAR PER ANNUM,

POSTAGE PREPAID BY THE PROPRIETOR.

All subscriptions will commence with the January number, unless otherwise ordered.

Dr. S. S. Rathvon, who has so ably managed the editorial department in the past, will continue in the position of editor. His contributions on subjects connected with the science of farming, and particularly that specialty of which he is so thoroughly a master—entomological science—some knowledge of which has become a necessity to the successful farmer, are what make much more than the price of this publication. He is determined to make "The Farmer" a necessity to all households.

A county that has so wide a reputation as Lancaster county for its agricultural products should certainly be able to support an agricultural paper of its own, for the exchange of the opinions of farmers interested in this matter. We ask the co-operation of all farmers interested in this matter. Work among your friends. The "Farmer" is only one dollar per year. Show them your copy. Try and induce them to subscribe. It is not much for each subscriber to do but it will greatly assist us.

All communications in regard to the editorial management should be addressed to Dr. S. S. Rathvon, Lancaster, Pa., and all business letters in regard to subscriptions and advertising should be addressed to the publisher. Rates of advertising can be had on application at the office.

JOHN A. HIESTAND,
No. 9 North Queen St., Lancaster, Pa.

$72 A WEEK. $12 a day at home easily made. Costly Outfit free. Address TRUE & Co., Augusta, Maine

ONE DOLLAR PER ANNUM.—SINGLE COPIES 10 CENTS.

THE LANCASTER FARMER

DEVOTED TO Agriculture, Horticulture, Domestic Economy and Miscellany

THE FARMER IS THE FOUNDER OF CIVILIZATION.—WEBSTER

Dr. S. S. RATHVON, Editor.   LANCASTER, PA. AUGUST, 1882.   JOHN A. HIESTAND, Publisher

Entered at the Post Office at Lancaster as Second Class Matter.

CONTENTS OF THIS NUMBER.

EDITORIAL.
A Chosen People 113
Green Corn Pudding 113
Kitchen Garden for August 113
Good Husbandry 113
How to Preserve Stable Manure 114
Gapes and Eels 114
Excerpts 114

QUERIES AND ANSWERS.
A Big Bug 116
Tomato Horn worm 116
Goldsmith-Beetle 116
The English Sparrows 116

CONTRIBUTIONS.
Gapes in Poultry 117
Lime 118
Tariffs and their Effects 118

SELECTIONS.
Silk Culture 118
  Nature of the Silk Worm—Enemies and Disease—Varieties of Eggs—Wintering and Hatching the Eggs—Feeding and Rearing the Worms.
Mineral at the Exposition 121
Diversified Farming in the South 121
The Mosquito 122

OUR LOCAL ORGANIZATIONS.
Lancaster County Agricultural and Horticultural
  Society 122
  Crop Reports—Wheat Crop of 1882—How should Manure be Applied—New Business—Miscellaneous Business.
The Poultry Society 123
  Prevention of Gapes.
Fulton Farmers' Club 124
  Exhibits—What is the Best Kind of Wheat?—The Best time to Sow—A Question of Plows—Russian Oats—Literary Exercises—A Farmers' Reunion Noxious Weeds.
Linnæan Society 124
  Museum—Donations to the Library.

AGRICULTURE.
Lying in Fallows 125
A Short-sighted View 125
Select Your own Seed Wheat 125
A Talk About Grasses 125

HORTICULTURE.
The Peach Crop 125
Value of Fruit 125
Shallow Cultivation for Fruits 125
The Vegetable Garden 125
Fig Culture 126

HOUSEHOLD RECIPES.
Cucumber Mangoes 126
Peach Mangoes 126
Veal a la Mode 126
Breast of Veal Baked with Tomatoes 126
Breast of Veal Braised 126
White Sauce 126

Boiled Tongue 126
Boiled Corned Beef 126
Boiled Ham 126
Pork Chops, Spanish Style 126
Roast Pork 126
Pork Tenderloins 126
Irish Stew 126
Persillade of Mutton 126
Fried Breast of Mutton 126
Breading 126
Ragout of Cold Beef and Vegetables 126
Roast Leg of Lamb or Mutton 126
Garlic Cloves 126

LIVE STOCK.
Advice of a Lancaster County Blacksmith on
  How to Shoe Horses 126
Training Horses 126
The Best Farm Horses 127
Draught Horses 127
Is Horseshoing Useless 127
Keep the Stable Clear of Flies 127
Remedy for Side Hole in Cow's Teat 127
Care of Horses 127
The Stock 127

POULTRY.
Poultry Gossip 127
Feather and Egg Eating 127
Geese 128
The Wonders of Incubation 128
A Meat Diet 128
Feed for Laying Hens 128
Literary and Personal 128

SILK-WORM EGGS.

Amateur Silk-growers can be supplied with superior silk-worm eggs, on reasonable terms, by applying immediately to
GEO. O. HENSEL,
No. 278 East Orange Street, Lancaster, Pa.
may-3m]

PENNSYLVANIA STATE COLLEGE,

FALL TERM OPENS AUGUST 29.

Located in one of the most beautiful and healthful spots of the entire Allegheny region. Open to students of both sexes and offers the following Courses of study:

1. A full Classical Course of four years.
2. A full Scientific Course of four years.
3. The following Technical Courses of four years each: In Agriculture; in Natural History; in Chemistry and Physics; in Civil Engineering.
4. A short Special Course in Agriculture.
5. A short Special Course in Chemistry.
6. A Classical and Scientific Preparatory Course.
Military drill is required. Expenses for board and incidentals very low. Tuition Free. Young ladies under charge of a competent lady principal.
For Catalogue or other information, address,
GEO. W. ATHERTON, President,
State College, Centre co., Pa.

$1000 Reward VICTOR
For any machine hulling as much clover and in 1 day as the
(Double Huller)
It has hulled
150
Bushels
In
ONE
DAY.
Illustrated Pamphlet mailed free,
Newark Machine Co.,
Newark, O., Formerly the Dayton Agricultural Mfg. Co., Dayton, O.
july-3m]

D. M. FERRY & CO'S
ILLUSTRATED, DESCRIPTIVE AND PRICED
SEED ANNUAL FOR 1882
Will be mailed FREE to all applicants, and to customers without ordering it. It contains five colored plates, 600 engravings, about 200 pages, and full descriptions, prices and directions for planting 1500 varieties of Vegetable and Flower Seeds, Plants, Fruit Trees, etc. Invaluable to all. Send for it. Address,
D. M. FERRY & CO., Detroit, Mich.
Jan-4m

$66 a week in your own town. Terms and $5 outfit free. Address H. HALLETT & CO., Portland, Maine.
jun-1yr]

WE WANT OLD BOOKS.
We Want GERMAN BOOKS.
WE WANT BOOKS PRINTED IN LANCASTER CO.
We Want All Kinds of Old Books,
LIBRARIES, ENGLISH OR GERMAN BOUGHT.
Cash paid for Books in any quantity. Send your address and we will call.
REES WELSH & CO.,
23 South Ninth Street, Philadelphia.

LIGHT BRAHMA EGGS
For hatching, now ready—from the best strain in the county—at the moderate price of
$1.50 for a setting of 13 EGGS.
I. RATHVON,
No. 9 North Queen st., Examiner Office, Lancaster, Pa.

WANTED.—CANVASSERS for the LANCASTER WEEKLY EXAMINER In Every Township in the County. Good Wages can be made. Inquire at
THE EXAMINER OFFICE,
No. 9 North Queen Street, Lancaster, Pa

$72 A WEEK. $12 a day at home easily made. Costly outfit free. Address TRUE & CO., Augusta, Maine.
jun-1yr]

SEND FOR SPECIAL PRICES
On Concord Grapevines, Transplanted Evergreens, Tulip, Poplar, Linden, Maple, etc., Tree Seedlings and Trees for timber plantations by the 1000 or 10000.
J. JENKINS' NURSERY,
3-9-79   WINONA, COLUMBIANA CO., OHIO.

MARBLEHEAD
Early Sweet Corn
Is the most profitable of all, because it matures before any other kind, giving farmers complete control of the early market. I warrant it to be at least a week earlier than Minnesota, Narragansett or Crosby, and decidedly earlier than Dolly Dutton, Tom Thumb or Early Boynton. Of sage of Minnesota, and very sweet. The original introducer, I send pure stock, postpaid, per package 15 cents; per quart, 30 cents; per peck, by express, $5.00. In my catalogue, (free to all,) are emphatic recommendations from farmers and gardeners.
JAMES J. H. GREGORY,
Marblehead, Mass.
apr-3t

PENNSYLVANIA RAILROAD SCHEDULE.

Trains leave the Depot in this city, as follows:

WESTWARD.	Leave Lancaster.	Arrive Harrisburg.
Pacific Express*	2:40 a. m.	4:05 a. m.
Way Passenger†	5:00 t. m.	7:30 a. m.
Niagara Express	11:30 a. m.	1:20 a. m.
Hanover Accommodation	11:35 p. m.	Col. 10:40 a. m.
Mail train via Mt. Joy	10:20 a. m.	12:40 p. m.
No. 2 via Columbia	11:25 a. m.	12:55 p. m.
Sunday Mail	10:50 a. m.	12:40 p. m.
Fast Line*	2:30 p. m.	3:25 p. m.
Frederick Accommodation	2:05 p. m.	Col. 2:45 p. m.
Harrisburg Accom	5:15 p. m.	7:50 p. m.
Columbia Accommodation	7:20 p. m.	Col. 8:20 p. m.
Harrisburg Express	7:25 p. m.	8:40 p. m.
Pittsburg Express	9:50 p. m.	10:59 p. m.
Cincinnati Express*	11:30 p. m.	12:45 a. m.

EASTWARD.	Lancaster.	Philadelphia
Cincinnati Express	2:55 a. m.	3:00 a. m.
Fast Line*	5:08 a. m.	7:40 a. m.
Harrisburg Express	8:05 a. m.	10:00 a. m.
Columbia Accommodation	9:10 p. m.	12:0 p. m.
Pacific Express*	2:40 p. m.	3:40 p. m.
Sunday Mail	2:00 p. m.	5:00 p. m.
Johnstown Express	3:05 p. m.	5:30 p. m.
Day Express*	5:35 p. m.	7:20 p. m.
Harrisburg Accom	6:25 p. m.	9:20 p. m.

The Hanover Accommodation, west, connects at Lancaster with Niagara Express, west, at 9:35 a. m., and will run through to Hanover.

The Frederick Accommodation, west, connects at Lancaster with Fast Line, west, at 2:10 p. m., and runs to Frederick.

The Pacific Express, east, on Sunday, when flagged, will stop at Middletown, Elizabethtown, Mount Joy and Landisville.

*The only trains which run daily.
†Runs daily, except Monday.

NORBECK & MILEY,

PRACTICAL

Carriage Builders,

COX & CO'S OLD STAND,

Corner of Duke and Vine Streets,

LANCASTER, PA.

THE LATEST IMPROVED

SIDE-BAR BUGGIES,

PHÆTONS,

Carriages, Etc.

THE LARGEST ASSORTMENT IN THE CITY.

Prices to Suit the Times.

REPAIRING promptly attended to. All work guaranteed.

79-9-

S. B. COX,

Manufacturer of

Carriages, Buggies, Phaetons, etc.

CHURCH ST., NEAR DUKE, LANCASTER, PA.

Large Stock of New and Second-hand Work on hand very cheap. Carriages Made to Order Work Warranted for one year. [tr-9-19

EDW. J. ZAHM,

DEALER IN

AMERICAN AND FOREIGN

WATCHES,

SOLID SILVER & SILVER PLATED WARE,

CLOCKS.

JEWELRY & TABLE CUTLERY.

Sole Agent for the Arundel Tinted

SPECTACLES.

Repairing strictly attended to.

ZAHM'S CORNER.

North Queen-st. and Centre Square, Lancaster, Pa.

79-1-12

E. F. BOWMAN,

AT LOWEST POSSIBLE PRICES,
Fully guaranteed,
No. 106 EAST KING STREET,
79-1-12] *Opposite Leopard Hotel.*

ESTABLISHED 1832.

G. SENER & SONS,

Manufacturers and dealers in all kinds of rough and finished

LUMBER,

The best Sawed SHINGLES in the country. Also Sash, Doors, Blinds, Mouldings, &c.

PATENT O. G. WEATHERBOARDING

and PATENT BLINDS, which are far superior to any other. Also best COAL constantly on hand.

OFFICE AND YARD:

Northeast Corner of Prince and Walnut-sts.,
LANCASTER, PA.

79-1-12]

PRACTICAL ESSAYS ON ENTOMOLOGY,

Embracing the history and habits of

NOXIOUS AND INNOXIOUS

INSECTS,

and the best remedies for their expulsion or extermination.

By S. S. RATHVON, Ph. D.

LANCASTER, PA.

This work will be Highly Illustrated, and will be put to press (as soon after a sufficient number of subscribers can be obtained to cover the cost) as the work can possibly be accomplished.
79-5-

 **$5 to $20** per day at home. Samples worth $5 free! Address STINSON & Co., Portland, Maine.

TREES

Fruit, Shade and Ornamental Trees.

Plant Trees raised in this county and suited to this climate. Write for prices to
LOUIS C. LYTE,
Bird-in-Hand P. O., Lancaster co., Pa.
Nursery at Smoketown, six miles east of Lancaster.
79-1-12

WIDMYER & RICKSECKER,

UPHOLSTERERS,

And Manufacturers of

FURNITURE AND CHAIRS.

WAREROOMS:

102 East King St., Cor. of Duke St.
LANCASTER, PA.

79-1-12]

Special Inducements at the
NEW FURNITURE STORE
OF
W. A. HEINITSH,
No. 15 1-2 E. KING STREET
(over Bursk's Grocery Store), Lancaster, Pa.
A general assortment of furniture of all kinds constantly on hand. Don't forget the number.
15 1-2 East King Street,
Nov-1y] (over Bursk's Grocery Store.)

For Good and Cheap Work go to
F. VOLLMER'S
FURNITURE WARE ROOMS,
No 309 NORTH QUEEN ST.,
(Opposite Northern Market),
Lancaster, Pa.
Also, all kinds of picture frames. nov-1y

GREAT BARGAINS.

A large assortment of all kinds of Carpets are still sold at lower rates than ever at the
CARPET HALL OF H. S. SHIRK,
No. 202 West King St.

Call and examine our stock and satisfy yourself that we can show the largest assortment of three Brussels, three plies and ingrains at all prices—at the lowest Philadelphia prices.

Also on hand a large and complete assortment of Rag Carpet.

Satisfaction guaranteed both as to price and quality. You are invited to call and see any goods. No trouble in showing them even if you do not want to purchase. Don't forget this notice. You can save money here if you want to buy.

Particular attention given to customer work.

Also on hand a full assortment of Counterpanes, Oil Cloths and Blankets of every variety. [nov-1y]

PHILIP SCHUM, SON & CO.,
38 and 40 West King Street.

We keep on hand of our own manufacture,

QUILTS, COVERLETS,
COUNTERPANES, CARPETS,

Bureau and Tidy Covers, Ladies' Furnishing Goods, Notions, etc.

Particular attention paid to customer Rag Carpet, and scouring and dyeing of all kinds.

PHILIP SCHUM, SON & CO.,
Nov-1y Lancaster, Pa.

THE HOLMAN LIVER PAD!

Cures by absorption without medicine.

Now is the time to apply these remedies. They will do for you what nothing else on earth can. Hundreds of citizens of Lancaster say so. Get the genuine at
LANCASTER OFFICE AND SALESROOM,
22 East Orange Street.
Nov-1yr

C. R. KLINE

ATTORNEY-AT-LAW,

OFFICE: 15 NORTH DUKE STREET,
LANCASTER, PA.
Nov-1y

The Lancaster Farmer.

Dr. S. S. RATHVON, Editor.   LANCASTER, PA., AUGUST, 1882.   Vol. XIV. No. 8.

EDITORIAL.

A CHOSEN PEOPLE.

"Those who labor in the earth are the chosen people of God, whose breasts he has made his peculiar deposit for substantial and genuine virtue."

The foregoing very pretty sentiment is one of the utterances of the "sage of Monticello"—the immortal Jefferson—and as we find it used as a motto under the title head of the *Farmers' Monthly Visitor*, an agricultural journal, conducted by Isaac Hill, at Concord, N. H., about forty years ago, we presume that the "workers in the earth" means farmers, although for the matter of that it might be so construed as to mean railroad excavating, canal digging, and perhaps also fence-making, at least as far as relates to the digging of post-holes. Of course, in Jefferson's time, although there may have been fence-making, yet there was no railroad or canal making, no excavating except an occasional tail-race to a mill, or cutting down a hill for a township road. We believe cultivators of the soil are entitled to an extra distinction above all other manual operators on earth, but we do not think they are "the chosen people of God," above all others who earn their bread by the sweat of their faces, *only because* they "labor in the earth." Cain labored in the earth, but he does not seem to have been chosen of God in as special a sense as Abel was, who *did* not labor in the earth, but was a herder. Many fine things are patronizingly said about farmers, and no matter how much they merit them yet there is reason to believe that many of them are said ironically. If the above sentiment, in its application to farmers, was popular in Jefferson's day, it is singular that farmers, as a class, should have had such a limited influence in the general make-up of the government. It is fifty-six years since Jefferson died, and yet it is only now beginning to be seen that agriculture ought to be represented in the national cabinet, for if the "people of God" are worthy of any position in the construction of civil government, it surely ought to be *there*. We may legitimately claim this position for *agriculture*, without setting up a special claim for agriculturists *per se*. Doubtless, like all other classes of men, they are "good, bad and indifferent." However this may be, it is unquestionable that *agriculture* should not only be a co-ordinate department of government, but should also outrank all other departments; because, if it were not for agriculture our government, our commerce, our manufactures, and our civilization itself, in no sense, would be much in advance of our aborigines or the clouted Patagonians.

Wherever there are thoroughly cultivated grain fields, orchards, gardens, including lawns, flowering vegetation, shrubbery, and even forests, *there* you may expect to find a corresponding civilization, and as such, they are in a large sense the "people of God" whose labors have produced a civilizing effect of this kind; not however to the total exclusion of others, who may be laboring as honesty, as faithfully, as effectually, and as usefully, in a different occupation.

It perhaps would not militate against any people to be considered the "people of God" provided they understood in what sense they were chosen to such a distinction—whether arbitrarily, or as the best instruments to effect certain ends.

The Israelites claimed to be the "chosen people of God," and this has also been accorded to them by christians, and yet according to the records of inspiration they were a dreadfully "stiff-necked, idolatrous and wicked nation;" but they were the best instruments in the hands of deity to accomplish His purposes among men on earth, just as under a stress in civil government, a thief may be chosen to catch a thief. Agriculturists certainly seem to have the advantage of all other occupations, because there seems to be no doubt about the legitimacy and usefulness of their calling, which is more than can be said about many others. And yet, there are many honest people in the world, who verily believe that the cultivation and sale of tobacco is a perversion of agriculture, and of course, are altogether unable to believe that any persons so employed are, by way of eminence, the "chosen people of God."

All this leads us to conclude that men often employ descriptive and explanatory terms of which they do not duly consider the import at the time they use them, and those in whose behalf they are employed, are astonished that they should have been so distinguished in a matter which they deem within the sphere of their bounden duty, and therefore entitled to no special distinction. The "people of God," are therefore those who are endeavoring to do the *will of God* on earth as it is done in heaven.

GREEN CORN PUDDING.

Take eighteen ears of green corn, split the kernels lengthwise of the ears with a sharp knife, then with a case knife scrape the corn from the cob; mix it with three or four quarts of rich sweet milk; add four eggs well beaten, two tablespoonsful of sugar, salt to the taste; bake it three hours. To be eaten hot, with butter.

We extract the above from the *Farmers' Monthly Visitor* of June 30, 1847, and we do so because we *know* it is a most toothsome dish, and have often wondered why housewives so seldom prepare it, for it is far superior to anything else that we know of of green corn—better than "corn starch." Of course the above quantities are too large for a small family, but it can be made proportionally with the one-half or one-quarter of those quantities. The corn should not be too ripe—just passing out of its milky state—when the inner pulp can be easily pressed out, by a moderately forced manipulation with the back of a common table-knife; leaving nothing but the empty shells of the grains adhering to the cob. When corn is too young and milky the operation may be facilitated by first scalding or parboiling it. As corn is just in season now (or soon will be) for this preparation we confidently recommend it to our worthy housewives. It, perhaps, may involve a little more time and labor than the usual modes of cooking corn, but the result will be ample compensation for all the additional trouble.

KITCHEN GARDEN FOR AUGUST.

In the Middle States the work in this month does not vary materially from the month of July. Cabbage for winter use may head if planted at once. Celery, earth up; plant for future use. Endive plant. Beans—Bush or Snap-plant; tender Snaps gathered in late autumn may be preserved in strong brine (salt and water) for winter use, and vary but little from those freshly gathered. Lettuce sow in drills to head. Peas sow. This vegetable is a delicacy in autumn, and should more frequently appear at table. Landreth's Extra Early, sown in latter end of this month and beginning of next, perfect before frost. Spinach sow for autumn use; for winter use sow in September. Radishes sow; the Spanish and China for winter; the Golden Globe and Red Turnip (rooted) for autumn. Ruta Baga sow without delay, if not already done. Should the ground be dry work thoroughly and sow in the dust; the seed may vegetate with the first shower. A roller to compress the soil sometimes promotes vegetation; but there is this disadvantage, if heavy dashing rains immediately ensue the ground packs and the seed is lost. Yellow Aberdeen, Pomeranian Globe and Amber Globe Turnips sow early in the month; also the German Sweet, don't forget it; the Early Dutch and Red-Topped Turnips—both strap-leaved varieties—may be sown until the first of September, although it is well to sow at least a portion earlier, as at a late day it is difficult to remedy a failure. Read remarks under the head of July.

Onion seed raised in that portion of Pennsylvania which surrounds Philadelphia unquestionably must be earlier than the New England seed, and still more so when compared with Western seed. The growth conclusively proves the assertion. This is an important feature, as the early marketed onions always bring the highest prices. Try the experiment and you will find that seed from this locality will make bulbs long before seeds from any other locality.—*Landreth's Rural Register.*

GOOD HUSBANDRY.

"A place for everything, and everything in its place," has passed into a domestic proverb, and doubtless many of those who most frequently and most earnestly use it, may suppose it had its origin in modern times, but this is a most egregious error. It is found frequently used in "The Science of Good Husbandry, or the Economics of Xenophon," and Xenophon died about 444 years before the beginning of the Christian Era. Xenophon records an interview between Socrates and Ischomachus, a rich and powerful Athenian, in which occur many of the economical maxims of the present day—not only secular or domestic maxims, but those also of a religious character, albeit both were Pagans.

For instance, when Socrates inquired whether Ischomachus had instructed his young wife in the things "which relate to the management of a house," he answered: "I did, but not before I had implored the gods to show me what instructions were necessary for her, and that she might have a heart to learn and practice those instructions to the advantage and profit of us both." And this noble Athenian also invoked the guidance of the gods in all his enterprises. No matter about the *quality* of his religion—it was the best then accessible to him—it was his *faith* we commend, a faith that puts to blush many of the pietetic practices of modern christianity, which have little regard to any influence outside of self.

"Husbandry," says Ischomachus, "is an honorable science, and the most pleasant and profitable of any other; it is favored by the gods and beloved by mankind."

Even the drawbacks to husbandry existed in ancient times very much as they do now. "There are many unforeseen accidents that happen in husbandry, which will sometimes destroy all our hopes of profit, though a husbandman has acted with the greatest skill and diligence. Sometimes hail, droughts, mildews, or continual rains, spoil our crops, or vermin will even eat up the seed in the ground." What period in the world's history can the husbandman point back to, when these contingencies did not exist? And yet many are fretful and dissatisfied with the calling of the husbandman, and are yearning after that of the artisan, as though these troubles alone were his, and belong to the evils of our modern times.

HOW TO PRESERVE STABLE MANURE.

First.—All urine should be gathered or made to flow into a well-puddled or cemanted cistern, covered and protected against currents of air, as experiments have proved that in one week four-fifths of the ammonia can be dissipated.

Second.—The stables should daily have a dusting of plaster-of-paris, and the solid manure when thrown out should have a slight sprinkling; [the quantity can be regulated by the number of animals, some idea of which can be formed in estimating, that to hold the 135 pounds of nitrogen from an animal of 1000 pounds weight during one year in the form of ammonia, would require 1000 pounds of plaster-of-paris, or 500 pounds of oil of vitriol.

Third.—The manure should frequently have some soil or turf thrown over it, especially if exposed to the sun in hot weather.

Fourth.—The urine should be frequently pumped over the manure heap, that the same may not become burned or dried out, and that the gypsum spread over it may combine with the ammonia generated. It will be well in the urine cistern to add 3 to 4 pounds of oil of vitriol to about every 100 gallons of urine, thereby preventing the escape of ammonia. The acid when applied should be thoroughly stirred in, otherwise it may sink to the bottom by its gravity, and lie inert, and in a cemented cistern prove injurious to the well by dissolving the cement.—*From What of Fertilizers.*

GAPES AND EELS.

All we have time and space to say, on this occasion, in regard to the theory of our contributor, W. J. P., on the origin of "gapes" and "hair worms," and the breeding habits of eels, is, that if his observations can be verified, or corroborated by any intelligent authority, he has made the greatest discovery, on these subjects, of the nineteenth century. We do not doubt his intelligence nor his integrity, but we think he may have based his conclusions on insufficient data.

EXCERPTS

INDIA has nearly 2,000,000 acres of land sown to wheat.

HOP-GROWERS are happy over the prospects of a heavy hop crop.

THE prospects for good crops in France, Germany and Holland are favorable.

CLOVER will be a short crop in Michigan this season, owing to winter killing and drouth.

APPLE trees in Bucks county, Pa., are said to be dying from the effects of last year's drought.

IN Geogia insects of all kinds are abundant and all kinds of crops are receiving their attentions.

CATTLE valued at $13,500,000 are calmly grazing in what was six years ago absolutely an Indian country.

A TOTAL of 85,160,866 fleeces were shorn in the United States in 1880, with an average weight of 4.42 pounds.

THE army worms, which are abundant in Lyons county, Ky., are being destroyed by miriads of small red ants.

A fruit-grower in California says that should the Chinese go the fruit interest in that State would suffer seriously.

CALIFORNIA takes the lead for heavy heads of wheat. Some stalks have been shown, six feet high, with heads six inches long.

Two hundred thousand head of sheep were driven from New Mexico recently to Texas, and 50,000 wethers to Nebraska.

THE silk trade of Switzerland gives employment to 70,000 hands. The yearly products of this industry amount to 130,000,000 francs.

THERE are over 150,000 orange trees in Florida, and the number is rapidly increasing annually. The product this year is put at 50,000,000 oranges.

THE large bean-raising districts of New York are afflicted by a worm called the bean weevil, which is doing great damage to the newly planted crops.

THE oleomargarine factories of New York have a producing capacity of 116,000,000 pounds annually, while the production of dairy butter in the State is only 111,000,000 pounds.

FIRE BRICK should be laid in a thin mortar made of fire clay, rather than in a lime and sand mortar, such as is used in ordinary brick-work. In laying up those portions of a boiler furnace requiring fire brick, provision should be made in the original wall for replacing the fire brick and without disturbing the outer brickwork.

WHEN corn on the ear is fed to horses they masticate it much more slowly than if the corn was shelled. As a consequence that on the ear is better digested. A horse requires more time to eat corn on the car than if fed either meal or shelled corn. If the horse can not have time to masticate a full feed of unshelled corn, then it is best to feed something else.

ROOTS OF GRASS.—The roots of grass being much shorter than those of cereals are less able to collect ash constituents from the soil. If, therefore, grass is mown for hay, manures containing potash, lime and phosphoric acid will generally be required. Like the cereal crops, grass is greatly increased in luxuriance by the application of soluble, nitrogenous manures.

BUTTER IN WINTER.—In Denmark in the management of the dairy rape cake, oats and wheat bran are reckoned as first-class butter foods, palm-nut cake and barley as second-class foods, while linseed cake, peas and rye are placed in the third class. By the employment of first and second-class foods, with out straw, hay and roots, an abundance of excellent butter is produced throughout the winter.

THE opinion has generally prevailed that a little bran mixed with meal would produce more pork than clear meal, but in some experiments lately tried it was found that clear meal made more pork than a mixture of bran and meal.

THE naturalists have found that trunks of trees undergo daily changes in diameter. From early morning to early afternoon there is a regular diminution, followed until twilight by an increase.

HOUSE-FLIES are found to be very frequently infested by parasitic worms, which suggests the possibility that they may also carry about the germs of infectious diseases.

DE VRIES believes that the true function of the resinous juices of plants is to serve as a balm for wounds, and that the resins, are not therefore excrementitious matter as some have thought.

INCLOSE a piece of ground adjacent to the hen-house with a high picket fence, and set out plum trees in it. Keep the hens in the inclosure during the curculio's ravages, and a crop of plums annually will be the reward of the pains and the outlay. The editor says the remedy is a good one, and has been used by plum growers for years. Whether it is as thoroughly efficacious as this correspondent thinks, is open to debate. However, in the vicinity of Detroit, two parties who have tried it for years declare it eminently satisfactory, as they have never had a curculio since they turned their fowls into their plum orchards.—*Michigan Farmer.*

STUDIES OF THE WIND IN JAPAN.—The India Bureau of statistics has received a report from the University of Tokio, Japan, on meteorology. Among other things, the movement of the wind for each day in the year is given. The total movement for 1880 was about 4,000 miles greater than for 1879. When States and countries make such a record of the wind as this, the law of its motion may be ascertained, and then the character of the seasons in every part of the world, perhaps, foretold. The bureau is making an effort to interest observers everywhere in this matter.

THE celebrated rose-bush at Hildesheim, in Hanover, reputed to have been planted by Charlemagne and therefore to be more than a thousand years old, has borne more blossoms this season than ever before, and is an object of much curiosity. The branches of the bush extend to about three feet and eight inches in height and three feet and four inches in width.

IT is not generally known that Fortress Monroe is the largest single fortification in the world. It has already cost over $3,000,000. The water battery is considered to be one of the finest pieces of military construction ever built.

HAVE your seed wheat perfectly clean. An hour spent in making seed clean will save a day or week in the future in eradicating weeds.

THE bull is half of the herd. Thus a bull of the best milking strain of blood used even in a small lot of dairy cows greatly and at once improves each of his get.

POULTRY.—As floors to poultry houses boards are not good, especially if chickens are to be brought up on them. Nothing is so good as deep, well pulverized, dry soil, which is really the least expensive of anything.

CARE OF SHEEP.—Keep sheep dry under foot. This is even more necessary than roofing them.

THE Castor bean is a special crop of increasing popularity in all the Western States, and in some counties in the West they are leading crops. They have proven a source of profit to the general farmer, as the cultivation and harvest are simple and require little or no outlay for machinery. As a crop they are nearly insect-proof, belong to the night-shade family; they also bid defiance to chinch bugs. For the last 15 years it is claimed they have not been below a profitable price but twice, and generally held at a most remunerative figure. Corn and other staples have been below this oftener a great deal.—*Louisville, Ky., Agriculturist.*

VENICE and Amsterdam are the cities of bridges. The first has 450, the last 300. London has 15, Vienna 20, Berlin will soon have 50. Altogether the most beautiful and striking bridge in Europe is that over the Moldau at Prague.

THE resident population of Great Britain in the middle of 1882 is estimated by the Registrar General at 35,280,299 persons ; that of England and Wales at 26,406,820 ; of Scotland at 3,85,400, and of Ireland at 5,088,079.

FARMERS do not be deceived by the cry, by our large city dailies, of such an immense crop of wheat in the West. This is gotten up for a purpose. While the crop is fair in some localities, large in others and poor in others, this cry is started for the purpose of stimulating the farmers of the States east of the Mississippi river to rush their wheat into the market at a beggarly price before the Western wheat is harvested. Once in the market the "cornering" and speculating business would commence in earnest. With the present poor prospects for corn there is no reason why farmers should crowd their wheat on the market.—*South Bend (Indiana) Era.*

HOW to FEED PIGS.—The great point in feeding pigs is to keep them growing. It is not a difficult matter to accomplish, but there are many who keep pigs that fail to grow them profitably. If our farmers would lay out and fit up clover pastures for their pigs there would be a great point gained towards economical feeding. Clover pasture with a little skim-milk, pure water and a little soaked corn will make pig pork at low cost.

M. TOUSSAINT has shown experimentally the serious dangers of eating meat nearly raw as is now so generally done. If the meat is unsound, the germs of disease must pass into the system. The most frequent and dangerous malady with which animals slaughtered for food were affected is consumption, and even if the animal is only slightly affected, persons eating the uncooked meat are liable to infection. The raw juice pressed from a slightly affected cow's lung was used to inoculate rabbits and young pigs, and all the subjects died in a short time from the disease. The experiment was repeated with a portion of the juice which had been partially cooked, and the result was the same. Thorough cooking of the meat at a temperature of 150 or 160 degrees, is recommended as a precaution unsafe to neglect.

THE *New England Homestead* says: "There is a right and a wrong time for everything. It certainly isn't the right time for a farmer to take a vacation before haying and hoeing are finished. But with these jobs (the greatest the New England farmer has to do) out of the way, there certainly ought to be a chance for farmers, like other folks, to get off for a few days' vacation. A day or two even of change, of absence from every-day cares, braces a man up wonderfully for future work. And if a farmer and his family who have toiled through the season till August don't deserve a little rest at least, then we don't know who does."

A LEADING farmer in Middle Tennessee states that a crop of 10 acres of amber cane was of more value to him for feeding hogs, cattle, and mules, than any 25-acre crop on his farm, and that it paid better than any other crop. Those who have had the most experience claim that the amber cane is twice as nutritious as common field corn, and yields nearly double the amount of the best varieties of the sweet corn usually sown for fodder.—*St. Louis Journal of Agriculture.*

THERE has lately been exhibited in the Botanical Garden of Berlin the biggest flower in the world—the great flower of Sumatra known in science as the Rafflesia Arnoldi,and peculiar to Java and Sumatra. It measures nearly ten feet in circumference, and more than three in diameter. Sir Stanford Raffles and Dr. Joseph Arnold were exploring in company when they discovered this champion plant.

DON'T kill the toads, the ugly toads that hop around your door. Each meal the little toad doth eat a hundred bugs or more. He sits around with aspect meek, until the bug hath neared; then shoots he forth his little tongue like lightning double-geared. And then he soberly doth wink and shut his ugly mug, and patiently doth wait until there comes another bug.—*Independent Farmer.*

MERINO sheep will yield from ten to twenty pounds of wool per head, and the Cotwold even more, while scrub sheep will give from three to six pounds. The fine sheep eat no more than scrub and produce more flesh, to say nothing of the superior quality of both wool and flesh. Therefore keep only good sheep.

WASHINGTON TERRITORY is now setting up its claims to distinction as a State. Two years ago the census of that Territory showed a population of 75,116. The people now claim a population of quite 150,000. Owing to the remoteness of the Territory from the East this increase is quite remarkable, and the completion of the Northern Pacific Railroad will result in the rapid filling up of that country.

WIDE TIRES.—Those who have learned to use wide tire wagons find great advantage in so doing. They could not be induced to go back to the narrow tire. The philosophy of this is readily observed. The broad tire does not cut through, either in mud or sand, thus making the draft much lighter ; besides this the roads are not cut up, but on the contrary the broad tire presses down the lumps and leaves a smooth track, thus bettering the roads, the advantage of which is easily understood. The tire which seems to meet with general favor is from three and a half to four inches wide.

IN consequence of the defective water supply there has been an increase of 25 per cent. in the price of fire insurance risks in the city of Galveston. The average under the old rates was 1¼ per cent. The new schedule of fire rates will make it 1½ per cent. There is about $20,000,000 regularly covered by insurance in Galveston.

FIFTY years ago the capital invested in cotton-factories was $40,000,000, and the amount of cotton used was 77,759,316 pounds; to-day the capital is $225,000,000 and the material used 793,210,500 pounds. Forty years ago the woolen factories used 50,808,524 pounds of wood, turning out products worth $20,696,999. In 1880 187,916,905 pounds of wool were manufactured into articles worth $234,587,671. In the last ten years our silk products have increased from a value of $12,210,662 to $34,410,462. Fifty years ago there were but few tanneries and no shoe factories. In 1870 4,247 tanneries, using 9,000,000 hides and 9,664,000 skins, produced leather worth $86,160,383; while the 3,151 shoe factories turned out articles worth $146,704,000. In 1830 the yield of the iron furnaces was 165,000 tons ; in 1880 that of iron and steel works was 7,265,000 tons, worth $296,557,685. In but twenty years the capital employed in making machinery has increased from $15,000,000 to $40,000,000, and the annual product is worth $20,000,000. In 1810 the value of paper made in the United States was $2,000,000 ; in 1870 it was $30,812,445. To quote the words of Commissioner Loring, from whose address these facts are collated, "the aggregate annual product of the manufacturing and mechanical industries of the United States is now more than $6,000,000,000. Of this vast product less than $200,000,000 are exported."

QUERIES AND ANSWERS.

A BIG BUG.

FAIRFIELD, July 31, 1882.

I send you herewith a bug, or something of the buggy nature, handed me by two of your readers—Messrs. Samuel Martin and Hiram Harvey, who would be pleased if you can persuade Dr. Rathvon to kindly classify it through your columns. It was found in Mr. Martin's tobacco patch, and the gentlemen are desirous of knowing the name of the "quar looking thing."

For myself, I am not much an admirer of

"Great ugly things,
All legs and wings,
With nasty long tails armed with
Nasty long stings,

nor have any inclination to be

"Poking and peeping
After things creeping,
Or eternally thinking
And blinking and winking,
At grubs,"

but one is forced to manifest an interest sometimes, and the learned doctor's previous kindness has spoiled us.—*Yours very truly, W. F. M., in Intelligencer.*

The insect referred to above is the *Belastoma Americana*, of naturalists, for which I know no common name other than the "American Belastoma."

It is a true "bug," and a "big bug" at that—indeed, it is distinguished as being the largest species of bug that exists in North America.

It is amphibious in its character, and during the larval and papal periods lives exclusively in the water; its cursorial or ambulatorial abilities are very defective, but it is a great swimmer and diver, and when its wings are fully developed it is to a limited extent a good flyer. But "it overrates its strength nor measures well the foe;" hence, it is often "brought to grief" when it ventures abroad; because, when it once falls to the earth it cannot rise again, and its legs being oar-shaped, and used as oars, it can make little progress on dry land.

It is entirely carnivorous in its habits, and although destitute of mouth and teeth it has a powerful sectorial apparatus with which it pierces the bodies of other animals and absorbs their fluids. It has long been known to destroy small fishes and sometimes infests fish ponds and preys upon the fry.

The flying abroad is said to be during the nuptial season, when the sexes meet and provide for the continuation of the species.

Its systematic position in insect classification is in the order *Hemiptera* (Half-wings) because its *elytra* or wing-covers are not wholly *coriaceous*, or leathery, but on the contrary the apical portions are membranaceous. This order includes the odoriferous "squashbug," the "chinchbug," and many other postiferous suctorials. Still, there are some individuals in it that prey upon many of our noxious species of insects, and hence they may be tolerated.

TOMATO HORN-WORM.

Miss S. S. L. and others.—The large green worm with diagonal whitish stripes along the sides, and a horn on the back near the hinder end, is the larva of *Sphinx* (*Macrosila*) *carolina*, the same that also infests the tobacco plant, the potato, the egg-plant, etc; and the little white follicles that stud the entire bodies of them, are cocoons of a small Hymenopterous parasite, that infests them (*Microgaster congregata*) and which has a very peculiar history. The parent, a small four winged fly—deposits her eggs on or *in* the worm, and as soon as they are hatched they penetrate the body of the worm and feed upon its substance. When they are mature they issue from the body of the worm, and each one spins a small white cocoon, that resembles a grain of rice. After a few days in the pupal state the fly is evolved, cuts off the upper end and issues forth a small fly, like the parent, and is soon ready to repeat the operation on some other worm. A worm so infested, rarely, if ever, has strength to effect its usual transformations. Even if it should be able to go into the ground and assume the pupal form it would be hardly able to change to a moth. The worm usually dies with its hooked feet firmly fixed in the plant, without the ability to disengage them, and as we have found the dried carcass of the worm with the cocoons on it so suspended during the winter season, it is very probable that those that evolve late in the season hibernate in that condition during the winter, and thus perpetuate the species the following season. It is possible that those which may be carried under ground in the body of the worm, would hibernate there and come forth the following spring. Be this as it may, it is very certain that they are in some manner carried from one season to another, for as often as the host occurs, the parasites are also present. These little insects may therefore be esteemed as friends, for, if the horn-worm is a female, they at least prevent the generation of as many horn-worms as the female would deposit eggs, which frequently amounts to two or three hundred if not more.

GOLDSMITH-BEETLE.

GERMANTOWN, July 24, 1882.

PROF. RATHVON—*Dear Sir:* A large bronze and black beetle has been sent to me, with a blackberry on which it was preying. The account is that these were on a cluster *eating the fruit*, and that they came out of the ground, mounted the stalk and began feasting. I know from this you have a very poor basis to say anything, unless the beetle is a familiar acquaintance. But if you have leisure please tell me in *brief* what you know, if anything.—*Very Truly, P. R. F.*

We have answered what we supposed the insect alluded to in the above, may have been, notwithstanding the paucity of the description; but, since that answer was written, Mr. John Thomas, of East Orange street, Lancaster city, sent us three specimens of the common "Goldsmith Beetle," (*Gymnetis nitida*) which he captured in the act of feasting upon his peaches. He found as many as five on a single peach, fairly wallowing in percican luxury, and never abandoning it until the whole pulp of the fruit was devoured except the "skin and bone," (stone).

This is nothing new for master goldsmith. We have observed it on many occasions during the last fifty years or more, but never in sufficient numbers to create any special alarm. We are inclined to believe also that the insect alluded to in the above note was either the goldsmith, or a nearly allied species; because we have found both, on rare occasions, feeding on the blackberry; and both are developed from a "white grub-worm" that lives under ground, and feeds upon the roots and tubers of vegetables.

About the *first* insect the average country or village boy observes, on *terra firma*, is the "Tumble-dung" (*Canthon lævis*) and the *second* is the "Goldsmith;" and, whatever may have been the character or quality of the sport, the boy that had not "dyed" a goldsmith, with a thread tied to one of his hind legs, would have been considered no "great shakes."

It is the absence of this positive recognition by the writer of the above note, which alone involves the subject in doubt.

In an editorial "Remark" the writer of the above states that the Goldsmith "fills the bill."

THE ENGLISH SPARROWS.

The papers complain of the injurious habits of the English sparrows, showing that they drive other small birds away. It is plain that they are becoming a nuisance. One farmer in Canada says there are about five thousand of them upon his farm, and have done great harm in eating up his corn and barley and other productions of the ground. Our good wife complains that they destroy many of her garden products. Shall this evil be permitted to continue? Is there a law to protect the little depredators? If there is, let it be repealed, and let the boys have the privilege of trapping and shooting them, and there will soon be a lessening of their numbers. I would suppose they would be a nice morsel for the breakfast table for those who have a love for little birds to eat.—*J. F. W., Lancaster, Aug. 10, 1882.*

Our views on the "English Sparrow" have been given at length in vol. 14, page 17 of the FARMER, and it is hardly necessary to repeat them here. They are doubtless protected under the laws protecting other birds, so far as they are *insectivorous*, but we do not consider the sparrow as legitimately belonging to this class; and therefore, its introduction and domestication here was a mistake.

MR. EDOUARD PERRIS, of France, says, "the peasants of Lombardy prepare nesting places for the sparrows and then destroy the nests." This might be a good plan to effect their ultimate extinction here, without a repeal of the laws. Prevent their procreation, and their sires and dams will eventually die of old age. The same authority says, "the sparrow is a pillager who carries on his depredations in the harvest-field, in the garden, in the granary, and among the ripe grapes on our trellises; and I cannot join in the kind of worship paid him by certain persons more credulous of his pretended utility than struck by his instinct of rapine and waste." As we may refer to this able essay on some future occasion again, it is only necessary to add at this time, that the above characteristics of the sparrow are being loudly echoed from various parts of the county. True, his character may sometimes be traduced, but there may be a well founded suspicion of the innocence of one, whom everybody deems *guilty*.

CONTRIBUTIONS.

For The Lancaster Farmer.

GAPES IN POULTRY.

So much have been said of late in regard to the gapes in poultry, and as nothing positive has yet been determined on, I feel as though I should say something on the subject, giving my experience. I find by close observations that they are neither a louse nor do they take them from the ground, nor yet is it contagious. It is nothing more or less than the pip, as it was called when I was a boy, and they come from the downy plumage which the chick is covered with when hatched; whilst picking themselves the down is sucked in the windpipe whilst breathing, and if the quill end enters first it invariably will work itself down, and it requires but a day or two until it is covered with a red fleshy substance and will move when disturbed, but it contains no ovas nor ever will. In time, if not removed it will dissolve and pass away.

They are generally double, one being a little longer than the other; they are not male and female as some suppose—far from it. If you will examine the down, on all small feathers, you will find them all double, the same as the gape worm. Young ducks do not moult their downy plumage, but on the contrary, it increases in length and in quantity, and adheres more firmly to the skin for the purpose of keeping the body dry. Anything of this kind, or hairs from anything, placed in a warm and wet place and receiving air, will become living animals. They are very common at this time of the year, where stock go to drink, in the footprints, containing water. They differ in size and length, depending on the part of the body from which the hair had fallen. I took the other day from the ditch below my pump a knot of hair that was all alive; it was just as it had been taken from the comb and wrapped around the finger, and a hairpin stuck through it. I removed the pin, shook them out in a basin of warm water; they appeared to enjoy their liberty very much. By drawing one through my finger nails, stripping off the red fleshy substance, the hair was then red. Just so with the gape worm. Chicks are more subject to gapes after a few days rainy weather. During this time they are cooped, and having no exercise, they pass the time in picking themselves; as after the first week they commence moulting the downy plume. The best remedy is the horse hair to remove them. The best preventive is dry food: wheat and cracked corn; nature's food is dry, let us not change it. A better remedy still is to grease the chicks with lard and salt mixed; this will kill the life of the down and most generally prevent gapes, as grease or salt, or both combined, is death to anything so delicate, or prevent any accumulation of it after passing into the pipe. I have made the chick my study for several years. My last essay on poultry I gave to the Farmer,'giving the contents of an egg and how it is made.

The Eel question is another puzzle to many and is still talked of through the papers. I will here give my experience in regard to their mode of breeding and where. I am a miller by trade and have lived near a mill pond for the past forty-five years. Some say that they descend down the streams until they reach salt water and there spawn, and whilst young ascend again hundreds of miles before they reach the head of the last mill pond. Wouldn't it be amusing to see a few hundred of young and old eels from three to twenty inches in length climbing up the breast of a mill pond, say ten to twenty feet high, and so on to the next. They in one respect are like other fish, they breed where they inhabit; their spawn is not round like other fish, they are more the shape of a hen's egg, and they are laid on sunken brush or any bits of wood under water; take from the same a splinter containing a few spawn, place it in a bucket of luke warm water, and in a few days they will hatch; now drop a few drops of melted grease free from salt, and as it spreads over the surface they will come up and feed upon it; continue this for a few days and you will be surprised to see how the little wigglers will grow. Knowledge derived from the closet in the way of book learning' in many instances is of but of little use or benefit. Self taught from close observations and experimenting, is knowledge beneficial and it is never forgotten.— Yours, truly, W. J. P.

For The Lancaster Farmer.
LIME.

Is Lime a Manure, or only a Stimulant?

Much has been published pro and con, on this subject without settling the question either way. I am inclined to the belief that it is a manure as well as a stimulant.

I well remember coming up from Baltimore to York in the stage, sixty or sixty-two years ago, on passing through what was called the "York county Barrens," to see very little cultivation, and the old fields without fences, or only two or three rails, and no vegetation except that the ground was entirely covered with daisies, as with a mantle of snow.

At a place where the stage stopped to change horses, a man got in the stage and took a seat along side of me. He at once began to "pump" me where I was going, when I resided, &c. I told him I had been to Baltimore, and was on my way home to Lancaster county. He said I was fortunate in living in so rich a county—that in this neighborhood the land was too poor to make a living on it—that they could hardly grow enough on it to keep the family in provision the year round—that if they sowed rye, they could get very little more than the seed—and corn would not produce enough to pay expenses. I said, why do you not manure the ground? Ah! that's the difficulty—we have no manure—no grass or feed, hay, &c., to keep stock, so we have to do as well as we can.

Now, that "barren locality" produces as heavy crops of corn, wheat, and grass as Lancaster co. A few years ago I was again in that locality and I saw better crops of corn than we had that season here—they having had more rain in that section. Clover fields too, so very rank as to lodge all over the field. What has brought about this wonderful change? Lime was the renovator, that like the alchemists of old, turned baser metal into gold, or money into the farmers' pockets!

The same may be said of parts of Lancaster county. I well remember hearing people talk of "poor Octorare." They said the soil was so poor that Kilderes could not live there—that they had to come over to Manor and Hempfield townships to get feed to live.'" It suited these birds very well for breeding purposes, as a few formed their nest on the ground, on a bare spot so they could see all around, thus guarding their nests from enemies, as polecats, possums, snakes, &c.—but no feed for their young in that section. I do not now know the locality of this poor spot, but it may have been in parts of Drumore, Fulton, or Martic townships. That country has changed quite as wonderfully as the "Barrens of York co." Land that could be bought sixty years ago for 4 or 5 dollars an acre, is now worth probably from 50 to 80 dollars, or more, according to improvements. Here too, lime was the forerunner of improvement, so this lime is evidently a manurial agent. I well recollect the time when lime was first being applied to land as a fertilizer, some 50 or 60 years since.

A farm less than two miles from me, on the river hills, produced nothing but chestnuts and garlic. An old field had been in corn, as the little hillocks proved, was thrown out as of no further use, no fences, and a public road passing on one side, to cut off a corner people traveled over the field. All the vegetation on it was cinquefoil and running blackberry vines. All the income the family had was from a crop of chestnuts. But when a new owner took possession, chestnut trees and garlic soon disappeared, the former for rails and the latter of no earthly use. Lime being liberally applied, corn, wheat and even tobacco took their place, and now that old farm has been rejuvenated, and produces as heavy crops of useful vegetation as any in Lancaster county.

A farm not a hundred miles west of Lancaster where corn grew to three feet high and rye in spots here and there where cattle had dropped their excrements, grew only in spots (no wheat was sown.) There was also a public house on the farm and teamsters to Pittsburg stayed over to rest and feed, their horses of course dropped considerable manure, yet the farm did not improve until lime was applied, then to see the change that soon became apparent was really wonderful. No heavier crops can now be grown anywhere than on that poor farm, poor no longer. Many yet living may remember the poor quality of the soil in Chester county sixty years ago where now such heavy crops are grown. All this I think goes to show that lime is a manurial agent of great power as a renovator of the soil.

Then we had none of the so called improved varieties of fruits, still such as we had, mostly seedlings, bore heavy crops of fruit. Apples every alternate year produced more than could be used, even after making cider, applebutter, vinegar and the cellar filled for winter use and wagon loads taken to the still-house for "apple-jack"—hogs having their fill for months and many bushels going to waste. Why is it that our trees are so barren now? There is a question in my mind as our lands becomes more productive for grass and grain by the use of calcareous manures, has it a contrary effect on fruit trees? It appears to me as if it might be so.

But I will not follow this train of reasoning at present as my sheet is full, but would wish to call the attention of farmers and fruit growers to the fact that our grain crops have greatly increased during the last half century —chiefly from the use of lime. and our fruit crops decreased from some cause to me unknown unless it is from the application of celearous manures.—*J. B. Garber.*

FOR THE LANCASTER FARMER.
TARIFFS AND THEIR EFFECTS.

The article in the June number of the FARMER entitled "change of prices," etc., in which I trenched on the tariff question, has brought out J. P. in the July number, who, by the by, handles the tariff question with a good deal of ingenuity and argument, and also at considerable length. I must confess, however, that I cannot clearly comprehend J. P's. argument, or rather, perhaps, J. P., does not take in my meaning in regard to the working the tariff. A country, or a government, comparatively stands in the attitude of an individual. If an individual has nothing to sell, he is, of course not considered a good manager of his affairs. Or rather, if he has to buy more than he sells, he is not considered a good housekeeper. But whether the community is injuriously affected, or better off, I am at a loss to know.

First. Taking the term tariff to mean the exacting or levying a tax, direct or indirect which should only be enforced to protect an individual or a government in particular emergencies, there ought to be no difficulty in reaching a conclusion, according to the very nature of the case. As times and people change in all things relating to human progress, so should our lawmakers change our laws. Our lawmakers should be selected from among such practical men as will be able to travel with the age in which they live, and the changes we are following. As I said before, discount and repudiate all pretended Statesmen who profess to act for the people, but are nothing but corrupt and and only are interested in official spoils.

The four hundred millions of dollars are collected in tariffs, revenues and various other sources, one million four hundred thousand of which is in the Treasury of the United States, and the four millions of dollars collected in our State of which over four hundred thousand dollars is now kept in the State Treasury, and the thousands of dollars collected in Lancaster county, etc., (all of which I am no advocate of collecting direct from the people) if the people would take the trouble to look into the matters it would have the effect of opening their eyes, and excite their inquiry as to what is becoming of all this money, *who* gets it, *how* do they get it, and *what* is given as an equivalent for it? The tariff, as I think it should be assessed and used, ought to be for reasonable protection—one nation to protect itself against another.

As a general thing a young nation, which is not yet firmly established in manufactures, and has not the hard money or specie, should be protected against the commercial innovations of older and more permanently established notions, having lower prices and greater manufacturing facilities, absorb-ing the pecuniary means of the younger nation, by the withdrawal of its specie, and an unequal competition.

Taking for one of our maxims—if perhaps, not the best that an individual or a nation can be governed by—that "that individual or nation which make economy and industry the enduring basis of individual, state and national prosperity, should be protected by the laws of the land, (always remembering to "first seek the kingdom of heaven, and all these things shall be added unto you.'")

We should follow, as I said before, time's changes; and, as we have had heretofore, from time to time, protective tariffs which worked well, so well as to enable us to compete with older nations, and brought us ample revenue, so, in my opinion, we should continue to keep the balance of trade in our favor, but at the same time protect economy and industry, and not luxury and extravagance.—*P. S. R.*

SELECTIONS.

SILK CULTURE.

We are one of those who believe in the ultimate feasability of silk culture in this county, and we desire to place on permanent record such information as may be useful to those who may desire to make this industry a specialty, among their secular occupations.

The following paper on this subject, by Prof. C. V. Riley, M. A. Ph. D. Entomologist of Department of Agriculture, we appropriate, from the columns of the *National Farmer*, Washington, D. C., as a proper introduction, because a history of the habits, the transformations and general character of the *worms*, is of paramount importance in the conduct of the silk industry; indeed the very foundation of the whole superstructure is involved in healthy, thrifty worms. Without *these* through whose bodies the tissues known as silk are elaborated, all else would be a mere inert and profitless skeleton, only fit to be consigned to the "valley of drybones," as an industrial enterprise.—[ED.]

Nature of the Silk Worm.

The silkworm proper, or that which supplies the ordinary silk or commerce, is the larva of a small moth known to scientific men as *Sericaria mori.* It is often popularly characterized as the mulberry silkworm. Its place among insects is with the *Lepidoptera*, or spinners. There are several closely allied species, which spin silk of different qualities, none of which, however, unite strength and fineness in the same admirable proportions as does that of the mulberry species. The latter has, moreover, acquired many useful peculiarities during the long centuries of cultivation it has undergone. It has in fact become a true domesticated animal. The quality which man has endeavored to select in breeding this insect is, of course, that of silk-producing, and hence we find that, when we compare it with its wild relations, the cocoon is vastly disproportionate to the size of the worm which makes it or the moth that issues from it. Other peculiarities have incidentally appeared, and the great number of varieties or races of the silkworm almost equal those of the domestic dog. The white color of the species; its seeming want of all desire to escape so long as it is kept supplied with leaves, and the loss of the power of flight on the part of the moth, are all undoubtedly the result of domestication. From these facts, and particularly from that of the great variation within specific limits to which the insect is subject, it will be evident to all that the following remarks upon the nature of the silkworm must necessarily be very general in their character.

The silkworm exists in four states—egg, larva, chrysalis, and adult or imago—which we will briefly describe.

Different States or Stages of the Silk Worm

THE EGG.—The egg of the silkworm moth is called by silk-raisers the "seed." It is nearly round, slightly flattened, and in size resembles a turnip seed. Its color when first deposited is yellow, and this color it retains if unimpregnated. If impregnated, however, it soon acquires a gray, slate, lilac, violet, or even dark green hue, according to variety or breed. It also becomes indented. When diseased it assumes a still darker and dull tint. With some varieties it is fastened to the substance upon which it is deposited, by a gummy secretion of the moth, produced in the act of ovipositing. Other varieties, however, among which may be mentioned the Adrianople whites and the yellows from Nouka, in the Caucasus, have not this natural gum. As the hatching point approaches, the egg becomes lighter in color, which is due to the fact that its fluid contents become concentrated, as it were, into the central, formiug worm, leaving an intervening space between it and the shell, which is semi-transparent. Just before hatching the worm within becoming more active, a slight clicking sound is frequently heard, which sound is, however, common to the eggs of many other insects. After the worm has made its exit by gnawing a hole through one side of the shell, this last becomes quite white. Each female produces on an average from three to four hundred eggs, and one ounce of eggs contains about 40,000 individuals. It has been noticed that the color of the albuminous fluid of the egg corresponds to that of the cocoon, so that when the fluid is white the cocoon produced is also white, and when yellow the cocoon again corresponds.

THE LARVA OR WORM.—The worm goes through from three to four molts or sicknesses, the latter being the normal number. The periods between these different molts are called ages, there being five of these ages including the first from the hatching, and the last from the fourth molt to the spinning period. The time between each of these molts is usually divided as follows : The first period occupies from five to six days, the second but four or five, the third about five the fourth from five to six, and the fifth from eight to ten. These periods are not exact, but simply proportionate. The time from the hatching to the spinning of the cocoons may, and does, vary all the way from 30 to 40 days, depending upon the race of the worm, the quality of the food, mode of feeding, temperature, etc.; but the same relative proportion of time between molts usually holds true.

The color of the newly hatched worm is black or dark gray, and it is covered with long stuff hairs, which, upon close examination, will be found to spring from pale-colored tubercles. Different shades of dark gray will, however, be found among worms hatching from the same batch of eggs. The hairs and tubercles are not noticeable after the first molt and the worm gradually gets lighter and lighter, until, in the last stage, it is of a cream-white color. It never becomes entirely smooth, however, as there are short hairs along the sides, and very minute ones, not noticeable with the unaided eye, all over the body.

The preparation for each molt requires from two to three days of fasting and rest, during which time the worm attaches itself firmly by the abdominal prolegs (the 8 non-articulated legs under the 6th, 7th, 8th and 9th segments of the body, called prolegs in contradistinction to the 6 articulated true legs under the 1st, 2nd and 3rd segments,) and holds up the forepart of the body, and sometimes the tail. In front of the first joint a dark triangular spot is at this time noticeable, indicating the growth of the new head; and when the term of "sickness" is over, the worm casts its old integument, rests a short time to recover strength, and then, freshened, supple, and hungry, goes to work feeding voraciously to compensate for lost time. This so-called "sickness" which preceded the molt, was in its turn preceded by a most voracious appetite which served to stretch the skin. In the operation of molting the new head is first disengaged from the old skin, which is then gradually worked back from segment to segment until entirely cast off. If the worm is feeble, or has met with any misfortune, the shriveled skin may remain on the end of the body, being held by the anal horn; in which case the individual usually perishes in the course of time. It has been usually estimated that the worm in its growth consumes its own weight of leaves every day it feeds; but this is only an approximation. Yet it is certain that during the last few days before commencing to spin, it consumes more than during the whole of its previous worm existence. It is a curious fact, first noted by Quatrefages, that the color of the abdominal prolegs at this time corresponds with the color of the silk.

Having attained full growth, the worm is ready to spin up. It shrinks somewhat in size, voids most of the excrement remaining in the alimentary canal; acquires a clear, translucent, often pinkish or amber-colored hue; becomes restless; ceases to feed and throws out silken threads. The silk is elaborated in a fluid condition in two long, slender convoluted vessels, one upon each side of the alimentary canal. As these vessels approach the head they become less convoluted and more slender, and finally unite within the spinneret, from which the silk issues in a glutinous state and apparently in a single thread. The glutinous liquid which combines the two, and which hardens immediately on exposure to the air, may, however, be dissolved in warm water. The worm usually consumes from three to five days in the construction of the cocoon and then passes in three days more, by a final molt, into the chrysalis state.

THE COCOON.—The cocoon consists of an outer lining of loose silk known as "floss," which is used for carding, and is spun by the worm in first getting its bearings. The amount of this loose silk varies in different breeds. The inner cocoon is tough, strong, and compact, composed of a firm, continuous thread, which is, however, not wound in concentric circles as might be supposed, but irregularly, in short figure of eight loops, first in one place and then in another, so that in reeling, several yards of silk may be taken off without the cocoon turning around. In form the cocoon is usually oval, and in color yellowish, but in both these features it varies greatly, being either pure silvery white, cream or carneous, green, and even roseate, and very often constricted in the middle. It has always been considered possible to distinguish the sex of the contained insect from the general shape of the cocoon, those containing males being slender, depressed in the middle, and pointed at both ends, while the female cocoons are of a smaller size and rounder form, and resemble in shape a hen's egg with equal ends. Mr. Crozier, however, emphatically denies this, and thinks it "next to impossible for the smartest connoisseur not to be mistaken."

THE CHRYSALIS.—The chrysalis is a brown, oval body, considerably less in size than the full-grown worm. In the external integument may be traced folds corresponding with the abdominal rings, the wings folded over the breast, the antennæ and the eyes of the inclosed insect—the future moth. At the posterior end of the chrysalis, pushed closely up to the wall of the cocoon, is the last larval skin, compressed into a dry wad of wrinkled integument. The chysalis state continues for from two to three weeks, when the skin bursts and the moth emerges.

THE MOTH.—With no jaws, and confined within the narrow space of the cocoon, the moth finds some difficulty in escaping. For this purpose it is provided, in two glands near the obsolete mouth, with a strongly alkaline liquid secretion with which it moistens the end of the cocoon and dissolves the hard gummy lining. Then by a forward and backward motion, the prisoner, with crimped and damp wings, gradually forces its way out, and when once out the wings soon expand and dry. The silken threads are simply pushed aside, but enough of them get broken in the process to render the cocoons from which the moths escape, comparatively useless for reeling. The moth is of a cream color, with more or less distinct brownish markings across the wings. The males have broader antennæ or feelers than the females, and may by this feature at once be distinguished. Neither sex flies, but the male is more active than the female. They couple soon after issuing, and in a short time the female begins depositing her eggs, whether they have been impregnated or not. Very rarely the unimpregnated egg has been observed to develop.

Enemies and Diseases.

As regards the enemies of the silkworm but little need to be said. It has been generally supposed that no true parasite will attack it, but in China and Japan great numbers of the worms are killed by a disease known as "uji," which is undoubtedly produced by the larva of some insect parasite. Several diseases of a fungoid or epizootic nature, and several maladies which have not been sufficiently characterized to enable us to determine their nature are common to this worm. One of these diseases, called *muscardine*, has been more or less destructive in Europe for many years. It is of precisely the same nature as the fungus (*Empusa muscæ*,) which so frequently kills the common house-fly, and which sheds a halo of spores. readily seen upon the window-pane, around its victim.

A worm, about to die of this disease, becomes languid, and the pulsations of the dorsal vessel of the heart becomes insensible. It suddenly dies, and in a few hours becomes stiff rigid and discolored; and finally, in about a day, a white powder or efflorescence manifests itself, and soon entirely covers the body, developing most rapidly in a warm, humid, atmosphere. No outward signs indicate the first stage of the disease, and though it attacks worms of all ages, it is by far the most fatal in the fifth or last stage, just before the transformation.

"This disease was proved by Bassi to be due to the development of a fungus (*Botrytis Bassiana*) in the body of the worm. It is certainly infectious, the spores, when they come in contact with the body of the worm, germinating and sending forth filaments which penetrate the skin, and, upon reaching the internal parts, give off minute floating corpuscles which eventually spore in the efflorescent manner described. Yet most silkworm raisers, including such good authorities as F. E. Guerin-Meneville and Eugene Roberts, (Guide a l'eleveur de vers a soie,) who at first implicitly believed in the fungus origin of this disease, now consider that the Botrytis is only the ultimate symptom—the termination of it. At the same time they freely admit that the disease may be contracted by the Botrytis spores coming in contact with worms predisposed by unfavorable conditions to their influence. Such a view implies the contradictory belief that the disease may or may not be the result of the fungus, and those who consider the fungus as the sole cause certainly have the advantage of consistency." Dr. Carpenter, an eminent microscopist, believes the fungus origin of the disease, and thinks it entirely caused by floating spores being carried in at the spiracles or breathing-orifices of the worm, and germinating in the interior of the body.

Whichever view be held, it appears very clear that no remedies are known, but that care in producing good eggs, care in rearing the worms, good leaves, pure, even-tempered atmosphere, and cleanliness are checks to the disease. The drawers, and other objects with which the diseased worms may have been in contact, should be purified by fumigations of sulphurous acid (S. O2), produced by mixing bisulphite of soda with any strong acid, or, better still, by subjecting them to a carbolic-acid spray from an atomizer. In this way all fungus spores will be destroyed. In fact it will be well to wash off the trays or shelves once in a while with diluted carbolic acid, as a sure preventive. It is the best disinfectant known to science. The cheapest kinds may be used with the same efficacy as the more expensive.

Another disease known as *pebrine*, has proved extremely fatal in Southern Europe, and for twenty years has almost paralyzed silk culture in France. It is a disease which, in its nature and action, except in being hereditary, bears a striking analogy to cholera among men. "The worms affected by pebrine grow unequally, become languid, lose appetite and often manifest discolored spots upon the skin. They die at all ages, but, as in muscardine, the mortality is greatest in the last age. The real nature of this malady was for a long time unknown. In 1849 M. Guerin-Meneville first noticed floating corpuscles in the bodies of the diseased worms. These corpuscles were supposed by him to be endowed with independent life, but their motion was afterwards shown by Pilppi to depend on what is known as the Brownian motion, and they are now known either by the name of panhistophyton, first given by Lebert, or by that of psorospermiæ. They fill the silk-canals, invade the intestines, and spread throughout the tissues of the animal in all its different states; and though it was for a long time a mooted question as to whether they were the true cause or the mere result of the disease, the praiseworthy researches of Pasteur have demonstrated that pebrine is entirely dependent upon the presence and multiplication of these corpuscles. He has analyzed the disease so clearly that not only do we see its nature, but we are able to point out the remedy. The disease is both contagious add infections, because the corpuscles which have been passed with the excrement or with other secretions of diseased worms have been taken into the alimentary canal of healthy ones in devouring the soiled leaves, and because it may be inoculated by wounds inflicted by the claws. It is hereditary on the mother's side, because the moth may have the germ of the disease and yet oviposit. Indeed, the eggs may be affected and yet look fair and good, the microscopic psorospermix not being visible, so that the only true test of disease or health is an examination of the parent moth; and by killing off all infected moths the disease can be controlled.

"Both the diseases mentioned are, therefore, in the strict sense of the word, silk-worm plagues; the one of a fungus and the other of an epizootic nature. Each may become epidemic when the conditions are favorable for the undue multiplication of the minute organisms which produce them, or when the checks to the increase of such organisms are removed by carelessness or ignorance." Cleanliness and purification are absolutely necessary in treating both these diseases, and in pebrine care must be taken that the eggs are sound by a microscopic examination of the moths. This may be done after the eggs are laid, and if the corpuscles be found in the mother, her eggs should be discarded.

Silkworms are subject to other diseases, but none of them have acquired the importance of those described. What is called gattine by older authors is but a mild phase of pebrine. The worms are apt to be purged by unwholesome leaves; too great heat makes them sickly, or they may become yellow, limp and die of a malady called grasserie or jaundice, which is almost sure to appear in large broods, and which is very common in those reared in this country. When the worms die from being unable to molt they are called lusettes, and such cases are most abundant at the fourth molt. All these different ailments, and others not mentioned, have received names, some local, others more general; but none of them warrant further notice here, as they are not likely to become very troublesome if proper attention and care be given to the worms.

Varieties of Races.

As before stated, domestication has had the effect of producing numerous varieties of the silkworm, every different climate into which it has been carried having produced either some changes in the quality of the silk, or the shape or color of the cocoons, or else altered the habits of the worm.

Some varieties produce but one brood in a year, no matter how the eggs are manipulated; such are known as Annuals. Others known as Bivoltins, hatch twice in the course of the year; the first time, as with the Annuals, in April or May, and the second, eight or ten days after the eggs are laid by the first brood. The eggs of the second brood are only kept for the next year's crop, as those of the first brood always either hatch or die soon after being laid. The Trevoltins produce three annual generations. There are also Quadrivoltins, and in Bengal, a variety known as Dacey which is said to produce eight generations in the course of a year. Some varieties molt but three times instead of four, especially in warm countries and with Trevoltins. Experiments, taking into consideration the size of the cocoon, quality of silk, time occupied, hardiness, quantity of leaves required, etc., have proved the Annuals to be more profitable than any of the Polyvoltins, although Bivoltins are often reared; and Mr. Alfred Brewster, of San Gabriel, Cal., says that he found a green Japanese variety of these last more hardy than the Chinese Annuals. Varieties are also known by the color of the cocoons they produce, as Greens or Whites or Yellows, and also by the country in which they flourish. The white silk is the most valuable in commerce, but the races producing yellow, cream-colored or flesh-colored cocoons are generally considered to be the most vigorous. No classification of varieties can be attempted, as individuals of the same breed exported to a dozen different localities, would, in all probability, soon present a dozen varieties. The three most marked and noted European varieties are the Milanese (Italian) breed, producing fine small yellow cocoons; the Ardeche, (French) producing large yellow cocoons, and the Brousse (Turkish) producing large white cocoons of the best quality in Europe. Owing to the fearful prevalence of pebrine among the French and Italian races for fifteen or twenty years back, the Japanese Annuals have come into favor. The eggs are bought at Yokohama in September, and shipped during the winter. There are two principal varieties in use, the one producing white and the other greenish cocoons, and known respectively as the White Japanese and the Green Japanese Annual. These cocoons are by no means large, but the pods are solid and firm, and yield an abundance of silk. They are about of a size, and both varieties are almost always constricted in the middle. Another valuable race is the White Chinese Annua which much resembles the White Japanese, but it is not as generally constricted.

Wintering and Hatching the Eggs.

We have already seen the importance of getting healthy eggs, free from hereditary disease, and of good and valuable races. There is little danger of premature hatching until December, but, from that time on, the eggs should be kept in a cool, dry room in tin boxes to prevent the ravages of rats and mice. They are most safely stored in a dry cellar, where the temperature rarely sinks below the freezing point, and they should be occasionally looked at to make sure that they are not affected by mold. If, at any time, mold be perceived upon them it should be at once rubbed or brushed off, and the atmosphere made drier. If the tin boxes be perforated on two sides and the perforations covered with fine wire gauze, the chances of injury will be reduced to a minimum.

The eggs may also, whether on cards or loose, be tied up in small bags and hung to the ceiling of the cold room. The string of the bag should be passed through a bottle neck or a piece of tin to prevent injury from rats and mice. The temperature should never be allowed to rise above 40° F., but may be allowed to sink below freezing point without injury. Indeed, eggs sent from one country to another are usually packed in ice. They should be kept at a low temperature until the mulberry leaves are well started in the spring, and great care must be taken as the weather grows warmer to prevent hatching before their food is ready for them, since both the mulberry and Osage orange are rather late in leafing out. One great object should be, in fact, to have them all kept back, as the tendency in our climate is to premature hatching. Another object should be to have them hatch uniformly, and this is best attained by keeping together those laid at one and the same time, and by wintering them, as already recommended, in cellars that are cool enough to prevent any embryonic development.

They should, then, as soon as the leaves of their food-plant have commenced to put forth be placed in trays and brought into a well-aired room where the temperature averages about 75° F. If they have been wintered adhering to the cloth on which they were laid, all that is necessary to do is to spread this same cloth over the bottom of the tray. If, on the contrary, they have been wintered in the loose condition, they must be uniformly sifted or spread over sheets of cloth or paper. The temperature should be kept uniform, and a small stove in the hatching-room will prove very valuable in providing this uniformity. The heat of the room may be increased about 2° each day, and if the eggs have been well kept back during the winter, they will begin to hatch under such treatment on the fifth or sixth day. By no means must the eggs be exposed to the sun's rays, which would kill them in a very short time. As the time of hatching approaches the eggs grow lighter in color, and then the atmosphere must be kept moist artificially by sprinkling the floor, or otherwise, in order to enable the worms to eat through the egg-shell more easily. They also appear fresher and more vigorous with due amount of moisture.

Feeding and Rearing the Worms.

The room in which the rearing is to be done should be so arranged that it can be thoroughly and easily ventilated, and warmed if desirable. A northeast exposure is the best, and buildings erected for the express purpose should, of course, combine these requisites. If but few worms are to be reared, all the operations can be performed in trays upon tables, but in large establishments the room is arranged with deep and numerous shelves, from 4 to 8 feet deep, and 2 feet 6 inches apart. All wood, however, should be well seasoned, as green wood seems to be injurious to the health of the worms. When the eggs are about to hatch, mosquito-netting or perforated paper should be laid over them lightly. Upon this can be evenly spread fresh-plucked leaves or buds. The worms will rise through the meshes of the net or the holes in the paper and cluster upon the leaves, when the whole net can be easily moved. In this moving, paper has the advantage over the netting, in that it is stiffer and does not lump the worms together in the middle. They may now be spread upon the shelves or trays, care being taken to give them plenty of space, as they grow rapidly. Each day's hatching should be kept separate, in order that the worms may be of a uniform size, and go through their different moltings or sicknesses with regularity and uniformity; and all eggs not hatched after the fourth day from the appearance of the first should be thrown away, as they will be found to contain inferior, weakly, or sickly worms. It is calculated that one ounce of eggs of a good race will produce 100 pounds of fresh cocoons; while for every additional ounce the percentage is reduced if the worms are all raised together, until for 20 ounces the average does not exceed 25 pounds of cocoons per ounce. Such is the general experience throughout France, according to Guerin-Melville, and it shows the importance of keeping them in small broods, or of rearing on a modern scale.

The young worms may be removed from place to place by means of a small camel's hair brush, but should be handled as little as possible. The best mode of feeding and caring for them is by continuing the use of the feeding net first mentioned. As the worms increase in size, the net must have larger meshes, and if it should be used every time fresh food is furnished, it will save a large amount of time and care. If entirely obviates the necessity of handling the worms to keep them thoroughly clean; for, while they pass up through the net to take their fresh food, their excrement drops through it and is always taken up with the old litter beneath. It also acts as a detective of disease; for such worms as are injured, feeble or sickly, usually fail to mount through the meshes, and should be carried off and destroyed with the refuse in the old net below. This placing on of the new net and carrying away of the old is such a great convenience and time-saver that, in France, for many years, paper, stamped by machinery with holes of different sizes suited to the different stages of the worms, has been used. The paper has the advantage of cheapness and stiffness, but a discussion as to the best material is unnecessary here, the aim being to enforce the principle of the progressive rise of the worms. Details will suggest themselves to the operator.

Where the nets are not used, there is an advantage in feeding the worms upon leaf-covered twigs and branches, because these last allow a free passage of air, and the leaves keep fresh a longer time than when plucked. In thus feeding with branches consists the whole secret of the California system, so much praised and advocated by M. L. Prevost. The proper stamped paper not being easily obtained in this country, mosquito-netting will be found a very fair substitute while the worms are young, and when they are larger I have found thin slats of some non-resinous and well seasoned wood, tacked in parallel lines to a frame just large enough to set in the trays, very serviceable and convenient—small square blocks of similar wood being used at the corners of the tray to support the frame while the worms are passing up through it. Coarse twine netting stretched over a similar frame will answer the same purpose, but wire-netting is less useful, as the worms dislike the smooth metal.

Where branches, and not leaves, are fed, the Osage orange has the advantage of mulberry, as its spines prevent too close settling or packing, and thus insure ventilation. It is recommended by many to feed the worms while in their first age, and, consequently, weak and tender, leaves that have been cut up or hashed, in order to give them more edges to eat upon and to make less work for them. This, however, is hardly necessary with Annuals, although it is quite generally practiced in France. With the second brood of Bivoltius it might be advisable, inasmuch as the leaves at the season of the year when they have attained their full growth are a little tough for the newly hatched individuals. In the spring, however, the leaves are small and tender, and nature has provided the young worms with sufficiently strong jaws to cut them.

MINERAL AT THE EXPOSITION.

Doctor W. T. Strachan, the superintendent of minerals for the New Mexico exposition is sending out the following circular to the miners of all the camps in the territory whose names have been given him:

I desire to call the attention of yourself and the miners of New Mexico, generally, to a rare opportunity for exhibiting to the world the mineral wealth of the territory, where it will do the most good, presented by our coming territorial exposition, commencing September 18th and ending September 28th, next. The association desire to make an especial success of the mineral exhibits of the territory, and it is hoped that all who are engaged in mining will render all the assistance in their power. Arrangements have been consummated for free return over the railroads of all exhibits; the association offers liberal premiums, and every arrangement will be made for the convenience of the exhibitors in the arrangement of their minerals. It is desired that each district prepare and send or bring a cabinet representing the different mines; but, when this can not be done, cabinets will be furnished, and, under the management of the superintendent of minerals, specimens will be exhibited to the best advantage. I also suggest that each specimen be labeled plainly with the name of the district, mine and owner. Will you not give this matter your careful attention, consult with your neighbors, and come here determined that it shall not be your fault if your district does not carry off the first premium?

The following is the list of premiums:

Best collection from any one mining district in New Mexico, first, $100; second, $50.
Best collection from the territory at large, $100.
Best collection from any one mine in New Mexico, $50; and a special premium of $100 for the best exhibit from any state or territory in the United States or Mexico.

All specimens forwarded and all communications directed to me here, will receive prompt and careful attention.

DIVERSIFIED FARMING IN THE SOUTH.

A few months ago, in an article designed to show the importance of diversified crops in the South, we presented some carefully compiled statistics pointing out that during the present crop-year the foodstuffs bill that the Southern States had to pay West footed up nearly $200,000,000, of which $75,000,000 was for wheat, $60,000,000 for corn, and 72,000,000 for provisions. These figures have been republished by the leading papers of the country, generally without any acknowledgment of their source, and sometimes credited to wrong papers, until probably a dozen or more different papers have received the credit for them.

It was natural that they should attract much attention; for while it was known that the South was largely dependent upon the North and West for its bread and bacon, few if any realized how great was this annual drain. When it is remembered that the total value of the cotton crop, the South's main source of money, averages only about $275,000,000 to $300,000,000 a year, fully two-thirds of which goes out for food, it is really a wonder that that entire section has not steadily decreased in material prosperity. With the entire profits on cotton culture in any one year forming but a very small percentage of what was paid for flour and provisions, it is somewhat of a mystery to know how the South could stand such an enormous annual loss. The New England States, as well as Great Britain, do not produce enough grain or provisions for their own consumption, and they are compelled to pay a good many millions of dollars to the West for these necessaries of life; but then they are not like the South in being dependent upon one staple, the profit upon which at the most are very small. They are largely engaged in manufactures, which produce hundreds of millions of dollars worth of goods.

That this unfavorable condition of affairs in the South is undergoing a rapid change is a matter of deepest interest. This of itself means a wonderful improvement in that section, and is fully as important as the increased attention given to manufacturing and mining by local as well as outside capitalists. The area in wheat in the South this year shows an increase of 800,000 acres compared with 1881, while in corn and oats there is a proportionately large gain. The changing condition of affairs is well illustrated by

Tennessee, which last year produced 6,400,000 bushels of wheat, and purchased from the West nearly 3,000,000 bushels at a cost of over $4,000,000; while the wheat crop of that State this year is about 2,000,000 bushels, which will provide for domestic wants and leave a surplus for sale of about 3,000,000 bushels. In Georgia, Kentucky, Texas, North Carolina, and other States, the change is equally as great.

The acreage devoted to corn and oats, as previously stated, is also much greater than in former years, and this means more home-raised bacon, and also less provender from the west for live stock. We think that we are fully within bounds in estimating that the decrease to be paid out by the South for foodstuffs during the crop-year 1882-83, as compared with 1881-82, will be not less than $35,000,000 to $40,000,000. A revolution in the affairs of the South so great as this will undoubtedly tell upon the future prosperity of that whole section, and a few years more of the same course of diversified farming will make the Southern States practically independent.—*Baltimore Journal of Commerce and Manufacturers' Record.*

THE MOSQUITO.

There is another little lady whom you have fed and regaled at your own expense and very unwillingly withal. She is by no means modest, but steals unbidden into your room. She generally heralds her coming with song that is anything but soothing, and she is so persevering that even the strong "bars" with which you protect yourself are not proof against her persecutions. You have all, no doubt, at times exercised a little strategy with the mosquito, and when the little torment was fairly settled, made a dexterous movement of the hand, and, with a slap, exclaimed: "I've got him this time." No such thing; you never got him in your life, but probably have often succeeded in crushing her, for the male mosquito is a considerate gentleman. In lieu of the piercer of the female he is decorated with a beautiful plume, and has such a love of home that he seldom sallies forth from the swamp where he was born, but contents himself with vegetable rather than animal juices. (I do not wish to make any reflection, but in the insect world it is always the females which sing.)

But to its history. The mosquito was not born a winged fly, and if you will examine a tub of rain water that has stood uncovered and unmolested for a week or more during any of the summer months, you may see it in all its various forms. You may see the female supporting herself on the water with her four front legs and crossing the hinder part like the letter X. In this support made by the legs she is depositing her eggs, which are just perceptible to the naked eye. By the aid of a lens they are seen to be glued together so as to form a little boat, which knocks about on the water till the young hatch. And what hatches from them? Why those very wrigglers (Fig. 14, f.) which jerk away every time you touch the water. They are destined to live a certain period in this watery element and cannot take wing and join their parent in her war song and house invasions, till after throwing off the ski an

few times, they have become full grown, and then with another molt have changed to what are technically known as papoe (g.) In this state they are no longer able to do anything but patiently float with their humped backs at the surface of the water or to swim by jerks of the tail beneath, after the fashion of a shrimp or a lobster. At the end of three days they stretch out on the surface like a boat, the mosquito bursts the skin and gradually works out of the shell which supports her during the critical operation. She rests with her long legs on the surface for a few moments till the wings have expanded and become dry, and then flies away to fulfill her mission, a totally different animal to what she was a few hours before, and no more able to live in the water as she did then than are any of us. Is it not wonderful that such profound changes should take place in such a short time? Even the bird has to learn to use its wings by practice and slow degree, but the mosquito uses her newly acquired organs of flight to perfection from the start.

In this transformation from an aquatic to an aerial life the mosquito has first breathed from a long tube near the tail; next through two tubular horns near the head, and finally, through a series of spiracles along the whole body.

From a calculation made by Baron Latour, the mosquito in flight vibrates its wings 3,000 times a minute—a rapidity of motion hardly conceivable.

Those who have traveled in summer on the lower Mississippi or in the northwest have experienced the torment which these frail flies can inflict. At times they drive every one from the boat, and trains can sometimes only be run with comfort on the Northern Pacific by keeping a smudge in the baggage car and the doors of all the coaches open to the fumes.

The bravest man on the fleetest horse dares not cross some of the more rank and dark prairies of Northern Minnesota in June. It is well known that Father De Smit once nearly died from mosquito bites, his flesh being so swollen around the arms and legs that it literally burst.

Mosquitoes have caused the rout of armies and the desertion of cities, and I would counsel all who desire to learn how the hum of an insignificant gnat may inspire more terror than the roar of the lion, to consult Kirby and Spencer's history of the former.

There are many species of the mosquito, all differing somewhat in habit and season of appearance, and doubtless also in mode of development, which, in fact, has been studied in but few. They occur everywhere, whether in the torrid or the arctic zone, and are nowhere more numerous or tormenting than in Lapland.

Both the fly and the mosquito are great scavengers in infancy, the one purifying the air we breathe, the other the water we drink. They perform, in this way, an indirect service to man which few perhaps appreciate, and which somewhat atones for their bad habits in maturity.

BREEDING from immature animals is a great mistake. It is the foundation of degeneracy.

OUR LOCAL ORGANIZATIONS.

LANCASTER COUNTY AGRICULTURAL AND HORTICULTURAL SOCIETY.

The Lancaster County Agricultural Society met statedly in their rooms on Monday afternoon August 7, 1882.

The following members were present: H. M. Engle, Marietta; Calvin Cooper, Bird-in-Hand, Jos. F. Witmer, Paradise; J. C. Linville, Salisbury; Johnson Miller, Warwick; Simon P. Eby, city; M. D. Kendig, Creswell; F. R. Diffenderffer, city; Peter H. Hershey, city; Henry Shiffner, Bird-in-Hand; J. Frank Landis, East Lampeter; Jacob B. Garber, Columbia; W. B. Paxson, Colerain; I. C. Hunsecker, Manheim; Ephraim H. Hoover, Manheim; Ezra B Engle, Marietta; J. M. Johnston and W. W. Griest, city.

On motion the reading of the minutes of the last meeting was dispensed with.

Crop Reports.

Henry M. Engle reported corn as rather irregular; some is excellent, but some rather poor. It may come to an average crop with good weather. The early set will make a full crop. Potatoes are hardly a full crop. Pears will be short. Pasture is pretty plenty, but a little short. Wheat never was of better quality not the average higher, perhaps. The oats are good—not quite as good as last year.

J. C. Linville said the wheat was very good; the oats had smut and rust; potatoes are an average crop. The tobacco is the poorest ever he saw. The grass is well set.

M. D. Kendig, of Manor, reported a very full wheat and straw crop, corn looks well; oats was medium; fruits are falling fast. Tobacco is growing well since the last rains. The green worms are very numerous. The rainfall for July was 1 2-5 inches for June it was 1¼.

Johnson Miller said the wheat crop was a remarkable one in quality and quantity. Oats about half a crop. Hay better than expected. Tobacco is now growing well; the prospects are encouraging; peaches are inferior, apples are dropping fast and are imperfect.

J. Frank Landis said that wheat was never better. A professional thresher reports an average of 30 bushels to the acre. Corn is doing well. Peaches are ripening immaturely. Grapes are rotting a little.

Calvin Cooper never saw a more promising crop of grapes but some are rotting. He was unable to account for it. Telegraphs and Hartford Prolifics seem exempt. The vines have been well cultivated. Rogers Nos 9 and 53 are afflicted mostly by the rot.

H. M. Engle said some growers hold cultivation is no preventive of rot. He did not know what the cause was.

J. C. Linville said in Ohio and New York the disease is very common, and their cultivation has in some cases been given up. Rose bugs have done much harm to his grapes. He has tried to put the clusters into paper bags.

H. M. Engle said it was believed by some that bagging was a remedy for rot. He proposes to do this himself.

S. P. Eby reported an unusual crop of grapes; very few have so far been injured. Peaches are very abundant on his trees. He has been obliged to thin out largely. Tobacco is West Hempfield is uneven.

W. B. Paxson said he had a peculiar experience with a grape vine. It died down to the ground. He applied bone, which seemed to nurse it, and this year it is very full. There are no peaches in his neighborhood this year. What is are incomparable in quality. Corn looks very well. Tobacco looks well in early planted fields and bad in the late.

Calvin Cooper placed tobacco stems around the trunks of peach trees, and there is not the sign of a borer in any of the trees so treated. He set the stems around the butts of the trees and tied them at the top.

H. M. Engle said the borer and the yellows are the two enemies we have to contend with in peach culture. We can head off the borer, but know of no remedy for the yellows. Have your tree clear of borers when you plant it; then put shmething around the stalk to keep off the insect that lays the borer egg, and thus prevent the embryo egg from reaching the roots, and your trees are safe. The danger is not so great when the tree is some years old.

J. C. Linville said bagging grapes was not so much of a job as some thought. One man can bag 500 bunches in a day. A hole must be made at the bottom of the bag to let out any water that may chance to get in. Bees, birds and bugs are all headed off in this way. Grapes can be left on the vines much longer in the fall.

Joseph F. Witmer reported for Paradise that corn is doing well, but is very uneven. He has always advocated late planting. This year his farmer planted part of a field of corn on April 28 and another on May 3, and it is to-day the best field of corn he knows of. Tobacco has improved very much, and will be a fair crop, with favorable weather.

Wheat Crop of 1882.

S. P Eby, Esq., read the following:

A few years ago the question was frequently asked this society: What was the probable cause of our poor and imperfect wheat crops.

Numerous reasons were assigned and remedies suggested, such as, we needed new seed; that wheat should be cultivated in rows, like corn, to strengthen the straw and prevent lodging. The most generally received opinion, however, seemed to be that something was wanting in our soil; that wheat had been grown for so long a time that the essential element for its production were exhausted, and that a chemical analysis of the soil should be made, the want ascertained and the deficiencies supplied by proper fertilizers; or, that we must turn our attention to the raising of other products, and leave the sow lands of the West to supply our markets with wheat.

This year has shown that our soil is still capable of producing as fine and perfect a crop of wheat as ever was harvested in the county, and all previous doubts on that point ought to be forever set at rest, and henceforth we must look in another direction for a solution of the question of imperfect wheat crops in this county.

The present season we were favored with weather such as had not been given us during the several previous years when the wheat crops were partial failures; namely, a cool and moist spring, thus re affirming the correctness of the old saying:

"A wet April and cool May
Bring much grain and make much hay."

How, then, can we secure cool and moist Springs, such as we had during the present season? This question is easily enough answered, but difficult of realization. The difficulty lies in the fact that it will again require extensive co operation and an outlay for which there is no immediate return. The remedy is no other than the partial restoration of our original forests. We must plant trees to secure to the county more frequent rainfalls, better retention of moisture, and, as a consequence, a more even and a lower temperature during the spring months.

H. M. Engle thought the essay had several good qualities—it was short, to the point, and full of good sense. It shows that in good seasons we have all grown good crops. But the good farmer is shown in raising good crops when his neighbors have poor ones. How can this be done? There are some good crops every year. Why are not all so? This is the fact that stares us in the face, for fact it is. The subject is one of much importance and ought to be thoroughly discussed. He had some doubts if all the hopes of the forest culture advocates would be realized, if their plans were carried out. It is very sure we should all plant trees. If we do not need them coming generations will. We cannot always look to tobacco to make our money out of. Much money has been realized in the West from forest culture, and perhaps some could be made here in the same way.

How Should Manure be Applied.

M. D. Kendig said this question was hard to answer. Some crops do better when manure is applied in one, and some when applied in another way. It should be kept near the surface. It does not matwhat kind of manure it is, so that it don't get down too deep.

J. C. Linville also believed in applying manure to the surface; a smaller amount will go further. It acts both as a mulch and a manure. In a few cases perhaps, it is better to plow down long manure, but the substance of the manure should be near the roots of the crop where it is most needed. Grass always does better on ground where the manure is applied on the top. Fruits also do better when manured in this way.

Mr. Eby believed when manure was applied to the surface it ought to be worked in as quickly as possible. If left untouched the ammonia evaporates and there is a loss. If placed under trees, the rootlets seek it near the surface and damage may result.

H. M. Engle said general sentiment is in favor of surface manuring, but he believes in working the manure under. Unless this is done there is a loss. Soil is a good absorbent and will take up all the essentials in the manure. We ought to use the powl less and the cultivator more.

Calvin Cooper said a neighboring farmer applied the manure to the surface of his fields and has better crops than any man in the township. But he applies only well-rotted manure—never long straw. He has got, as it were, one year ahead with his manure pile, and therefore it is always rotted.

H. M. Engle said there was no necessity to keep manure over a year. If the manure pile is turned over two or three times in a season it will become thoroughly decomposed and as fine as need be.

J. Frank Landis gave his assent to this theory and related his experience, which confirmed its benefit.

New Business.

H. M. Engle alluded to the lack of interest in our meetings. He thought we ought to make an effort to overcome this. Lectures he believed would do it. We should have some well known man lecture at least quarterly. He made a motion that the secretary be instructed to procure some one.

Calvin Cooper thought we ought to procure a large room and advertise the lectures, so that a full attendance could be secured.

The motion was adopted and the name of Thomas J. Edge, Secretary of the State Board of Agriculture, was mentioned in connection therewith.

C. L. Hunsecker believed the meetings of the society were held in too obscure a place. So long as we continue to meet here our audience will be small. He gave three members an overhauling who came to town and neglected to attend the meetings. The keeping up of the society was left to half a dozen active members.

A good deal was said about the propriety of procuring a more accessible room.

Miscellaneous Business.

W. B. Paxson was named as the essayist for the next meeting.

The following subjects for discussion at the next meeting were named:

What is the cause of "streaks" in butter? Referred to J. Frank Landis.

Will it not pay the farmer to cut his corn fodder before feeding it to his cattle? Referred to Peter Hershey.

Is it best to sow timothy seed before or after the drill? Referred to Levi S. Reist.

What is the best method of preparing and seeding corn ground with wheat? Referred to Joseph F. Witmer.

Ought early potatoes, that are intended for winter use, be taken up when mature, or should they be left in the ground until cold weather? Referred to H. M. Engle.

The President appointed Messrs. Cooper and Kendig to report on some seedling apples, sent in by Mr. J. B. Lichty, of Lancaster. They reported the apple of good size and pleasant flavor, but recommended that on account of its toughness and the prevalence of better sorts, it be not recommended for general introduction.

There being no further business, on motion the society adjourned.

POULTRY ASSOCIATION.

The regular monthly meeting of the Poultry Association was held Monday morning, August 3, 1882. The following members were present: Isaac H. Brooks, Marticville, and J. B. Lichty, Charles Lippold, W. W. Greist, John A. Schom, Charles E. Long, F. R. Diffenderfer and W. A. Schoenierger, all from the city.

The minutes of last meeting were read and approved

On motion of Mr. Trissler, Mr. Milton Evans, of this city, was proposed and elected to membership.

F. R. Diffenderfer read the following letter, received from a correspondent:

Prevention of Gapes.

I perfectly agree with the assertion in the article Gapes in Chickens, that "It is not a remedy we want so much as a preventive," hence I give you my experience in the matter. Ever since I commenced raising poultry in 1873, I have been troubled with gapes. A few years ago Mrs. B. gave my a hint on the subject, but I never gave it a fair trial until the past spring. On all other occasions I lost so many young chickens that I concluded last spring to see how well I could succeed in preventing the gapes and how many of my young chicks I could raise.

Early in March I gave 39 eggs to three hens and from these eggs got 34 young chicks, all of them out a few days before the last of March. On the first of April one of them was overcome by cold and from the effects of this died a few days after. Another was hurt by one of the mother hens and also died about the same time. The balance—being 32 out of 34—are still living and are doing well. I have had no gapes and no sickness among them of any kind whatever. If I live I expect to try the same p an next year, and I have full confidence in the success of the experiment.

My recipe is "keep the young chicks off from the ground." I have the hens and young chicks in boxes facing the south, with all the openings for sun-light and air on that side possible, and then an outside pen for the young chicks, with board floor, and the sides and top of ordinary plastering laths, so as also to let in plenty of sunlight and air. I tried to be governed by three rules :

1. Give them plenty of sunlight and fresh air.

2. Good food and plenty of it—cracked corn, dry, and wheat screenings and grass.

3. Cleanliness. I clean the pens and boxes often and give them air slaked lime, coal ashes, etc.

Perhaps some of you can give us a better plan, but until then I expect to pursue the course so successful during the past spring. I give you this statement, because it may be of interest to you to know how others do, and with what success they meet, and because you may be able to use the facts some time.

I must add that I keep my chicks in their boxes and pens until they were two months old, and then in a small yard by themselves until to day (July 8), when, for the first time, I turned them out to run with the other chickens.

The Secretary read a number of letters from poultry fanciers in different parts of the country, offering special premiums on certain classes of birds entered at the next exhibition. Some of these are quite valuable and will no doubt attract exhibitors.

The secretary also stated that he had met with good success so far in procuring advertisements for the new catalogue.

Mr. Schum stated that he was more successful in raising pigeons than ever before. He sent at least 150 pairs to New York, Washington and other cities.

Charles Lippold reported good success with some varieties of pigeons, but poor with some of the rest. There being no further business, the society adjourned.

FULTON FARMERS' CLUB.

The Fulton Farmers' Club met at the residence of William King in Fulton township, August 5, and all the members and several visitors were present.

Exhibits.

J. R. Blackburn exhibited a sample of Deiges's prolific wheat, raised from seed received from the Patent Office. The quantity planted was too small to enable him to decide on the productiveness of the variety.

Wm. King exhibited samples of his wheat and oats. The wheat was thrashed from the rakings of his field, and was a rather inferior article. The oats weighed 29 pounds per bushel, and produced only at the rate of about 19 bushels per acre.

Sallie Hambleton exhibited some home-made hard soap made from Lewis's prepared lye. She recommends it for all kinds of washing and scrubbing, and it requires no boiling while being made.

John Coates exhibited some cheese from the Boyd creamery near Parkesburg. It was made on the 10th of June last from partly skimmed milk. Some considered it good while others thought it rather inferior to what their mothers used to make.

What is the Best Kind of Wheat?

Simpson Preston asked: "Do the members consider the Fultz wheat the best for us to sow?" The members replied as follows: J. R. Blackburn said he had tried the Fultz wheat two years and abandoned it. Josiah Brown likes the Fultz; he believes that on an average it yields the best for him. Day Wood has raised the Fultz for five years and likes it well enough to continue sowing it. S. L. Gregg has sowed the Fultz for several years, and last fall he sowed Fultz, Key's Prolific, Davis Brown wheat and Italian; he has not thrashed yet but is satisfied that the Fultz is the best and Italian second. Thomas P. King considers the Fultz the best under very favorable circumstances. Montillion Brown had found the Fultz to do the best for him, although it does not stand up long after it is ripe; but before that time it stands well. It seemed to be generally believed that on strong land the Fultz was pretty certain to do well, but where the land is thin some other varieties are likely to do better.

The Best Time to Sow.

Joseph Brown asked "What time will the members sow wheat this fall?" The answers to this showed that nearly all were in favor of sowing between the 10th and 20th of September. T. P. King said he was in favor of sowing either quite early or else not till late in September. Last year S. L. Gregg sowed an acre after tobacco on the 10th of October, and it was the best wheat he had.

A Question of Plows.

E. H. Haines said it is now several years since the chilled plows were introduced into this neighborhood and wished to know if the members now considered them better than the common Wiley plow. Joseph R. Blackburn said he bought a chilled plow last fall and after repeated trials and returning to the intervals to his old plows, he considers the new plow rather an improvement on the old. His new plow is the "Advance," and he uses the slip point.

Day Wood, S. L. Gregg, Montillion Brown, Simpson Preston and C. L. Gatchel, are each using some pattern of the new plows and all are pleased with their work. Some think they run hard and nearly all consider them more expensive than the Wiley plows for points. John Coates said he had a Syracuse chilled plow and found it so expensive for points that he put the jointer on the Wiley and it did just as good work.

Russian Oats.

Montillion Brown asked, "How many have tried Russian oats and how do they like it?" Several had sown small parcels of it and found it to ripen late, but the trials were on too small a scale to decide upon its merits.

Literary Exercises.

John Coates made some remarks on Creameries. He thinks that if people had facilities to attend to dairying it would pay, but if not, it was best to sell the milk at a creamery.

Essay on Noxious Weeds, by Wm. King.

Carrie Blackburn recited "A Doctor's Story."

Mary Hoopes read, "Sookey's Appeal."

A Farmer's Reunion.

A committee was appointed to make arrangements for a meeting to be held in the Hon. James Black's grove at Black Barren Springs, Sept. 9, 1882. It will be a farmers' reunion and all persons interested in agriculture are invited to be present and bring fruits, flowers and vegetables for exhibition. There will be tables arranged for that purpose. Any person engaged in the manufacture of machinery of any kind is also invited to come and bring machinery. Several public speakers are expected and the occasion no doubt will be one of great interest as well as benefit to our farming community. The following persons were appointed: J. R. Blackburn, Mary Blackburn, Montillion Brown, Wm. King, Rebecca King, George A. King and Martha Brown.

Noxious Weeds.

When quite a small boy we well remember seeing our father's hired man reach out both hands and with one grasp an ox eye daisy and with the other a wild carrot saying, "Pink and carrot, two of the worst things that can grow on a man's farm." This remark was made in accordance with the general sentiment of the time and to this day the two weeds are looked upon as the greatest pests that infest our farms. But while we would not willingly encourage the growth of pestiferous weeds we will say that in a life of more than half a century that has been wholly passed upon a farm, we have never been able to see why the two weeds above mentioned should have been singled out from the myriads of their companions as objects that were to be regarded as especially troublesome. With concerted effort on the part of farmers in any neighborhood the carrot could easily be exterminated as it is altogether propagated from seed and but for the fact that the adjoining farms are polluted with it any farmer could eradicate it from his farm in a few years if he should think it worth the effort. The daisy having a perennial root, and propagating itself from it, and also the seed, is a far more troublesome plant, but even it will give place to a rank growth of grass: but is almost sure to put in an appearance again if the grass dies off and becomes thin on the ground. The two plants take up room that might produce something better. This is the worst and about the only thing that can be said against them. We have several others that give us more trouble, but they are here and are likely to stay, so we will at present make no effort to point them out. It is the pests that are coming that we wish, at this particular time, to call attention to. While we live in a part of the country that is free from the scourge, we have the Canada thistle on every side of us, and only a few miles off. In any year we may find it growing in our fields, for the seed has been known to travel for miles on the crest of the snow. Fortunately it produces scarcely any seed and with a little attention can be kept from overrunning our fields. Not so with the horse nettle; its progress is ever onward. But little more than a decade ago it was extremely rare; at the present time it is quietly peeping up along our roadsides and spreading at a rate that but a few of the most observing have any idea of. It is brought here by Virginia cattle and can be found in almost any place where droves of them are kept or pass along.

Along the road from Rock Springs to Oak Hill there is scarcely a half mile of road where it has not taken hold in some place, while from the Baptist church to Conowingo it grows almost continuously. In any kind of soil or under any circumstances it is showing itself; and it is coming to stay. Once well rooted it is no boy's play to eradicate it. If allowed to spread to any extent in our fields it will seriously detract from the value. It is therefore the interest of every farmer to see that it is confined to its present limits, if it cannot be eradicated. How this is to be done we will not at present attempt to show. Our purpose has been simply to call attention to the fact that we have an enemy advancing, and that we cannot be too prompt or energetic in preparing to meet it.

LINNÆAN SOCIETY.

The society met on Saturday afternoon, July 29, 1883, in the ante-room of the museum; in the absence of the executive officers, S. M. Sener, Esq., was called to the chair, and Mrs. L. N. Zell was appointed secretary pro tem.

After the usual opening order the following donations and additions were made to the museum and library:

Museum.

Dr. M. L. Davis donated a large specimen of *Vanadium*, which he obtained from Mr. Hathaway, the owner and discoverer of the mine from which it was obtained, at Tioga, Pa. Mr. Hathaway is a blacksmith by occupation, and experimenting with the ore he found that when melted with iron and copper it rendered the former as hard as steel, and the latter a few degrees softer. He had a razor blade made from an old stove grate mixed with this ore, and on melting them together in a crucible, then beaten into shape on an anvil, the metal became firm in texture, and admitted of a very high polish. *Vanadium* was discovered in the year 1830. Some authors have attributed it to Del Rio in 1801, but the former, by whom it was named after Vanadis, a Scandinavian deity, was the original discoverer. The metal is found in nearly all clays in small quantities, but its most abundant source is the *Vanadiate of Lead*, which has been found in Scotland, Mexico, and some of the South American States. The metal may be chemically obtained by the reduction of Vanadic acid, in the form of a brilliant powder, having a silvery lustre.

It is not acted upon by sulphuric or nitric acids, but nitro-muriatic acid dissolves it, the solution resembling an aqueous solution of sulphate of copper.

Dana describes *Vanadinite*, or Vanadate of Lead (Vanadioblelera) as crystallizing hexagonally, but mostly occurring in implanted globules or incrustations; he is also one of the authorities who attributes its discovery at Zimpana, in Mexico, by Del Rio. This ore has a dark brown or brownish black color, and is generally observed only in an earthy state, much like a ferruginous clay. It is an interesting fact that it is now found in the State of Pennsylvania.

An interesting little fresh-water fish, donated by Dr. M. L. Davis. This is the *Boleosoma tessellatum*, locally called the "Sand-Perch," but it belongs to the family *Etheostomidæ*, which is only remotely related to the true *Perciæ*. This fish is remarkable for being destitute of an air-bladder, hence it is always observed lying upon the bottom of the pool in which it is found, and never swims with the graceful buoyant motion of other fishes, but changes its location by a sudden darting motion. The whole family to which it belongs are small fishes. The late Prof. S. S. Haldeman, described two new species from the Susquehanna belonging to an allied genus; and the late Jacob Stauffer discovered a third one from the Conestoga, which was described by Prof. Cope. Perhaps there were few boys who had access to a stream of water to whom this little fish was not familiar. It was quite abundant in my boyhood in the Susquehanna, and I have often succeeded in angling for it with a small hook, but it was more frequently taken with a "dip net," and used as a bait for larger fishes.

A bottle of insects taken at and in the vicinity of York Furnace spring, during the encampment of the Tucquan club, in the present month, consisting mainly of the general *Calasoma*, *Prionus*, *Orthosoma*, *Chrysoeus*, *Languria*, *Tereopes* and *Eriphus Suturalis*. The last named occurred in tolerable abundance on the bloom of a species of *Solidago*, along the river, from the York Furnace station to the mouth of the Tucquan. Four specimens of "shell rock," found on Bair's island by Mr. Wm. L. Gill, differing very much from each other. These were found in large water-worn pebbles, and must have been borne down the stream from remote localities above, as no locality of such a rock has yet been discovered in Lancaster county. For more than forty years, small boulders and various sized pebbles have been found on the beaches, bars and islands of the Susquehanna, but no locality has yet been found where this rock exists *in mass*. Some of these "fossil remains" appear to belong to the *Radiates*, some to the *Articulates*, and others to the *Mollusks*, and perhaps also to different geological periods.

Donations to the Library.

Proceedings of the *American Philosophical Society* from January to June 1882, 207 pp. octavo.

Geodetic and United States Coast Survey, for 1878, 404 pp., quarto, with 39 maps and illustrations from the Department of the Interior.

Parts 1, 2, 3 and 4 of the Official Patent Office Gazette, vol. 22, from the Department.

LANCASTER FARMER for July, 1882.

Ten Catalogues of Historical, Biographical and Scientific Books.

Six circulars of interesting publications.

Two envelopes containing eighteen historical and biographical selections.

No new business was brought before the Society, and the meeting was small.

After some deliberation it was voted to hold a recess for two months, after which the Society adjourned to meet in the ante room of the Museum, on the last Saturday in September next (30), with a hope that the members would not forget it.

AGRICULTURE.

Lying in Fallows.

That there is a wonderful progress in agriculture, a comparison of the practices of the present with the not very remote past abundantly shows. There is little doubt but that considerably more profit is derived from the same space of ground than even men not very old used to obtain. In these increased productions consist the most encouraging of progressive features. Not thirty years ago, a year of idleness was an essential feature in the regular rotation of an English farm. The summer "fallows" almost invariably preceded the wheat growing. But now the laying down of land to rest as a preliminary to the sowing of grain is rarely thought of. Still it continues in other countries, where the free communion of mind with mind, through the means of agricultural papers, has not been brought about. In France, especially, it seems that the practice of summer fallowing is as common as it ever was. It is quite likely that the change in the practice in England is due indirectly to the writings Liebig and others, who, about the time we refer to, created much thought by their writings. Although some of their views failed in time to secure the attention hoped for them, there is no doubt that we owe them much. It is, indeed, not always that as much direct good flows from the work of a great genius as is expected, but the indirect good, not so often recognized, is often much greater than all. In this case we have not derived as much benefit from soil analysis as the great agricultural chemist hoped for; and yet, what they told us about the elements of nutrition and the nature of plant food was no doubt the entering-wedge which ultimately broke up a very absurd and wasteful system.

That land will slightly improve by being "rested" there is, of course, no doubt. It was part of the old wasteful system, or no system, of Southern agriculture. A crop of cotton or of corn was taken successively from the same land, and then it was left to grow to weeds and briers, until after a few years it was taken in hand for the same crops again. But in this case it was as much the decaying matter formed by the weeds, if not wholly by them, as from any imaginary principle of rest. In this then our generation has gained one great advantage. We need not every fifth year or so give a fifth of all we possess as a sacrifice to the recuperative powers of nature. She need never rest, in the sense that human minds understand. Give fuel to the fire and it will burn forever; and, with the proper plant food, continually and intelligently applied, there is no reason why the same piece of land would not bear annual crops to the end of the world.

A Short-Sighted View.

Since the beginning of the Egyptian troubles many writers have been predicting that should the war prove a serious one, it would redound to the benefit of the United States by increasing the demand for our foodstuffs, and that the benefit would be still greater should the present trouble lead to a general European war. Ignoring the desire to build up our trade through the misfortunes of others, the predictions themselves are false. It is true that a foreign war might for the time being stimulate our commerce and increase the price of our foodstuffs, but there would surely come a reaction in the future. The various nations of the world are so closely interwoven in their trade relations that one cannot suffer without the others feeling it. Under the stimulus of a war, England might pay more for our grain than she otherwise would; but she would simply be impoverishing herself and thus be less able to purchase in the future. We, as the seller should desire from pecuniary motives, if from no other, to see our best customer, Great Britain, in the full enjoyment of prosperity, knowing that the greater prosperity of her people the greater will be the consumptive wants and their ability to supply them.

Select Your Own Seed Wheat

Our best varieties of wheat were produced by the careful selection of the best heads and plumpest grains to be found in the field. These being sowed by themselves the succeeding season, and the best again preserved, a great improvement has uniformly been the result. Any farmer can do this, and it is not necessary to depend on some scientist or particular seedsman for an explanation of the proper method. On this subject Prof. A. E. Blount, of the State Agricultural College, Fort Collins, Colorado, states that farmers generally permit their wheat to retrograde one eighth, whereas if proper care should be bestowed upon its improvement by *selection* alone, not one would ever find it necessary to procure better seed. It only takes two years to make wheat No. 1 and pure by selection, and from three to ten to make a successful hybrid. Wheats raised upon the soil of any locality are better than those from other points for seeding. To prove this fact Prof. Blount says:

"I have sown—and am nicely growing now—this spring 181 different varieties, the seed of which I obtained from every country in the world. Many I received are winter wheats, which I have converted into spring wheats. Of all the samples received not one was as good, or begin to be as good, *as the poorest I now have*." In other words, by careful selection the poorest wheat now is better than the best of the original lot, and this improvement has been made in one year. If it is desired to improve wheat try this method: Go over the field and select the largest heads for the best stalks. Spread the grains on a table and examine each one separately, discarding all but the best formed and fullest. Next season make a seed bed, putting one seed to a hill, one foot apart each way. Hoe well and keep clean. The result will be suprising, as the yield will be larger, the grains better and the seed clean. When, by doing this on a small piece of ground, the seed is perfected it can be sowed for a crop. Practice this annually, as there is no limit to improvement.—*Philadelphia Record.*

A Talk About Grasses.

The Deerfield Valley Agricultural Society had the following to say about grasses: James S. Grinnell, of Greenfield, says that grasses was the foundation of our success in farming, and it is of the greatest importance that we sow the best varieties of grass and cut it at the right time. Mr. Johnson, of Greenfield, said that although raising grass is the foundation of farming, it is astonishing that farmers take so little pains with it, in fact do not know even the names of the grasses they cut. The principal grasses he would recommend are blue grass, red top bent-grass and red clover. There are other varieties that may be cultivated to some extent, as the soil and circumstances allow, such as the sweet scented vernal, white clover and orchard grass. Farmers should study their nature, and be sure to sow varieties that would ripen at the same time. J. N. Abbott, of Buckland, exhibited ten different kinds of grass. His favorite is orchard grass, which he would sow with red clover, but it requires strong, moist land. It ripens about the same time as red clover, and he considered these two varieties mixed together the best hay he cuts. E. C. Harris found his hay more increased when he sowed a variety of grasses. Mr. Grinnell said we must wage war on the weeds and subdue them. Part of his meadows had gone to sorrel and it grieved him very much.

HORTICULTURE.

The Peach Crop.

Superintendent Mills, of the Delaware Railroad estimates the coming peach crop in the districts traversed by railway at 4,000,000 baskets. The heaviest yield will be in the district between Middletown and Clayton, and, with continued favorable weather, the crop may reach 5,000,000 baskets. This is exclusive of the sections above, which are dependent upon water transportation. The peach growers of the peninsula will meet in convention at Dover to-morrow to discuss transportation rates and facilities and other matters of interest.

Value of Fruit.

It is a fact that fruit is a great regulator of the human system. It will keep the blood in order, the bowels regular, tone up the stomach, and is positively a specific in many diseases. It is said of a doctor who became largely interested in peach growing, that he recommended peaches to his patients on all occasions. The story was told to illustrate the man's meanness, but if he was mean it was a meanness that benefited his patients. If men were wise they would spend two days in a vineyard or orchard to every five minutes in a drug store when anything is the matter with them. If you have dyspepsia eat fruit. Did you ever think what a doctor gives for dyspepsia? He gives an acid. Fruit will furnish better acid than the drug store will. Do you know what the doctors dose you with when your liver is out of order? With acids. Then why not supply the remedy yourself from your own garden? Why continue to have your medicine done up in such a repulsive mixture when nature furnishes it in so palatable a shape. Every home should have at least one grapevine. Once in possession it would be almost above price.

Shallow Cultivation for Fruits

Fruit growers must be reminded that their hoes, cultivators and ploughs may do more damage to plants than good if not used with discretion. The small fruits—berries, currants, grapes, also dwarf pears, quinces, etc., root near the surface. Here are found the best roots, those that provide the most nourishment. Nature designed these to be mulched by the dead leaves, and in our fields mulching would be the best treatment if it were possible. As it is, the best we can do is to give frequent but shallow cultivation. I have seen intelligent men ploughing deep furrows alongside of their raspberries, currants and grapes, well satisfied that they were doing thorough work that would secure an abundant harvest. Let such men dig up one plant before thus ploughing and one after and see what butchery they have committed. There are no top-roots stretching far down into the subsoil, but simply a few laterals branching out, say from two to four inches below the surface, and more than half of these have been sacrificed by the ploughshare. When we set green hands hoeing strawberries and newly set raspberries we know what they will do if not watched—they will destroy half their roots and loosen the hold for life that the struggling pets have secured by chopping close about them.

The Vegetable Garden.

In these days of a scant supply of labor and high prices for it, economy which does not approach meanness, is one of the fine arts. For instance, it is economy to hire only good, well-trained farm hands, but it is not economy to set them down to a meanly furnished table. It is economy to feed your people well, but it is far from economical to have to purchase all they eat. A farmer is expected to have something besides salt pork and potatoes on his table, and a variety is more economical than meat, wheat flour, canned goods and store truck generally. Vegetables are so healthful, so economical and so indispensable that it is a sign of a poor farmer if his table lacks them at any season of the year. There is no land on the most productive farm in the country which pays one-third as well as a well laid out vegetable garden, properly planted and cultivated. As hand labor is costly, it is both wise and economical to dispense with it whenever possible. My experience in market gardening has proved that all kinds of garden truck can be grown and cultivated by a man or boy, with a one-horse single shovel cultivator far more successfully and at far less expense of time and wages than by the employment of any three good men with hoes. A garden for vegetables should have no bushes, trees or perennial plants in it, but should be convenient for the plow and teams at all times. It should be long, to save time in turning, and wide enough to supply the family demand. It should be covered with good, old, well-rotted manure six inches deep, plowed under twelve inches.

Fig Culture.

The subject of our heading is attracting much attention, both South and North. The people are waking up to the idea that the fig is of great commercial value, and the Press are stimulating the people to go ahead in the introduction and preparation for home use and the market.

In former years Virginians cultivated this delicious fruit quite extensively, but as it would not grow without some little care and attention, its old time, general cultivation has been (unwisely) discontinued.

"Eternal vigilance is the price" of fruit of all kinds. The grub destroys the peach tree, the curculio robs us of our plums, the blight kills our pear and quince trees, and yet we replant and fight the foe, and enjoy as our laurels the wholesome fruits of summer and autumn.

There are very many reasons why fig culture should become a fixed fact, with every one having a few rods of land.

1. The trees give two crops annually.
2. They commence bearing early—say at two years from the cutting.
3. The fruit is very delicious and healthful and better than medicine in malarial and febrile diseases.
4. The cultivation is at once simple, easy and profitable.

Now for the other side.

For the Middle States and, say, north of 36°, winter protection is necessary. And as a matter of fact, the same is true of all the Southern States.

In the autumn of 1879, fig trees were cut down to the ground by the frost in November; and in the winter of 1880–81 the cold made a clean sweep all over the South.

I have said that "a little care" is necessary, but the fig must have *that*; and that care is *protection* from the cold of winter.

Well, really, there is no other trouble; there is neither blight nor insect trouble.

I could wish that our American women would find "a field of labor" in this pure fruit, for to them it would be a "joy forever!"

HOUSEHOLD RECIPES.

CUCUMBER MANGOES.—Select the largest sized pickling cucumbers and put them into a strong brine for two weeks; then take them out, drain well, and heat the brine to a boiling point. Pour it over the pickles immediately, and let them stand until the next day. Repeat this process nine times and after they are cold the last time throw them into pump water for eight hours. Drain them well, and dry each one upon a coarse towl, then with a sharp knife make a slit in the side and remove the seeds. Make each one perfectly dry, and fill with the following mixture: To three dozen cucumbers take six large onions, chop them very fine, and add half an ounce celery seed, one ounce turmeric, one pound white mustard seed, one ounce chopped mace, half an ounce powdered nutmeg, half a pound grated horse radish and a quarter of a pound of ground mustard. Mix all into a paste with a cup of salad oil. Tie a string around each pickle in order to keep the filling in, and pack them down into a stone jar. Take as much vinegar as will be necessary to cover them, and let it boil up once with a handful of cloves, a head of garlic, and a tablespoonful of sugar. Pour this over the pickles while it is hot, and tie the jar closely. Do not open for five or six weeks and they will be ready for use.

PEACH MANGOES.—Take one peck of large Morris white peaches, or large firm yellow freestone peaches and cover them with brine for twenty-four hours. Take them out and remove the seed by making a slit in the side and partially opening the peach, then throw them back into the brine and let them remain another 24 hours. Drain them, wipe dry inside and outside, and fill them with the following mixture: One pint of chopped onions, a teaspoonful of chopped green ginger, half an ounce of celery seed, one ounce of white mustard seed, half an ounce of tumeric, and one ounce of black mustard seed. Tie a string around each peach, pack them in a strong jar, and cover them with cold vinegar, adding a tablespoonful of oil, head of garlic, and a tablespoonful of sugar. Ready for use in three weeks.

VEAL A LA MODE.—Wipe with a wet cloth a solid piece of lean veal weighing five or six pounds; make half a dozen holes in it by running the knife steel through it, parallel with the fibre of meat, and working it about to make holes large enough to admit the forefinger, fill the holes with a forcemeat made as directed in the recipe for roast chicken, omitting the chicken liver, lay the veal in a saucepan just large enough to hold it, pour over it boiling water, sprinkle in a teaspoonful of salt, and half a saltspoonful of pepper, cover it steam tight, and stew it gently for two hours; if the gravy is not thick enough stir in a little flour mixed with cold water, and boil it for two minutes; remove the meat before adding the flour.

BREAST OF VEAL BAKED WITH TOMATOES.—After wiping a breast of veal with a wet cloth lay it in a small dripping-pan, and brown it quickly in a hot oven. Meantime peel and slice a pint of tomatoes; or use those which have been canned; when the veal is brown season it highly with pepper and salt, pour the tomatoes over it, and bake it until the meat is well done. Serve it with the tomatoes on the same dish.

BREAST OF VEAL BRAISED.—After wiping a breast of veal with a wet towel remove the bones with a sharp knife, season it with salt and pepper, roll it and tie it compactly; put it over the fire in boiling water enough to cover it, with a small onion and turnip peeled, a small carrot scraped, a dozen whole cloves, half a teaspoonful pepper-corns, and a teaspoonful of salt; fasten the cover of the sauce-pan with a thick paste of flour and water, and gently cook the veal for two hours. Then take it up, remove the string, and keep it hot; strain the broth, and use it to make a white sauce as follows:

WHITE SAUCE.—For each pint desired mix together over the fire a tablespoonful each of butter and flour until they bubble; then gradually stir in a pint of boiling broth or water, stirring constantly until all lumps are removed, season palatably with pepper and salt, and let the sauce boil before using it.

Before taking up the meat peel a pint of potatoes, cut them in half inch dice, throw them into salted boiling water and boil them until tender, then mix them with the white sauce, put them on a platter, and serve the veal on them.

VEAL WITH BROWN SAUCE.—Cut cold veal in two inch pieces, brown them over the fire in sufficient hot butter to prevent burning, dust flour over them, about a tablespoonful for two pounds of meat; when the flour is brown cover the meat with boiling water, season the stew highly with salt and pepper, add to it sufficient nice table sauce to flavor it, and when it has boiled five minutes serve it on toast.

BOILED TONGUE.—Proceed according to the directions given in the recipe for boiling salt meats. When the tongue is done the skin can easily be stripped off and the rough parts about the root trimmed away; these parts, freed from bone and gristle, make excellent hash.

BOILED CORNED BEEF.—Follow the directions for boiling salt meats. When vegetables are to be served with corned beef they may be boiled with it until tender, allowing them to cook only long enough to make them tender. Cabbage is usually boiled several hours, when it will sometimes boil tender in a quarter or half an hour; by unnecessary boiling it becomes watery, and emits an unpleasant and penetrating vapor.

BOILED HAM.—Follow the directions for *Boiling Salt Meats*. When the ham is done, if it is to be served hot, take it up, strip off the skin, dust it with fine bread-crumbs or cracker dust, and brown it in a quick oven.

PORK CHOPS, SPANISH STYLE.—Trim off nearly all the fat, chop it and put it into a hot frying pan over the fire until it is brown; then fry the chops brown in the same fat, season them with salt and pepper, squeeze over them the juice of a sour orange, and keep them hot while some eggs, one for each chop, are being fried in the same fat; when the eggs are cooked to the desired degree lay them on the chops, pour the fat over them, and serve at once.

ROAST PORK.—Use the chine or loin of fresh pork; cut out the bone, replace it with a stuffing of stale bread soaked soft in cold water and seasoned highly with salt, pepper, powdered sage, and a little chopped onion; sew up the cut to keep in the stuffing, and bake the pork in a moderate oven half an hour to each pound; season it when brown with salt, pepper, and powdered sage. More than all other meats pork requires thorough cooking.

PORK TENDERLOINS.—Stuff and roast them according to the preceding recipe; or split them open, and fry, or broil them very brown; season them with salt, pepper, and powdered sage; or in place of the sage, when they are fried, mix some chopped pickle with gravy, and pour it over them. Cook them thoroughly.

IRISH STEW.—Cut three pounds of breast of mutton in two-inch pieces, put it into a saucepan with a quart of boiling water, two teaspoonfuls of salt, and a saltspoonful of pepper, and stew all together gently for an hour; then add a pint of onions peeled and sliced, and a quart of potatoes, peeled and cut in inch pieces, and again stew gently for an hour; the stew should be kept closely covered while cooking.

PERSILLADE OF MUTTON.—Slice cold mutton, lay it on a dish which can be sent to the table, sprinkle the surface thinly with salt and pepper, cover it with cold gravy, dust the surface with bread or cracker crumbs, and brown it in the oven; serve it at once.

FRIED BREAST OF MUTTON.—Boil a breast of mutton, according to the directions for boiling meat, until it is tender enough to permit the bones to be pulled out; lay it between two platters, under a weight, until it is quite cold; then roll it in bread-crumbs and fry it whole in a dripping-pan large enough to hold it flat, in sufficient smoking-hot fat to cover it; or cut it in small pieces, before breading it, and fry it in an ordinary frying-kettle. Fried onions or tomatoes may be used to garnish this dish.

BREADING.—The "breading" of any article consists of simply rolling its moist surface in sifted bread crumbs or cracker dust; if the crumbs are unlikely to stick the article is next dipped in beaten egg, and then again rolled in crumbs. The crumbs should always be sifted so that they may be of one size, and as fine as possible, or they will be apt to fall off during frying. They are made by drying stale bread, rolling it fine, and sifting it through a fine sieve; the coarser crumbs may again be rolled and sifted, or kept for stuffing or puddings. Crackers can be rolled in the same way; cracker dust is sold ready for use.

RAGOUT OF COLD BEEF AND VEGETABLES.—Cut cold beef in inch squares, brown it in hot drippings, sprinkle it with flour and let it brown, cover it with boiling water, and season it with salt and pepper; add to it any cold vegetables cut in similar pieces, heat them, and serve the stew.

ROAST LEG OF LAMB OR MUTTON.—Wipe a leg of lamb or mutton with a wet cloth; run a sharp thin-bladed knife between the skin and flesh where the leg is thickest, in such a manner as to form a pouch for the stuffing; into this pouch put the flesh of a red herring, highly seasoned with pepper, and pounded to a paste, forcing it as far as possible under the skin; roast the leg according to the directions given for roasting all kinds of meat. French cooks put a clove of garlic into the flesh close to the end of the shank bone of a leg of mutton before roasting it.

GARLIC CLOVES.—Garlic, when marketed, looks somewhat like a dried tuberose root; it divides when broken into many small lobes called "cloves"; each clove is covered with an inner skin which must be removed before the clove is used for flavoring.

LIVE STOCK.

Advice of a Lancaster County Blacksmith on How to Shoe Horses.

A Lancaster county subscriber sends to the Germantown *Telegraph* the following statement from a noted horseshoer in his vicinity, as to his mode of shoeing. It strikes the *Telegraph* as being about as nearly perfection as it could be, and it therefore commends it to the attention of every shoer. The shoer some years ago was asked to write down his mode, and did so, and thinks that it might have been published in a Western paper where he formerly resided:

"The way I shoe a horse is this: First, see that the animal stands in a natural position, so that it is perfectly upright. If so, level the wall of the foot to receive the shoe, and nothing more. Never cut the frog, braces or sole, nor the heels; let nature do its own work. If let alone, once in six weeks or two months, the frog-braces and sole will shed. Make the shoe light at the toe, heavy at the heels, (for the heels are the tenderest part of the foot). Put the nails well forward of the quarters. Use light nails. Concave the shoe until you get to the quarters to protect the soles. Then convex the heels a little, and you cannot make narrow heels, corns, quarter cracks or contracted feet. Fit the shoe cold and fit it to the foot and not the foot to the shoe. Follow the wall carefully; fit the shoe as broad as you can; bring the heels around to the frog, but do not touch it. Never make the shoe longer than the foot; never use bar-shoes, as in every case they are wrong and hurtful. Never rasp the outside wall, for you destroy the glass or enamel that protects it against decay, and prevents it from growing rough and ridged. This improper method of shoeing is the cause of more lameness than any one thing. I causes corns, quater-cracks and contracted feet. Whoever pares ol allows to be pared, a horse's sole, brace or frog, and burnes the foot with a shoe, or puts clips on the shoes in front or sides, is by such useless, harmful and outrageous mutilation guilty of cruelty to the noble beast. *No frog, no foot; no foot, no horse.*

Training Horns.

If it is desirable to straighten a horn you may frequently scrape with a piece of glass, or a knife, the hollow side, which will cause it to grow faster on that side; but in that case it must not be scraped deeply, for then it becomes weaker on that side, and will be turned toward the weaker side. Some scrape the side toward which they wish to turn the horn quite thin, and then scrape the opposite side just enough to make it grow faster, and that will turn it toward the thinly scraped side. If you wish to turn a horn up, scrape on the under side just enough to make it grow faster on that side. A very barbarous way to turn a horn is sometimes practiced, by searing with a hot iron on the side toward which the horn is to be turned This prevents the growth of the horn on that side and the growth upon the other side turns the horn. The horns may be polished by rubbing them with fine sandpaper, and then with pumice-stone and oiling them. But this artificial manipulation of horns is seldom necessary. The horns of well-fed cattle will generally grow in comely shape if let alone. The hair is sometimes oiled to give it a glossy appearance, but the best gloss is put upon the hair by rich and appropriate feeding. Nature, under proper conditions, does this work best.

The Best Farm Horses.

What class will be the most profitable to raise is a question of interest to nearly every farmer. Trotters may be set aside. The care and skill required in training, even when the colt has all the advantage of pedigree is such as would make serious inroads upon the time and patience of all but a fortunate few. So the trotter may be set down as not a profitable horse for the farmer to breed; but carriage and heavy draft horses are. Both of these kind are scarce in all our large cities, and the demand for them greater than the supply; hence prices are always renumerative. For some years to come no sort of farm stock will be more profitable than these two classes of horses. The carriage horse requires a good share of thoroughbred blood in him, else he will show a deficiency in style, spirit, action and endurance, qualities that constitute the chief value of that class. For draught horses the native breed is entirely too small. To remedy this defect we must employ the best types of imported stallions. The writer has watched with interest the importations of foreign stock, as telegraphed over the country for the last few years, and gives it as his judgment that the Clydesdale has been the favorite, and represents the best type of imported stallions. What seems singular the heaviest shipment of these horses have almost invariably been for the Eastern or Western States, where they seem to be in high favor. Rarely has a shipment for Ohio been recorded. Skipping the details of their anatomy, appearance and peculiarities, we can say that no breed of heavy draught horses is more valuable on the farm, either as pure breeds or to improve our native horses, and this has been a rendered verdict in both the East and the West.

Draught Horses.

There has been such a demand made upon Western Pennsylvania for draught horses the past few years that farmers who have been fortunate enough to breed heavy horses have found it decidedly to their advantage when their stock was brought to market. The supply is not yet up to the demand for heavy draught horses, and we see as a result that there is a tendency on the part of breeders to meet this demand. The question with the average farmer and breeder is not so much as to what he may prefer but what is the most advantageous, the most profitable horse to breed. For ordinary farming purposes in a comparatively level country a horse weighing 1,200 pounds is perhaps in most respects the most desirable animal. But as farmers raise four or five times as many horses as they themselves use, the question of breeding simply is, what is the most marketable horse? In selling cattle to the butcher, he pays according to the weight—the heavier the steer the higher the price. Two pounds of beef are worth twice as much as one pound. In draught horses this rule does not hold good, the advantage being on the side of the heavier horse. When a 1,200 pound horse sells at $150 or 12½ cents a pound, a 1,700 pound horse sells at $300, or nearly eighteen cents a pound, making a market of forty per cent. premium over the lighter in favor of the heavier horse. It is not difficult to see, therefore, which is the most advantageous horse to breed for market. It is simply a question of dollars and cents, and is readily seen by any one.—*Pennsylvania Farmer.*

Is Horseshoeing Useless.

A recent issue of *Frazer's Magazine* contains an article by Sir George W. Cox, in which he estimates that the English custom of horseshoeing costs the nation as much as $44,000,000, which might be saved if the horses were allowed to go unshod. He quotes the authorities from Xenophon, who marched his horses from Cunaxa over the Armenian highlands to the walls of Trebizond, down to the "free lancers" of the present day, and contends that it is safer, cheaper and better to let horses go unshod over the hardest roads, and especially in the slippery streets of London. He estimates that over twelve million dollars would be saved in farriers' bills alone. And he calculates further that the working life of a horse would be trebled by the change so that a horse which is now worn out at twelve would live to twenty-six. The figures seem somewhat startling, and have hardly been sufficiently proved to be trustworthy. Meanwhile it is said that a medical man in Waterbury, Conn., has not put shoes on his horses for two years, driving them them winter, summer, spring and autumn with bare feet without any trouble. The doctor's theory is that nature has provided for the horse; that a horse can travel over all kinds of roads; that the hoof will be moist, and that the frog coming to the ground keeps the hoof properly spread, and free from founder and other diseases.

Keep the Stable Clear of Flies.

One of the greatest hindrances to thrift during hot weather is the annoyance caused by flies. This is true both in the field and stable. In the former we cannot, in any considerable degree, control them, but in the latter we can. The better class of stables should be provided with screens. By this mode, fumigation being practised to drive the flies out, the stock may be quite well protected. The placing of small vessels of chloride of lime about the ceiling will sometimes answer the purpose of keeping them out of the building. If a decoction of *lycopodium* (sometimes called wolf's claw), which is the largest of the European mosses, be placed in a bladder, the neck being supplied with a quill nozzle, by means of which the liquid can be sprinkled where the flies accumulate; early in the morning, the effect upon the flies will soon be seen; as it quite promptly destroys them. This article is also used to destroy vermin.

Remedy for Side Hole in Cow's Teat.

Make the edges of the opening raw with a sharp knife, or cauterize with a pointed stick or nitrate of silver. The hole may then be closed with strips of adhesive plaster, or better yet by a coating of "collodion," which can be obtained of any photographer, if the nearest druggist does not keep it. In milking be careful not to displace the dressing—and it will perhaps be better to draw the milk with a tube for several days. If the opening in the teat is not quite small, it may be necessary to close it by a stitch just through the skin with a fine thread. In most cases the scratching of the edges of the opening with a knife and the application of collodion will however be sufficient.

Care of Horses.

The following abridged observations of a French writer are deserving the attention of all who have horses under their care: The same quality of oats given to a horse produces different effects, according to the time they are administered. I have made experiments on my own horses, and always observed matter not digested, when I purposely gave them water immediately after a feed of oats. There is decidedly, then, a great advantage in giving horses water before grain is fed to them. There is another bad practice I observe, that of giving grain and hay on their return to the stable, immediately after hard work.

The Stock.

Give all the stock a bedding, and especially the working oxen and horses. The cows will prove the better for it, so will the yearlings; to the swine a warm place and dry bed are indispensable to profit. Remember the zero weather we have in winter, and how much comfort we can bestow upon the animals dependent upon us, by a little timely care.

POULTRY.

Poultry Gossip

An Indiana man has a bronze turkey cock, nine months old, which weighs 31½ pounds.

An occasional, or rather, even a frequent whitewashing of the hen house will make the air there sweeter and purer than would otherwise be the case.

The Board of Health of New York city will probably take a hand in the question of undrawn poultry. It is to be hoped they will show more sense than the aldermen.

The art of caponizing roosters does away with the worry over the sex of eggs. Experts can do this work for ten cents a piece, and capons are of more value than hens.

Four pairs of prize ducks at a recent English fair weighed as follows: 1st, 19 pounds 5 ounces; 2d, 19 pounds 1 ounce; 3d, 18 pounds 15 ounces; 4th, 18 pounds 10 ounces.

Hens that lay few eggs, or eggs that will not hatch, are sometimes very earnest and persistent sitters. Perhaps they comprehend their weakness and desire to make amends in a useful way.

A healthy hen, sitting early in the season, can be made to keep right on and incubate a second batch of eggs if her first hatching is removed in time and given to another hen. Sometimes this is very desirable when sitters are scarce.

A New Jersey man recommends keeping eggs in whitewash. We should think that this would be hard on the shell. If they are to be kept away from the air and cool, why not keep in water without lime or anything else? Has any one tried that?

Some deodorizer under hen roosts is a very important matter, both for health and economy. Cover the droppings every morning with sawdust, road dust, dry muck, plaster, dried clay, or anything that will act as an absorbent, and the dangers of cholera will be slight indeed.

A defender of undrawn poultry claims that it is a full crop which spoils dressed poultry, and not the intestines. Chickens should be made to fast at least twelve hours in advance of butchering. No doubt that is good advice, but to clean out all the uneatable "innards" is better.

Poultry is not safe in a house infested with rats. These creatures when hungry—and they seem to be hungry most of the time—will even pull chickens and small fowls from the roosts at night and kill them. They will also rob hatching hens of their young Judicious care will make a hen house rat proof.

Poultry does not sell according to its low price in market, but, like good butter and fresh eggs, according to its quality. Still so many people in the great cities are so accustomed to poor, flavorless, insipid poultry, that the first-class article, freshly killed and drawn, would not be recognized at first. There is education in such matters as well as in the fine arts.

Somebody in the interest of incubator manufacturers avers that city people can raise chickens (in the back parlor probably) by means of incubators by getting eggs "from the grocers and farmers near by," and that even two hundred chickens can be raised "by special care in a room fifteen feet square." All right; let them try it. The parties sending out this advice must have had experience as lightning-rod agents.

Ducks can be raised with more certainty than chickens and turkeys. Sink a tub for them in some out of the way place and keep water in it, and that is enough as regards water. Or a sort of basin may be excavated in the ground, and this can be cemented easily so that it will hold water until the next winter's frost cracks it. But ducks are great gormandizers and will destroy more flowers and flower beds and grass about the lawn than chickens, so this trouble must be guarded against.

Feather and Egg Eating.

Fowls in confinement are apt to contract vicious habits, chief among which is feather eating. It is often necessary to confine fowls in certain portions of the season, if not the whole year, and during this period of inactivity they learn this bad trick, which they seldom give up. One teaches another, and they soon denude the bodies of the cocks, and then begin to pluck one another. Feather-eating begins, in the first place, from a lack of something better to do, and at length an appetite is acquired. Cocks are rarely or never guilty of it. Fowls that are confined should be well supplied with vegetable and animal food. This prevents much mischief. When milk may be had, a basin given daily is of great benefit in supplying the lack of vegetable and animal food, and at the same time giving occupation. When fowls are at large they gather innumerable insects and other wild food. When confined they are shut off from this, and feather-eating is learned.

Another habit equally bad, if not worse, is that of eating eggs. The fowls learn this in confinement by scratching in the nests, from a lack of something better to do. After the eggs are once broken they, of course, eat them, and thus the taste is formed. Any nest material like hay or straw, invites the hens to scratch, which is second nature to the fowl. To avoid this, give plenty of occupation outside of the building where they roost and lay. The fowls should have yards, and be allowed to run there. Let them labor a little for food. Give them fresh food each day, and allow them to pick their corn from the cob. This they will do if the grain is dry and shells easily, thus giving employment. The eggs, however, should be brought in two or three times a day, so that the breakage may be avoided. By 2 o'clock in the afternoon the most of the hens have deposited their eggs, and since they have not the privilege of going abroad, they look for something else to do. Almost anything that offers they are willing and ready to do.

Habits thus formed are seldom forgotten. The better way is to prevent the formation. I have seen fowls so given to feather-eating that nothing short of death would cure. Any preparations applied to the feathers has no effect whatever. I think sometimes it may be brought on through the means of vermin on the body, the presence of which is discovered by the hens when at rest. I do not think that it is always the case, but cocks, as a general thing, are more infested with vermin than the hens. They do not wallow in dust like the hens. A pretty sure way to teach the fowls how to eat feathers is to leave the plucked ones from the dead birds about where the living ones have access to them, more especially if they are pen feathers. This habit is acquired by all breeds, but the rapid layers and non-sitters are the worst. They posses an almost irresistible appetite for animal food, and it is this appetite and the gratifying of it that gives us so many eggs.

That fowls require a great deal of care when made thourougly profitable is readily admitted by all that have once had the experience. The practice of feeding green vegetables is a good one if persevered in. Of this they are fond. Above all, do not neglect fowls that are in confinement; give them their rations at regular intervals, and all the occupation that may be afforded. They will pick bones and pluck the greenness from a fresh sod in a short time, and afterwards scratch among the fresh earth for a length of time. If the yard is ample a portion of it may be dug over, and the fowls find some insects in the turned up earth, and will hunt for more, which gives natural employment and exercise. It is useless to plant any seeds for green food in their yards, as nothing will grow beneath their constant tread.—*Country Gentleman.*

Geese.

We think more attention should be paid to the rearing of geese than is usual. One may often travel half a day's journey in the country without seeing a flock of geese. Their flesh is by no means to be despised when the birds are young, and their feathers always command ready sale.

It is an erroneous opinion, and one without doubt generally prevalent, that geese cannot be successfully raised away from ponds and streams of water. Persons may soon disabuse themselves of this idea by visiting the suburbs proper of the city of Chicago, when the green patches about the tenement houses and shanties are converted into "geese farms," flocks averaging from the progeny of a single goose to that of half a dozen. One of the best flocks we ever knew, and which, for the years we knew it, ran from forty to fifty goslings each year, had for its nearest water a brook three quarters of a mile away, and which the geese never saw. Our own flock, when we kept geese, had ample water facilities in a river close at hand. Our friend used to beat us and laugh at us, when visiting each other, at my losses. His standing joke was: "I only have weasels and skunks to look after, and you, in addition, have catfish and snappers (turtles)." We think he was right.

If a pool of water is near, it is desirable and an advantage. If not, a shallow tub in which they can plunge, dabble, and drink, will really fill all the absolute necessities of the case.

As to varieties, we think it lies between the Embden and the Toulouse geese—both of them descended from the gray-legged goose (*Anser ferus*) of the north of Europe. Either of the varieties are of the largest size, growing to the extreme weight of twenty two to twenty six pounds.

The true Embden, called also the Bremen goose, should be pure white with brick-red legs, and heavily feathered. The Toulouse goose is gray, but darker and more uniform in color. Both are round-bodied, compact, short-legged, with large abdominal development, are quiet, lay plenty of eggs, fatten readily, and have excellent flesh. A cross of the Embden and Toulouse is said to make better birds than either of the pure breeds. This we cannot vouch for, but we have found the Toulouse rather better able to take care of themselves than the Embden. On the other hand, the excellent white feathers of the Embdens are more valuable than those of their relatives.

Geese are not difficult to manage. They want a dry, warm place to huddle under in winter, and which, in summer, may be given plenty of air. This must be cleaned regularly, and often enough to be sweet and wholesome. In the summer they will pretty much supply themselves with food, grass, worms, and various insects—but what grain they will eat should be also supplied every night, and it is always better that this be supplied to them at the bottom of a vessel filled with water. In winter this food may be corn and the screenings of small grain, in connection with cabbage leaves or other greens, or else chopped root, daily.

The Wonders of Incubation.

It is wonderful to trace the development of the chicken during the process of incubation, from the day in which the mother hen begins her tedious term of setting to the moment when the downy biped bursts the shell and enters on life as an animate and independent existence. In the pursuit of science and the interest of learning no seeming destruction of material is of any moment, and we trust no economical poultry raiser will accuse us of extravagance if we remove each day or oftener of the twenty one days required for the perfection of the chicken, a single egg, and show you (as far as we can understand the principles of creation) how the feathered tribes of our barnyards are made.

Of course the germ of life is in the egg from the beginning, as no amount of warmth and quiet will produce a bird from a sterile egg, but with this fact assured, the hen has sat on her eggs hardly twelve hours before we find some lineament of the head and body of the chicken. The heart may be seen to beat at the second day and the aspect or shape is that of a a tiny horseshoe. Blood vessels appear at the end of the second day and their faint pulsation is distinguishable, one being the left ventricle and the other the rudiment of the great artery. About the fifteenth hour one auricle of the heart appears, resembling a loop folded down upon itself. At the end of seventy hours symptoms of the wings are apparent and on the head five bubbles are seen, two of the incipient brain, one for the bill and the other two for the front and back of the head. At the end of the fourth day the auricles, already visible, approach nearer to the heart, and the liver appears towards the fifth day.

At the end of seven hours more we see the lungs and stomach, and, with wonderful rapidity, are developed; four hours afterwards the intestines, the loins and the upper jaw. At the 144th hour two ventricles are visible, and two drops of blood instead of a single drop which we had seen previously.

The seventh day the brain begins to have some consistency; and at the 119th hour of incubation the bill opens and flesh appears on the breast. Four hours after the breast bone is seen, and in six hours after this the ribs appear, forming the back of the chicken; and the bill is distinctly visible, as well as the gall bladder. The bill becomes green at the end of 236 hours, and if remove the chick from the shell it evidently moves itself. At the 200th hour the eyes appear, and 38 hours after the ribs are perfect. At the 331st the spleen draws near the stomach and the lungs to the chest. About the fifteenth day the bill frequently opens and shuts; and a careful listener can catch the smothered cry of the imprisoned chick at the end of the eighteenth day.

For the remaining three days it grows continually, developing the finishing touches to its various organs and to the silken color of down which envelopes the tiny creature from glossy beak to tender drumstick. Strength comes with all the accelerated forces of quickening life, and a few strokes of the powerful bill sets the pretty prisoner free, and his after life and prosperity is something with which we as its owners have more or less connection.

A Meat Diet.

It is generally conceded by the majority of poultry breeders that a meat diet is essential during cold weather, when worms, bugs and insects are not to be found by the birds. But though considered necessary to atone for the lost insect food it should be used sparingly and not fed too often to young fowls.

In winter and early spring to keep up egg production, the fowls must have something to work on. The best way to supply them if there is not enough of waste meat scraps from the breeder's table to meet the required demand, is to get scraps from the butcher or slaughter house. The waste meat, offal and the bloody pieces which are unsalable can be bought for a cent or two a pound.

The best way to utilize these scraps and to render them more digestible and nutritious is to cut them into fine pieces, put them into a boiler with plenty of water and boil them until the bones separate from the flesh. Then stir cornmeal into it until it makes a thick mush, season with salt and pepper, and cook till done. Feed this when cold to the poultry and they will eat it with evident relish, and you have a most excellent food which will keep during cold weather.

Our experience is in favor of cooking the meat. It goes further, is more nourishing and less injurious if over fed than in a raw state. Sheep's heads, shanks, livers and bone pieces can be utilized in this way and the soup mixed in with meal or scalded wheat and seasoned to suit. Young fowls should be fed sparingly with flesh; meat, grain and cooked vegetables is the best staple food when properly varied.—*Poultry Monthly.*

Feed for Laying Hens.

Fat hens rarely lay. If hens are fed so much or so often that they begin to fatten rapidly, they will soon stop laying. No food is better than Indian corn or ground corn (Indian meal), to fatten hens, and of course it should be fed sparingly to laying hens. If hens do not lay and are fat, feed them but once a day—at evening, just before they go to roost—giving wheat screenings, buckwheat and oats, in such proportions as you judge best. Throw the feed upon clean ground only so fast as they pick it up. Stop just as soon as you see any of the flock begin to wander away. Let them forage all day for weed seeds, grass, insects, etc. They must have warm quarters, well ventilated at night, and a sunny run by day in winter. After a while begin to feed them sparingly a little meat scrap chopped fine, broken bones, oyster shells, etc., and they will probably soon begin to lay.

LITERARY AND PERSONAL.

THE SHAKER MANIFESTO, an octavo of thirty pages, published by the United Societies, Shaker-village, New Hampshire, monthly, at sixty cents a year, devoted to moral and miscellaneous literature, poetry, domestic economy, farm and garden, household affairs, etc. Its general *utilitarian* spirit may be illustrated in the following description of "A Minister of the Olden Times":

"There was once a minister of the gospel
Who never built a church;
Who never preached in one;
Who never proposed a church fair to buy the church a new carpet;
Who never founded a new sect;
Who never belonged to any sect;
Who frequented public houses and drank wine with sinners;
Who never received a salary;
Who never asked for one;
Who never wore a black suit, nor a white necktie;
Who never used a prayer-book;
Or a hymn-book;
Or wrote a sermon;
Who never hired a cornet soloist to draw souls to hear the 'word;'
Who never advertised his sermons;
Who never even took a text for his sermons;
Who never went through a course of theological study;
Who was never ordained;
Who never was 'converted';
Who never went to conference.
Who was he?
Christ." —*N. Y. Graphic.*

If the foregoing should not be deemed sufficiently radical and conclusive, it might be added that
He never wore a hat or cap;
Never wore boots or shoes;
Who never was married;
Never wore breeches;
Never used a fork;
Who ate his meals "lounging";
Who never used coercion, except to drive people *out* of the Temple, instead of driving them *in*;
Who never traveled by railway or canal.

But, it must be remembered, *He* lived nearly nineteen hundred years ago, and said with emphasis,

"BEHOLD, I MAKE ALL THINGS NEW."

REGULATIONS and premiums list of the "Frederick County Agricultural Society." Twenty second Exhibition, 1882. $8,000 are offered in premiums. Lists of all the officers are carefully given, and the regulations and general arrangements elaborately set forth, and the whole exhibition divided up into thirty-five classes, embracing all the objects, products, implements, machinery and industries usually included in such fairs; but, except that the books of entrance will be closed at 10 o'clock on Tuesday, October 10th, it is no where stated explicitly on what days of the month the exhibition will be held.

DEPARTMENT OF AGRICULTURE Special Report No. 45, upon the area and condition of corn, the condition of cotton, and small grains, sorghum, tobacco, etc., July 1882, 33 pp. octavo, Washington, D. C.

The returns for July indicate an increased area planted in corn exceeding 4 per cent, or fully 2,500,000 acres. The general condition on July 1 gives an average of 85 against 90 a year ago; although in eleven States it was over 100, notably in Georgia, which is registered 108. Late planting, could and wet weather, and planting after floods, is the principal cause. Winter wheat averaged 104, and spring wheat 96—14 per cent above 1881, indicating an aggregate crop of 500,000,000. Rye, similar to that of wheat, 100 and upwards. Oats in a high condition, with a persentage of 103. Barley averages 100. The general average of cotton is 92. Seven per cent. increase in the area of potatoes, averaging 102. Acreage of tobacco same as 1881—condition high southward, but low north. Southern increase, and northern decrease in Sorghum. Apples and peaches fairly abundant. Delaware and Maryland will exceed 4,000,000 baskets. The report contains many valuable tabulated statistics.

THE SIDEREAL MESSENGER, conducted by Wm. W. Payne, Director of Carleton College Observatory. No. 5, vol. 1, of this interesting astronomical journal has reached our table, and we are pleased to see that it not only maintains the excellence with which it started out a few months ago, but that it very perceptibly improves. It is an octavo of 32 pages, exclusive of the tinted covers, and is published at $2.00 for ten numbers, Northfield, Minnesota. The material, typographical execution and the literary contents are of a high order of excellence, the contributions being able, scientific and practical. In a personal remark, the editor says: "C. Piazzi, Astronomer Royal of Scotland, was the first foreign subscriber to the *Sidereal Messenger*, which at least indicates that it is appreciated by learned astronomers abroad, if it should find no recognition at home. It is fortunate in having an able corps of contributors, and is well posted in the current astronomical literature and discoveries. Surely our "Star Club" would become vitalized, under the most discouraging circumstances, by the perusal of such an able publication.

THE SUGAR BEET. Third year, number 3, has been received. This handsomely illustrated quarto abates not in the least in its advancement of the utilization of the sugar beet. The production of sugar and of silk in this country, sufficient for the needs of the country alone, involves industries that must ultimately redound to its more perfect independence, and the wonder is that their progress has been so slow.

The cultivation and utilization of root crops in general have an immense bearing upon the quantity and quality of other productions than sugar, that seem to be but faintly apprehended by agriculturists.

THE WORLD OF NATURE

The world of animated nature is more splendidly represented under the canvas of Forepaugh's Great Show than in any zoological collection existent. Not since the day Noah lifted his hawser off the snubbing post have so many distinct varieties of rare animals been collected under one charge. This important fact should not be lost sight of by schools and parents. Boys and girls can learn more in an afternoon of natural history, in the great Menagerie of Forepaugh's Show, than by months of book study. Recognizing this, Mr. Forepaugh makes reduced rates to schools, and admits all children to orphan asylums free of charge. This Great Show will exhibit in Lancaster, Monday, April 24.

THE
LANCASTER EXAMINER

OFFICE

No. 9 North Queen Street,

LANCASTER, PA.

THE OLDEST AND BEST.

THE WEEKLY
LANCASTER EXAMINER

One of the largest Weekly Papers in the State.

Published Every Wednesday Morning,

Is an old, well-established newspaper, and contains just the news desirable to make it an interesting and valuable Family Newspaper. The postage to subscribers residing outside of Lancaster county is paid by the publisher. Send for a specimen copy.

SUBSCRIPTION:

Two Dollars per Annum.

THE DAILY
LANCASTER EXAMINER

The Largest Daily Paper in the county.

Published Daily Except S nday.

The Daily is published every evening during the week. It is delivered in the City and to surrounding Towns accessible by railroad and daily stage lines, for 10 cents a week. Mail Subscription, free of postage—One month, 50 cents; one year, $5.00.

JOHN A. HIESTAND, Proprietor,

No. 9 North Queen St.,

LANCASTER, PA.

Important to Grocers, Packers, Hucksters, and the General Public.

THE KING FORTUNE-MAKER.

OZONE
A New Process for Preserving all Perishable Articles, Animal and Vegetable from Fermentation and Putrefaction, Retaining their Odor and Flavor.

"OZONE—Purified air, active state of Oxygen."—*Webster.*

This preservative is not a liquid pickle, or any of the old and exploded processes, but is simply and purely OZONE, as produced and applied by an entirely new process. Ozone is the antiseptic principle of every substance, and possesses the power to preserve animal and vegetable structures from decay.

There is nothing on the face of the earth liable to decay or spoil which Ozone, the new Preservative, will not preserve for all time in a perfectly fresh and palatable condition.

The value of Ozone as a natural preserver has been known to our abler chemists for years, but, until now, no means of producing it in a practical, inexpensive, and simple manner have been discovered.

Microscopic observations prove that decay is due to septic matter or minute germs, that develop and feed upon animal and vegetable structures. Ozone, applied by the Prentiss method, seizes and destroys these germs at once, and thus preserves. At our office in Cincinnati can be seen almost every article that can be thought of, preserved by this process, and every visitor is welcomed to come in, taste, smell, take away with him, and test in every way the merits of Ozone as a preservative. We will also preserve, free of charge, any article that is brought or sent prepaid to us, and return it to the sender, for him to keep and test.

FRESH MEATS, such as beef, mutton, veal, pork, poultry, game, fish, &c., preserved by this method, can be shipped to Europe, subjected to atmospheric changes and return to this country in a state of perfect preservation.

EGGS can be treated at a cost of less than one dollar a thousand dozen, and be kept in an ordinary room six months or more, thoroughly preserved; the yolk held in its normal condition, and the egg as fresh and perfect as on the day they were treated, and will sell as strictly "choice." The advantage in preserving eggs is readily seen; there are seasons when they can be bought for 8 or 10 cents a dozen, and by holding them, can be sold for an advance of from one hundred to three hundred per ct. One man, with this method, can preserve 5,000 dozen a day.

FRUITS may be permitted to ripen in their native climate, and can be transported to any part of the world.

The juice expressed from fruits can be held for an indefinite period without fermentation—hence the great value of this process for producing a temperance beverage. Cider can be held perfectly sweet for any length of time.

VEGETABLES can be kept for an indefinite period in their natural condition, retaining their odor and flavor, treated in their original packages at a small expense. All grains, flour, meal, etc., are held in their normal condition.

BUTTER, after being treated by this process, will not become rancid.

Dead animal bodies, treated before decomposition sets in, can be held in a natural condition for months, without puncturing the skin or mutilating the body in any way. Hence the great value of Ozone to undertakers.

There is no change in the slightest particular in the appearance of any article thus preserved, and no trace of any foreign or unnatural odor or taste.

The process is so simple that a child can operate as well and as successfully as a man. There is no expensive apparatus or machinery required.

A room filled with different articles, such as eggs, meat, fish, etc., can be treated at one time, without additional cable or expense.

☞ In fact, there is nothing that Ozone will not preserve. Think of everything you can that is liable to sour, decay, or spoil, and then remember that we guarantee that Ozone will preserve it in exactly the condition you want it for any length of time. If you will remember this it will save you asking questions as to whether ozone will preserve this or that article—it will preserve anything and every thing you can think of.

There is not a township in the United States in which a live man can not make any amount of money, from $1,000 to $10,000 a year, that he pleases. We desire to get a live man interested in each county in the United States, in whose hands we can place this Preservative, and through him secure the business which every county ought to produce.

A FORTUNE
Awaits any Man who Secures Control of OZONE in any Township or County.

A. C. Bowen, Marion, Ohio, has cleared $2,000 in two months. $2 for a test package was his first investment.

Woods Brothers, Lebanon, Warren County, Ohio, have made $6,000 on eggs purchased in August and sold November 1st. $2 for a test package was their first investment.

P. K. Raymond, Morristown, Belmont Co., Ohio, is clearing $2,000 a month in handling and selling Ozone. $2 for a test package was his first investment.

D. F. Webber, Charlotte, Eaton Co., Mich., has cleared $1,000 a month since August. $2 for a test package was his first investment.

J. H. Gaylord, 80 La Salle St., Chicago, is preserving eggs, fruit, etc., for the commission men of Chicago, charging 1½c. per dozen for eggs, and other articles in proportion. He is preserving 5,000 dozen eggs per day, and on his business is making $3,000 a month clear. $2 for a test package was his first investment.

The Cincinnati Pearl Co., West 6th Seventh street, is making $3.50 a month in handling brewers' malt, preserving and shipping it as feed to all parts of the country. Malt impreserved soars in 24 hours. Preserved by Ozone it keeps perfectly sweet for months.

There are instances where we have asked in the privilege of publishing. There are scores of others. Write to any of the above parties and get the evidence direct.

Now, to prove the absolute truth of every thing we have said in this paper, we propose to place in your hands the means of proving for yourself that it we have not claimed half enough. To any person who doubts any of these statements, and who is interested sufficiently to make the trip, we will pay all traveling and hotel expenses for a visit to this city, if we fail to prove any statement that we have made.

How to Secure a Fortune with Ozone.

A test package of Ozone, containing a sufficient quantity to preserve one thousand dozen eggs, or other articles in proportion, will be sent to any applicant on receipt of $2. This package will enable the applicant to pursue any line of tests and experiments he desires, and thus satisfy himself as to the extraordinary merits of Ozone as a Preservative. After having thus satisfied himself, and had time to look the field over to determine what he wishes to do in the future—whether to sell the article to others or to continue it to his own use, or any other line of policy which is best suited to him and to his township or county—we will give exclusive territorial privileges to the first responsible applicant who orders a test package and desires to control the business in his locality. The man who secures control of Ozone for any special territory, will enjoy a monopoly which will surely enrich him.

Don't let a day pass until you have ordered a Test Package, and if you desire to secure an exclusive privilege we assure you that delay may deprive you of it, for the applications come in to us by scores every mail—many by telegraph. "First come first served" is our rule.

If you do not care to send money in advance for the test package we will send it C. O. D., but this will put you to the expense of charges for the return money. Our correspondence is very large; we have all we can do to attend to the shipping of orders and giving attention to our working agents. Therefore we can not give any attention to letters which do not order Ozone. If you think of any article that you are doubtful about Ozone preserving remember we guarantee that it will preserve it, no matter what it is.

REFERENCES.

We desire to call your attention to a class of references which no enterprise or firm based on any thing but the soundest business success and highest commercial merit could secure.

We refer, by permission, as to our integrity and to the value of the Prentiss Preservative, to the following gentlemen: Edward C. Hoyer, Member Board of Public Works; F. O. Eshelby, City Comptroller; Amor Smith, Jr., Collector Internal Revenue; Malcim & Worthington, Attorneys; Martin H. Harrell and B. F. Hopkins, County Commissioners; W. S. Cappeller, County Auditor; all of Cincinnati, Hamilton County, Ohio. These gentlemen are each familiar with the merits of our Preservative, and know from actual observation that we have without question

The Most Valuable Article in the World.

The $2 you invest in a test package will surely lead you to secure a township or county, and then your way is absolutely clear to make from $2,000 to $10,000 a year.

Give your full address in every letter, and send your letter to

PRENTISS PRESERVING COMPANY. (Limited,)

S. E. Cor. Ninth & Race Sts., Cincinnati, O.

Nov-8m

WHERE TO BUY GOODS IN LANCASTER.

BOOTS AND SHOES.

MARSHALL & SON, No. 12 Centre Square, Lancaster, Dealers in Boots, Shoes and Rubbers. Repairing promptly attended to.

M. LEVY, No. 3 East King street. For the best Dollar Shoes in Lancaster go to M. Levy, No. 3 East King street.

BOOKS AND STATIONERY.

JOHN BAER'S SONS, Nos. 15 and 17 North Queen Street, have the largest and best assorted Book and Paper Store in the City.

FURNITURE.

HEINITSH'S, No. 15½ East King st., (over China Hall) is the cheapest place in Lancaster to buy Furniture. Picture Frames a specialty.

CHINA AND GLASSWARE.

HIGH & MARTIN, No. 15 East King st., dealers in China, Glass and Queensware, Fancy Goods, Lamps, Burners, Chimneys, etc.

CLOTHING.

MYERS & RATHFON, Centre Hall, No. 12 East King St. Largest Clothing House in Pennsylvania outside of Philadelphia.

DRUGS AND MEDICINES.

G. W. HULL, Dealer in Pure Drugs and Medicines, Chemicals, Patent Medicines, Trusses, Shoulder Braces, Supporters, &c., 15 West King St., Lancaster, Pa.

JOHN F. LONG & SON, Druggists, No. 12 North Queen St. Drugs, Medicines, Perfumery, Spices, Dye Stuffs, Etc. Prescriptions carefully compounded.

DRY GOODS.

GIVLER, BOWERS & HURST, No. 25 E. King st., Lancaster, Pa., Dealers in Dry Goods, Carpets and Merchant Tailoring. Prices as low as the lowest.

HATS AND CAPS.

C. H. AMER, No. 39 West King Street. Dealer in Hats, Caps, Furs, Robes, etc. Assortment Large. Prices Low.

JEWELRY AND WATCHES.

H. L. RHOADS & BRO, No. 4 West King St. Watches, Clock and Musical Boxes. Watches and Jewelry Manufactured to order.

PRINTING.

JOHN A. HIESTAND, 9 North Queen st., Sale Bills, Circulars, Posters, Cards, Invitations, Letter and Bill Heads and Envelopes neatly printed. Prices low.

Thirty-six Varieties of Cabbage; 26 of Corn; 28 of Cucumber; 6 of Melon; 33 of Peas; 2 of Beans; 17 of Squash; 25 of Beet and 20 of Tomato, with other varieties in proportion, a large portion of which were grown on my five seed farms, will be found in my **Vegetable and Flower Seed Catalogue for 1882**. Sent FREE to all who apply. Customers of last season need but write for it. All seed sold from my establishment warranted to be fresh and true to name, so far, that should it prove otherwise, I will refill the order gratis. The original introducer of Early Ohio and Burbank Potatoes, Marblehead, Early Corn, the Hubbard Squash, Marblehead Cabbage, Phinney's Melon, and a score of other New Vegetables, I invite the patronage of the public. New Vegetables a specialty.

JAMES J. H. GREGORY,
Marblehead, Mass.
Nov-5mo]

EVAPORATE YOUR FRUIT.
ILLUSTRATED CATALOGUE
FREE TO ALL.
AMERICAN DRIER COMPANY,
Chambersburg, Pa.
ap1t-tf

FARMING FOR PROFIT.

It is conceded that this large and comprehensive book, (advertised in another column by J. C. McCurdy & Co., of Philadelphia, the well-known publishers of Standard works,) is not only the newest and handsomest, but altogether the BEST work of the kind which has ever been published. Thoroughly treating the great subjects of General Agriculture, Live-Stock, Fruit-Growing, Business Principles, and Home Life; telling just what the farmer and the farmer's boys want to know, combining Science and Practice, stimulating thought, awakening inquiry, and interesting every member of the family, this book must exert a mighty influence for good. It is highly recommended by the best agricultural writers and the leading papers, and is destined to have an extensive sale. Agents are wanted everywhere. jan-tf

LANDRETH'S BLOOMSDALE SWEDE, OR RUTA BAGA,

Is the result of critical selection, and has proved to be unquestionably the most desirable of all known strains of

PURPLE TOP YELLOW RUTA BAGA.

The foliage is not superabundant, the shape is nearly globular, the crown deep purple, and the flesh a deep yellow. The illustration conveys a good idea of the shape assumed by this strain.

Also, strap-leaved Garden Ruta Baga Turnip, white leafed, Purple top Ruta Baga Turnip, Hanover Long French or Sweet German Turnip, Yellow Aberdeen, or Scotch Yellow Turnip, Pomeranean White Globe (strap leaved) Turnip, Amber Globe (strap leaved) Turnip, Yellow Stone Turnip, Early Flat Dutch (strap leaved) Turnip, the Flat Red, or Purple Top (strap leaved) Turnip, Cow Horn Turnip, Early White Egg Turnip, Large Early Red Top Globe Turnip, White Norfolk Globe Turnip, Seven Top Turnip.

Bloomsdale Swede or Ruta Baga.

Every farmer should sow Turnip Seeds. A good stock of turnips is the best and most economical food for cattle during the winter and early spring months. Also, turnips grown on the ground, and plowed in, make very valuable manure.

Descriptive and Illustrated Catalogue free on application.

D. LANDRETH & SONS,
AGRICULTURAL AND HORTICULTURAL IMPLEMENT
AND SEED WAREHOUSE.

Nos. 21 and 23 South Sixth Street,
BETWEEN MARKET AND CHESTNUT STS.,
AND S. W. CORNER DELAWARE AVENUE, AND ARCH ST.,
apr-6m   PHILADELPHIA.

MERCHANT TAILORING.

1848 (The Oldest of All.) 1881

S. S. RATHVON,

MERCHANT TAILOR AND DRAPER,

respectfully inform the public that having disposed of their entire stock of Ready-Made Clothing, they now do, and for the future shall devote their whole attention to the CUSTOM TRADE.

All the desirable styles of CLOTHS, CASSIMERES, WORSTEDS, COATINGS, SUITINGS and VESTINGS constantly on hand, and made to order in plain or fashionable style promptly, and warranted satisfactory.

All-Wool Suits from $10.00 to $38.00.
All-Wool Pants from 3.00 to 10.00.
All-Wool Vests from 2.00 to 6.00.

Union and Cotton Goods proportionately less.

Cutting, Repairing, Trimming and Making, at reasonable prices.

Goods retailed by the yard to those who desire to have them made elsewhere.

A full supply of Spring and Summer Goods just opened and on hand.

Thankful to a generous public for past patronage they hope to merit its continued free gratias in their "new departure."

S. S. RATHVON,
PRACTICAL TAILOR,
No. 101 North Queen Street,
LANCASTER, PA.
1848   1881

GLOVES, SHIRTS, UNDERWEAR.
SHIRTS MADE TO ORDER,
AND WARRANTED TO FIT.
E. J. ERISMAN,
56 North Queen St., Lancaster, Pa.
79-1-12]

A HOME ORGAN FOR FARMERS.

THE LANCASTER FARMER,

A MONTHLY JOURNAL,

Devoted to Agriculture, Horticulture, Domestic Economy and Miscellany.

Founded Under the Auspices of the Lancaster County Agricultural and Horticultural Society.

EDITED BY DR. S. S. RATHVON.

TERMS OF SUBSCRIPTION:

ONE DOLLAR PER ANNUM,

POSTAGE PREPAID BY THE PROPRIETOR.

All subscriptions will commence with the January number, unless otherwise ordered.

Dr. S. S. Rathvon, who has so ably managed the editorial department in the past, will continue in the position of editor. His contributions on subjects connected with the science of farming, and particularly that specialty of which he is so thoroughly a master—entomological science—some knowledge of which has become a necessity to the successful farmer, are alone worth much more than the price of this publication. He is determined to make "The Farmer" a necessity to all households.

A county that has so wide a reputation as Lancaster county for its agricultural products should certainly be able to support an agricultural paper of its own, for the exchange of the opinions of farmers interested in this matter. We ask the co-operation of all farmers interested in this matter. Work among your friends. The "Farmer" only one dollar per year. Show them your copy. Try and induce them to subscribe. It is not much for each subscriber to do but it will greatly assist us.

All communications in regard to the editorial management should be addressed to Dr. S. S. Rathvon, Lancaster, Pa., and all business letters in regard to subscriptions and advertising should be addressed to the publisher. Rates of advertising can be had on application at the office.

JOHN A. HIESTAND,
No. 9 North Queen St., Lancaster, Pa.

$72 A WEEK. $12 a day at home easily made. Costly Outfit free. Address TRUE & Co., Augusta, Maine

ONE DOLLAR PER ANNUM.—SINGLE COPIES 10 CENTS.

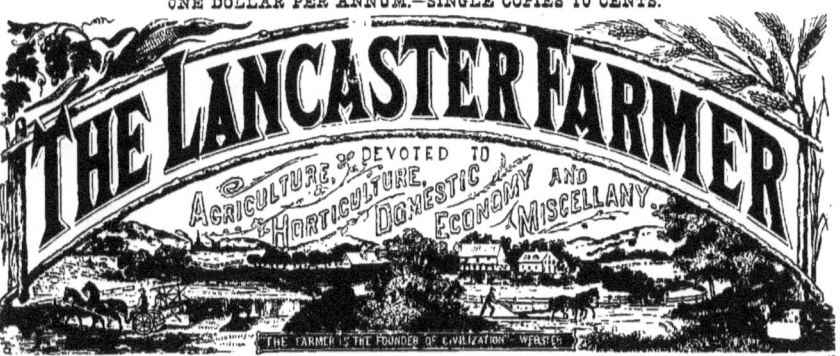

Dr. S. S. RATHVON, Editor.   LANCASTER, PA. SEPTEMBER, 1882.   JOHN A. HIESTAND, Publisher

Entered at the Post Office at Lancaster as Second Class Matter.

CONTENTS OF THIS NUMBER.

EDITORIAL.
State and County Fairs of 1882 129
Kitchen Garden for September 129
  Seed Purchasing a Matter of Confidence.
Insect Migrations 129
The Wheat Crop of 1882 130
  Luck Good Management Manure.
Tobacco Worms—Curious Facts Concerning them 130
Excerpts 130

QUERIES AND ANSWERS.
The Royal Horned Caterpillar 133

CONTRIBUTIONS.
Gapes in Poultry 133
Shallow Cultivation 133
Not the Tariff Question 133
The Eel—Its Habit and Growth 133

SELECTIONS.
A Grand Harvest 134
Occupation and Longevity 134
The War in Egypt 135
The Climate in different parts of the Union 135
Pure and Wholesome 135
Temperature and Rainfall 136
Barn Yard Manures 136
Preserving Fence Posts 136
Some Wheat Statistics 136
Importance of Having a Good Queen 137
Draining of Land 137
The Practical and the Scientific in Agriculture..138

OUR LOCAL ORGANIZATIONS.
Lancaster County Agricultural and Horticultural Society 138
  Crop Reports—Selecting and Breeding Dairy Stock
  —The Question Discussed—How to Secure Better Meetings—Cutting Corn Fodder for Feed
  —Streaks in Butter—How to Sow Timothy Seed
  —The Best Variety of Wheat—Following Corn with Wheat—Forestry—Fruits on Exhibition—Herbs County Fair.
The Poultry Society 140
State Board of Agriculture 140
Fulton Farmers' Club 140
  Articles Exhibited—Address of Welcome—Seth L. Kinsey—"Manual Labor and How it May be Encouraged."

AGRICULTURE.
Pasture Grasses 141
Experiments with Green Manuring 141
Wheat Raising 142
What of the Future as Regards Grain 142
What Manure Loses by Heating 142
Good Crops in Alabama 142
Magnesia for Wheat 142

HORTICULTURE.
Keeping Grapes Fresh 142
Beneficial Effect of Mulching on Berries 142
Taking in Fall Flowers 142

Save the Peach-tone 143
A Hint for Window Gardening 143

HOUSEHOLD RECIPES.
Fig Pudding 143
To Whiten Scorched Linen 143
To Cook Turnips 143
Almond Cake 143
Pan Dowdy 143
Smothered Chicken 143
Pumpkin Pie 143
Sheep's-Head Soup 143
Pickled Onions 143
Lemon Pudding 143
Ready-Made Glue 143
Apple Jelly 143
A Remedy for Diptheria 143
Household Hints 143
Health Hints 143
Dry Curing Pork and Beef 143
Stewed Corn 144
Brown Sauce 144
Boiled Sweet Corn 144
Stewed Corn and Tomatoes 144

LIVE STOCK.
Improved Sheep 144
Management of Pigs 144
A New Cattle Disease 144
Literary and Personal 144

SUBSCRIBE FOR THE

LANCASTER WEEKLY EXAMINER

$2.00 PER YEAR.

The Largest Weekly Paper in the State.

☞ GIVE IT A TRIAL.

JOHN A. HIESTAND, Proprietor,

No. 9 North Queen Street.

$72 A WEEK. $12 a day at home easily made. Costly Outfit free. Address TRUE & CO., Augusta, Maine.
jun-1yr*

LANDRETH'S
BLOOMSDALE SWEDE, OR RUTA BAGA,

Is the result of critical selection, and has proved to be unquestionably the most desirable of all known strains of

PURPLE TOP YELLOW RUTA BAGA.

Bloomsdale Swede or Ruta Baga.

The foliage is not superabundant, the shape is nearly globular, the crown deep purple, and the flesh a deep yellow. The illustration conveys a good idea of the shape assumed by this strain.

Also, nine-leaved Garden Ruta Baga Turnip, white fleshed, Purple top Ruta Baga Turnip, Hanover Long French or Sweet German Turnip, Yellow Aberdeen, or Scotch Yellow Turnip, Pomeranean White Globe (strap leaved) Turnip, Amber Globe (strap leaved) Turnip, Yellow Stone Turnip, Early Flat Dutch (strap leaved) Turnip, the Flat Red, or Purple Top (strap leaved) Turnip, Cow Horn Turnip, Early White Egg Turnip, Large Early Red Top Globe Turnip, White Norfolk Globe Turnip, Seven Top Turnip.

Every farmer should sow Turnip Seeds. A good stock of turnips is the best and most economical food for cattle during the winter and early spring months. Also, turnips grown on the ground, and plowed in, make very valuable manure.

Descriptive and Illustrated Catalogue free on application.

D. LANDRETH & SONS,
AGRICULTURAL AND HORTICULTURAL IMPLEMENT AND SEED WAREHOUSE,

Nos. 21 and 23 South Sixth Street,

BETWEEN MARKET AND CHESTNUT STS.,
AND S. W. CORNER DELAWARE AVENUE AND ARCH ST.,
apr-6m   PHILADELPHIA.

We mail seeds to all applicants, and to customers without ordering it. It contains five colored plates, 600 engravings, about 200 pages, and full descriptions, prices and directions for planting 1500 varieties of Flower and Vegetable Seeds, Plants, Fruit Trees, etc. Invaluable to all. Send for it. Address,
D. M. FERRY & CO., Detroit, Mich.
jan-6m

$66 a week in your own town. Terms and $5 outfit free Address H. HALLETT & Co., Portland, Maine.
jun-1yr*

WE WANT OLD BOOKS.
We Want German Books.
WE WANT BOOKS PRINTED IN LANCASTER CO.
We Want All Kinds of Old Books.
LIBRARIES, ENGLISH OR GERMAN BOUGHT.
Cash paid for Books in any quantity. Send your address and we will call.

REES WELSH & CO.,
23 South Ninth Street, Philadelphia.

THE LANCASTER FARMER.

PENNSYLVANIA RAILROAD SCHEDULE.
Trains LEAVE the Depot in this city, as follows:

WESTWARD.	Leave Lancaster.	Arrive Harrisburg.
Pacific Express	2:40 a. m.	4:05 a. m.
Way Passenger	5:00 a. m.	7:50 a. m.
Niagara Express	11:00 a. m.	11:24 a. m.
Hanover Accommodation	11:05 p. m.	Col. 10:40 a. m.
Mail train via Mt. Joy	—	12:20 p. m.
No. 2 via Columbia	11:25 a. m.	12:50 p. m.
Sunday Mail	10:00 a. m.	12:40 p. m.
Fast Line	2:30 p. m.	3:25 p. m.
Frederick Accommodation	2:35 p. m.	Col. 7:45 p. m.
Harrisburg Accom	5:45 p. m.	7:40 p. m.
Columbia Accommodation	7:20 p. m.	Col. 8:30 p. m.
Harrisburg Express	7:30 p. m.	8:40 p. m.
Pittsburg Express	8:50 p. m.	10:10 p. m.
Cincinnati Express	11:30 p. m.	12:45 a. m.

EASTWARD.	Lancaster.	Philadelphia
Cincinnati Express	2:55 a. m.	5:00 a. m.
Fast Line	5:04 a. m.	7:40 a. m.
Harrisburg Express	8:05 a. m.	10:00 a. m.
Columbia Accommodation	9:10 p. m.	12:01 p. m.
Pacific Express	10 p. m.	3:40 p. m.
Sunday Mail	2:00 p. m.	5:00 p. m.
Johnstown Express	3:05 p. m.	5:30 p. m.
Day Express	5:35 p. m.	7:20 p. m.
Harrisburg Accom	6:25 p. m.	9:30 p. m.

The Hanover Accommodation, west, connects at Lancaster with Niagara Express, west, at 9:35 a. m., and will run through to Hanover.
The Frederick Accommodation, west, connects at Lancaster with Fast Line, west, at 2:10 p. m., and runs to Frederick.
The Pacific Express, east, on Sunday, when flagged, will stop at Middletown, Elizabethtown, Mount Joy and Landisville.
*The only trains which run daily.
†Runs daily, except Monday.

NORBECK & MILEY,

PRACTICAL
Carriage Builders,
COX & CO'S OLD STAND,

Corner of Duke and Vine Streets,
LANCASTER, PA.

THE LATEST IMPROVED

SIDE-BAR BUGGIES,
PHÆTONS,
Carriages, Etc.

THE LARGEST ASSORTMENT IN THE CITY.
Prices to Suit the Times.

REPAIRING promptly attended to. All work guaranteed.
79-2-

S. B. COX,
Manufacturer of
Carriages, Buggies, Phaetons, etc.
CHURCH ST., NEAR DUKE, LANCASTER, PA.

Large Stock of New and Second-hand Work on hand very cheap. Carriages Made to Order. Work Warranted for one year. [79-9-12

EDW. J. ZAHM,
DEALER IN
AMERICAN AND FOREIGN
WATCHES,
SOLID SILVER & SILVER PLATED WARE,
CLOCKS,
JEWELRY & TABLE CUTLERY.
Sole Agent for the Arundel Tinted
SPECTACLES.
Repairing strictly attended to.
ZAHM'S CORNER,
North Queen-st. and Centre Square, Lancaster, Pa.
79-1-12

E. F. BOWMAN,
Watches & Clocks
AT LOWEST POSSIBLE PRICES,
Fully guaranteed.
No. 106 EAST KING STREET,
79-1-12] *Opposite Leopard Hotel.*

ESTABLISHED 1832.

G. SENER & SONS,
Manufacturers and dealers in all kinds of rough and finished
LUMBER,
The best Sawed SHINGLES in the country. Also Sash, Doors, Blinds, Mouldings, &c.
PATENT O. G. WEATHERBOARDING
and PATENT BLINDS, which are far superior to any other. Also best COAL constantly on hand.
OFFICE AND YARD:
Northeast Corner of Prince and Walnut-sts., LANCASTER, PA.
79-1-12]

PRACTICAL ESSAYS ON ENTOMOLOGY,
Embracing the history and habits of
NOXIOUS AND INNOXIOUS
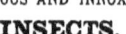
INSECTS,
and the best remedies for their expulsion or extermination.
By S. S. RATHVON, Ph. D.
LANCASTER, PA.
This work will be Highly Illustrated, and will be put in press (as soon after a sufficient number of subscribers can be obtained to cover the cost) as the work can possibly be accomplished.
79-9-

$5 to $20 per day at home. Samples worth $5 free! Address STINSON & Co., Portland, Maine

---

SEND FOR
SPECIAL PRICES
On Concord Grapevines, Transplanted Evergreens, Tulip, Poplar, Linden Maple, etc. Tree Seedlings and Trees for timber plantations by the 10,000.
J. JENKINS' NURSERY,
3-9-79 WINONA, COLUMBIANA CO., OHIO.

$1000 Reward **VICTOR** (Double Buller)
For any machine hulling as much clover seed in 1 day as the
It has hulled
150 Bushels in ONE DAY.
Illustrated Pamphlet mailed free.
Newark, O. Formerly the Hagerstown Ag. Imp. Mfg. Co. Hagerstown, Md.
July-3m)

THE
LANCASTER EXAMINER
OFFICE
No. 9 North Queen Street,
LANCASTER, PA.

THE OLDEST AND BEST.

THE WEEKLY
LANCASTER EXAMINER
One of the largest Weekly Papers in
the State.
Published Every Wednesday Morning,

Is an old, well-established newspaper, and contains just the news desirable to make it an interesting and valuable Family Newspaper. The postage to subscribers residing outside of Lancaster county is paid by the publisher.
Send for a specimen copy.

SUBSCRIPTION:
Two Dollars per Annum.

THE DAILY
LANCASTER EXAMINER
The Largest Daily Paper in the county.
Published Daily Except Sunday.

The daily is published every evening during the week. It is delivered in the City and to surrounding Towns accessible by railroad and daily stage lines, for 10 cents a week.
Mail Subscription, free of postage—One month, 50 cents; one year, $5.00.

JOHN A. HIESTAND, Proprietor,
No. 9 North Queen St.,
LANCASTER, PA.

The Lancaster Farmer.

Dr. S. S. RATHVON, Editor.   LANCASTER, PA., SEPTEMBER, 1882.   Vol. XIV. No. 9.

EDITORIAL.

STATE AND COUNTY FAIRS OF 1882.

There are 47 State Fairs in the United States and Canada—the latter holding *six*, and several of our States holding *two*—for the year 1882.

Of county fairs notices of 671 being held in 21 States have been published, and these are confined to the Northern and Border States. Of what the South is doing in this direction we have not been yet advised. Maine holds 17 county fairs; Massachusetts, 20; Connecticut, 20; New Hampshire, 1 Vermont, 3; Rhode Island, 2; New York, 48; New Jersey, 10; Pennsylvania, 79; Illinois, 89; Indiana, 58; Iowa, 83; Michigan, 34; Ohio, 69; Kansas, 33; Minnesota, 1; Wisconsin, 12; Kentucky, 12; Maryland, 8; Delaware, 1; Virginia, 1; West Virginia, 1. The largest number are held in Illinois, only 10 of the counties holding "no fair," but many others holding *two*. Pennsylvania is third on the list, 10 of her counties holding no fairs, namely, Perry, Monroe, Mifflin, Huntingdon, Franklin, Adams, Cambria, Cameron, Centre and Lancaster. Lancaster, perhaps equals, if she does not exceed all the other non-holding counties put together in population, wealth and agricultural resources—too rich, too populous, and too prominent, perhaps, to *need* such an adjunct as a fair. If it were possible for one wide-awake and observant individual to visit all these 718 fairs, what a multitude of life's phases would be brought under his notice, and what a variety of local productions. There may be much labor, expense and vexation of spirit attending these enterprises, but surely there must be some compensation or they would not be continued.

KITCHEN GARDEN FOR SEPTEMBER.

In the Middle States many and varied are the duties which devolve on the gardener at this season. Not only do the growing crops demand attention, but seeds are to be sown to provide the necessary plants for the ensuing spring. Roots are to be divided and reset Strawberry beds planted, &c. Cabbage, Jersey Wakefield, and Landreth' Large York, sow, to plant out in autumn, where the locality admits, or box up in cold frame, to keep till planting time in spring; the latter end of the month will be time enough to sow in the latitude of Lancaster county; especially sow the newly introduced sub-variety Bloomsdale; also Bloomsdale Brunswick, as a succession. Turnips, the early Dutch and Red-topped may be sown within the first half of this month, if failure has attended earlier efforts. In some sections the fly devours the early sowing. They are less voracious after the nights become cool and the dews heavy. Celery, earth up. Corn salad, scurvy grass and chervil, sow for winter salad. Lettuce, sow for spring planting; the plants to be kept during winter in cold frames. The better sorts for autumn sowing are the Dutch Butter, Royal Cabbage, Bloomsdale Early Summer, and India. Spinach sow early in the month for autumn use; later winter and spring. Turnips and Ruta Baga, cultivate.

Seed Purchasing a Matter of Confidence.

It is *entirely* so. The man who buys drygoods, groceries, corn or cotton, can, to a very considerable extent, judge of the quality and value of the article. This is not the case with *seeds*. Simply because a dealer *says* a certain cabbage seed he holds in his hand is "large late flat Dutch" it does not follow that it *is* so; he may have been deceived himself. No one can tell till valuable time and labor has been expended on the crop. No other commodity but drugs is so entirely a matter of confidence. It behooves every one to get his supplies from dealers of recognized repute; men who have a reputation at stake which they value. Cheapness at once is sufficient to raise a doubt both as to vitality and quality. Good seeds have a *value*—they cannot be *cheap*, in the common acceptation of the word.—*Landreth's Rural Register.*

Of course the foregoing, in relation to seeds, is not intended as a reflection upon any one engaged in the seed business, except such as pursue it fraudulently. A man who possessed a great reputation as a seedsman would also possess a great opportunity to perpetrate a fraud, but he would soon be found out. Retailers of seeds may intend no deception whatever, and yet may most egregiously deceive, because they may have been deceived themselves. The best plan is either to buy from the seedsman himself, or from his accredited agent. Landreth's sealed packages we believe can be safely recommended to our patrons.

INSECT MIGRATIONS.

Nothing seems to be more indisputable, or more fully authenticated, than the migratory habits of some species of insects—indeed, the great African Locust (*Locusta migratoria*) has received its specific name from that very habit; but, it must be borne in mind, that insects do not migrate in the same sense that birds do. Birds, excepting a few local species, at the end of every summer season, migrate to a warmer region of the earth than the one in which they have passed the summer and reared their broods, and this is especially the case with *insectivorous birds*. In the northern temperate zone, at least, they migrate southward in the autumn of the year, and return again to their old haunts in the spring, and it is on record that the same pair have occupied the same nest for different periods, covering from five to fifteen years, or more. And, we may infer, *a priori*, that those that pass the summer in the south temperate zone, at the end of the season, migrate northward, and return again to their old haunts in the spring. Although some of the birds that visit the northern zones in the summer may leave the continent altogether, and pass our winter season in the West India Islands, yet the larger number only remove to our Southern, or the Mexican States, seemingly all the while lingering on the verge of spring. About five and forty years ago we passed a winter in Oldham county, Kentucky, and we were rather surprised, during a few warm days in the first half of January, to find the woods and the fields numerously inhabited by Robins, Bluebirds, Red-headed Woodpeckers, Flickers (Golden-winged Woodpeckers) Black-birds, Wrens, Sapsuckers, and a number of other familiar examples. After a week of balmy spring weather there followed a sudden change in the temperature; a snow fell sufficient to afford tolerable sleighing for two or three days, clearing up cold and freezing; after which not a single bird could be seen. A similar warm spell occurred again about the 10th of February, when the birds returned with greatly augmented numbers, but retired again before the cold blasts that ushered in the month of March. We then left the State and cannot say how soon the birds returned again, but according to our observation they seemed to be all the while "waiting and watching" for the advent of spring and summer. It is not so with *insects*. When we say "not so " in regard to insects, we mean that it is not so in the same sense or degree, for there are some approximations among some insects to the migratory habits of birds.

Again, among mammals and among fishes we find abundant testimony to this habit, and especially in reference to the latter. Those persons residing on or near the Susquehanna river, in our own county, are well aware of the upward migrations of the adult shad in the spring, and the downward migrations of the young shad in the fall. The seine fisheries, for a long series of years, have proved the former, and the fish pots, or "baskets," have borne lamentable evidence of the latter. This has also been evinced to a considerable degree in regard to rock-fish, carp, several species of perch, and last, not least, the eels; but in this last instance the migratory periods are reversed—that is, eels migrate towards the head-waters of the streams when they are young—from three to five or six inches in length—and in the late spring; and migrate downward, in the adult state, in the fall. Both of these positions have been established by testimony as incontrovertible as that relating to shad, although it has some exceptionable or modifying phases. But then, it will be observed, in both of these cases, that it is not the same individuals that go and return again in either case, in which they greatly differ from migratory birds.

As to mammals, from our early boyhood we were impressed with the stories of the western migrations of squirrels, and especially in the States of Ohio and Indiana. These occurrences were frequent for a long series of years afterwards—the squirrels even swimming across the Ohio and other rivers in passing from one locality to another. These animals were not only a nuisance, but also a source of considerable destruction to the ripe corn crops standing in the fields, and their migrations were doubtless governed by a question of food. Organized parties annually slaughtered thousands of them, so that now perhaps such migrations do not occur. It is also on record that rats, mice, &c., for the same reasons, occasionally migrate.

As to migratory insects, neither in the great African locust, nor yet in the "Rocky Mountain Locust" of our own country, is it the same individuals that return towards the locality from whence their progenitors migrated, but an entirely new generation, in which they very materially differ from birds. After the locusts have devoured all the herbage in the locality where they have been bred, or may have subsequently located themselves, it is but natural to suppose that when they make their departure they go in quest of new feeding grounds; but, what should induce a subsequent brood to return to the point of their ancestral departure, must remain an enigma, unless we may suppose that their movements are merely incidental, governed by "wind and tide," and without any of that instinctive judgment so manifest in some of the higher animals. We only know for certain that they come and go again. The "Armyworms," we may legitimately suppose, in their various migrations, are in quest of new pasture, having devoured all that is desirable in the fields they abandon.

The "Colorado Potato Beetle," in its migrations, from the time it left its Rocky Mountain wilds, and perseveringly bent its course eastward, was doubtless mainly governed by gastronomical considerations, and yet it may have been under a migratorial impulse that it could not disobey. During one of our visits to the Atlantic coast, in the State of Delaware, we found, for five or six miles along the beach, many thousands of these beetles (and also other species), and almost every returning wave dashed others on the beach, some of which were yet living. Delaware that year was seriously infested by them; but why they should leave the green potato fields, fly two or three miles across a sandy flat, destitute of succulent vegetation, and out into the Atlantic Ocean, only to drop in and become food for fishes, or be cast upon the shore again, it would be difficult to find out, unless they were driven out by the wind, or, in obedience to that characteristic impulse to be ever on "the go," which has been such a distinguished feature of their advent.

In August and September, 1839, the common "Red-legged Locust," (Caloptenus femurrubrum) was more abundant in Lancaster county than it has ever been since, or perhaps than it ever had been before. Its destructive character was plainly visible in many of the corn fields, and amongst other species of vegetation. Towards the end of September they began to migrate; and, to a considerable height, the air was full of them. They did not seem to manifest any special purpose as to direction, but merely rose to fifty or a hundred feet in height, and apparently submitted themselves to the control of the prevailing wind. What astonished every beholder was, that they had the power to rise so high. They were mainly carried southeast by the winds coming from the northwest.

But, perhaps, the more marked insect migrations have been among the diurnal Lepidoptera —Butterflies. In our very first entomological readings, over forty years ago, we were impressed with the extraordinary migrations of the "Painted Lady Butterfly," (Vanessa cardui) from the continent into England across the straits of Dover. If we have not this identical insect in this country, we have one so near like it as to be indistinguishable and it is just as likely to have migrated hither by various stages, as to have been brought here by other means, and it is now almost a cosmopolitan. But, our own "Milk-weed," or "Wild-cotton" butterfly, (Danais archippus), is a more familiar example of these migrations than any that has yet been recorded, for according to papers published in the Canadian, and also in the American Entomologist, these butterflies have gathered together in large flocks and have migrated to Florida, where the trees have been "literally festooned" with them. And this is the more curious from the fact that there is far less milk-weed there than there is in the valley of the Mississippi, from whence they departed. Of course the milkweed is of no account to them as a butterfly, (only the larva feeding on it,) and therefore they must instinctively have gone South, as a safe place of hibernation. Except a very few straggling, gravid females, they never get back again to their native valley, and it is questionable if ever these do personally, therefore their migrations differ from that of birds.

THE WHEAT CROP OF 1882.

Luck—Good Management—Manure.

Our last wheat crop was one of the best we have had for many years; the yield was from 15 all the way up to 40 bushels per acre, averaging about 27 bushels for the entire county. Those persons who fed their corn into stock cattle, or bought stable manure, brought from Philadelphia or Pittsburg, got their 40 bushels from the acre; and those who kept on farming in the old way, pasturing close in the summer and feeding no cattle in the winter, where the ones who got only from 15 to 20 bushels to the acre. Feeding and making stock cattle fat depends a great deal on good judgment in buying and selling, and requires the best attention during the winter season.

Last spring will be long remembered as an extraordinary one for both good and bad luck in fattening cattle. Cattle were bought in the fall of 1881 for from 3 to 5 cents per pound. For 5 cents you could buy steers nearly fat, weighing from 1,000 to 1,300 pounds. In the early part of 1882 small steers were sold, when fat, at from 4½ to 5 cents per pound, and gradually advanced in price until June, when the best brought from 8 to 9 cents per pound. Farmers got well paid for their corn, realizing, according to good luck, from 50 cents up to $2 00 per bushel.

Farming is like everything else. "Whatever is worth doing at all is worth doing well," hence the success of farming depends largely, and in many cases entirely on good management. I have a neighbor who has a forty acre farm, and he feeds five or six steers, and he yearly got as much as 400 bushels of wheat from 12 acres. He also sells from $300 to $400

worth of tobacco from his place, and is improving it all the time, but he is one of the "come boys" stamp. We are getting too many of the "go boys" farmers, and they are generally among the unsuccessful.

Our lands will be made to increase in fertility and value, through our cattle feeding, and from manures brought from the cities of Philadelphia and Pittsburg, or elsewhere, when we have not a sufficient supply of our own making.

A good coating of barnyard manure will make a good crop of wheat, and will be followed by a good crop of grass and corn. I am strongly in favor of enriching our soil from its own drafts—have more faith in good stable manure than in all your forcing fertilizers, and lime thrown into the bargain.

Cattle and corn are both high in price, and things may look a little demoralized just now, and if beef should fall as suddenly as it rose then there may be some danger of small profits in feeding stock. Our compensation will then be in the manure.—L. S. R., Oregon, Sept., 1882.

[We have taken the liberty to italicise the words "luck" and "good management" in our contributor's otherwise excellent paper, because it seems to involve a contradiction. Does not good luck depend on good management? If so, then the converse must be governed by a similar rule or its absence.]

TOBACCO WORMS—CURIOUS FACTS CONCERNING THEM.

We have before us a large specimen of a green "Horn worm"—two inches and a-half long, and an inch and a-half in circumference—which was brought to us as a "great curiosity." It is wonderful that the phenomenon which we shall attempt to describe should be still regarded as a great curiosity, especially since we first noticed it fully forty years ago, and hardly a year has passed since then in which we have not noticed it, and often half a dozen times in the same season. This worm is the larva of one of the great "Hawk-moths," or "Humming Bird Moths," known to entomologists under the names of Macrosila carolina or quinque maculata—two species that have a close specific alliance, and the larvæ of both of which feed upon the tobacco plant; also, upon the tomato, the potato and the egg plants, and perhaps on other solanaceous vegetation. Before the introduction of the tobacco plant so generally in the county of Lancaster, we found this worm mainly on the potato or tomato plants. The moths can easily be distinguished, the first-named being rather small in size and lighter in color than the last named, but the larvæ to us, at least, are not readily distinguishable.

Perhaps the curiosity did not consist so much in the worm itself, as in the fact that it was covered over its entire body—except the underpart—from the head to the very last segment, almost hiding the posterior horn, with a compact coating of small white spindle-shaped cocoons, resembling small grains of rice attached to the skin of the worm by one end, and so close together that the body of the worm could not be seen between them. We had never seen so many on one worm before, and we were astonished that the host possessed so much vitality under such depleting circumstances. The worm was brought to us, with a part of the tomato plant on which it was found, on the 9th of August, and on the morning of the 12th most of the cocoons were deftly cut off at the upper ends and fully three hundred small four-winged flies had issued forth, and were vainly trying to make their escape from the glass jar in which we had confined the worm. The worm still continued to crawl over the plant but was evidently much weakened. It seemed also to be annoyed by the pressure of the flies, and feebly struck about with both ends, as though it desired to get rid of something very disagreeable.

The worm belongs to the crepuscularian *Lepidoptera* (Twilight-flying moths) and the little four-winged flies belong to their parasitic *Hymenoptera*, of which there is a very large family (*Ichneumonidæ*)—the *Microgaster congregata*, or a specific closely allied. Supposing the worm to be about dying we attached it to a cork and suspended it in a small jar of alcohol; and, although it suffered us to run a needle and thread through its head, and by a similar process attach a small leaden weight to its tail, with almost entire impunity, yet, when we suspended it in the alcohol, it writhed vigorously for five minutes, and detached about twenty-five of the follicles from the thoracic segments. The worm and all the flies are carefully preserved in alcohol. There cannot be much less than three hundred and fifty of them, but we shall base our estimates on the round number of three hundred in our remarks. This phenomenon surely furnishes food for practical reflection in its connection with insect economy. Suppose this worm to have been a female, which, had she developed the moth, would have been able to have deposited at least three hundred eggs (on one occasion we counted more than that number) the possibilities, therefore, existed to have reproduced three hundred horn-worms at least. Now, we will suppose that one-half of that number would have been females endowed with the same reproductive powers, and the result would have possibly been an increase of forty-five thousand horn-worms for next season, and all the legacy of a single worm. Perhaps out of these forty-five thousand not five thousand would have fallen victims to the most efficient remedies, nor yet that number have reached maturity. What then would have become of the other thirty thousand. Let us see. It would be going too far beyond the pale of probability to suppose that only one-half of the parasites would have been females, for the females among these insects always greatly exceed the males in number—a hundred, a thousand, and often ten thousand to one. This has especially been the case among the gregarious gall-insects which are parasitic on plants. Out of these three hundred little *microgasters* we must, therefore, claim two hundred and seventy females at least, capable of reproducing eighty-one thousand parasitic enemies to the three hundred horn-worms, or over seven millions against the forty-five thousand horn-worms, perhaps not as probabilities, but as ultimate possibilities, all other things being equal.

Now, we advise tobacco-growers, gardeners, fruit-growers and agriculturists in general, that, whenever they discover a horn-worm, or any other kind of worm, infested with these or similar parasites, they "severely let it alone," and allow nature to take its course. There is no danger of such worms ever doing any further damage, and by crushing them or trampling them under foot, they may be only destroying a multitude of little insect friends. During the past forty years we have had at least one hundred worms of different kinds infested by parasites brought under our observation, and we never knew a single instance in which any of them survived. All eventually perished. Therefore, the wisest economy is not to disturb them, but to permit the parasites to develop, and when developed they will find another or scores of other worms that will serve us as nidusses for future generations. They will find those, perhaps, that eluded the utmost vigilance of the tobacco cultivators, who generally relax their watchfulness as soon as the crop is harvested, after which hundreds of worms, left in the field, are permitted to mature and go into winter pupation in the ground. Should any of these late worms go into the ground, carrying in their bodies the eggs or immature larvæ of the parasites, although they might be able to effect their pupal transformations, yet the moths will never be evolved.

Perhaps the naked caterpillars (those quite or nearly destitute of hair) are more liable to these parasitic infestations than those that are protected by long stiff hairs, but even some of *these* will develop a few. We have found on opening the tough follicle of the "sack worm" (*Thyridopteryx ephemeriformis*) on several occasions, that the inner cavity was packed nearly full of the small cocoons of a hymenopterous parasite. Now, no bird can dislodge the larvæ of this insect from its strong cocoon, which it always carries with it wherever it goes. We have seen both chickens and birds attempt it, but they always have abandoned it without accomplishing their object. But it seems these parasites can circumvent those almost otherwise unapproachable worms, which affords an ample illustration of the superiority of parasitic infestation over all other known remedies, either natural or artificial. Of course, there are different genera and different species even among those that affect worms in a similar manner. Those before us now constructed white cotton-like cocoons, but there are others more silky. They are, however, not all white; some are different shades of yellow, some drab-colored, and some brownish.

Of course, in one sense, parasitic infestations may be classed among natural remedies, but in this paper we wish them understood as entirely distinct. Natural remedies may therefore be interpreted to mean those animals that naturally or incidentally feed on insects themselves or provide them for their young—such, for instance, as birds, poultry, skunks, moles, snakes, etc., and may also include such predaceous insects as capture or feed upon other insects for their own sustenance, such, for instance, as dragon-flies, tiger-beetles, ant-lions, wheel-bugs, camel-crickets and many others. But all these are either spasmodic or indiscriminate, or both, in their antagonism to the insect world. They destroy friendly and innoxious insects, as well as those that are noxious, and some of them only devour insects when they can obtain nothing better; others survive only for a brief season, and others again desist when they are surfeited, so that they are inconstant in their antagonism with insects.

Nothing could be more spasmodic than the application of artificial remedies for the destruction of our insect pests, for these are mainly governed by the caprice of man. They seldom if ever anticipate the efflux or influx of noxious insects, and are generally only applied when the enemies of vegetation have been augmented and have become destructive. The remedy is then liable to be applied at the wrong time and place, or the quality of the material used may be inferior, or the quantity may be excessive or insufficient, or it may not come in contact with the subjects intended to be destroyed. Many people use a remedy as a patient takes a pill—shuts his eyes, swallows it at random, and then lets it work its way through the stomach and bowels as best it may. We by no means intend to disparage either natural or artificial remedies, for often contingencies arise when it is absolutely necessary to do something, and that promptly, too, in order to rescue a crop from destruction; but when the evil is overwhelmingly present, it is very seldom that artificial remedies ever amount to anything. The Chinch Bug, the Rocky Mountain Grasshopper and the Colorado Potato Beetle have not been exterminated beyond a peradventure, and may become abundant whenever meteorological and other combinations are favorable.

Parasitic infestation, however, occupies a very different ground in its relations to the noxious tribes; it is not merely spasmodic in its operations, but on the contrary, it is constant and continuous, and in obedience to an unalterable dictate of nature. We might almost as reasonably expect a fish to cast its spawn in a hay-mow, or a robin to build its nest on the bottom of a mill dam, as for a parasitic insect to find any other *nidus* for its eggs than the body of a living caterpillar, grub or worm; and there is sufficient analogical evidence to lead to the conclusion that there is not an insect species on earth that has not one or more parasitic enemies. In the general equipoise of nature's economy the bane and antidote go hand in hand, but through human intervention this equilibrium is disturbed or entirely destroyed. This is especially the case where the noxious insects of one realm have, through commercial intercourse, been introduced into another of a similar climate. The bane may be introduced and not the antidote, and hence an insect that was not specially destructive in its native country may become an unmitigated scourge in a foreign one. The "white cabbage *butterfly*," when first introduced into the United States, increased more rapidly, and consequently was more destructive than it is now. This was supposed to be owing to the fact that its parasite (*Pteromalus puparum*) had not yet been introduced. But that parasite is in the country now, and through its intervention the butterfly has, in some localities, become almost exterminated.

Three years ago we received from Franklin

*Pieris rapæ.

county, Pa., twenty-one chrysalids of this butterfly, out of which we only bred four of the flies, but from the other seventeen we bred over fifty of the parasites. Two years ago, out of twelve army† worms which he had confined, eight were infested by parasites—from one to three in a single worm. Only ten days ago S. P. Eby, Esq., gave us the fragments of a cocoon containing the dead body of what appeared to be *Orgya leucostigma*, or "vapor moth," which contained fully a dozen larvæ of a hymenopterous parasite.

In our early entomological experiences we were surprised when we bred a large wasp (*Trogus fulvus*) from the chrysalids of the "Parsnip worm" (*Papilio asterias*), but it soon became a common occurrence, nor longer excited surprise.

But the foregoing is not "*the all*" of parasitic insects. In addition to a large number of *gregarious* species, there is still a greater number that may be regarded as solitary in their infestations—that is, only one individual occurring in the body of its host, the magnitude of the host being just large enough to fully develop the parasite. Perhaps the reason that only *one* individual is found in such a nidus is because there would not be sufficient aliment to develop *two;* and often the parasite itself is infested by one that is smaller still. *Aphids*, or "plant lice," are very liable to be infested by parasites, and it frequently occurs that a whole colony is extinguished by parasitic infestation. The Aphids are found firmly adhering to a twig or leaf of tree or plant, the body swollen or bleached, with a small aperture on the side or back of the abdomen, through which the parasite (*a hymenopter*) has made its escape. But there are also cases in which two or more may occupy the same host. Nor is this the limit of parasitic infestation, for the very *eggs* of insects becomes infested by parasites, and, small as the infested eggs are, they find sufficient aliment within the shell to complete their larval development.

All of the foregoing relates to those parasites that penetrate the body of their hosts, and live upon their substances, manifesting no external sign of their presence until they arrive at maturity; they, however do not all spin a cocoon; some pass also the pupa state within the body of the host. But there are also external parasites that affect insects the same as lice do other animals. The "dor-beetles," (*Copris*) and the "chick-beetles," (*Eluter*) are especially subject to them. On one occasion we continued seven or eight large white "grubworms" in a box of decayed wood and earth. We raised from them one mature specimen of *Pelidnota punctata* (great grape beetle.) All the remainder of them fell a prey to a large voracious white "maggot," from which we reared a specimen of *Mydas flata*, a large two-winged black fly, with an orange band near the base of the abdomen. We found the maggot among the grubs, but can't say how many grubs it destroyed before we obtained it. It however only seemed to live on the fluids of the grub, and hence it killed more than it consumed. When a grub became putrid it abandoned it and attacked a fresh one. It also fed on a specimen of *Lumbricus* or fish-worm, given it.

†Leucania unipuncta.

Thus, the silent work of nature is ever pertinaciously working onward towards its ultimate ends. It may be, and often is, thwarted, partially defeated, or turned aside from its legitimate purposes by contingent interventions; but when its freedom is restored it will gradually converge towards its accustomed channel. Perhaps the forces of nature encounter no greater barriers to their harmonious progress than those imposed through human ignorance. Many years ago we noticed a man in a "potato-patch" with his brows knit and his lips compressed running along the rows, and engaged in a most vigorous manipulation. Curious to know upon what he was exercising himself, we drew near him, and as we approached he assumed an attitude of triumph, exclaiming: "There, I have just smashed the last d——d ladybug in the patch." When we desired to know his reasons for smashing them, he replied that they laid the eggs from which the plant lice bred, for his potato vines were seriously infested by a species of *Aphis*. He did not trouble himself about the Aphids—there were too many of them—but felt sure that they would not long survive their progenitors. Our adverse views had no effect whatever; he knew all about them; he was raised among them. Now, the "lady-birds" (*Coccinillidæ*) are so distinctly the enemies of the Aphids that the group including them is called *Aphidiphaga*—"aphid-eaters." Manual effort alone will not accomplish the destruction of noxious insects; it must be intelligent effort, discriminating effort, persevering effort; the intelligence, the discrimination and the perseverance of the little *microgaster*, which is the joint subject of these reflections. It would not deposit its eggs in a piece of putrid flesh, in a decayed fruit, or in animal fæces, but only in a living, noxious worm. The progress of improvement on the earth's surface may necessarily disturb the equipoise of nature, and where this is the case, it will impose additional vigilance, additional labor, and additional intelligence, in order to insure additional compensation. In the matter of willing and doing, it is of some moment that we know what not to will and do, else we may be standing in our own light, and knowing what not to do is a progressive step towards knowing what we ought to do.

EXCERPTS.

BOILING water will remove tea stains and many fruit stains; pour the water through the stain, and thus prevent it from spreading over the fabric.

RIPE tomatoes will remove ink and other stains from white cloth; also from the hands.

A TEASPOONFUL of turpentine, boiled with white clothes, will aid the whitening process.

BOILED starch is much improved by the addition of a little spermaceti, or a little salt, or a little gum arabic dissolved.

BEESWAX and salt will make flatirons as clean and smooth as glass; tie a lump of wax in a rag, and keep it for that purpose; when the irons are hot rub them with a rag, and then scour with a paper or rag sprinkled with salt.

KEROSENE will make the tea-kettles as bright as new; saturate a woolen rag and rub with it; it will also remove stains from clean varnished furniture.

KEROSENE will soften boots or shoes which have been hardened by water, and render them as pliable as when new.

AGRICULTURE is the financial barometer of the United States.—*London Paper*.

WE would not advise the sowing of white clover in lawns. It exterminates other grasses and does not stand heat drought.

GIVE the laboring class 10 or even 12 hours work a day, with plenty or good newspapers and no strong drink, and the country will soon become prosperous and its men enlightened.

THE value of poultry in the United States amounts to over $100,000,000. This large sum would be increased if poultry received the same attention as is bestowed on sheep, cattle or horses.

THE winter wheat crop of Illinois this year exceeds 50,000,000 bushels, and it is the largest, except that of 1820, ever harvested in that State. The spring wheat aggregates over 52,000,000 bushels, a little under the average of 1879, but the quality is much better.

THE Herefords in the London market are always worth more per pound than Shorthorns. We supposed that the Short-horn editors and advocates had conceded this fact, but in conversation with a prominent editor a few days since he denied it was true. If he will place this in some direct and positive form we will produce the proof of it, and will offer this much now. The Chamber of Agriculture *Journal*, in its issue of June 19th, speaking of the Smithfield market of London, says : "The Herefords range with Scott cattle at 5s. 10d. to 6s. as the topping current rate of the morning trade, the Hereford cattle ranging up to 5s. 8d. to 5s. 10d. These prices are for the stone of eight pounds weight. The Canadians sold at 5s. 4d. to 5s. 8d.; Danish at 5s. 4d. to 5s. 6d." The editor he referred to has all the means at his disposal to inform himself, and he ought to know that Hereford beef is always at the top of the market, and that this is especially true with the Herefords from grass.

THERE are but few circumstances that will justify the burning of straw as it comes from the machine. Upon all uplands, or soils inclined to be light, and which are deficient in vegetable mold, it is better to rot the straw and apply the same to the most unproductive portions. Where the soil is of a heavy clay character, and fall plowing can be done for the growing of a crop the following summer, the plowing under of a heavy coating of straw will render the soil lighter by reasons of the drainage afforded, and richer by the partial decay of the straw. Occasionally it may be of advantage to burn the straw, especially if weed seeds or the eggs and larvæ of insects are unusually abundant. Consider the matter well before burning the straw, for when rotted it forms a rich mold, which is the "one thing needful " on all our clay uplands, East, West, North and South.

THE MEXICAN DOG.—Of the hairless Mexican dog, which is the shepherd dog of that country, the *Texas Siftings* has this to say :

The Mexicans call him pelon, the Americans refer to him as no-hair dog, while the stranger from the North who sees him for the first time calls him a cast-iron dog, for that is what he looks like at first glance. Although not particularly intelligent the no-hair dog is susceptible of a high polish, for his hairless hide shines in the sun as if it had been recently touched up with stove-polish. His body is about the size and somewhat the shape of a watermelon—that is, of one of those small watermelons that is about the size of a pelon dog. He differs, however, from the melon in that his tail is adorned with a tuft of blonde hair, which is never the case with a watermelon. He wears a tuft of hair—another tuft of course, not the same one at all—on his head, which gives him a very striking appearance. The pelon dog is found in Austin, in San Antonio, and in tamales, the latter being a Mexican dish, the ingredients of which are as uncertain as those of hash.

WHY 1900 IS NOT A LEAP YEAR.—The year 1900, although it is divisible by 4 without a remainder, is not leap year, and it comes about in this way: Under the "Julian period" the solar year was considered to consist of three hundred and sixty-five days and a quarter of a day, but as the actual or civil year could not be made to include a quarter of a day, an additional day was inserted in the calendar every fourth year to make up for four lost quarters, and this is the 29th of February. But the Julian method of intercalation made the year too long by eleven minutes, ten and one-third seconds. This put the calendar ahead of solar time one day in 129 years; so to balance this, in the adjustment of the calendar known as the "Gregorian" after Pope Gregory the XIII., now universally adopted in Christian countries except Russia, one of the leap years is dropped at the close of every century, except when the figures of the centurial year, leaving out the two cyphers at the end, can be divided by four without a remainder. Thus 1,600 was a leap year, and 2,000 will be, but 1,700, 1800 and 1900 are not.

QUERIES AND ANSWERS.

THE ROYAL HORNED CATERPILLAR.

CONEWAGO, Lancaster co., Pa.,
August 25th, 1882.
EDITOR LANCASTER FARMER:

Dear Sir: With this mail I send you a tin box inclosing a large worm or caterpillar, which I found on a walnut tree. Please describe it in the next number of the FARMER.
J. F. B.

The box and caterpillar came safely to hand, and is the *larva* of the "Regal Walnut Moth," (*Cratocampa regalis*) but it does not confine itself to the walnut, for we have found it on the hickory, and Prof. Riley has received it from correspondents who found it on the persimmon. It is also found on the butter-nut, on the cultivated "Dutch-nut," and occasionally on the sumac. It is better known, and is oftener found in the caterpillar state than in the moth state. In Virginia it is called the "Hickory Horned Devil," and by many people is as much dreaded as a venomous snake. There is something repulsive in its looks, but it is entirely harmless, as far as our experience goes. This specimen is over five inches in length, when it is crawling, and fully two inches in circumference. The whole body is green in color; the head and feelers are orange, and eight large spines on the first three segments are of the same color, tipped with black. There are five black spines on all the segments, and the dorsal spine on the last segment, but one, is much larger than the others. The larva goes into the ground in September, forms a cell therein and changes to a black pupa, and comes forth a large and beautiful moth in the month of June of the following year. The body of the moth is fully two inches in length (female specimens) and expands six inches from tip to tip of the front wings. The front wings are fawn colored, broadly latitudinally lined with orange, and two large lemon-colored spots near the anterior margin and the tips. The hind wings and abdomen are orange colored, with a few inconspicuous ashen and other yellow markings.

This insect is solitary in its habits, and one brooded; and, as it is usually found on wild forest trees, it is not generally considered noxious.

CONTRIBUTIONS.

For the LANCASTER FARMER.
GAPES IN POULTRY.

Croup, would be a very appropriate name, as it has much the same effect as croup with children; believing that young poultry is as liable to take cold, as anything else, therefore a phlegm, or a roapy substance accumulates in the windpipe, with a slight discharge of blood from the lungs, and the same enclosing the downy plumes, stop up the air passage. I have no doubt but there are many who will doubt my theory of the gape worm. But as I said in my last, if you will examine the down on the young chick, you will find them all double, the connection being the same, and at the same place as the gape worm; you will also find the end of the worm nearest the connection the hardest, this being the quill of the feather. If the worm is not full grown, it will be quite hard, and with a strong glass you will see fine threads in the worm, always in color the same as the chick (not intestines), then; after the threads of down; on mashing it, they will mingle with the blood and water, as they have become soft by decomposition. Probably if the chick did not take cold, there would be no accumulation of worms; otherwise the down would pass away without any hindrance of breathing; although anything so fine, like hair, in a warm and moist place, is liable to become a living animal (the hair worm). If the chick is well greased soon after hatching, with lard and salt, and kept in a warm and dry place, and fed on dry food, there will be little or no gapes. Yours truly—*Wm. J. Pyle, Aug. 30, 1882.*

For the LANCASTER FARMER.
SHALLOW CULTIVATION.

The remarks in the August Lancaster FARMER, on "Shallow Cultivation for Fruit," reminded me of the case of farmer Wm. Klinger, at Weishampletown, Schuylkill co., Pa., who a few years ago was gravely advised by a stranger to plow deep around the apple trees in his fine young orchard, if he wanted to see the trees prosper well. He did so, ripping up the roots on both sides of the trees effectually, leaving very few untouched. The result was that some of the trees died and all the rest suffered for several years from want of succors in the ground. It took about three years before the orchard recovered from the injury. The farmer was very much vexed at his own folly for following the advice of that wiseacre.—*J. F. H.*

NOT THE TARIFF QUESTION.

For the LANCASTER FARMER.

My respected opponent, P. S. R., writing in the last number of THE FARMER, while professing to answer my communication in the July number, gets entirely away from the question we were discussing and favors us with an essay on "The Tariff"—a subject that has no necessary connection with the one at issue between us, and enters upon new ground where I feel no call to follow him. The only question was respecting the so-called Balance of Trade. My opponent assumed that if in trading with other countries we import more value than we export it was proof of a balance against us and that we were doing a losing business. This I denied, and gave my reasons for my opinion—reasons which it is unnecessary to inform those who have read both articles, Mr. R. has not even attempted to confute.—*J. P., Lancaster Sept. 8, 1882.*

For the LANCASTER FARMER.
THE EEL—ITS HABITS AND GROWTH.

The following sketch on the habits and growth of the eel, has been prepared from an article on "Eels and Eel-sets" which appeared in the January number of *Blackwood's Magazine*. As the article was much too lengthy for publication in the NEW ERA, it was very much cut down, and only the part relating to their habits and mode of reproduction are given:

The eel has puzzled many naturalists, and is destined to puzzle many more. As to the natural history and habits of the eel, naturalists generally agree that there are three sorts indigenous to this country (England) namely, the sharp-nosed or silver-bellied eel, the grig or surg, and the broad-nosed eel.

The grig is a yellowish eel with a projecting under-jaw; the broad-nosed eel is stated to be an uglier-looking eel, with a broader head, and, according to Pennell, fierce and voracious to its habits; while the silver-bellied eel is a firm, fine-flavored eel, with a dark, almost black back, a silvery belly, and a fine sharp head. This is the eel which migrates seaward in the autumn, and is the eel by which eel-setters live.

Mr. Pinkerton says the grand distinction between the sharp-nosed and broad-nosed eel is, that the sharp-nosed species is a migratory fish while the other is not. He admits that the latter has its summer and winter quarters, for eels are very susceptible of the effects of cold and electricity, and it wanders about a good deal at night, in search of prey; but it does not migrate to the sea in large shoals, as the sharp-nosed species usually does. It is about the middle of autumn that the annual migration commences, the eels moving in the night, and always choosing a dark night for the purpose. A change of wind, a clap of thunder, a cloudy night becoming clear and starry, will at once stop the movement.

No one has ever seen the eels returning, but in the spring of the year the young eels come up by millions, keeping close to the banks, and swimming in almost solid columns. They will surmount almost any obstacle, creeping wherever there is any moisture, oftentimes through grass and over stones and timber. This "eel fare" lasts several days; and the tiny elvers, something like darning needles in size, can sometimes be scooped out by the bucketful, and applied to the land for manure, baked into cakes for the men, or given to the pigs for food.

But to return to the movements of the eels in the migrating season. Big and little, old and young, start on this singular voyage; and big and little, old and young, remain and "bed" themselves. Thousands of bubbles rising to the surface show where they work far down into the soft mud. This bedding is to escape the cold winter to which eels are very sensitive, and is easily intelligible. But why do they migrate ? For one reason, the brackish water, estuaries and harbors, is warmer than either sea or river. The admixture of fluids of different densities causes a rising temperature, and fresh and salt water are daily mixed by the tides, and lessen the cold. Thus while some eels prefer to seek the warmth of the mud, others seek heat in brackish water. But sooner or later, all eels of the silver-bellied species go down to the sea, and none of those that go down return. This is spoken of so positively by all eel-fishers, that it cannot be doubted; and in such rivers as the Severn, there is no room for doubt, because of the facilities there are for observation. Then how is the supply kept up; and how is it that eels are always found in the river of a large size?

The answers to these questions are, that young ones are produced; and the eels are so that although immense numbers leave the rivers each year, yet equally immense numbers remain. Now comes the curious part of it, so far as Norfolk rivers are concerned. In other rivers the procreation takes place largely in the estuaries or sea, and the elvers return to stock the rivers. In the lower part of the Norfolk river the elvers are not noticed in spring, or any other time of the year; and so continually are these eel-men on the river day and night that such a phenomenon could hardly escape their attention. Neither could they fail to detect the return of the old eel, supposing they came back singly or in small detachments, for seeing that the cold weather does not end until March, and that the eels begin to descend in July, and continue until the end of November, only three months would be allowed for their ascent, so that if they did ascend they must come up in droves.

We have all heard of the notion that chopped horse hairs thrown into the water turn into eels, and the many other ideas accounting for their breeding in equally absurd ways. Some of the more intelligent, however, believe that the young ones are produced in the river in the spring, and have been stated positively that they have cut eels open in February and found them full of young eels.

It is only at the first obstacles on the rivers Yare and Bure—the flouring mills on the upper reaches—that the elvers are ever noticed; and here they appear in large numbers. In the "New Mills" in the city of Norwich is a building which completely spans the stream. There are brick walls on each side of the river, and no means of access save through the sluices, and by the floats of the wheels. Here the tiny elvers force their way in countless thousands, wriggling through every crevice, on their upward march. But it is positively stated that no adult eels ever ascend, yet between "hay harvest" and November the eels descend in thousands, and of all sizes. Now, although the silver-bellied eel is undoubtedly a fast grower, yet eels of the size caught in the nets at the New Mills must be several years old, and must have passed all their days, since elverhood, above the mills. Can it then be reasonably supposed that these eels have passed so much of their lives without procreation of their species? This can scarcely be; and it is therefore a fair conclusion that the procreation of a large number of eels takes place in fresh water. This leads then to the question, What is the object of the yearly migration of the silver-bellied eel ? If the above suggestions are correct, it cannot be for breeding purposes alone; and it is more than probable that, as eels multiply as fast as other fish and probably grow faster, and as they bring forth their young alive, and so are not subject to so many chances of destruction as the spawn of other fish; their numbers are so incredibly large that the rivers must get over-crowded. Therefore each year a certain portion "swarms" off and is lost in the sea.

It will have been noticed that the eel has been alluded to as being viviparous. Naturalists affirm that the eel deposits in spawn as other fish do, and state that the microscope reveals the presence of spawn and milt in the eel. This is so much opposed to all the statements and experience of eel-fishers and eel-setters that it cannot be accepted as a fact, and after listening to so many eel-fishers who stoutly affirm that they have constantly opened eels in February that have been full of minute living eels (not parasites), and that in a tub of eels young ones have been found in the morning that were not there over night, we strongly lean to the theory that eels are viviparous.

The young fry are contained in a membranous sac, as long and thick as one's finger, and eyes and back-bones of the fry are distinguishable. When the sac is cut open, the fry unbend themselves and wriggle about. Eels are found in this State during February, March and April.

SELECTIONS.

A GRAND HARVEST.

From all parts of the country we have concurrent reports of the abundance of nearly or quite all the staple crops produced by our agriculture. Not only is the prospect good, but already the receipts at the leading commercial centres of the seaboard and the West far exceed those of last year at the corresponding periods. What is still more remarkable is the fact that the European harvests are rather short, especially those in the British Isles; while the troubles in Ireland interfere very sadly with the harvest there, and the Egyptian war threatens that country with famine. Thus financiers and speculators appear at the present time to agree in regarding the condition of things commercially as highly favorable to another large increase of the balance of trade in favor of this republic in her commerce with the world at large. It is true that nominally this balance has still been in our favor during the whole of the year, notwithstanding the shortness of the crops of last season; but nevertheless, the current of gold turned towards Europe and has caused us a net loss of at least $20,000,000 of our stock of gold; while at the time our foreign debt has again been increasing in consequence of the enormous railway building movement in this country, and the excessive demand for large loans of money for that purpose abroad and at home.

But it is now considered quite probable that the American harvest will be so vast as to stimulate a tremendous export trade in all our food-products and agricultural staples, and thus to compel the European bankers to make exchange either by shipment of gold hither or by the return of masses of American stocks and bonds. These alternative would be quite acceptable to the country, for although the gold would of course be preferred, as under all circumstances the best and most substantial consideration, yet the liquidation of the whole of the foreign debt is the greatest possible desideratum known to American finance.

It will be perceived that this present prospect is wholly due to the blessings of Providence upon the labors of our husbandmen in those arts which some politicical economists have foolishly discouraged as fit only for barbarians. We have several times spoken of last year's crops as having been unusually short, but we must beg our readers to bear in mind that those crops supplied all the wants of our own people with a large margin for export and a considerable surplus to carry us safely through the year until the new harvest of the present season could be garnered. We consider this as a remarkable illustration of the safety of this country from the dangers of famine, since it was precisely a similar state of affairs which in 1880 forced France to ship to the United States $10,000,000 in gold to buy food to supply the deficiencies of her own crops.—*Germantown Telegraph.*

OCCUPATION AND LONGEVITY.

"Woe to them that are at ease!" says Carlyle, but his anathema does not prevent the English village parson from outliving every other class of his countrymen, not excepting the British farmer, whose peace of mind cannot always be reconciled with high rents and the low price of American wheat. Where agriculture is what it would be—a contract between man and nature, in the United States, in Australia, and in some parts of Switzerland—the plow furrow is the straightest road to longevity; in Canada where nature is rather a hard taskmaker, the probabilities are in favor of such half-indoor trades as carpentering and certain branches of horticulture—summer farming as the Germans call it. Cold is an antiseptic, and the best febrifuge, but by no means a panacea, and the warmest climate on earth is out and out preferable even to the border lands of the polar zone. The average Arab outlives the average Esquimaux by twenty-five years.

The hygienic benefit of sea voyages, too, has been amazingly exaggerated. Seafaring is not conducive to longevity; the advantage of the exercise in the rigging is more than outweighed by the effluvia of the cockpit, by the pickle diet, the unnatural motion and the foul weather misery, and, from a sanitary standpoint, the sea air itself is hardly preferable to mountain and woodland air. The cozoon may have been a marine product, but our Pliocene ancestor was probably a forest creature.

"For what length of time would you undertake to warrant the health of a seaman?" Varnhagen asked a Dutch marine doctor. "That depends on the length of his furlough," replied the frank Hollander, and it will require centuries of reform to redeem our cities from the odium of a similar reproach. In victuals and vitality towns consume the hoarded stores of the country, and only the garden suburbs of a few North American cities are hygienically self-supporting. Permanent in-door work is slow suicide, and between the various shop-trades and sedentary occupations the difference in this respect is only one of degree. Factories stand at the bottom of the scale, and the dust and vapor generating ones below zero; the weaver's chances to reach the average age of his species have to be expressed by a negative quantity. In France, where the tabulation of comparative statistics is carried further than anywhere else, the healthfulness of the principal town trades has been ascertained to decrease in the following order: Housebuilding, huckstering, hot-bed gardening (florists), carpenter and brick mason trades, street paving, street cleaning, sewer-cleaning, blacksmiths, artisansmiths (silver, copper and tin concerns), shoemaking, paper making, glass-blowing, tailor, butcher, housepainter, baker, cook, stonemasons and lapidaries, operatives of paint and lead factories, weavers, steel grinders—the wide difference between brick and stonemasons being due to the lung-infesting dust of lapidary work, which, though an outdoor occupation, is nearly as unhealthy as steel grinding. Lead paint makers have to alternate their work with jobs in the tin shop, and after all can rarely stand it for more than fifteen years. Needle-grinders generally succumb after twelve or fourteen years.—*Popular Science Monthly.*

THE WAR IN EGYPT.

The prospect of a speedy termination of the Egyptian difficulty does not improve. Indeed, it looks now as though England has on hand a serious war which is not likely to be brief, even if no general European complication arises from it.

Meantime the industries of Egypt are grievously deranged; trade is at a stand-still, and all manufacturing operations are suspended, and agriculture is largely interrupted.

The geographical and the social characteristics of Egypt are peculiar, and of such a nature that war affects the country far more disastrously than would be possible in any other land.

The Europeans who have been driven out furnished most of the capital for all commercial and industrial enterprises, filled most of the positions requiring scientific knowledge or mechanical skill, and controlled the majority of the means for making productive and profitable the labor of the native masses. In their absence a speedy revival of prosperity is impossible, even if the war should end at once.

Within the past twenty years the agricultural products of Egypt have been nearly trebled by means of the capital and machinery introduced from Europe. The irrigation and consequent cultivation of vast areas of sugar and cotton and corn land have been made possible by the introduction of steam pumps and other modern irrigation machinery. Were the natives able to operate such machinery they cannot now do so for lack of coal, and so to a serious extent they cannot produce the crops on which their prosperity depends.

The cotton-ginning factories and steampresses, by means of which the cotton crop of Egypt has been made fit for profitable exportation, were introduced by Europeans and largely operated by them. The same is true of the sugar mills and the railways and other means of rapid and economical transportation. The natives themselves are incapable of operating the railways or of conducting an export trade, were such trade possible in Egypt in time of war. As a consequence the gathered crops are lying in the interior unsold; cultivation is largely suspended and thousands of native workpeople are threatened with starvation.

The commercial and industrial arrangements incident to the war are not confined to Egypt. Even if no harm befalls the Suez Canal, and there is no suspension of traffic through it, England cannot but suffer severely, though indirectly, in her commercial and manufacturing interests.

Fully two-thirds of the cotton crop of Egypt, averaging 280,000,000 pounds, has hitherto gone to England. In the Bolton district alone 5,000,000 spindles are employed on Egyptian cotton; and in the whole of England some 26,000 work people are employed upon this staple. The stoppage of the supply cannot but affect them disastrously.

The large dependence of English industry upon Egyptian products is further illustrated in the case of cotton-seed, about $9,000,000 worth of which is imported annually. Last year Hull alone took 120,000 tons, and in its crushing 2,500 men and boys were employed. Still more serious will be the effect of the stoppage of the supply of Egyptian cotton seed upon English agriculturists, who depend very largely upon cotton-seed oil-cake for feeding their cattle. The English soap boilers use about 50,000 tons of Egyptian cotton-seed oil a year, and must likewise severely feel a cutting off of the supply from that region. England also draws from Egypt annually $6,000,000 or $7,000,000 worth of wheat and beans, $3,000,000 worth of sugar, and more than $2,000,000 worth of wool, ivory, gums, and other native products.

In return for all these, Egypt has taken manufactured goods, machinery, coal, and cotton fabrics, the producers of which cannot but lose heavily by the ruin which has fallen upon Egypt.

How far these English losses will react upon American trade it is impossible to foresee. The deficiency in cotton and corn can be made good from this side, but it is doubtful if any marked advantage will accrue to American producers unless the war should involve other powers than Egypt and Great Britain.

The first effect anticipated by our shipping merchants is an advance on ocean freight and in marine insurance, through the withdrawal of first-class steamers for transport service to the seat of war, and the substitution for them of second and third-class freighters in the regular carrying trade.—*Scientific American.*

THE CLIMATE IN DIFFERENT PARTS OF THE UNION.

Figures gleaned from the observation points of forty-nine States and Territories show that the hottest places in the Union are Florida, Louisiana and Arizona, the mean annual temperature of which is 69. Texas ranks next at 67, Alabama 66, Mississippi 64, Arkansas 63, South Carolina 62, Indian Territory 60, North Carolina 59, Georgia and Tennessee stand on a par at 58, Virginia 57, Kentucky 56. The mean temperature of 56 prevails in California, Missouri and the District of Columbia; 54 in Maryland and Pennsylvania, 53 in Delaware, Ohio and Oregon; 52 in Idaho, Utah and West Virginia. 51 in Indiana, Kansas, New Mexico and Washington Territory; 50 in Connecticut, Illinois Nevada and New Jersey; 49 in Iowa and Nebraska; Massachusetts ranks with Rhode Island, New York and Colorado at 48; Michigan and Dakota are equal at 47. Alaska is not the coldest part of the Union, as is commonly supposed, but stands with New Hampshire at 46; colder than these are Maine and Wisconsin at 45, Montana and Vermont at 43, Minnesota at 42, and coldest of all Wyoming at 41.

PURE AND WHOLESOME.

Nearly all the American cotton seed oil shipped to Europe is christened "olive" oil, and re-exported to this country, where we consume it with the greatest gusto, as "real extra Lucca." This suggests to our mills the importance of securing a market here at home, where they can sell their oil to much greater advantage, since they will not have to pay double freight to Marseilles and back. Our people have been putting cotton seed oil on their salads as olive oil for years; why continue this practice any longer? Why not confess what is well known that cotton seed oil is not like glucose or oleomargarine—an adulturant—but as good and as pure as the product of Italy in every respect? It is true that when we first began to manufacture it many persons pretended to find about it a somewhat bitter taste. But this taste has latterly been completely eradicated, and now our factories turn out as fine a salad oil, and chemically and gastronomically exactly the same as the best farms of Tuscany and Lucca. This oil should supplant lard in the Southern household; it is cleaner, better, cheaper and in every way superior to lard. Its use instead of lard has become quite common in the southern Atlantic and Gulf States during the past few years, and everybody who has tried it has been delighted with it. Here is a field for our mills more promising than Italy, that needs only to be properly worked up to make rich returns. It is something that will benefit the whole South, for the cotton planters are all interested in finding a market, and thus giving a value to a product which, a few years ago, was waste, and which they were then anxious to get rid of at any price.

TEMPERATURE AND RAINFALL.

The following tables of temperature and rainfall for June have been received from the Signal Service Bureau, prepared under direction of General W. B. Hazen.

Average Temperature for June, 1882.

Districts.	Average for June, Signal Service observations.	For several years.	For June 1882.	Comparison of June, 1882, with the average of several years.
New England		61.8	65.0	3.2 above
Middle Atlantic States		70.7	70.8	0.1 above
South to Atlantic States		77.4	77.4	Normal
Florida Peninsula		80.8	81.4	0.7 above
Eastern Gulf States		79.2	79.1	0.1 below
Western Gulf States		79.1	79.1	Normal
Rio Grande Valley		85.1	84.3	0.8 below
Tennessee		76.4	76.2	0.2 below
Ohio Valley		72.8	70.9	1.9 below
Lower Lake Region		66.0	64.1	1.6 below
Upper Lake Region		62.6	60.6	2.0 below
Extreme Northwest		62.7	61.2	1.5 below
Upper Mississippi Valley		70.1	68.0	2.1 below
Missouri Valley		70.9	71.3	0.4 above
Northern Slope		63.4	62.7	0.7 below
Middle Slope		71.1	69.0	2.1 below
Southern Plateau		70.9	70.1	0.8 below
Northern Plateau		61.6	65.5	0.9 above
Middle Plateau		67.0	65.4	2.2 below
Southern Plateau		79.0	75.1	3.8 below
North Pacific Coast		60.7	61.0	0.3 above
Middle Pacific Coast		60.1	66.9	2.3 below
South Pacific Coast		71.2	69.1	2.1 below
Mount Washington, N. H.		45.6	41.8	4.8 below
Pike's Peak, Colorado		33.7	30.8	6.4 below

Average Precipitation for June, 1882.

Districts.	Average for June, Signal Service observations.	For several years.	For 1882.	Comparison with the average of several years.
New England		2.75	3.61	0.14 deficy
Middle Atlantic States		3.81	2.80	1.01 deficy
South Atlantic States		4.90	4.45	0.45 deficy
Florida Peninsula		4.76	5.59	0.65 excess
Eastern Gulf States		4.48	2.85	1.64 deficy
Western Gulf States		4.11	2.24	1.54 deficy
Rio Grande Valley		1.77	1.12	0.65 deficy
Tennessee		4.31	4.80	0.63 deficy
Ohio Valley		4.54	5.41	0.87 excess
Lower Lake Region		3.51	3.62	0.11 excess
Upper Lake Region		4.37	5.76	1.39 excess
Extreme Northwest		4.13	4.48	0.35 excess
Upper Mississippi Valley		5.05	7.01	1.96 excess
Missouri Valley		4.91	6.29	1.25 excess
Northern Slope		3.00	2.81	0.11 excess
Middle Slope		1.88	3.21	1.33 excess
Southern Slope		2.50	2.78	0.28 excess
Northern Plateau		1.51	0.11	0.99 deficy
Middle Plateau		0.58	1.70	1.12 excess
Southern Plateau		0.41	0.40	0.12 excess
North Pacific Coast		1.38	1.17	0.21 deficy
Middle Pacific Coast		0.21	0.10	0.11 deficy
South Pacific Coast		0.01	0.13	0.12 excess
Mount Washington, N. H.		8.71	11.10	3.16 excess
Pike's Peak, Colorado		1.80	3.10	1.30 excess

BARN YARD MANURES.

In the system of agriculture practiced in the United States, barn yard manure, from its cheapness and efficiency, must for a long time constitute the staple fertilizer under ordinary conditions of practice.

Dr. J. B. Lawes, in his valuable pamphlet on "Fertility," says:

In the district where I live the land is cultivated on a four course shift, and the crops which are grown and sold off the land would cost more to produce by the means of purchased artificial manures than the sum which the tenant, under the above system of cultivation, pays for them in rent, or in other words, as far as regards the production of the crop, the landowner sells his fertility cheaper than the manufacturer of manure could supply it.

The principle that underlies this statement, startling as it may appear, applies with twofold force to successful farm practice in this country.

On the average American farm, with its cheaper land, and soils that have been under cultivation for a comparatively short time, the natural stores of fertility that have been accumulated in past ages must be the leading element in determining the profits of grain production at low prices; and when this natural source of profitable cultivation is properly reinforced with the barn-yard manure that can readily be made, under a fairly good system of management, to retard and diminish the exhaustion that is unavoidable in a paying system of husbandry, the commercial fertilizers, which are too often urged upon farmers as the essential basis of good farming will find their true place as supplemental manures that are desirable for special purposes.

Aside from the fact that barn-yard manure is a complete fertilizer, supplying, as it does, the potash, phosphoric acid, and nitrogen, which are considered the only valuable constituents of purchased manures, it seems to have a specific action on the soil that cannot be obtained with any combination of chemical fertilizers.

In the Rothamsted experiments with drainage waters, from the plots which had been under continuous cultivation with the same crop for more than thirty years, it was observed that "whilst the pipe-drains from every one of the other plots in the experimental wheat-field run freely, perhaps four or five times or more annually, the drain from the dunged plot seldom runs at all more than once a year, and in some seasons not at all."

Dr. Voelcker remarks that "this result is interesting and important, for it illustrates in a striking manner the beneficial effects of barn-yard manure on the soil in ameliorating its texture, and, generally speaking, its mechanical or physical condition, in consequence of which the growing crops will suffer less during seasons of drought."

After a careful investigation as to the causes of the small discharge of water by the drain of the dunged plot, Drs. Lawes and Gilbert concluded that "the result was due to the greater power of absorption and retention of moisture by the dunged soil near the surface."

The power of retaining a large amount of moisture, in an available form, and without making the soil wet, seems, therefore, to be increased by the application of barn-yard manure, and this, with the increased porosity which renders the water of the lower strata of soil available for plant growth, explains the effects of excessively dry or wet seasons.

The advantages of the barn-yard manure, under the unfavorable conditions of a wet, backward spring, followed by a severe drought, were decidedly marked in the crop of 1881 throughout the entire season.

From the first appearance of the plants above the surface to the time of harvest, the barn-yard manure plots could be clearly distinguished, even at a distance, by the vigorous and rapid growth of the crop, and when the tassels and ears were forming, the stalks were not only much larger, but they gave indications of a mature development that was not observed on the other plots.—*Manly Miles, Houghton Farm.*

PRESERVING FENCE POSTS.

Several plans have been tried for increasing the natural durability of the poplars, elms, etc., when used for posts.

Of these, the most effective has been immersing them in hot coal tar, where they are kept at a boiling temperature for thirty or forty minutes. Vats are used for the work built on the principle of the old sorghum evaporators. The posts are put in and taken out of the hot tar with large nippers made for the purpose.

Pots of willow, cottonwood, white elm, etc., thus treated have proven more durable than white oak posts set green. With the cottonwood it has been found that the tar would penetrate into the pores of the wood better when green than after they become dry.

It has been also found that elm boards were very strong, durable and free from warping when treated to a bath of the boiling gas tar.—*Western Farmer.*

SOME WHEAT STATISTICS.

It is to be regretted that a journal of such good standing as *Bradstreet's* should fall into such errors regarding wheat statistics as are to be found in its last issue. Commenting on wheat, *Bradstreet's* says:

Two years ago, on August 1, 1880, it was estimated that there were 50,000,000 bushels of wheat left over out of the crop of 1879; add to this the large crop of 1880, which was stated by the Agricultural department at 490,000,000 bushels, and the crop of 1881, which may be estimated at not over 460,000,000 bushels – giving a total supply of 940,000,000 bushels of wheat for the two years to August 1, 1882. Out of this we exported in the form of wheat and wheat flour in the year to July 1, 1881, 198,828,581 bushels of wheat, and in the year to July 1, 1882, 121,523,246 bushels of wheat, making a total of 320,351,-827 bushels.

On August 1, 1882, there was practically no wheat of the old crop left in the country; consequently the balance over what was exported was consumed, say, 619,648,173 bushels, or at the rate of, say, 320,000,000, which, with an increasing population but by immigration and natural increase, would probably require 340,000,000 for the year ending August 1, 1883.

The estimate of 50,000,000 old stock on hand August 1, 1880, is much too high, as it was a noticeable fact that stocks of old wheat were at that date unusually small. Instead of the crop of 1880 being 490,000,000 bushels it was 8,000,000 bushels larger, while the crop of 1881 was 380,000,000 bushels, or 20,000,000 bushels less than *Bradstreet's* figures. For the year ending July 1, 1881, we exported of wheat and flour 186,321,464 bushels, or 12,-000,000 bushels less than what *Bradstreet's* gives. In speaking of the stock on hand in 1880 and 1882, *Bradstreet* takes August 1 as the date, but when giving the exports it uses the fiscal year ending July 1 or rather June

30, which causes a serious error—as July 1882 is thus lost sight of, when it ought to have been stated that at July 1, 1882, the stock of old wheat, including flour, was probably not less than 40,000,000 bushels. Certainly it was fully as large as the old stock on hand July 1, 1880. So that the amounts for the two dates will balance each other, leaving the exports and consumption to be provided for solely out of the two crops. Another mistake of *Bradstreet's* is making one-half of 619,000,000 bushels 320,000,000 bushels, instead of in round numbers 310,000,000 bushels. Thus in the few lines devoted to the subject there were some six or seven serious blunders, well calculated to mislead.

The crop of 1880 was 498,000,000 bushels and that of 1881, 380,000,000 bushels, or a total of 870,000,000 bushels. From this we exported 305,000,000 bushels, leaving a balance of 573,000,000 bushels for two years' consumption, or an average of 287,000,000 bushels a year for seed and bread, which just corresponds with what we have always claimed. The amount required for bread and seed this year will be a little over 300,000,000 bushels and no more.—*Baltimore Journal of Commerce*.

IMPORTANCE OF HAVING A GOOD QUEEN.

In every season the queen must be accompanied by worker bees sufficient to produce and retain an increased temperature in the hive. As the queen is not assigned to build the breeding cells, or furnish the brood with food, the workers attend to that for her. She is rightly called the "mother of bees" because she gives life to all the young bees that exist in the colony, by producing the eggs which develop into the future workers and drones. The success of the colony and its perpetuation depends upon the fruitfulness of the queen. If a weak colony be given a prolific queen, it will quickly increase to a strong one and the strongest colony will soon be reduced to weakness if the queen produce few or no eggs, either on account of advanced age or other defect. Being aware of these facts the apiarist should tenderly care for his queens and especially to winter only such colonies as have very fruitful, faultless and not too old queens. Many queens are nearly useless, even when young, and others still prolific in old age, but the latter are very liable to lose their strength and fruitfulness at a very inconvenient time, when they should be depositing the most brood and when substitution is very uncertain on account of the scarcity of drones. In consideration of this it is very advisable to supersede a queen in about the third summer; and the most favorable time is when the bees are swarming.

The first part of this operation will be the most difficult, especially if the colony is very numerous or has gathered much honey. One method which has been recommended is to allow the colony from which you wish to remove the queen to become quiet, then quickly remove her, place the young queen in the hive and the superseding is over—before the colony fairly realizes her presence. But there is no surety of success in this method, for the bees are often so attached to the old queen that they will not brook substitution, and immediately destroy the intruder.

There are so very many plans given for catching the queen that the operation has become so simplified that it can easily be accomplished in the strongest colony. It is not necessary to look all over the combs and in every corner of the hive to find the queen, but you can easily locate her upon a comb, in any part of the hive. This is not done by inserting combs of honey from the cells but is fed by the bees—but by giving her an opportunity to deposit eggs without disturbance, especially drone eggs, which occupation best pleases her majesty. For this purpose choose empty brood-combs, or such as are partially filled, for the queen will be in haste to occupy all space and fill these cells with eggs in order to close the brood. If you will examine that hive in twenty-four hours, without creating disturbance, you will, in nearly every instance, find the queen on this comb.

To get a queen out of a box hive, about the only way is to drum the bees out and allow the queen to pass out with them. There will be no difficulty in discovering an Italian queen from her golden color, for she excels the worker bees in brightness. The astronomer does not have to search the heavens when seeking Venus, Jupiter, or Mars, for they so far surpass the surrounding planets in brilliancy that they catch the eye at a glance. No more does the bee-keeper seek in vain his Italian queen, and in queen-rearing this is quite an object. After the queen is captured and the colony becomes fully aware of its loss, the bees will build queen cells and rear a successor. We may also expect some "after swarms;" and the first one will probably appear in about fourteen days, the time being varied by the strength of the colony.

But to those bee-keepers who are not seeking an increase of colonies but rather depend upon the honey harvested for their profits, the method we have given would be of no value. Such bee-keepers must immediately place a young queen in the colony from which the queen has been removed, in order to prevent after swarming and cause as little disturbance among the honey gatherers as possible. The new queen must be caged at least 24 hours, when introduced; some prefer placing a queen cell in the hive that is nearly developed, but this requires skill and patience. I have recently tried—and with much better success—hanging the entire comb containing the queen-cell in the hive which contains no queen. Queen cells are not scarce in the swarming season; every colony which has produced an early swarm will contain several queen cells which must be used at just the proper time—that is, when nine or ten days old, for if delayed longer, some may have fully matured, and if the bees are not inclined to swarm these new queens may destroy those remaining undeveloped, by biting through the cells. The bees usually place the queen cells upon one or two combs; attention is necessary to distribute them sufficiently, that every queenless colony may be supplied with comb containing one or more queen cells. This method of superseding queens is certainly very simple and practical, as well as expeditious. Very little disturbance is created among the bees, and scarcely any interruption of labor. The young queen will soon become fertilized and commence depositing eggs. Should she

by any means be lost or destroyed during the wedding flight, a new queen cell should be immediately inserted, and care should be taken to select one nearly matured, that the bees may not become too much excited.

DRAINING OF LAND.

Notwithstanding all that has been said and written upon the subject of underground drainage, it has not yet become a popular operation on our farming lands. There are various obvious reasons for this. Many persons have doubts regarding its value. The expense of thorough drainage is considerable, although the labor of digging and cost of materials is often exaggerated, yet the difficulty of procuring either is, in most localities, sufficient to deter those who have no practical experience in the execution of the work, and who cannot avail themselves of intelligent supervision, for in this, as in all other practical operations, very much depends upon the economic application and direction of labor.

The experience of practical drainers, both in this and other countries, proves beyond all controversy the great advantages which accrue from the thorough drainage of all soils. Even in lands not particularly retentive of water the effect of underground ventilation or aeration is evidenced by the increased capacity for production. With a drained soil the cultivator is prepared either for a wet or dry season, for it is well established that draining increases the capacity of the soil for retaining moisture or moist air, which is precisely what the roots of plants require. It is a mistake to suppose that draining actually has the effect of drying land to the extent of depriving it of all available moisture. The reverse is nearer the truth, that there is more available moisture for plants in drained than there is in undrained land. Every description of soil has its relative degree of porosity or power for retaining moisture. Peaty or mossy soils, mainly composed of partially decayed organic matters, are the most porous, and consequently are the greatest absorbents of water, while compact clayey soils have this capacity in a very limited degree. Draining a peaty soil will not deprive it of porosity; it may be likened to a sponge which will retain all the water that may be poured on it until its pores are filled, but no more; so draining relieves the soil of superfluous moisture that cannot be retained or held in suspension by air, and which, if not removed by percolation can only be removed by the slow process of surface evaporation.

Clay soils cannot produce to the full extent of their ability unless underdrained. The ordinary routine operation of plowing has a tendency to form a compacted strata immediately below the cultivated or plowed portion, which acts as a basin in the retention of water. Such soils are cold and late, because the water prevents the heat of the sun from warming the soil until the water has been removed by evaporation, which produces cold; so that in addition to the impracticability of early spring cropping of such soils, every summer shower cools the earth surrounding the roots of the growing plants, which thus sustain a series of checks in their progress to maturity.

These evils are removed by draining. Even the strongest clays are more or less permeated by veins of sand or gravel, sometimes by a layer of vegetable matter which has collected in a crack or fissure, but so long as there is no outlet beneath, these conducting veins are inert, but when underlaid with drains their action is at once apparent; the subsoil that previously held water like a basin now transmits it like a filter, and as the water sinks the air follows; the rains descend freely through the soil, carrying to the roots the nutritive elements with which it is charged; the absorbing property is increased, it holds more moisture in suspension, and crops remain luxuriant even in seasons of drouth, and superfluous water being removed from below, the heat of the sun is economized in warming the soil, instead of being expended in the evaporation of surface water.

Briefly it may be stated that some of the advantages of underdraining consist in the removal of stagnant water from the surface, and excess of moisture from heavy rains; the temperature of the soil is increased, which allows early planting of crops, and hastens their maturity; it equalizes the temperature of the soil during the growing season; it equalizes the moisture of the soil; so that crops are in a great measure exempt from the evils resulting from excess of rainfall on the one hand, or from a deficiency of rain-fall on the other; the roots of plants are supplied with soluble food carried down by rains, as well as that which is rendered available by the decomposing influences of air and moisture on the surrounding soil, and on such manures as are applied for additional fertilization; the land is more economically worked, and cultivation suffers less interruption at all seasons, and as a consequence crops are increased to their maximum production, at least so far as they are dependent upon the physical condition of the soil, a factor of equal importance with that of its chemical constitution.—*Wm. Saunders.*

THE PRACTICAL AND THE SCIENTIFIC IN AGRICULTURE.

The earnestness and zeal with which agricultural investigations are conducted in our day, indicate not only great intellectual activity in the agricultural community, but they indicate also a deep desire to ascertain the best fountains of agricultural knowledge. The work of the college is becoming more and more acceptable, but the work of agricultural associations is becoming more and more useful. The experimental science of farming, that science which, without exercising undue curiosity with regard to the laws of nature, observes and collects all the facts which may guide us in such an observance of these laws as will secure our prosperity and success in farming. And ;this is the science we most need; "a science founded upon the accumulation of facts and the accumulation of experiments." For, as the Duke of Argyle recently said, "we can never have agricultural science unless we know the facts with which we have to deal. So long as we want a system of agricultural statistics, we are deficient in one of the very bases upon which an agricultural science can be founded."

Of the value of this kind of science to the farmer the most enterprising agriculturists have long been aware. The foundation of all knowledge of agriculture is the accumulation of fixed facts, suggested perhaps by accident, discovered perhaps by science; but, however obtained, proved or confirmed by the practical farmer on the land. A theory which bears this test may become a law at once for the farming community, and until it has borne such a test it is theory still, no matter what its origin may have been, whether college or farm yard. While, therefore, an agricultural school may be devoted to science as a guide to agriculture, and may be engaged in cultivating a single farm according to the best known principle, it must depend upon a widespread community of farmers for the last grand process of proving and diffusing its theories. And when we remember that agriculture is not an exact science, and cannot be until the skies and the seasons are subdued by man, and that the facts discovered in the field by the diligent cultivator are often of more practical value than those laid down by the student in his closet, we shall not be surprised at the success which associated farmers have met with in the work of advancing agricultural education. In fact the most substantial and useful literature of agriculture goes to prove this.

The books to which the farmer turns most eagerly for knowledge are those which contain the facts which now constitute the treasury of his library: Arthur Young, traversing all England for the materials out of which to write his admirable volumes, Jethro Tull toiling with his own hands to extract from the soil itself the doctrines of horse-hoeing and drill husbandry with which to enrich his native island; Mr. Cully, devoted to the improvement of cattle as the best college in which to learn how to discuss their breeding and feeding; Fitzherbert, who, although justice of common pleas, was as he tells us "an experienced farmer of more than forty years," and wrote the "Books of Husbandries;" and so the admirable writers of modern days all write from the great standpoint of experience, What richer fountains of agriculture knowledge can be found than the transactions of our agricultural societies? Where can a better lesson be read than is contained in those modest volumes issued annually and containing the recorded experience of successful farmers? We turn to this fountain of knowledge with confidence, and we turn from it with new light and courage for the pursuit of farming. What a treatise on sheep-husbandry might be written by sitting at the firesides or roaming over the pastures of our great wool-growing States and taking notes of the experiences and labors of the farmers there? What fund of information upon the cultivation of crops, the management of orchards, the use of manures, the conduct of the dairy, lie concealed in the farm-house everywhere? It is a combination of this practical and economic science which should be the desire and motto of every farmer's association, and is the foundation of the farmer's best knowledge.

Let the example thus set be followed always and everywhere. Let our scientific teachers learn to respect the practical knowledge of the farmer, and let the farmer lay aside his jealousy of the learning of the schools. To this just and proper combination of mental forces how would the earth unfold her secrets! how would the fields rejoice under welldirected cultivation; how would the whole animal economy of the farm be developed and improved; how would the whole business of agriculture be brought into subjection to systematic laws. Without this combination, deprived of this accumulation of facts, science in agriculture becomes powerless; with it, it becomes a most important ally to the farmer; in fact, it is reduced to one mode of practice itself, and meets with the highest success. For in whatever the farmer does he is obliged to recognize an influence which the hand of man cannot reach, which no investigation can fathom, no human power guide. Agriculture obeys the laws of nature; science endeavors to ascertain and explain them. Science may attend upon agriculture as a guide and stimulus to the best exertion; but it is the patient and prudent and experienced farmer who knows what land he needs, what crops he can raise, what fertilizers he requires, and what labor he can best apply. It is the union of practice and science which makes farming perfect.—*Hon. Geo. B. Loring.*

OUR LOCAL ORGANIZATIONS.

LANCASTER COUNTY AGRICULTURAL AND HORTICULTURAL SOCIETY.

The Lancaster County Agricultural and Horticultural Association held a stated meeting on Monday afternoon, September 4, 1882, in their room in City Hall.

The following named members were present: John C. Linville, Salisbury; S. P. Eby, Esq., city; James Wood, Little Britain; W. B. Paxson, Colerain; Daniel Smeych, city; F. R. Diffenderffer, city; C. H. Gast, city; Peter Hershey, city; J. M. Johnston, city; J. Frank Landis, East Lampeter; Johnson Miller, Lititz; Levi S. Reist, Oregon; Phares Buckwalter, East Lampeter; Eph S. Hoover, Manheim; W. H. Brosius, Drumore.

The president being absent James Wood was called to the chair.

John C. Linville stated that Henry M. Engle had informed him that he had corresponded with Prof. Thomas M. Edge, of the State Board of Agriculture, and that Prof. Edge had consented to deliver a lecture before the society at its next stated meeting in October.

Crop Reports.

Levi S. Reist, of Oregon, reported that the crops in all sections of Lancaster county were good; the wheat and hay had been garnered and produced bountifully; corn and potatoes promised equally well, and even the tobacco, which only a short time ago it was feared would be almost a failure, is turning out unexpectedly good. Within the past ten days it has grown wonderfully; the leaves are large and clean, of good color and apparently good quality. The peach, pear and apple crops in his neighborhood are quite fair, and taken all in all, the farmers have to be thankful for as prolific a harvest of all kinds as he has any recollection of.

J. Frank Landis, of East Lampeter, reported that there would be about three fourths of a full crop of corn and potatoes in his section; the late tobacco is growing finely and will yield much better than was expected; apples are scarce and imperfect, and grapes are rotting on the vines.

John C. Linville, of Salisbury, agreed that most of the crops were good, as represented by Mr. Reist, but in his neighborhood the oats was not a good crop. He does not grow tobacco, but he has seen it in his neighbors' fields, and never before saw such an improvement as there has been within the past ten or fifteen days. Notwithstanding the long drouth, the grass and clover fields look well—the young clover looking better than he has seen it look within the past four years. Apples are diseased and are dropping off—scarcely a perfect one is to be found on the trees.

W. E. Paxson, of Colerain, had never seen the prospect of the corn crop better; the clover and pasture fields also look well and the late rains have greatly improved the tobacco, which looks very well; apples, with the exception of the russets, are dropping off.

Johnson Miller, of Lititz, said the late rains have greatly improved the corn and tobacco, and the latter now looks very well; the wheat crop was excellent; the apples are rapidly dropping off; the peaches are being attacked on the tree by bees and wasps: the grapes are rotting on the vines in some places; the young clover looks better than he has ever seen it at this time of the year; some farmers are cutting their tobacco, which looks as if it would be about an average crop.

Peter Hershey said he thought the tobacco in Salisbury was better than that nearer Lancaster, though everywhere there has been wonderful improvement within the last ten days; the corn is healthy, but the ears are short; grass, timothy and clover look very well; apples are scarce and falling off, except the Baldwin, Smith's Cider and a few others, which contain some fine fruit; wheat turned out very well and oats poorly.

James Wood, of Little Britain, said the wheat was excellent, and believes the corn will be equally good; of oats there was not more than half a crop; his own yielded 21 bushels to the acre; not many apples are grown in his neighborhood, and no peaches; he don't grow tobacco, but sees some very good crops in the lower end; the hay was good and potatoes will yield over an average crop. Take it all in all this is one of the very best seasons the farmer has ever had.

Selecting and Breeding Dairy Stock.

Mr. W. E. Paxson, of Colerain, read the following essay:

In respectful obedience to your request I will present as briefly as may be some remarks based on the subject of "selecting breeding and dairy stock." The milking qualities of our domestic cows are to some extent artificial, the result of judicious care and breeding. In the natural or wild state the cow yields only enough to nourish her offspring for a few weeks and then goes dry for several months or during the greater part of the year. There is therefore a constant tendency to revert to that condition which is prevented only by judicious treatment, designed to develop and increase the milking qualities so valuable to the human race. If this judicious treatment is continued through several generations of the same family or race of animals the qualities which it is calculated to develop become more or less fixed, and capable of transmission; and instead of being exceptional or peculiar to an individual they become the permanent characteristics of a breed. A knowledge of the history of the different breeds, and especially of the dairy breeds, is of manifest importance and will aid the farmer perhaps in making an intelligent selection with reference to the special object of pursuit. In selecting any breed, therefore, the farmer must select that breed or herd which is best adapted to that branch of dairying which he pursues. An intimate acquaintance with the various breeds of cattle known among us has led us to distinguish the most prominent breeds—especially those adapted to the dairy; and the importations of these famous breeds have been so frequent and extensive in the United States within the last few years that they are now pretty generally diffused over the country and within the reach of every farmer. If the dairyman is selling his milk the cow that will yield the most milk will be the profitable one. If he is making butter then he must have a large yield of cream of the best quality, no matter what the flow of milk; and the cow that will be the profitable one. How careful then should he be in his selection, and to breed from that stock with the hope of improvement.

With the dairyman the cow is the machine that manufactures the dairy goods. Then should he not be as careful in selecting his herd of cows as the manufacturers would be in selecting machinery to work out his fabrics? And yet the manufacturer gives greater attention to his machinery than the dairyman to his herd of cows. Nearly one third of all the cows kept by dairymen in the county produce less milk than will pay their keep. They are simply a clog upon the business, and the sooner they are disposed of the better. Does not this important matter behoove every one engaged in the dairy business to set himself at work in weeding out the poor cows so that they may be able to reap larger products from their dairies? Test each cow by herself and see how many do pay profit enough to retain them, and get rid of these small profit ones as soon as possible. It is necessary that we should endeavor to improve our dairy stock; and how can this be done? Not surely by indiscriminate crossing. This is why we have so many mongrel herds, and why so large a share of their progeny is so poor. Our object in crossing should be improvement, and we ought to know when we make a cross whether we are likely to get it. We must be careful not to in any way deteriorate the form and health of the animal, and the quality of the milk, as well as not to lessen the flow. As far as possible it is best to breed our herds from pure blood. A great part of the art of breeding lies in the principles of judicious crossing, for it is only by attending properly to this that success is to be attained. All eminent breeders know that ill-bred animals are unprofitable, and the old saying still holds good that "like begets like." If this be true, which doubtless it is, then how careful we should be not to breed from an inferior animal; and I hope the time is gradually passing away when the intelligent, practical farmer will be willing to put his cows to any mere runt of a bull simply because his services may be had for 25 cents. A calf sired by a pure-bred bull, particularly of a race distinguished by firmness of bone, symmetry of form, and early maturity, will bring a much higher price at the same age than a calf sired by a scrub.

In closing, let me make one remark in regard to the treatment of our cows. There is an old adage among the Germans that "the cow milks only through her throat." Never was better said. Alas, how many of us forget this, and instead of giving them the treatment and care which they deserve, they are neglected. The productiveness of the cow does not depend on her breed so much as on her food and management. Proper shelter and good nutritious food should be provided for her during the winter months and then we may expect good results.

It is fortunate, indeed, that wiser and more humane ideas prevail with regard to the care of stock of all kinds, now, than that treatment which it received in the early history of our country, when many thousand perished from exposure and starvation. And I hope that the idea which was so prevalent among our farmers once who styled themselves "practical farmers," is thoroughly rooted out never to be practiced again that cows and young stock should remain outdoors exposed to the cold winter days in order that they might be toughened. Is not this an erroneous idea? No thrifty farmer will subject his stock to such treatment with such an object in view.

The Question Discussed.

J. C. Linville thought the essay much to the point and furnished much information that would be of value to the farmer as well as to the dairyman. We don't pay enough attention to our dairy stock; the margin of profit on dairy products is small at best, and the difference between a good and a bad cow is the difference between a moderate profit and an absolute loss. We should dispose of all cows that do not pay for their feed; if we have good butter cows we should raise our stock from their calves and not buy our cows from the West.

S. P. Eby thought the essay contained a number of valuable suggestions. The proper care of cattle is important; the thoroughbred is, as the essayist says, an artificial animal and will run back to its natural state unless judiciously treated and kindly cared for. We should devote as much attention to our cows as we do to our horses. The one should be curried and kept clean as well as the other. They should not be merely well sheltered in winter, but their stalls should be kept clean and well ventilated.

In answer to the question as to what is a thoroughbred, Mr. Linville said it is an animal that has been carefully bred for many generations until its type is fixed—whether that type be Alderney, Jersey, Guernsey, Devon, Durham or other breed. Some of our native cows are as good or better than the thoroughbreds, but the difficulty is the type is not fixed, and three times out of four the calves will not be of the type of the cow. In answer to another question, Mr. Linville said the Jersey and Guernsey breeds were much alike, and that no Alderneys are imported into this country.

Mr. Paxson said the Guernseys were rather larger than the Jerseys, and of not quite such fine points or color; for milk and butter production they are much alike. He thought farmers should raise their own stock from carefully-selected bulls and cows.

Levi S. Reist said that notwithstanding the great advantages resulting from improved breeds of cattle, Lancaster county farmers have always been slow in introducing them. They stick to the old, common breeds, saying that cattle-raising don't pay anyhow, and that our land is too dear to devote to stock-raising. He agreed, however, with what the essayist had said.

On motion, the thanks of the society were voted to Mr. Paxson.

How to Secure Better Meetings.

Johnson Miller said some plan should be adopted by which a better attendance at the regular meetings of the society could be attained. But very little progress has been made in this direction during the past ten years. He suggested that if the meetings, instead of being held in the city every month, were held at the residences of the members in different parts of the county, the attendance would be much larger and meetings more attractive. This plan has been adopted by the Octoraro Farmers' Club, with good results. He had attended one of their meetings at which a large number of farmers were gathered, and a pleasant and profitable time followed. Our meetings are held on Monday, always a busy day, on which those members who do get to town have a great deal of work to do and rarely reach the meeting before half past two o'clock, and some of them, to reach home by the cars, have to leave before the adjournment, as was his case now. He had no motion to make, but asked the society to consider the suggestion.

S. P. Eby moved that the consideration of the matter be postponed until next meeting so that members could have time to think over it. The present meeting was a small one, and it would not be advisable to act hastily. Mr. Eby's motion was agreed to.

Cutting Corn-Fodder for Feed

The question, "Will it not pay the farmer to cut his corn fodder before feeding it to his cattle?" was answered by Peter Hershey, who said that his experience was that it did not pay to cut the fodder; he had practiced cutting it for several years but had quit it. It is true that the fodder is more easily handled when cut and when it finds its way to the manure pile it rots more rapidly and makes shorter manure, but the labor and expense of cutting it over balances the advantages gained.

John C. Linville said he had a good fodder cutter and used it two or three years and liked it; the cornstalks have not much nutriment in them, but they help to fill the stomach and will do to eke out a short hay crop; but the labor of cutting the fodder costs more than it comes to, if you have to hire help to do it; labor is expensive and unless you have plenty of spare time of your own he would not recommend cutting the fodder.

Streaks in Butter.

"What is the cause of streaks in butter?" was answered by J. Frank Landis. He said the streaks resulted from imperfect working and washing of the butter, and they might be avoided by care in this respect. The butter should be made of good cream from good cows, the cream should be at a temperature of 55 or 60 degrees when churned, and after being churned should be carefully worked so as to remove all the buttermilk and the best quality of salt should be used. Mr. Landis read from an agricultural journal a paragraph which substantially agreed with his own views.

How to Sow Timothy Seed.

"Is it better to sow timothy seed before or after the drill?" was answered by Levi S. Reist, who said he would prefer sowing it after the drill, and before a fall of rain. Unlike some other farmers he could not grow timothy from seed sown in the spring.

The best Variety of Wheat.

"What kind of wheat should we sow this fall?" was next discussed.

J. Frank Landis said a majority of the farmers in his neighborhood who have been growing the Fultz variety are giving it up and going back to the old Mediterranean or Lancaster variety, which brings a better price at the mills than the Fultz wheat.

Levi S. Reist said the farmers in his neighborhood were also giving up the Fultz wheat. In Mount Joy they are introducing a new variety, the name of which he had forgotten. Samuel Hossler has the seed and it is said to be very prolific.

James Wood said there would be very little change of seed in his neighborhood. The farmers were about ready to give up the Fultz, but it has done so well this year that they are going to give it another trial. There is a new variety being introduced by some farmers called the Russian wheat, for which $5 per bushel is asked. It was introduced into Lancaster county by New York seedmen who were so well pleased with it that they bought up for seed all the crops that were sown.

J. C. Linville said the shumaker wheat, being introduced by some farmers, has red chaff and red grain, and is better for milling purposes than the Clauson or Fultz; It stands the winter better than either, but is more liable to be attacked by the Hessian fly; the straw is tall and liable to lodge.

J. Frank Landis said the Clauson did not do well on heavy soil; it ripened too late and the seeds did not mature as well as other varieties.

Following Corn with Wheat.

"What is the best method of preparing corn ground for wheat?"

Levi S. Reist answered, cut the corn stubbles off close to the ground and drill in the wheat without plowing.

J. Frank Landis endorsed this plan : he cultivated his corn as late as August, kept the ground as level as possible and after the corn is off drill in the wheat without plowing.

J. C. Linville said that no answer will apply to all cases. He had not succeeded in getting good rye or wheat on corn ground without plowing. His soil is too heavy. He would plow the ground, roll it, harrow it and roll it again. If the soil is loamy it is not necessary to plow it, but in heavy soils inside plow by all means.

James Wood harrows in the seed without plowing, but has known both plans to succeed and both to fail. Much no doubt depends on the soil and the season.

Forestry.

Levi S. Reist announced that he had received from the Lieutenant Governor of Canada an invitation to attend the Forestry Convention, and in connection read from a paper some startling statistics showing the wonderful consumption of wood in the construction of our railroads—one fact being that the ties alone of the railroads in the United States if placed end to end would reach to the moon and back again.

Fruits on Exhibition.

F. B. Diffenderffer, Levi S. Reist and J. Frank Landis were appointed a committee to test and report on the fruit brought to the meeting. Their report was as follows :

The fruit on exhibition consisted of one plate of seven peaches, large in size and fully ripe ; three plates of seedling peaches of the Sener variety, all resembling the parent in general appearance, although not quite so large ; the one marked No. 2 was rather better flavored than the Sener itself. All are worth cultivating, and are very handsome in appearance. These were all exhibited by Daniel Smeych of this city.

Mr. Smeych also had a plate of Telegraph and Champion grapes, very fine in appearance and well flavored. Also, a large foreign plum of a fine yellow color and handsome appearance.

S. P. Eby Exhibited one of Rogers grapes, but not being fully ripe the particular variety is unknown. Also, some Bartlett pears from a tree planted in 1870; also, a Benoni apple from a tree planted in the fall of 1875; also, a pear of the Brandywine variety, and some early Crawford peaches of large size; also, Clapp's Favorite pears from a tree planted in 1879; also some very handsome trumpet flowers of reddish color and growing abundant clusters.

Mr. Reist had two seedling apples for name, but your committee are unable to pronounce definitely upon this. They are of medium size, reddish in color, and of a pleasant taste. They are very juicy and would make fine cider apples.

Berks County Fair.

Joseph F. Witmer, Eph. S. Hoover and Calvin Cooper were appointed delegates to attend the Berks county fair, commencing on the 24th of September.

POULTRY ASSOCIATION.

The Lancaster County Poultry Association met statedly in their room, in the City Hall, at half-past ten o'clock on Monday morning, September 4, 1882, with the following members present : George A. Geyer, Spring Garden ; J. B. Lichty, city ; Charles Lippold, city ; J. B. Long, city ; J. M. Johnston, city; J. E. Schum, city ; C. A. Gast, city ; and F. R. Diffenderffer, city.

The minutes of the last stated meeting were read and approved.

Peter Brunner, of Mount Joy, and Mrs. Theodore H. Patterson, of Safe Harbor, were elected to membership in the society.

Secretary Lichty reported that he was receiving much encouragement in securing advertisements for the catalogue. A number of special premiums were also being offered, among which was a silver cup valued at $25.

Several members reported that they had sent a number of pigeons and chickens to the exhibition of the State Agricultural Society, at Pittsburg.

After an informal discussion on the subject of ducks, chickens and pigeons, at which nothing of interest to the public was developed, the meeting adjourned.

STATE BOARD OF AGRICULTURE.

The next meeting of the Pennsylvania Board of Agriculture will be held in Washington, Pa., beginning on Wednesday, October 18, and continuing several days. A large number of delegates from different parts of the State will be present. These meetings are open to everybody, and, as the subjects all interest our farmers, it is expected that they will be in attendance in large numbers. Each subject will be treated by a person who knows what he is talking about. Among the essayists is Henry M. Engle, who will read a paper on "Ice Houses for Farmers."

FULTON FARMERS' CLUB.

In accordance with their established custom, the Farmers' Club, of Fulton township, held their annual fair and pic-nic on Saturday, September 2, at Black Barren Spring, in Fulton township, one of the prettiest places in the lower end of the county. Though it threatened rain all morning the sturdy farmers with their wives and daughters began to come in at an early hour and by noon fully four hundred persons were on the grounds, which were prepared for the occasion. A grand-stand had been erected, which was handsomely decorated with flowers and greens and in front of it were placed rows of benches for the accommodation of the audience. Near the stand were long tables, which had been placed under the trees, and these were soon made attractive by the large variety of articles placed on exhibition by the farmers as they came in. In another part of the grounds were the fanning implements and the live stock. Though no premiums are offered, those who attend these gatherings take enough interest in them to make a very creditable display as the following list of

Articles Exhibited

with the names of the exhibitors will show :

William King, of Kirk's Mills, exhibited three varieties of apples, two varieties of grapes, corn and Worster's Cabinet Creamery.

May H. Stubbs, Wakefield.—Three varieties of grapes, plums, double squash, box of honey and a cup of crab apple jelly.

Josiah Brown, Lyle.—Corn, cucumbers, Jerusalem cucumber and tomatoes.

Lyman C. Blackburn, Pleasant Grove.—Watermelons.

May Morgan, Goshen.—German wax beans and corn.

Grace A. King, Lyle.—Two varieties of grapes, tomatoes, cabbage, apples, peas, beets and canned peaches.

J. R. Blackburn, Pleasant Grove.—Peas, sweet potatoes and wheat.

Rebecca D. King, of Kirk's Mills. Jelly, canned plums, and canned gooseberries.

Rachel Gibson, Little Britain.—Preserved tomatoes.

Jacob Moore, Lyle.—Peas and beets.

Deborah Jackson, Wakefield.—Can of apples.

Lindley King, Wakefield.—Two varieties of apples and one of sweet potatoes.

W. P. King, Wakefield.—Lima beans, acme tomatoes.

Kirk & King, Wakefield.—Canned tomatoes and canned corn.

Lauretta A. Kirk, Wakefield.—Large beet.

Thos. P. King, Wakefield.—Corn.

Gilpin Reynolds, Wakefield.—Watermelons, citron and beet.

Philena Reynolds, Wakefield.—Canned and preserved peaches.

Joseph Brown, Wakefield.—Early Rose potatoes.

George Balderson, Colora, Cecil county, Md.—Forty two varieties of apples, cut flowers, and potted strawberry plants.

S. L. Gregg, Greene.—Four varieties of apples, five varieties of pears and grapes.

Haines, Brown & Bro., Lyle.—Green fox grapes, potatoes, short-horn bull and four short-horn cows, and corn on which the ears grew fully eight feet from the ground.

Neal Hamilton, Goshen—Buckeye cultivator, Davis swing churn, four varieties of corn egg plant, three varieties of tomatoes, lima beans, nasturtions, and Hartford prolific grapes.

Sallie Hamilton, Goshen.—Cabbage and hard soap.

Wm. Ingram, Pleasant Grove.—Bartlett pears and sweet potatoes.

Willie N. Hamilton, Goshen.—Oddities in potatoes.

Jonathan Pickering, Little Britain—Seven varieties of grapes and Fultz wheat.

S. S. Herr, Pleasant Grove.—Potatoes.

J. W. Thompson, Pleasant Grove.—Odessa or Russian white wheat and Root's corn planter.

Melissa Tucker, Harford county, Md.—Leaves of Egyptian corn and twenty varieties of cut flowers.

Alvin King, Wakefield.—Five varieties of potatoes, apples and tomatoes.

Jos. C. Stubbs, Peters Creek.—Two varieties of wheat and pen of Southdown sheep.

Annie Hamilton, Goshen.—Corn husk doll and tea cake.

Ella Tucker, Harford county, Maryland.—Concord grapes.

C. H. Stubbs, M. D., Wakefield.—Christine grapes and Rhubarb.

Wm. A. Johnson Oxford.—Estey Organ.

Isabella Reed, Pleasant Grove.—Two coops Plymouth Rock chickens.

Emmor Smedley, Wakefield.—Short-horned bull.

A. C. Jenkins, Rock Springs, Cecil county, Maryland.—Missouri grain and fertilizer drill.

Jas. C. Bird, Rising Sun, Maryland.—Spangler fertilizer attachment for grain drills and the Success pump.

A. M. Brown, Pleasant Grove.—Wheat and oats.

Watson Reeder, Rising Sun, Md.—Penn Revolving harrow.

Howard Coates, Little Britain—Acme barrow two varieties of potatoes, lima beans, and Livingston tomatoes.

Jos. A. Roman, Colora, Cecil county, Md.—Mangel wurzel beets.

R. L. Flaherty, Pleasant Grove.—Peerless potatoes.

Harry Reed, son of George K. Reed, of this city, who is now boarding with J. Wesley Thompson, the lessee of the farm, had on exhibition two of Mr. Black's Alderney cows, "Belle" and "Maggie," which were cared for by him and are in good condition.

The morning passed very quickly, all present seeming to enjoy themselves as only farmers and their families can when thrown together. They examined each other's exhibits and compared notes, each being benefited by the experience of the other, and the implements were constantly surrounded by small crowds of men eagerly listening to the explanations made by the owners.

Address of Welcome.

After dinner an hour or more was spent in social converse, when the president, William King, called the meeting to order and introduced James Black, Esq., of this city, who on behalf of the Fulton Farmers' Club, extended thanks and a cordial welcome to all pressent and more than all to those who have tried to contribute to the success of the fair, by bringing with them the products of the neighborhood. The face of the members of the club are not known to the citizens of Lancaster, but the published reports of their proceedings are read with great interest, and the little society is held in high regard. This club and like associations are of value to the public as well as socially, and the country is better in every way for them. They will be found to exist only where there are educated, moral and intelligent people. The speaker said he might talk on the common theme that agariculture, farmers and farms are the basis of wealth, but the main thought with him was that these organizations are making better farmers, better men and better Christians and that they can thus band together and destroy every foe to their homes and happiness. Therefore he hoped that if they met together again in another year the club will be enlarged in numbers, and that the exhibit will be as much larger as this exhibit is larger than last year's. Addressing the young men and women he asked why they could not have in that place a counterpart of the Oxford fair or the county fairs. He believed they could have there exhibited the finest specimens of all that is produced on the farm, and standing on that platform men to address them who have given the science of agriculture the attention of their lifetime. He closed by thanking his hearers for their close attention.

Seth L. Kinsey,

of Harford county, Md., an intelligent young farmer was the next speaker introduced by the president to address his "friends and fellow tillers of the soil," and in the beginning he spoke of the purpose for which this assemblage was held as being most worthy and commendable; not a meeting to further political schemes but one where there exists only a generous ambition to excel in the condition, quality, and perfection of the articles placed on exhibition and a general desire to pass a day of social enjoyment, instruction and improvement. How great is the delusion of those persons who suppose that farm life is rude and uncultured, and attended only with toil and weariness. Who can look on this assembly, on the evidences of civilization and advancement we see about us in the fertile fields the grazing herds, the handsome yet substantial and useful buildings and remembering the wonderful art and skill there represented yet say that farm life is rude and uncultured. But we need not stop here for greater and more wonderful and beautiful than all, the works of Nature surround us, offering us ideas, teaching us lessons, and leading us into paths that tend ever to our moral, intellectual and spiritual advancement. There are two ways of farming, as perhaps of doing everything else—a right and a wrong way. The first leads to abundant success, and the second to discouragement and ultimate failure. Farming is not as many appear to suppose, merely a laborer's occupation, in which animal force is alone required; it is a science, and the man who makes farming a success has as good a right to be proud of what he has done as the most worthy and eminent of our merchants or professional men. It is not for us now to attempt to suggest how farming is to be made a success, but one thing is very certain—different soils need different kinds of fertilizers to enable them to produce satisfactory yields, except in the case of manures, which are suited to all soils, and upon which too much value cannot be placed or too much care exercised to collect as large a quantity as possible. They may be justly regarded as the farmer's savings banks, while chemical fertilizers, which we cannot do without at present, are surely a drain through which passes much of the profits of farming. The man who expects to succeed at farming must not only work, but also study and observe, and here comes in the advantage of establishing farmers' clubs in order that the knowledge gained by individuals may be imparted to others, for by this mutual exchange of ideas, opinions, etc., much good must inevitably result. The speaker had read with pleasure and profit the reports of the meetings of the club as published, and by no means the least interesting parts of those reports is that the young women of the community are represented at these meetings and aid in their success by their literary offerings. Woman's presence and influence must always add to the interest and success of every enterprise where business and social life can be combined. In farming communities the wives and mothers who teach their daughters to respect work and consider a competent knowledge of housekeeping as one of the essential parts of their education, will be given them knowledge that is of more vaule then gold, as it will insure to them such happiness as they could never know were they ignorant of these subjects, but knowing are able to fill the positions of wife and mother. The speaker closed by congratulating the management on the success of their meeting and with many wishes for the welfare and success of the club and success of its members.

"Manual Labor and How it May be Encouraged"

was the subject chosen by the next speaker, Washington B. Paxson, of Colerain, and he said he knew of no subject that was of greater interest and importance to the agriculturist. Labor is the bone and sinew of every nation. It was labor that has laid the foundation of our republic; that has cleared away the forest and cultivated the soil and caused it to bring forth the golden harvest; that has built our towns and cities; that has constructed our railroads; in short, it feeds, clothes and defends us. All that man possesses or may expect to possess is acquired by incessant toil. Every path that leads from the great highway of labor is cut out by human invention to shirk duty and is leading and training the rising generation to dislike labor, instilling into their minds that it is a disgrace to work. Such a training is ruinous, and these ideas must be eradicated before the laboring class is appreciated as it should be. Labor would become more honorable if we could do away with those absurd class distinctions which make the occupation of a man the standard of his worth. The man who spends his life in clearing and cultivating a farm, provides for his family and the community, but lives and dies almost unnoticed. Think of his exalted position, his unrivaled industry and frugality, discharging his duties to his family serving his country and honoring his God and tell me that such an occupation is degrading. We esteem too highly the man that gains a fortune through trickery, and moves through the world making a grand display in society. How little the world admires yet how heroic the resolution which prompts the young man to clear land and make a home for his family. How can manual labor be elevated in the scale of employment and encouraged to take a higher rank? In the first place honest labor should receive more attention in our domestic circles, and, secondly, it should be introduced into our schools. It is evident that there is something wrong in our educational system, for to educate a child now is to wean it from manual labor. To educate a young man now is sending him forever from the farm. Education and manual labor are not working harmoniously together and the speaker feared they never will until that aversion and prejudice is removed. How can this be done! Only by a proper system of training. In some of the leading countries of Europe, industrial schools have been established, where agricultural and the industrial arts are taught, not only in the higher but in the primary schools, and these have proved highly successful.

It is now a recognized truth that the successful cultivation of the soil is both a science and an art. The idea that of all vocations in life the tilling of the soil requires the least education and training must be eradicated, for experience has demonstrated that a high degree of knowledge is requisite to determine what kinds of crops are adapted to different kinds of soil and to preserve the the fertility of that soil, and great skill is required in planting, cultivating and securing the crops. What our country wants to-day are practical and intelligent agriculturists: there is a great demand for skilled farm help. If our farmers' sons could be taught to believe that laboring on the farm is as honorable as any other employment, and as much desirable, then this lack of skilled labor on the farm would in some degrees be mitigated.

In conclusion the speaker advocated the formation of an agricultural society in every community where farmers and their families can mingle together and discuss the various modes of farming and exchange ideas. Such associations are instructive and those who belong look forward with bright anticipations to the meetings. Farmers as a rule have too little acquaintance with each other, know too little of what others are doing even in their own neighborhood. Absorbed each in his own affairs, they do not consider the ways in which they might be mutually helpful to each other.

The above are but short abstracts of the addresses which occupied several hours in their delivery. Mr. Dickey, of Oxford, played several fine selections on the organ. and the audience again dispersed through the ground to pass the rest of the day as suited their fancy.

AGRICULTURE.

Pasture Grasses

Pastures should not consist of one kind of grass only, because (1) stock prefer a variety, going from one to the other thus keeping their appetites whetted; (2) because the grasses having different periods at which they mature, one kind having passed its best stage, another comes to its best, and takes its place, and (3) because grasses vary in the degree of standing wet and drouth, hence if one sort is injured, by vicissitudes of the weather, another may be to an equal degree benefited. It should be more the practice to stimulate pastures with special manures. This is as necessary a thing to do as to feed a particular animal freely, because it is falling off in flesh. Among the best stimulants to tardy-growing grass, is nitrate of soda; and this may be used freely on pastures without great outlay, and with prompt and beneficial results.—*National Live Stock Journal.*

Experiments with Green Manuring.

Mr. J. C. Chadbourne, of Vassalboro', has been experimenting with green manuring on a small scale, and with very satisfactory results. He had a piece of land containing about two acres which had been neither ploughed nor dressed for fifteen or twenty years, and was producing not more than five hundred pounds hay to the acre. A year ago last spring he ploughed it, and after thoroughly pulverizing the sod, he sowed upon it at the rate of four bushels of western corn to the acre. The corn grew well and when it was at maturity of growth, he ploughed it under. It was estimated that there was from forty to fifty tons of green fodder per acre.

In April last he sowed the field to clover and Timothy, and harrowed it in; and the last of July he made from the two acres, three tons of excellent hay. When ploughed, on a portion of the field, the plough turned up white sand; on another, black mould, and on the balance coarse gravel. On a part of the field Timothy was in full bloom the last of July and very handsome. Mr. Chadbourne says it was the finest hay he ever cut upon his farm. He proposes to continue his experiments with green manuring and is very much encouraged in his past success. Other parties in Vassalboro' are moving in the same direction, and are making anxious inquiries for the best methods of fertilization by green manuring.

Wheat Raising.

A great strike towards successful wheat-raising was made when the drill was brought into use, and a much greater strike could be made if the drill-hoes were twice as far apart and were made to sow not more than thirty to forty pounds to the acre.

In order to make the greatest possible quantity on a given area, the wheat plant (or any other plant) must have room to carry out its habit and develop according to its nature. One grain of wheat cannot do this on less than sixteen square inches. One kernel should make on an average all over a field at least twenty good heads, and every head should produce at least forty grains, every pound should be made to produce its bushel all over the world. One pound has been made to produce from sixty to one thousand fold. These facts are from thin-sowing. No instance is on record where ,thick-sowing ever produced more than seventy bushels per acre.

Sowing much wheat "to get a good stand" is the worst kind of economy. The farmer loses his seed, and never, in any instance, can make as large a yield as by thin sowing.

The greatest enemy wheat has to contend with is wheat. Instead of giving each kernel about an inch square, as most farmers do, they should in every case reduce the quantity per acre and sow thin enough to give it sixteen. Thousands of instances are on record where one grain has produced from ten to one hundred and eighty good stalks and as many heads without dividing. Last year from seventy six kernels ten and one fourth pounds of good, plump grain was raised. On fifty-one square rods this year I sowed just twenty-eight ounces of picked seed in rows one and two feet apart, and what I have now to show as the product is 19 1-2 bushels of as nice grain as the sample enclosed.

To thick-sowing in every State and locality I can offer many objections, but to thin sowing and cultivation there is not one that can be made tenable.—*A. E. Blount in Germantown Telegraph.*

What of the Future as Regards Grain.

As the decline in wheat has attracted so much attention, and farmers are reported to be holding back for better prices, it may be well to examine a few statistics upon the matter. To us the decline seems only natural, though from its suddenness there may possibly be a temporary reaction. We have a yield of wheat of not less than 500,000,000 bushels and the very superior quality of it will increase its bread making properties to much above the average. For a population of 54,000,000 we need at the outside for bread 243,000,000 bushels and for seed not over 57,000,000 bushels, or a total for all of our home wants of 300,000,000 bushels, leaving a surplus of 200,000,000, all of which is available for export, as the stock of old wheat and flour in the country is fully 40,000,000 bushels—and this is ample for reserves.

Now, with 000,000,000 bushels surplus, what are we to do with it? The highest amount of wheat ever exported was 180,000,000 bushels (flour included) for the year ended June 30, 1881, and the average export value for the whole year was $1.11 a bushel while for the year ended June 30, 1879, the average value was $1.06 a bushel. During the year ending June 30, 1881, we had almost the world for our customers. Nearly every European country needed large imports, both France and England having smaller crops than they are promised this year. The requirements of Europe were greater than they will probably be for the next twelve months, and then we had but little competition. Russia had short crops and consequently a very small surplus, while India had not begun to ship wheat to any considerable extent. Thus, we have the prospect for a smaller demand than in 1880-'81, and more competition from other countries than in that year, while in the face of all this our surplus is 14,000,000 bushels greater than the amount we then exported. In view of these facts the part of wisdom would seem to be to ship our wheat out as freely as possible.—*Baltimore Journal of Commerce.*

What Manure Loses by Heating.

It is not always true that a pile of manure steaming with heat and smelling strongly is losing ammonia. Ammonia is a very volatile and pungent gas and might be known by its peculiar scent, which is freely given off by close, ill-ventilated horse stables, or by the coat of ill-cleaned horses. But it is not often that this peculiar scent escapes from manure heaps; on the contrary it is a more disagreeable odor, similar to that of rotten eggs. This is sulphureted hydrygen, and not ammonia, and occasions no loss to manure except the sulphur. If, in making a manure pile, some plaster is mixed in the heap, all the ammonia will be caught and held by it, and all water contained in the manure will also contain a large quantity (700 times its bulk) of it, and will not give it off at a heat that can be raised in a manure pile. If the manure is left to heat and get dry and "fire fang" or slowly burn to a white, dry, light stuff, then the ammonia is lost and the manure seriously injured.

Good Crops in Alabama.

Never in the country were better crops made than this season, and we may reasonably calculate a brisk business with the business men. All kinds of crops are a certainty, except cotton, and the cotton is forming, growing, and all the bolls found up to the 20th of September will make good cotton before frost ; with an average season for the next 50 days, and notwithstanding the small acreage, the yield will be larger at Selma by 20,000 bales than it was this year. Farmers have very generally produced their own substances, and by next Christmas we predict the farmers of south Alabama will be in a condition they have not been since the war—out of debt and their cribs and smoke-houses full, and of substances the result of their own abor and economy.—*Mobile, Ala., Gazette.*

Magnesia for Wheat.

The author ranks magnesia along with nitrogen, phosphoric acid, lime, and potash. The proportion of nitrogen and of phosphoric acid increases in wheat from time of blossoming to maturity. Lime, on the contrary, decreases, and does not seem to play a very important part in the production of the grain, but along with potash serves chiefly in the development of the straw. Magnesia is more important than lime in the formation of grain. The mean requirements of wheat in order to produce 40 hectoliters per hectare are: Nitrogen, 92.6 kilos; phosphoric acid, 37; lime, 25.2; magnesia, 12.2, and potash, 116.2. The laying of wheat and other corn is not due to deficiency of silica in the stalks, but to a diseased condition, consequent on excessive moisture and deficient sunlight.—*H. Joulie.*

HORTICULTURE.

Keeping Grapes Fresh.

Particularly at this season, when grapes are ripening, the discussion is generally started as to the best method of preserving them through the winter. Some of these methods involve a great deal of labor and after all are seldom successful and rarely worth the labor and expense. Besides, who cares about keeping grapes all winter? Every fruit has its season, and when that comes to an end the desire for it passes. Apples can be kept until July in a very good condition and with very little labor; but who cares for them after April ? It is so with pears—the relish for them disappears at the end of January, about as long as they can be easily kept. Peaches, the season of which is very brief, are canned, and when well done they are liked by a great many people, though we do not think they are growing in popularity with those of a rather fastidious taste—many families never using them at all.

As to grapes, they can very easily be kept in good condition until Christmas, and beyond this are very little cared for. And the best method to put them up is that pursued by the grape-growers of New York and Michigan—the clean, dry pine box, packing them, after removing all the decayed or over-ripe berries, firmly, without the addition of any substance as a protection—putting in the boxes, holding from two to four pounds, only the pure bunches of grapes.

We know of no better way; there may be: and if any of our readers possess it, we shall be glad to make room for telling us what it is.—*Germantown Telegraph.*

Beneficial Effect of Mulching on Berries.

Among the more intelligent horticulturists of this country the plan of mulching the surface for a part of the summer months with some cheap material has long ago been accepted as a wise and economical method for fruit growers to adopt. That such a system will keep the surface soil moist in time of drought, and the soil loose and open during a wet season, there can be no doubt, as any fruit grower who has tried the experiment can testify. While talking to a successful small fruit grower about mulching a short time ago, he said : "If I could find no material to mulch my berries with, I would abandon the business." Another person remarked : "I covered my acre of Kittantony blackberries last year with a heavy coating of salt hay, and the effect was magnificent—large berries and plenty of them—while some of my neighbors who did not mulch suffered severely from the drought." This kind of testimony could be given without limit, as the experience of practical men who have given the subject careful thought and practically tested the value of mulching. Until quite recently strawberries seemed to be the only fruit that was benefitted by mulching, and that more on account of the mulch keeping the berries clean and free from dirt or gravel than anything else. But the usefulness of a mulch is by no means confined to the strawberry ; but where material can be had cheap, there is no question but it would pay well to mulch raspberries, blackberries, currants, gooseberries and pears. Nor is there any question but what the size of fruit would be increased, and growth more uniform. Where the surface is covered with loose hot weather sets in, the mulch will serve a threefold purpose when put on heavy enough. As stated, it keeps the surface soil moist and of uniform temperature during the growing months, the crop of fruit is not checked, nor the growth of wood retarded by an excessive drouth. Again, under a mulch, the surface never becomes compact, no matter how much rain strikes the hay or straw, and then filters through gradually giving the best condition for plant growth. Even on clay ground, where the surface has been mulched for three or four consecutive years, it is difficult to compact the surface.

Taking in Fall Flowers.

The time is approaching when we must do what we can to secure the floral beauties that have been with us the latter part of summer and the first part of autumn. But how many of them will we have to give up to die ? We really needed them only for the summer decoration of the grounds; and we have no place to keep them over winter, and besides this it is certain that in the case of many young plants it will be better in every way than those things which we covered, even if all things suited to that end. Still there are some which we will save "anyhow," and it may be as well to say a few words as to the proper way to go about it.

Of course the leading difficulty is that the plants are so likely to wither up and die away after taking up and potting, and we have therefore to direct our energies to prevent this very thing. The kind of plants will decide the treatment. Some things, like carnation or sweetwilliam, have a mass of small roots in a close bunch, and with this comparatively small tops. These rarely wither, even under rather poor hands. On the other hand, a geranium has very few roots. It seldom comes up but all the dirt falls away, and in an unskillful hand all the leaves would fall, and for the whole winter the plant presents a sorry sight. To prevent such leaves from withering and dying away is the point. Much may be done with these sprawly-rooted things by "watering them well before beginning to lift them, and they should have a thorough soaking. Then, some of the younger and softer leaves should be picked off, for it is these which are the most reckless in drawing on the plant's liquid supplies. Of course the plants must be put into their pots or tubs at once on lifting, to keep them from drying, and the whole thoroughly soaked with water on completion. Then the pots should be set into shade and shelter, where neither sun nor wind can get at them, and where air without the loss of moisture can be given to them. Some plants will not much "miss their move," as the gardeners say, and may be put in the full light after a day or so, while some may need this sort of protection for a week. The rule is to put them into the full light as soon as they show no disposition to wither under a moderate sun.

Save the Peachstones.

Now is the time to be putting these away under ground, so as to have them in good condition to be planted out in beds of rows in the spring. Nor is it well to put it off until next season, thinking to do it then, as it may find you without any to put away. So, now, when they are to be had in abundance, it is better to attend to it at once. If some of the seed of the best varieties are left to grow without being budded there will be a fair chance of their producing fruit of as good, and perhaps better, quality than those that you bud, but generally speaking it is not advisable to depend too much upon seedlings.

A Hint for Window Gardening.

A recent English writer gives the following, which suggests a way in which hardy wood-climbers might be made available for window decoration in winter or early spring:

"Some years ago, as I was passing through a room used only occasionally, I perceived an odor of fresh flowers that surprised me, as none was ever kept there. On rising the curtain of the east window, I saw that a branch of Dutch honeysuckle had found its way between the two sashes at one corner, while growing in the summer, and had extended itself quite across the window; and on the branch inside there were three or four clusters of well-developed flowers; with the usual accompaniment of leaves, while on the main bush outside there was not yet a leaf to be seen. The flowers inside were just as beautiful and fragrant as if they had waited until the natural time of blooming. Since then I have tried the experiment purposely, and always with the same result."

A heavy covering of the ground over the roots of the plants with leaves, and sufficient protection of the stem outside, would allow this method to be practiced in quite severe climates.

HOUSEHOLD RECIPES.

FIG PUDDING.—An excellent pudding can be made of figs, and I think it will be generally liked if well made, as every thing ought to be. Let the figs be cut up and mixed with eggs, flour, suet, milk etc., in the usual current method, and that is all.

TO WHITEN SCORCHED LINEN.—If a shirtbosom or any other article has been scorched in ironing, lay it where bright sunshine will fall directly upon it, it will remove it entirely.

TO COOK TURNIPS.—Pare and slice and boil in as little water as possible. When almost done and almost dry, add an even tablespoonful of sugar to each quart of turnips, and salt to make palatable. When dry and tender mash, add two or three spoonfuls of thick sweet cream and serve hot.

ALMOND CAKE—One and one-half cups sugar, half cup butter, four eggs, half cup milk, two cups flour, two teaspoonfuls baking powder; bake in sheets. Icing: White of three eggs beaten stiff, three tablespoonfuls white sugar, one cup chopped nut meats; flavor to taste and put these between and on top of layers.

PAN DOWDY—Fill a pudding pan with apples—pared, quartered and cored; cover the top with a crust rolled out of light bread dough; make a hole in the lid and set the pan in a brick oven. After it has cooked lift the crust and add molasses or brown sugar, a little powdered cinnamon and nutmeg to taste, also one tablespoonful of butter; stir it well, cut the crust into square bits, mix altogether, cover it with a large plate, return it to the oven for three or four hours. Serve hot. A pan dowdy may be baked in a stove oven, in which case the apples had better be stewed and the crust baked separately, then mix altogether and bake two hours.

SMOTHERED CHICKEN.—Cut a good sized chicken open on the back and spring the breast bone back so that it will lie flat; wash it well in salt and water; lay it in the baking pan with the outside of the bird up; rub it over with butter and sprinkle well with flour; cross the legs and tie them and cramp the wings; pour over it a quart of water and set it into the oven to bake, dripping the gravy over it occasionally. When well browned turn it over and sprinkle a little flour over the inside surface and set it back in the oven. About ten minutes before it is needed for the table turn it over again, so as to have the outside of the chicken a bright yellow brown when placed on the platter. It is very delicious cooked in this style.

PUMPKIN PIE.—Cut the pumpkin into as thin slices as possible, and in stewing it the less water you use the better; stir so that it shall not burn; when cooked and tender stir in two pinches of salt; mash thoroughly, and then strain through a sieve; while hot add a tablespoonful of butter; for every measured quart of stewed pumpkin add a quart of milk and four eggs, beating yolks and whites separately; sweeten with white sugar and cinnamon and nutmeg to taste, and a saltspoon of ground ginger. Before putting your pumpkin in your pies it should be scalding hot.

SHEEP'S-HEAD SOUP.—Cut the loins and lights into small pieces, and stew them in four quarts of water with some onions, carrots and turnips, one cup of rice, pepper and salt, a few cloves, a little parsley and thyme; stew until nearly tender, strain, and when cold remove the fat; when used thicken with flour and butter.

PICKLED ONIONS.—Peel the onions and let them lie in strong salt and water nine days, changing the water each day; then put them into jars and pour fresh salt and water on them, this time boiling hot; when it is cold take them out and put them on a hair sieve to drain, after which put them in wide-mouthed bottles and pour over them vinegar prepared in the following manner: Take vinegar and boil it with a blade of mace, some salt and ginger in it; when cool pour over the onions.

LEMON PUDDING.—Put in a basin one-quarter pound of flour, same of sugar, same of bread crumbs and chopped suet, the juice of one good sized lemon, and peel grated, two eggs, and enough milk to make it the consistency of porridge; boil in a basin for one hour; serve without sauce.

READY-MADE GLUE.—A good glue ready for use is made without the application of heat by dissolving the glue in common whisky instead of water. Both are put together in a bottle, which is then corked tight and allowed to stand for three or four days. If prepared in this way, it will keep for years and always be ready for use, except in extremely cold weather, when it will be necessary to set it in warm water before using. A strong solution of isinglass made in the same manner is an excellent cement for leather.

APPLE JELLY.—Put three quarts of water into your stew-kettle and pare one dozen large apples and slice them into the water; when all are cut, boil until soft, then pour into a jelly-bag. Let drain and press out all you can. To one pint of juice add one pound of white sugar and boil moderately for half an hour, stirring occasionally.

A REMEDY FOR DIPTHERIA.—Dr. Setserich for children of one year gives a remedy, for internal use every one or two hours, as follows: Natr. benzoic. pur, 5 0 solv. in aq. distill at aq. menth. piper. ana 40.0 syr, cort. aur. 100. For children from one to three years old he prescribes it from seven to eight grammes for 100 grammes of distilled water with same syrup; for children from three to seven years old he prescribed ten to fifteen grammes and for grown persons from fifteen to twenty-five grammes for each 100 grammes. Besides this he used also with great success the insufflation on the diptherial membrane through a glass tube, in serious cases every three hours, in light three times a day of the natur. benzoic pulver. For grown people he prescribes for gargling a dilution of ten grammes of this pulver for 200 grammes of water. The effect of the remedy is rapid. After twenty-four or thirty six hours the feverish symptoms disappear completely and the temperature and pulse become normal. This remedy was used also with the same success by Draham Braun and Professor Klebs, in Prague; Dr. Senator Cassel, and several others in Russia and Germany.

HOUSEHOLD HINTS.—To determine the quality of silk, take ten fibres of the filling in any silk, and if on breaking they show a feathery, dry and lack lustre condition, discoloring the fingers in handling, you may at once be sure of the presence of dye and artificial weighting. Or take a small portion of the fibres between the thumb and forefinger, and very gently roll them over and over, and you will soon detect the gum, mineral, soap and other ingredients of the one and the absence of them in the other. A simple but effective test of the purity is to burn a small quantity of the fibres; pure silk will instantly crisp, leaving only a pure charcoal; heavily-dyed silk will smoulder, leaving a yellow, greasy ash. If on the contrary, you cannot break the ten strands, and they are of a natural lustre and brilliancy, and fail to discolor the fingers at the point of contact, you may well be assured that you may have a pure silk that is honest in its make and durable in its wear.

HEALTH HINTS.—Flaxseed tea, which is good for cough and sore throat, is made as follows: Put two tablespoonfuls whole flaxseed in a pint of boiling water, boil fifteen minutes. Cut up one lemon and put in a pitcher, with two tablespoonfuls of sugar. Strain the tea boiling hot through a wire strainer into the pitcher and stir together. Medical men claim that a pound and a quarter of oatmeal will supply as much nitrogen and almost as much nitrogen and almost as much fat to the body as one pound of uncooked meat of ordinary quality. A man gets three times as much nourishment at the same cost in oatmeal as he does in meats. One pound and a half of Indian meal is equal to one pound of uncooked meat in nitrogen, and surpasses it in fat. One who has tried it communicates the following about curing sore throats: Let each one of your readers buy at any drug store one ounce of camphorated oil, and five cents' worth of chlorate of potash. When any soreness appears in the throat, put the potash in a half tumblerful of water, and with it gargle the throat thoroughly; then rub the neck thoroughly with the camphorated oil at night before going to bed, and also pin around the throat a strip of woolen flannel. This is a simple, cheap and sure remedy.

DRY CURING PORK AND BEEF.—Mr. Gillette informed us that he had for a number of years practiced, with entire success and great satisfaction, a method of dry curing, which supplied far better and sweeter bacon and ham than the usual brining process, After killing the carcasses, dry and thoroughly cool 24 hours or so. The sides and hams are then rubbed over thoroughly with molasses—he used the Porto Rico. Salt is heated in an iron vessel to a dry fine powder, and almost "red hot," when it is spread quickly over the smeared pork, and when cool enough is thoroughly mixed with the hand. After three days the same process is repeated. They they then lay in a dry and cool place for a couple of weeks, when they are ready for smoking. No brine is used to toughen the pork or hams or affect the flavor. The smoking is continued at intervals, with care not to get up a heat by a continuous fire. Two fires a day are make with corn cobs, or dry oak or hickory. The total smoking that is the time the meat is totally surrounded with smoke, 100 to 120 hours in all. After smoking enough, the bacon or hams are packed in barrels, and covered over with with a thick layer of dry wood ashes. He says he has never lost a pound, and never failed to have bacon and hams sweet and delicious to the taste, and commanding the highest price in the market. We should add, that in curing very large hams by the process, as a safety precaution, he makes a small opening down to the bone joints, and fill them with the hot salt.

He cures beef in the same way, but only puts it through the salting process. Indeed, some of his neighbors give ham only the first salting, but he deems the second application, as above described, as better, and insuring perfect success always.

STEWED CORN.—Into a pint of nicely-prepared brown sauce put in a pint of sweet corn cut from the ear, and cook it slowly for half an hour. Serve it hot.

BROWN SAUCE.—Make a brown sauce as follows: Put over the fire in a sauce-pan one tablespoonful each of butter and flour, and stir them constantly until they are light brown, and then stir in very gradually a pint of boiling water, a teaspoonful of salt, and a quarter of a saltspoonful of pepper, let it boil two minutes, and use it hot.

BOILED SWEET CORN.—Remove the husk, except the inner layer, from short, plump ears of sweet corn; turn this layer far enough to permit the removal of the corn silk, then replace it, and tie a short string around it to hold it in place; boil the corn in boiling water without salt until the milk is opaque white, ten to twenty minutes; then remove string, husk, cover corn with napkin, and serve it with a dish of melted butter, pepper and salt.

STEWED CORN AND TOMATOES.—Stew together for half an hour one pint of corn cut from the ear, one pint of tomatoes peeled and sliced, one tablespoonful of butter, one teaspoonful of sugar, one saltspoonful of salt, and quarter of a saltspoonful pepper; serve hot on toast.

LIVE STOCK.

Improved Sheep.

The Alabamains have given but little attention to growing stock, and more especially to sheep, because the negroes must have a hog for every negro child, and the dog's appetite is very injurious to the health of the sheep. We hope hereafter Alabama will determine to have more sheep and fewer dogs. All the sheep that we see are of the commonest and coursest grades. It is just as easy to grow a fine Southdown, Cotswold or Merino as it is a scrub, and the fine animal is doubly profitable. We intend to appeal to the farmers of our State to improve their breed of sheep until they hear us and follow our advice. The next two months is the period for looking after fine wooled rams, to cross on the common ewes that are now here. This is the cheapest and most successful way to improve the common sheep, and at the same time keep them strong enough to endure the climate.

Sheep taken from Kentucky or Ohio and carried to Southern Alabama or Florida, lose at first in the weight of their fleece, and finally in its quality, wool becoming courser and mixed with hair. To keep up the quality of their flocks, the owners of sheep ranches in those sections must therefore bring in new blood from Northern flocks, or their wool will rapidly deteriorate in value. The Northern sheep man, therefore, while not able to compete with them in raising wool cheaply, owing to the difference of climate and cost of land, yet has the advantage of being able to dispose of young stock at good prices to his competitors in wool-growing. It is as necessary with them to procure this stock, and they can never breed it for themselves.—*Montgomery (Ala.) Southern Agriculturist.*

Management of Piggs.

The greatest danger to which young pigs are subject to overfeeding. A pig at weaning has a very small stomach and very limited powers of digestion, and yet these young animals are permitted to gorge themselves with sour milk and meal slops as soon as they are weaned until their sides are swollen. This over-feeding produces indigestion, with disorder of the brain, or so-called staggers, nervous disorders, with paralysis or epilepsy: the growth is arrested, the breath is fetid, the teeth become black, and some people ignorantly believe that black teeth are doing it all. The teeth are knocked out with a stone or a bolt in a rough manner, and the mouth made so sore that the pig refuses to eat for a while, and then recovers from the abstinence. So that the removal of the teeth is claimed to be the real cause of the recovery. Black teeth do not cause disease: they are a symptom of it only, and when the health is good the teeth are all right. Half a pint at a time of sweet skimmed milk is a sufficient meal for a weaned pig.

A New Catte Disease.

Persons who have just returned from a tour of 18 miles through North Heidleberg and Jefferson townships bring the most alarming reports concerning the deaths of cattle from a new and mysterious disease. Cattle have been known to drop dead 15 minutes after they were first attacked. Two cows of Harrison Haak were driven into pasture early in the morning. They were apparently well, but in 20 minutes they dropped dead. The rest of the herd commenced bellowing and pawing the earth, and pranced about the dead carcasses that were rapidly swelling. In a short time six more of the same drove were dead. The owner had the swollen bodies carefully limed and buried in the woods. In this way some 35 head of cattle perished on different adjoining farms. Some died in the stable. One farmer found two cows dead in the barnyard. Among the other losers are Levi Moyer, Moses Schaeffer, John Snyder, Henry Zerbe, Gabriel Lutz, Benjamin Haas, Widow Klopp, John Lutz, William Umbenhower, Joseph Ernest and others.

When the cattle are first attacked they refuse to eat or drink. They seem to be seized with a chill and breathing becomes difficult. Some moan and appear to be in great pain. In a short time they lie down and die in great agony. Their bodies swell out of proportion and a very foul odor is emitted. A hasty examination of one of the bodies shows that the blood of the dead animals turns completely black.

Benjamin Lutz, a veterinary surgeon, has been kept very busy for the past few days, and at present is working day and night. He says the disease starts in the head, and he has become deathly sick while boring the horns of sick cattle. He says that the cows are dying from appoplexy of the spleen, and his opinion is concurred in by Drs. Owens and Collins, who are busily engaged in the work of attending to various herds now in quarantine. The spleen of some of the dead carcasses is found to be quite putrid. The bodies of dead animals are very poisonous and one man has already died from lockjaw and blood poisoning. His name was Harrison Haag. He undertook to skin a carcass for its hide and also to perform a post mortem. Some of the poison of the animal got into his system through a wound on the hand, and in a few hours his entire system was poisoned. His body, arms and limbs became fearfully swollen and covered with black blotches. He was then attacked with lockjaw and died in terrible agony. Two others who assisted him narrowly escaped death. Their blotches were burned with caustic Since then no attempts have been made to skin animals or examine them. They are buried in a hurry and the balance of the herd quarantined. All barnyards and stables are being thoroughly cleaned, and farmers are strictly quarantining all their cattle. The disease is contagious and said to be worse than rinderpest or pleurophenumonia.

THE latest returns of live stock and fresh meat importations from the United States and Canada into England at the port of Liverpool shows large increases. For a single week in August the quantity of live stock was double the quantity for the week preceding it, and in fresh meat there was considerable advance, particularly in beef. The totals were: Cattle, 1,808; sheep, 2,800; quarters of beef, 4,748; carcases of mutton, 453. No hogs whatever were landed.

LITERARY AND PERSONAL.

WALT WHITMAN'S "LEAVES OF GRASS."—Years ago, we had a copy of this work, on loan, for a week or ten days, and gave it an ordinary perusal, and whatever opinion we may have entertained or expressed in regard to it, we certainly never would have thought of classing it with obscene or immoral literature. There may be some passages in it that are repulsive or impure to the "immodestly modest," but there are also such passages in the Bible.

If "to the pure all things are pure" can be at all predicated of intelligent beings, then we may view humanity in all its aspects, and from any standpoint in which it may be presented, with moral, philanthropic and philosophic incontamination.

SIMULTANEOSLY with the announcement of the reappearance on Thursday July 20th, of Walt Whitman's "Leaves of Grass" from the presses of Rees Welsh & Co., book publishers of this city, comes the statement that the Philadelphia Society for the Suppression of Vice and Immorality are preparing to anticipate the issue of the much maligned book by an endeavor to have it placed under the category of obscence literature, thus to prevent its circulation through the mails.

THE REV. MR. MORROW'S VIEWS.

Speaking yesterday on the subject of the Vice Society's measure of interference and his own position in regard to the Philadelphia publisher's request that he would review the book Mr. Morrow said: "The members of this society are my friends, and its work is my work, but in this particular instance I think they have made a mistake. I deprecate the attempt to suppress the circulation of a book of this character, and fear that it will not have the good effect intended. In Europe where Whitman's "Leaves of Grass" is looked upon as one of the highest types of the American classic, the endeavor to suppress the book is regarded with astonishment. As an exponent of a peculiar form of thought, it is entitled to a place in American literature, and as such its publication should be unobstructed." I have no wish at the present time to become identified with either side in the fight which is now going on over Mr. Whitman's work, as it would probably interfere with certain movements with which I am myself connected. But if you ask me what I think of the book from a moral point of view, I would say that in my opinion it is neither lewd nor obscene. Nor but that to the minds of many of our readers it may appear so and be so—that depends largely upon the purpose for which it is read. The obnoxious poems, I believe, were not written in a spirit of lewdness.

"Walt Whitman is robust, virile, but not obscene. In his poetry he tries to carry out certain ideas of his own, ideas that may not be consonant with accepted notions of morality, but which with him are convictions. He believes that the human form in all its parts and functions should be made a commonplace theme in social intercourse, and one or two of his poems are exponents of this belief.

"'Leaves of Grass' should be read in the same spirit as that in which it was written, and not as an encouragement to immorality. I give Whitman credit for attempting to formulate thoughts which are to him earnest convictions. His doubtful passages differ from those in Shakespeare and other classical poets, in that the latter are expressions of the current notions of the morality of the day, while Whitman's are exponents of his own ideas, and are at variance with present conceptions of morality. I should not, however, for my own use, want one line in his book expurgated. But if I wanted my daughter to read it I would expurgate many passages. Now, you can understand my reasons for deeming it unwise to suppress the work. It is a book which can only circulate where the contents will be digested by mature minds, and where its capabilities for moral injury are null."—*From the Philadelphia Press, July 15th,* 1882.

DEPARTMENT OF AGRICULTURE.—Special Report No. 47. Climate, soil and agricultural capabilities of South Carolina and Georgia. By J. C. Hemphill, Washington. D. C., 1882. 65 pages octavo, in which is ably discussed a multitude of matter relating to the Sea Islands of the State; how the soil is fertilized; the preparations for planting; the planter's profit on long staple cotton; the rice growing region; the central cotton region; the methods of cultivation; the cultivation of corn; the upper and middle country; labor and wages; small farms in South Carolina; number of farms; farms occupied; the colored people of the State; the lien law and its operations; farming on advances; official figures as to the extent of the system; cotton manufacturing in the State; profits of the mills; compression of cotton; the phosphate industry of South Carolina; trade in fertilizers; agriculture in Georgia; Upper Georgia; Southern or lower division; market gardening in the State, with many instructive tabulated statistics.

PREMIUM LIST OF THE NEW MEXICO EXPOSITION, and driving Park Association. Second annual fair, to be held at the city of Albuquerque, September 18; 19, 20, 21, 22 and 23, 1882. A demi eight vol. of forty-eight pages, divisions A to G, subdivided into twenty-four classes, including rules and regulations, etc., etc. We observe that our friend, Dr. W. L Strachan has a "finger in the pie," being the superintendent of minerals, which will doubtless be a prominent feature in the exhibition. The premiums are very liberal, and there *ought* to be and doubtless *will* be a successful fair.

JUSTICE. A weekly newspaper devoted to anti monopoly principles. New York, one dollar a year. No. 1, vol. 1, (of an indefinite series) has reached our table, an imposing folio with an imposing title, and the following inscription decorating its banner: 'Our principles anti-monopoly. We advocate, and will support and defend the rights of the many as against privileges for the few. Corporations, the creation of the State shall be controlled by the State. Labor and capital allies, not enemies; justice for both."

Published by the "Justice Publishing Company," a very fair and well gotten up journal, the proclaimer of principles which we have heard announced by differnt political parties these very many years—principles endorsed by the majority of educated mankind, and doubtless also practical until they become incorporators themselves, or possess the means to become bondholders thereof, then

"Whiteman berry unsartin,
And nigga neber sure"

is too often most lamentable realized.

One would suppose that *Justice* is a too self evident factor in the moral ultimation of man, to need a corporate effort to effect its illumination in practical life. If we thought it essential we might say that we are in sympathy with above principles in *spirit and in truth*—and much more of the same sort, but, having heard them "trumpeted" these fifty years or more, we have concluded that they never will be practiced until each individual begins the work of reform in his own person. There may be other modes and the newspaper may be an auxilary in their development—we bid it "God-speed."

THE PENN HARROW

BEST IN THE WORLD

IT HAS NO EQUAL

Patented April 13, 1880.

The above cut represents the Penn Harrow complete, with all its combinations of Five Harrows and a sled for each Harrow; and each succeeding change is made from this Harrow without the least additional expense. By hooking the beam in other points, D or C, the center revolves and gives pulverizer for the ground. Two Strokes and Two Crossings in one pass over the field. This Harrow is the most effective pulverizer in the market.

THIS HARROW HAS ONLY TO BE USED TO BE APPRECIATED.

See it before purchasing and you will buy no other.

The Penn Harrow
CHANGED TO A THREE-CORNER ROTARY HARROW.

Indispensable for Orchards, as the revolving wheel harrows right up to and all around the trees without barking them.

The Penn Harrow
CHANGED TO SINGLE "A" HARROW.

PHILIP SCHUM, SON & CO.,
38 and 40 West King Street.

We keep on hand of our own manufacture,
QUILTS, COVERLETS,
COUNTERPANES, CARPETS,
Bureau and Tidy Covers, Ladies' Furnishing Goods, Notions, etc.
Particular attention paid to customer Rag Carpet, and scouring and dyeing of all kinds.

PHILIP SCHUM, SON & CO.,
Nov-1y
Lancaster, Pa.

THE HOLMAN LIVER PAD
Cures by absorption without medicine.

Now is the time to apply these remedies. They will do for you what nothing else on earth can. Hundreds of citizens of Lancaster say so. Get the genuine at
LANCASTER OFFICE AND SALESROOM,
22 East Orange Street.
Nov-1yr

By removing the wing and wheel from the original you have a complete one-horse "A" Harrow.

The Penn Harrow
CHANGED TO DOUBLE "A" HARROW.

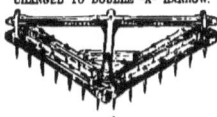

Remove the wheel from the original, reverse the wing, and it makes the most complete Double "A" Harrow in the market.

The Penn Harrow
CHANGED TO A SQUARE HARROW.

By removing the wheel from the original you have a Harrow with three points to lead to. By hooking to B or C you can use the Harrow as a Harrow, rail harrow the bottom and both sides, or over a ridge and furrow the top and both sides, or can lift either point and have three points on the ground—something that cannot be done with any other harrow.

The Penn Harrow
ON ITS SLED.

It has always been a great inconvenience to get the Harrow to and from the field. The Penn Harrow obviates this, as no matter which Harrow you wish to use in the combination, it has its own sled to haul it on.

The Penn Harrow
is made of the best white oak, with steel teeth, well painted, in every way first-class. Formerly a harrow was the most unhandy implement on the farm; with our improvement it is the most convenient, will do double the work of any other harrow and save the farmer half his labor, and is warranted to do all we represent or money refunded. ORDER AT ONCE AND BE CONVINCED.

Price of the light draft Combination Penn Harrow, $30. Send for a Catalogue and see what farmers say.

AGENTS WANTED IN EVERY COUNTY.

PENN HARROW MANUFACTURING CO.
CAMDEN, N. J.
sep-3m

DISSOLUTION OF PARTNERSHIP
The co-partnership in the merchant tailoring business heretofore existing under the firm of Rathvon & Fisher, is this day dissolved by mutual consent. All persons in any manner indebted to said firm, are respectfully solicited to make immediate payment to S. S. Rathvon, who is hereby authorized to receive the same, and those having claims against said firm, will please present them for settlement.
S. S. RATHVON,
M. FISHER,
101 North Queen Street, Lancaster, Pa.

Until further announcement, the business, without interruption, will be conducted by the undersigned, who solicits a continuance of the patronage heretofore bestowed upon the firm, and which is hereby greatfully acknowledged.

S. S. RATHVON,
PRACTICAL TAILOR,
No. 101 North Queen Street,
LANCASTER, PA.

TREES

Fruit, Shade and Ornamental Trees.

Plant Trees raised in this county and suited to this climate. Write for prices to
LOUIS C. LYTE,
Bird-in-Hand P. O., Lancaster co., Pa.
Nursery at Smoketown, six miles east of Lancaster.
79-1-12

WIDMYER & RICKSECKER,
UPHOLSTERERS,
And Manufacturers of
FURNITURE AND CHAIRS.
WAREROOMS:
102 East King St., Cor. of Duke St.
LANCASTER, PA.
79-1-12]

Special Inducements at the
NEW FURNITURE STORE
OF
W. A. HEINITSH,
No. 15 1-2 E. KING STREET
(over Burk's Grocery Store), Lancaster, Pa.
A general assortment of furniture of all kinds constantly on hand. Don't forget the number,
15 1-2 East King Street,
Nov-1y]
(over Burk's Grocery Store.)

For Good and Cheap Work go to
F. VOLLMER'S
FURNITURE WARE ROOMS,
No. 309 NORTH QUEEN ST.,
(Opposite Northern Market),
Lancaster, Pa.
Also, all kinds of picture frames.
nov-1y

GREAT BARGAINS.
A large assortment of all kinds of Carpets are still sold at lower rates than ever at the
CARPET HALL OF H. S. SHIRK,
No. 202 West King St.

Call and examine our stock and satisfy yourself that we can show the largest assortment of these Brussels, three ply and ingrain at all prices—at the lowest Philadelphia prices.
Also on hand a large and complete assortment of Rag Carpet.
Satisfaction guaranteed both as to price and quality.
You are invited to call and see my goods. No trouble in showing them even if you do not want to purchase.
Don't forget this notice. You can save money here if you want to buy.
Particular attention given to customer work
Also on hand a full assortment of Counterpanes, Oil Cloths and Blankets of every variety. [nov-1yr

---

C. R. KLINE
ATTORNEY-AT-LAW,
OFFICE: 15 NORTH DUKE STREET,
LANCASTER, PA.
Nov-1y

SILK-WORM EGGS.
Amateur Silk-growers can be supplied with superior silk-worm eggs, on reasonable terms, by applying immediately to
GEO. O. HENSEL,
may-3m)
No. 254 East Orange Street, Lancaster, Pa.

LIGHT BRAHMA EGGS
For hatching, now ready—from the best strain in the county—at the moderate price of
$1.50 for a setting of 13 Eggs.
I. RATHVON,
No. 9 North Queen st., Examiner Office, Lancaster, Pa.

WANTED.—CANVASSERS for the
LANCASTER WEEKLY EXAMINER
In Every Township in the County. Good Wages can be made. Inquire at
THE EXAMINER OFFICE,
No. 9 North Queen Street, Lancaster, Pa.

WHERE TO BUY GOOD IN LANCASTER.

BOOTS AND SHOES.

MARSHALL & SON, No. 12 Centre Square, Lancaster, Dealers in Boots, Shoes and Rubbers. Repairing promptly attended to.

M. LEVY, No. 3 East King street. For the best Dollar Shoes in Lancaster go to M. Levy, No. 3 East King street.

BOOKS AND STATIONERY.

JOHN BAER'S SONS, Nos. 15 and 17 North Queen Street, have the largest and best assorted Book and Paper Store in the City.

FURNITURE.

HEINITSH'S, No. 15½ East King st., (over China Hall), is the cheapest place in Lancaster to buy Furniture, Picture Frames a specialty.

CHINA AND GLASSWARE.

HIGH & MARTIN, No. 15 East King st., dealers in China, Glass and Queensware, Fancy Goods, Lamps, Burners, Chimneys, etc.

CLOTHING.

MYERS & RATHFON, Centre Hall, No. 12 East King St. Largest Clothing House in Pennsylvania outside of Philadelphia.

DRUGS AND MEDICINES.

G. W. HULL. Dealer in Pure Drugs and Medicines Chemicals, Patent Medicines, Trusses, Shoulder Braces, Supporters, &c., 15 West King St., Lancaster, Pa.

JOHN F. LONG & SON, Druggists, No. 12 North Queen St. Drugs, Medicines, Perfumery, Spices, Dye Stuffs, Etc. Prescriptions carefully compounded.

DRY GOODS.

GIVLER, BOWERS & HURST, No. 25 E. King St., Lancaster, Pa., Dealers in Dry Goods, Carpets and Merchant Tailoring. Prices as low as the lowest.

HATS AND CAPS.

C. H. AMER, No. 39 West King Street, Dealer in Hats, Caps, Furs, Robes, etc. Assortment Large. Prices Low.

JEWELRY AND WATCHES.

H. Z. RHOADS & BRO., No. 4 West King St. Watches, Clock and Musical Boxes. Watches and Jewelry Manufactured to order.

PRINTING.

JOHN A. HIESTAND, 9 North Queen st., Sale Bills, Circulars, Posters, Cards, Invitations, Letter and Bill Heads and Envelopes neatly printed. Prices low.

FARMING FOR PROFIT.

It is conceded that this large and comprehensive book, advertised in another column by J. C. McCurdy & Co., of Philadelphia, the well-known publishers of Standard works, is not only the newest and handsomest, but altogether the BEST work of the kind which has ever been published. Thoroughly treating the great subjects of general Agriculture, Live-Stock, Fruit-Growing, Business Principles, and Home Life; telling just what the farmer and the farmer's boys want to know, combining Science and Practice, stimulating thought, awakening inquiry, and interesting every member of the family, this book must exert a mighty influence for good. It is highly recommended by the best agricultural writers and the leading papers, and is destined to have an extensive sale. Agents are wanted everywhere. jan-li

GLOVES, SHIRTS, UNDERWEAR.

SHIRTS MADE TO ORDER,

AND WARRANTED TO FIT.

E. J. ERISMAN,

56 North Queen St., Lancaster, Pa.

79-1-12]

Thirty-Six Varieties of Cabbage; 26 of Corn; 28 of Cucumber; 41 of Melon; 33 of Peas; 28 of Beans; 17 of Squash; 23 of Beet and 60 of Tomato, with other varieties in proportion, a large portion of which were grown on my five seed farms, will be found in my **Vegetable and Flower Seed Catalogue for 1882**. Sent FREE to all who apply. Customers of last Season need not write for it. All Seed sold from my establishment warranted to be fresh and true to name, so far, that should it prove otherwise, I will refill the order gratis. The original introducer of **Early Ohio** and **Burbank Potatoes**, **Marblehead, Early Corn**, the **Hubbard Squash**, **Marblehead Cabbage**, **Phinney's Melon**, and a score of other New Vegetables, I invite the patronage of the public. New Vegetables a specialty.

JAMES J. H. GREGORY,
Marblehead, Mass.

Nov-5mo]

EVAPORATE YOUR FRUIT.

ILLUSTRATED CATALOGUE
FREE TO ALL.

AMERICAN DRIER COMPANY,
Chambersburg, Pa.

Apl-lf

LANDRETH WHEAT.

Under this name we offer to Merchants and Farmers a

NEW WINTER WHEAT

of superior excellence. Not till this year had the stock increased sufficiently to offer it for sale—the strain all being derived from one stool selected five years ago. We control every bushel and expect to distribute it widely, feeling sure that it is an acquisition of value, being Hardy, Vigorous, Early, Stiff in Straw, very Prolific, entirely from Rust, and making Flour of the Highest Quality. This Wheat is far superior to the Clawson, and those who sow it this Autumn will be able to sell to their neighbors for Seed all the resulting crop at good prices. We do not expect any will be sent to mill.

We append a few sworn testimonials showing the estimation in which it is held by well known millers in the State of New York.

PRICES, including bags: **$1.50 per Peck, $5.50 per Bushel, $10.00 2 Bushels.**

DAVID LANDRETH & SONS,
SEED GROWERS, Philadelphia.

OVID STEAM MILLS,
George W. Jones & Bro., Prop's.
Having ground and baked some of the flour made from the "Landreth" White Wheat, we find the Wheat to be A No. 1 White, and a first-class wheat for grinding. The flour being very white, the bran thin and light. We regard the "Landreth" Wheat much superior to the Clawson variety. We saw it before it was harvested, the heads were very large, the straw bright and stiff, and think it will become one of the leading wheats.
August 14, 1882.   GEO. W. JONES & BRO., Millers.

STATE OF NEW YORK, } ss.
COUNTY OF ONTARIO, }
Richard H. Wilmot, of Phelps, in said county, being duly sworn deposes and says, I have used the flour made from the New White Wheat known as "Landreth," from the grist I ground for H. S. Bonnel, and I have no hesitation in saying that in my long experience in milling I have never seen or had such nice sweet and spongy bread.
R. H. WILLING, Miller.
Subscribed and sworn to before }
me August 5, 1882. }
LYSANDER REDFIELD,
A Justice of the Peace in and
for the County of Ontario N. Y.

OVID, August 14.
I have ground trial samples of the New Wheat "Landreth," and find it excelling the Clawson and equal to any variety I have ever seen. The berry is large, white, with thin skin and light bran. The flour makes unusually white bread.   M. MAXWELL,
Miller.

STATE OF NEW YORK, } ss.
COUNTY OF ONTARIO, }
Ezra A. Hibbard, and Fanny Hibbard, his wife, of the town of Phelps, in said county, being duly sworn, depose and say : We have used in our family flour made from the "Landreth white Wheat," grown by H. S. Bonnell, and we can say that it makes the sweetest and best bread and pastry that we have ever had or used.
E. A. HIBBARD,
FANNY HIBBARD.
Subscribed and sworn to before me, }
August 5th, 1882. }
LYSANDER REDFIELD,
Justice of the Peace in and for Ontario co., N. Y.

sep-lt

A HOME ORGAN FOR FARMERS.

THE LANCASTER FARMER,

A MONTHLY JOURNAL,

Devoted to Agriculture, Horticulture, Domestic Economy and Miscellany.

Founded Under the Auspices of the Lancaster County Agricultural and Horticultural Society.

EDITED BY DR. S. S. RATHVON.

TERMS OF SUBSCRIPTION:

ONE DOLLAR PER ANNUM,

POSTAGE PREPAID BY THE PROPRIETOR.

All subscriptions will commence with the January number, unless otherwise ordered.

Dr. S. S. Rathvon, who has so ably managed the editorial department in the past, will continue in the position of editor. His contributions on subjects connected with the science of farming, and particularly that specialty of which he is so thoroughly a master—entomological science—some knowledge of which has become a necessity to the successful farmer, are alone worth much more than the price of this publication. He is determined to make "The Farmer" a necessity to all households.

A county that has so wide a reputation as Lancaster county for its agricultural products should certainly be able to support an agricultural paper of its own, for the exchange of the opinions of farmers interested in this matter. We ask the co-operation of all farmers interested in this matter. Work among your friends. The "Farmer" is only one dollar per year. Show them your copy. Try and induce them to subscribe. It is not much for each subscriber to do but it will greatly assist us.

All communications in regard to the editorial management should be addressed to Dr. S. S. Rathvon, Lancaster, Pa., and all business letters in regard to subscriptions and advertising should be addressed to the publisher. Rates of advertising can be had on application to the office.

JOHN A. HIESTAND,

No. 9 North Queen St., Lancaster, Pa.

$72 A WEEK. $12 a day at home easily made. Costly Outfit free. Address TRUE & Co., Augusta, Maine.

ONE DOLLAR PER ANNUM.—SINGLE COPIES 10 CENTS.

THE LANCASTER FARMER

DEVOTED TO Agriculture, Horticulture, Domestic Economy and Miscellany

"THE FARMER IS THE FOUNDER OF CIVILIZATION."—WEBSTER

Dr. S. S. RATHVON, Editor.   LANCASTER, PA. OCTOBER, 1882.   JOHN A. HIESTAND, Publisher

Entered at the Post Office at Lancaster as Second Class Matter.

CONTENTS OF THIS NUMBER.

EDITORIAL.
The Stanwich Nectarine................................145
Luscious Grapes.....................................145
Something about "Hair-Worms" and Eels...145
Kitchen Garden for October...................146
   How not to apply Stable Manure.
Necrophore..146
Seedling Peach.....................................146

CONTRIBUTIONS.
The Value of Clover on Land..................147
The Leaves...147
Save the Peach Stones..........................147

SELECTIONS.
Fighting the Phylloxera in Europe..........147
Protecting Plants during Winter.............148
Self Dependence..................................148
The Preservation of Forests from wanton Destruction, and tree planting.......148
  A Universal Mine of Wealth—War Against Trees and Its Effects—Calling a Halt—Forests in the Territories—Forestry Laws—Judicious Thinning—A Unmophonsor of Woods and Forest—Tree Planting—Planting Trees on Public Roadsides—Tree Planting on Farms—Stony Ground—A Forestry Commissioner—The Dominion.
Cultivation of Peppers............................150
How to Bottle Wine................................150
Practical Forestry Illustrated..................151
Summer...152
  Autumn—Winter—Spring.
How to Keep Houses Healthy................152
The Coming Fence................................152
The Trade in Nuts.................................152
Work and Leisure..................................153
Stable Cleaning....................................153
Worthless Dogs....................................153
The Black Walnut.................................153

OUR LOCAL ORGANIZATIONS.
Lancaster County Agricultural and Horticultural Society.................................154
  Crop Reports—The Value of Clover Land—Should There be Less Fencing—Hay as a Fertilizer—Going to the York Fair—The Next Meeting—Fruit Report
The Poultry Association.........................155
Fulton Farmers' Club..............................155
  Questions and Answers—Crops—Literary Exercises.
Linnean Society....................................156

AGRICULTURE.
Wheat Growing.....................................156
An Excellent Fertilizer...........................156
How to Remove Stumps.......................156
The Telephone on the Farm.................156
Octagonal Barns...................................156

HORTICULTURE.
York Imperial Apple..............................157
Keeping Apples....................................157
Apple Notes...157
Root Pruning..157
The Cherry and Apple...........................157

HOUSEHOLD RECIPES
Chow, Chow..157
Stuffed Tomatoes..................................157
Pancakes..157
Rissole Soup...157
Lamb Chops..157
Potato Mound..157
Ladies' Cabbage...................................158
Damson Tart..158
Potato Porridge.....................................158
Roasted Sweetbreads............................158
Boil and Blanch the Sweetbreads...........158
Potato Croquettes.................................158
Rice Pudding Cold................................158
Breakfast Cakes....................................158
Cream Nectar..158
Potatoes au Maitre d'Hotel....................158
Stewed Tomatoes and Onion.................158
Stewed Pears with Rice.........................158
Ox-Cheek Soup.....................................158
Stewed Calfs' Hearts.............................158
Apple Souffle Pudding...........................158

LIVE STOCK.
Raising a Colt..158
Hints on Raising Stock..........................158
Swine Raising—A Different System Desirable..158
More Frequently Milking.......................158
Jersey Cows and their Records............158
Facts About Horses..............................159
Overloading Cows' Stomachs...............159
Quarantined Cattle................................159

APIARY.
Some Information About the Queen Bee...159
Twelve Facts for Beginners...................159
A System for Wintering.........................159
Preparing for Winter..............................159

POULTRY.
Guinea Hens..160
Care of Fowls..160
Ducks..160
Which is the More Profitable.................160
Fattening Turkeys..................................160
Farm and Workshop Notes...................160
Literary and Personal............................160

NONPAREIL FARM & FEED MILLS
The Cheapest and Best.
Will Crush and Grind Any thing
Illustrated Catalogue FREE.
Address **L. J. MILLER,** Cincinnati, O.
oct-2m

Agents Wanted. The Culminating Triumph.

HOW TO LIVE!

A complete Cyclopedia of household knowledge for the masses; now ready. Nothing like it. Going fast. Low priced, illustrated, unequalled in authorship. Send for Prospectus and full particulars now. Outfit and instructions how to sell, free to actual agents. Success guaranteed faithful workers. State experience, if any. Salary desired. W. H. THOMPSON, Pub., 654 Arch St., Phila.

LAND LETH'S FIELD SEEDS,
LANDRETH'S FLOWER SEEDS.
Agricultural Implements in great variety.
Horticultural Tools in great variety.
Requisites for Lawn and Green House.
Red and White Clover, Alsike Clover, Lucerne, Blue Grass, Grass, Orchard Grass, Herds Grass, Perennial Rye Grass.
Mixed Lawn Grass Seed, very best quality.
Plaster of Paris for use, Plaster.
Bone Meal of the finest quality.
Peruvian Guano, Land Plaster.
Farm Salt, Flaxseed Meal.
Carbolic Soap, Paris Green.
London Purple, Paris Purple.
Insect Powder, Tobacco Dust.
ILLUSTRATED CATALOGUES FREE. PRICES LOW. CAREFUL ATTENTION GUARANTEED.
Founded 1784. 1500 acres under cultivation growing Landreth's Garden Seeds.

D. LANDRETH & SONS,
Nos. 21 and 23 South Sixth Street,
BETWEEN MARKET AND CHESTNUT STS.,
AND S. W. CORNER DELAWARE AVENUE AND ARCH ST.,
oct-6m   PHILADELPHIA.

Garmore's Artificial Ear Drum.
As invented and worn by him perfectly restoring the hearing. Entirely deaf thirty years, he hears with them even whispers, distinctly. Are not uncomfortable, and remain in position without aid. Descriptive Circular Free. CAUTION: Do not be deceived by bogus ear drums. Mine is the only successful Artificial Ear Drum made.
JOHN GARMORE,
Fifth & Race Sts., Cincinnati, O.
oct-3m

D. M. FERRY & CO'S ILLUSTRATED DESCRIPTIVE AND PRICED SEED ANNUAL FOR 1882

Will be mailed to all applicants, and to customers without ordering it. It contains four colored plates, engravings, about 200 pages, and full directions for planting all varieties of Vegetable and Flower Seeds, Plants, Fruit Trees, etc. Invaluable to all. Send for it. Address,
D. M. FERRY & CO., Detroit, Mich.
Jan-9

$66 a week in your own town. Terms and $5 outfit free. Address H. HALLETT & Co., Portland, Maine.
jan-1y*

WE WANT OLD BOOKS.
We Want German Books.
WE WANT BOOKS PRINTED IN LANCASTER CO.
We Want All Kinds of Old Books.
LIBRARIES, ENGLISH OR GERMAN BOUGHT.
Cash paid for Books in any quantity. Send your address and we will call.
REES WELSH & CO.,
23 South Ninth Street, Philadelphia.

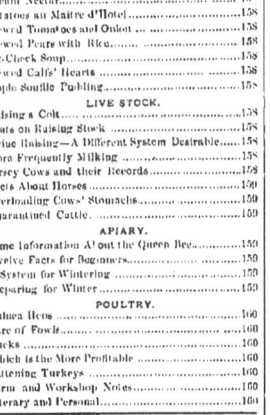

THE LANCASTER FARMER.

PENNSYLVANIA RAILROAD SCHEDULE.
Trains LEAVE the Depot in this city, as follows:

WESTWARD.	Leave Lancaster.	Arrive Harrisburg.
Pacific Express*	2.40 a. m.	4.05 a. m.
Way Passenger*	5.04 a. m.	7.50 a. m.
Niagara Express	11.00 a. m.	11.20 a. m.
Hanover Accommodation	11.05 p. m.	via Col. 10.40 a. m.
Mail train via Mt. Joy	10.20 a. m.	12.40 p. m.
No, 2 via Columbia	11.25 a. m.	12.55 p. m.
Sunday Mail	10.15 a. m.	12.50 p. m.
Fast Line*	2.30 p. m.	3.25 p. m.
Frederick Accommodation	2.35 p. m.	Col. 2.45 p.m.
Harrisburg Accom.	5.15 p. m.	7.40 p. m.
Columbia Accommodation	7.30 p. m.	Col. 8.20 p. m.
Harrisburg Express	7.30 p. m.	8.40 p. m.
Pittsburg Express	8.50 p. m.	11.55 p. m.
Cincinnati Express*	11.50 p. m.	12.45 a. m.

EASTWARD.	Leave Lancaster.	Philadelphia
Cincinnati Express*	2.55 a. m.	5.00 a. m.
Fast Line*	5.08 a. m.	7.40 a. m.
Harrisburg Express	8.05 a. m.	10.05 a. m.
Columbia Accommodation	9.10 a. m.	12.01 p. m.
Pacific Express*	1.30 p. m.	3.40 p. m.
Sunday Mail	2.00 p. m.	5.00 p. m.
Johnstown Express	3.05 p. m.	5.50 p. m.
Day Express*	5.35 p. m.	7.20 p. m.
Harrisburg Accom.	6.25 p. m.	9.30 p. m.

The Hanover Accommodation, west, connects at Lancaster with Niagara Express, west, at 9.35 a. m., and will run through to Hanover.

The Frederick Accommodation, west, connects at Lancaster with Fast Line, west, at 2.10 p. m., and runs to Frederick.

The Pacific Express, east, on Sunday, when flagged, will stop at Middletown, Elizabethtown, Mount Joy and Landisville.

*The only trains which run daily.
†Runs daily, except Monday.

NORBECK & MILEY,

PRACTICAL
Carriage Builders,

COX & CO'S OLD STAND,

Corner of Duke and Vine Streets,
LANCASTER, PA.

THE LATEST IMPROVED
SIDE-BAR BUGGIES,
PHÆTONS,
Carriages, Etc.

THE LARGEST ASSORTMENT IN THE CITY.

Prices to Suit the Times.

REPAIRING promptly attended to. All work guaranteed.
79-9-

S. B. COX,
Manufacturer of

Carriages, Buggies, Phaetons, etc.

CHURCH ST., NEAR DUKE, LANCASTER, PA.

Large Stock of New and Second-hand Work on hand very cheap. Carriages Made to Order Work Warranted for one year. [ly-6-19

EDW. J. ZAHM,

DEALER IN

AMERICAN AND FOREIGN
WATCHES,

SOLID SILVER & SILVER PLATED WARE,

CLOCKS.

JEWELRY & TABLE CUTLERY.

Sole Agent for the Arundel Tinted

SPECTACLES.

Repairing strictly attended to.

ZAHM'S CORNER,

North Queen-st. and Centre Square, Lancaster, Pa.
79-1-12

E. F. BOWMAN,

AT LOWEST POSSIBLE PRICES.
Fully guaranteed.

No. 106 EAST KING STREET,
79-1-12]   *Opposite Leopard Hotel.*

ESTABLISHED 1832.

G. SENER & SONS,
Manufacturers and dealers in all kinds of rough and finished

LUMBER,

The best Sawed SHINGLES in the country. Also Sash, Doors, Blinds, Mouldings, &c.

PATENT O. G. WEATHERBOARDING

and PATENT BLINDS, which are far superior to any other. Also best COAL constantly on hand.

OFFICE AND YARD:
Northeast Corner of Prince and Walnut-sts., LANCASTER, PA.
79-1-12]

PRACTICAL ESSAYS ON ENTOMOLOGY,
Embracing the history and habits of

NOXIOUS AND INNOXIOUS
INSECTS,

and the best remedies for their expulsion or extermination.

By S. S. RATHVON, Ph. D.
LANCASTER, PA.

This work will be Highly Illustrated, and will be put in press (as soon after a sufficient number of subscribers can be obtained to cover the cost) as the work can possibly be accomplished.
79-9-

$5 to $20 per day at home. Samples worth $5 free! Address STINSON & Co., Portland, Maine

---

SEND FOR
SPECIAL PRICES

On Concord Grapevines, Transplanted Evergreens, Tulip, Poplar, Linden Maple, etc. Tree Seedlings and Trees for timber plantations by the 10,000
J. JENKINS' NURSERY,
3-2-79   WINONA, COLUMBIANA CO., OHIO.

$1000 Reward
For any machine hulling as much clover seed in 1 day as the

VICTOR (Double Huller)

It has hulled **150 Bushels** in **ONE DAY.**

Blue print & Pamphlet mailed free.
Newark Machine Co., Newark, O. Branch Office Hagerstown Agr. Imp. Mfg Co. Harrisonburg, Va.

July-3m]

THE
LANCASTER EXAMINER

OFFICE
No. 9 North Queen Street,
LANCASTER, PA.

THE OLDEST AND BEST.

THE WEEKLY
LANCASTER EXAMINER

One of the largest Weekly Papers in the State.

Published Every Wednesday Morning,

Is an old, well-established newspaper, and contains just the news desirable to make it an interesting and valuable Family Newspaper. The postage to subscribers residing outside of Lancaster county is paid by the publisher.
Send for a specimen copy.

SUBSCRIPTION:

Two Dollars per Annum.

THE DAILY
LANCASTER EXAMINER

The Largest Daily Paper in the county.

Published Daily Except Sunday.

The daily is published every evening during the week. It is delivered in the City and to surrounding Towns accessible by railroad and daily stage lines, for **10 cents a week.**
Mail Subscription, free of postage—One month, **50 cents;** one year, **$5.00.**

JOHN A. HIESTAND, Proprietor,

No. 9 North Queen St.,
LANCASTER, PA.

The Lancaster Farmer.

EDITORIAL.

THE STANWICH NECTARINE.

Mr. Samuel W. Taylor, of North Mulberry street, Lancaster, Pa., has placed us under obligations for a small but handsome donation of this luscious fruit, and what increases its value is the fact that it was of his own cultivation, an achievement in which he stands alone in Lancaster county, if not in the State of Pennsylvania; for it was not professed, even by the nurseryman from whom he obtained it, that it could be successfully raised in the open air. Such a contingency might occur, but it was by no means an assured quality of the fruit. Mr. Taylor has, however, successfully accomplished it for the last two or three years. The present season was the most successful one, the tree yielding a fair crop, and some of the fruit measuring 7¼ inches in circumference, of a greenish tinge, with a bright vermilion cheek, and finely flavored. The tree is very healthy looking, being deep green in the color of its foliage; stands near the middle of his lot; has a southern exposure, and a western and northern protection, but not entirely shielded from the wintry blasts. Of course, our fruit-growing patrons will be aware that the nectarine (*Persica lævis* of De Candolle) is nearly allied to the peach; indeed, in flavor and in form, it is essentially a peach, with the smooth skin of a plum. When fully ripe the skin of the fruit is thin and tender, and need not be removed in eating it. It, however, lacks the color and the peculiar flavor of the best varieties of the peach; but for canning there is less waste, because the skin need not be removed, and its presence does not impair the quality of the fruit.

Mr. Taylor thinks in another season, he can improve the size and quality of the fruit by judicious thinning. The present season when this work should have been attended to, he was ill, and confined to his room, and did not recover until it became too late. It is much more of a peach than the Apricot is, and the seed differs very little from the appearance of a peach-stone. On the whole, we think the result is very much in favor of its general cultivation.

LUSCIOUS GRAPES.

We gratefully acknowledge the receipt of nine very fine clusters of grapes, from our esteemed friend and clever fruit-grower, H. M. Engle, of the Marietta nurseries. If these fairly represent the quantity and quality of his crop the present season they most certainly indicate an unqualified success—at least as far as quality is concerned. The clusters, according to the respective varieties, were large, full, and the berries remarkably perfect and uniform in size. The *Merrimac* (Rogers No. 19,) large berry and color black. *Wilder* (Rogers No. 4), of the same color, but less in the berry. *Rogers* No. 33, nearly the same size and color as the *Wilder*.

The *Israella* and the *Eumelan* are medium and small in size and dark in color, and the clusters of a fair size. The *Lindly* (Rogers, No. 9), a large red grape; and the *Iona*, similar in color, but small in berry and fair cluster. The *Martha* and the *Croton*, both green in color, but differing much in size of berry. This is by no means intended as a technical description of the above-named fruit, but only an external glance at it. As to flavor, or edible qualities, of course the respective varieties more or less differ, but we don't know that we are able just now to make that difference plain in its details. Moreover, tastes also differ. Let any novice take a dozen varieties of grapes, in order to test their qualities, and by the time he gets to the bottom of the list his taste will be so much demoralized that he can hardly tell "'tother from which." It may be different with an expert, but we confess we are like the unsophisticated Hibernian, who when asked which of his dozen children he liked the best, very promptly replied, "Faith, then, I like them all the best." Our gratitude, however, is not based up on the quality or external appearance of the fruit. In its donation is manifested a recognition of our humble and long continued labors as editor of the FARMER; and, although we have not in any sense demanded such recognition, yet on all occasions when it has been vouchsafed, it has loomed up as a green spot in the desert of our editorial days, and we have felt duly grateful.

SOMETHING ABOUT "HAIR-WORMS" AND EELs.

In the article of our esteemed contributor, W. J. P., on the Origin of Gapes and the Generation of Eels, published in the August number of the *Farmer* (p. 117), he makes use of the following unqualified declarations on the first-named topic. "Anything of this kind" (that is. the down of young chickens), "or hairs from anything, placed in a warm wet place, and receiving air, will become living animals. They are very common at this time of the year, where stock go to drink, in the foot-prints, containing water. They differ in size and length, depending upon the part of the body from which the hair has fallen. I took the other day from the ditch, below my pump, a knot of hairs that was all alive; it was just as it had been taken from the comb and wrapped around the finger, and a hair-pin stuck through it. I removed the pin, shook them out in a basin of warm water; they appeared to enjoy their liberty very much."

We don't dispute the *facts* of our contributor's observations in regard to his "knot of hairs;" for hair-worms are usually found in such knots, and, from that very circumstance, the generic name—*Gordius*—which has been given to them is derived, in allusion to the "Gordian knot" cut by Alexander the Great. But we can't believe that that knot of hairs ever belonged to a human head (even if it *had* a "hair-pin stuck through it"). And we would suggest that before he comes to an unqualified conclusion on the subject, he should try the experiment of animating human hairs, or the "hairs from anything." Forty years ago we devoted six months to the experiment, and, signally failed; but we experimented with horse-hair, for we were assured that our living Gordians were only animated horse-hairs. We have before us an old illustrated work on natural history, published one hundred and twenty-five years ago, in which the author combats the idea that horse-hairs become animated, very successfully. Although the light of science seems to have exploded the theory over and over again, it has still its followers, and probably always will have, simply because mankind is disposed to judge from *appearance* only. Finding a saddle and bridle under a patient's bed, is no evidence that he has eaten a horse.

We have found these Gordians or Hair-worms in roadside pools; in enclosures where no domestic animals were allowed to intrude; in the bodies of beetles and grasshoppers; in the seed cavities of apples; in the centre of cabbage-heads; and they, or species allied to them, have been found in the eyes of horses, in the brain of birds, in the bodies of calves, pigs and sheep, and in the trachea of fowls; and they have generally been found knotted or tangled, except when found very small. Nor is that all. In the museum of the Linnæan Society we have a female Hair-worm that has a string of eggs externally attached to her, and another specimen in which the eggs can be seen within her body. Still, we do not consider that our observations are final, but we think that our contributor has taken a too narrow and hasty view of the subject.

We are not now, and never have been, a Gallinicultturist, and pretend to no special knowledge on the subject of the diseases that are common to the feathered tribes. We are always willing to accord all the merit we possibly can to the theories and opinions of those who profess an experimental knowledge of the life, habits, qualities and diseases of the "poultry world," but we confess we are not prepared to receive the doctrine that heat and moisture will transmute the down of little chickens into *strongylidæ*, or "gape-worms." We are compelled to doubt it, even if such doubt should consign us to the company of those who in aforetime were wont to ask, "Have any of the doctors believed on him?" We mean no disrespect, but we think the doctrine needs revision and further confirmation. No doubt our contributor is sincere, and the *appearances* may be quite satisfactory to him, for we observe in an explanatory paper on the same subject, published in the September number of the FARMER (p. 133), he reiterates the same views with the qualification that, "probably if the chick did not take cold there would be no gape-worm." A view, which, instead of simplifying only complicates the question. We supposed that those who were practically connected with poultry culture, and who ought to have some knowledge of the diseases that are contingent to that business, would have given their views upon the subject.

As to the question of the breeding of eels, we are confident, from personal observation, that the young ascend the streams in the spring and descend them in the fall in very large numbers; but, at the same time, we do not deny that there are individuals, if not species, that become local in their habits. Indeed, some thirty years ago, we visited a lake in Lycoming county, into which eels had been long previously introduced, and in which they became very large, weighing as much as ten pounds, and out of which they could not have migrated without going some distance over land; nor could there have been any migration into it, for it was at least thirty miles distant from the river, and the discharge of water from the lake was small and precipitous, and as their presence was continuous and the fish very large, the inference is that they must have inhabited the lake for many years. In former numbers of the FARMER we have given our observations on the migration of eels in the Susquehanna, and these have been corroborated by intelligent authority.

It seems a little singular that the most experienced pisciculturists and naturalists have never placed on record the easy and certain mode of eel generation that our contributor does. In the September number of the FARMER, commencing on page 133, is an article, abridged from a larger article published in *Blackwood's Magazine*, in which the writer puts a different face on the matter. He reiterates the doctrine that eels are not *oviparous*, but on the contrary viviparous, or perhaps *ovo-viviparous*. As some other fishes, and also some snakes possess these different characteristics, it is not impossible they may be extended to some species of eels. Under any circumstances, it shows that the simple process of eel-generation is not a definitely settled question, however confident different observers may be in the finality of their theories.

We speak *that which we do know* on the subject, and it is, that *eels do ascend* the Susquehanna river in the spring, and *descend* in the fall; whether the habit is universal or not, it at least *exists*.

KITCHEN-GARDEN FOR OCTOBER.

In the Middle States, the labors of the gardener are varied, as during the preceding month; but he who then neglected duties necessary to be done, has lost time not to be regained—the autumn is upon him. Seeds of a few varieties may still be sown. The principal labors are, however, the protection of crops, already grown, transplanting others, and setting out *trees* and shrubs. Asparagus-beds dress. Cabbage plant out in light land for next season's use. Beets and carrots store now, or early next month. Lettuce plant out for next spring. Potatoes dig. Spinach sow at once, if not sown last month. Vacant ground, trench. (*Landreths' Rural Register.*)

And when you sow spinach, don't forget the extra curled Bloomsdale, an improvement on the best former stock. This edible plant was introduced and named by Landreths many years ago, and is conceded by all gardeners to be the very best under cultivation in this country. Very productive in leaf, thick foliage, and deeply corrugated, very like a Savoy cabbage.

How not to Apply Stable Manure.

"The worst method of applying manure is to place the same in small piles all over the field, and allow it to remain to be leached by every rain storm. The spots thus covered soon become saturated, and, in loose, open soil, the soluble soaks into the subsoil, or is carried away in drains (if any exit), thus forcing an overgrowth on these spots, and a baldness in other places. Manure when carried to the field, should be at once spread and ploughed in, if not intended as a top dressing."

Good wholesome advice, but one would hardly suppose that any intelligent farmer of the present period needed it; and yet, now and then, these neglected piles of manure may be seen for weeks after they have been hauled on the land. Some other work has been considered of more importance, and the manure-spreading has been compelled to await a more convenient season. Like a good many other manual operations of the household, that which is last ought absolutely be the first.

NECROPHORE.

The Burying Beetle, Alias Clarion Beetle.

The word Necrophore is derived two Greek words signifying a *dead body* and a *carrier*; and is applied to a genus of beetles named *Necrophorus*, of which there are about twenty species in the United States, that have been described, named and catalogued. They belong to the family SILPHIDAE, called after *Silpha*, the typical genus of the family group, an arbitrary term, perhaps, of which the derivation is not clear. In plain English the different genera of the family are usually denominated "scavenger beetles," and they perform an important function in the economy of nature. This mild October weather revives them and brings them abroad; and to-day (6th) one was captured by the senior reporter of the INTELLIGENCER and placed in our possession. This is *Necrophorus marginatus*, about an inch in length, in color black, with transverse orange marks on the wing covers. The individuals of this genus are commonly called "sexton beetles," or "burying beetle," from a singular habit they have of burying the substances in which they deposit their eggs. Perhaps I ought also to state that the name of "carrion beetle" has been very generally applied to the family, and wherever there is a putrid carrion exposed there you may find them, their engeners and their more remote family relatives. But the genus *Necrophorus* has a peculiar habit that does not pervade the whole family. If the carcass is large, for instance that of a horse, ox, sheep, swine, dog, &c., these beetles will make no attempt to bury it; except, perhaps, a small detached portion of it. But when it finds the dead carcass of a toad, a rat, a mouse, or a small bird, it forwith commences to bury it, in which labor the sexes unite. They dig out the earth from under it, and when the carcass sinks down a little below the surrounding level, they cover it with the earth removed in making the excavation. Into this nidus the eggs are deposited, the larvæ are hatched and matured. Exposed to the hot sun and the rapid decomposition, assisted by the various species of "Blow-flies," their larvæ could not become fully developed, and hence would perish. There are districts in Mexico and South America, where carrion-beetles are not known; the air is so pure and dry that carcasses become "jerked" or dried, before decomposition takes place. Our beetle buries the carcass to prolong decomposition. The name of "Buzzard-beetle" suggested by the aforesaid reporter, would be appropriate; because as a scavenger and for its size it will compare with the "Buz."

SEEDLING PEACH.

EPHRATA, Pa., October 12th, 1882.

PROF. S. S. RATHVON : Herewith we send you a late seedling peach. It is a medium sized peach. Having noticed it two years ago we then thought it merited further notice. This year the tree is full again of very fine peaches. We do not think that there is a finer peach out. We want your opinion whether you think it is worth propogating; or if anything is cut similar to it let us know what it is. You may report in the FARMER or by mail, as you think proper. Yours respectfully,
S. R. HESS & SON.

Your peach was duly received, and almost immediately submitted to the practical test of Messrs. H. M. Engle, J. C Linville and W. L. Hershey—three members of the Lancaster County Agricultural and Horticultural Society—and they unhesitatingly concluded that it is worthy of propagation and distribution, in which we heartily coincide. Good late peaches are very desirable, and this seems to "fill the bill." There seems to be some indecision as to the variety, but the preponderance of opinion is that it is a "Salway," or a cross between that variety and "Kieffer's Smock."[*] Mr. Engle happened to have a Smock with him, so that we were able to compare. In external appearance it closely resembles the Smock, but is more acid—not so sweet. The Smock is a foreign peach, and was originated by the celebrated Mr. Rickard, of England. We had not a Salway to compare it with, but if it differs from that variety in appearance flavor, texture and habit, of course, you will have the privilege of naming it as a new variety. Therefore, we would say, "go ahead," "give it a wide circulation, and you will be doing a good work for the late peach crop of Lancaster county. There is, however, a later peach cultivated in the county, Mr. Windolp, of Marietta, we think, has a variety that will not ripen before the 20th of October. Probably some of yours may continue until that date.

ONE writer on ensilage truly says: "It will assist enormously in making mankind independent of the weather, for the constant use of the plow and the cultivator, and the raising of strong, growing crops will greatly obviate the difficulties from drought while the serious loss and expense of harvesting crops in wet seasons will be greatly diminished by this method of preserving food."

[*]Not the "old Smock," of twenty or thirty years ago.

CONTRIBUTIONS.

For The Lancaster Farmer.
THE VALUE OF CLOVER ON LAND.

The grass crop is said to be one of the three greatest crops of the country, and we may safely say that clover is the most valuable of all the grass crops.

After our land has lost its virgin soil from the cultivation of crops we must use some means to renovate it again, to bring it back to its virgin soil, which to accomplish we must necessarily use some artificial means.

I presume the object of the question was to discuss the value and benefits of clover, and its effect on our lands over other grasses. No grass and no hay, no manure; no manure, no hay.

One among many other ways to bring our worn out lands to their virgin soil again is to let it rest with a crop of grass, and I could recommend no better than clover. It may make an inferior hay to some others, such as timothy and blue or meadow grass, but it comes up quicker and endures more pasturing, keeps the soil covered more, draws more from the atmosphere than other grasses, makes more milk, and lastly, keeps a better mulch or top dressing, which we all know to be so essential to keep up the soil. This mulch serves the same as the foliage of trees, which serves to feed the soil.

Another good effect we must not forget, namely: the turning down of this grass, which keeps the soil loose and mellow to retain the rain.

I think about the latter end of March or the beginning of April a good time to sow clover seed, providing the weather is favorable; about six quarts to the acre or one bushel on five acres, on wheat or rye. After harvest good judgment and discretion ought to be used to treat it to its best advantage. If dry weather, do not pasture too close, notwithstanding turning cattle on to tramp it solid has a good effect. The drouth can not scorch it so severely. Keep it covered all the time. It is decidedly the best crop of grass for the land. Timothy may be more valuable, at least in the market; it may bring more from the acres, but will exhaust the land more, and is admitted to leave the soil more compact and hard than clover. Now, I would say, if you try clover and it has not the effect you think it ought to have, sow more.

In another thing, which will be adopted in the near future, clover will have the advantage over all other grass-soiling. A farmer can cut his grass and begin again at the beginning when he is through, and so repeat a number of times; more so than any other grass.—*P. S. R.*

For The Lancaster Farmer.
THE LEAVES.

It is said that the leaves are the lungs of trees, through which the sap passes before it is taken up by the growing fruit, and that by this process it (the sap) gets its supply of vitality from the air, as the blood of the animal does in passing through the lungs of the body. That leaves are essential to the growth of the fruit is evident to my mind from the condition I find my grapes in this season. Early in the summer I observed that the leaves were attacked on the underside by a little fly,whose depredations gave the leaf a speckled appearance, and many of them withered and died. I once had a mind to use a hose, and drench the vines from underneath with an abundance of water from the hydrant, but I neglected to do it. Now (Sept. 8) nearly all the leaves are half dead, and many of the grapes are wilting and falling off; and, although the vines are full, I fear there will be but a scant crop of grapes fit for use. Some remedy ought to be provided to prevent these "what-do-you-call-'ems" from destroying the grape fruit.

Please call and see my vines, and then say in your excellent FARMER what you think.—*J. F. W., Lancaster, Pa.*

For The Lancaster Farmer.
SAVE THE PEACH STONES.

Mr. Rathvon, Ed.: In this month's issue of your Journal, an article (Save the Peach Stones) is timely, and to show the importance of it I will give a few instances of success, without even trying.

From an orchard not far from here La Grange Peaches grew, and a party taking some with him home, there came up a tree somewhat resembling the parent, but larger, better and later, and which I named Steadily, that is now flourishing from New Jersey to the middle of Texas, and giving general satisfaction. But for my discovering it, it would never have been known outside of the immediate neighborhood. A few years after I got it the original tree was carried down the Missouri river, and now steamboats pass where it stood. Another instance is, a Miss Ellen Fauner took some fine late yellow clings to one of my daughters, in Hermann, Texas, some years ago, who, being a horticulturist, told me they were so fine that I ought to procure buds. It was then too late for that season; but the year following I sent for buds, but the tree was on the decline, and nothing but a few feeble twigs were got. Thinking it would possibly recover, the buds were set, and, to my surprise, in the following season they made fine trees, apparently healthy and quite vigorous. In the same year the old tree died, and would have been gone irretrievably had I not saved it.

Now there are three trees here bearing fruit, and fruit it is. About as late as a peach can ripen. The largest size specimens measuring 12 inches in circumference, and weighing 14 ounces. Some on the trees yet, which may even exceed the above.

Form nearly round, deep yellow, with a red check. Flesh yellow next the skin; but nearly blood red at the stone, which red extends fully half way out, and grows paler as it recedes from the stone. Highly aromatic, but not so sweet as the heath cling, a cling stone, named Ellen Fauner. But my own experience with early ones is what was intended more particularly in this article. About six years ago about 200 stones of Hale's Early were planted in nursery beds for the purpose of budding early varieties upon, and perchance something new in case some were left unbudded.

This season about fifty of them bear fruit. Some identical with the Hale, and rotted just as bad; some very inferior, while about twelve were really valuable. But only three were noticed particularly, and the trees notched as they were discovered, Nos. 2, 3 and 5. They came in within a few days of Anjou's June, and lasted until that variety was gone. Nos. 2 and 4 are larger, handsomer and better than the Anjou, and No. 3 not quite as large as the others, but as handsome as anything could be. An extensive nurseryman told me that he could sell trees of them at fifty cents apiece if his agents had a painting of them and my recommendation of them. But I only budded for my own use, not wishing to add to the already numerous list of peaches, although I added some twenty more varieties to my collection this season.

In planting the stones of choice varieties I think one can count upon at least one-half not worth growing. I have always such coming on, and if they don't suit me, bud them, frequently budding on limbs two inches in diameter, and which will grow large enough to bear half a bushel on one limb. We fruited about forty varieties this season; and in all my time of observation, fifty years, never saw them so fine, nor ever expect to see such again.—*Samuel Miller, Bluffton, Missouri, Sept. 25, 1882.*

SELECTIONS.

FIGHTING THE PHYLLOXERA IN EUROPE.

The destruction to which the grape crop of France and Germany is subjected on the appearance of the phylloxera in the vineyards has been the cause of much study and investigation by wine-producers and scientific men, to find out the most practical and economical means of preventing these pestilential insects from pursuing their course of devastation.

In order to form an idea as to what extent the grape crop of France suffered by the plague of 1881, it is only necessary to refer to a report made to the Department of State by our able consul at Bordeaux. He reports that there were 100,000 hectares of vineyards destroyed in 1881, against 37,000 hectares in 1880.

Of the various remedies recommended for destroying the phylloxera, the results obtained from the use of *bisulphide of carbon* and *potassium sulpho-carbonate* have been most satisfactory. By comparing the condition of the vines treated respectively with bisulphide of carbon and potassium sulpho-carbonate, it was found that in the former case they grew stronger and remained green for a longer period than those under the influence of the latter, or the potassium sulpho-carbonate treatment, but the yield in grapes was not so good.

It appears that the use of bisulphide of carbon in the vineyards of France in 1881 did not have the desired effect which Pasteur, from personal observations made, was attributed to the extreme amount of moisture contained in the soil during that season. The same authority, having obtained good results from its use in his vineyards, makes the following suggestions as his experience :

1. Only those vines should be treated whose diseased condition has not been of long standing.

2. The treatment to be continued during

the winter only by normal temperatures, and discontinued as soon as the temperature begins to fall too low.

3. The soil must be good and dry.

4. If the vines are in a compact soil and the phylloxera is old, the number of holes in the soil should be increased and only small doses of the bisulphide of carbon administered.

5. The manuring must be strong, and potash salts applied in conjunction with stable manure.

6. The condition of the vines, must be watched closely, and if, after one or two years, the insects have been entirely destroyed, the treatment is to be discontinued; but if there are a few still remaining unkilled the bisulphide of carbon should be resumed only in very moderate doses.

As a preventive against the plague, Avignon recommends the use of common tar. Gayon has carried on investigations with the view of finding out such organisms that could live as parasites on the phylloxera, but as yet his efforts have not been very successful. Thurmann writes concerning the proposed method of Gouillond Depret in the use of bromine against the phylloxera that, besides its being an expensive one, the injurious effects it produces on the health of the workmen renders it quite dangerous.

With reference to the treatment of phylloxerated vines with potassium sulpho-carbonate, P. Mouillo, who has had eight years' experience with its use, makes the following observations:

1. The application of sulpho-carbonate before or in the beginning of the phylloxera invasion will prevent its increase in the vineyards and allow to the vines the power of producing their normal yield of grapes.

2. From its use the destruction of the vines will be lessened and the sickliest ones restored under all climatic influence.

3. Potassium sulpho-carbonate is a manure of the first quality.

Mouillefert claims that by the judicious use of sulpho-carbonate it is a complete means of fighting against this terrible enemy to the grape crop.

In conclusion, I would say that the prospects of this year's grape crop in Germany are most favorable, and an abundant crop is predicted. The season thus far has been extraordinarily propitious for all other crops.— *Wm. D. Wamer, Commercial Agent of Dusseldorf.*

PROTECTING PLANTS DURING WINTER.

The utility of protecting plants during winter is not sufficiently appreciated; even those of reputed hardiness in any given climate will well pay the expense of partial protection from the severity of low temperatures. It is sometimes remarked that a plant to be fitted for general cultivation must have, among its good qualities, the faculty of taking care of itself at all seasons; but it must be remembered that the majority of plants, grown for the sake of their products, have been removed from their natural conditions, by change of climate, selection, crossing, hybridizing, &c., to such conditions as are found most conducive towards realizing the purposes for which they are grown; protection from extremes of temperature, therefore, becomes a part of culture routine, and in many instances it is one of much importance.

The degree of cold that plants will resist without being injured cannot be definitely ascertained short of actual experiment; their powers of resistance depend upon many contingencies. A plant will sometimes be destroyed by exposure to a temperature not lower than it had previously encountered without sustaining any apparent injury. It is not to be supposed that this seeming anomaly is due to any change in the laws of nature; but it is to be traced to causes that influence the resisting power, and upon the knowledge of these causes depends our ability to aid, by culture processes and appliances, this power of resistance in plants which form the objects of special culture and care.

The exact process by which cold destroys plants is a matter upon which there is yet room for conjecture. The mechanical action of frost on vegetable tissue is undoubtedly a cause of injury; fluids expand while freezing, and the expansion of the sap while undergoing this process lacerates and disrupts the tissue, interrupts the convection of the sap vessels and hastens destruction and decomposition, especially in delicate and succulent growths. When, therefore, a plant has reached a degree of maturity which has converted the fluid matter into woody fibre, its power of resisting cold is much greater than when its tissue is highly charged with watery matter, so that it is a well established axiom that plants resist cold in the inverse ratio of the quantity of water which they contain, or in proportion to the viscidity of their fluids.

But it is also well known that the mere thermometric degree of cold does not indicate the extent of the injury that plants suffer during winter. The hygrometric condition of the atmosphere is at least of equal importance.

Plants that pass with safety through a zero cold in December, will frequently be destroyed by the cold dry winds of March, although the thermometer may not indicate more than ten degrees of frost. The intense acidity of these cold winds act in a similar manner as the hot dry winds of summer.

The moisture of the plant is extracted by evaporation, and the resulting injury will depend upon the amount thus evaporated.

It follows, therefore, that whatever tends to render tissue moist and prevents its solidification, increases its susceptibility to injury from cold; and whatever tends to reduce its humidity and hasten the conversion of fluid matter into woody fiber, increases its power of resisting cold; and upon the recognition of these principles all protecting appliances should be based.—*William Saunders.*

SELF-DEPENDENCE.

No alliance with others can ever diminish the necessity for personal endeavor. Friends may counsel, but the ultimate decision in every case is individual. As each tree though growing in the same soil, watered by the same rains, and warmed by the same sun as others, obeys its own law of growth, preserves its own physical structure and produces its own peculiar fruit; so each person, though in the closest communion and intercourse with others, and surrounded by similar influences, must be himself, and must do his own duties, contest his own struggles, resist his own temptations, and suffer his own penalties.

There is too much dependence placed upon co-operation for security from evil, and too little reliance upon personal watchfulness and exertion. There are some who seem to feel in a great measure released from obligations if they do not receive such aid, and some will plead the shortcomings of others as an excuse for their own.

We would by no means disparage the effect of influence or discourage in the slightest the generous assistance which we all owe to one another, or undervalue the important effect of a worthy example. These are vital elements of growth, and their results can never be fully estimated. But they should not usurp the place of a proper self-reliance, nor diminish the exercise of individual powers. Moral force must be a personal possession. It can never be transferred; and while we gladly welcome whatever is good from all sources, it can only be food that must be digested before it can truly nourish us. Material benefits may be conferred by simple gift, but mental and moral activities can only be sustained by their own exercise. Thoughts may be changed but not thought powers; moral help and encouragement may be given, but virtue cannot be transferred; responsibility cannot be shifted.

The most permanent good we can do to others is to nourish their individual strength. To aid the physically destitute most effectively, food, fuel and clothing are not nearly so valuable as steady remunerative employment. To educate a child is not half so important to instill large amounts of information as to set his mind to work, to bring out his mental powers, to stimulate his thought and quicken his faculties. And in moral life, especially in cities, where massed together, and men inclined to lean upon each other, the best lesson to enforce is, that virtue, to exist at all, must be strictly individual. That which cannot stand alone, but depends on props and supports, which needs the constant spur of fear and the bribe of reward, to insure its activity, is but the semblance of virtue and will crumble before temptation. A well developed body ever excites admiration, but a well-developed and self-reliant spirit is a nobler thing. It is calm, modest and unassuming, yet firm in conscious integrity of purpose and steadiness of aim. Inflated by no vanity it is at once humble yet courageous; helpful to the tempted, and yet resolute in assailing evil.

THE PRESERVATION OF FORESTS FROM WANTON DESTRUCTION, AND TREE PLANTING.

The greater part of the North American continent was covered with forests when first invaded by Europeans. These forests had stood for many years undisturbed, except by the slow decay of one generation of trees, if we may so speak, and the slow growth of another. These operations had been going on simultaneously since the creation, or since the last great convulsion of nature, and the annual falling of leaves and the gradual decay of branches and trunks had covered the earth with a vegetable mould of considerable depth.

A UNIVERSAL MINE OF WEALTH.

This mould, possessing all the elements of fertility, was an immense treasure, everywhere abounding, and tempting the settler to clear away the trees and reap the benefit of the virgin soil. When trees were cut down, a crop, which had probably required several hundred years to grow, was reaped in a few weeks or years, thereby leaving the earth bare, and the vegetable mould was used up in a few years by continued cropping in wheat, corn and potatoes. The writer knew an excellent bush lot which produced great crops at first to be reduced in less than ten years to mere rocks and stones. And this process of exhausting the vegetable soil went on everywhere as fast as settlements advanced. Of course where the subsoil was good and was turned up in part to mix with the vegetable mould, fertility continued much longer, but, in course of time, all, except prairie lands, were reduced so much in fertility as to require the application of fertilizers at great expense. Had the soil at first required these fertilizers the progress of settlement would have been exceedingly slow or more probably there would have been no progress at all.

War Against Trees and Its Effects.

The labor of cutting down great trees, cutting them into short logs and piling them up in log heaps to burn, was, however, so great, that a feeling of dislike to trees as the settler's natural enemy became general, and the vengeance against them was so great that in extensive regions the land was completely bared, and rendered thus not only unsightly but unsheltered. Bleak winds had full play and droughts parched the earth. What was even worse, the clearing away of trees on the hills and mountains by the settlers, the lumbermen and forest fires, left the snow of winter exposed to the spring sun ; and the sudden melting and running off of this accumulation of frozen water made dangerous floods in the streams in early summer and left those streams nearly dry in the hot season.

Calling a Halt.

At length the evil results of the indiscriminate cutting down of trees began to be perceived. The improvidence of previous generations was lamented, and efforts to conserve what forests were left and to plant trees, gradually became popular. The first class of efforts was directed to preserving a few acres of the original forest in each farm where that still could be done, and merely thinning the trees for firewood, fencing, &c., thus leaving the smaller trees room to grow more rapidly. The grove thus preserved became one of the most necessary and valuable portions of the farm, and that without any labor of plowing, sowing or cultivating. It also afforded a delightful shade in hot weather for man and beast.

Forests in the Territories.

The preservation of the vast forests in the Territories belonging to the nation attracted attention also, and laws were enacted to protect them from wanton waste. Secretary of the Interior Schurz distinguished himself for endeavoring to enforce these laws, which are very difficult of execution on account of the opportunities lumbermen have in an almost uninhabited region for cutting trees on Government land, and the frequency of forest fires kindled by careless Indians, hunters, trappers and lumbermen settlers. These fires often do more damage to forests in a few days than lumbermen could do in as many years, and how to prevent them is yet an unsolved problem.

Forestry Laws.

The only remedy, and that only a partial one, that can be suggested, for the wanton destruction of forests, is a national system of forestry laws somewhat similar to those of France, Germany, Austria, Norway and other European countries, which prohibit, under severe penalties, the injury or destruction of trees by unauthorized persons ; and also the kindling of fires, or even smoking in the woods. A forest police was created to see to the execution of these laws, and at the same time providing for the utilizing of forests by gradual thinning out and selling the largest trees, so as to leave more room for the smaller ones. In this way the public forests are an annual source of revenue, and after centuries of such management they are in as good condition as they were at first.

Judicious Thinning.

In passing through Plattsburgh, N. Y., once the writer saw the Saranac thickly covered with sawed lumber, and he asked an old gentleman if that river was not yet lumbered out. The reply was, "I have known it for 60 years, and the quantity of lumber coming down has been pretty much the same all the time. There is as much now as there was 60 years ago." This shows the result of a judicious system of thinning forests.

A Commissioner of Woods and Forest.

If the United States, and each State had a department of woods and forests, with a suitable head and the necessary subordinates, much could be done, not only for the preservation of forests belonging to the public, but to persuade settlers to leave a suitable portion of their farms in wood, and to counsel from time to time in public documents, not only care in husbanding present forests, but some general system of tree planting by States, corporations, and individuals, so as to provide a supply of timber for the future.

Tree Planting.

The second branch of this great subject is tree planting, and here credit must be given to the United States Government for its encouragement of this necessary work in the prairies. The law giving 160 acres to anyone who will plant and maintain for a few years 40 acres in trees, has had a great effect already in providing for a future supply of timber in the prairie States ; those groves will also break the terrible prairie "blizzards," and, probably, to some extent, attract rainclouds to mitigate prairie droughts. A fine spirit of tree-planting has also been manifested in many cities and villages ; and "Arbor Day," or a day set apart in spring for tree planting, has become, in some parts of the country, an institution for the purpose of beautifying streets and public and private grounds.

Planting Trees on Public Roadsides.

The public roads should be lined on each side with trees, which, when grown, would do something toward sheltering and beautifying the country everywhere ; but along railroads there should be something more than isolated trees. There should be a rather broad belt on the windy side, thickly planted with the various kinds of trees needed for repairing the roads. This belt would shelter the railway from storms, catch and retain the winter's snows which gives us so much trouble, and, before many years, supply much useful timber when the supply from other sources might be exhausted.

Tree Planting on Farms.

Every farm should have a belt of timber planted all along its windy side ; this belt, not less than fifty feet wide, should be planted thickly with the various kinds of trees that grow best and fastest in the neighborhood, the thinnings of which for useful purposes would soon be valuable, whilst the shelter it would give from prevailing winds would be invaluable. All swamps not covered with trees should be planted with white and red cedar and tamarac, all of which grow best in damp ground, and produce most excellent timber for various purposes. The leaves also of these trees would absorb the unwholesome air which swamps generate.

Stony Ground.

There is on many farms more or less of ground so rocky that it will not repay the expense of cultivation, and all of such spots should be planted with trees. These may be got out of the woods or farm nurseries ; or what would be easier, cheaper and probably much more effectual, the seeds of various kinds of trees could be sown, imitating as nearly as possible the natural processes which have produced all the forests of the country. The seeds of the different trees should be gathered in the woods just at the time that they fall naturally, and they should be immediately planted in little shallow holes among the stones and covered with a little earth. There the rains of autumn, the snows of winter, and the sunshine of spring would bring up quite a crop of young trees, which should be fenced in from cattle and left to themselves. They would require no labor after the first sawing and fencing except subsequent thinning out from year to year of those that were too crowded or most valuable for economic purposes. If hickory nuts, black walnuts, butternuts, chestnuts and the seeds of sugar maples, pines and spruces were any of them or all of them sown every here and there over the place intended for a grove the most valuable kinds and those that thrive best could be ultimately left to become great trees. After ten years the annual thinnings of this grove for firewood, fencing, hop-poles, railroad ties, etc., would probably make it as valuable a part of the farm as any other, and when the black walnut and butternut trees became large enough to be sold to cabinet-makers the value of the grove would be very great. The present race of farmers may say they would not live to see the trees become fit for the cabinet-makers, but none the less would the growth of that grove increase the value of the farm every year, and that whether the owner sold it or left it to his children.

A Forestry Commissioner.

What is very much needed as a preliminary o covering of a considerable portion of the land with these groves is the advice of scientists and experts as to the kind of trees suitable for different soils, the rapidity of their growth and the relative value of their wood. This information should be collected and scattered by a judicial commissioner of woods and forests in each State, just as the fish commissioners now give information about fishes. To plant or sow millions of trees is just as necessary as to hatch and distribute millions of food fishes.

The Dominion.

With respect to the Dominion of Canada there is great need for tree planting in the fertile valley of the St. Lawrence for a considerable distance around Montreal, and still more need in the prairies of the northwest. In the latter region of vast capabilities, to which much attention is now turned, a system of granting land on condition of planting trees might be most advantageously introduced now, as every year will render such an arrangement more difficult. The other provinces of the Dominion are still well supplied with timber, and the system of selling "timber limits" to lumbermen is conservative of the forests, but there is need for great precaution against forest fires or wasteful usee of valuable timber. A capable commissioner of woods and forests for the Dominion would therefore prove a valuable functionary, if he were not only an expert, but an enthusiast in forestry, as otherwise his appointment would merely add another salary to the expenses of Government.

CULTIVATION OF PEPPERS.

Black and white pepper grow on the same vine; and green pepper-berries, just before maturity, after gathering, turn black and make "black pepper," while " white pepper" is obtained by gathering the berries—fire-red in color—when fully ripe, and, through long soaking in water and subsequent stirring and slaking, relieve the berries of the outer skin; whereafter the latter, on being dry, become "white."

In what country the pepper-vine originated I am unable to say; but Eastern history shows that the northern half of Sumatra, the once mighty old sultanate of Atjeh, (pronounced Aijeh, by the Malays), when the Portuguese, Dutch, and British (in rotation) came to that country, was far-famed for that spice, which drew, at Acheen Busar in North Sumatra (near the entrance of the Straits of Malacca), the native traders from many Eastern countries and islands, who there exchanged the products of their countries or purchased for cash. And during the reign of Queen Elizabeth and her successor, King James, a British naval squadron visited Acheen and met with a hearty reception from the Sultan; succeeded in making a treaty of peace and commerce with him, and, as a result, in establishing factories for and of the British East India Company.

After the British East India Company, during the last century, acquired the island of Penang from the Rajah of Queddah, a Siamese sozerain (much interesting history is attached to this acquisition,) so favorably situated for commercial purposes, and made it a very important factory and place of residence for a sub-governor (the famous Sir Stanford Raffles resided there for a long time) the great Acheen trade gradually drifted to Pulo-Penang (Prince of Wales Island,) and with it the pepper trade, principally. At that time Singapore had not been acquired by the British, and not before 1819, when the island was covered with a dense trackless jungle.

After the acquisition of Penang the natives on the peninsula of Malacca, especially in the Province of Prang, a Siamese sozerain province, commenced to plant pepper, and with excellent success; and now it is extensively planted by Malays and Chinese in many places on the peninsula of Malacca, also in Siam, Cochin China, and in Sarawak, Borneo. That grown in the southern part of the peninsula and on the island of Singapore, known in the market as "Singapore pepper," is by far the best, commanding a higher price than Acheen pepper.

Penang maintained the Acheen pepper trade until the Dutch commenced their war of conquest in Acheen in 1873, blockading the coast and preventing the exports of all Acheenese products. At that time, owing to the spread of wild rumors about the destruction of the pepper gardens in Acheen, etc., pepper reached the figure of $14 for a short time. It was feared the supplies from Acheen being cut off, that the spice would become scarce, and as a consequence many Chinese planters increased its cultivation; in fact, to such extent that the Acheen war was no longer looked upon as the cause of influence in prices. Later some of the chiefs of certain Acheenese provinces, having submitted to Dutch rule, were allowed to send pepper to Penang on vessels having a permit from the Dutch consul in Penang to supply them with rice and other needed goods. Then it happened that some of the rajahs who had submitted to the Dutch, after having been pretty well supplied with the necessaries of life, turned truant again, and, as a consequence, their coasts were again blockaded, or it was discovered that some of them had received contraband of war, and were, therefore, put under close restrictions. At any rate, the "old Acheen trade at Penang has become seriously crippled since that war commenced, but contractors in Penang profitted heavily by supplying the Dutch army in Acheen, owing to the proximity, with beef and many other necessaries of life.

The Dutch are now making efforts to make "Ole-Seh," the old port of Acheen Busar, in fact well protected by a fort and man-of-war, a trading port, and to export "Acheen pepper from Acheen" themselves; at least so the Java and Singapore papers inform, and I presume they will. As I shall in a future report write about Sumatra and her provinces again, I will not now enter upon any further details about "Atjeh."

As to the pepper vine, it presents a very handsome appearance; a pepper garden at a distance looks like a "hop-yard. Some planters, however, trellis the vine, and I think myself that it is the best plan. It grows everywhere round about here very easily and luxuriantly on fair upland soil, and, like the grape-vine, needs occasional pruning, weeding and fertilizing. With a little care and attention it yields abundantly and proves a good source of income. The quantity of pepper exported annually from the Malay peninsula and ports in Dutch India is simply immense, and is almost exclusively planted, gathered and brought to market by natives, Malays and Chinamen chiefly.

HOW TO BOTTLE WINE.

Every wine-owner, wishing to bottle his wine, should analyze it to ascertain what proportion of alcohol, sugar and neutral constituents it contains, for the production of effervescence, which is a very important operation; and the breakage which might arise from an excess of saccharine matter would lead to very serious losses.

To help the formation of effervescence, if the wine has not enough sugar in it, more is given to it by means of a preparation called "liqueur," or sirup, which is nothing but pure wine containing candied sugar in solution, and is composed of 100 kilos of candied sugar to 100 liters of wine. The analysis of the wine having been made beforehand, it is easy to calculate the amount of sirup to be inserted. An almost exact measure can be taken with the help of a vinometer.

At this time much care must be taken in the choice of bottles, to the regularity of shape, to their color, and especially to their superior strength. It is worse than useless to use poor bottles. The bottles are thoroughly rinsed, perfectly dried, and before being filled are carefully inspected, both inside and outside.

When the wine and bottles are ready the workmen proceed to drawing or bottling, an operation which is effected by means of taps with six, eight, or ten spouts.

The bottles when filled pass into the hands of the corker. The methods of corking most in use are done with a mallet.

When the bottle is corked the cork must be secured. This third operation is the "wiring," to effect which short pieces of wire are looped in the middle and fitted under the ring at the top of the neck of the bottle, and for which a wiring machine is used.

The bottles are then stored in the wine-vaults, or left in the cellars, as the case may be, where they are stacked with rods and laths. In hot years fermentation sets in sooner; the sugar is decomposed by the acids in the wine; carbonic acid is set free; alcohol is produced, and a deposit; effervescence takes place, and the weaker bottles break. They are then removed as soon as possible to the vaults, after being marked on the upper sides, so that they may be placed in the same position as before, the marking being made with white chalk. These are piled in the vaults in stacks, as in the cellars.

Often when this effervescence occurs the deposit mentioned is of more or less tenacious character. As a rule it presents dark traces, and on the side a thin, white substance, which appears to adhere to the bottle, and which sometimes cannot be removed by shaking the bottle.

In the month of July, when the bottling is over, the workmen find employment in hooping, renewing the defective hoops, replacing them by new ones, &c. These are the operations included in the word "hooping," and they, with the vintage, lead us to the time of early frost.

At this time some houses bring what they have bottled up again to the upper cellar. This is a good plan, as the cold helps the deposit to dry. Others content themselves with changing their position, removing the leaky and broken bottles, and making new piles of bottles. It is at this time also that attention is paid to the *masque*, the name given to the deposit on the side of the bottle, and which must be removed. This deposit is removed by means of a machine which consists of a box, into which are placed two bottles having this deposit in them; by means of a handle a rotary motion is imparted to the bottles, which are further subjected to continual blows from two little hammers. These continued shocks produce a shaking which is sufficient to detach the adhesive deposit. The removal by hand requires much more attention. The workman is supplied with an iron implement, and has to take care to hit only just hard enough to detach the crust. If he were to hit too hard it might give rise to accident.

When the bottles are entirely cleared of deposit they are placed neck downwards, either on tables or on racks. The latter having certainly the advantage of taking up less room than rows of tables.

After being kept for some time in this position, the wine should be shaken, so as to make all the deposits fall on the cork. This is an important operation, and great care is taken in the selection of workmen to do it. It is done by very slightly lifting the bottle, and giving it a shake or two in that position. To bring it to a successful issue requires a month or six weeks, or even more, the bottle being moved every day. When the deposit has altogether settled on the cork, the good bottles are placed in stacks, with necks down, at a sharp angle, to await the time when they are again uncorked. The rest are replaced and "worked" a second time on the racks.

The uncorking is also a difficult and delicate operation. It is necessary to remove the cork and wiring with the least possible loss of wine, the bottle being all the time kept neck downward. To do this the workmen watches the bubble of air which is in the bottle, and so removes the cork that only the deposit is ejected by the rush of gas. When the froth comes he uses a part of it to wash the neck of the bottle, and then inserts a small cork prepared for the purpose, which prevents too great a loss of gas.

The bottle then passes into the hands of a man who takes out enough wine to admit the necessary amount of sirup. The wine is now very "dry" and would not be very drinkable, although some countries, especially England, will have nothing but dry wines. This dryness is corrected by the addition of what is called the "export sirup," which differs from what is put in at the time of the bottling; it is composed of 150 kilos of candied sugar to 100 liters of wine, and three quarts of alcohol, which is added to increase the strength. As the sugar dissolves, the wine becomes thick, and must be filtered to make the liquid perfectly clear.

The bottles, when opened and emptied to a certain depth, are taken to the "mixer." The mixing consists in putting into each a certain equal quantity of sirup, the precise portion differing for each country. When a bottle is not full enough it is filled with a little wine. The bottle is then placed on a revolving table, and as it revolves all the bottles in turn come to the corker.

This second and final corking requires more care than the first. The corks used are of Spanish cork, soft and strong, hard and full, or red corks, according to the country the wine is to be sent to; they are soaked for a few days before in cold water to soften them, and they bear the name of the manufacturing firm branded on the end which enters the bottle.

For the final corking before dispatch, the corking machine is often a mallet machine, but others are also used. The cork is put into the tube, pressed and made to come level with the lower end of the tube, and with a clean sponge the few drops of water which have resulted from the compression of the cork are wiped off, and then the bottle is filled as much as is wanted and is corked, the cork being driven in to a greater or less extent, according to the destination of the wine.

The tying up is then proceeded with, oiled string being used, this being prepared for exportation, and which lasts longer in cool vaults. A new stringing machine has been invented which has been of much use. Before the invention of this machine the working of a "stringer" could only be performed for a few hours at a time, it being so hard and tiring, but now a young man from 16 to 18 years of age can easily string bottles all day.

After the string is put on the wire is added; the kind most in use now is galvanized wire. It is at this time that the bottle is often shaken once or twice to mix the "sirup" thoroughly with the wine. Then the bottles are arranged in piles, always on ends and are left still for a month or two, being examined then to see if the cork is in order, or if it shows marks of leaking.

Then comes the packing. This is done in boxes or hampers. The bottles are in straw, or wrapped up in straw covers, which are manufactured beforehand in different manners; tin or golden leaves; cratings of wax of different colors; leaden covers; labels with the name of the house, &c. The baskets are closed with flexible twigs or willow. Besides these fastenings some houses use a wire all around the basket; its ends sealed together with a leaden seal, so that if the receiver finds any loss when it is delivered to him, he cannot claim anything from the carrier unless the seal or fastening has been broken. The cases are closed by means of nails. They are bound with wooden or metallic bands, and some are also sealed.

When the time comes to send the wines away, the senders should take care that the wines are in their right positions, *i. e.*, recumbent. When the wine has reached its destination, it should be taken into a cool place and laid down horizontally. It should not be used for a full fortnight, or even for a longer time, as the traveling injures it, and it would be unfair to judge of the wine on immediate arrival.—*John L. Frisbie.*

PRACTICAL FORESTRY ILLUSTRATED.

A quartette of our Lancaster disciples of Blackstone, composed of W. R. Wilson, Simon P. Eby, A. F. Hostetter and Andrew M. Frantz, Esqs., visited our well-known rural friend of Warwick, Mr. Levi S. Reist, who is more than any other man in Lancaster county identified with the planting and rearing of new forests. Mr. Reist was honored lately with a personal invitation to attend the American Forest Convention, sitting at Montreal, in the British dominion. The new forest, comprising about twenty-five acres, has been named as above in commemoration of an aboriginal settlement of American Indians occupying the same spot in the early history of the country, called the " Le Roy " settlement. The ground occupied by the new forest was under the plough as common farm, and as late as twenty years ago. It is situated about one mile from the residence of Mr. Reist, on a high point of Gravel Hill, presenting a view for extent and variety of scenery unsurpassed. Mr. Reist owns most, if not all, the land along the roads from his residence to the forest. These roads are continuous avenues planted with trees on either side, making the passage one of great interest and pleasure to the pedestrian as well as the carriage goer.

Before reaching the new forest you pass a place devoted by Mr. Reist to horticultural and vegetable culture, where now are in full ripeness a large quantity of rare grapes and the popular Hartford in perfection. An order was left with the Superintendent to pick a basket for the party and hand it over upon the return.

The forest cannot be described in detail. Its tree growth is rank and healthy-looking, most of which was planted and reared from the seed, consisting of oak, chestnut, locust, poplar, maple, etc. There is a circuitous drive in, around and through the forest, along which the green foliage is constantly brushing the horses and carriages, up and down hill, over ravines and past sparkling springs. Aside from the utilitarian view, it is one of the most romantic and pleasurable spots in the county, which must be seen and its inspiration felt to be properly appreciated.

The order with reference to the Hartfords was well executed. The superintendent handed over a basket of rich fruit, not by struck measure, but heaped full. It is needless to say that the party, under the sharpening influences of a free mountain air, pitched in pretty freely.

In view of the reckless destruction of the old forests, too much credit cannot be given to the few leading men who devote themselves to the matter of raising new forests for the use of future generations. The party was highly pleased with their visit and would suggest that others follow their example. Few men have more to show to interest the visitor than our friend Mr. Reist, and still fewer are so heartily disposed to make a visit instructive as well as pleasant. The party of visitors feel that it is due to their best that they should make this public acknowledgment of their satisfaction with their visit personally to Mr. Reist as well as in recognition of the noble enterprise which he has undertaken. The forest now could furnish fencing material, telegraph poles, etc., in considerable quantities. The home of Mr. Reist and its surroundings illustrate the character of the owner for planting and renewing original forest productions, the beautiful Golden rod, trees loaded with the poppy fruit, etc., adorning the lawn.—*New Era.*

SUMMER.

Summer, astronomically, includes the period between the vernal and autumnal equinoxes, or from June 21, 8.08 a. m., lasting 93 days, 14 hours and 22 minutes.

In the United States we call June, July and August the summer months. In England, May, June and July are known as the summer months. Between the tropics there is, properly speaking, no summer, the hottest periods being when the sun passes to the zenith at noon, corresponding at the equator to our equinoxes.

Autumn.

In the northern temperate zone it begins when the sun, in its apparent descent to the south, crosses the equator September 22, 10.30 p. m., ends at the time of the sun's greatest southern declination, December 21st, 10.52 a.m., lasting 89 days, 18 hours and 15 minutes.

In the United States, September, October and November are known as the autumn or fall months; in England, August, September and October are so called. In the Southern Hemisphere they have their autumn when we have our spring.

Winter.

Winter begins, astronomically, on the shortest day, December 21, at 10.52 a. m., and lasts 89 days, 1 hour and 4 minutes (March 21.) In the U. S. winter months are commonly reckoned December, January and February; in England, November, December and January. In the Southern Hemisphere, by the American style, the winter months are June, July and August; by the English style, May, June and July.

Spring.

The passage of the sun across the equator, when the days begin to be longer than the nights, is the vernal equinox. In the Northern Hemisphere this occurs March 20, at 11.56 a. m., when spring begins. Spring lasts 92 days, 20 hours and 12 minutes. In the United States, March, April and May are popularly known as the spring months; in England, February, March and April.

HOW TO KEEP HOUSES HEALTHY.

The custom of working or exercising horses directly after eating; or feeding after hard work, and before they are thoroughly rested, baiting at noon when both these violations of a natural law are committed; these are the predisposing causes of pinkeye, and of most diseases that effect our horses. Keep the horse quiet, dry, warm, and in a pure atmosphere, the nearer out'door air the better, and stop his feed entirely at the first symptom of disease, and he will speedily recover. It has been demonstrated in tens of thousands of cases in family life that two meals are not only ample for the hardest and most exhaustive labors, physical or mental, but altogether best. The same thing has been fully proved in hundreds of instances with horses, and has never in a single instance failed, after a fair trial, to work the best results. An hour's rest at noon is vastly more restoring to a tired animal, whether horse or man, than a meal of any sort, although the latter may prove more stimulating.

The morning meal given, if possible, early enough for partial stomach digestion before the muscular and nervous systems are called into active play; the night meal offered long enough after work to insure a rested condition of the body; a diet liberal enough, but never excessive; this is the law and gospel of hygienic diet for either man or beast. I have never tried to fatten my horses, for I long ago learned that fat is disease; but I have always found that if a horse does solid work enough he will be fairly plump if he has two sufficient meals. Muscle is the product of work and food; fat may be laid on by food alone. We see, however, plenty of horses that are generously—too generously—fed, that still remain thin, and show every indication of being under-nourished; dyspepsia is a disease not confined exclusively to creatures who own or drive horses. But for perfect health and immunity from disease, restriction of exercise must be met by restriction in diet. Horses require more food in cold than in warm weather, if performing the same labor. In case of a warm spell in winter I reduce their feed, more or less, according to circumstances, as surely as I do the amount of fuel consumed. I also adopt the same principle in my own diet. The result is, that neither my animals nor myself are ever for one moment sick,—*Medical and Surgical Journal.*

THE COMING FENCE.

Farm-fencing has been discussed year after year with increased interest ever since fencing was used to divide fields and farms and to keep out of fields and crops the roving cattle which formerly filled the highways and did immense damage to the honest, hard working farmer. The original "worm fence" is still in existence, and so is the "stump-fence" in the wilder parts of the State where fences are used at all. Then came the "post-and-rail," which in most improved sections continues to be the most popular and we may say the most efficient fence; but lumber is getting scarce, and some other material than wood must be substituted. The "hedge fence" in the Western States has, within the last ten years, been most extensively introduced, and many believe that is *the* fence. Next came the iron fence, the common wire fence, followed by the "barbed wire fence," which just now seems to claim the most popularity. But there is still another just tried in the West, which is coming in for a full share of popular favor. This is simply a wire fence without barbs, woven together similar to a fishing seine, with a large heavy top and bottom wire. This fence, it is said, will completely withstand all kinds of cattle, with no possibility of injury, while it is "no more expensive than the ordinary board fence." As to the real truth of this statement we cannot say, but we should fear that from the lightness of the wire, unless well galvanized, it would succumb to the effects of the weather. One thing, however, seems to be well-established, that iron, in some form, must eventually be the "coming fence" to stay. Wood has become too expensive, but we cannot bring our mind to believe that the live-fence, however it may be esteemed by some, will ever be a fixture in this country.

THE TRADE IN NUTS.

In the past few years, says *The World*, the trade in foreign and domestic nuts has developed largely, and in New York, with its widespread facilities for distribution, and its local wants, has become an important center of trade. The old traditions as to the indigestibility of nuts has evidently lost its terrors. Wholesale fruit dealers now regard this stock as a steadily selling commodity, and count surely upon an increased demand at the winter holiday season. For peanuts, the South has become famous; and Africa, which used to send whole ship cargoes of peanuts here, is almost swept out of the market by their cultivation in Virginia, North Carolina and Tennessee. The crop for these States this season was 1,110,000 bushels for Virginia, 550,000 bushels for Tennessee, and 120,000 bushels for North Carolina. The peanut is well known to be of a mild nature, which permits its sale the year round. It is estimated that this city handles nearly one-half of the Virginia crop.

The pecan of Texas is increasing in favor at the North, and especially in the Eastern States. A few barrels or boxes made up the consignments a few years ago; now car-loads and invoices of one to two hundred barrels are not uncommon. This nut is of the family of hickory nuts, but has a much softer shell and a richer flavor. The local crop of hickory nuts, or shell-barks, is scanty this year. The West, however, meets the deficiency—half a dozen car-loads a week, if they are needed. Wild chestnuts are getting scarcer at the North, and there is difficulty in obtaining sound lots. They will not disappear, though, for they can be successfully cultivated, and, in a few years more, there will be a full supply of larger and better quality that will compete with the expensive Italian chestnuts. Black walnuts are mainly used by the confectioner; they and butternuts are apt to be found too rich and oily for table use. West Virginia and Pennsylvania furnish the chief supply. The hazelnuts of this country are too insignificant for commerce, especially as their noble cousin, filbert, is always to be had in plenty. A few samples of English hazelnuts, in their outside "shuck," occasionally arrive here for show. California growers promise, in a few years, to make additions to the list of domestic nuts of sorts that come from Italy or Spain, and of what is known as Mediterranean stock. An immense trade is doing in foreign nuts. A trustworthy dealer assured *The World* reporter that the demand is fully three times as great as it was five years ago. Almonds have always a steady sale, since large quantities are used in fine baking. What are called English walnuts come mainly from France and Spain, and were formerly called Maderia nuts. The English crop is used at home. The best received here are the Grenoble and Marbeau. English walnuts are successfully raised in the Pacific Coast States. Brazil nuts never sell largely. They are peculiarly fat, and a few go a long way. For cocoanuts there is a steady and large demand. The process of desiccating them has widened their family use. One purveyor in this city buys by the hundred thousand. The season for the delivery of foreign nuts here begins about the first of November and ends early in the spring. They are largely transhipped from English ports.

WORK AND LEISURE.

Old-fashioned, routine farmers are afraid sometimes that they will get out of work, and look upon such an occurrence as a sort of calamity. We have known farmers to refuse to buy threshing machines for the avowed reason that to thresh out all their grain all at once would leave them nothing to do in the winter except to feed stock, and to pound the rye out with flails, and the oats, wheat and corn by driving horses over it would furnish them and the boys with the employment desired, and keep all hands so employed that ennui need not be apprehended. That was not the word used, but that was what was meant.

Employment is good, excellent indeed, but there can be too much of it sometimes, and particularly of certain kinds. If one wants a son to hate farming and to determine to leave it at first chance, it is only necessary, at least if he is a bright, busy thinker, to make a drudge of him in this way. Give him no rest. Make it appear that you think work, chiefly for its own sake, is the chief end of life, and if he don't leave at the first opening he isn't much of a boy. Work is good, so is medicine, but both are means to an end, not usually the end itself. Work is honorable and necessary, but people who work merely because leisure makes them lonely, and who cannot find entertainment in reading, or visiting other farms to learn new methods, or in some of the thousand ways by which intellectual progress may be promoted, tells a poor story of himself. He is carrying to excess a thing which is good in its place, and is as much off the track in this respect as if he were intemperate in some other form.

On the other hand there is no virtue in indolence. The farmer can justly be busy all the year, because farming is his business. When not actually engaged in manual labor he can be busy at planning work for the next day, the next week, or the next season, and finding out all he can as to the best methods through books, agricultural papers, discussions with neighbors or at agricultural societies, or observations on the work of others. If owning a large farm and at all "forehanded" there may be little necessity for him to work at all, as many consider work; he can employ himself more profitably at supervising, planning and keeping his employees at work—not in any mean or offensive sense, but to see that they work to the best advantage. It is the true principle in farming that the sagacious and clear-head thinker shall be at the head of the farming operations, and those who cannot plan shall be in subordinate positions to carry out orders. The general in the ranks is out of place, because any good or strong soldier can do just as well there as he can, but at the head of a great army his abilities are worth more than many thousands of soldiers. So of the first-class farmer ; the real farmer is more than a laborer, and should not be classed in the same category.

Leisure is valuable for review, rest and recuperation. The man who has none is badly off. But usually this is a matter of habit ; devotion to business never should be so great that there is no leisure for anything else. It is excess, and excess means an early wearing out and premature death. Work and recreation or leisure should be so blended that no sense of continuous weariness shall be felt, and it usually can be if people are inclined to use forethought, and not give way to imagined necessities. The necessity—or desire rather—to amass a fortune has killed thousands ; and thousands of farmers and others are shortening their lives and losing the chance for enjoying old age by devotion to manual labor, with nobody to be a real gainer in the end. The necessity of giving their children "a good start" in the world is often made the excuse for this sort of sacrifice ; but it is a poor one. The best start for children is to bring them up properly with well-balanced minds and the capacity to judge of things by their real merits, and they can be depended on to start themselves quite as well as they should. The lesson has been taught too often that the child "started" with abundance of money at his command is, on the average, even more likely to fail wretchedly than the one who has the discipline of adverse circumstances to encounter.

STABLE CLEANING.

Forty to fifty years ago, and we are sorry to say that the evil still exists at the present time at points far away from towns and cities and dense populations, there was nothing so much neglected as the keeping of cow stables clean. As a common rule they were cleaned out once a week—on Saturday—and then it was not so much on account of the comfort and health of the animal and the convenience and tidiness of the milker, as simply because the pile of manure must be gotten out of the way to allow of the putting up and letting loose of the cattle mornings and evenings. We have seen the manure in the cattle stalls two feet deep of almost clear dung, with the hinder part of the animal at least one foot higher than the front part, and the cattle being driven out the pasture field with quantities of fresh dung hanging to their flanks, which from day to day received layer upon layer until it was one disgusting mass, and was left there until it became dry and hard and fell off in flakes of its own action. The litter—about a fourth of the quantity cattle now receive—consisted of the stalks of cornfodder which could not be eaten, the weeds left in the hay, the rakings of dirty straw lying about the outbuildings, and sometimes mixed with a few leaves from the woods. The food of the dairy cows consisted of musty corn-fodder, second crop clover and orchard-grass, badly cured, chaff from the winnowings of the threshed grain, oats straw, &c. The cows were of course as thin almost as skeletons, and their product of poor milk was about one-half of what would have been obtained from properly-fed cattle. The fact is that the farmer took no pride in his live stock. The idea of giving them clean stalls, good ventilation and nourishing food, never entered his thoughts, and if it did would have been regarded as an utter waste of money, without any return.

But look at the stables now of the dairy-stock ! Their stalls are wide, clean and fresh, the cattle themselves are bright and sleek, with no projecting ribs, and pleasant to handle, well-fed, comfortable in every-way, and giving two or three times the quantity of milk and as rich as it is abundant. The butter from such cows commands twice the price from its careful manufacture and uniform excellence. At the present time also the farmer feels more pride in his dairy stock than in anything else upon his farm. He finds that they give a double return for all the extra care and cost of their improved treatment, and that he has nothing upon his premises that pays him so well in every respect as they.

Finding so satisfactory a return from this part of his stock, he extends this extra care in his purchases of sheep and swine, and after a few years of trial he discovers that they pay equally well in proportion as his cows. And in this way his improved system of husbandry progresses from year to year, and his methods are patterns for his neighbors, until a whole district is revolutionized and the old harum-scarum ways are utterly abandoned.

WORTHLESS DOGS.

Some heartless wretch has been putting out poison again, and this time Mr. Ferguson's dog Daisy is the victim. We have no doubt that any one of the poor brutes thus destroyed were of infinitely more value to the community than the cowardly carcass of the destroyer.—*Potaba Spirit*.

This is drawing it pretty strong, and we should judge that the writer was an ardent lover of dog-flesh or entertained a desired spite at some suspected party. In any event he is a little too harsh. If the truth was known we should learn, no doubt, that the lamented curs were molesting some poor, hard-working farmer's sheep, who, becoming exasperated, put out poison in self-defense. It would be a blessing to the country at large if there were more men with nerve enough to do just what this one has done. True, a good and valuable dog might occasionally be brought to an untimely end, but the country would soon be relieved of one of the greatest, if not the greatest, drawbacks to sheep husbandry. In Ohio it has been estimated by competent authority, that the loss sustained by sheep raisers from the depredations of dogs alone would fully reach the handsome sum of $60,000 annually. This estimate might safely be applied to the Pacific Coast in proportionable ratio—as no country can produce so many worthless curs to the population. A prominent sheep raiser once told us that if people would take care of their dogs he could take care of the wolves, it having been argued that were it not for the dogs (bounds) the wolves would destroy all the sheep in the community. A good dog is a valuable animal, but he should be looked after by his owner just as much as a herdsman looks after his herd, it will be better for all concerned. We cannot blame the half-starved dogs that run about seeking something to eat ; it is the masters who are to blame and should be made to stand the damage done.

THE BLACK WALNUT.

An address delivered last winter by W. H. Ragan, secretary of the Indiana Horticultural Society, on cultivating the black walnut for profit, contains so much that is valuable that we are induced to refer on the present occasion to some of the facts which it presents, and to add a few further suggestions. Mr. Ragan thinks the black walnut the most valuable of all trees for artificial plantations and timber belts. He states that a man in Wisconsin planted "a piece of land" twenty-three years ago with this tree. We are not informed the extent of the land covered with it, but that the trees, sixteen to eighteen inches in diameter, were sold for $27,000. He adds that walnut lumber now commands from $75 to $100 per thousand feet in the cities, for parlor decoration and other purposes. The tree bears nuts at an early age, and annually thereafter, which have an important commercial value.

In raising the trees, it is of utmost importance to do everything in the best manner. Those who carelessly plant the nuts, especially after they have dried for a long time, will probably fail to get trees; or if any grow, and the owner expects the young trees to take care of themselves, he will be greatly disappointed. Mr. Ragan's directions are, therefore, to the point, when he says the ground should be prepared in the best manner in the autumn. Furrow the ground off each way as for corn, except that the rows should be seven feet apart. Take the nuts, fresh from the tree, and plant two at each crossing. They are to be covered shallow, just enough to hide them. So much for planting. Then next spring furrow the seven-feet spaces intermediate between the rows, and plant with corn or potatoes. The corn and young trees will be all cultivated alike, and young trees must be kept clean. The second spring thin out the trees to one in a hill. The thinnings will fill any vacant spaces where needed. Corn or potatoes may be planted the second, or even the third year, and after that the trees must be cultivated and kept clean until they occupy the whole ground so fully as to keep down by their shade all weeds and grass. Standing so near as seven feet, the trees will not require trimming, but will thus trim themselves. But when they begin to suffer from crowding, take out every alternate tree in each row, and in a few years another thinning may be made by taking out every alternate tree in the rows at right angles to the first, leaving them fourteen feet each way. If the trees are to stand until they become quite large, additional thinning may be necessary. But they should always be thick enough to obviate the side trimming of branches. The thinnings will always possess considerable value.

At fourteen feet apart there would be over 200 trees to the acre, and these should sell for five dollars each in a quarter of a century, or at $1,000 an acre. It is not likely that the timber will become cheaper in future years. If the good cultivation and management here described are given, there will be little or no failure of a full, even growth. If the work is carelessly performed, and the trees neglected, they will be poor and scattered. The regular planting in rows, and the continued cultivation until they wholly shade down all other growth, are indispensable to success, and they are equally necessary in raising plantations of any other trees, as chestnuts, locusts, or catalpas.

OUR LOCAL ORGANIZATIONS.

LANCASTER COUNTY AGRICULTURAL AND HORTICULTURAL SOCIETY.

The society met in their rooms on Monday, October 2d, 1882, when the following persons made their appearance: Wm. H. Brosius, Liberty Square; H. M. Engle, Marietta; W. B. Paxson, Colerain; Levi S. Reist, Oregon; Peter S. Reist, Lititz; M. D. Kendig, Creswell; Robert Patterson, Coleraln; C. L. Hunsecker, Manheim township; John M. Clark, Chestnut Level; J. Hoffman Hershey, Rohrerstown; Daniel Smoych, city; S. P. Eby, city; H. M. Mayer, Rohrerstown; J. M. Johnston, city; H. W. Stein, city; W. W. Griest, city.

Vice President Engle called the meeting to order at three o'clock, and M. D. Kendig was elected temporary secretary.

On motion of Wm. H. Brosius, Wm. T. Clark, of Liberty Square, was elected a member of the society.

Crop Reports.

Peter S. Reist reported for Warwick township, that the wheat is about an average crop; corn three fourths; oats one-half; grass three-fourths; apples three-fourths; tobacco three-fourths, but of very excellent quality. The chestnut crop promises to be extraordinary. Sowing is about three-fourths finished.

Washington B. Paxson, of Colerain, said that grass is green and growing; seeding three-fourths done; wheat looks well; corn is as good as usual; grapes abundant; apples generally dropped off; tobacco generally pretty good, though a great deal was destroyed by hail.

Down in Drumore, according to the report of Wm. H. Brosius, grass is good, corn a full crop, tobacco about all harvested, and seeding pretty much done; apples falling off.

M. D. Kendig noticed that more cattle are full fed in his section of Manor township than usual, probably by reason of the great crops of straw and grass. The tobacco leaves are longer than they have been in many years, which augurs well for a good curling. It is about all cut. Corn is a fair crop. All the apples are doing poorly, most of them falling off, except Smith's cider and the Fallowater, which hang well and are in fine condition. Rainfall for September 4 6-10 inches.

The seeding is not advanced so much in the river corner of East Donegal, reported Henry M. Engle, as it is in the other sections heard from. Corn is a full crop. The ground is in excellent condition; young grass hasn't looked so well for years; apples are poor and falling off. Rainfall for September 2 inches, for August 2 inches.

H. M. Mayer, of East Hempfield (Rohrerstown), reported for that section very good wheat; a good average corn crop; apples not falling off so much as reported elsewhere; peaches poor; seeding three-fourths finished; young grass in excellent condition; tobacco a good crop, it having been topped low and then obtained magnificent growth.

The Value of Clover Land.

Peter S. Reist read a paper on the "Value of Clover Land." See page 147.

Wm. H. Brosius asked whether in the light of the fact that timothy is supposed to exhaust the soil, it would do to stop sowing it?

Mr. Reist thought not, as his experience was that you would be unable to grow clover alone for two or three years.

Robert Patterson inquired what is the best time to cut clover?

Mr. Reist answered: When the blossoms are just drying off, and before the second crop has much of a start.

M. D. Kendig said that ten years ago he stopped sowing timothy, and let the clover go it alone. In three years he couldn't get his clover to grow, and was compelled to mix them again, after which the clover flourished.

C. L. Hunsecker acknowledged that timothy is injurious, but contended that the best farmers for years have been sowing timothy and clover half and half, and down along the Conestoga, where many of the farmers are large timothy raisers, he failed to notice any particularly deleterious effect. Oats is also more or less injurious, but the finest wheat comes after it, and oats clears the land of weeds, etc. Beans also injure the soil. Clover is first-rate to repair the damage, and so is lime. Jacob B. Garber, who probably knows as much if not more about agricultural affairs than any other man in the county, writes to the *Agriculturist* that lime is both an excellent manure and a first-class stimulant.

Mr. Patterson asked if lime is of any advantage without manure or vegetable matter mixed with it?

Mr. Hunsecker hadn't studied the matter scientifically and didn't know.

Henry M. Engle thought the clover question a very important one. Clover is undoubtedly the best and cheapest manure for poor land, although lime may bring the answer in less time. He would advise every farmer to avoid timothy on poor land, but, if the soil is in a fertile condition, we can sow it with some degree of impunity. Farmers are too saving of clover seed; they ought to sow more than they do, and they ought to be careful when they cut it. It is objected to as horse feed, and the reason is that it is over-ripe. Cut it at the proper time—when the heads are all fresh—gather it damp, and it makes the very best horse feed. The only trouble in sowing both timothy and clover on rich land is that they don't reach maturity at the same time, and consequently when the clover is ready for harvest the timothy is not, and *vice versa*.

Should There be Less Fencing?

S. P. Eby, Esq., read the following paper, entitled, "Could the farmer do with less fencing?"

So long as the laws of Pennsylvania, relating to fences and cattle remain unaltered, the farmer will be obliged to inclose his farm with a fence "at least five feet high, of sufficient rails or logs, and closed at the bottom." Failing in this, he has no redress for damages that may be done by cattle or other stock running at large. And he will be liable for any hurt or damage he may do to live stock in driving them out of his grounds.

His neighbor may put up the line fence between their adjoining properties, in case he refuses to do so, and make him pay the one half of the cost thereof.

There seems, therefore, to be no escape from the expense of keeping up fences surrounding his farm, except through means of the Legislature, and a change of our fence and cattle law.

As to the interior of his farm, it becomes a question of economy with himself. While he continues to rotate his crops, and pastures all his fields alternately every fourth and fifth year, he will be obliged to have some barrier, either temporary or permanent, between his fields to protect the crops from his own cattle.

A few farmers have adopted soiling as a substitute for fencing. Instead of pasturing they cut the grass and feed it to the stock in the stable or barnyard.

This practice is well spoken of by some who have tried it. They allege it saves feed, increases the manure pile, and keeps the stock in good flesh alternately, but it has some objections, however; it adds greatly to the labor of the farmer in his most busy season. The attendance must be regular and unremitting, and the grass be newly cut; otherwise the stock will suffer. And it may be a question whether with the best attendance cattle thus confined will keep in as healthy condition as if allowed free range of the field, to crop the grass at will freshly from the sod.

Another mode is to fence off and keep a certain part of the farm for exclusive and continuous pasturage. This, if managed with proper care, not too closely cropped in dry weather, and treated to a coat of manure occasionally, can be brought into a thick growth of natural grass, very nutritious and greatly relished by cattle. It is the mode practiced in many parts of Chester and Delaware counties, and is well thought of.

Of course since rainfall is on the decrease the lowest and moistest ground on the farm should be selected for this purpose, and it should have shade trees for the benefit of the stock in hot seasons, and if possible, water for them to drink.

Another mode is to keep portable fence, and inclose with it such of the grass land as is desired to pasture. This will involve the cost of the fence itself and the trouble of moving it as often as the pasture is changed.

The last remaining manner, which can profitably be adopted when the stock consists of only one or two cows, is that of staking.

This will require some training to accustom the animals quietly to submit to the restraint of rope and stake.

Mr. Broslus thought that there were two things to consider: The expense of fencing and the value of land. He favored the soiling system and argued that all farm land could be enriched more by cultivating than by pasturing.

Mr. Paxson inquired whether it would be advisable to cut and haul grass from meadow land rather than pasture.

Mr. Broslus replied that he had referred to upland, to agricultural land.

Peter S. Reist found there was a great deal of difference between the theory and practice on the fence question. It was hard to talk about, but some fences were an absolute necessity to the farmer. He had seen many try to abolish them, and he had seen them rebuild their fences. He, however, denounced the barbed wire fence and said that some day a law would be enacted forbidding its use. In fine, he thought that no farm should be divided into fewer than four fields.

Hay as a Fertilizer.

W. B. Paxson answered the question : "Can the farmer sell his hay and maintain the fertility of the soil?" He said that the most prominent question with every tiller of the soil is, how can the farmer preserve and increase the fertility of his land, or how can he restore exhausted soil, and if possible increase its productive power ? This is a question of vast importance. Experience has demonstrated the fact that barnyard manure does return to the soil all the fertility that the crop takes from it. If the farmer sells his hay then he will have less barnyard manure than otherwise, and in order to restore the exhausted soil he must apply artificial fertilizers, which, in my opinion, should not be used as a substitute for, but to supplement barnyard manure. Therefore I answer the question in the negative.

Going to the York Fair.

The following were appointed a committee to attend the York fair : Wm. H. Broslus, M. D. Kendig, C. L. Hunsecker.

The Next Meeting.

At the next meeting Wm. H. Broslus will speak on agriculture.

Fruit Report.

There was a fine display of grapes on hand, and the following committee, L. S. Reist, Robert Patterson and S. P. Eby, reported as follows concerning all the fruit exhibited :

Goethe Rodgers, No. 1, very nice ; Devon Rodgers, No. 15, very fine ; Wilder Rodgers, No. 4, large and fine ; Massasoit Rodgers, No. 3, very good ; Lindley Rodgers, No. 9, sweet ; Salem Rodgers, No. 53, very good ; Rodgers No. 33, large ; Euncline Rodgers, small and very fine ; Clinton, a good grape ; Martha, sweet and very good ; Concord, a fine grape ; Paxton, a good grape ; Isabella, very fine ; Croton, small but good; Telegraph, very fine ; Maxatawney, white and good ; Franklin, small and tartish ; Harford, over ripe ; Ives' Seedling, small and luscious ; Iona, beautiful and sweet.

Mr. Smeych exhibited the following : Four plates seedling peaches, very fine, Grapes: One plate Black Hamburg, large and fine ; one plate Bordeaux, fine ; one plate Diana, very sweet ; one plate Rodgers No. 1, good ; one plate Rodgers No. 28, fine ; one plate, no name, good ; one plate, good and sweet. Two Springer plums, very fine and sweet.

Mr. Levi S. Reist exhibited some very fine York Imperial apples.

POULTRY ASSOCIATION.

The meeting of the Poultry Association met on Monday morning, October 31, 1882, and was attended by the following persons : President, G. A. Geyer, Secretary, J. P. Lichty, city ; Charles Lippold, city ; John E. Schum, city ; Dr. E. H. Witmer, Neffsville ; Harry Stein, city ; W. W. Griest, city ; Dr. H. D. Longaker, city ; I. Hoods, West Willow ; Joseph Triesler, city ; J. M. Johnston, city ; Wash. Hershey, Chickies.

Mr. Lichty, as chairman of the Executive Committee, reported that the premium list had been arranged in part as follows :

For poultry $2 for first, $1 to second, highly commended to third ; for pigeons, $1 to first, fifty cents to second, highly commended to third ; entrance fee, seventy-five cents for single bird for poultry and thirty-five cents per pair for pigeons. Of the cash specials offered by the association the society will charge ten per cent. The premium for breeding pens is $5.00 each, and entrance fee $2.00.

A breeder's stake will be made up, and every bird is charged one dollar; the purse will be divided thus : Sixty per cent. to first, 30 to second, and 10 to third ; the birds must be raised by the breeder.

The offer of T. B. Dorsey, of Delaware, of a $25 cup for the best bantam on exhibition—he to contribute $15 and the society $10 of the required $25—was accepted. The entrance fee $2. A Polish cup with same conditions was also accepted.

Mr. W. A. Jeffrey, of Ashland, Ohio, editor of the *National Poultry Journal*, offered a twenty-five dollar silver cup for the best collection of white crested black Polish by any one exhibitor, and his offer was accepted with thanks.

The above report of the Executive Committee was unanimously adopted.

Charles E. Long, by reason of his inability to attend to the duties of a member of the Executive Committee, tendered his resignation, which was accepted. Charles Lippold was nominated to fill the vacancy, and he will, under the rules, come up for election at the next meeting.

Dr. Longaker said he was enlarging his hatching house, and Mr. J. M. Johnston, of the *Intelligencer*, precipitated an adjournment by telling how, " when I was down in Georgia," a spring chicken no larger than his fist swallowed a snake as big as his fist !

FULTON FARMERS' CLUB

The regular monthly meeting of the club was held at the residence of Day Wood, on Saturday, October 7th, 1882.

The following members were present : Day Wood, Josiah Brown, J. H. Blackburn, Lindley King, Montillion Brown, S. L. Gregg, Wm. King, John Cauffman, Joel King and E. H. Haines.

Visitors : Dr. S. T. Roman, J. J. Carter, Harvey Hower, Neal Hamilton, Isaac Bradley and Clifford Cook, all of whom were accompanied by portions of their families, making a company that filled to overflowing the capacious and elegant parlor of the host.

The members all having exhibited specimens of their farm products at their last meeting made the exhibit to-day quite small, a single apple, brought by E. H. Haines to be named, was the sum total, and it was decided to be a King of Tompkins county.

Questions and Answers.

Day Wood—I have heard it said that it pays best to feed cattle when corn is high in price, is this the case ? This was a question that most of those present seemed to have thought but little about, but all seemed to agree that as corn is the material out of which winter beef is made, a scarcity, and consequently high price of the former must of a necessity be followed by a scarcity and high price of the latter. If corn can be fed to steers or hogs with a certain per cent. of profit on its market value, whether the price be low or high, then the production of beef or pork is much more profitable to the farmer in times of high prices than in times of low prices, just upon the same principle that a miller who tolls the grain he grinds gets twice as much for grinding a bushel of wheat when it is worth two dollars per bushel as when it is worth only one dollar per bushel.

Josiah Brown—In this neighborhood does it pay better to put in wheat after corn than to put the stalk ground in oats ? Several were in favor of sowing the wheat provided the land was rich, but on this land past experience did not recommend the plan much. Some objected to putting in wheat after corn on account of the labor required to remove the corn, while others did not consider this any great obstacle if the right plan was pursued, which is to cut two rows of corn where you wish the shocks to stand and lay one on each side, then cultivate and drill in these two rows, after which cut and shock the corn on the ground thus planted. In this way

the corn need not be carried more than the usual distance, and the shocks will not kill the wheat if not suffered to stand long.

Montillion Brown—Has any one present a variety of grape that does better for him than the Concord ? Some of the members were cultivating as many as eight or ten varieties, but no one had found any of them to do better than the Concord.

William King—Is it better to put apples in the cellar immediately after picking or barrel them and leave them out until cold weather ? J. J. Carter was in favor of leaving them out of doors in a dry, cool place. Day Woo l preferre l to put them in the cellar. Dr. Roman said to get barrels that were perfectly tight or make them so by pasting muslin over the joints, put the apples in these, head them up and leave the barrels under the trees, but on something to keep them off the ground, until freezing weather.

Josiah Brown said if you leave the apples out until the cold weather, they should be put in barrels, but if put in the cellar, they should be put on shelves.

Martha Brown—When is the best time for storing away cabbage for winter ? Answered—Just before Hallow-E en.

Josiah Brown cuts the stalks off his cabbage and packs the heads in a barrel planted in the ground ; they keep well and are easily got when wanted.

After dinner the usual inspection of the farm and stock was made. The condition of the farm, the crops and stock were all considered good.

The host made the following report of his farming operations for 1881 :

Crops.

26 acres of wheat produced	467½ bushels
17½ " corn	1,050 "
10 " oats	263½ "
5 " tobacco	6,241 pounds
30 " hay produced	30 4-horse loads
	Hungarian hay ....5 "
18 acres clover seed	6 bushels
½ acre potatoes	34 "

Receipts between Jan. 1, 1881 and Jan. 1, 1882.
For 4½ acres of tobacco (8,052 lbs) .... $ 877 21
" 6 cows and heifers, ... 201 00
" 26 lambs (a 4.50 .... 134 20
" 139 pounds wool (a 20c 27 80
" 28 head old sheep $4.00 .... 112 00
" 1,796 pounds butter 572 80
" grain sold (wheat and clover seed) 514 36
" 494 dozen eggs 82 30
" poultry, calves, pork and lard 199 23

Total sales.............................$2,719 80

Literary Exercises.

Montillion Brown read from THE LANCASTER FARMER an article on "The Practical and Scientific in Agriculture," of which the following is an extract : ."The foundation of all knowledge of agriculture is the accumulated of fixed facts, suggested by accidents discovered perhaps by science ; but, however obtained, proved or confirmed by the practical farmer on the land. A theory which bears this test may become a law at once for the farming community, and until it has borne the test, it is theory still, no matter what its origin may have been, whether college or farmyard. While, therefore, an agricultural school may be devoted to science as a guide to agriculture, and may be engaged in cultivating a single farm according to the best known principle, it must depend upon a widespread community of farmers for the last grand process of proving and diffusing its theories. And when we remember that agriculture is not an exact science and cannot be until the skies and seasons are subdued by man, and that the facts discovered in the field are often of more practical value than those laid down by the student in his closet, we shall not be surprised at the success which associated farmers have met with in the work of advancing agricultural education."

Sadie Brown read from the Lancaster *Examiner* an editorial on " Farmers' Societies and Festivals."

Rebecca D. King read a temperance story, "A Strong Temptation."

Neal Hamilton delivered an amusing stump speech he had learned when a boy.

Day Wood read an article on "Ants," by Mark Twain.

Resolution for discussion at the next meeting: *Resolved*, That the experience of a farmer is of more benefit to him than the writings of others.

Montallion Brown, J. R. Blackburn, Phœbe King and Sadie Brown were appointed a committee to have some literary exercises for the next meeting, which will be held at the residence of C. C. Cauffman on the first Saturday in next month.

LINNÆAN SOCIETY.

The Society met in the ante-room of the Museum on Saturday afternoon, Sept. 30th, at 2 o'clock P. M. Prof. J. S. Stahr in the chair, and S. M. Sener, Esq., secretary *pro tem.*

The donations, contributions, and additions were the following to the museum : two specimens of parrots were donated by Mr. N. B. Fondersmith, bird tancier of East Orange street, and stuffed and mounted by Mr. George O. Hensel, taxidermist and florist. One of these birds was an ash-colored, or gray parrot, and the other green and red, and both had died in their cages, a fate to which foreign birds are extremely liable in our northern climate. The former is Psittacus erythacus, and the latter seems to be an immature specimen of Psittacus festivus, of South America. An alcoholized specimen of a "Hornworm," Macrosila Carolina, infested by about three hundred insect parasites, a description of which had been published in the *New Era* and the LANCASTER FARMER during the month of September. Also a specimen of the larva of Dryocampa imperialis, donated by Maj. J. R. Wimloiph of the Cornwall farms. Also a mature female specimen of the "oil-beetle," Meloe angusticollis, captured about a mile north of Lancaster, about a week ago. If this cannot be regarded as a rare insect, on the other hand it is never found very abundantly in this locality. When captured, it exudes its oil very freely, has an unpleasant odor, and from the fact that all the other insects in the bottle died, and the oil-beetle alone survived, it may be inferred that it was poisonous to them. Mr. Milton Wike, of Columbia, donated a very extraordinary cranium of a "ground hog—Arctomys monax—found in Martic township, near McCall's Ferry, at a place called the "Pinnacle." One of the front teeth (incisors) in the upper jaw grew round in a circle and entered the jaw again near the base of the first molars and the other in a similar form, grew out of the side of the mouth. The animal could not have possibly brought its molars together within an inch, and as this animal is a rodent, and lives exclusively on vegetable food, it must certainly have starved to death. The incisor teeth of rodents are mainly used for cutting, and the question is, how could these teeth possibly have grown so rapidly, as to prevent the animal from bringing their ends together, and wearing them down as is usually the case? for there is hardly room for the inference that they grew in that condition after the death of the animal. If it starved, it would be an interesting fact to know how long it had survived in that condition before death ensued. Two double peach stones, from double peaches, in both cases growing from one stem, and inferentially from one blossom; donated by Messrs. Thomas and Fondersmith. This phenomenon occurred quite frequently the present season and in one instance, at least, *three* were found growing from a single stem. A specimen of bituminous coal from Vancouver's Island, British America, donated by Mr. Washington L. Hershey, Chiques farm. Specimens of coffee, and a piece of the keel of the ill-fated vessel S. S. Pliny, which was wrecked in May last, on the coast of New Jersey, between Deal Beach and Elberon, N. J., donated by Messrs. C. A. Heinitsh and Joseph Steinhauser. A prepared specimen of Erythræa comælestina, var. pulchella; or, as it is some times called, E. mublenbergia; was added to the Herbarium of the society by Prof. J S. Stahr. This plant is new to the flora of Lancaster county, and was found by Prof. Stahr in a small ravine in the western part of Lancaster city in July last. Since

then it has also been found at Media Hill, by Mr. Vetscher, a student of Dr. M. L. Herr, of the graduates of Franklin and Marshall College. It belongs to the family Gentianaceæ, and is nearly related to Sabatia angularis, commonly called "Centuryplant," a favorite bitters among the Pennsylvania Germans. A collection of prepared plants and flowers was exhibited by Mrs. P. E. Gibbons, also specimens of the water chestnut, used as a food in France. The plants were collected in Huntingdonshire, England, by Mrs. Gibbons on her late visit there.

To the library, first series of the official records of the Union and Rebel armies, in the war of the rebellion; from the Department of the Interior. This series includes four volumes, royal octavo, comprising in all 3,400 pages. United States Coast and Geodetic Survey for 1879, a quarto of 214 pages and fifty-three progressive sketches and illustrations, also from the Department of the Interior. Nos. 6 to 12 of the "Official United States Patent Office Gazette," Vol. XXII., from the same. THE LANCASTER FARMER for August and September, Part 2, Vol. III, Proceedings of the Davenport Academy of Natural Science, from January, 1879, to December, 1881, 193 pp. Royal 8 vo. and 4 plates ; from the corresponding secretary. This report contains a large amount of interesting Western mound-lore, and announces the death of its late president, Joseph Duncan Pitman, in the prime of life, an industrious and progressive scientist, and one who had already made his mark in the scientific world. Three envelopes containing thirty historical and biographical selections. A number of book catalogues and circulars.

Prof. Stahr read a paper entitled "Botanical Notes," in which he referred specifically to the plant he donated ; also to Viola lanceolata, provisionally, found by Mr J. C. Foltz, in Drumore township ; also, to some peculiarities in several specimens of the night-blooming cacti of Lancaster city.

Prof. Buehrle, City Superintendent of the public schools, was proposed for active membership by Prof. Rathvon, which, under the rules, lies over until the next meeting.

S. M. Sener was unanimously elected assistant secretary.

After a short session of science gossip, the society adjourned to meet at the office of Dr. Knight, North Queen street, in October, of which due notice will be given by the secretary.

AGRICULTURE.

Wheat Growing.

The success in growing wheat in Pennsylvania the last few years should stimulate us to raise a greater average per acre than has been the case in many portions of the State. We notice that as much as an average of thirty bushels has been obtained this year in some of the Western States; and we are well aware that the yield has been increased this year in Pennsylvania—in some special instances over forty bushels per acre have been obtained. Of course there are various causes influencing success. That which might be an aid at one point might be an injury at another. But there are one or two matters that wheat growers are apt to forget. The first is, that as a general thing it is well understood that manure must be liberally applied to induce a good crop; but many persons plow it under, hence, it is not until the plant has set its roots deep down into the soil that it derives much benefit from the manure. But if the manure is so placed that the young rootlets could push at once into it on germinating, it would get an early start on its vital course, which would establish it firmly against any future drawbacks.

In the second place few persons have any idea how manure operates in making roots. If we bury a shovelfull of manure some distance from a thrifty tree in early spring, and examine it again the ensuing fall, we find the lump of dung a complete mass of roots, while the earth in other parts contiguous

has but a few struggling ones. Some people think that the roots are attracted to the spot by the manure; but it is not so. They are actually created by the manure. A leading root sucks into the rich mass, and finding plenty to eat, at once sets to work to increase and multiply. Contact with the manure, therefore makes roots; and the principle in successful wheat culture should place the grain and the food as close together as possible, if we would encourage it to root out well and get a good start. We all know very well how this is done with corn. Manuring in the hill is quite a universal practice; but where it is not, the result is well known. We repeat, therefore, give the crops an early start. It has a wonderful influence in its efforts to in after life to come out.

An Excellent Fertilizer.

A German farmer once told the writer that every year he prepared a heap of manure which, when applied to his soil, made it produce marvelous yields. His mode of preparing it was as follows, to use his own language. "I have but one horse, one cow and about two dozen fowls. I save every particle of their droppings and place them under a shed which has a cemented floor; upon this I spread a layer of forest mold, and in order to preserve the ammonia in it I cover the dung with another layer of mold, taken from the woods close to my house. I continue this system of layering each time the stable, cow and hen houses are cleaned out. I also save the urine of the animals and that from my house, and pour it upon the heap; sometimes I also add a small quantity of litter from the stable, and, when not too busy to collect them, a lot of leaves. By attending to the heap in person, and seeing that all the manure is rigidly saved, I find on hand by the early spring a large quantity of the richest fertilizer I have ever used. During the few winter months it has thoroughly rotted, and when needed to spread upon my garden it resembles a heap of ashes, so completely is it pulverized. My garden consists of five acres of ground, which receives this valuable manure. On a farm where twenty or thirty horses, mules and cattle are kept, and a flock of sheep and a fair number of fowls, besides their combined droppings, if treated as above, enough of this excellent fertilizer could be saved each year to thoroughly manure fifteen or twenty acres of land, and no farmer should be at a loss to have what manure he requires, for this is the foundation of successful farming."

How to Remove Stumps.

The *Scientific American* gives the following receipt for getting rid of stumps : " In the autumn of every winter bore a hole one or two inches in diameter, according to the girth of the stump, and about eighteen inches deep. Put into it one or two ounces of saltpeter, fill the hole with water and plug it close. In the evening spring take out the plug and ignite it. The stump will moulder away without blazing, to the very extremity of the roots, leaving nothing but the ashes "

The Telephone on the Farm.

A French farmer uses a portable telephone to carry on the work of his farm without going away from his house. His plan is simply to have a tripod carrying a movable roller, on which is wound a conducting cable composed of two insulated wires. Below this on a movable board is a small box, in which is placed a telephone and bell. The system allows the current to pass from the bell to the telephone without using a commutator. Thus, the telephone being at rest, the bell is in connection with the line, and when the telephone is in use the bell is cut out of the circuit. Another telephone and bell are fixed in the house of the farmer, with a commutator.

Octagonal Barns.

If a barn is wanted to accomodate a certain number of animals, the proper space is better and more cheaply obtained in the octagonal form, for this gives an equal space in every direction, and requires the least outside wall. It would require a 66 foot octagon to accommodate forty head of cattle and give six box stalls (9x10 feet,) with plenty of room for calves besides. This form of barn might also be enlarged by building a wing on when wanted. Four wings would look well on such a barn, and might be built wide enough for two rows of cattle. On a larger octagonal centre eight wings might be built, increasing the room to almost any extent. On the size above given, two wings, in the direction to extend the feeding floor, and the rows of cattle in the octagon, might be built without injuring the appearance of the barn—the octagonal centre relieving the long line by the appearance of the elevated dome. And when wanted, two more wings could be added, still improving its appearance. This form of barn is certainly the most convenient, and is least expensive according to space inclosed. The octagon gives a wider space, which can be laid out more conveniently than in a long narrow barn, and all parts being equi-distant from a centre, such a barn requires less travel in doing the daily work.

HORTICULTURE.

York Imperial Apple.

If ever a fruit did better in Eastern Pennsylvania than the York Imperial apple in the few years it has been tested, then it must be as near perfect as we can expect. It is as regular in bearing as the return of the seasons; as large as the favorite old Pennock, and as handsome in color as was the variety in "the good old times" of our fathers; has no imperfections to speak of mar its glossy red surface; and in quality just that nice commingling of acid and sugar sure to please the majority of judges of good fruit. It is not so rich as the Smokehouse, and yet it is by no means deficient in flavor; nor so spicy as the Newtown Pipin, although it possesses a fragrance peculiarly its own. In the orchard the outline of the tree is not to be commended, and yet it is a remarkably healthy and vigorous grower, with rich dark green foliage. It will not produce so many apples as Smith's cider, but there will be more bushels per tree; and as regards value, the York Imperial is immeasurably its superior, and always commands much better prices.

Keeping Apples.

As the time is at hand when the work of picking and putting up the apple crop for the winter and spring will have to be attended to, it is well that the methods of preserving this valuable fruit should be considered. We have hitherto on frequent occasions discussed them, and pointed out what we conceived to be the best method to pursue. In brief, we would, therefore, repeat in substance, as follows: 1st. The apples must be good keepers, free from bruises or blemish. 2d. They must be spread out on shelves or packed in barrels, and kept in an atmosphere of from forty to fifty degrees, better from forty to forty-five, and at a temperature as equable as possible. Some cellars are just the thing and preserve them beautifully. Others are too moist. Where this is the case a few bushels of stone lime should be used. Sliding shelves, six inches apart, latticed bottom with a single layer of fruit, are extremely convenient, as they allow of constant examination without disturbing the fruit. A friend informed us some years ago that with a large stand of these shelves in his cellar, with a few inches of lime on the bottom of the cellar, he kept his apples into May in perfect condition and good flavor.

A vault in the cellar, kept closed, but with some ventilation, frequently answers admirably, as we know from personal experience. If carefully packed in clean, naked barrels, the head forced down in order that the fruit may be quite solid, and the barrels placed under an open shed until late in November, but before hard freezing comes on, and then be removed to a dry cellar, where the temperature will auge about what is stated above, there will be litt danger of the fruit not keeping through the winter and late into the spring. Indeed, we have known it to keep until June.

Apple Notes.

Apple exhibitors at the Southern Illinois fair report the Nickajack as worthless for this latitude and the Lawyer as a very shy bearer. There were some monstrous specimens of the Buckingham show, and all growers united in declaring this variety to be one of the very best for Southern Illinois, as a late summer and fall apple. Winesaps are reported as falling off, and some growers have already begun to harvest them. The St. Lawrence was reported as one of the very best table and market summer apples. It is a great bearer of beautiful red striped, good flavored fruit. Growers united in commending the Benoni as the very best and first good apple in the market. It is of fine color and flavor, an enormous bearer, early in coming into bearing, and brings more money than any other early apple. It is said to be a far better variety every way than the Red June. The Red June is apt to run small and badly shaped, whereas the Benoni is uniformly perfect in form and increases in size as the crop is thinned. Its one fault is that of occasionally being water cored.

Root Pruning.

The experiments were made on the apple and pear. A vigorous apple tree, eight or ten years old, which had scarcely made any fruit buds, has done best when about half the roots were cut in one season and half three years later, by going half way round on opposite sides in one year and finishing at the next pruning, working two feet underneath to sever downward roots. It has always answered well also to cut from such trees all the larger and longer roots about two and a half feet from the stem, leaving the small and weaker ones longer, and going half way round, as already stated. The operation was repeated three or four years later by extending the cut circle a foot or two further away from the tree. By this operation unproductive fruit trees be come thickly studded with fruit spurs, and afterward bore profusely. This shortening of the roots has been continued in these experiments for twenty years with much success, the circle of roots remaining greatly circumscribed. The best time for the work has been found to be in the latter part of August and beginning of September, when growth has nearly ceased, and while the leaves are yet on the trees, causing greater increase of bloom buds the following year than when performed after the leaves had fallen.—*London Garden.*

The Cherry and Apple.

S. F. Larkan, of Delaware county, contributes the following to the *Germantown Telegraph:*

Various letters of inquiry as to the profitableness of the cherry as a market fruit, having reached me at various times, and not having been as fully answered as their importance demands, I appeal to the reliable old *Telegraph* for a more complete answer to all. We are cultivating the cherry, the apple, the pear, the grape, and several berries for market—having abandoned the plum long since, and lately the peach, as being impractical. About three acres in all may be occupied by the cherry, and fully seven by the apple. I will first compare these. Taking an average of seasons we can receive annually six hundred dollars for cherries to three hundred for apples. We can raise twice as much other produce from a cherry orchard as we can from an apple orchard. The cherry harvest lasts about a month; the apple six.

You, Mr. Editor, can speak from experience that we can make a cider from the apple that is equal to three-fourths of the famous champagne sold, and it is really superior for invalids. We have never attempted to make wine from the cherry, profitable as it might be. Then, again, the apple is not subject to the losses occasioned by the gambling middlemen, the cherry market is. Yet, with all, the cherry beats the apple in profit two to one. I think this experience ought to satisfy any one that there is no risk in planting the cherry as a profitable market fruit. Our canned cherries take premiums, and the dried fruit is unsurpassed by any other dried fruit in the markets. The cherry tree requires less care to propagate than the apple, though neither should ever be set in what is known as an "orchard," which is too much of a good thing together.

Pine-Apples.

Pine apple culture will in a short time become one of the best paying businesses in South Florida. The success that has attended it the past two years has encouraged a number of people to turn their attention to this fruit. The Indian-river country as low down as Lake Worth is admirably adapted to it and large sums of money are being made. Around this settlement almost every one has his patch, and pine-apples have been selling in town all the summer. Dr. Voorhis, Messrs. O. P. Terry, W. B. Wood, F. Norris, H. B. Austin, and others have done well. The apples grown here range in size from 6 to 8 pounds, and are well-flavored; the Red Spanish and Sugar Loaf have been those hitherto cultivated, but finer varieties are now receiving attention, those sold here have brought 10 to 35 cents each.—*Florida Agriculturist.*

HOUSEHOLD RECIPES.

CHOW CHOW.—Two quarts of tomatoes, two white onions, half dozen green peppers, one dozen cucumbers, two heads of cabbage, all chopped fine; let this stand over night; sprinkle a teacup of salt in it. In the morning drain off the brine and season with one tablespoonful of celery seed, one ounce of turmeric, half teaspoonful of cayenne pepper, one cup of brown sugar, one ounce of cinnamon, one ounce of allspice, one ounce of black pepper, one quarter ounce cloves, vinegar enough to cover, and boil two hours.

STUFFED TOMATOES.—Take six large, well-shaped tomatoes; cut a slice off the stem end and take out all the pulp and juice, being careful not to break the tomatoes; then sprinkle them inside with a little salt and pepper; have a pound of cold cooked veal, beef or chicken, a slice of boiled ham or fried bacon, chop very fine, and add the pulp and juice of the tomatoes; chop fine and fry to a light brown, half an onion, and mix with the meat a teacupful of fine bread crumbs, two eggs, a teaspoonful of salt, a salt-spoonful of white pepper, and a pinch of cayenne; fill the tomatoes with the force-meat, piling it quite high, and bake for an hour.

PANCAKES.—Beat up three eggs and a quart of milk; make it up into a batter with flour, a little salt, a spoonful of ground ginger, and a little grated lemon peel; let it be of a fine thickness and perfectly smooth. Clean your frying pan thoroughly, and put into it a good lump of dripping or butter; when it is hot pour in a cupful of batter, and let it run all over of an equal thickness, shake the pan frequently that the batter may not stick, and when you think it is done on one side toss it over; if you cannot, turn it with a slice, and when both are of a nice light brown, lay it on a dish before the fire; strew sugar over it, and so do the rest. They should be eaten directly, or they will become heavy.

RISSOLE SOUP.—Take the fat from the top of your cold stock. Pick out some of the best pieces of meat—about a cupful—and set aside. Add a pint of boiling water to the stock, and boil slowly, with the bones and the rest of the meat, for nearly an hour. Chop the meat reserved from the stock; make into force-meat with fine crumbs, seasoning with onion, parsley, pepper, nutmeg, and binding with beaten egg. Flour your hands and make this into round balls. Roll them in flour; set in a floured pie-dish, not touching each other, and leave in a quick oven until crusted over. Let them cool. Strain your soup; add such seasoning as you desire; heat to a boil; drop in the force meat rissoles, and heat without boiling three minutes.

LAMB CHOPS.—Trim off fat and skin it, leaving a bare bit of bone at the end of each. Broil quickly over a clear fire; butter, salt, and pepper each, and stand them on the larger ends, just touching each other, around your mound of potato.

POTATO MOUND.—Mash smooth, with butter, milk, salt, and pepper; make into a smooth mound upon a hot dish, and arrange the chops around it.

LADIES' CABBAGE.—Boil a firm cabbage in two waters. When done, quarter it and let it get perfectly cold. Chop fine; add two beaten eggs, a tablespoonful of butter, pepper, salt, and three tablespoonfuls of milk. Stir all well; pour into a buttered pudding-dish, and bake, covered, until very hot; then brown. If your dish has been well-buttered, turn the cabbage upon a hot dish, and pour over it a cupful of drawn butter.

DAMSON TART.—Fill a pie-dish, lined with good paste, with ripe, sound damsons; sweeten very plentifully; cover with crust and bake. Brush with beaten egg when done, and return to the oven one moment, to glaze.

POTATO PORRIDGE.—Twelve potatoes, peeled and sliced; 1 large onion, also pared and sliced; 2 quarts of boiling water; 1 cup of hot milk; 3 beaten eggs; 3 tablespoonfuls of butter rolled in flour; salt, pepper, and 1 teaspoonful celery essence; chopped parsley.

Fry potatoes and onions light brown in a little butter. Put into a soup pot with the boiling water, and cook gently until soft. Rub through a colander to a smooth *puree*. Add the water in which they were boiled, and return to the fire. When the *puree* begins to bubble, stir in the buttered flour, pepper, salt, and chopped parsley, and simmer five minutes. Heat the milk in another vessel; pour upon the eggs; cook one minute, and pour into the tureen. Add the *puree;* stir in the celery essence, and it is ready.

ROASTED SWEETBREADS.—Three fine sweetbreads; 1 cup of gravy—a cup of your soup will do; 1 beaten egg; cracker-dust; 1 teaspoonful mushroom catsup; 1 small glass wine; a very little minced onion put into the gravy; 2 tablespoonfuls melted butter; fried bread.

BOIL AND BLANCH THE SWEETBREADS.—Wipe perfectly dry, roll in egg, then in the pounded cracker. Lay in a baking-pan; pour in the melted butter slowly over them, that it may soak into the crumbs. Set in the oven, cover and bake 45 minutes, basting freely, from the time they begin to brown, with the gravy. Dish upon crustless slices of fried bread. Strain the gravy; add catsup and wine; boil up, and pour over the sweetbreads.

POTATO CROQUETTES.—Mash the potatoes, and beat in a raw egg, butter, milk, nutmeg, a little grated lemon-peel, with pepper and salt. Heat in a saucepan, stirring constantly, for three minutes. The saucepan should be buttered first. When cool enough to handle with comfort, make into croquettes, roll in flour, or dip in egg and cracker-crumbs, and fry—not putting too many into the pan at once—in boiling lard, or dripping. Drain in a hot colander, and serve.

RICE PUDDING COLD.—Two quarts of milk, one gill of rice, one teacup brown sugar, one stick of cinnamon about three inches long; wash the rice in a colander to remove the floury particles, which are so much loose starch and spoil the pudding; put it in the baking dish, scattering in a quarter of a pound of rasins; cook very slowly for two hours. Keep a cover over the dish until the last half hour, when the upper skin may be allowed to brown; do not stir it, as this breaks up the rice; it ought to look like rich yellow cream when done. A large piece of thick paper or a large plate can be used to cover up the pudding dish.

BREAKFAST CAKES.—To make warm weather breakfast cakes take one cup of brown sugar, nearly one cup of butter, or lard and butter mixed, one cup of sour milk, four cups of flour, four teaspoonfuls of soda (not heaping, but even full), one teaspoonful each of cinnamon, salt and ginger, one egg; bake in gem tins. These will keep well for a week.

CREAM NECTAR.—Two pounds of lump or granulated sugar, two ounces of tartaric acid, juice of one lemon, half a cup of flour mixed smooth in a little water and three pints of water. Boil five minutes. When cold stir in the whites of three eggs beaten to a stiff froth and a half ounce of wintergreen or any other flavoring one may fancy. Bottle and keep in a cool place. When wanted put a fourth teaspoonful of soda into a glass of ice water, and then add two tablespoonful of this syrup.

POTATOES AU MAITRE D'HOTEL.—Slice cold boiled potatoes rather thick. Have ready in a saucepan four or five tablespoonfuls of milk, a good lump of butter, with salt, pepper and minced parsley. Heat quickly; put in the potatoes; and stir until almost boiling. Stir in a little flour, wet with cold milk; cook a moment to thicken it; add the juice of half a lemon, and pour out into a deep dish.

STEWED TOMATOES AND ONION.—Peel, slice and stew a dozen tomatoes ten minutes. Then add a small parboiled onion, cut up small; stir in sugar, salt and pepper, with a good spoonful of butter rolled in flour. Simmer five minutes and pour out.

STEWED PEARS WITH RICE.—Pare and halve eight large pears. Put into a saucepan with eight tablespoonfuls of sugar and a cup of claret—or if you prefer, clear water. Stew slowly until tender and clear. Take out the pears and boil down the syrup to one-half, flavoring, then, with essence of bitter almond. Have ready two cupfulls of boiled rice, cooked in milk and sweetened. Spread out upon a flat dish; lay the pears upon it, and pour on the syrup eat very cold.

OX-CHEEK SOUP.—Two ox cheeks, three onions, two carrots, two turnips, twelve whole black peppers, six cloves, salt, five quarts of water, one half cup of German sago. Break the bones of the cheeks, and wash well with salt and water. Cover with cold water; bring to a boil, and throw off the water. Fry the sliced onions, and put into the pot with the meat, also the sliced carrots, onions, and spice. Cover with a gallon and a quarter of water. Bring to a slow boil, and keep this up, skimming often, for four hours. Strain off the liquor; pick out the meat and bones, salt highly, put into your stock-pot with nearly half the broth. Set in a cold place for to-morrow. Pulp the vegetables into that meant for to-day; let it cool; take off the fat, and put back over the fire. Season to your liking; add the sago, which should have been soaking for two hours in a little water, and simmer until it is clear.

STEWED CALFS' HEARTS.—Wash two fresh calf's hearts; stuff with a force-meat of crumbs, chopped salt pork, a little thyme, sage, and onion. Tie up snugly in clean mosquito netting; put into broad saucepan; half cover with broth from your soup from yesterday or to-day. Cover and stew an hour and three-quarters gently, turning several times. Take out the hearts, and keep them hot, while you thicken the gravy with a tablespoonful of butter cut up in flour. Boil up, add pepper, salt, a little grated lemon-peel, and the juice of half a lemon, with a small glass of wine. Pour over the hearts.

APPLE SOUFFLE PUDDING.—Seven or eight juicy apples; four eggs; one cup fine crumbs; one cup of sugar; two tablespoonfuls of butter; nutmeg and a little grated lemon-peel. Pare, core and slice the apples, and cook tender in a covered farina-kettle without adding water to them. Beat to a smooth pulp, and stir in butter, sugar, and seasoning. When cold whip in the yolks of the eggs; then the frothed whites, alternately with the crumbs. Beat to a creamy batter; put into a buttered pudding dish, and bake, covered, fifty minutes. Then brown quickly. Eat hot with custard sauce, or cold, with cream and sugar.

LIVE STOCK.

Raising a Colt.

A colt is regarded as an incumbrance, because he is useless until he arrives at a suitable age for work, but it really costs very little, compared with his value, to raise a colt. When the period arrives at which the colt can do service, the balance sheet will show in his favor, for young horses always command good prices if they are sound and well broken. One of the difficulties in the way is the incumbrance placed on the dam, which interferes with her usefulness on the farm, especially if the colt is foaled during the early part of the spring. Some farmers have their colts foaled in the fall, but this is open to two objections. In the first place, spring is the natural time, for then the grass is beginning to grow, and nature seems to have provided that most animals should bring forth their young in a season beyond the reach of severe cold, and with sufficient time to grow and be prepared for the following winter.

Again, when a colt is foaled in the fall he must pass through a period of several months' confinement in the stable, without exercise, or else be more or less chilled with cold from time to time. Should this happen, the effect of any bad treatment will be afterward manifested, and no amount of attention can again elevate the colt to that degree of hardiness and soundness of body that naturally belong to a spring colt. Besides a colt foaled in the spring will outgrow one foaled in the fall. An objection to spring colts may be partially overcome by plowing in the fall, or keeping the brood mares for very light work, with the colts at liberty to accompany them always. A colt needs but very little feeding if the pasture is good, and there is water running through it. He needs then only a small feed of oats at night—no corn—and if he is given hay it is not necessary to give him a full ration. What he will consume from the barn will not be one-third his value when he is three years old, and if he is well bred the gain is greater.

When a farmer raises his horses he knows their disposition, constitution, and capacity. It is the proper way to get good, sound, serviceable horses on the farm. It should not be overlooked that a colt must be tenderly treated from birth, and must be fondled and handled as much as possible. He should never hear a harsh word, but should be taught to have confidence in everybody he sees or knows. This is an easy matter if his training begins from the time he is a day old. He can be thus gradually broken without difficulty, and will never be troublesome. No such thing as a whip should be allowed in a stable that contains a colt. Colts should not be worked until three years old, and then lightly at first, as they do not fully mature until they are six years old, and with some breeds of horses even later. Mares with foals at their side should be feed on the richest and most nourishing food.—*Dixon, Ill., Western Farmer.*

Hints on Raising Stock.

Every farmer who raises his own cows knows full well that their future value depends largely upon their first year's growth when calves. If the calf is stunted, half-starved and ill-used there is not one chance in ten of its ever becoming a good cow on reaching the proper age. The calf must be supplied with an abundance of the proper food for securing the best conditions of growth. In fact, the same attention is necessary with yearlings and two-year olds.

Among the most desirable foods are good hay, linseed meal and cotton-seed meal. Shorts are also excellent for growing animals. The highest farm economy demands the rapid and early fattening of all steers, as well also as such heifers as are not wanted for breeding purposes. When an animal is grown it has taken up all the phosphate and nitrogen it is likely to require, which elements are the most expensive to supply. On the other hand, the fattening animal abstracts from its food nothing except fat, starch and sugar, the nitrogen, phosphate and potash contained therein being returned to the soil through the manure heap.

While the farm of the breeder is likely to grow poorer without the extensive use of commercial fertilizers or purchased farmyard manure, yet the lands of the feeder are always gaining and growing richer. The farmer who sells lean stock is robbing his farm of its vital and most valuable elements, while he who purchases lean stock for fattening on his own lands will prove a successful cultivator. Progressive farmers should always strive to produce only good stock, thus insuring remunerative rewards as well as maintaining the value and fertility of their farms. —*American Cultivator.*

Swine Raising—A Different System Desirable.

Pure air helps to make pure blood, which, in the the course of nature, builds up healthful bodies. Out-of-door pigs would not show as well at the fairs, and would probably be passed over by judges and people who have been taught to admire only the fat and helpless things which get the prizes. Such pigs are well adapted to fill lard kegs, whereas the standard of perfection should be a pig which will make the most ham with the least waste of fat, the longest and deepest sides, with the most lean meat; it should have bone enough to allow it to stand up and help itself to food, and carry with it the evidence of health and natural development in all its parts. Pigs which run in a range or pasture have good appetites—the fresh air and exercise give them this—hence they will eat a great variety of food and much coarser than when confined in pens. Nothing need go to waste on the farm for lack of a market. They will consume all the refuse fruit, roots, pumpkins, and all kinds of vegetables, which will make them grow. By extending the root patch, and planting the fodder corn thinner, so that nubbins will form on it, and by putting in a sweet variety, the number of pigs may be increased in proportion. A few bushels of corn at the end of the season will finish off the pig. The pig pasture will be ready the next year for any crop, and ten times the advantage accrue to the farm than if the pigs are confined in close pens, for, as pigs are usually managed on the farm only little manure is ever made from them.

More Frequently Milking.

Mr. L. T. Hawley, of the Onondaga Farmers' Club, lately reported an experiment in more frequent milking, which we quote from the Syracuse *Journal:*

"The cow with which he experimented dropped the calf when twenty two months old, in February, 1881, and gave thirty-two pounds of milk per day with two milkings, ten days after the calf was born. A change to three milkings a day was made, with an increase in ten days to forty-two pounds. The milk was set by itself for fourteen days, and from the cream twenty one pounds of well worked butter was obtained. The feed was corn stalks from which the ears had been taken and green cut hay, timothy and clover well cured in the cock, cut and mixed together and fed three times a day, together with one pound of linseed meal and four pounds of Indian meal. Water tempered to 65 degrees was given three times a day. He added that Professor Arnold has stated that increasing the milkings from two to three times per day will increase the percentage of cream from 12½ to 18½."

Jersey Cows and their Records.

In view of the heavy prices paid at various public sales of Jersey cattle, the Live Stock Journal comments as follows: If anybody had predicted ten years ago that the mild eyed little Jerseys would have their $1,000 boom on their butter records, he would have been considered on the borders of lunacy. The breeders of fancy Short-horns have seldom considered the butter or milk record as worthy of note. They ignored the most valuable characteristic of any breed of cattle for use in a highly civilized country—their milk and butter production. These yield more annual profit than beef production; and every breed that maintains a permanent foothold in the United States must meet this test or stand aside. Happily, the Short-horn with its magnificent beef form, can also point proudly to its achievements in the dairy. Its temporary eclipse in this line, through some of its noblest strains has resulted from the fault of the breeder, and not from the capacity of the breed. But the little Jersey is having her boom upon her honest merit in producing very large yields of golden-colored and nutty-flavored butter. Perhaps her admirers are somewhat extravagant in their valuation of these records. They may not always scan them as closely as they should. As these extreme prices must be based upon a confidence in the truth of these records, the records themselves should be well attested. Tests for a year must also be a safer reliance than for a shorter time. The tests of milk and butter yield for a few days are open to so many errors, that they cannot form a basis for calculating the annual yield. The variability in the yield of some cows in different parts of the season of lactation is very great, while other cows are very uniform through three-fourths of the season, only decreasing gradually during the last two or three months. The circumstances, then—all being favorable—may produce a very large yield for a few days, when the annual yield would be only respectable. If the short test is given, several important points should also be given to assist in forming a correct estimate—such as the length of time from calving, the season of the year, the food before and at the time of trial—all these are necessary elements for determining the value of a test.

Facts About Horses.

The horse's stomach has a capacity of only sixteen quarts, while that of an ox has 250. In the intestines this proportion is reversed, the horse having a capacity of 190 quarts against 100 of the ox. The ox and most other animals have a gall bladder for the retention of a part of the bile secreted during digestion. The horse has none, and the bile flows directly into the intestines as fast as secreted. This construction of the digestive apparatus indicates that the horse was formed to eat slowly and digest continually bulky and innutritious food. When fed on hay it passes rapidly through the stomach into the intestine. The horse can eat but five pounds of hay in an hour, which is charged during mastication with four times its weight of saliva. Now the stomach, to digest it well, will contain but about ten quarts, and when the animal eats one-third of his daily ration, or seven pounds, in one and one half hours, he has swallowed at least two stomachfuls of hay and saliva, one of these having passed to the intestines. Observation has shown that the food is passed to the intestine by the stomach in the order in which it is received. If we feed a horse with six quarts of oats it will just fill his stomach; and, if as soon as he finishes this, we feed him with the above ration of seven pounds of hay, he will eat sufficient in three quarters of an hour to have forced the oats entirely out of his stomach into the intestine. As it is the office of the stomach to digest the nitrogenous parts of the feed, and as a stomachful of oats contains four or five times as much of these as the same amount of hay, it is certain that either the stomach must secrete the gastric juice five times as fast, which is hardly possible, or it must retain this food five times as long. By feeding with the oats first, it can only be retained long enough for the proper digestion of hay; consequently, it seems logical, when feeding, a concentrated food like oats with a bulky one like hay, to feed the latter first, giving the grain the whole time between the repasts to be digested. The digestion of a horse is governed by the same laws as that of a man; and as we know that it is not best for a man to go at hard work the moment a hearty meal is eaten, so we should remember that a horse ought to have a rest after his meal, while the stomach is most active in the processes of digestion.—*American Cultivator.*

Overloading Cows' Stomschs.

When cows are changed from scanty to flush feed it often happens that the benefit of the more liberal supply is neutralized for some time by allowing them to gorge themselves to the extent of uncomfortable fullness. An excessive distension of the stomach produces inflammatory action and impedes digestion, and tends to diminish the flow of milk and to impair its quality. Overloading a cow's stomach invariably gives a strong and disagreeable odor to her milk that injures it for butter or for cheese-making, and also its healthfulness for food. Such an overloading is always indicative of a double loss—a loss from failing to utilize as fully as might the flush feed, and a previous loss from a supply of food insufficient to enable the cows to give as much milk as they are capable of giving. When cows are fed with a liberality that develops a full flow of milk, they will not overload with a food so little concentrated as green grass. The fact that they do overload is an evidence that their previous food was too scanty for profit, and consequently that loss has been endured on account of it. But when such a course of feeding has existed, and a change is to be made to a better one, loss from over-eating may be prevented by admitting the herd gradually to the new feed and supplying them with all the salt and water they desire. The increase in the new ration should never be so great as to change the flavor of the milk.

Quarantined Cattle.

The Governor of Illinois issued a proclamation prohibiting the importation of cattle into the State from Philadelphia and adjacent localities, unless the shipment is accompanied by a certificate of health signed by a duly-authorized veterinary inspector. The proclamation states that there is good reason to believe that the pleuro pneumonia exists as an epidemic among the cattle in the eastern part of this State, as well as in Maryland, Delaware, New York and Connecticut, rendering such action necessary. This action on the part of the State of Illinois does not cause alarm to the Philadelphia beef-cattle dealers, but is considered a wise measure.

City Treasurer Martin, who is President of the Philadelphia Stock Yard Company, says this establishment of a cattle quarantine by the State of Illinois cannot affect the regular trade. He says that the disease is caused by the importation of fancy stock from England, and is, in his opinion, confined to dairy cows exclusively. Mr. Martin also stated that the inspection of cattle in this vicinity is very rigid. Dr. J. W. Gadsden, United States Cattle Inspector for Pennsylvania, considers that this measure on the part of Illinois is one that should have been taken long ago. He says, however, that in his opinion the proclamation is too sweeping. The Doctor thinks that this proclamation will have the effect of causing the passage of a bill to eradicate pleuro-pneumonia in this country. The disease has been stamped out in England and in Massachusetts, and it can be done everywhere. Several other physicians, whose opinion of the proclamation was asked, concurred in the judgment of Dr. Gadsden, that it is a wise measure and that it will have a beneficial effect.

APIARY.

Some Information about the Queen Bee.

There is an impression prevailing among the uninitiated that the queen of a hive leads off the swarm, but this is by no means the case with first issues, for, as a rule, the queen does not come forth from the hive until the greater part of the bees are on the wing. Another erroneous idea in existence is that the queen bee is the first to alight upon a branch or a bush, and that the bees congregate about her, but the reverse of this is the fact. When a swarm begins to issue, if the bee-keeper will place himself on the shady side of the hive and watch the stream of bees which pour forth like an army through a gateway, he may see the queen come out, and, if inclined to prove our assertions, he may capture and cage her, and put her in his pocket while he watches the proceedings of the bees. When the throng is circling in the air he may imagine that the bees are searching for her, and will perhaps conclude that, as they cannot find her, they will return at once to the hive; but no, they will first congregate near a convenient tree or bush, and make a great noise sufficient to attract the attention of her majesty, if she were abroad, and then they will alight and form a cluster, and wait for some minutes to give her an opportunity of joining them. If now she be taken to them, she will join the mass and all will be well; if not, the bees after a short time will disperse and return to the hive. Now this kind of experiment has been so often proved that it may be taken for granted when a swarm of bees has alighted, and afterwards returned to the hive, that the queen was not able to join them, or she would assuredly have done so.—*British Bee Journal.*

Twelve Facts for Beginners.

MR. EDITOR: I will offer for publication a few facts which every beekeeper ought to know:

1 That the life of a worker bee, during the working season, is only from six to eight weeks' duration, and that a large majority of them never live to see seven weeks.

2. That a worker is from five to six days old before it comes out of the hive for the first time to take an airing, and that it is from fourteen to sixteen days old before it begins to gather either pollen or honey.

3. That all swarms engaged in building comb, when they have not a fertile queen, build only drone comb, and that all the comb in the lower or breeding apartment should be worker or brood comb, except a very small quantity of drone comb, four inches square being amply sufficient.

4. That the more prolific the queen is the more young bees you have, and the more surplus honey will be gathered, other things being equal.

5. That you ought never to cut moldy combs out of the hives, for the reason that you should never allow it to become moldy.

6. That you ought never to double swarms or stocks of bees in the fall, because you ought to attend to that and make them strong during the summer by taking the brood from the strong stocks and giving it to the weaker.

7. That a drone laying queen should be taken away and one producing workers be put in her place, else the colony will soon come to naught.

8. That as a rule, as soon as an Italian queen shows signs of old age or feebleness, the bees themselves will supersede her.

9 That all colonies should be kept strong in order to be successful.

10. That every hive should contain about two thousand cubic inches in the breeding department.

11. The beginners in bee keeping should be very cautious about increasing the number of their swarms or stock rapidly until they thoroughly understand the business.

12. That the hive itself, if well constructed, is all the bee house you need.—*Bee-Keepers' Review.*

A System for Wintering.

We have long thought that there is as much need of a system for the management of bees as there is for a system of penmanship, and it is quite likely that without the principles and forms in writing to day, there would be confusion in penmanship equal to the confusion which exists among bee-keepers. Our operations and extensive experiments give us the impression that a system of wintering on summer stands might be based on the following directions which involve principles:

1st. It must be determined by weighing that each hive contains twenty pounds or more of honey, November 1st.

2d. The hive must be perfectly tight, so that, if inverted and filled with water, there will be no leakage.

3d. The bottom-board on which the hive is placed must have an opening through it of fifty square inches, covered with wire cloth, and elevated from the ground about six inches.

4th. The hive must be protected on all sides with dry saw dust or clover-chaff, six inches thick. When double-walled hives are used four inches of packing will do.

5th. When the thermometer indicates zero or below, bank the hives with snow. If there is no snow, use straw. When the thermometer indicates 30° above zero, and there are prospects of a thaw, remove the snow or straw from around the hives, and allow the sun to shine under the bottom board, if possible.

6th. Colonies arranged along the south side of a tight-board fence, running east and west, are more secure than if set in an open yard.

7th. Examine all stocks on the first warm day in April, and, if any are wanting in stores, feed enough at once to suffice until fruit blossoms appear. Remove winter protection the first day of May.

Preparing for Winter.

When the month of October has arrived in this latitude the fall honey yield ceases, and during the month all the brood hatches and the queen lays but sparingly. This is the desirable condition for the hive to be in when weighing and feeding is done. That winter preparations should not be delayed after the first of November, we have had strong intimations from the weather of previous seasons. Colonies could not be fed to any advantage after the last day of November, 1880, to pass through the coming severe winter. For the first six years no colony in our apiary has consumed over twenty pounds of stores from November 1st to April 1st, and we are satisfied that this amount makes all colonies safe. A hive with its combs and bees weighs about twenty pounds, so that forty pounds is the standard weight of a colony with plenty of stores to winter on, and it is our practice to make every colony reach this weight. We make up all the deficiencies by feeding extracted honey, and when the supply is exhausted standard A. sugar is used. The feeding should be done as rapidly as possible. There is no gain in weight if only a pound or two is fed each day, but if ten pounds be fed to one stock in a day there will be a gain in weight of eight or nine pounds. Now is the time for all Northern apiarists to make their reports, which should include every stock in their apiaries. We are confident that a lack of system results in many disasters. We would be glad to publish a large number of these reports. Last fall we had a larger amount of work on hand than usual, and help was scarce. Had it not been for our habit of making a report each year on a certain day, and our extra effort just at that time to keep the report unbroken, we would have lost heavily.

A badly worn or broken down farm implement of any kind is a bad investment. The loss of time from stoppage when work should be hurried is usually more expensive than the money cost of repairs.

POULTRY.

Guinea Hens.

Objection is made to the guinea fowl in domestic quarters because its voice is hardly less musical than "the controversy of two ripsaws;" because it bullies less pugnacious poultry and because its lays are in stolen nests. Still its eggs are of what an Englishman call a "a good flavor" and its flesh "most delicious, resembling that of the pheasant." He tells in *The Illustrated Book of Poultry* how to ameliorate the birds by kind acquaintance above: "By setting eggs under common hens, and rearing them at home, they grew up much tamer, and will flock round the person who feeds them, and even allow themselves to be taken up petted, like other poultry. When reared thus kindly, and secluded nests are provided, they will generally lay in the house; and if perches are placed high for them, and they are regularly fed every night, will roost at home also. So far domesticated, they will pay to rear in places where they can have ample range."

Care of Fowls.

Frost-bitten combs and wattles not only injure the looks of fine fowls, but affect the health as well. To guard against this, warm houses must be furnished. A house that is partially underground is well adapted in our climate for a winter habitation. The northern portion, particularly, should be sheltered, while the southern should have an exposure of glass. Oftentimes the fowls will pass through the winter unharmed, and become frosted in March. Fowls suffer much from the cold if not frosted, and should have a generous protection from severe weather. Fowls should never be removed or changed in the spring until the weather is warm and settled.

The difference in flavor of the eggs of the different breeds of hens is a matter of fact which few people appreciate, although it is as distinct as the difference in flavor of the different kinds in potatoes, and may be varied to some extent by the kind of food—although the dark color of the shells almost invariably indicates the rank egg flavor while the most delicate flavor is found with white or slightly tinted shells. The every-day layers seldom produce as rich eggs as those which lay every second day; and for hatching the latter are by far the best in point of strong shells, fertility and strength of chickens. These advantages can also be varied by feeding and other influences; the most natural conditions of food, exercise &c., producing the best results. It is the *whites* of the eggs that generally determine their richness (although the pastry cook tells us the "light yellow yolks do not color cakes and custards sufficiently to make them look rich,") and the kinds of feed which stimulate hens to produce the greatest number of eggs are the most deficient in the albumen which constitutes their richness, not only making the yolks light yellow, but the whites watery.

The food value of an egg can be easily tested by breaking it into water just before boiling heat. If the white draws up around the yolk, and covers it thickly the egg is rich in albumen; but if the white spreads through the water in stringy flees, leaving the yolk uncovered, or slightly covered, the egg is proportionately poor, though the yolk be ever so dark in color. Feeding hens broken boiled-lobster shells, will make the yolks of their eggs dark color, at any season, and there is no kind of shell-forming substance which they more crave or more eagerly eat.

Ducks.

We are occasionally led into wondering why more ducks are not bred and marketed among our poultry-breeders in America. We have now in this country three or four a varieties of imported ducks, at the head of which the Pekins stand to day, without question, for size, early maturity, hardiness, and thrift. The Aylesbury (pure white, like the Pekin), the Rouen (brown or parti colored), and the Cayuga (black) are notable, and of good quality. Each of these varieties, within our knowledge, has been successfully bred in New England, upon a country place where there was neither pond nor rivulet for their amusement on the farm.

The ducklings were hatched under hens, and the ducks were raised with the other poultry and fowls on the estate, with similar feed and care, the owner claiming that for marketing purposes ducks can be reared, like any other fowls, upon dry land; and he has found no perceptible difference in their proportionate thrift during the season, though in his experiments, in the last two years, his ducks never had access to any body of water. A pleasanter kind of poultry we do not know of.

Which is the More Profitable?

"Do we derive a better profit from the non sitter than from the sitter?" is as yet an unsettled question. Poulterers are prone to give more credit for the time being rather than to enter into a closer examination of facts. The best of the non-sitters do not average over 180 eggs during the year; but the sitter is equally as sure for 132 or more. This is a difference of four dozen eggs, and as the non-sitters lay more eggs in the summer season than in winter the monetary value will not exceed seventy-five cents as the measure of difference. Using the Leghorn as a sample of the non-sitting breeds and the Brahma to represent the sitters, it cannot be denied that the latter, being better winter layers, are nearly if not quite equal to the Leghorns in monetary product, even if the number of eggs laid is smaller, owing to the enhanced price of eggs during the cold terms. But the young chicks should be taken into account; and on an average of only five to the brood, after deducting loss and the low price of 20 cents per pound, at two pounds each, the account will be \$2. Estimating yet lower in the price, making it 12½ cents per pound, and we still have an advantage in favor of the sitter. The expense will correspond with the ratio of sale, and the gain cannot be offset. The Leghorn matures earlier than the Brahma, and gains time in that respect. They take resting spells, however, from laying, and if they are not at work bringing forth chicks the time is lost. With all that may be said in favor of the non-sitter, it must be remembered that the young chicks count in value as well as eggs.

Fattening Turkeys.

An old turkey raiser gives an account of an experiment in fattening turkeys as follows: Four turkeys are fed on meal, boiled potatoes and oats. Four others of the same brood were also at the same time confined in another pen and fed daily on the same articles, but with one pint of very finely pulverized charcoal mixed with their food—mixed meal and boiled potatoes. They had also a plentiful supply of charcoal in their pen. The eight were killed on the same day, and there was a difference of one and one-half pounds each in favor of the fowls which had been supplied with charcoal, they being much the fatter, and the meat being greatly superior in point of tenderness and flavor.

FARM AND WORKSHOP NOTES.

There is more smut in corn this season than usual.

The Bermuda onion always does best from imported seed.

A Maryland farmer produced a 26 pound cantalope this season.

It is said that sheep in orchards will annihilate the codling moth.

Hogs in the neighborhood of Reading are dying of some unknown disease.

At a recent cattle show the Polled Angus bulls captured the premiums over the Short-horns.

It is conceded that crossing breeds of poultry promotes laying, and gives better results in hatching.

At an Indiana Fair the Holsteins won the sweepstakes on bulls, the Shorthorns on cows and the Herefords on steers.

Mr. H. G. Mumma, of Washington county, Md., raised 452 bushels from 9¾ acres of land, or about 46 bushels to the acre.

Farmers should be cautious in destroying unknown insects, as they often turn out to be kinds that prey on those that injure crops.

The Boston *Cultivator* says that new varieties of corn are produced from the small nubbins that grow on the end of the tassel.

J. A. Dodge, in the *Journal of Agriculture*, says he cures hog cholera by giving half a teaspoonful of carbolic acid in a gill of milk. It is administered from a long-necked bottle.

Samuel T. Earle's cow, Valma Hoffman, of Queen Anne's county, Md., has produced in thirty days over eighty seven pounds of butter, or nearly three pounds a day. She is valued at \$5,000.

Celery will be much better if allowed to grow until checked by cool weather at the end of the season, and then placed in boxes or trenches and blanched for future use.

LITERARY AND PERSONAL.

THE POULTRY AND STOCK JOURNAL.—A chronicle for country gentlemen, breeders and fanciers. Published at the National Capital on the 15th of each month, at one dollar a year. Grant Parrish, editor and proprietor. Address, Capital Hill P. O., Washington, D. C. This is a first-class illustrated quarto, printed in clear type on heavy calendered paper, and in embellished tinted covers. Although largely devoted to poultry, in its pet and fancy phases, it also includes a general view of stock and literature appertainining thereto, and discusses the various subjects with ability. To a person desiring variety in his literary repast we can recommend nothing better than this *Poultry Review*.

APPEAL of Mrs. Elizabeth Thompson to the American people, on the subjects of "Intemperance and Ignorance," together with the poem, "What Right?" as a preface to 14 pages of the "Congressional Record," containing the speeches of Hon. Henry W. Blair, of New Hampshire, on the "Manufacture and sale of intoxicating liquors;" and on "Aid to Common Schools." Time and perseverance may render the last named measure possible, if not probable, because it belongs to the intellectual in man; and, as the intellect becomes enlarged and elevated, he may be enabled to see and acknowledge the wisdom of the measure, and finally yield his assent. But the first named seems to be more nearly related to the affectional in man. It strikes at his animal nature and its loves: a domain in which it is difficult to get him to see what is best for his moral and physical welfare, or even to acknowlege it, and bring himself under its reforming influences, when he does see it. Mrs. Thompson's appeal is but a reiteration of similar appeals which we have been hearing and reading these fifty years or more. The task is a herculean one, and it is not at all surprising she should be pained at the little advance that has been made. So far as prohibitory laws *per se* are concerned it would make little difference to us, personally, what the punishment for their violation might be—whether fine, disqualification or imprisonment—if the prohibition was accompanied by a rational discrimination; and until all men take a similar view of the subject it would be vain to look for an honest execution of such a law, even if it were enacted.

SILOS AND ENSILAGE, a Record of Practical Tests in several States and Canada. Special Report No. 48 of the Deparment of Agriculture. The Commission of Agriculture sent out a schedule of questions to the number of 26, and received statements from 92 persons who had built silos and tested ensilage, in reply from different parts of the United States, and the Dominion of Canada, and in their details they are generally very favorable to ensilage. On the profitableness of ensilage "there is hardly a doubt expressed—certainly not a dissenting opinion." "The general use of ensilage must depend largely on its cheapness. Costly silos and expensive machinery must always be insurmountable obstacles to a majority of farmers. For this reason, experience tending to show what is *essential* to the preservation of fodder in silos, is of the first importance."

TWO LEGENDARY POEMS: "The Botanist checkmated. By A. G. P. and "The Plague of Flies." By T. G. P., from the annals of the P—— Family. Printed for private distribution. Express Publishing co., Easton, Pa., 1882. 28 pp. 12 mo. A clever satire, written in the Hiawatha measure.

FREE TRADE BULLETIN, issued monthly at 50 cts. per annum. G. U. Wing, Publisher, Nos. 27 and 29 West 23d street, New York.

This is a demi-folio, and is a zealous, and able advocate of the views of Freetraders, a subject which, we must confess, has never occupied much of our attention, and even those who have made commercial intercourse a specialty do not seem to have come to a harmonious conclusion. It would perhaps require some self-concession to arrange a system of duties on merchandize, that would be acceptable to people in general. There are people who demand the very highest prices for their own manufactures, especially when they *know* they can get it, but when they are in need of the manufacture of others they go where they can obtain them at the very lowest price. The former position may be regarded as *high tariff*, and the latter, *low tariff*. These individualities carried into the enactment of general laws give them a similar taint.

THE AGENTS' HERALD.—L. Lum Smith, editor, publisher and proprietor; 50 cents a year, monthly; a demi-folio of 16 pp. in the interest of legitimate agents and agencies, and the exposure of frauds. It is wonderful what a power agents have become in the world of business. Where there was *one* twenty or thirty years ago there seems to be a *thousand* now; therefore it is not surprising they should need a representative organ. This journal has a very characteristically embellished "title-head." On the right hand an agent, with his traps under his arm, is sitting on the "cow-catcher" of a locomotive, with a bright expectant look towards the station the train is approaching. On the left hand an agent is standing as far forward as he possibly can get on the bow of a boat approaching a wharf, and straining his gaze over a town; whilst just below them in the foreground an agent with a long (gum-elastic) arm is presenting his card to a victim on the right, who is receiving it with a similarly elongated member. The countenance of the former illustrates a dogged, yet bland, importunity, whilst the latter exhibits a jaded affability, such as *business* men sometimes manifest when they wish their pertinacious importuners at the devil.

Since the times and the fashions are such as characterize the period in which we live, it is, perhaps, essential that all parts of human industry, energy and enterprise should assume an organized form, in order to effect their successful progress. "Times ain't as they used to was;" and "we can't do as we used to did." Hence, provision must be made to sustain a class that, in our earlier days, had not "a local habitation and a name," but now is becoming "a power in the land."

EVERY lady should send 25 cents to Strawbridge & Clothier, Philadelphia, and receive their *Fashion Quarterly* for 6 inos. 1,100 illustrations and 4 pages new music each issue.

A MANUAL OF ELOCUTION AND READING, embracing the Principles and Practice of Elocution. By Edward Brooks, Ph. D., Principal of the State Normal School, Millersville, Pa. Philadelphia: Eldridge & Bro. Price, 1.50. To teachers, for examination, $1.00.

PHILIP SCHUM, SON & CO.,
38 and 40 West King Street.

We keep on hand of our own manufacture,

QUILTS, COVERLETS,
COUNTERPANES, CARPETS,

Bureau and Tidy Covers, Ladies' Furnishing Goods, Notions, etc.

Particular attention paid to customer Rag Carpet, and scouring and dyeing of all kinds.

PHILIP SCHUM, SON & CO.,
Nov-1y   Lancaster, Pa.

THE PENN HARROW

BEST IN THE WORLD

IT HAS NO EQUAL

Patented April 13, 1880.

The above cut represents the Penn Harrow complete, with all its combinations of Five Harrows and a sled for each Harrow; and each succeeding change is made from the Harrow without the least additional expense. By hooking the team to either point, B or C, the center revolves and gives the ground Two Strokes and Two Crosses in passing over it once, making it the most effective pulverizer in the market.

THIS HARROW HAS ONLY TO BE USED TO BE APPRECIATED.

See it before purchasing and you will buy no other.

The Penn Harrow
CHANGED TO A THREE-CORNER ROTARY HARROW.

Indispensable for Orchards, as the revolving wheel harrows right up to and all around the trees without barking them.

The Penn Harrow
CHANGED TO SINGLE "A" HARROW.

DISSOLUTION OF PARTNERSHIP.—
The co-partnership in the merchant tailoring business heretofore existing under the firm of Rathvon & Fisher, is this day dissolved by mutual consent. All persons in any manner indebted to said firm, are respectfully solicited to make immediate payment to S. S. Rathvon, who is hereby authorized to receive the same, and those having claims against said firm, will please present them for settlement.

    S. S. RATHVON,
    M. FISHER,
101 North Queen Street, Lancaster, Pa.

Until further announcement, the business, without interruption, will be conducted by the first signed, who solicits a continuance of the patronage heretofore bestowed upon the firm, and which is hereby greatfully acknowledged.

S. S. RATHVON,
PRACTICAL TAILOR,

No. 101 North Queen Street,
LANCASTER, PA.

By removing the wheel from the original you have a complete one-horse "A" Harrow.

The Penn Harrow
CHANGED TO DOUBLE "A" HARROW.

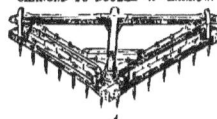

Remove the wheel from the original, reverse the wing, and it makes the most complete Double "A" Harrow in the market.

The Penn Harrow
CHANGED TO A SQUARE HARROW.

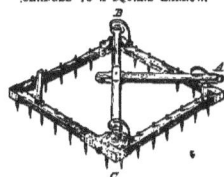

By removing the wheel from the original you have a complete Square Harrow. By looking to B or C you can harrow as a harrow, and harrow the bottom and both sides, or scarify either end. By looking to the top and both sides, or you can lift either point and have three point; on the ground—something that cannot be done with any other Harrow.

The Penn Harrow
ON ITS SLED.

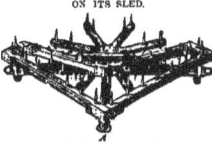

It has always been a great inconvenience to get the Harrow to and from the field. The Penn Harrow obviates this, as no matter which Harrow you wish to use in the combination, it has its own sled to haul it on.

The Penn Harrow

Is made of the best white oak, with steel teeth,well painted, in every way first-class. Formerly a harrow was the most unhandy implement on the farm, with our improvement it is the most convenient, will do double the work of any other harrow and have the farmer half his labor, and is warranted to do all we represent or money refunded. ORDER AT ONCE AND BE CONVINCED.

Price of the Right draft Combination Penn Harrow, $30. *Send for a Catalogue and see what farmers say.*

AGENTS WANTED IN EVERY COUNTY.

PENN HARROW MANUFACTURING CO.
    CAMDEN, N. J.
sep-3

TREES

Fruit, Shade and Ornamental Trees.

Plant Trees raised in this county and suited to this climate.
Write for prices to

    LOUIS C. LYTE
Bird-in-Hand P. O., Lancaster co., Pa.
Nursery at Smoketown, six miles east of Lancaster.
79-1-12

WIDMYER & RICKSECKER,
UPHOLSTERERS,
And Manufacturers of

FURNITURE AND CHAIRS.
WAREROOMS:

102 East King St., Cor. of Duke St.
LANCASTER, PA.
79-1-12]

Special Inducements at the
NEW FURNITURE STORE
OF
W. A. HEINITSH,
No. 15 1-2 E. KING STREET
(over Bursk's Grocery Store, Lancaster, Pa.
A general assortment of furniture of all kinds constantly on hand. Don't forget the number.

15 1-2 East King Street,
Nov-1y] (over Bursk's Grocery Store.)

For Good and Cheap Work go to
F. VOLLMER'S
FURNITURE WARE ROOMS,
No. 309 NORTH QUEEN ST.,
(Opposite Northern Market),
    Lancaster, Pa.
Also, all kinds of picture frames.   nov-1y

GREAT BARGAINS.

A large assortment of all kinds of Carpets are still sold at lower rates than ever at the

CARPET HALL OF H. S. SHIRK,
No. 202 West King St.

Call and examine our stock and satisfy yourself that we can show the largest assortment of these Brussels, three ply and ingrain at all prices—at the lowest Philadelphia prices.

Also on hand a large and complete assortment of Rag Carpet.

Satisfaction guaranteed both as to price and quality.
You are invited to call and see my goods. No trouble in showing them even if you do not want to purchase.
Don't forget the notice. You can save money here if you want to buy.

Particular attention given to customer work. Also on hand a full assortment of Counterpanes, Oil Cloths and Blankets of every variety. [nov-1,yr.

C. R. KLINE

ATTORNEY-AT-LAW,

OFFICE: 15 NORTH DUKE STREET,
LANCASTER, PA.
Nov-1y

SILK-WORM EGGS.

Amateur Silk-growers can be supplied with superior silk-worm eggs, on reasonable terms, by applying immediately to

    GEO. O. HENSEL,
may-3m] No. 278 East Orange Street, Lancaster, Pa.

LIGHT BRAHMA EGGS

For hatching, now ready—from the best strain in the county—at the moderate price of

$1.50 for a setting of 13 Eggs.
    L. RATHVON,
No. 9 North Queen st., Examiner Office, Lancaster, Pa.

WANTED.—CANVASSERS for the LANCASTER WEEKLY EXAMINER

In Every Township in the County. Good Wages can be made. Inquire at
    THE EXAMINER OFFICE,
No. 9 North Queen Street, Lancaster, Pa

WHERE TO BUY GOOD IN LANCASTER.

BOOTS AND SHOES.

MARSHALL & SON, No. 12 Centre Square, Lancaster, Dealers in Boots, Shoes and Rubbers. Repairing promptly attended to.

M. LEVY, No. 3 East King street. For the best India-rubber shoes in Lancaster go to M. Levy, No. 3 East King street.

BOOKS AND STATIONERY.

JOHN BAER'S SONS, Nos. 15 and 17 North Queen Street, have the largest and best assorted Book and Paper store in the City.

FURNITURE

H. S. SHIRK, No. 15 East King st., Cover China Hall, is the cheapest place in Lancaster to buy Furniture. Parlor Frames especially.

CHINA AND GLASSWARE.

HIGH & MARTIN, No. 15 East King st., dealers in China, Glass and Queensware, Fancy Goods, Lamps, Burners, Chimneys, etc.

CLOTHING.

MYERS & RATHFON, Centre Hall, No. 12 East King st., Largest Clothing House in Pennsylvania outside of Philadelphia.

DRUGS AND MEDICINES.

G. W. HULL, Dealer in Pure Drugs and Medicines, Chemicals, Patent Medicines, Trusses, Shoulder Braces, Supporters, &c., 15 West King St., Lancaster, Pa.

JOHN F. LONG & SON, Druggists, No. 12 North Queen St. Drugs, Medicines, Perfumery, Spices, Dye Stuffs, Etc. Prescriptions carefully compounded.

DRY GOODS.

GIVLER, BOWERS & HURST, No. 25 E. King St., Lancaster, Pa., Dealers in Dry Goods, Carpets and Merchant Tailoring. Prices as low as the lowest.

HATS AND CAPS.

C. H. AMER, No. 30 West King Street, Dealer in Hats, Caps, Furs, Robes, etc. Assortment Large. Prices Low.

JEWELRY AND WATCHES.

H. Z. RHOADS & BRO., No. 4 West King St., Watches, Clock and Musical Boxes. Watches and Jewelry Manufactured to order.

PRINTING.

JOHN A. HIESTAND, 9 North Queen st., Sale Bills, Circulars, Posters, Cards, Invitations, Letter and Bill Head and Envelopes neatly printed. Prices low.

FARMING FOR PROFIT.

It is conceded that this large and comprehensive book, (advertised in another column by J. C. McCurdy & Co., of Philadelphia), the well-known publishers of Standard works, is not only the newest and handsomest, but altogether the first work of its kind which has ever been published. Thoroughly treating the great subjects of general Agriculture, Live-Stock, Fruit-Growing, Business Principles, and Home Life; telling just what the farmer and the farmer's boys want to know, combining Science and Practice, stimulating thought, awakening inquiry, and interesting every member of the family, this book must excel in adding influence far good. It is highly recommended by the best agricultural writers and the leading papers, and is destined to have an extensive sale. Agents are wanted everywhere. jan-tf

GLOVES, SHIRTS, UNDERWEAR.
SHIRTS MADE TO ORDER,
AND WARRANTED TO FIT.

E. J. ERISMAN,
56 North Queen St., Lancaster, Pa.
79 1-12]

GREGORY SEED CATALOGUE

Thirty-Six Varieties of Cabbage; 26 of Corn; 28 of Cucumber; 17 of Melon; 35 of Peas; 28 of Beans; 17 of Squash; 23 of Beet and 40 of Tomato, with other varieties in proportion, a large portion of which were grown on my five seed farms will be found in my Vegetable and Flower Seed Catalogue for 1882. Sent free to all who apply. Customers of last Season need not write for it. All Seed sold from my establishment warranted to be fresh and true to name, so far, that should it prove otherwise, I will refill the order gratis. The original introducer of Early Ohio and Burbank Potatoes, Marblehead, Early Corn, the Hubbard Squash, Marblehead Cabbage, Phinney's Melon, and a score of other New Vegetables, I invite the patronage of the public. New Vegetables a specialty.

JAMES J. H. GREGORY,
Marblehead, Mass.

[Nov-6mo]

EVAPORATE YOUR FRUIT.
ILLUSTRATED CATALOGUE
FREE TO ALL.

AMERICAN DRIER COMPANY,
Chambersburg, Pa.
Apl-tf

LANDRETH WHEAT.

Under this name we offer to Merchants and Farmers a

NEW WINTER WHEAT

of superior excellence. Not till this year had the stock increased sufficiently to offer it for sale—the strain all being derived from one stool selected five years ago. We control every bushel and expect to distribute it widely, feeling sure that it is an acquisition of value, being **Hardy, Vigorous, Early, Still in Straw, very Prolific, entirely from Rust, and making Flour of the Highest Quality**. This Wheat is far superior to the Clawson, and those who sow it this Autumn will be able to sell to their neighbors for Seed all the resulting crop at good prices. We do not expect any will be sent to mill.

We append a few sworn testimonials showing the estimation in which it is held by well known millers in the State of New York.

Prices, including bags: **$1.50 per Bushel, $5.50 per Peck, $10.00 2 Bushels.**

DAVID LANDRETH & SONS,
SEED GROWERS, Philadelphia.

OVID STEAM MILLS,
George W. Jones & Bro., Props.
Having ground and baked some of the flour made from the "Landreth" White Wheat, we find the Wheat to be a No. 1 White, and first-class wheat for grinding. The flour being very white, the bran thin and light. We regard the "Landreth" Wheat much superior to the Clawson variety. We saw it before it was harvested, the heads were very large, the straw bright and still, and think it will be come one of the leading wheats.
August 11, 1882. GEO. W. JONES & BRO.,
Millers.

STATE OF NEW YORK,
COUNTY OF ONTARIO, } ss.
Richard B. Willing, of Phelps, in said county, being duly sworn deposes and says, I have used the flour made from the New White Wheat known as "Landreth," from the grist ground for H. S. Bonnel, and I have no hesitation in saying that it has long experience in milling I have never seen or had such nice sweet and spongy bread. R. H. WILLING, Miller.
August 6, 1882.
Subscribed and sworn to before me,
August 5th, 1882.
LYSANDER REDFIELD,
A Justice of the Peace in and
for the County of Ontario N. Y.

OVID, August 14.
I have ground trial samples of the New Wheat "Landreth," and find it excelling the Clawson and equal to any variety I have ever sown. The berry is large, white, with thin skin and light bran. The flour makes unusually white bread. M. MAXWELL,
Miller.

STATE OF NEW YORK,
COUNTY OF ONTARIO, } ss.
Emery A. Hibbard, and Fanny Hibbard, his wife, of the town of Phelps, in the said county, being duly sworn, depose and say: We have used in our family flour made from the "Landreth white Wheat," grown by H. S. Bonnell, and we can say that it makes the sweetest and best bread and pastry that we have ever had or used.
E. A. HIBBARD,
FANNY HIBBARD.
Subscribed and sworn to before me, }
August 5th, 1882.
LYSANDER REDFIELD,
Justice of the Peace in and for Ontario co., N. Y.

sep-lt

A HOME ORGAN FOR FARMERS.

THE LANCASTER FARMER,

A MONTHLY JOURNAL,

Devoted to Agriculture, Horticulture, Domestic Economy and Miscellany.

Founded Under the Auspices of the Lancaster County Agricultural and Horticultural Society.

EDITED BY DR. S. S. RATHVON.

TERMS OF SUBSCRIPTION:

ONE DOLLAR PER ANNUM,

POSTAGE PREPAID BY THE PROPRIETOR.

All subscriptions will commence with the January number, unless otherwise ordered.

Dr. S. S. Rathvon, who has so ably managed the editorial department in the past, will continue in the position of editor. His contributions on subjects connected with the science of farming, and particularly that specialty of which he is so thoroughly a master—entomological science—some knowledge of which has become a necessity to the successful farmer, are alone worth much more than the price of this publication. He is determined to make "The Farmer" a necessity to all households.

A county that has so white a reputation as Lancaster county for its agricultural products should certainly be able to support an agricultural paper of its own, for the exchange of the opinions of farmers interested in this matter. We ask the co-operation of all farmers interested in this matter. Work among your friends. The "Farmer" is only one dollar per year. Show them your copy. Try and induce them to subscribe. It is not much for each subscriber to do but it will greatly assist us.

All communications in regard to the editorial management should be addressed to Dr. S. S. Rathvon, Lancaster, Pa., and all business letters in regard to subscriptions and advertising should be addressed to the publisher. Rates of advertising can be had on application at the office.

JOHN A. HIESTAND,
No. 9 North Queen St., Lancaster, Pa.

$72 A WEEK. $12 a day at home easily made. Costly Outfit free. Address TRUE & Co., Augusta, Maine.

ONE DOLLAR PER ANNUM.—SINGLE COPIES 10 CENTS.

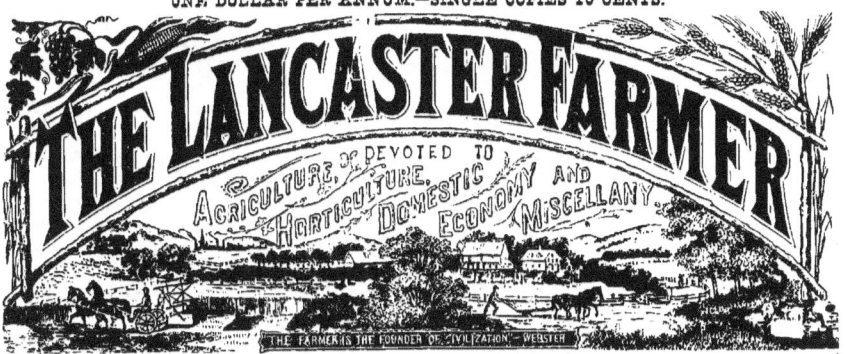

Dr. S. S. RATHVON, Editor. LANCASTER, PA. NOVEMBER, 1882. JOHN A. HIESTAND, Publisher

Entered at the Post Office in Lancaster as Second Class Matter.

CONTENTS OF THIS NUMBER.

EDITORIAL.
The History of the Tomato..................161
"Leaves"...................................161
Kitchen-Garden for November................161
Insects Injurious to Forests and Shade-trees.....161
A Plea for Trees............................162
The Farmer's Creed..........................162
Excerpts....................................163

CONTRIBUTIONS.
Balance of Trade............................164

SELECTIONS.
Trees, Climate and Soil.....................164
Relation of Forests to Rainfall.
Heavy Manuring, and How?..................165
Artificial Incubation.......................166
Indian Corn in Kansas—Its Value and Importance..166
The Effect of a Good Silo..................167
Agricultural Prosperity Should Benefit the Farmer.................................167
Tree-Planting in Streets and Grounds........168
The Fair Season............................168
Italian Bees and How to Italianize the Common Black Bees...........................168
Preventable Losses on the Farm.............169
Yield and Condition of Crops...............169

OUR LOCAL ORGANIZATIONS.
Lancaster County Agricultural and Horticultural Society..................................170
The Poultry Association....................171
Linnæan Society............................171
Fulton Farmers' Club.......................171
Exhibits—Questions and Answers.

AGRICULTURE.
The Use of the Roller......................172
Progressive Farmers........................172
Effect of Draining.........................172
Fall Plowing...............................172

HORTICULTURE.
Pear Raising...............................172
The Effect of Dry Weather on Apples........172
Saving Cabbage till Spring.................173
The Fruit Supply...........................173
Bananas and Plantains......................173

HOUSEHOLD RECIPES.
Graham Bread...............................173
Indian Cake................................173
Crullers...................................173
Doughnuts..................................173
Buns.......................................173
Roast Mutton...............................173
Mashed Potatoes............................173
Mashed Turnips.............................173
Baked Potatoes.............................173
Apple Pudding..............................173
Spanish Cream..............................173

Boiled Flank of Beef.......................173
Meat Hash..................................173
Veal Loaf..................................173
Tomato Sauce...............................173
Steamed Beef Steak Pudding.................173
Stewed Lobster.............................174
Boiled Rice................................174
Boiled Cider...............................174
Steamed Pudding............................174
Nice Griddle Cakes.........................174
Cottage Pudding............................174
Griddle and Indian Cakes...................174
Escalloped Mutton..........................174
Mock Oyster Soup...........................174
Excellent Cold Cake........................174
Lemon Cake.................................174
Fried Chicken..............................174
Plain Fruit Cake...........................174
Boiled Rice Pudding........................174
Okra Soup Equal to Turtle Soup.............174
Steamed Brown Bread........................174
Rhubarb Pies...............................174

LIVE STOCK.
Cattle-Raising in Montana..................174
To Utilize Jersey Bulls....................174
The Shropshire Sheep.......................174
Rearing Sheep for Their Milk...............174
Making Good Pork...........................175
The Coming Sheep...........................175

POULTRY.
Moulting...................................175
How to Be Rid of Them......................175
A Poultry House............................175
Literary and Personal......................176

Queen OF THE South.
PORTABLE
FARM MILLS
For Stock Feed or Meal for Family use.
10,000 IN USE.
Write for Pamphlet.
Simpson & Gault M'f'g Co.
Successors to Straub Mill Co.
CINCINNATI, O.
nov-3t

NONPAREIL
FARM & FEED MILLS
The Cheapest and Best.
Will Crush and Grind Any thing
Illustrated Catalogue FREE.
Address L. J. MILLER, Cincinnati, O.
oct-2m

Agents Wanted. The Culminating Triumph,
HOW TO LIVE!
A complete Cyclopedia of household knowledge for the masses; now ready. Nothing like it. Going fast. Low priced, illustrated, unequalled in authorship. Send for Press notices and full particulars now. Outfit and instructions how to sell, free to actual agents. Success guaranteed faithful workers. State experience, if any, and territory desired. W.H.THOMPSON,Pub.,604 Arch St.,Phila.
oct-1t

LANDRETH'S
GARDEN SEEDS

LANDRETH'S FIELD SEEDS,
LANDRETH'S FLOWER SEEDS.
Agricultural Implements in great variety.
Horticultural Tools in great variety.
Requisites for Garden and Green House.
Red and White Clover, Alsike Clover, Lucerne, Blue Grass, Gray Grass, Orchard Grass, Herd's Grass, Perennial Rye Grass.
Mixed Lawn Seed, very finest quality.
Plants, Pot Plants and House Plants.
Bone Meal of the purest quality.
Peruvian Guano, Land Plaster.
Farm Salt, Flaxseed Meal.
Carlisle Soaps, Paris Green.
London Purple, Paris Purple.
Insect Powder, Tobacco Dust.
ILLUSTRATED CATALOGUES FREE. PRICES LOW. CAREFUL ATTENTION GUARANTEED.
Founded 1784. 1500 acres under cultivation growing Landreth's Garden Seeds.

D. LANDRETH & SONS,
Nos. 21 and 23 South Sixth Street,
BETWEEN MARKET AND CHESTNUT STS.,
AND S. W. CORNER DELAWARE AVENUE AND ARCH ST.,
oct-6m PHILADELPHIA.

HEAR
YE
DEAF.
Garmore's Artificial Ear Drum.
As invented and worn by him perfectly restoring the hearing. Entirely deaf for thirty years, he hears with them even whispers, distinctly. Are not observable, and remain in position without aid. Descriptive Circular Free. CAUTION: Do not be deceived by bogus ear drums. Mine is the only successful artificial Ear Drum manufactured.
JOHN GARMORE,
Fifth & Race Sts., Cincinnati, O.
oct-3m

D.M.FERRY&CO'S
ILLUSTRATED, DESCRIPTIVE AND PRICED
SEED ANNUAL FOR 1882
Will be mailed FREE to all applicants, and to customers without ordering it. It contains four colored plates, 600 engravings, about 200 pages, and full descriptions, prices and directions for planting 1200 varieties of Vegetable and Flower Seeds, Plants, Fruit Trees, etc. Invaluable to all. Send for it. Address,
D. M. FERRY & CO., Detroit, Mich.
Jan-4m

$66 a week in your own town. Terms and $5 outfit free Address H. HALLETT & Co., Portland, Maine.
jun-1yr*

WE WANT OLD BOOKS.
WE WANT GERMAN BOOKS.
WE WANT BOOKS PRINTED IN LANCASTER CO.
We Want All Kinds of Old Books.
LIBRARIES, ENGLISH OR GERMAN BOUGHT.
Cash paid for Books in any quantity. Send your address and we will call.
BEES WELSH & CO.,
23 South Ninth Street, Philadelphia.

THE LANCASTER FARMER

PENNSYLVANIA RAILROAD SCHEDULE.
Trains LEAVE the Depot in this city, as follows:

WESTWARD. — Lancaster, Harrisburg.
Pacific Express*, Way Passenger*, Niagara Express, Hanover Accommodation, Mail train via Mt. Joy, No. 2 via Columbia, Sunday Mail, Fast Line*, Frederick Accommodation, Harrisburg Accommodation, Columbia Accommodation, Harrisburg Express, Pittsburg Express, Cincinnati Express*.

EASTWARD. — Lancaster, Philadelphia.
Cincinnati Express, Fast Line, Harrisburg Express, Columbia Accommodation, Pacific Express*, Sunday Mail, Johnstown Express, Day Express*, Harrisburg Accommodation.

The Hanover Accommodation, west, connects at Lancaster with Niagara Express, west, at 9:35 a. m., and will run through to Hanover.
The Frederick Accommodation, west, connects at Lancaster with Fast Line, west, at 2:10 p. m., and runs to Frederick.
The Pacific Express, east, on Sunday, when flagged, will stop at Middletown, Elizabethtown, Mount Joy and Landisville.
*The only trains which run daily.
†Runs daily, except Monday.

NORBECK & MILEY,

PRACTICAL
Carriage Builders,
COX & CO'S OLD STAND,

Corner of Duke and Vine Streets,
LANCASTER, PA.

THE LATEST IMPROVED

SIDE-BAR BUGGIES,
PHÆTONS,
Carriages, Etc.

THE LARGEST ASSORTMENT IN THE CITY.

Prices to Suit the Times.

REPAIRING promptly attended to. All work guaranteed.

S. B. COX,
Manufacturer of
Carriages, Buggies, Phaetons, etc.
CHURCH ST., NEAR DUKE, LANCASTER, PA.

Large Stock of New and Second-hand Work on hand very cheap. Carriages Made to Order. Work Warranted for one year.

EDW. J. ZAHM,
DEALER IN
AMERICAN AND FOREIGN
WATCHES,
SOLID SILVER & SILVER PLATED WARE,

CLOCKS,
JEWELRY AND TABLE CUTLERY.

Sole Agent for the Arundel Fluted

SPECTACLES.

Repairing strictly attended to

ZAHM'S CORNER,
North Queen-st. and Centre Square, Lancaster, Pa.

E. F. BOWMAN,

Watches & Clocks
AT LOWEST POSSIBLE PRICES,
Fully guaranteed.
No. 106 EAST KING STREET,
Opposite Leopard Hotel.

ESTABLISHED 1832.

G. SENER & SONS,
Manufacturers and dealers in all kinds of rough and finished

LUMBER,
The best Sawed SHINGLES in the country. Also Sash, Doors, Blinds, Mouldings, &c.

PATENT O. G. WEATHERBOARDING
and PATENT BLINDS, which are far superior to any other. Also best COAL constantly on hand.
OFFICE AND YARD:
Northeast Corner of Prince and Walnut-sts.,
LANCASTER, PA.

PRACTICAL ESSAYS ON ENTOMOLOGY,
Embracing the history and habits of

NOXIOUS AND INNOXIOUS
INSECTS,
and the best remedies for their expulsion or extermination.

By S. S. RATHVON, Ph. D.
LANCASTER, PA.

This work will be Highly Illustrated, and will be put in press (as soon after a sufficient number of subscribers can be obtained to cover the cost) as the work can possibly be accomplished.

$5 to $20 per day at home. Samples worth $5 free! Address STINSON & Co., Portland, Maine.

SEND FOR
SPECIAL PRICES
On Concord Grapevines, Transplanted Evergreens, Tulip, Poplar, Linden, Maple, etc. Tree Seedlings and Trees for timber plantations by the 100,000.
J. JENKINS' NURSERY,
WINONA, COLUMBIANA CO., OHIO.

$1000 Reward

For any machine hulling as much clover seed in 1 day as the **VICTOR** (Double Huller)
It has hulled **150 Bushels** in **ONE DAY.**
Illustrated Pamphlet mailed free. Newark Machine Co., Newark, O. Formerly the Hagerstown Agr. Imp. Mfg. Co., Hagerstown, Md.

THE
LANCASTER EXAMINER
OFFICE
No. 9 North Queen Street,
LANCASTER, PA.

THE OLDEST AND BEST.

THE WEEKLY
LANCASTER EXAMINER
One of the largest Weekly Papers in the State.

Published Every Wednesday Morning,

Is an old, well-established newspaper, and contains just the news desirable to make it an interesting and valuable Family Newspaper. The postage to subscribers residing outside of Lancaster county is paid by the publisher.
Send for a specimen copy.

SUBSCRIPTION:
Two Dollars per Annum.

THE DAILY
LANCASTER EXAMINER
The Largest Daily Paper in the county.

Published Daily Except Sunday.

The daily is published every evening during the week. It is delivered in the City and to surrounding Towns accessible by railroad and daily stage lines, for 10 cents a week.
Mail subscription, free of postage—One month, 50 cents; one year, $5.00.

JOHN A. HIESTAND, Proprietor,
No. 9 North Queen St.,
LANCASTER, PA.

The Lancaster Farmer.

Dr. S. S. RATHVON, Editor.　　　LANCASTER, PA., NOVEMBER, 1882.　　　Vol. XIV. No. 11.

EDITORIAL.

THE HISTORY OF THE TOMATO

It is just fifty-five years since the first tomato was grown in this county. In the spring of 1827 a traveler who tarried over the night at a village inn at Frankstown, Pa., presented the landlord's little daughter with a few seeds, which she carefully planted and from which the first ripe tomatoes ever grown in this county were produced. The little girl was so proud of her little tree full of bright red "love apples" that she carried it around every house in the village, and as the fruit was supposed to be poisonous the children were cautioned not to touch or handle it. The following year the little girl's mother, against the protest of all the other members of the family, prepared, cooked and ate some of the fruit. As she pronounced it good and palatable and did not sicken, the tomato soon became a favorite dish in the family. The young girl, whose name was Matilda Brotherline, is still living, the wife of Mr. John Barr, of East Hollidaysburg, and mother of S. B. Barr, of Hays City, Kansas.—*Altoona Tribune*.

The foregoing historical item is interesting, and the more so because it allows other people, and other places, to put in their claims in regard to the introduction, and culinary preparation of the tomato by claiming to account for its introduction only in "this county." Some months ago a similar paragraph was going the rounds of the press to the effect, that about half a century ago, a prisoner, discharged from the York jail, had presented some seeds to the wife or daughter of the jailor, who had planted them in an enclosure attached to the jail, and had grown the beautiful red "love-apples", which were also deemed poisonous, but which were subsequently discovered to be edible, and this was claimed as the first introduction of the tomato into York county. There is no reason why both accounts should not be true; and, if the *whole* truth were known, perhaps a dozen other places might establish a similar claim. It is certain, from a record made by Thomas Jefferson in his domestic diary, while he was President of the United States, that tomatoes were sold in Washington market for edible purposes. Their cultivation and sale may have been limited, but they were used as a culinary preparation on the table of the President, at least. We, ourself, saw the tomato and the white egg-plant, under ornamental cultivation, more than sixty years ago, and, overshadowed, coincidentally, by the same notion that it was poisonous, indeed was readily impressed by its peculiar odor. About the summer of 1831 or 1832, we very dubiously tasted the first stewed tomatoes, and we were by no means prepossessed in their favor; we learned to appreciate bull-frogs and lobsters in a shorter time, and with far less repulsion. Philip Miller, in his *Gardener's Dictionary*, published about the year 1731, speaks of tomatoes being introduced into England from Spain, where they long had been cultivated for edible purposes, but were chiefly used as a condiment. Miller, himself, spoke disparagingly of them—considered them too "watery" ever to come into general use. This objection obtains in some of the varieties cultivated at the present day. But, a wonderful improvement has been made, both in the quality and quantity of this fruit. The best we have ever eaten were supplied abundantly in our markets the present season, large, solid all though, comparatively dry, granular, and finely flavored. They have become a culinary necessity.

"LEAVES."

We regret that through an oversight, discovered too late, we failed to respond in anywise to the queries of J. F. W., in his contribution published in the October number of the FARMER on the subject of "Leaves." He is physiologically correct that the function of leaves in plants is analogous to the function of the lungs in the animal body; and that the premature destruction of the leaves will mar, or totally prevent the development of the fruit.

The "little fly" that infested his grape-vines was doubtless the "Grape-leaf Hopper" (*Tettigonia vitis*), an insect belonging to the order HOMOPTERA; but, we do not think that "drenching the vines from beneath with abundance of water from the hydrant" would have had any permanent effect. It might have driven them off for an indefinite time, but as soon as the operation would have ceased, and the water had evaporated, they would have been all back again. A few of them might have been washed down and have perished, but the larger number would only have hopped off to safer quarters, and there awaited the first opportunity to return. In consequence of this peculiar characteristic, they are difficult of access with a drench of any kind, although if applied when the insects are quite young (June and July) and before the development of their wings, even water would prove more or less destructive, especially if applied with force, and from beneath. But, if a saponaceous solution, an alkalinous dilution, or a tobacco, capsicum, or pyrethrum infusion, were applied with a garden syringe, every morning early, or late in the evening when they are in repose on the undersides of the leaves, the remedy would be very effectual. The transformation of these insects, like that of the *Orthoptera* (grasshoppers, etc.,) and the *Hemiptera* (bugs) is what is termed "incomplete"—indeed rather a *transition* through successive stages of development, in which there is very little difference between the young and the adult, save in the development of the wings and coloration. In their earlier stages they are feeble, and when disturbed merely shift their positions from the lower to the upper sides of the leaves, or *vice versa*, but as they advance in life they are given to flying or hopping, hence called "Leaf-hoppers."

SEND in your subscriptions early.

KITCHEN-GARDEN FOR NOVEMBER.

In the Middle States, the season for gardening is drawing to a close; indeed, it is limited to the preservation of roots, and the hardier vegetables for winter use, and such operations as may be preparatory to another season. Now is a good time to transplant fruit and ornamental trees, shrubbery, &c. On loamy and light land we prefer, decidedly, fall planting; on heavy soil, or where the subsoil is clay, thus retaining the moisture near the surface, spring may be a more favorable season; and it is also generally esteemed the best for evergreens. Asparagus beds winter dress. Beets dig and store. Cabbages place in safe quarters. Carrots dig and store. Celery earth up finally. Drain vacant ground if needful. Horse-radish dig and store for convenience. Onions, in store, examine. Parsnips dig for convenient access. Salsify ditto &c. From 1st to 20th of this month, according to locality, the winter supply of turnips should be cared for.—*Landreth's Rural Register*.

The proper time for transplanting fruit, ornamental, and other trees—that is, whether the spring or autumn of the year is best—is a question of very long standing; indeed, we can remember it from an early boyhood, at least sixty years ago. Both periods had their staunch friends and advocates then, as they have now, and both could point to numerous cases, by way of illustration, where the one had succeeded and the other had failed, and neither party would plant out of their favorite season, "hit or miss." Perhaps it never occurred to either party that "circumstances alter cases." Soil and meteorological conditions have certainly much to do in determining the question.

We recall a circumstance that occurred fully fifty years ago, when an extraordinary drought prevailed from the middle to the end of summer, when nearly all the late potato vines were burnt brown and crisp, and the tubers were about the size of marrowfat peas. One cultivator, either through indolence or indifference, failed to weed his "patch," feeling that there would be no potatoes anyhow. Well, it transpired that he had the best crop of potatoes in the district—indeed a fair crop. The weeds shaded the vines and kept them green, and matured the tubers. This, however, would not have been successful as a rule—*circumstances alter cases*. Both periods of transplanting have their advantages and disadvantages, according to the conditions above stated.

INSECTS INJURIOUS TO FORESTS AND SHADE-TREES.

Kaltenbach, in his work entitled "*Die Pflanzenfiende aus der Klasse der Insekten*," has enumerated, in a closely-printed volume of 848 pages, the species of insects preying upon the different trees and plants of all sorts in Central Europe. The number of insects found upon some kinds of forest-trees is astonishing, though it is to be remembered that all kinds are not equally destructive, the most injurious and deadly forms being comparatively few.

The above named author enumerated 537 species of insects injurious to the oak, and 107 obnoxious to the elm. (Some of these species, however, may number millions of individuals.) The poplars afford a livelihood to 204 kinds of insects; the willows yield food to 396 species; the birches harbor 270 species; the alder, 119; the beech, 154; the hazelnut, 97; and the thornbeam, or "iron-wood," 88. The junipers supply 33 species, while upon the pines, larch, spruce, and firs, collectively, prey 299 species of insects. In France, Perris has observed over one hundred species, either injurious to, or living upon, without being especially injurious to, the maritime pine; these he describes in an octavo volume of 532 pages with numerous plates." On this same subject, Ratzeburg, of Germany, published three beautifully-illustrated quarto volumes, of about 500 pages each, over thirty years ago, and it is a standing regret to us that we did not secure the work when we could have done so on the most liberal conditions.

Dr. A. S. Packard, Jr., of the Entomological Commission, has written a work, noticed in our Literary and Personal columns, in which he brings together a large number of American species destructive to forest and shade trees, and, although little more than the technical names of many of them are given, and extended descriptions to only a few, still the work is a needful one, and comes before the public as the probable forerunner of a more ample work on the subject on some future occasion, when the subject of forestry, and the preservation of trees of all kinds shall have taken a deeper hold upon the minds and the hearts of those who are materially interested therein. Indeed, it is a subject in which all are more or less interested, whether they possess large landed estates, or an acre, or nothing.

Of course Dr. Packard does not profess to have enumerated *all* the destructive species of insects that infest fruit, shade and forest trees, but even the few he *does* describe is suggestive of what remains to be described. The Oak is infested by 214 species, only 59 of which have appended descriptions. Although all these insect species live upon the substances of the various oaks, yet it is not manifest that the larger number are injurious—indeed, it is possible that some of them may in some way be even beneficial. Still, there are so many of such a decidedly injurious character, that it would not be wise to permit any of them to increase, if it can possibly be prevented; insect injury is a matter that depends much on numbers. A little blood taken from an animal being may do no harm, or even be beneficial, but the result would be quite different if *all* the blood were tapped.

regretted because this tree has been largely relied on to furnish a forestry for the Western prairies. The *Poplar* is infested by 36 species, and the *Linden*, 23; the *Birch*, 19; the *Beech*, 15, and the *Tulip tree*, 7. The Wild Cherry, 22; the *Choke-cherry*, 4; the *Red Wild Plum*, 6; the June Berry, 4; the *Mountain Ash*, 19; the *Sweet Gum*, 5; the *Persimmon*, 3; *Gum*, 1; the *Laurel*, 19; *Sassafras*, 6; *Sycamore*, 9; Hazel, 8; Hornbeam, 12; Water Beech, 2, and the *Alder*, 19. The willow has 99 species to feed; the *Pine*, 102; the Spruce, 24, and the *Fir*, 19. The Hemlock, 10; the Juniper, 12; Larch, 8; the Cedar, 3; the *Sequoia gigantea*—the great California mammoth, 3, and the Cypress, 1.

Of course, it is not to be understood that all these insects confine themselves to the particular kind of tree mentioned in connection with them, for many of them are either indiscriminate, or readily adapt themselves to different kinds. Notwithstanding this, there are *one thousand* species of insects enumerated that are injurious, or may become injurious to forest and shade trees in the United States, and known to be such at the present day; and this does not include the many that infest fruit and other species of vegetation. This little work makes no attempt to deal in insect remedies; that is not its object; it is merely a synoptic compilation of the knowledge accessible to the author in a special department of practical entomology, and partially illustrates the immensity of the labor yet to be performed in this rapidly developing field. We have hardly more than entered the vestibule of practical entomology, and yet it is possible the government may eventually relax its aid, simply because it may be unable to appreciate the progress that has been made, and what yet remains to be accomplished.

A PLEA FOR TREES.

The High Commissioner of Cyprus attributes its chief curses, droughts and floods, to the reckless destruction of the forests. As the woods disappeared, so did the soil that covered the hills; that soil was washed down to the plains, choked the rivers, and formed marvelous swamps, the hills became bare rocks, incapable of growing a blade of vegetation, and the locust at once took possession of the barren ground, while the absence of trees deprived the earth of its annually fertilizing agent—leaf mould. The same process is going on upon the higher hills, and Sir R. Biddulph believes it is no exaggeration to say that Cyprus is in a critical state on this account, from which, however, there is reason to hope that it may yet be recovered. There are districts of this country, too, where these remarks may be pondered over with advantage.

How often do we meet with paragraphs that any one whose opinion is entitled to respect, should gainsay them. The theory itself constitutes a convenient little species of "thunder," wielded by agricultural writers and speakers, from the township club up to the halls of Congress, or *vice versa*, if that form of expression would be putting the matter more correctly. In one lifetime of seventy years we have been cognizant of several local changes on the earth's surface, which, if not caused by the removal of the trees that once occupied places made bare and arid apparently by their removal, then the phenomena seem absolutely inexplicable; and yet Prof. Isaac Bassett Choat, of Cambridge, Mass., in a communication to the editor of the New York *Tribune*, under date of September 30, 1882, goes very far towards exploding the theory, the arguments it relies on, and the presumed results. As the subjects of forests, rain-falls, droughts and floods seem now to be eliciting more than ordinary attention, it may subserve a useful end to to place both sides of the question before the thoughtful reader, and, if he can, enable him to make up his mind thereon; hence, in another column we insert Prof. Choat's paper, for the edification of our interested readers. It is not to be inferred, however, that it expresses our sentiments, or that we unqualifiedly endorse it simply because we re-publish it. "It mought be so, but then again it moughtn't." It is just as likely that the con side of the question may be influenced by appearances as much as the *pro* side is. It would require twenty years of thorough observation and experience, perhaps, to affirm or overthrow the theory effectually, and our time seems a little too short to begin such a labor now. We will have to view it from other stand-points for the present.

THE FARMER'S CREED.

"Let this be held the farmer's creed:
For stock, seek out the choicest breed;
In peace and plenty let them feed;
Your land sow with the best of seed;
Let it not dung nor dressing need;
Inclose, plough, reap with care and speed;
And you will soon be rich indeed."

THE FARMER'S FIEND.

A tender young potato bug
Sat swinging on a vine,
And sighed unto a maiden bug:
"I pray you will be mine."
Then softly spake the maiden bug:
"I love you fond and true,
But oh! my cruel-hearted pa
Won't let me marry you!"
With scorn upon his buggy brow,
With glances cold and keen,
That haughty lover answered her:

Our joys, when extended,
  Will always increase ;
Our griefs, when divided,
  Are hushed into peace."

THE FARMER'S HOPE.

"Hope springs eternal in the human breast ;
Man never is, but always to be, blest ;
The soul, uneasy and confined from home,
Rests and expatiates on a life to come."

THE FARMER'S PROVIDENCE.

"All nature is but art, unknown to thee ;
All chance, direction, which thou canst not see,
All discord, harmony not understood,
All partial evil, universal good ;
And, spite of pride, in erring reason's spite,
One thing is clear—whatever is, is right."

EXCERPTS.

THE rye crop will probably reach 20,000,-000 bushels.

OF buckwheat, Pennsylvania produces nearly one-half the entire crop. The total yield will be over 11,000 000 bushels.

THE potato crop covers an area approaching 2,000,000 acres, with a yield of about 80 bushels per acre. A short crop is foreshadowed in New York State.

RETURNS to the Department of Agriculture from all the 1700 counties of the United States indicate a wheat product slightly exceeding 500,000,000 bushels, or an average yield per acre of about 13.5 bushels.

IN cotton, an unusual size and vigor of plant, with capacity for a large production, is reported. The general average of condition is higher than in any October for ten years, with the exception of 1875 and 1878.

MORE than one-half of all the barley produced in the United States is raised in New York, California, and Wisconsin. The average yield is 23.5 bushels per acre, and the total product will reach 45,000,000 bushels.

OATS are an immense crop. The average yield is higher than that of last year, Kansas ranks among the highest, as it does in wheat. The total product in oats of all the States will probably be 480,000,000 bushels.

THE six principal winter wheat States will aggregate 244,000,000 bushels. There is a reduction in the acreage of the spring wheat area of the Northwest, but the yield may reach 113,000,000 bushels. The Pacific coast will probably yield 45,000,000 bushels, the Middle States 40,000,000 bushels, and the Southern States a trifle more than 50,000,-000 bushels.

KANSAS holds its reputation for large returns to the toiler, with the extraordinary average yield of 19.5. The country north of the Ohio river, in the great wheat belt, averages nearly 16 bushels. Kentucky and Missouri promise about 14 bushels, and California 13 bushels.

THE average yield of wheat the country over has never fallen quite to 10 bushels, and it has never quite reached 14 bushels in years of greatest abundance. This season it is unusually high in New York—18.7 bushels. In the New England States, except Vermont, it runs as low as 14 bushels. In nearly all the Southern States the average is low, ranging from 7 to 10 bushels. Texas and Arkansas are exceptions.

THE yield of corn cannot yet be accurately estimated. Much of it is still standing in stock in the fields. It is believed, however, that there will be at least 1,680,000,000 bushels, or an average yield of 25 bushels, or an average yield of 25 bushels to the acre, against 28 in 1879, and 18 in 1881. Of this total the States north of Tennessee and west of Virginia and Pennsylvania produced 1,250,-000,000 ; the Southern States, 340,000,000 ; Middle States, 82,000,000, and New England over 7,000,000. The total product will be more than four hundred millions greater than last year.

INDIGESTION IN HOGS.—When pigs do not thrive and try to eat gravel or earth it is a symptom of indigestion. They are probably overfed. Reduce their food one-half. Give the two pigs half a pint of sweet-oil or linseed-oil in the food daily for two or three days, and as they recover gradually give them a little dry corn in addition to their other food. Some charcoal would be of service and may be given frequently.

DIARRHŒA IN A MARE.—When a horse is in good health and condition while feeding upon grass, and when changed to hay is affected by diarrhœa, it is doubtless something in the hay that causes it. Give the mare half a pint of linseed oil once a day for a few days ; cut the hay and wet it, and add to it a quart of bran and linseed-oil cake meal in equal parts, and add little salt to it. This will probably remove the trouble.

REMEDY FOR FLIES.—As a remedy for flies of all kinds in houses, stables, and greenhouses, it has been recommended to boil tobacco in water until the juice has been nearly all evaporated. As it is the essential oils of the tobacco which are effective for this purpose when it is burned and these are evaporated with the steam in the boiling, all the effect is produced without the disagreeable smoke of the burning.

THE CABBAGE-WORM.—There is no doubt the cabbage-worm can be destroyed by using some very soluble substance that is poisonous to it, but not hurtful to persons if it is wholly washed off—but that may easily be done. Nitrate of soda, glauber salts, and muriate of potash have each been tried the past season, and each one killed the worms. The different salts were dissolved in water, half an ounce to a quart, and sprinkled over the cabbages.

REDUCING BONES WITH POTASH.—The waste potash from the muriatic acid makers can be used to soften bones in the following manner : Pack the bones in a tank or pit or heap, with the potash and quicklime in proportion of 25 pounds of each to 100 pounds of bone, or even double that quantity, as they are of value for the fertilizer. When the heap is complete wet it until the lime begins to slake ; then cover it with earth and leave it exposed to the rain during the winter. In the spring it may be shoveled out and mixed.

THREE and one-tenth pounds of corn will produce, when fed to a hen, five-sixths of a pound of eggs ; but five-sixths of a pound of pork requires about five pounds of corn for its production. Taking into account the nutriment in each and the comparative prices of the two on an average, the pork is about three times as costly a food as the eggs, while it is certainly less healthful.—Hartford (Conn.) Farmer.

A SALMON was caught in the Penobscot, near Bucksport, Me., the other day, that was 31½ inches long and weighed 16½ pounds. It was " tagged " as follows: " Salmon No. 1,135. This was a female tagged Oct. 28, 1880, and dismissed a few days later, weighing 7½ pounds and measuring 30 inches in length. She had just yielded 1 pound 15 ounces of spawn, which would make her weight before spawning 9 pounds 7 ounces." Another, recently caught, was marked Nov. 13, 1880, and then weighed 8½ pounds, Recaptured June 23, 1882, it weighed 14½ pounds, and was in good condition.

ONE hundred and fifty thousand pounds of wool were purchased by one firm in Washington county, Pa., in one week.

Cows cannot be cheated into giving liberal quantities of milk. That which they give is in proportion to what they receive.

A GOOD farmer is better than a poor doctor, and a good horse-shoer is better than a bishop who preaches sermons nobody wants to hear.

ALWAYS have a place where your chickens can be sheltered from the storms, and be kept comfortable. It is the lack of this that kills so many chickens.

THERE are two things that every farmer must have—things that subserve like purposes and are of equal importance—a grindstone and a newspaper.

A VERY successful farmer once remarked that " he fed his land before it was hungry, rested it before it was weary, and weeded it before it was foul."

COTTON-SEED FOR FOWLS.—We do not know if fowls would eat cotton-seed ; it is hardly probable, unless the husk were free from lint. The writer has fed the meal to fowls mixed in equal portions with ground corn and oats and wet with hot water, or mixed with sour milk, and they thrived exceedingly well upon it. There is no doubt the meal will be a very useful feed for poultry, but as regards the whole seed it is doubtful. If some of our Southern readers would try it and report we should be glad to publish the facts.

GIVE the steers about two quarts of grain every day.

STABLE the horses at night, if they are worked. In rainy weather, work them as little as possible, and rub dry.

PLENTY of night feed for the milch cows. Soft corn, corn meal, corn fodder, hay, bran, beet tops, cabbage leaves and pumpkins, are what they ought to have.

PENS for farrowing sows should have a rail round the interior, about a foot high and six inches from the sides. Feed warm slops after farrowing, and increase the quality of the food as the pigs grow. After three weeks' growth, feed the little pigs in a separate trough. Don't delay shutting up the pigs intended for fattening.

FOR early lambing, choose the best ordinary Merino ewes and a ram of pure breed.

About half a pound of grain daily, and your best pasturage for a few days, and they will take the ram readily. Sheep intended for winter-fattening should be well fed now. Don't attempt to winter-fatten Merinoes until they have had three years' growth. Keep the lambs in a separate flock. Use carbolic soap as a wash for ticks. Shelter the lambs—and indeed all sheep—well in severe weather. Be on the alert for dogs that kill sheep.

FEED the poultry well. Don't let them get the habit of roosting on trees or utensils. The greater the number of eggs produced by a fowl, the less vitality there will be in each; therefore, the first only of a laying should be set. Early chickens are the most certain to live, because force is stored up in the parent, before laying commences, sufficient to endow the first eggs or chickens with plenty of vigor. The chickens being hatched and assigned quarters, see that animal food is artificially provided for them, for they cannot thrive upon grain and vegetables alone.

CONTRIBUTIONS.

For THE LANCASTER FARMER.
BALANCE OF TRADE.

Mr. Editor:—I have been considerably interested in the discussion that has been going on for some time in THE FARMER on "The Balance of Trade." The articles over the signature "J. P." seem to me so unreasonable in argument and so false in statistics that I would like to make some reply. The idea that a nation, any more than an individual, can grow rich by buying and consuming more than it produces and sells would seem only to need statement to meet with ridicule. I suppose it to be axiomatic that all production is gain and all consumption is loss; and as our imports are for consumption, and are actually consumed in our country, they measure our consumption over our production, and our exports measure our production over our consumption, and therefore the excess of imports (consumption) over exports (production) must necessarily be loss, and *vice versa*. It is very plain when we apply it to a farmer or manufacturer: what he produces and sells in the market are his exports; what he buys and consumes are his imports. If the former exceed the latter, he gains; if the latter the former, he loses. It cannot be different with a nation. If we import only necessary articles, of course the more we get for a given amount of exports the better; for, being necessary, we have to have them, and the cheaper we get them the better. But we import and consume hundreds of millions of dollars worth of luxuries every year, that, so far as increasing the material wealth of our country is concerned, amount to nothing whatever. Take the single article of wine. We import and consume about $12,000,000 worth of foreign wines, champagnes, etc. annually. Would any one but J. P. say that this $12,000,000 worth of liquor that is used almost entirely as a beverage, enriches our nation as much as though we had brought back, for our exports bills of exchange or gold coin, that would stand as solid, reserved wealth in our country, to be used in time of need, and not vanish, leaving us poorer than before, as the liquor does. Gold coin being transmitted to pay the balance of trade, and not to be counted as imports or exports in this argument. Necessaries we must have, or fail in our production, and of course it is best to get them where they can be obtained the cheapest; but J. P.'s argument rests on the fallacy that all imports that have the same money value are of equal worth, to the nation that consumes them. No one would think of applying this reasoning to an individual, and say that a farmer who sells (exports) his wheat crop for $100 and brings back (imports) its value in whisky, and drinks up, or rather down, would be just as well off at the end of the year when the liquor is all gone as though he had invested the $100 in State or Government bonds.

Take the case that J. P. supposes, of a miller exporting $50 worth of flour, and importing for it salt worth here $75. Salt being a necessary article, the country may have gained as the miller did ; but suppose he had imported the same value in rum and drank it, how would the country have been benefited, and how would it have helped the matter if he had sold it to his neighbor for consumption. J. P. seems to think that if the importer makes a profit, our country is that much richer. As well say that all the lottery dealers, stock-jobbers and gamblers, who grow rich by filching from the pockets of others, are adding wealth to the country. The supposition about the salt being lost at sea and hence not being set down at the custom house as imports is only trifling, as it is the *purchase* of the imports, and not their actual passage over the water, of which we complain. We might as well suppose the salt spilled or destroyed in the street on its way from the ship to the warehouse after it had passed the books of the collector's office. Then the imports would exceed the exports, and according to J. P. there should be a gain, but neither the importer nor the country could be easily convinced of it. Or suppose the salt should prove worthless and would bring nothing in the market; here again the imports would exceed the exports, for is it the cost of the merchandise abroad that is put down at the custom house as the value of our imports, as it should be, and not what they may sell for here. Where would be the gain in this case? Or, how much richer would our country become by consuming seventy-five dollars worth of good rum than thirty sacks of worthless salt. If our prosperity is to be measured by the excess of imports over exports, all we have to do to get rich is to trade our goods abroad for merchandise at enormous prices; for, the higher the price we pay the more would our imports exceed our exports. If this theory is correct, why is it that all our stocks and securities go up in the market when our exports are in excess of our imports and gold is coming into the country, and go down when it is going out, to pay the balance of trade against us?

But it is J. P.'s statistics that I wish more particularly to correct, his arguments not being very dangerous. One would suppose that a theory so false could hardly have many facts to support it ; but quite a row of figures is presented. Without pretending to follow him through "England, Denmark, Austria and Hungary," the prosperity or adversity of which countries few of us know much about, let us see how near he comes to the truth in our own country, at a time that most of us can remember to our sorrow. I mean the period through our late civil war and the extravagance of our nation for nearly a decade after its close; namely, from 1861 to 1873. J. P. is compelled to acknowledge this to be a period of great depression and loss, and to make his theory hold good he has the exports during these years of great and extravagant consumption and comparatively small production, exceeding the imports by nearly a billion of dollars. I was astounded on seeing these figures, for if they were correct all my ideas of political economy must be given up. We all know our nation was a losing one from 1861 to 1873, through that terribly destructive and unproductive period of our civil war, and the extravagantly consumptive period since, till our financial panic in 1873 compelled us to stop in our ruining course. If our exports during this time were exceeding our imports, why, black was white, and white black, and all the old rules about industry and economy leading to wealth, were false.

I got the American Almanac, compiled by Mr. Spofford, the Congressional librarian, who is the very best authority on these subjects. It puts the imports in excess of the exports in every year of these 12 except 1862, and the excess of imports in the whole 12 years was $1,190,103,171. Not satisfied with this, I wrote to Joseph Nimmo, Jr., Chief of Bureau of Statistics at Washington, and got his reports from 1861 to 1879. His figures agree almost precisely with Mr. Spofford's, making our imports exceed our exports from 1861 to 1873 by over a billion of dollars. It thus appears that J. P.'s statistics are exactly reversed. He had the right figures but got them on precisely the wrong sides of the account.

After our extravagantly inflated balloon burst in 1873, and let us down so hard that some of our bones are aching yet, and we were taught by severe adversity, that we must go to work and practice economy, our exports began to exceed our imports, and from 1874 to 1879, our exports, according to the authority just quoted, exceeded our imports by $657,206,961, or about 130 million dollars annually. Does any one pretend to say we were not gaining during these latter years? or, that we could have resumed specie payments, as we did in 1879, if our imports had exceeded our exports, as they had done, during the previous destructive decade? According to J. P. our country must have been losing at a terrible rate from 1874 to 1879, when our exports were exceeding our imports 130 millions annually, and gold was pouring into our country. John Sherman didn't think so.—*S. P., Lincoln, Del., Oct. 23, 1882.*

SELECTIONS.

TREES, CLIMATE AND SOIL.

Relation of Forests to Rainfall.

SIR : The idea has long prevailed that the removal of forests is accompanied with a diminished rainfall. As a matter of course the converse of this would be held as widely and with equal confidence. Such, no doubt, is the popular belief at this time. While in the Eastern States during the excessive drought of the past summer men were constantly attributing the failure of rain to the removal of the forests from our older States; we read in the public journals that Nebraska was favored with a more copious and more equally distributed supply of rain than in any former year within the history of that State. It was claimed that the increased amount of moisture was due to the planting of trees, and that a sufficient breadth of forest growth was now planted to secure the State against drought for all coming time. This tradition is that the growth of trees favors the increase of rain, and that their removal is followed by drought. It is only within the present year that this theory has been combated by Professor J. D. Whitney, in his recently published monograph on "Climatic Changes." The views therein set forth have not as yet been widely disseminated. They are so radically opposed to the opinions commonly held that even if made familiar through the public prints, they are not likely soon to gain general acceptance.

Against the popular notion that the certain drying-up of the lands is the result of removing the forests, the Professor claims that "the question of desiccation is one essentially removed from the domain of man's influence." He would prove this to be the case by showing that the process began in geological epochs, long before man was on the earth to interfere with any of the operations of nature, that it has been continued down into historic times, and that it is now going on in the same general way, neither hastened nor retarded by the intervention of human agency. He believes that "the human race is no way responsible for the changes which have brought and are bringing ruin upon those countries which, once prosperous, have now sunk into comparative decay." Egypt and the countries north of the Mediterranean are instanced as showing decay from a drying-up of the land and an increasing absence of moisture from the atmosphere. "As a rule, these nations have reached a stage of decadence from which they can never rise to occupy again the position which they have lost." No efforts of man are of the slightest avail to restore the former conditions of climate by planting forests or by any other means. There has been a loss rather than a gain in the quantity and frequency of rain in Egypt since the beginning of this century, despite the vigorous measures of the Government in planting forest trees.

Professor Whitney shows from many instances observed in our own land that the removal of forests has nothing to do with the falling-off in the amount or the frequency of rain, neither does the planting of trees occasion any increase in this, for while the Great Salt Lake has been rising for some years since the Mormons began planting trees upon its shores, in neighboring States, the Winnemucca and the Pyramid lakes have been rising in an equally rapid ratio although their shores "were being stripped of their trees with the greatest rapidity." "There is just as much reason for inferring that the rise of these latter was produced by disforesting the country as that the similar increase of Great Salt Lake was the result of tree-planting by the Mormons; in other words, there is no truth in either statement." So plainly and so boldly are set forth these scientific principles! It is not the wasteful destruction of forests by man in a wholesale slaughter and burning which has brought the desolation of dryness upon so many desert regions of the globe, but rather the failure of rain from wholly natural causes which has led to the disappearance of the woods. There is left no ground to hope that civilization will ever reappear in lands it has once forsaken if the operations of nature continue constant. The loss of forests and the deterioration of soils are but examples of the difficulties which man meets in his struggles against unfavorable physical conditions, and are proofs of his inability to overcome them. Doctrines like these seem tinged with the despairing thought of fatalism.

All that Professor Whitney has to say upon the cultivation of forests comes in incidentally to help out the discussion of "Climatic Changes." Since that work was published in the spring some statistics representing forestry have been published by the Census Bureau, affording much interesting and timely information. These statistics were collected by Professor Sargent, who has made this matter a subject of special study. In his contribution to *The North American Review* for October, on "The Protection of Forests," he presents views identical with those of Professor Whitney, and is no less clear and emphatic in expressing them. "The popular belief that forests affect the rainfall has too long," he says, "confused the discussion of the forest question, and carried it far beyond its legitimate limits." He is positive that trees have no power to increase the quantity of rain. He manifests not the slightest faith in the endeavors of Government and of individuals to overcome the natural dryness of soils and of climate by planting forests. On the contrary, he looks upon this dryness as the cause and not the effect of the lack of trees. Rain he regards as the agency which will clothe the treeless regions of the interior with woods. Indeed, he declares that "the position of the forests and plains of North America can be explained upon no other theory." From this it will naturally be inferred that the density of the original forests varies directly with the rainfall.

Here is a point which seems not well established. To the unscientific observer rain does not seem to be the one sole thing essential to the growth of forests on the plains in the Mississippi Valley. Even the casual visitor to that section must have noticed peculiarities in the growth of timber which climatic conditions will not account for. The character of the soil seems to have much to do with the kind of growth that covers it. Let the peculiar soil of the Illinois prairies be met with in the timber region of Wisconsin, as it will at times be met with even north of the Fox River, and it will be found covered with luxuriant grass just as would be the case with a similar piece of ground located in southern Illinois. On the other hand, a deposit of drift, coarse in texture, and mineral in its composition, occurring as it sometimes does in Illinois, will be found covered with a growth of trees which not even the assaults of fire from the surrounding prairies have been able to exterminate. And yet these wooded gravel-beds are often higher than the grassy lands about them, and their loose texture lets the rains run through much faster than the water-drains off the level lands around.

Again, in those regions which are designated on the maps as treeless, wherever the ledge crops out along the borders of ravines, many varieties of trees, as cedars and crab-apples, take root in the crevices of the rock and flourish there. It may be said that these owe their existence to the water oozing from the ledge or trickling down its sides. That this is not the case, however, will be seen from the fact that where the debris-piles of loose chips and fragments of the rock lying heaped against the base—becomes sufficiently pulverized to support trees under our New-England climate, it bears them just as naturally there; and in general, wherever in the West the soil is formed from the underlaying ledge, whether that be lime or sandstone, slate or granite, there its natural covering will be a luxuriant growth of forest trees. We are indebted for the extensive pineries of northern Wisconsin and Michigan to the circumstances that all that region is overspread with drift similar in its character to the drift which abounds in those parts of Maine where pine is the native growth. Conditions of soil appear to have as much to do with determining the kind of growth upon it, whether trees or grass, as do conditions of climate. That trees are not born of the copiousness and frequency of rains is evident from this, that when it became desirable some generations ago to convert old fields and pasture lands in the west of Scotland into timber, it was found necessary to plant young trees, since these did not spring up on the abandoned farms as they would do under similar circumstances on the hillsides of New England. And yet Scotland has a climate proverbially moist. Again, here is our own country, of all the lands once covered with trees, none are slower to renew their forest growth than some of the rocky pastures about Cape Ann, where excessive drought is much less frequent and less severe than in the well-wooded interior. May it not be the case that on lands long kept in grass and where the dampness of the sea-air maintains a well-matted sod, the seeds of trees fail to germinate, or, if they do, have the young life choked out of them by the all-engrossing grasses? And may it not with good reason be supposed that the treeless condition of the prairies of the West is largely owing to the fact that there, too, the grasses have assumed and maintained the right of eminent domain?—*Isaac Bassett Choate, Cambridge, Mass., Sept. 30, 1882, in New York Tribune.*

HEAVY MANURING, AND HOW?

Probably very few men in the West spend so much for fertilizers upon an equal area of land as I do. I am cultivating about forty-five acres, and, although I get fertilizers at a small cost as compared with prices about Eastern cities, yet their cost upon that surface this season will not be less than $2,000. A large share of this amount has already been returned to me, and unless the final result this season belies all present indications the whole amount will come back with at least 50 per cent. added to it. My main reliance is upon the compost heap, stable and barnyard manures and wood ashes. How shall they be used to be of the greatest advantage to us? My early teachings were all in favor of plowing them under, and I remember some of the teachers saying the deeper the better.

Nearly twenty years ago I had occasion to break up a piece of very heavy soil. Believing in deep plowing, as well as heavy manuring, two yoke of oxen were put to the plow, and it was turned over not less than eight, and most of it ten inches deep. Upon a portion of it I put some very rich manure, putting it in the bottom of the furrows. It was of course buried very deep; but if my theory of deeply covering the manure was correct it would be all right, and I should in due time receive my reward. I watched and waited with a good deal of interest for the result. I am waiting yet, but not watching with any great amount of interest any longer. So far as any benefit to the subsequent crops was concerned I believe the manure might as well been buried under the Pyramid of Cheops as to have been put where it was.

This set me to thinking as well as to studying and experimenting in other directions. The result has been that my system of fertilizing my land has been almost completely revolutionized within the last ten or twelve years. At present my rule is to have my compost heaps (which, by the way, are my main reliance) as fine and thoroughly rotted as possible, and then spread upon the top of the ground after plowing and harrowing it. I still plow under some manure, but never so deep as I did years ago. I will not pretend that my present plan is the best for others and under all circumstances, or that I have reached perfection as regards my own land; but I am certainly receiving much better returns for my manures than I did years ago when they were buried much deeper. When I can find a better method than my present one I shall not hesitate to adopt it.—*J. M. Smith, N. Y. Tribune.*

ARTIFICIAL INCUBATION.

There is not the slightest reason to believe that when the ancient Egyptians invented a method of artificially hatching eggs they were influenced by any desire to lessen the labor of hens. Their sole object was to produce more chickens than the hens produced. Although we may give a setting hen credit for the best possible intentions, it must be admitted that she is a very clumsy bird. She will tread on her eggs and will leave more or less of them out in the cold. Besides, her capacity to hatch eggs is limited by her size. There are very few hens who can hatch out more than a dozen chickens, and, of course, if a man wishes to raise chickens on a large scale he must supply himself with an immense number of hens. Artificial incubation obviates all these difficulties. As invented by the Egyptians, and extensively practiced in our own day, a thousand eggs can be hatched at one time in a single incubator, and not one of these runs any risk of being broken or chilled.

The immense success which has attended the artificial incubation of chickens in France recently attracted the attention of Dr. Tavernier, a learned and ingenious physician. He was attached to a hospital for foundlings, and although the position gave him an admirable opportunity for experimenting with new medicines, he was a humane man, and he was annoyed at the large number of foundlings who died within the first six months of their life. The majority of those admitted to the hospital were weak and sickly, but in that respect they did not differ from the majority of all sorts of French infants. Dr. Tavernier felt that it was a reproach to medical science that French infants could not be cultivated with as much success as French chickens, and he resolved to try what artificial incubation—if it may be so called—would accomplish if applied to infants.

The doctor constructed a child incubator on precisely the model of the ordinary chicken incubator. It was a box covered with a glass side furnished with a soft woolen bed and kept at the temperature of 86° Fahrenheit by the aid of hot water. He selected as the subject of his first experiment a miserably made infant, one, in fact, that had rashly insisted upon beginning the world at an injudiciously early period. This infant was placed in the incubator, provided with a nursing bottle, and kept in a dark room. To the surprise of the doctor it ceased to cry on the second day after it was placed in the incubator, and although it had been a preternaturally sleepless child, it sank into a deep and quiet sleep. The child remained in the cubator for about eight weeks, during which time it never once cried, and never remained awake except while taking nourishment. It grew rapidly, and when, at the expiration of sixty days, it was removed from the incubator it presented the appearance of a healthy infant of at least a year old.

Delighted with the success of this experiment, Dr. Tavernier next selected an ordinary six-months-old infant addicted to the usual pins and colic, and exhibiting the usual fretfulness of French infants. This child conducted itself while in the incubator precisely as its predecessor had done. It never cried; it spent its whole time in sleep, and it grew as if it had made up its mind to embrace the career of a professional giant. After a six weeks' stay in the incubator it was removed and weighed. During this brief period it had doubled its weight. It had become so strong and healthy that it resembled a child of three years old, and it could actually walk when holding on to a convenient piece of furniture.

These two experiments satisfied Dr. Tavernier of the vast advantages of artificial child incubation. He immediately proceeded—with the permission of the authorities of the hospital—to construct an incubator of the capacity of four hundred infants, and in thus he placed every one of the three hundred and sixty infants who were in the hospital on the 10th day of February last. With the exception of one who died of congenital hydrocephalus and another who was claimed by its repentant parents, the infants were kept continuously in the incubator for six months, when they were removed in consequence of having outgrown their narrow beds. The result will seem almost incredible to persons who are unfamiliar with the reputation of Dr. Tavernier, and have not seen the report made to the French government on the subject by a select committee of twelve. The average age of the infants last February was 3 months and 3 days—the youngest being less than 12 hours old and the oldest not more than 11 months. Their average weight was 16 pounds, only one of the entire 360 having attained a weight of 32 pounds. At the end of six months of artificial incubation the average weight of each infant was 84 pounds, and there was not one who would not have been supposed by a casual observer to be at least eight years old.

In other words, six months of artificial inculation did as much in the way of developing Dr. Tavernier's foundlings as eight years of ordinary life would have done. The infants were strong and healthy, as well as big; they walked within a week after leaving the incubator, and most of them have since learned to talk. These results surpassed Dr. Tavernier's most enthusiastic expectations, and there can be no doubt that his system of artificial child incubation will be adopted not only in every child's hospital in France, but in every private family throughout the civilized world.—*N. Y. Times.*

INDIAN CORN IN KANSAS—ITS VALUE AND IMPORTANCE.

The crop of Indian corn is one of the most important and valuable in the United States. The crop of 1880 was estimated at 1,717,000,000 bushels; the wheat crop of the same year was estimated at 498,000,000 bushels. It must be considered as being the staple crop of the Western and Southwestern States. In 1880, Illinois produced 240,000,000 bushels, as against 60,000,000 bushels of wheat. The acreage of corn in Kansas the same year was 2,165,070 acres, and the product 108,704,927 bushels, against an acreage of 1,520,639 acres of winter wheat, with a product of 17,500,259 bushels.

On land as well adapted to cultivation and production of corn as the prairie and bottom lands of the West, it has the advantage of any other crop of grain.

The cost of an acre of corn, put in the crib, is as follows:

Plowing	$1.00
Planting and seed	35
Harrowing twice	25
Plowing three times	1.00
Husking	1.00
Total	$3.60

The average yield of corn for 20 years is 35 bushels per acre, and the price has averaged, for the same time, 30 cents per bushel, giving a product per acre of $10.50, at a cost of $3.60.

But the great advantage of the corn crop is, that it can be fed out at home, and taken to the market in the shape of beef and pork. Where this is done, there is hardly a year in Kansas that will not return more than 30 cents per bushel. With a good stock of hogs, and pork at $3.50 per 100 pounds gross, 40 cents per bushel can be realized for corn, or when fed to good grade steers, at $4.25 per 100 pounds, will make 40 cents per bushel, besides the waste picked up by the hogs following them.

Corn is not as exhausting to the soil as wheat or oats. There is but a very small percentage of our prairie land that will not produce corn successfully from 10 to 15 years in succession. I am not in favor of raising the same crop on the same land year after year, and would consider such a course poor farming.

There is crop that feels the effect of good land more than corn. The application of 20 loads of manure to the acre on land planted to corn will increase its yield from 8 to 10 bushels for three years, and its effect will be seen after that time.

We do not, at the best, make the most that can be made out of our corn crop. It pays well to grind corn for horses, beef-cattle, milk-cows, and partly for hogs. By grinding corn there is a saving of one-third. The farmer who feeds from 1,500 to 2,000 bushels per year can well afford to invest $200 in a mill and horse-power.

We do not fully utilize the crop of cornfodder. The fodder on an acre of corn yielding 40 or 50 bushels is worth as much for feed as a ton of timothy hay, which is about an average yield of timothy on prairie land. The cost of cutting up an acre of corn is $1.25. Of course there is some value to the food left standing in the field until the corn is gathered and stock turned in upon it, but 10 acres of corn-fodder cut in good season is worth as much as 50 acres left standing.

We need to say but little about the cultivation of corn. The means for planting and cultivation of corn have greatly improved during the last 20 years. With the implements of to-day, two men with good teams can plant and well cultivate from 80 to 100 acres of corn. The lister is of modern invention, but there is a question whether it is really an improvement. No better corn can be raised by its use than is raised where the ground is well plowed, and planted with a horse-planter with check-row attachment. In very wet springs there are many objections to the use of the lister. This spring, corn planted with the lister is not as good as that planted with a planter. The heavy rains have done much damage to listed corn; some fields washed out entirely. If a little saving of labor is effected by the use of the lister for three or four years, damage may be done the fifth year by heavy rains to more than overbalance the gain. If we wish to grow corn with success in Kansas, we had better settle down on good plowing, thorough culture, and keep our lands in good heart by application of manure, turning under clover and other grasses, feed our corn at home, and in this way every farmer will make a success of raising corn.—Hon. Joshua Wheeler, Atchison county, Kansas.

THE EFFECT OF A GOOD SILO.

Last year I built a silo of 230 tons capacity, wholly of stone and Rosendale cement, with a flume and roof for cover. It is a good one (I believe in no other), no water can get in; no sap from the corn can get out, as so many complain of when their silos are not half built, or made from stale cement, or any poor material. On account of the long-extended drouth in this part of New Jersey, I was able to scrape together of good, bad and indifferent, half-dried, wilted, grown and half-grown corn, some 30 tons of ensilage after cured. This, however, was enough to satisfy my mind on this subject, if there had ever been any doubts. I used it as food for cows 110 days continuously, until all was fed out. Within a week from the time we began feeding hay, and though with an addition of grain, the cows lost at least 25 per cent. of milk; the cream did not make as much butter, and the butter was not of as good color or flavor. During the time of feeding ensilage we were unable to discover any other than the most satisfactory taste to milk, cream or butter. The cows were in the most perfect state of health, and kept in fine condition.

I raised a Jersey calf dropped in September, which had all it wanted of ensilage, and I will show it any day beside any man's calf six months older. I fed for 40 days eight Western steers, which averaged a gain of over 1½ pounds per day. The ration for cows and oxen was 22 pounds of ensilage morning and night, and 5 pounds of cut cornstalks at noon. The cows had three quarts of cornmeal and two quarts of wheat bran per day, and the steers had four quarts of cornmeal for 45 days, and five quarts for the last 45 days. Our success with the steers quite astonished my neighbors, who feed in the old way. The butcher says the cattle slaughtered well, and the meat was remarkably fine, and gave him every satisfaction. The use of poor ensilage, made from corn half ripe, or frost-bitten, I have reason for believing, would not give such satisfactory results.' I am one who believes that to make good ensilage the corn should be cut at the right time, cut the right length, put away in a good silo, and covered over nicely, and then well and thoroughly weighted down.—W. W. M., in Country Gentleman.

AGRICULTURAL PROSPERITY SHOULD BENEFIT THE FARMER.

The future of farming in the United States has never, in the history of the country, been so propitious as this season. Two of the great staple products of American agriculture, the two which are most to be relied upon for the general prosperity, to wit, hay and wheat, have never been so abundant. It is not unlikely that the corn crop also may exceed that of any year that has preceded this. The average of all other staple crops is good. Already these facts of great power are beginning to be felt. "This is the tide in the affairs of men —farmers—which taken at its flood will lead to fortune." The farmer, living in his retirements little knows the influences which his general prosperity exerts over the commercial relations of the country. The manufacturer knows, the importer and exporter knows, the railroad king knows, the politician knows, and all appreciate this influence. Unfortunately for the farmer there are, among all classes of men, ambitious speculators whose their knowledge of the above facts not only to increase their legitimate operations, but to study how they can best take advantage of the ignorance of farmers to appropriate the prosperity of the latter to their own advantage. To this end there are "stock exchanges," "corn exchanges," "bankers' and brokers' boards," "boards of trade," "railroad syndicates," etc., etc., the business of which is, by organization, to control the transportation, purchase and sale of farm products, not for the benefit of the farmer at home, nor for the benefit of the consumer of such products, but for the sole benefit of the jobbers and traffickers in these commodities, and the farmer in the country home, without organization, is the victim of these combinations and machinations.

It is a serious problem how farmers can so unite and combine as to protect their own interests against the organized outrage and extortion of these other combinations.

Running through most of the grain and grass-growing portions of the country there are net-works of railroads, the original stock of which was largely subscribed by farmers, in the belief that proximity to railroad transportation would advance the profits of farming, but unfortunately railroad corporations and syndicates so completely monopolize the operations of railroads that the farmer never sees the benefits of the roads, nor the color of the money he invested in them.

Farmers generally know these facts and deplore their inability to remedy the evils.

The organization of "Patrons of Husbandry" was conceived by and created in the belief that a solid combination of farmers with their individual intelligence and combined strength, could control legislation, and through it regulate transportation and purchase and sale, so that the producers would, at least, be able to divide with the operators in the profits of farm products.

But this organization, in its extreme caution to keep out of it "politics and religion," and avoid dangers within itself, carried its caution to the extent of inability to guard against danger from without, and after years of labor finds its organization outgeneraled by combined corporations, and out-witted by wily politicians, until its very power is turned against itself, and made the instrument of those outside combinations which it was designed to protect the farmer against.

If the founders of the Patrons of Husbandry, instead of excluding politics from its deliberations had made it a political, not partisan organization, the purpose of which was to protect the interests of the farmer by electing legislators who would enact laws in the interest of agriculture, and by electing executive officers who would preserve those interests, that organization might, to-day, instead of being managed and manipulated in the interests of capital and monopolies, be the dictating power in State and National Legislative bodies. Capital in the United States to-day controls legislation, and legislators control the industries of the country, more especially the agricultural industry. This order of things should be and can be reversed, so that farmers, being in the majority over any other class of men, should control legislation—State and national and honest legislation should control capital. In a Republican form of government majorities should rule. The farmer is the majority. Forewarned is to be forearmed. Farmers should and can protect their industry, and reap the fortune from this tidal wave of prosperity.

TREE-PLANTING IN STREETS AND GROUNDS.

The whole of good farming and gardening does not consist alone in raising good stock and growing crops; the farm and home need ornament and pleasant appearances.

Trees and shrubbery, well chosen and properly arranged, constitute the means to give charm to the homestead, as also to give beauty to the town and park; their shade is also a comfort.

Taste, knowledge, and skill are as necessary to secure highest satisfaction in tree and shrubbery planting, as in drawing fine landscapes and designing elegant buildings; being simply a good civil and topographical engineer does not qualify a person to be a successful and tasteful landscape gardener; landscape gardening, in its completeness, is a high order of profession in itself, requiring talent and experience to attain high efficiency. True, a thorough study of surveying and engineering is considerable help in that direction. It takes fine talent and varied experience to make such a happy landscape gardener as Wm. Saunders has proved himself to be.

Besides knowing the character and habits of trees, their climatic requirements, the best mode of planting, and the right adaptation of soil for their successful growth, an eye and judgment for pleasant effects are equally requisite; in fact, both taste and experience are indispensable in producing the most delightful results in lawn, park, and street planting of trees and shrubbery.

In addition to various evergreens — as cedars, firs, pines, arborvitæs, and the like — the ashes, catalpa, elms, lindens, maples, oaks, and some others are both useful and handsome, and are adapted to a wide range of climate and diversity of soils.

The best experience proves that the planting of young trees should not be too deep; in this, as in most operations with plant-growing, the methods or habits of nature are safe guides to tree-planters.

An experienced writer in this matter says: "Large, round holes for tree-planting are better than square ones; the bottom of the hole should be elevated towards the center, with rich, pulverized soil, upon which the tree should be placed in planting; then the roots should be carefully spread out in all directions toward the circumference of the hole, and carefully covered with rich, fine soil, the tree to be gently shaken and slightly lifted while this is being done in order that soil will settle around all the roots."

Full and complete guide and instructions cannot be given nor expected in a single short article, on this beautiful and too much neglected subject; but only a few suggestions are thrown out on the kinds of trees to be chosen and the rudiments of planting. In future articles I will give more specific details, in regard to the natures and necessities of various trees and shrubs, as also in regard to the natures and necessities of various trees and shrubs, as also in regard to adapting different ones to various locations and soils. This is to awaken interest only.

The planting of various fruit-trees, as well as their adaptation to soil and locality, will also be considered in these plain and brief articles, with the aim of being perfectly practical rather than fanciful.

Some varieties of fruit-trees are decidedly ornamental as well as useful for fruits, especially when interspersed among cedars and pines, and in many cases drawing among evergreens, particularly cedars, is known to prevent, to a good degree, the ravages of many insects upon the fruit-trees.

The pear tree, the apple and the cherry are many of them very fine and graceful in their habits and forms. We have known plum-trees growing among walnut and hickory trees to be preserved from insects.—*D. S. Curtiss.*

THE FAIR SEASON.

We are in the midst of the season for fairs and expositions. Whoever has produced anything of more than usual excellence in his own estimation and that of his friends, whether in the farming, mechanical or art line, brings it forth for public examination and approval. Mankind thus puts its best foot forward. The scope of the agricultural fairs has enlarged of late years, until they have become comprehensive industrial symposiums, with a little extraneous entertainment included in the shape of horse-racing and occasional other diversions. In many localities there is a department of art, and where circumstances will not permit of that dignity in its high sense, there are exhibitions of fancy needlework and what not from fair fingers, that help to give an agreeable coloring to the whole and widen the field of interest. The motto seems to be, "something of everything for everybody," and it is a very good one.

The animating principle of fairs is competition, and the benefits they bestow come in the shape of the advertisement of new and practical ideas, comparisons, and a relaxation of the humdrum round of daily life. The farmer and the mechanic are brought together in a mutually profitable way. If the latter has an improvement in the way of saving agricultural labor, and is interested in finding customers for it, the former is equally interested in finding it out. Tests between the different machines are frequently made before discriminating witnesses, and, by a process of natural selection, the good are established in the market, while the inferior are numbered with the infinity of human failures upon which progress is built. The results of different methods of cultivation are brought together and discussed. The merits of various kinds of stock are illustrated, and that species of intelligence is disseminated among farmers which is of the greatest service to them.

The aggregate influence of fairs upon the advancement of the agricultural interests must be very great. Formerly they were mainly places of bargain and sale; now they have a more direct educational influence, and they are developing in accordance with the demands of the age, until something like a universal system has been evolved.

The county fairs supply the want of local interchange of ideas and comparisons; then come fairs, representing larger sections; then State fairs, and so on up to the world's fairs, which have now, it may be said, become established institutions held at a comparatively regular intervals every few years, and in the sustentation of which the civilized nations have spontaneously and, in a manner, instinctively united.

These fairs, large and small, are great levelers, but they level up. Their effect is to raise the low places, not to cut down the substantial heights. They do not strike an average, but push the inferior out of existence altogether, and when all the world's excellence and advances are to choose from, the effect is a compact partnership of civilized forces in the work of progress. In thus regarding the world's fairs, the smaller ones are not to be despised. They are as important in their sphere as the larger ones are in theirs. It is through local endeavor and the inspiration of local competition that the marvels of ingenuity and of careful labor are produced. Usually each separate locality possesses advantages in some particular direction that others are deficient in. The local competition they all represent is not only an incentive to the best effort, but it is instructive. Many heads can furnish more valuable hints than one can. The more the general subject is regarded the more it will appear that the fair system is a very important one, and bears a little short of vital relation to the various industries. It supplies them with a nervous circulation that they would advance very sluggishly and unevenly without.

ITALIAN BEES AND HOW TO ITALIANIZE THE COMMON BLACK BEES

After having tested the Italian bees for ten years we can say very truly that they are far superior to the black or native bees. First, they are more energetic and resist the attack of robbers and the bee-moth; never had a strong colony of Italian robbed or destroyed by the bee-moth. Second, they are better honey gatherers and can gather honey from flowers that the black bees cannot. Our Italians, during a dry spell, the fall of 1881, were busy working on red clover while there could not be a black bee seen. Third, they will gather at least one-third more honey than the black bees, to take one year with another. Fourth, and last, they are more quiet and better to handle, the bees stick close to the combs.

A pure Italian should have three distinct, yellow bands or rings across the lower part of the abdomen, and a bright yellow hair over the body. The so-called Albino bees are a strain of Italians, having white bands and hair; they are the finest workers of the two and very nice to handle; they are of American origin, and are distinguished in scientific bee culture as (*Apis America.*) We got our first queen of this strain of Italians, October, 1879. The next year, 1880, this colony gave us two swarms and 110 pounds of one-pound sections of honey, and last year the same queen's colony gave us 63 pounds of one-pound sections of honey. The honey of 1880 brought us $16.50, while that of 1881 brought us $12.60.

Our average last year was 42 pounds per colony (Italians,) when the average per colony black bees, last year, fell below par.

How to Italianize.

First, procure a good queen from a reliable breeder, and when the queen arrives, if in movable frame hive, commence on one side of hive and take out one or two frames and shake off the bees so as to be sure that the black queen is not on them. Now have a new hive; put the two frames in and set in place of the old hive, and carry the old hive and remaining bees some six or eight rods away, then examine each frame carefully until the black queen is found, then kill her or make a new colony by giving her about half of the frames, and set it some distance from where it first stood. Queens are mostly sent in a cage one inch thick and two inches square. Take this cage and lay it on a frame of brood, near the top bar, and with a sharp knife cut out a piece of comb, just as large as the cage and no larger. Now remove the two tacks that hold the tin gate, but do not let the gate slip out of place, slip the cage in the hole cut in the comb with the gate down, be sure the gate is in the right place so it will be impossible for the queen to get out, place the frames in the hives, just as they were, and leave it for 36 to 48 hours, then remove the tin gate, but leave the cage in position, and with a sharp, thin knife give two or three cuts just below the opening, but do not remove any comb ; now close the hive and the bees will know their way out, but before closing the hive be careful to destroy all queen cells. In about five days open the hive and see if all is right and remove the cage. The above plan is intended for those who have had but little experience at the business, and not for the practical apiarian.—*Maryland Farmer.*

PREVENTABLE LOSSES ON THE FARM.

It is a "penny wise and pound foolish" system, to breed from scrub stock. There is not a farmer in this region who has not access to a pedigreed Shorthorn bull, by a payment of a small fee of two to five dollars, and yet we find only one animal in ten with Shorthorn blood. It is a common practice to breed to a yearling, and as he is almost sure to become breechy, to sell him for what he will bring the second summer. Many farmers neglect castrating their calves until they are a year old. I think ten per cent. are thus permanently injured—must be classed as stags, and sold at a reduced price. Fully half the calves so stunted never recover.

With many, the starving process continues through the entire year. They are first fed an insufficient quantity of skim milk; then in July or August, just at the season when flies are at their worst, and pastures driest, they are weaned, and turned out to shift for themselves, and left on the pastures until snow fall, long after the fields yield them a good support. They are wintered without grain, spring find them poor and hide-bound, and the be st grazing season is over before they are fairly thrifty.

The keeping of old cows long past their prime is another thing which largely reduces the profits of the farmer. We have found quite a large per cent. of cows, whose wrinkled horns and generally run-down condition, show that they have long since passed the point of profit. A few years ago, these cows would have sold at fall prices for beef, now they will only do for bologna at 2 cents per pound. Thus cows have, in a majority of cases, been kept, not because they were favorites, or even because they were profitable, but from sheer carelessness and want of forethought. Another fruitful cause of loss to the farmer, is attempting to winter more stock than he has feed for. Instead of estimating his resources in the fall, and knowing that he has enough feed even for a hard winter, he gives the matter no thought, and March finds him with the choice of two evils, either to sell stock, or buy feed. If he chooses the former, he will often sell for much less than the animals would have brought four months earlier, and if the latter, will usually pay a much higher price for feed than if it had been bought in autumn. Too often he scrimps the feed, hoping for an early spring, and so soon as he can see the grass showing a shade of green around the fence rows, or in some sheltered ravine, turns his stock out to make their own living. This brings one of the most potent causes of unprofitable cattle raising ; namely, short pastures. The farmer who is overstocked in winter, is almost sure to turn his cattle on his pastures too early in the spring, and this generally results in short pasture all summer, and consequently the stock do not thrive as they ought, and in addition, the land which should be greatly benefited and enriched, is injured, for the development of the roots in the soil must correspond to that of the tops, and if the latter are constantly cropped short, the roots must be small. The benefit of shade is lost, and the land is trampled by the cattle in their wanderings to fill themselves, so that it is in a worse condition than if a crop of grain had been grown on it. From all these causes combined, there is a large aggregate of loss, and it is the exception to find a farm on which one or more of them does not exist, and yet without exception, they may be classed as "preventable," if thought and practical common sense are brought to bear in the management.—*American Agriculturist.*

YIELD AND CONDITION OF CROPS.

The October returns include the entire area of nearly seventeen hundred counties of the United States, representing nearly all of the breadth in cereals, potatoes, cotton, tobacco and sorghum. They give direct estimates of the yield per acre of the small grains, all of which are harvested, based on thrashers' records as far as obtainable. Errors have been carefully eliminated, and unreasonable estimates examined for correction. The result of this test of production gives the largest figures of the official series of tests, from the involuntary impulse of farmers to think and speak well of their acres; so that, on completing the direct comparison, by counties, with the product of last year, and the adjustment of possible discrepancies by further investigation, the outcome may possibly be lower than is indicated by the figures of yield per acre.

The crops not yet generally harvested, corn, potatoes, and buckwheat, and cotton also, make a final report of condition, the rate of yield to follow in November.

WHEAT.

The October returns of yield per acre of wheat, estimated from results of thrashing, foreshadows a product slightly exceeding 500,000,000 bushels. The average yield per acre will not much exceed an average of 13.5 bushels, on an acreage slightly under 37,-000,000. There is a reduction of area in the spring-wheat region, and a large yield in the great winter-wheat-growing belt of the West. The six principal winter-wheat States will aggregate about 214,000,000 bushels, or nearly half the crop of the United States. The spring wheat of the Northwest may make 113,000,000 bushels. The Pacific-coast crop, which has been persistently exaggerated in commercial estimates, may possibly reach 45,000,000 bushels. The Middle States have produced about 40,000,000 bushels, and the Southern States slightly in excess of 59,-000,000. Slight modifications may come from further investigation as the results of the harvest are more closely tested ; but the total cannot be much changed, and certainly cannot be expected to enlarge the aggregate above, which requires nearly as large a yield per acre as has ever been reported in this country by census or official estimate. The average yield has never fallen quite to 10 bushels (though very near ,it last year), and never has quite touched 14 bushels in years of greatest abundance. It was 12.9 in the census year, and the crop of 1880 was estimated at 13.1.

The yield in New England varies from 14 bushels in Maine to 18.7 in Vermont. It is unusually high in New York, 18.7 bushels; in Pennsylvania not quite so high, 15.5 bushels, Delaware and Maryland secure good yields ; but the South, from Virginia to the Mississippi River, though yielding better than usual, ranges 7 to 10 bushels ; Arkansas and Texas do better.

Coming to the winter wheat belt of the Ohio Valley, the country north of that river averages nearly sixteen bushels. Michigan and Illinois stand highest in this belt. Kentucky and Missouri promise about 14 bushels; Kansas reports the extraordinary yield of 19.5, a crop of about 34,000,000 bushels. The yield of California is apparently about 13 bushels, while Oregon and Washington are higher and more uniform in local areas.

The quality of wheat is generally good ; high in the Eastern and Middle States and approximating 100 in the South. In Illinois the average is 99 ; in Indiana, 97 ; in Ohio, 96. Some loss of quality in Michigan from heating in the stack, reducing the average to 90. In West Virginia it fails to reach perfection by nine points. Iowa, in the spring wheat belt, makes lowest returns, averaging 87. Further west, and on the Pacific coast, quality is reported uniformly good.

OATS.

The average yield of oats will be somewhat higher than last year or 1879, and the product will be nearly as large as that of wheat, probably about 480,000,000 bushels. Illinois, Iowa, New York, Wisconsin, Missouri, Pennsylvania, Ohio, Indiana, and Kansas are States of highest rank.

RYE.

The indicated average yield of rye is 14.7, making a crop of 20,000,000 bushels, or nearly the same as that reported by the last census. The quality ranges, with few exceptions, from 95 to 100.

BARLEY.

The indicated average yield of barley is 23.5 bushels per acre, aggregating 45,000,000 bushels. California, New York and Wisconsin, together produce more than one-half, or 27,000,000 bushels. The product in 1879 was 44,000,000.

BUCKWHEAT.

The prospect for buckwheat is good for a nearly average product, eleven to twelve million bushels. Pennsylvania produces nearly half of the crop, and reports 95 as the average of condition, 100 representing a full normal yield, and not an average of good and bad seasons. New York makes an average of 95. No other States produce half a million bushels.

CORN.

The yield per acre of corn will be reported in November. Condition averages 81, being very high in the South, and comparatively low in States of largest production. In Illinois, with 8 per cent. decrease of area, condition is only 72; it is 70 in Iowa, and 87 in Ohio; these States produced 40 per cent. of the crop of 1879. A careful comparison of changes in area and condition indicates an average yield of 25 bushels per acre, against 28 in 1879, and 18 in 1881. The average of a series of years is between 26 and 27 bushels. New England will produce, according to these returns, seven to eight millions; the Middle States, 82,000,000; the Southern States, 340,000,000; those north of Tennessee and west of Virginia and Pennsylvania, 1,250,000,000; an aggregate of 1,680,000,000. Later returns of product may slightly reduce, but cannot materially increase this result. The 1,800,000,000 product predicted by the corn buyers is a myth, which has been so persistently assumed that the public may be misled. The increase in the South, where ten to fifteen bushels may be considered a large yield, cannot make good the reduction in Illinois alone. It is gratifying to know that the product is more than four hundred millions greater than last year, and ample for a liberal supply for domestic wants and exportation; a supply never exceeded, with two exceptions, 1879 and 1880, notwithstanding a later and more unpropitious planting season than has occurred in many years.

The injury to corn in New England, by drought, was somewhat serious. The desired rains did not fall anywhere out of Vermont, except locally. Oxford county, Maine, "has the largest and best crop for ten years."

Burlington and Gloucester counties, New Jersey, report falling off in condition during September, but the reverse is true of the State.

Much "soft" corn is mentioned in Pennsylvania, but no fears of injury by frost had yet been realized, and the fair weather has taken much of the crop out of danger.

While corn in Richmond county, Virginia, was "seriously injured by drought," and in Botetourt "suffered materially from too much rain," in Craig "some fields will yield 70 bushels per acre," and Fauquier reports "25 per cent. above an average." The general report from the State is not unsatisfactory.

In the South the large promise of the season fails but slightly by reason of severe storms of wind and rain in September. Damage from these causes are mentioned in Transylvania, Stanley, Davie, Cabarrus, Davidson, Henderson, Beaufort, Camden, Gaston, Iredell, McDowell, Rowan counties, North Carolina; Union, Chester, Edgefield, South Carolina; Decatur, Harris, Banks, Gordon, Talbot, Carroll, Floyd, Wilkes, Baldwin, Putnam, Early, Dawson, Wilcox, and Habersham, in Georgia.

In Alabama and Mississippi rain and floods are mentioned in some localities, but an abundant crop is harvested in both States.

But little discontent with the harvest is evinced in Texas. Hardeman county mentions clinch bugs, and more rains would have been acceptable in limited portions of the State.

West Virginia made no improvement during the last thirty days. Several counties suffered too much rain.

No complaint comes from any section of Kentucky. Returns indicate no abundant crop in Hickman. Best yield in twenty years in Kenton.

September's considerate weather did all that could be done for corn in Ohio, and "out of danger" is the report from all parts of the State. Cutting is in progress, and frost did not injure even the latest fields.

The danger from frost has passed in Indiana; there is but slight local variation in condition, and the improvement by reason of fair weather in September is general.

Frost caught some fields in Illinois, and its damage is mentioned in Kankakee, Edgar, Kendall, Henry, Fulton, Winnebago, Boone, De Kalb, Grundy, and Livingston counties, yet the late grain received most of the injury, and the condition improved wonderfully during September.

Wisconsin corn suffered from frost in a few counties. In Dodge it was cut prematurely for its protection, and the quality was lowered. Many counties report an average crop of good quality—among them, Waushara, Milwaukee, Waupaca, and Juneau.

In Watonwan county, Minnesota, corn was "badly injured" by frost. Other counties mention it, while many escape wholly.

POTATOES.

The average condition of the potato crop is 81, indicating a yield of about 80 bushels per acre on an area approaching 2,000,000 acres. In New York the average is 70, foreshadowing a short crop in a State of large production. In Maine, 85; Vermont, 84; and less in other parts of New England. In Michigan the prospect is very flattering, and throughout the Ohio Valley, Missouri, and Kansas, and in the Southern States condition is unusually high. In the northwest it is somewhat reduced.

Potatoes in New England shared in the disastrous influences of the drought. In New York and New Jersey a poor condition exists. Pennsylvania reports large crops in many counties, and rot in others. August rains did good in Delaware.

Large crops are mentioned in Williams, Allen, Franklin, Knox, and Geauga counties, Ohio, and little complaint comes from any part of the State. Rot is mentioned in Monroe, Delta, and Houghton counties, Michigan. A fine crop is reported in Indiana. In Illinois, rot is mentioned in Du Page, Kendall, Jo Daviess, Carroll, and Boone. Shelby has an immense crop. The Wisconsin product is large, but rotting in Kewanee, Washington, Pierce, Fond Du Lac, Racine, and Dodge. Local variations occur in Iowa. A fair yield is reported in Missouri and Kansas, especially of the early planted.

COTTON.

The cotton returns of the Department of Agriculture for October indicate unusual size and vigor of plant, and a capacity for a large production. The late development of fruitage, and the reported indications of a small top crop, limit the otherwise extraordinary prospect. The coincidence appears of the same general average of condition in 1881 and 1882 for June, July and August, 89, 92, and 94, respectively. During August and September, in 1881, condition fell from 94 to 66, but in the same period of this season to 88 only. This is higher than in any October for ten years with two exceptions, 1875 and 1878.

Compared with the August returns, there is a loss of one point in Florida and Texas; two in Alabama; three in North Carolina and Georgia; four in Virginia, Mississippi, and Arkansas; five in Tennessee, and six in South Carolina. The figures for Virginia are 86; North Carolina, 85; South Carolina, 89; Georgia, 86; Florida, 82; Alabama, 88; Mississippi, 82; Louisiana, 82; Texas, 100; Arkansas, 96; Tennessee, 84.

Rains have been abundant throughout the belt, with a few local exceptions in the southwest. Severe storms are reported generally, with occasional injurious consequences, while some correspondents claim a benefit in partial breaking of roots, stopping growth, and hastening maturity.

Rust is slight and not injurious.

The caterpillar is present in the Gulf States, but no appreciable damage is reported east of Mississippi. The partial loss of leaves where the worm exists is favorable to development of the boll. Slight damage is reported in Madison and Caddo, in Louisiana, and in a few Texas counties.

The boll worm is doing some injury in bottom lands of Russell county, Alabama; in Dallas, Denton, Eastland, and Stephens, Texas; in Pope, Arkansas, and in Fayette, Tennessee. This pest has perhaps done more injury than the caterpillar, but the losses from all insects will be insignificant.

The range of possibilities between early frost and a long and favorable season for maturing and picking is much wider this season than usual, owing to the present rank growth and greenness of the weed, and later ripening.

OUR LOCAL ORGANIZATIONS.

LANCASTER COUNTY AGRICULTURAL AND HORTICULTURAL SOCIETY.

The Lancaster County Agricultural and Horticultural Society met stately in their room, Monday afternoon, Nov. 6, with the following members present: Joseph F. Witmer, Paradise; H. M. Engle, Marietta; John C. Linville, Salisbury; M. D. Kendig, Manor; John H. Landis, Manor; W. H. Bollinger, Warwick; C. A. Gast, city; H. G. Rush, Pequea; F. R. Diffenderffer, city; M. Hershey, Salunga ; Calvin Cooper, Bird-in-Hand ; C. L. Hunsecker, Manheim ; E. S. Hoover, Manheim ; E. W. Eshleman, Manheim ; S. P. Eby, city.

On motion the reading of the minutes of the last meeting were dispensed with.

Mr. M. D. Kendig, who was one of the committee to visit the York County Fair, reported that he found a large exhibit. The attendance was also large, and the display of cattle, horses, sheep, etc., was good. The fruit display, however, was not so good, and what was on exhibition was of a very poor quality.

J. C. Linville stated that it had been understood that Mr. Edge had agreed to address the society at its October meeting, but on account of illness he was unable to do so. He was unable to promise when he would be allowed to do so, on account of ill-health and a great amount of other work on hand.

Mr. Linville, of Salisbury, reported an unusual growth of grass this fall. The stubble fields have an extraordinary growth of clover. The wheat sown in October looks well. The corn crop is about an average. He never knew the corn to be put in the crib so green as it was this fall, and he was afraid this would prove a disadvantage.

H. M. Engle, of Marietta, said young clover was doing remarkably well. He had made young hay from it, something he had never done before. It was a little coarse, but the cattle appeared to like it. The rainfall for last month was one inch.

M. D. Kendig reported the rainfall in his section to be 1 1-5 inches.

These reports were corroborated by other members of the society.

Mr. Kendig inquired as to the probable tendency of the wheat market.

Mr. Diffenderffer said he was of the opinion that the price of wheat would remain about as it is at present. He said there could not be any argument which would lead him to believe that it would advance in price, for the reason that the late crop was larger than it was for some years. The Chicago Board of Trade put the crop much higher than it ever was.

Mr. Engle asked when it was most advisable to plow clover sod for corn, fall or spring? He said he was of the opinion that fall was the best time, especially if the season was dry. It should then be thoroughly cultivated in the spring.

Mr. Bollinger's practice was fall plowing. His reasons for fall plowing were because he had more time ; another was because in spring the ground was too wet, as a general rule ; the corn should be thoroughly cultivated in the spring ; this year he was expecting 65 or 70 bushels per acre, although it was planted in May.

Mr. Resh advised manuring heavily in the fall, and then plowing it thoroughly in the spring. He had had tried this plan and always found it to work well, especially if there was a good set of clover.

Mr. Hoover could be held the same idea, to a great extent, as that held by the gentleman who had preceded him. If he had level ground he thought the plan would work well, but if the land was sloping he would not advise it. He was in favor of fall plowing as a rule, but he was of the opinion that in the fall there would be a searcity of manure, and then some of the land would not get any manure at all. He, therefore, advocated doing it by degrees, just as he secured his manure.

Mr. Engle was of the opinion the plan of exposing the plowed surface, during the winter, would prove beneficial. The general sentiment throughout the country tended towards surface manuring. He did not, however, like to leave his manure exposed to the elements during the summer season. He was in favor of having his manure in good condition, then he would have a good cultivator and properly apply it.

Mr. Bollinger said the only objection he had to top dressing in the fall was that then the manure would not as a rule be in proper condition. He thought we would destroy the insects, especially the cut-worm, if we would plow in the fall.

Mr. Eby instanced a case where the cut worms were most numerous in a tobacco patch which had been plowed in the fall, but Mr. Engle thought this was was an exception.

Mr. Linville plowed in the spring, because he then could have his manure in proper condition. He was of the opinion that land plowed in the winter and exposed to the elements, would become very poor. It may kill the cut-worm, but he thought it did not pay in the end.

Mr. Engle stated that J. B. Garber contributed several specimens of the Kieffer hibrid pear, and he moved that a committee be appointed to examine the fruit.

The motion was carried, and Messrs. Engle Cooper and Eby were appointed.

The following questions were referred : At what age should stock cattle be put up for feeders ? to E. S. Hoover. At the present prices of corn and bran, what constitutes a profitable ration for beef cattle ? to H. G. Resh. What is a profitable ration for milk cows ? J. H. Landis. Is there any truth in the oft repeated assertion that farmers eat too much ? F. R. Diffenderffer. Adjourned.

POULTRY ASSOCIATION.

The Lancaster County Poultry Society met statedly in their room in the City Hall building, Monday morning, Nov. 6, with the following members present : Geo. A. Geyer, Spring Garden ; J. B. Lichty, city ; Dr. E. H. Witmer, Neffsville ; S. G. Engle, Marietta ; W. A. Schoenberger, city ; S. P. Eaby, Esq., city ; F. R. Diffenderffer, city ; C. A. Gast, city ; Edward Brackbill, Strasburg ; Peter Bruner, Mount Joy.

The minutes of the previous meeting were read and approved.

Philip Rorogresser, city, was elected a member of the society.

The Secretary, Mr. Lichty, stated that he had written to Mr. J. W. Bicknell, of Buffalo, and Charles Becker, of Baltimore, to act as judges, the former on poultry and the latter on pigeons, as in the coming exhibition of the society. These men are said to be among the best judges in the country. As it is the intention of the managers of the show to keep the man selected to judge the poultry exhibit at the show during tha entire time, one person, it was thought, would be sufficient.

The question of a suitable room for the exhibition was then brought up, and the Executive Committee was instructed to secure, if possible, Excelsior Hall.

The present season was reported to have been a profitable one for poultrymen, as eggs and chickens have been unusually high

The Secretary said he had received up to the present time nearly $60 for advertising space in the catalogue, and he had no doubt a great deal more would be paid for advertisements than would be necessary to pay for the printing and mailing of the catalogue.

After an informal talk on the subject of "gaps" in chickens and the various remedies for the same, the society adjourned.

LINNÆAN SOCIETY.

The regular monthly meeting of the Linnæan Society was held in Dr. Knight's office on Thursday evening, October 26, at 7½, P.M. The following members were in attendance : Dr. Dubbs, Dr. M. L. Davis, Dr. Knight, Prof. Rathvon, S. M. Sener, W. L. Gill, Mrs. Zell and Miss Lefevre. Dr. J. R. Dubbs, the Vice President occupied the chair in absence of the President Dr. S. T. Davis donated to the museum a fine specimen of opalized wood obtained by him in Dakota, while on a visit there recently. Dr. H. B. Knight donated a specimen of the hermit crab, occupying its stolen habitation, which was procured on Long Island. The Patent Office Gazette, Nos. 14, 15 and 16, and yearly index of the same were donated to the library. Prof. R. K. Buehrle was unanimously elected a member of the society. After passing an hour in scientific gossip of an interesting character the society adjourned to meet on Saturday, November 25th, at 2½ P.M., in the museum room.

FULTON FARMERS' CLUB.

The Fulton Farmers' Club met statedly on the 4th inst., at the residence of Christopher C. Cauffman, at Wakefield. There was a full attendance of members and their families, together with a large number of visitors.

Exhibits.

S. L. Gregg exhibited an apple to be named, and it was pronounced to be a Tewkesbury Winter Blush. Day Wood exhibited a Spanish chestnut. C. C. Cauffman, who had recently been on a Western trip, exhibited some Odessa wheat from Iowa, some it was said, and a specimen of paper manufactured from poplar wood. The wheat was the product of 1881, and was said to be better than the wheat of the present year ; but here such wheat would be considered of extremely poor quality, and it speaks but poorly for Iowa as a wheat producing State. Wm. King exhibited Pomegrine, Nottingham Brown, and an unknown variety of apples.

E. H. Flatues reported having received from John B. Landis ten copies of the report of the State Agricultural Society for distribution among the members of the club, but said he was unable to have them present.

Questions and Answers.

Montillion Brown—How is the corn yielding the present season? Most of those present could only guess at the number of bushels per acre. One or two had made an estimate by the number of wagon loads, and the whole seemed to indicate an average of from 60 to 65 bushels per acre. The lowest being 40 and the highest 75 bushels per acre.

Day Wood—Does it pay to shell corn to grind for feed? Nearly all preferred to have it ground in the ear until toward spring, when the cobs get hard and then they shell it.

Montillion Brown feeds his without grinding as long as the cobs are soft and after that he shells it before grinding. He cuts the ears into small pieces before feeding them.

Josiah Brown—Does it pay to buy bran to mix with corn for cows? Day Wood said he believed a portion of cow's feed should be bran ; he had known it to increase the amount of butter when there was no perceptible increase in the amount of milk. Lindley King also prefers to feed bran. Montillion Brown thinks that the principal benefit to be derived from feeding bran is in the health of the animals. He had known serious results from confining animals to corn alone.

After dinner the men inspected the buildings, farm and stock, and after reassembling at the house all expressed themselves well pleased with what they had seen.

Montillion Brown read from the LANCASTER FARMER an article on "Gapes in Chickens." This started quite a discussion on the cause of the gapes. Edwin Stubbs said that at three different places where he had lived he found that when the chickens were allowed to drink from the ditch that carried the slops and water from the kitchen they were always affected with the gapes, and that when they were not allowed access to this ditch, or when it was cleaned out every few days, the chickens were free from this disease. His experience confirms him in the belief that if the coops are kept in dry, warm places, and the vessels out of which the young chickens drink are kept scrupulously clean there will be no trouble with the gapes.

Robt. Gibson dissented from this view of the matter, and said that at his place the chickens had the gapes much worse than at some of his neighbors, where there was much more impure water standing about.

Montillion Brown said the gapes were worse in some locations than others. He had on several occasions divided broods of chickens, taking a part each time to a tenant house of his, where they invariably escaped the gapes, while those left at home were badly affected with the disease, and yet at the tenement there was no aparent advantage either in regard to pure water or healthy location.

8. L. Gregg said that at his place the waste water was all conducted away under ground to a sink, and chickens were regularly supplied with clean water and that during a residence of ten years at his present place the chickens had never suffered from the gapes.

After some unimportant discussion on the comparative value of one's own experience, and agricultural reading, the club adjourned to meet at the residence of E. Henry Haines, on the second Friday in December, and there discuss the question of the advantage of keeping up inside fences on farms on this section.

AGRICULTURE.

The Use of the Roller.

The *New England Farmer* has a timely article on this subject. Indeed it is almost always timely to talk about the good effect of rolling land. The roller will not make moisture, but it will tend to retain some of it that is already in the soil, and its use may make the difference between a crop and no crop on land that is to be seeded down during a dry period. In a soil made compact by the roller, a light shower may afford sufficient moisture to the surface to germinate the seeds and give them a healthy start, while in an ever-mellow soil they would lie dormant or merely sprout and then dry up and die. The iron roller is far better than a wooden one in every respect. It turns easily, being made in short sections; it is heavy according to its size, and bears harder on the soil it covers. The weight of a large wooden roller is distributed over too much surface at once. The roller is often useful in the spring for compacting the surface of newly-seeded mowing or grass fields, sown the previous autumn, and which the frosts of winter have loosened up or torn to pieces. If cloverseed be sown on such land the roller becomes almost indispensable, and some farmers practice covering their grass seed with a roller in place of a harrow or brush, which is an excellent method where the soil is sufficiently moist. Another good use of the iron roller is upon mowing lands recently top-dressed with stable manure. The weight is needed to press the manure down close to the surface, where it will keep moist, and all the sooner help start the new growth, at the same time leaving the surface smooth for the scythe or mowing machine. It is also used by gardeners to break up lumpy soil, and with alternate harrowings to render it fit for receiving the seeds of tender garden vegetables.

Progressive Farmers.

The true farmer does not stop to count the cost of improvement, for his reason prompts him to believe that he cannot go wrong by endeavoring to improve. Every acre of his farm is cultivated to its highest capacity, and his soil never deteriorates in quality. He rotates his crops with a view to increased fertility, and he estimates his profits by the amount of expense created in securing that profit. The failure to realize immediate results does not discourage him, for he knows that, through his judicious system of cultivation, the realization is but deferred for a little while longer. He farms for profit and he speaks for profit. He knows nothing of stinted economy, which saves to-day and robs to-morrow. The farm is his bank, his workshop and his occupation, no stone being left unturned, and no portion slighted at the cost of another part.

A good farm means good stock. The squealing hog has no place on it, but must be superseded by the quiet thoroughbreds. The tangle-fleeced, small-carcassed sheep cannot be allowed where only the Merino, the Cotswold and the Oxford Down are adapted. The scraggy bovines of the past are seen

no more, for the deep milking Holstein, the creamgiving Jersey and the beef-producing Hereford have occupied their places. The thoroughbred and the Clydesdale plow the fields that formerly yielded to the wind broken plugs, and the wagons and implements are of the most approved labor-saving patterns. All this means capital and is expensive; but when we consider the fact that it costs no more to keep the best than the bad, and that expense means profit in the end, the cost is not so formidable as it seems.

But the manure heap is the most important of all. A good farmer can be selected by the manner in which he keeps his manure. The manure is the wealth—the bank on which the check is drawn—and it is imprudent to neglect it. Drenching rains and scorching suns carry upward and downward the soluble and volatile constituents of the unprotected heap, and often great ditches are dug to allow the black liquid riches to pass off and away forever. But the good farmer works differently. He makes his manure fine, attends personally to the process of decomposition, protects it from the weather and endeavors to make it a ready food for the crops when hauled to the fields. Farming pays well—to good farmers.

Effect of Draining.

First. It removes the surplus water and prevents ponding in the soil. It should be noted that if the drains are used, they should be of sufficient size to remove the surplus water in twenty-four hours. Second. It prevents the accumulation of poisons in the soil which result from stagnant water, either above or under the surface. Third. The ammonia is carried down into the soil by the descending rain, stored for the plant food instead of stopping on the surface and passing off by evaporation, or borne away with the surface waste. Fourth. It deepens and enriches the soil by opening the ground, allowing the roots of the plant to go deeper into the earth; decaying after harvest, they form this subsoil into surface soil, providing resources for the plant more reliable, and making the same ground better for cultivation for a greater length of time. Fifth. It avoids drought, by enabling the plant to thrust its roots deeper into the soil. Sixth. The drainage increases the temperature of the soil. In some cases the average has been increased as much as ten degrees. Seventh. By securing the uniformity of condition for plant growth, it hastens the maturing of the crop from ten days to two weeks. Eighth. It enables the farmer to work his land in wet or dry seasons, and insures a return for the labor bestowed. With our land thoroughly drained we can carry on the operation of farming with as great success and as little effect from bad weather as any business which depends on such a variety of circumstances. We shall have substituted certainty for chance, as far as it is in our power to do so, and make farming an art rather than a venture.—*Prairie Farmer.*

Fall Plowing.

Any one who has seen the best European farming knows how important it is to thoroughly prepare the soil for the seed. The working of the soil adds nothing, but it helps in changing the form of the plant food compounds, and thus plowing and harrowing becomes indirectly a source of nourishment. The soil is a vast storehouse of plant food, which it holds by virtue of its insolubility. Furthermore, it is only through the action of the air and all three processes, chemical and otherwise, which are covered by such terms as weathering, nitrification, etc., that these essential elements are brought into a soluble form and made available for the use of the growing plant. The chemistry of the soil, as it becomes better understood, teaches in every line the importance of a frequent stirring of the surface of the cropped field. With this in mind it is to the purpose to urge the importance of fall plowing. For other than chemical reasons the stubble or sod may be turned under this fall. Not only will the air circulate more freely, and the processes of reducing

the insoluble substance go on more rapidly, but the mechanical texture of the heavy soil especially will be improved. Should insects or their larvæ, or "worms," abound in the earth they will be turned out of their winter quarters and destroyed. Aside from these advantages there is a lull in the farm work at this season, and any plowing or other labor with the soil will help materially to lessen the rush and hurry that otherwise comes with the busy months of spring. The thoughtful and successful farmer so plans his farming operations that one season helps the next in more ways than one.—*American Agriculturist.*

HORTICULTURE.

Pear Raising.

It has often been said by those in a position to know, that more money can be made from an acre of ground planted in choice fruit trees than out of any other crop, and after seeing what Mr. William Weidle, of No. 542 East Orange street, has cropped from his comparatively small lot, we are ready to believe it. His small fruits, such as plums, grapes and raspberries, are over, but his pears are still in his cellar and show what his product has been in that line. He has no fewer than thirty-six varieties of this fruit, beginning with the earliest, the Giffard and Bloodgood, and closing with the Clout Morceau and Winter Nellis, which come into season any time from December until April. Between these early and late kinds come the Bartlett, Seckel, Beurre Bosc, Buffum, Louise Bon de Jersey, Flemish Beauty, Lawrence, Sheldon, Beurre Diel, Vicar of Winkfield, Urbaniste and many others, all of the most approved varieties. Mr Weidle put into his cellar about one hundred bushels of these luscious pears. He has several plans of keeping them. Some are put on trays and these are fixed on stands specially constructed for this purpose. Others are wrapped in paper and put into boxes, while still others have strings attached to their stems and are then hung to nails driven in the joists of the floor above, where they hang in huge masses from end to end. It requires much attention to look after this fruit. As some of it is ripening daily, the boxes must be examined every few days and the ripe fruit removed. In warm weather it ripens much faster than in cold. There is a market for all he has. Not only do hotels and grocers buy them, but private individuals take more or less every day. The price varies with the kind and quality ; fine fruit now sells from sixty to seventy cents per half peck. It was a fine sight to see all these pears strung along the joists, in the trays and in the numerous boxes, and we were not long in reaching the conclusion that next to being a newspaper reporter, the most delightful occupation in the world was growing pears and eating them.

The Effect of Dry Weather on Apples.

The effect which a protracted drought has on the fruit of an apple orchard depends on location, condition and the treatment of the trees. If the orchard be on high land, and is kept in grass cut at the usual time, even a short drought will affect the trees and the fruit. The first indication of injury will be the turning of the leaves to a lighter color, followed by the shedding of a considerable portion of them ; the fruit stops growing, or grows very slowly, and finally a considerable portion drops off. But if the land be kept well cultivated, no ordinary drought will affect either the trees or the fruit, though the land be quite high and dry. When it is not convenient to cultivate the land, the trees can be protected by mulching quite as well, if not better, than by cultivation. An orchard should never be set on high land unless it be kept cultivated or mulched.

In many places this year the drought has been so protracted that even trees on what is usually quite moist land have suffered, and the fruit is much below the usual size, and within the past week or two a considerable portion has fallen off.

A season like this teaches us how important it is to not only make a good selection of land upon which to set an orchard, but also to keep the trees in that high state of cultivation that will insure good fruit when the weather is unfavorable, thus enabling the grower to go into the market with an abundance of good fruit when it will bring good prices. The successful fruit grower and the successful farmer usually learn to succeed where the ordinary cultivator fails. In a good season almost any one can grow fair crops, but in a year like this, it is only those who work intelligently that are able to overcome the unfavorable conditions of the natural elements.

When the farmer becomes thoroughly acquainted with the best methods of protecting crops from the dry weather, we shall fear droughts much less than we now do, and our losses will be very small in comparison to what they are now.

Saving Cabbages Till Spring.

We know no better way to preserve cabbages through the winter than that which we have recommended for a number of years. It is to plant or set them up in rows as they grow—that is, with the roots down—fill in with soil pretty freely, than make a covering by planting two posts where there is a fence to rest on, or four where there is not, allowing for a pitch to carry off the water; lay bean-poles opposite the way of the pitch, and cover with cornfodder, or straw, or boards. In going through the winter, avoid as much as possible the sun-side and close up again. We have not found that setting the cabbage upside-down in the rows, as many do, of any advantage, as we have kept ours for more than twenty years in the way we mention in a sound, perfect condition, through the winter into the spring, and could even up to the first of May if desirable. We see other methods recommended, and they may answer just as well, but as to our own we speak from a long experience.—*Germantown Telegraph.*

The Fruit Supply.

In all of the principal markets of the Northern States there has been an under supply of home-grown fruit this season. Last year it was better, but for three or four years it has been evident that the demand was increasing faster than the supply. With a flourishing state of horticulture throughout the country, why is it there should be so steady an advance in the price of these products? In the first place, the population in cities and large villages has been rapidly increasing, causing a proportionately increased consumption of fruit. Then the principles of diet and hygiene that have for many years been disseminated among the people through the press have been accepted as true, and practically applied, until every one considers as a necessity a certain amount of fruit. Again, new methods of preserving, such a canning, bottling and drying, have been learned, so that immense quantities are used for this purpose, keeping a supply all through the year and foreign markets consume increasing quantities.

Even if this immense increased demand could have been properly appreciated considerable time must elapse before it could be met by fruits like the apple and the pear. The cherry, the plum and peach, that give returns quicker, would respond more alertly, but those fruits can be raised to advantage only in particular sections, and probably only a few of those persons who may be favorably situated to raise these crops will avail themselves of the opportunity, since their attention is directed to some other branch of industry. The small fruits, strawberries, etc., will turn sooner, but for the reason probably that they are most available for canning, the proportionate supply of them diminishes apparently more than that of the large fruits. It is then quite clear that it is safe for those having suitable soils in favorable localities to plant fruit trees. With proper attention they will be sure in time to yield handsome returns. Again, we could afford to give our orchards and fruit grounds more attention than has been the custom. They should have the best care, with the expectation of receiving ample payment for it in return. More of the small fruits can be cultivated, and a ready market will be found for all the products.—*Farmers' Review.*

Bananas and Plantains.

A pound of bananas contains more nutriment than three pounds of potatoes, while as a food it is in every sense of the word far superior to the best wheaten bread. An acre of ground planted with bananas will return, according to Humboldt, as much food material as thirty-three acres of wheat, or over a hundred acres of potatoes.

The banana (it should be called plantain, for, until lately, there was no such word as banana) is divided into several varieties, all of which are used for food. The platino maxinito is a small, delicate fruit, neither longer nor stouter than a lady's forefinger. It is the most delicious and prized of all the varieties of the plantain.

El plantino guinea, called by us the banana, is probably more in demand than any other kind. It is subdivided into different varieties, the principal of which are the yellow and purple bananas that we see for sale in our markets; but the latter is so little esteemed by the natives of the tropics that it is seldom eaten by them.

El plantino grande—known to us as simply the plaintain—is also subdivided into varieties, which are known by their savor and their size. The kind that reaches our market is almost ten inches long, yet on the isthmus of Darien there are plantains that grow from 18 to 22 inches. They are never eaten raw, but are either boiled or roasted, or are prepared as preserves.

HOUSEHOLD RECIPES.

GRAHAM BREAD.—Make a stiff batter of half a pint of warm water thickened with graham, flour and add to it a third of a cupful of yeast. Let it rise over night, and in the morning add a little piece of butter, half a cupful of sugar, and wheat flour enough to mould. Let the bread rise in pans, and bake an hour.

INDIAN CAKE. One pint of Indian meal, a cupful of flour, half a cupful of sugar, one-third of a cupful of butter, a teaspoonful of soda, one of cream of tartar, an egg, and some salt. Mix in enough sweet milk to make a soft batter.

CRULLERS.—These dainties are easily and quickly made. A piece of butter about the size of an egg, a nutmeg, a cupful of sugar, and three eggs are to be made stiff with flour, cut in fancy shapes, and fried in boiling lard.

DOUGHNUTS.—One and a half cupfuls of milk, the same quantity of sugar, two eggs, a scant teaspoonful of soda, a teaspoonful of salt, and half a nutmeg. Very toothsome doughnuts are made by this rule.

BUNS.—Half a cupful each of yeast, sugar and butter, one and a half cupfuls of milk, half a nutmeg, and a little salt. Mix together at night, and in the morning add half a cupful of sugar, and some currants.

ROAST MUTTON.—Wipe the mutton with a damp cloth; then dredge with salt, a little pepper, and generously with flour. Place on a meat rack in the baking pan before dredging, see that the bottom of the pan shall be covered with flour. Place in a hot oven, and as soon as the flour in the pan is brown (which will be in about five minutes), pour in hot water enough to cover the bottom of the pan. Baste every fifteen minutes. Cook a leg weighing six pounds, one hour and a quarter, and give ten minutes for every additional pound. This cooks it rare. If it is to be well done, roast one hour and a half, with fifteen minutes for every pound over six. When the meat is done, pour all the fat from the gravy and add a cupful of boiling water to what remains in the pan. Thicken this with a smooth paste made of a tablespoonful of flour and a little cold water. Stir well, and boil two or three minutes. Season with salt and pepper. Strain and serve. All the dishes must be very warm for a mutton dinner.

MASHED POTATOES.—Pare and boil for thirty minutes. Mash light and fine with a wooden masher. To every twelve potatoes add one teaspoonful of butter, half a cupful of boiling milk, and salt to taste.

MASHED TURNIPS.—Pare, and cut into slices. If the white turnips be used and they are fresh, they will cook in forty minutes, but if they be the yellow kind they must boil for two hours in plenty of water. Mash and season with butter, salt, and pepper.

BAKED POTATOES.—Wash, nip good sized potatoes and bake in a moderate oven forty-five minutes. They are spoiled by being over-cooked.

APPLE PUDDING.—Pare and chop fine six large apples. Put in a pudding-dish a layer of grated bread crumbs, one inch deep, then a layer of apple. On this put bits of butter, sugar, and a slight grating of nutmeg. Continue as before, and finally pour on a teacupful of cold water. Bake half an hour. Use in all two tablespoonfuls of butter and a small cupful of sugar.

SPANISH CREAM.—One quart of milk, three eggs, one cupful of sugar, one-third of a box of gelatine, one generous teaspoonful of vanilla flavor. Put the gelatine in a bowl with half a cupful of cold water, and when it has stood an hour add it to a pint and a half of the milk, and then place the sauce pan in which it is to be cooked (it should hold two quarts), into another of boiling water. Beat the yolks of the eggs with the sugar and one fourth of a teaspoonful of salt. Beat the whites to a stiff froth. Add the half pint of cold milk reserved from the quart to the yolks and sugar, and stir all into the boiling milk. Cook five minutes, stirring all the time; then add the whites and remove from the fire. Add the vanilla, and pour into moulds. Place on ice to harden.

BOILED FLANK OF BEEF.—Wash the flank, and make a dressing as for turkey, with spread over it, first having salted and peppered it well; then roll up and tie. Wind the twine around it several times, to keep it in place; then sew into a cloth kept for that purpose. Put a small plate in the pot, and put in the meat; then your on it boiling water enough to cover and boil gently six hours. When done, remove the cloth, but not the twine until stone cold; then cut in thin slices, and you will have alternate layers of meat and dressing. This is a nice dish for breakfast or tea.

MEAT HASH.—Dredge with salt and pepper any kind of cold meat, and chop it fine. This is always the best manner of seasoning hash, as all parts will be seasoned alike. If you have cold potatoes, chop fine and mix with the meat; if they are *hot*, mash. Allow one pint of meat to two of potato. Put this mixture in the frying pan with a little water or soup stock to moisten it, and stir in a spoonful of butter; or if you have nice beef dripping, use that instead of butter. Heat slowly, stirring often, and when warmed through, cover and let it stand on a moderately hot part of the stove or range twenty minutes. When ready to serve, fold as you would an omelet.

VEAL LOAF.—Three pounds of veal or fresh beef, half a pound of salt pork chopped fine, two beaten eggs, one teacupful of cracker crumbs, three teaspoonfuls of salt, two teaspoonfuls of pepper. Mix and press hard into a tin. Bake one and a half hours.

TOMATO SAUCE.—One pint of stewed tomato, one tablespoonful of butter, one of flour, four cloves, a tiny bit of onion. Cook the tomato, clove and onion together ten minutes. Heat the butter in a small pan and stir the flour into it. Cook, stirring all the time, until smooth and a light brown; then stir into the tomato. Cook two or three minutes longer. Season with salt and pepper, and strain.

STEAMED BEEF STEAK PUDDING.—One quart of flour, one large teaspoonful of lard, two teaspoonfuls of cream of tartar, one teaspoonful of soda, two cupfuls of milk or water, a little salt, one and a half pounds of beef steak. Roll out the crust and line a deep earthen dish; then lay in part of the steak, with a few pieces of butter, a little salt, and a few whole cloves ; then lay on the rest of the steak, with seasoning as before. Turn the crust up over the whole. Steam two hours.

STEWED LOBSTER.—Open a lobster weighing two and a half pounds and cut the meat into little dice. Heat two tablespoonfuls of butter, and add the dry flour, stirring until perfectly smooth ; then gradually add the water, stirring all the while. Season to taste. Add the lobster, and heat thoroughly.

BOILED RICE.—Wash in two waters one cupful of rice. Put it to boil in two quarts of boiling water and one tablespoonful of salt. Boil rapidly, with the cover off the sauce pan, for twenty-five minutes. Turn into a colander to drain, and place where it will keep warm while the steak is broiling. The water in which it was boiled may be used to starch prints.

BOILED CIDER.—Take four gallons of cider and boil it to one gallon.

STEAMED PUDDING.—One cupful of molasses, one of sweet milk, one of raisins, half a cupful of butter or two-thirds of a cupful of chopped suet, one teaspoonful of mixed spice, one of soda, half a teaspoonful of salt, four cupfuls of flour. Dissolve the soda in the milk. Mix all the ingredients thoroughly, and steam three hours in a buttered mould. To be eaten with lemon sauce.

NICE GRIDDLE CAKES.—Two quarts of flour, a handful of Indian meal, two eggs, a teaspoonful of salt, one of soda, one quart of milk.

COTTAGE PUDDING.—One cupful of sugar, two of flour, one of milk, one egg, butter the size of an egg, one teaspoonful of soda, two of cream of tartar. Beat the sugar and butter together ; then add the egg, well beaten, then the milk, and finally the flour, in which the soda and cream of tartar have first been well mixed. Bake in a pudding dish for half an hour in a moderate oven. To be eaten with sauce. The lemon sauce is good with it.

GRIDDLE AND INDIAN CAKES.—For the griddle cakes use two coffee cupfuls of sour milk or buttermilk, one teaspoonful of saleratus dissolved in a little hot water, and flour enough to pour. Grease the griddle with a piece of fat salt pork, and fry the cakes a light brown. Indian cakes are made in much the same way, save that half flour and half Indian meal is used, and also a teaspoonful of salt. They require a somewhat longer time to fry.

ESCALLOPED MUTTON—Chop some cold mutton rather coarse and season with salt and pepper. For one pint of meat use half a cupful of gravy and a heaping cupful of grated bread crumbs. Put a layer of the meat into an escallop dish, then some gravy, then a thin layer of crumbs. Continue in this way until the dish is full. The last layer must be a thick one of crumbs. Cook fifteen minutes in a hot oven.

MOCK OYSTER SOUP.—Peel twelve good sized tomatoes, and boil in a little water until quite soft. Let two quarts of milk come to a boil, and thicken with two large crackers that have been rolled fine. Add one teaspoonful of soda to the tomatoes. When these are well broken up, season with salt, pepper and three tablespoonfuls of butter. Add to the milk and serve immediately. The tomato may be strained if you prefer.

EXCELLENT GOLD CAKE.—A cupful of sugar, half as much butter, half a cupful of milk, one and three-fourths cupful of flour, the yolks of three eggs and one whole egg, one-fourth of a teaspoonful each of soda and cream of tartar, half a teaspoonful of lemon flavor. Mix together the sugar and butter, and add the eggs, milk, lemon extract and flour, in this order. Bake for half an hour in a moderate oven.

LEMON CAKE.—The rind and juice of a lemon, a teaspoonful of cream of tartar, as much saleratus, a teacupful of butter, one of sweet milk, three of sugar, four and a-half of flour, and five eggs—the yolks and whites beaten separately. Bake in two ovens for forty-five minutes in a rather quick oven.

FRIED CHICKEN.—Cut the chicken into six or eight pieces, and season well with salt and pepper. Dip into beaten egg, and then into fine bread crumbs, in which there is a teaspoonful of chopped parsley for every cupful of crumbs. Dip once more in the egg and crumbs, and fry two minutes in boiling fat.

PLAIN FRUIT CAKE.—Half a cupful each of milk and butter, one and a half cupfuls of sugar, two and a half cupfuls of flour, two eggs, half a teaspoonful of soda, spices and fruit.

BOILED RICE PUDDING.—Pick and wash one cupful of rice and pour in one quart of boiling water for fifteen minutes ; then drain dry. Wring a pudding-cloth out of boiling water, and spread in a deep dish, and turn the rice into it. Sprinkle in one cupful of raisins, and a tablespoonful of salt ; tie the cloth loosely, that the rice may have room to swell, and boil two hours. Serve with lemon sauce, or sugar and cream. Or, apples may be used in place of the raisins.

OKRA SOUP EQUAL TO TURTLE SOUP.—One leg of beef, quarter of a package of okra, two carrots, eight tomatoes, two onions, cut fine, nine quarts of water. Boil six and a half hours. Cut the meat off the bone in small pieces. Take the most glutinous parts of the leg and a little of the flesh, and mix with the soup when it is made. Cut the okra in small pieces. Boil steadily but not hard.

STEAMED BROWN BREAD.—Two cupfuls of new milk, two of Indian meal, one and a half of flour, one of molasses, one teaspoonful of soda. Steam three hours.

RHUBARB PIES.—Do not cut the rhubarb until the morning it is to be used ; or, if you have to buy it, keep it in a cool place. Strip off the skin and cut the stalk into pieces about an inch long, and stew in water just enough to prevent burning. When cold, sweeten to taste. Cover the pie-plates and roll the upper crust about half an inch thick ; cut into strips, an inch wide, and, after filling the plate with the rhubarb, put on four cross pieces and the rim. Bake half an hour.

Live Stock.

Cattle-Raising in Montana.

To assert that Montana is the best grazing country in the world, writes a correspondent of the St. Paul Pioneer Press, is merely to report the deliberate verdict of hundreds of practical stockraisers who of late have visited this region and made it the subject of cautious investigation. For some time to come the eastern half of the Territory is likely to stand foremost among the beef and wool producing sections of America. It is now known that Montana cattle make better beef than the average stock of other beef-producing Territories and States, and this is largely due to the nutritious quality of the native perennial grasses.

Unlike cultivated grasses, these prolific wild products have firm, solid stocks, and their heads are full of seeds, a combination whose merits are aptly described in the assertion that "to pasture an animal on bunch grass is like giving him plenty of good hay with regular and liberal feeds of grain." Before the frost has left the ground the grass appears above the soil, covering the face of nature with brilliant emerald verdure. At this time of year, however, its freshness has nearly all gone. The period of moisture has passed, and the plains present a yellow and withered appearance for the rest of the year. The fact is that the grass has been converted, on the stock, into hay, upon which the sheep and cattle pasture and fatten throughout the coldest winters sheep require greater care than cattle, but if successfully handled the profits are considerably larger. The average increase of the herd is about 75 per cent. The production is the measure of the profit in sheep-raising, as the sale of wool, which is always in eager demand, defray the whole expense of maintaining the herd, and sometimes exceeds it, to the extent of $1 or $1.25 per head.

It is evident that Montana stock-raising rests on a solid basis as a legitimate field of enterprise widely separated from the character of wild-cat speculation in which many capitalists regard it. It should be stated that Western cattle-raisers are no longer the uncouth half-civilized beings that they were ten years ago. A majority of them are men of education and enterprise from the older states, who have come out here and invested their capital in cattle and sheep raising. With few exceptions, those who have realized large profits or are in a sure way to do so. Such men, of course, do not pass their whole time upon the ranches, but live chiefly in towns. Here, where I write in Billings, a number of them are taking up their residence. I have met many of these cattle-kings, and I have found them all, without exception, well-informed, generous, enthusiastic, hospitable men. Many of them have planned great improvements for Billings, notably the case in regard to the construction of the stock yards.

To Utilize Jersey Bulls.

A correspondent of The Rural New Yorker suggests that it will be well to utilize Jersey bulls as working oxen rather than nip them in the bud for veal, and he cites this illustrative instance : "A neighbor has a pair of four-year-old full bred Jersey steers at work now on his farm, which are as strong and useful as a pair of good 15-hand horses. Their natural walk is at least four miles per hour, and they are fair brothers, reminding me to all their movements and work of the admirable Connecticut working oxen. They are also very hardy, and do not mind the hot sun at all. Considering the small cost of their harness—a simple yoke and a pair of bows—and the quickness and ease with which they can be attached to and detached from the cart, plough or harrow, these cattle are more economical than either of the three pairs of work horses which are kept on the same farm with them."

The Shropshire Sheep.

The development of great industries in iron and coal in the districts of Shropshire, at the beginning of the century, gave rise to a large and increasing demand for mutton. To meet this demand, the farmers of that part of the country turned their attention to the raising of mutton sheep. Breeding ewes were sought for from the midland and southern counties, and in time Shropshire became not only a leading sheep raising region, but also the home of an important breed, the parentage of which it is difficult to state, for the reason that it is derived from and combines a number of the best mutton breeds. The Shropshire is more strictly speaking a cross breed, in which "natives" of the districts, the Cotswold, and later the Leicester and Southdown have been combined. On account of this complex admixture of blood, the Shropshire is one that varies somewhat in character. The original sheep was hornless, black or brown faced, hardy and free from disease, producing 44 to 50 pounds of mutton to a carcass, and a fleece of two pounds of moderately fine wool. The present Shropshires are without horns, the legs and face dark or spotted with gray, the neck thick, the head well shaped, ears neat, breast broad, back straight, barrel round, and the legs strong. They are easy keepers, hardy, fatten quickly, and at the age of two years give 100 to 120 pounds of excellent flesh. She fleece is longer, heavier, averaging 7 pounds, and more glossy than that of the Southdown. The Shropshire is a valuable sheep for the American farmers.—Dr. Byron D. Halstead, in American Agriculturist for November.

Rearing Sheep for Their Milk.

In the south of France, where the climate is hot and the country mountainous, rearing sheep for their milk to produce cheese (Roquefort) is largely extending. The best milking ewes ought to have four or six teats, the udder voluminous, the wool rare and secreting much grease, ears long, head small and without horns. Sheep with four teats ought to be sought. In the Agricultural College of Montpelier there is a ewe with two lambs and yielding milk from six teats. So far the experiments have not succeeded in obtaining an animal producing much milk and a good fleece at the same time. Counting milk, lamb, and wool, a ewe produces net about 48 francs yearly. Six quarts of milk yield one pound of cheese. The Chillians, to obtain special skins much sought after, cross the sheep with the goat. Experiments are being conducted to the end of similar crossing for improving the milking capacities of ewes. Goat farming does not pay. The animal is destructive, its flesh held in little repute, and its offal of no value.—*New England Farm.*

Making Good Pork.

The first thing in order to make a letter A pork product is to secure the right breed of porkers. Tastes differ on this point. We like the small breeds, such as the Suffolks, Yorkshires, and Essex. The old-fashioned ambition to make a hog weigh 500 pounds at 18 months or 2 years old was not profitable to the producer, and the consumer certainly had "too much pork for his shillings." If a pig can be made to weigh 250 or 300 pounds at 8 months, as the Suffolks usually do, there is a saving of a year's keeping, and the pork is of a much better quality. We have eaten none other than pig pork for four years, and desire to eat no more of the big, strong sort. The Western producers are finding the best market for the small breeds, the spring pigs of which are fit for slaughter before Christmas, weighing, when dressed, 250 pounds on an average, and furnishing hams of about fifteen pounds weight. The early maturity of the small breeds gives them a great advantage over the larger kinds. We have known Suffolk pigs to weigh 300 pounds at seven months. To secure this result they must be fed with skimmed milk when first weaned, mixing with it a little bran and oat meal, and gradually increasing the ration of oats till the pigs have attained such a size that it will answer to put on fat, when corn meal may be substituted gradually for the bran and oats. There is nothing equal to milk for young pigs, but for inducing the growth the skimmed is fully as good as the pure article.—*New York Times.*

The Coming Sheep.

The philosophy of evolution and development appears to be supported by the history of our live stock. Those who have traced out the rise and progress have also had to recount the decadence and the fall of races of cattle and sheep. The old Long-horn, brought to perfection under the skillful management of Bakewell, waned and vanished under the superior qualities of the Short-horn. It would indeed be touching upon delicate ground so hint that this pet of the great ones of the earth could be displaced from her temple. All things, however, must come to an end, and exorbitant sums of money given by individuals for no special excellence except what exists, or is supposed to exist, potentially in the mysterious virtues of pedigree, savours of that luxury which precedes decay and dissolution.

The history of our chief breeds of sheep affords more than one instance of improvement and abandonment. Take, for example, the Leicester. Fifty years ago this breed might appropriately have been said to "rule the roost." Now, except in very few counties and among a small minority of farmers, the Leicester has been superseded. The Cotswold sheep is said to be going out, even upon his own hills, and does not seem to be spreading rapidly in any other locality. The Southdown was to the shortwooled races as the Leicester was to the longwools. Scarcely a breed was not improved by his touch, and for this reason alone the southdown will always hold a high position in the history of British flocks. Still, it must be confessed that the Southdown has ceased to be a rival for popularity with larger and more profitable, if less shapely, breeds of sheep.

One of the greatest advances in sheep breeding was made by Mr. Druce, of Eynsham, when he successfully crossed the Hampshire Down and Cotswold, and thereby produced the Oxford Down. The rise of this remarkable breed has been rapid, and it seems likely to extend further in its geographical distribution. An unfortunate predisposition to foot lameness is one of the weakest points in the favorite breed of the midlands, and a slowness in coming to maturity may possibly be also recorded as a frequent mark against him.

The last breed we have to mention is one which deserves very special mention. He has not as yet attracted a large share of public notice. Columns of show reports have been lavished upon Leicesters and Southdowns, but scant notes have been usually thought enough for the Hampshires. They have not been pushed up by the great. They have, however, been long carefully bred by a large number of first class tenant-farmers around Salisbury, and tended by a good and faithful race of shepherds. We venture to assert that the Hampshire sheep is not sufficiently known and appreciated. There is no race in England, or in the world, which can vie with it in the production of large sized lambs of from six to eight months old. Shropshire lambs are simply "nowhere" to them. Let any unprejudiced person attend the ram sales in July, near Salisbury, and if he has never before seen a Hampshire lamb, he will be astonished. Then he will see lambs which present you with a pound weight per quarter from the day they were born. No one thinks of using shearing rams, as they would be too heavy and unwieldy if not used as lambs. As yet the Hampshire breed has been insufficiently represented by our show-yards, but we expect soon to see a change in this particular. Such a breed cannot be comparatively kid from public notice, but must come out. His hardihood, size, and quality of mutton are unsurpassed. He thrives between hurdles and never asks for greater liberty. He is extraordinarily docile and intelligent, and can be brought into such perfect training that a word from the shepherd suffices to guide and control his movements. In the district in which this splendid race of sheep are found in greatest perfection it is not uncommon to realize as much as 60s. or even 65s. per head for lambs of from seven to eight months old. It is in those parts customary to sell off the wether lambs and retain the ewe lambs and ewes as winter stock. If instead of selling the lambs at the autumn fairs they were kept on through the winter and sold out, as is the case with most other breeds of sheep at ten or thirteen months old, they would make prices which we are confident in maintaining that no other race of sheep could touch. These are strong points in favor of the Hampshire sheep, insuring him a brilliant future, and, in a certain sense, the title we have placed at the head of these remarks.—*Agricultural Gazette (English).*

POULTRY.

Moulting.

As this is the time of year for fowls to moult (cast off their feathers and put on new ones), there must be greater attention given to them than usual. It matters not how well a bird looks when commencing to moult, or how well it feels, in two or three days there is so much change in its system and in its feelings and looks that one would not recognize that it was the same one. The bright red combs become pale and wilt down to quarter their usual size; their heads ;[hat were carried so stately, are now dropped, and the bird walks as if it was weary ; it appears weak, as it really is, and if ever an extra feed is given to fowls, it is now that it should be given. Quantity is not the only requisite, though it is something, but quality is the main object—something strong and in good proportion, such as a loaf of baked middlings (or rather a mixture of shorts, cornmeal, or buckwheat), with plenty of boiled potatoes, and a good seasoning of salt, red pepper or ginger. When kneading this add a few drops of tincture of iron, say half a teaspoonful for a two-pound loaf, which loaf given to a flock of twenty fowls will be sufficient for one day, and whole corn (old, not new) wheat screenings, peas, boiled oats or boiled barley, may be given in such quantities as will be eaten up clean without wasting.

In England many poultry breeders confine their fowls in small apartments and give a teaspoonful of camphor to each five in its drinking water, which assists in casting off the feathers, and they are not allowed to get any other water to drink but this for a week.

I think the process of moulting is the least understood, has the least care bestowed, and is the most neglected of anything belonging to the poultry-yard, whether fancier or farmer. During September and October—the times when birds are at their most critical period of health during the year—farmers generally have to rough it, and little or no attention is given them ; they are permitted to roost in wet lofts, or exposed to draughts of wind and sometimes in apple trees. Now, this should not be so ; they should be given not only the same care as other farm stock, but a little more just now, and when eggs are wanted in winter, and when good fat turkeys are wanted for that time, they will be forthcoming in plenty, or according as they have had attention. Those hens that have been kindly treated, and have had the best attention, will start to lay the first after moulting, as a hen will never lay while in this stage of nature's development.—*R. A. Brown,* in *Farmer's Advocate.*

How to Be Rid of Them

Is a question which is very apt to come if care is not constantly exercised. We mean the mites, jiggers or hen spiders, call them whatever name you please. They are the little lice that swarm everywhere, like the frogs of Egypt, unless kept out of the fowl-house. Don't fear to use plenty of whitewash with a little carbolic acid, and perhaps a solution of potash.

Stamp them out. Clear the fowls of this pest. Destroy the young broods of insects, now just coming forth. Keep the parasites at bay. Fumigate the closed houses with a pot of burning sulphur and crude rosin, shut the smoke in five or six hours. Then ventilate the premises thoroughly before roosting time. Wash the perches with kerosene—all over, underneath, edges and top. Destroy these annoying depredators, before they get old enough, strong enough and numerous enough to kill your young chickens, and devour the flesh of your adult stock.

We can not too often impress this important work upon the attention of good breeders, who entertain a disposition to render their domestic fowls comfortable. Especially is this advice needful to be observed in the hot weather we are at present in the midst of. And so we repeat it, if you would have your birds healthy and happy, drive off the lice from amongst them.

Last winter we saw a pile of boards lying in a farmer's barnyard which looked gray. On closer examination the gray shade proved to be the same "little insects" of which we are speaking.

The owner had built a new fowl-house last fall and threw these old boards out of doors to notice the effects on the lice. By taking one of these boards into a warm room, the multitude began to march, which shows pretty fully that cold will not kill, although it may paralyze them. From this trial we see that it is not safe to result the washing even in winter. Poison these parasites. Suffocate them. Do anything rather than lose a fowl from lack of care.—*Hartford, Conn., Poultry World.*

A Poultry House.

"How, when and where shall we go to work to build us a poultry house?" is an old refrain to the song, "We want to keep poultry right off." And we are expected to be able to stand up and give a satisfactory reply under any and all conditions. A person about to build should, if possible, observe and investigate some fowl house already erected that gives its owner satisfaction, and by practical consultation with some one near at hand can learn much more than one who knows nothing of his situation can inform him. A few general hints, however, may also be given.

For aspect, the glazed front should face the east and south. This affords you the sun's rays from the earliest morning, to late afternoon, as a rule, and it is the early hours of sunlight and warmth that fowls mostly covet in winter and chilly spring time. The glazing should be entire upon one or two sides of the house, whatever may be its size or length. If the sashes are tightly placed, it is amply protective as a wall upon these two warm sides, while the cost is no more than ceiling or battens, and clap boarding. The birds will enjoy both the light and the warmth thus afforded them; and if the other two walls are banked up, or are made impervious to wind and weather by a double boarding (four inches between the inner and outer walls), packed to the eaves with straw daubed with coal tar to keep off vermin, you may thus have a cheap, comfortable house that your early spring chickens will thrive in, and your adult birds will appreciate from December to April.

LITERARY AND PERSONAL.

STOCK AND POULTRY INDEX.—This is a neat 16 page octavo monthly, devoted exclusively to the breeding and management of stock and poultry, and filled with the choicest matter for every one interested in its specialties. It is not merely a magazine of advertisements, but contains 15 pages of good reading matter pertaining to stock and poultry. As its advertising patronage increases, extra pages will be added. No. 3, vol. 1, of this spicy little journal has found its way to our table, and although entirely unpretentious, and lacking the embellishments of more pretentious publications, we find it solid and sensible. Only 50 cents a year with clever premiums. Address *Stock and Poultry Index*, Waynesburg, Greene county, Pa., Lock Box 16.—*W. E. Robinson, Publisher and Editor.*

THOROUGHBRED STOCK JOURNAL.—A demifolio of 16 pages, only two of which are advertisements, Published by the L. S. P. Co., Philadelphia, Pa. U. S. I. Hunt, proprietor; W. S. Webster and Joseph Barbiere, editors; at $1.50 per annum; single copies 15 cents. Good material, good print, and good illustrations. This is an entirely new enterprise in Pennsylvania stock journalism, the copy before us being vol. 1, No. 1, October, 1882. The contents of this number, although in the main able and appropriate, yet it contains three lengthy papers on subjects foreign to stock-breeding, or any other subject relating thereto, which perhaps may be acceptable to many readers. These are "Egypt—past and present"—"The Christian Religion," by Col. Robert G. Ingersoll, and a "Reply," by Hon. Jeremiah S. Black. Ingersoll's statements, arguments and denials, are only a rehash of the "stuff" we met with and read fifty years ago, only they do not seem to possess the same ability, for he certainly can have but a limited knowledge of the contents of the Old and New Testaments, or he would not make the assertions he does. In our opinion, if he does not live to see the day to regret these utterances, it will be for want of opportunity.

Of course, Mr. Black had to meet Col. Ingersoll on his own plane—the merely natural plane—for he does not recognize the Christian's God, the inspiration of the Scriptures, nor the existence of the spiritual world; still, in our view, and we think in any rational or common-sense view, Mr. Black utterly vanquished him. Mr. B. is however, not a theologian, and does not seem to possess a knowledge of the more advanced views of the present day on that subject, or he might have made some things more clear upon which he confessed himself uninformed. But then it might have had no weight at all with that school of philosophy which acknowledges nothing less tangible than "buckwheat cakes and sausages." Poor Col. Ingersoll, in his tirade against Christianity, he reminds us of the "crone" who shook her fist at the rear-guard of Bonaparte's army, when he invaded Italy. "Tell your General," said she, "I have a mean contemptible opinion of him." The soldier replied, "I will madam; but only think how hard he'll take it." The comparative relation between Robert G. Ingersoll and the Christian religion "hath this extent, no more."

CITY AND COUNTRY.—An illustrated literary and agricultural journal, 20 pages, monthly, at the very low price of 50 cents per annum. No 10, Vol. 1. of this paper is before us: it is the same in size as the immediately preceding, but has about ten pages of advertisements: but is more exclusively agricultural and domestic, with a moderate sprinkling of general literature, and notices of the general topics of the day. Published by the City and Country Co., Columbus, Ohio. Will C. Turner, editor and general manager, A. W. Lincoln, associate. We don't know another similar journal in the country that contains so much, nor of a better quality, at so low a price. We notice that it devotes at least one of its large pages to matters relating to physical health, through a regular physician of large experience, which we deem a commendable feature.

DEPARTMENT OF AGRICULTURE.—Special report. No. 46, on the condition of corn and cotton, of spring wheat, fruits, etc., also freight rates of transportation companies, August, 1882. No. 47, climate, soil and agricultural capabilities of South Carolina and Georgia, by J. C. Hemphill, Government office, Washington, D. C., comprising jointly 120 pages octavo. The value of these bulletins, of course, depends upon the value of the information communicated to the Department from local reporters. The means are commendable, and the fountain will be more and more useful as the stream flows from reliable and practical sources. The severely criticised, and often much maligned department, is certainly making an effort to impart information to the agriculturists of the country, commensurate with its abilities.

UNITED STATES ENTOMOLOGICAL COMMISSION.—Bulletin No. 7, *Insects injurious to Forest and Shade Trees*, by A. L. Packard, Jr., M. D. This bulletin bears the imprint 1881, but it has only come into our possession within the past two or three weeks—too late to notice it in our October issue. It is an octavo of 275 pages, uniform in size with preceding bulletins, and contains many appropriate illustrations. We are indebted to the Department of the Interior for a belated copy of the work for which we are exceedingly thankful. To protect our forest trees from insect infestations is hardly second in importance to protecting them from the (fell) destruction of the woodman's axe. Of course, such a work cannot fall into the hands of everyone in the country, who has an interest in forest and shade trees, but it is safe to say that it will fall into the hands of as many as are likely to study it and make a practical use of it. The insects destructive to forest and shade trees are "legion," and a description of each one in detail would involve a book or books, too formidable for any ordinary man to look into; hence, in many instances only the technical name is given, and this too, only because they have not yet received a specific common name, and perhaps never will.

LADIES' FLORAL CABINET, *a Monthly Home Companion*. L. F. C. Publishing Co., 22 Vesey street, New York. A beautifully illustrated quarto of 30 pages, with embellished tinted covers. The October number (Vol. XI., No. 10.) of this handsome publication has been laid on our table, and in addition to first-class material and superior mechanical execution, it contains that variety in its able literary contents which relieves it from the monotony that distinguishes many journals devoted to a single specialty. Any lady at all interested in floriculture, poetry, gardening, horticulture, domestic economy, and general literature, would find this journal an appropriate *vade mecum*.

ANNUAL wholesale and retail list of the *Ephrata Nursery* and Green House. Fruit and ornamental trees, grape vines, small fruits, etc. S. R. Hess and Son, proprietors, 8 pp, 12 mo.

GOODWIN'S IMPROVED BOOK-KEEPING AND BUSINESS MANUAL. Synopsis of contents, 32 pp, 16mo.

PREMIUM LIST of the New Mexico Exposition and driving Park Association. Second annual fair held at the city of Albuquerque, September 18, 19, 20, 21, 22 and 23, 1882. 48 pages, demi-octavo. This catalogue reached us too late to receive a notice either in our September or October issues. It is a very liberal one, and so far as we have been enabled to learn, the fair was a complete success.

PERSONAL.—Notably among the superintendents of departments, we observe the name of Dr. Wm. T. Strachan, a nativ of Lancaster county, and formerly a resident of Lancaster city. The doctor is a resident of New Albuquerque; is extensively engaged in mining, and it was appropriate that he should have been appointed superintendent of the mining department.

FARMER AND MANUFACTURER.—A journal devoted to the farming and manufacturing interests of the country, published by the *Farmer and Manufacturing Company*, Cleveland, Ohio, at 50 cents a year in advance, postage included. The November number of this excellent publication has reached our table, and it ought to be welcome anywhere in the world, where the English language is spoken, and, "you had better believe it." Among the multitudes of "sorts and sizes" of journals now issued from the printing press of the country, we are a little puzzled as to whether we should style this a royal quarto or a demi-folio—it is 10 by 20—with four columns to the page. In addition to farming, manufacturing, domestic and polite literature, it also includes history, philosophy, poetry and fiction, but these are of the most practical and instructive character, possessing just that brevity, diversity, and moral quality, which go to make up the most interesting and useful daily reading. If any of our patrons desire to try it, "just to see," we will furnish a copy of the *Lancaster Farmer*, and the *Farmer and Manufacturer*, at $1.25 a year, in advance, and have no hesitation in assuring our readers that the arrangement will be satisfactory to them. The material and "make up" is equal to the average of our very best serial publications.

"SEED-TIME AND HARVEST," which has for the past three years been published as a quarterly, has now entered the field as a 24-paged monthly magazine, and is filled to overflowing with notes and illustrations of the most popular new fruits, flowers and vegetables of American origin. It has among its contributors some of the best writers of the day upon horticultural subjects. Every page is made interesting and instructive. It is published at La Plume, Lack'a Co., Pa., by Isaac F. Tillinghast, at the low price of 50 cents per year.

Mr. A. G. Tillinghast, a brother of its editor, recently started on a trip across the continent to California and thence up the coast to Washington territory. One of the attractions of *Seed-Time and Harvest* for the next few months will be the publication of letters giving daily reports of the incidents of this journey, which will prove very interesting to every one who is interested in the subject of emigration, as he will let you know how an emigrant is treated, how fast he travels, what he sees and what it costs him to see it.

All the principal agricultural and horticultural papers in the country are taken and read by the editor of *Seed-Time and Harvest*, and he will endeavor to give monthly under the head of "Notes and Gleanings," all the new ideas of interest which may spring up anywhere; thus its readers will for 50 cents per year get an epitome of the agricultural and horticultural worlds.

A noticeable feature of *Seed-Time and Harvest* is the offer of its editor of $50 in gold to the person who sends him the most perfect list of the different words to be found in one number of the magazine, having eight or more letters, and no letters repeated. This exercise will be repeated in the January number and competition is free to all subscribers. Full and precise rules will be given in the number containing this offer.

This excellent and most practical journal and the LANCASTER FARMER, will be furnished to subscribers at $1.25 a year.

THE LANCASTER FARMER

EVERY lady should send 25 cents to Strawbridge & Clothier, Philadelphia, and receive their *Fashion Quarterly* for 6 mos. 1,100 illustrations and 4 pages new music each issue.

A MANUAL OF ELOCUTION AND READING, embracing the Principles and Practice of Elocution. By Edward Brooks, Ph. D., Principal of the State Normal School, Millersville, Pa. Philadelphia: Eldridge & Bro. Price, 1.50. To teachers, for examination, $1.00.

PHILIP SCHUM, SON & CO.,
38 and 40 West King Street.

We keep on hand of our own manufacture,

QUILTS, COVERLETS,
COUNTERPANES, CARPETS,

Bureau and Tidy Covers, Ladies' Furnishing Goods, Notions, etc.
Particular attention paid to customer Rag Carpet, and scouring and dyeing of all kinds.

PHILIP SCHUM, SON & CO.
Nov-ly   Lancaster, Pa.

THE PENN HARROW

BEST IN THE WORLD
IT HAS NO EQUAL

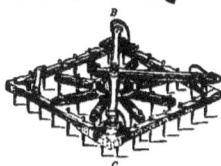

Patented April 13, 1880.

The above cut represents the Penn Harrow complete, with all its combinations of Five Harrows and a sled for each Harrow; and each successive change is made from the Harrow without the least additional expense. By hooking the beam in other point B or C, the cross pieces and axles can possibly rest on ground. Two Crossings in passing over it once, making it the most effective pulverizer in the market.

THIS HARROW HAS ONLY TO BE USED TO BE APPRECIATED.

See it before purchasing and you will buy no other.

The Penn Harrow
CHANGED TO A THREE-CORNER ROTARY HARROW.

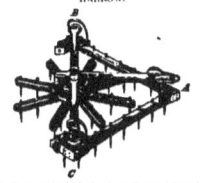

Indispensable for Orchards, as the revolving wheel harrows reach up to and all around the trees without barking them.

The Penn Harrow
CHANGED TO SINGLE "A" HARROW.

DISSOLUTION OF PARTNERSHIP.—The co-partnership in the merchant tailoring business heretofore existing under the firm of Rathvon & Fisher, is this day dissolved by mutual consent. All persons in any manner indebted to said firm, are respectfully solicited to make immediate payment to S. S. Rathvon, who is hereby authorized to receive and receipt for the same, and those having claims against said firm, will please present them for settlement.

S. S. RATHVON,
M. FISHER,
101 North Queen Street, Lancaster, Pa.

Until further announcement, the business, without interruption, will be conducted by the undersigned, who solicits a continuance of the patronage heretofore bestowed upon the firm, and which is hereby greatfully acknowledged.

S. S. RATHVON,
PRACTICAL TAILOR,

No. 101 North Queen Street,
LANCASTER, PA.

By removing the wire and wheel from the original you have a complete one-horse "A" Harrow.

The Penn Harrow
CHANGED TO DOUBLE "A" HARROW.

Remove the wheel from the original, reverse the wire, and it makes the most complete Double "A" Harrow in the market.

The Penn Harrow
CHANGED TO A SQUARE HARROW.

By removing the wheel from the original you have a Harrow with three points to look to. By looking to D or C you can harrow in a furrow, and harrow the bottom and both sides, or never rider and harrow the top and both sides, or you can lift either point and have three points on the ground—something that cannot be done with any other Harrow.

The Penn Harrow
ON ITS SLED.

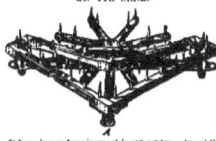

It has always been a great inconvenience to get the Harrow to and from the field. The Penn Harrow obviates this, as no matter which Harrow you wish to use in the combination, it has its own sled to haul it on.

The Penn Harrow

is made of the best white oak, with steel teeth, well painted, in every way first-class. Formerly a harrow was the most unhandy implement on the farm, with our improvement it is the most convenient, will do double the work of any other harrow and save the farmer half his labor, and is warranted to do all we represent or money refunded. **ORDER AT ONCE AND BE CONVINCED.**

Price of the light draft Combination Penn Harrow, $30. *Send for a Catalogue and see what farmers say.*

AGENTS WANTED IN EVERY COUNTY.

PENN HARROW MANUFACTURING CO.
CAMDEN, N. J.
sep-3

TREES

Fruit, Shade and Ornamental Trees.

Plant Trees raised in this county and suited to this climate. Write for prices to

LOUIS C. LYTE.
Bird-in-Hand P. O., Lancaster co., Pa.
Nursery at Smoketown, six miles east of Lancaster.
79-1-12

WIDMYER & RICKSECKER,
UPHOLSTERERS,
And Manufacturers of

FURNITURE AND CHAIRS.
WAREROOMS,

102 East King St., Cor. of Duke St.
LANCASTER, PA.
79-1-12]

Special Inducements at the
NEW FURNITURE STORE
OF
W. A. HEINITSH,
No. 15 1-2 E. KING STREET
(over Bursk's Grocery Store), Lancaster, Pa.
A general assortment of furniture of all kinds constantly on hand. Don't forget the number,
15 1-2 East King Street,
Nov-ly] (over Bursk's Grocery Store.)

For Good and Cheap Work go to
F. VOLLMER'S
FURNITURE WARE ROOMS,
No. 309 NORTH QUEEN ST.,
(Opposite Northern Market)
Lancaster, Pa.
Also, all kinds of picture frames.   nov-ly

GREAT BARGAINS.

A large assortment of all kinds of Carpets are still sold at lower rates than ever at the

CARPET HALL OF H. S. SHIRK,
No. 202 West King St.

Call and examine our stock and satisfy yourself that we can show the largest assortment of their Brussels, three plies and ingrain at all prices—at the lowest Philadelphia prices.

Also on hand a large and complete assortment of Rag Carpet.

Satisfaction guaranteed both as to price and quality. You are invited to call and see my goods. No trouble in showing them even if you do not want to purchase. Don't forget the letter. You can save money here if you want to buy.

Particular attention given to customer work.

Also on hand a full assortment of Counterpanes, Oil Cloths and Blankets of every variety.   [nov-ly.

C. R. KLINE
ATTORNEY-AT-LAW,

OFFICE: 15 NORTH DUKE STREET,
LANCASTER, PA.
Nov-ly

SILK-WORM EGGS.

Amateur Silk-growers can be supplied with superior silk-worm eggs, on reasonable terms, by applying immediately to
GEO. O. HENSEL.
[may-3m]   No. 258 East Orange Street, Lancaster, Pa.

LIGHT BRAHMA EGGS

For hatching, now ready—from the best strain in the county—at the moderate price of

$1.50 for a setting of 18 Eggs.
L. RATHVON,
No. 9 North Queen st., Examiner Office, Lancaster, Pa.

WANTED.—CANVASSERS for the
LANCASTER WEEKLY EXAMINER
In Every Township in the County. Good Wages can be made. Inquire at
THE EXAMINER OFFICE,
No. 9 North Queen Street, Lancaster, Pa.

WHERE TO BUY GOOD IN LANCASTER.

BOOTS AND SHOES.

MARSHALL & SON, No. 12 Centre Square, Lancaster, Dealers in Boots, Shoes and Rubbers. Repairing promptly attended to.

M. LEVY, No. 3 East King street. For the best Dollar Shoes in Lancaster go to M. Levy, No. 3 East King street.

BOOKS AND STATIONERY.

JOHN BAER'S SONS, Nos. 15 and 17 North Queen Street, have the largest and best assorted Book and Paper Store in the City.

FURNITURE.

HEINITSH'S, No. 15½ East King st., (over China Hall) is the cheapest place in Lancaster to buy Furniture. Picture Frames a specialty.

CHINA AND GLASSWARE.

HIGH & MARTIN, No. 15 East King st., dealers in China, Glass and Queensware, Fancy Goods, Lamps, Burners, Chimneys, etc.

CLOTHING.

MYERS & RATHFON, Centre Hall, No. 12 East King St. Largest Clothing House in Pennsylvania outside of Philadelphia.

DRUGS AND MEDICINES.

G. W. HULL, Dealer in Pure Drugs and Medicines Chemicals, Patent Medicines, Trusses, Shoulder Braces, Supporters, &c., 15 West King St., Lancaster, Pa.

JOHN F. LONG & SON, Druggists, No. 12 North Queen St. Drugs, Medicines, Perfumery, Spices, Dye Stuffs, Etc. Prescriptions carefully compounded.

DRY GOODS.

GIVLER, BOWERS & HURST, No. 25 E. King St., Lancaster, Pa., Dealers in Dry Goods, Carpets and Merchant Tailoring. Prices as low as the lowest.

HATS AND CAPS.

C. H. AMER, No. 30 West King Street, Dealer in Hats, Caps, Furs, Robes, etc. Assortment Large. Prices Low.

JEWELRY AND WATCHES.

H. Z. RHOADS & BRO., No. 4 West King St., Watches, Clock and Musical Boxes. Watches and Jewelry Manufactured to order.

PRINTING.

JOHN A. HIESTAND, 9 North Queen st., Sale Bills, Circulars, Posters, Cards, Invitations, Letter and Bill Heads and Envelopes neatly printed. Prices low.

FARMING FOR PROFIT.

It is conceded that this large and comprehensive book, (advertised in another column by J. C. McCurdy & Co., of Philadelphia, the well-known publishers of Standard works.) is not only the newest and handsomest, but altogether the BEST work of the kind which has ever been published. Thoroughly treating the great subjects of general Agriculture, Live-Stock, Fruit-Growing, Business Principles, and Home Life; telling just what the farmer and the farmer's boys want to know, combining Science and Practice, stimulating thought, awakening inquiry, and interesting every member of the family, and the leading papers, and is destined to have an extensive sale. Agents are wanted everywhere. Jan-tf

---

GLOVES, SHIRTS, UNDERWEAR.

SHIRTS MADE TO ORDER,

AND WARRANTED TO FIT.

E. J. ERISMAN,

56 North Queen St., Lancaster, Pa.

79-1-19j

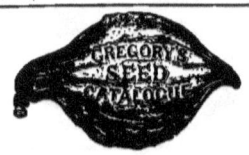

Thirty-Six Varieties of Cabbage; 26 of Corn; 28 of Cucumber; 41 of Melon; 35 of Peas; 28 of Beans; 17 of Squash; 31 of Beet and 40 of Tomato, with other varieties in proportion, a large portion of which were grown on my five seed farms, will be found in my Vegetable and Flower Seed Catalogue for 1882. Sent Fruit to all who apply. Customers of last Season need not write for it. All Seed sold from my establishment warranted to be fresh and true to name, so far, that should it prove otherwise, I will refill the order gratis. The original introducer of Early Ohio and Burbank Potatoes, Marblehead, Early Corn, the Hubbard Squash, Marblehead Cabbage, Phinney's Melon, and a score of other New Vegetables, I invite the patronage of the public. New Vegetables a specialty.

JAMES J. H. GREGORY,
Marblehead, Mass.

Nov-6mo]

EVAPORATE YOUR FRUIT.

ILLUSTRATED CATALOGUE
FREE TO ALL.

AMERICAN DRIER COMPANY,
Chambersburg, Pa.

Apl-tf

THE COOLEY CREAMER

Raises all the cream between the milkings. Saves two-thirds the labor. Increases yield of butter. Improves the quality. Quadruples the value of skim milk. Will pay for itself twice or more every season. The Cooley System is the only uniform dairy method in existence. Requires no spring house, or milk room. May be placed in a shed, cellar, or any place that cold water is handy.

The Best Hired Girl.

In the fall of 1879 I bought a Cooley Creamer. I have used it ever since with entire satisfaction. It makes more butter, of better quality, without ice, and half the labor, than the old process. A lady friend who has used one for about six months says it is "the best hired girl" she ever had. I have also used the Davis Swing Churn for the last 16 months, and am highly pleased with it. It churns the cream at a higher temperature and brings the butter in a better condition than any other churn. I have given the Eureka Butter Worker a fair trial, and am happy to recommend it to others. I can work twenty pounds of butter with it in five minutes, and thus save a half hour's work.

Yours truly,
SAMUEL S. OCLKITT,

Mt. Holley, Burlington County, N. J., August 22, 1881.

☞ Send for Circular free to

D. LANDRETH & SONS,
Sole Agents, Philadelphia Pa.

A HOME ORGAN FOR FARMERS.

THE LANCASTER FARMER,

A MONTHLY JOURNAL,

Devoted to Agriculture, Horticulture, Domestic Economy and Miscellany.

Founded Under the Auspices of the Lancaster County Agricultural and Horticultural Society.

EDITED BY DR. S. S. RATHVON.

TERMS OF SUBSCRIPTION:

ONE DOLLAR PER ANNUM,

POSTAGE PREPAID BY THE PROPRIETOR.

All subscriptions will commence with the January number, unless otherwise ordered.

Dr. S. S. Rathvon, who has so ably managed the editorial department in the past, will continue in the position of editor. His contributions on subjects connected with the science of farming, and particularly that specialty of which he is so thoroughly a master—entomological science—some knowledge of which has become a necessity to the successful farmer, are alone worth much more than the price of this publication. He is determined to make "The Farmer" a necessity to all households.

A county that has so wide a reputation as Lancaster county for its agricultural products should certainly be able to support an agricultural paper of its own, for the exchange of the opinions of farmers interested in this matter. We ask the co-operation of all farmers interested in this matter. Work among your friends. The "Farmer" is only one dollar per year. Show them your copy. Try and induce them to subscribe. It is not much for each subscriber to do but it will greatly assist us.

All communications in regard to the editorial management should be addressed to Dr. S. S. Rathvon, Lancaster, Pa., and all business letters in regard to subscriptions and advertising should be addressed to the publisher. Rates of advertising can be had on application at the office.

JOHN A. HIESTAND,
No. 9 North Queen St., Lancaster, Pa.

$72 A WEEK. $12 a day at home easily made. Costly Outfit free. Address TRUE & Co., Augusta, Maine.

ONE DOLLAR PER ANNUM.—SINGLE COPIES 10 CENTS.

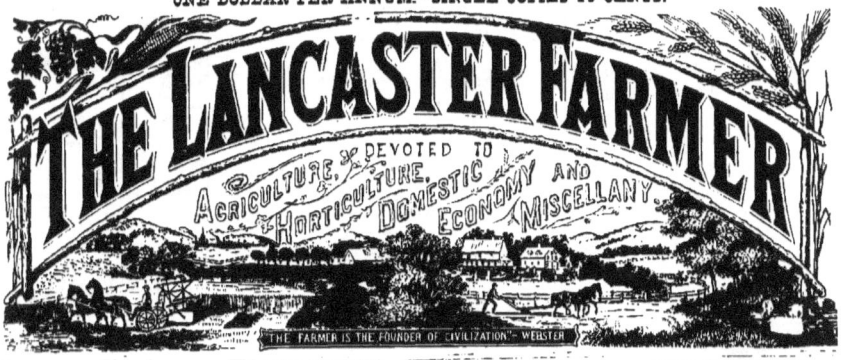

Dr. S. S. RATHVON, Editor.   LANCASTER, PA. DECEMBER, 1882.   JOHN A. HIESTAND, Publisher

Entered at the Post Office at Lancaster as Second Class Matter.

CONTENTS OF THIS NUMBER.

EDITORIAL.
Volume Fourteen 177
Myriapoda 177
The Tariff and Free Trade 178
The Turkey 179
Kitchen Garden for December 179
Excerpts 180

CONTRIBUTIONS.
"The Farmer's Friend" 181
A Sure Preventive of Chicken Cholera 182
The Balance of Trade Delusion 182

SELECTIONS.
The Virtues of Coffee 183
Feeding Stock in Winter 183
The Rational Method of Tree-Pruning 184
Letter from the Mother of Bayard Taylor to
  Prof. R. V. Riley 185
Soiling Cattle 185
Smoke house at Small Expense 186
The Sugar-Beet 186

OUR LOCAL ORGANIZATIONS.
Lancaster County Agricultural and Horticultural
  Society 187
  Crop Reports—Experimenting with Fertilizers
    —Answering Questions.
The Poultry Association 187
Linnæan Society 188
  Donations to the Museum.

AGRICULTURE.
Ivory Wheat and Millo Maize 188
Economy on the Farm 188
Rule Adopted by the Hay Trade 188
Effects of Broom-Corn on the Soil 188
The Agricultural Interests of the Country 188
Small Potatoes 188

HORTICULTURE.
Winter Flowers in the Window 189
Preserving Garden Flowers 189

HOUSEHOLD RECIPES.
Roast Turkey Garnished with Sausages 189
Mashed Turnips 189
Canned Corn Pudding 189
Cranberry Sauce 189
Orange Snow and Snow Drift Cake 189
Oyster Soup 189
Boiled Chicken 189
Browned Potatoes 189
Baked Sweet Potatoes 189
Scalloped Squash 189
Baked Custards 189
Simple White Soup 189
Stewed Fillet of Veal 190
Spinach 190
Boiled Beans 190
Mashed Potatoes 190
Queen's Toast 190

Brown Giblet Soup 190
Minced Turkey and Eggs 190
Stewed Potatoes 190
Celery 190
A Plain Rice Pudding 190

LIVE STOCK.
Cotton-Seed Meal as Stock Feed 190
Dry Food for Hogs 190
Lincoln Sheep 190
Pasturing and Soiling Hogs 190
Growth of Colts 190
Sheep 191
Cattle Range of Wyoming 191
Training Horns 191

POULTRY.
Poultry Nonsense 191
Poultry 191
Women as Poultry Raisers 191
To Fatten Fowls or Chickens in Four or Five
  days 191
Winter Rations for Hens 191
Pekin Ducks 192
Literary and Personal 192

We continue to act as solicitors for patents, caveats, trade-marks, copyrights, etc., for the United States, and to obtain patents in Canada, England, France, Germany, and all other countries. **Thirty-six years'** practice. No charge for examination of models or drawings. Advice by mail free.

Patents obtained through us are noticed in the **SCIENTIFIC AMERICAN**, which has the largest circulation, and is the most influential newspaper of its kind published in the world. The advantages of such a notice every patentee understands.

This large and splendidly illustrated newspaper is published **WEEKLY** at $3.20 a year, and is admitted to be the best paper devoted to science, mechanics, inventions, engineering works, and other departments of industrial progress, published in any country. Single copies by mail, 10 cents. Sold by all newsdealers.

Address, Munn & Co., publishers of Scientific American, 261 Broadway, New York.
Handbook about patents mailed free.

Queen OF THE South.
PORTABLE
FARM MILLS
For Stock Feed or Meal for Family use.
10,000 IN USE.
Write for Pamphlet.
Simpson & Gault M'f'g Co.
Successors to STRAUB MILL CO.
CINCINNATI, O.

nov:3t

WEBSTER'S UNABRIDGED.
In Sheep, Russia and Turkey Bindings.

"A LIBRARY IN ITSELF."

GET
THE
BEST

the latest edition with 118,000 Words, (3000 more than any other English Dictionary,) Biographical Dictionary which contains gives brief histories concerning 9700 noted persons, in Illustrations—3000 in number, (a least three times as many as found in any other Dict'y.)

HOLIDAY GIFT.
Most acceptable to Pastor, Parent, Teacher, Child, Friend; for Holiday, Birthday, Wedding, or any other occasion.

It is the best practical English Dictionary extant.—*London Quarterly Review.*
It is an ever-present and reliable school master to the whole family.—*N.Y. Herald.*
G. & C. MERRIAM & CO. Pub'rs, Springfield, Mass.

LANDRETH'S FIELD SEEDS.
LANDRETH'S FLOWER SEEDS.
Agricultural Implements in great variety.
Horticultural Tools in great variety.
Requisites for Garden and Green House.
Red and White Clover, Alsike Clover, Lucerne, Blue Grass, Orchard Grass, Herds Grass, Perennial Rye Grass.
Mixed Lawn Grass Seed, very finest quality.
Pure Food for House Plants.
Bone Meal of the Purest quality.
Peruvian Guano, Land Plaster.
Farm Salt, Flaxseed Meal.
Catholic Soaps, Paris Green.
London Purple, Paris Purple.
Insect Powder, Tobacco Dust.
ILLUSTRATED CATALOGUES FREE. PRICES LOW. CAREFUL ATTENTION GUARANTEED.
Founded 1784. 1500 acres under cultivation growing Landreth's Garden Seeds.

D. LANDRETH & SONS,
Nos. 21 and 23 South Sixth Street,
BETWEEN MARKET AND CHESTNUT STS.,
AND S. W. CORNER DELAWARE AVENUE AND ARCH ST.,
oct-6m   PHILADELPHIA.

Garmore's Artificial Ear Drum.
HEAR
YE
DEAF
As received and worn by him perfectly restoring the hearing. Entirely deaf for thirty years, he hears with them ever whispers, distinctly. Are not observable, and remain in position without aid. Descriptive Circular Free. CAUTION: Do not be deceived by bogus ear drums. Mine is the only successful artificial ear Drum manufactured.

JOHN GARMORE,
Fifth & Race Sts., Cincinnati, O.

oct-3m

THE LANCASTER FARMER.

PENNSYLVANIA RAILROAD SCHEDULE.
Trains leave the Depot in this city, as follows:

WESTWARD.	Leave Lancaster.	Arrive Harrisburg.
Pacific Express*	2:40 a. m.	4:05 a. m.
Way Passenger	5:00 a. m.	7:50 a. m.
Niagara Express	11:00 a. m.	1:20 a. m.
Hanover Accommodation	11:45 p. m.	Col. 11:40 a. m.
Mail train via Mt. Joy	10:50 a. m.	12:40 p. m.
No. 2 via Columbia	11:25 a. m.	12:55 p. m.
Sunday Mail	10:50 a. m.	12:40 p. m.
Fast Line*	2:30 p. m.	3:25 p. m.
Frederick Accommodation	2:35 p. m.	Col. 2:45 p. m.
Harrisburg Accom.	5:45 p. m.	7:40 p. m.
Columbia Accommodation	7:30 p. m.	Col. 8:20 p. m.
Harrisburg Express	7:30 p. m.	8:40 p. m.
Pittsburg Express	8:50 p. m.	10:10 p. m.
Cincinnati Express*	11:50 p. m.	12:45 a. m.
EASTWARD.	Lancaster.	Philadelphia
Cincinnati Express	2:55 a. m.	3:00 a. m.
Fast Line*	5:04 a. m.	7:40 a. m.
Harrisburg Express	8:05 a. m.	10:00 a. m.
Columbia Accommodation	9:10 p. m.	12:30 p. m.
Pacific Express*	2:40 p. m.	3:40 p. m.
Sunday Mail	2:00 p. m.	5:00 p. m.
Johnstown Express	3:05 p. m.	5:30 p. m.
Day Express*	5:15 p. m.	7:20 p. m.
Harrisburg Accom.	6:25 p. m.	9:30 p. m.

The Hanover Accommodation, west, connects at Lancaster with Niagara Express, west, at 9:35 a. m., and will run through to Hanover.
The Frederick Accommodation, west, connects at Lancaster with Fast Line, west, at 2:10 p. m., and runs to Frederick.
The Pacific Express, east, on Sunday, when flagged, will stop at Middletown, Elizabethtown, Mount Joy and Landisville.
*The only trains which run daily.
Runs daily, except Monday.

NORBECK & MILEY,

PRACTICAL
Carriage Builders,
COX & CO'S OLD STAND,

Corner of Duke and Vine Streets,
LANCASTER, PA.

THE LATEST IMPROVED
SIDE-BAR BUGGIES,
PHÆTONS,
Carriages, Etc.

THE LARGEST ASSORTMENT IN THE CITY.
Prices to Suit the Times.
REPAIRING promptly attended to. All work guaranteed.
79-2-

S. B. COX,
Manufacturer of
Carriages, Buggies, Phaetons, etc.
CHURCH ST., NEAR DUKE, LANCASTER, PA.
Large Stock of New and Second-hand Work on hand very cheap. Carriages Made to Order. Work Warranted for one year.

EDW. J. ZAHM,
DEALER IN
AMERICAN AND FOREIGN
WATCHES,
SOLID SILVER & SILVER PLATED WARE,
CLOCKS.
JEWELRY & TABLE CUTLERY.
Sole Agent for the Arundel Tinted
SPECTACLES.
Repairing strictly attended to.
ZAHM'S CORNER.
North Queen-st., and Centre Square, Lancaster, Pa.
79-1-12

E. F. BOWMAN,
Watches & Clocks
AT LOWEST POSSIBLE PRICES.
Fully guaranteed.
No. 106 EAST KING STREET,
79-1-12   Opposite Leopard Hotel.

ESTABLISHED 1882.

G. SENER & SONS,
Manufacturers and dealers in all kinds of rough and finished
LUMBER,
The best Sawed SHINGLES in the country. Also Sash, Doors, Blinds, Mouldings, &c.
PATENT O. G. WEATHERBOARDING
and PATENT BLINDS, which are far superior to any other. Also best COAL constantly on hand.
OFFICE AND YARD:
Northeast Corner of Prince and Walnut-sts.,
LANCASTER, PA.
79-1-12

PRACTICAL ESSAYS ON ENTOMOLOGY,
Embracing the history and habits of
NOXIOUS AND INNOXIOUS
INSECTS,
and the best remedies for their expulsion or extermination.
By S. S. RATHVON, Ph. D.
LANCASTER, PA.
This work will be Highly Illustrated, and will be put in press (as soon after a sufficient number of subscribers can be obtained to cover the cost) as the work can possibly be accomplished.
79-2-

$5 to $20 per day at home. Samples worth $5 free! Address STINSON & Co., Portland, Maine.

SEND FOR
SPECIAL PRICES
On Concord Grapevines, Transplanted Evergreens, Tulip, Poplar, Linden Maple, etc. Tree seedlings and Trees for timber plantations by the 100,000
J. JENKINS' NURSERY,
3-2-79   WINONA, COLUMBIANA CO., OHIO.

$1000 Reward For any machine hulling as much clover seed in 1 day as the **VICTOR** (Double Huller). It has hulled **150 Bushels in ONE DAY.** Illustrated Pamphlet mailed free. Newark Machine Co., Newark, O. Formerly the Hagerstown Agl. Imp. Mfg. Co., Hagerstown, Md.
july-3mj

THE
LANCASTER EXAMINER
OFFICE
No. 9 North Queen Street,
LANCASTER, PA.

THE OLDEST AND BEST.
THE WEEKLY
LANCASTER EXAMINER
One of the largest Weekly Papers in the State.
Published Every Wednesday Morning,

Is an old, well-established newspaper, and contains just the news desirable to make it an interesting and valuable Family Newspaper. The postage to subscribers residing outside of Lancaster county is paid by the publisher. Send for a specimen copy.

SUBSCRIPTION:
Two Dollars per Annum.

THE DAILY
LANCASTER EXAMINER
The Largest Daily Paper in the county.
Published Daily Except Sunday.

The daily is published every evening during the week. It is delivered in the City and to surrounding Towns accessible by railroad and daily stage lines, for 10 cents a week.
Mail Subscription, free of postage—One month, 50 cents; one year, $5.00.

JOHN A. HIESTAND, Proprietor,
No. 9 North Queen St.,
LANCASTER, PA.

The Lancaster Farmer.

EDITORIAL.

VOLUME FOURTEEN

Of the LANCASTER FARMER is completed by the issue of this number. We have endured longer than any of the French Governments since the Revolution of '93, although we may not have made as great a noise in the world as the worst among them. We believe, too, that our mission has been and will be of as permanent benefit to mankind as the mission of any of the heads of the governments aforesaid, or, perhaps any other merely nominal government. We believe this, because our lot has been cast in the peaceful walks of life, and we have not come to destroy, but to the end that men may enjoy physical life, and enjoy it more *abundantly;* for we know that men cannot be spiritually comfortable so long as they are not physically so. *Want,* stern want, whether natural or morbid, is the parent of many misdemeanors and crimes; and we have for fourteen years been gathering and distributing such items of information as are calculated, if appropriated, to elevate men above the plane of want and crime.

We have no special promises to make here for the future, nor any reproaches to offer for the past. We can give no other guarantee of what we *may* be than that which we *have* been : but at the same time, whilst manifesting our thanks to our old patrons, we would admonish them to reflect whether each one of them ought not feel it his duty to add at least one additional subscriber to the list of the FARMER.

In conclusion, when this meets your eye, you will doubtless be absorbed in thoughts pertaining to the approaching *Christmas holidays.* Hence, we would admonish all to rationally enjoy themselves, but not to forget worthy objects apart from themselves; for there are situations and circumstances under which what you freely *give* is really the only thing you truly *have* and never lose. In this spirit we commend our patrons to the festal customs of the season.

MYRIAPODA.

(Acetabula.)

The specimens of "centipedes" before us suggest some remarks upon the different orders belonging to the class *Myriapoda,* and the very marked distinction in their appearance, their economics, and their characters.

The term, *Myriapoda,* is a compound of two Greek words, namely, *murias,* ten thousand; and *pous,* foot. Of course, no subject of this class has ten thousand feet, although there are some species that have one or more hundreds. Formerly they were classed with insects, but have now been erected into a distinct class, divided conspicuously into three orders. The common names of "centipedes" and "millipedes" have been rather indiscriminately applied to them; but for the sake of simplifying their study, I would suggest that these names be applied to the two most prominent orders that distinguish the class. These animals differ from insects, in that they are excluded from the egg with two, three, or four pairs of feet, or come forth without any feet at all, and, as they are developed by age, the number of segmental rings and feet increases, sometimes running beyond hundreds. Indeed, it requires two years, according to authors, to complete their development, so far as to enable them to continue their species.

The 1st order, CHILOPODA, which is from two Greek words meaning *lip* and *foot,* because the anterior pair of feet approximate to *papi,* and perform the functions of those organs, consists of four families, namely, *Cermatiidæ, Lithobiidæ, Scolopendridæ,* and *Geophilidæ.* The subject before us is *Scolopendra heros,* and may be regarded as the type of that family. The term "centipede," I think, should be restricted to this order, not because the individuals belonging to it possess one hundred feet, any more than millipedes possess a thousand feet, or one of our city squares contains one hundred houses. But the distinction in their habits, their forms, and in their organs of locomotion are so great that, in common parlance as well as in scientific nomenclature, there is room for a different appellation.

In this order,—namely, centipede—the head and the body are depressed or flattened, and there is but one pair of feet attached to each segment. The *Antennæ* are long and in some instances twenty-five or thirty jointed; the feet are five jointed, terminating in a sharp heel spine, and their cursorial powers are extraordinary, if they choose to exercise them, which they generally do when exposed to the light, and the temperature is warm. When interrupted, they instinctively hide themselves, seemingly annoyed by light, and preferring darkness.

They are carnivorous in their gastronomical habits, and I have on several occasions observed them with wood-boring larva in their possession, which they refused to relinquish, even when captured themselves. There is a specimen in our collection (or was in it), which refused to release his captive when immersed in alcohol, and in its death only grasped its prey the closer. I do not *know* that they make burrows for themselves in decayed wood, but I know that they are frequently found in the burrows of wood-boring insects of the smaller species, especially those that make excavations immediately under the bark. The larvæ of small *Elateridæ* and of *Breathus* I have seen in their jaws. The one that I "bottled" had a small specimen of the larva of *Passalus cornutus* in his jaws. These animals, therefore, may be classed among the beneficial kinds, especially in relation to forest trees, and the destructive insects that bore into them.

But lest too much credit may attach to them on account of their antagonism to the insect world, I must here state that I am experimentally cognizant of another fact in regard to them, and that is, that they are really poisonous. They may not be poisonous to all persons and under all circumstances, but on one occasion they were very poisonous to me. In one of my excursions with our late Secretary, Mr. Stauffer, to Manheim township, I captured a specimen of *Scolopendra* about two or two and a half inches in length, which immediately inflicted a wound on the middle finger of my left hand, from which I had some difficulty in releasing it without decapitation. When it withdrew its mandibles two small drops of blood followed. The pain at first was pungent, but I sucked the wound and endeavored to forget it, but " it would not down." I had nothing to apply but alcohol. This only gave a momentary relief. The pain continued up the finger to the wrist, from thence to the elbow, and from thence nearly to the shoulder, and continued half a day. Crossing a small rivulet, I alighted and applied a clay poultice, which afforded relief. After the first twenty minutes the pain was no longer acute, but a continuous, dull, aching sensation, seemingly affecting the muscles and the nerves. After I reached home I applied ammonia, and in half an hour I felt little pain, but the hand was somewhat benumbed, and the following morning this feeling was also removed, but a hard tubercle remained for a fortnight. I have often reflected that if a small, scarcely more than two-inch centipede can inflict so much pain, what might be expected from one that measures ten or fifteen inches, under similar circumstances.

The second order includes the DIPLOPODA, and means twofold in allusion to the double pair of feet on each of the segmental divisions of the body, and these animals may be embraced under the common name of "Millepedes," some species of which have over two hundred feet, although none that have a thousand. This order includes the single family JULIDÆ, composed of the genera *Spirobolus* and *Julus. Spirobolus marginatus* occurs in Lancaster county, and is our largest species, measuring from two to three inches in length. I, on several occasions, detected this species feeding on a fungus belonging to the genus *Agaricus,* and from the fact that smaller species feed upon turnips, radishes, cabbages, strawberries, potatoes, and other vegetables and fruits, we may infer that they all prey upon vegetation. The bodies of these are tubular or cylindrical, the antennæ short and bent, the feet short, and two pairs attached to each segment immediately in the centre of the body beneath. Their locomotion is very slow, and when disturbed they make no attempt to escape, but merely coil themselves up spirally and turn over on their sides. Their pedal members, their locomotion, their gastronomical habits, and the conformation of their bodies, are entirely distinct from the other, although they and the centipedes form one class, and are similar in organic structure and development. Between the CHILOPODS and the DIPLOPODS, however, according to some systematists, is a sub-order named CHILOGNATHA, a name derived from the Greek words signifying *lip* and *jaw*. This order is composed of the families LYSIOPTALIDÆ and the POLYDESMIDÆ. They have the flattened bodies of the Chilopoda, but the antennal and pedal arrangements of the Diplopoda. Their locomotion is more rapid than the latter, but not nearly so rapid as the former. The members of this sub-order are presumably also vegetarians, and at least one species of the genus *Polydesmus* is known to have been exceedingly destructive to the young tobacco plants, in the bed, early in the spring. They have also been detected preying on other species of vegetation, and especially on small fruits, when near, or in contact with the earth. Their bodies are shorter and proportionately broader than either of the two orders, to which they are mutually related, and their legs are articulated at the sides. Of course, in the present advanced state of science, there are many minor divisions, not essential to the objects of this brief paper.

Belonging to the first family of the first order—namely, CERMATIDÆ, is a species that is very abundant in the city of Lancaster, but I have no recollection of ever having observed it in any other locality in the county. This is *Cermatia forceps*, or a species very nearly related to it, and is the typical genus of the family to which it belongs. This animal is endowed with the most remarkable cursorial powers, and unlike most of the chilopods, its *locale* is not merely or mainly in moist places, nor yet only in dark places. I have observed it not only at nearly all hours of the day, but also at nearly all hours of the night, and under a brilliant gas light as well as in dark corners. Intensely shy as it is, it frequently emerges from a cover of pamphlets and papers on my desk, and occupies a prominent position within twelve or fifteen inches of the hand with which I am writing, manipulating its long filiform antennæ, seemingly canvassing the area around it, in search of prey, or guarding against possible danger. The body of this animal, in proportion to other subjects of the order, may be called short, and the feet and the antennæ very long and slender. I do not think that I have noticed one more than two inches in length, and yet with the anterior and posterior feet extended, some individuals seem to be three or four inches in length. When quite young and small they are nearly colorless, but the adults are dorsally tinged with blue or green, and whitish beneath, the feet and antennæ being also whitish.

When Mr. James Thackara was express agent, he witnessed a deadly conflict between a large specimen of *Cermatia* and a *Blatta*, or cockroach, which ended in the death of the latter. Indeed, the former appeared to be reconnoitering and deftly approaching the latter, whilst the roach made no attempt at offense or defence, being solely occupied in avoiding a rear or lateral attack. The roach raised itself up on its feet as high as it possibly could, its body turning around as if on a pivot, with its head towards its enemy. Finally, in an unguarded moment, *Cermatia* sprang upon it, and in another moment the roach was on his back, with no signs of life, save in his trembling limbs. Then, at Mr. Thackara's approach, *cermatia* fled.

As a general rule, articulated animals possessing swift running powers, are predaceous in their habits, and carnivorous in their appetites. It is true, some carnivorous insects are slow in their locomotion, and lie in wait for their prey—like the MANTIDÆ and REDUVIDÆ—but these are generally provided with a largely developed pair of anterior feet, and are raptorial in their habits; but as a general thing, swiftness is not meaningless, but is to aid the possessor of it in capturing its prey. On the other hand, all, or nearly all, slow moving insects are noxious, and feed on plant food, or bore into living or decaying wood; and this is especially the case with those that are mandibulated, that is, possessing jaws. Some of the haustellated species—those that pierce vegetation, and suck out its sap—are very nimble and quick in flight, but the larger number of even these are merely a sort of "dodgers," and do not rely on their running speed. Slugs, grubs, caterpillars, maggots, mites, worms, etc., etc., are slow in their locomotion, and do not seem to depend upon it to secure their escape from danger. Some of them, seemingly conscious of an unfriendly presence, will relax their hold, fall to the earth and hide therein; but the larger number manifest no consciousness of present danger, and hence allow themselves to be destroyed with impunity. No animals, however, are more conspicuous for this characteristic trait than the *millepeds*. Occasionally we may find a caterpillar, or an array of them, moving with about as much haste as such animals are capable of making, but it is not because they are afraid of any person or thing, but because they may be in search of food or a proper place to undergo their metamorphoses.

As a general rule the order *Orthoptera* may be excluded from this rule, and yet there at least is one family in that order that is strictly predaceous, but it does not depend upon its cursorial powers, but on its raptorial powers in securing its prey. The genus *Blatta*, or cockroaches, are swift runners, but they are not strictly vegetarian in their feeding habits. Crickets, grasshoppers, and locusts, can get out of the way quick enough to rank with predaceous insects, but they depend more upon their salutatorial than their cursorial powers. If they were predaceous in their habits their leaping powers would be of no assistance in capturing prey, for their leaps are most heedless, and they never seem to know where they are going to alight; it may be against a fence or wall, in a hole, a pond of water, or in a fire. The Cicindelans and Carabidans, among the *coleoptera*, are all predaceous and all swift runners, in that respect, possessing the characteristics of the Centipedes.

Crickets, however, are not purely vegetarian; they have also carnivorous habits. Some years ago, during a few very warm days in the month of November, I found on the lacerated carcass of a calf that appeared to have been recently killed, about one hundred field crickets, in company with a large number of *Necrophore* feeding greedily on the flesh of the calf. So stupidly voracious were they that they made no attempt to escape, but allowed themselves to be captured with seeming indifference. This characteristic in noxious insects is very fortunate to us, but not to them.

THE TARIFF AND FREE TRADE.

It would be a great pity if the party that has seemingly been so largely triumphant throughout the country in the late elections should so far misinterpret its mission as to unduly and mistakenly tamper with the tariff laws when it comes into power. It is true, there may be occasion for a partial revision of these laws, and the very fact that a congressional commission had been previously appointed to elicit testimony on the subject seems to imply that there is room for such revision, but the result of the ballot box is by no means an intelligent expression of the popular sentiment, so far as to commit the country against either unqualified protection or free trade; for the masses of the people in no section of the country have a clear and practical understanding of the questions involved. We have only to follow up the commission in its various sittings at various points, and the nature of the testimony brought before it, to learn that the manufacturing and producing interest of the entire country are largely biased by sentiments of self. That is, each particular interest would have such an adjustment of the tariff laws as would inure to its own pecuniary benefit without any particular regard to the benefit of others. It is a question upon which, as yet, there cannot possibly be either a national or a party issue; and laws, whether friendly or unfriendly, must necessarily benefit or injure both political parties. An amicable adjustment can only be effected through mutual concessions at the present period, and it may take years yet before the country will be ripe for a radical change. We are not yet "out of the woods," contingent upon our late war and our great national debt, and the affairs of the country still require judicious management; there is no necessity in destroying or abolishing our revenue system, although it may need to be purified or purged.

The following from the columns of the *Thoroughbred Stock Journal*, of Philadelphia, Pa., may be added in further illustration of this interesting subject:

"There is no political enigma as puzzling to the brain of the average politician as that of the proper distribution of tariff and free trade, for unless these national economics can be so adjusted as to fit the necessities of the different sections of the country, they fail in their purpose, and the statesman who allows himself to remain in the old ruts of a worn-out policy will find himself so snugly ensconced in his selected groove, that even his best friends cannot extricate him. The tariff question of to-day is not that of a decade or a quarter of a century ago; economical positions have changed. The South, whose slogan in *ante bellum* days was free trade, has reversed her economics. She is to-day in her infancy, perhaps, but on the way to a full maturity of manufacturing development. This of course means the presentation of her claim for protection, and a falling into line with similar interests of the North and East—her former competitors in political ethics, her present rivals in economical practice.

"The South, i. e. the New South, will soon be clamorous for a protective tariff as of yore she was belligerently clamorous for free trade, and knowing the South as well as we do, we can safely venture the opinion that she will occupy no middle ground; that she will not accept any sophism, or be cajoled by the sophistry of a "Revenue Tariff," or a "Tariff for Revenue," so called, because they will accept as readily as do the great manufacturers of the Middle and New England States the theory of proper protection, knowing that a failure to de so places them in a subordinate position to foreign manufacturing interests, no matter how valuable her natural resources may give her advantages not possessed by the North and East; and she well knows that what will place in jeopardy the interests of her neighbors, will assail with as much force the interests involved in the looms and spindles of Georgia, and other manufacturing centres of the Southern States.

"It were well, indeed, that the political economist study these facts, and remember in connection that issues and occasions present themselves from time to time in the national economy that puzzle the most astute economist, and of all the questions, none require the attention of wise, profound, and skillful consideration more than the complex theories of tariff and free trade. Which of these is the more important factor in the economical administration of the government is a mooted question; but that one must yield to the other is as irrepressible a fact as was the same theory of irrepression presented by Mr. Seward on one of the most vital questions ever presented to the thought and action of the country.

"There can be no compromise—ever an evidence of weakness—but one or the other of these issues must submit to the power of the other; which will yield, it is not in the province of this article to determine. We cast the thought upon the waters of public opinion, and will be pleased to answer any and all correspondents who may desire our views as journalists, not as partisans."

THE TURKEY.
(*Melcagris gallo-pavo.*—Linn.)

"Man, cursed man, on turkeys preys,
And Christmas shortens all our days.
Sometimes with oysters we combine;
Sometimes assist the sav'ry chine,
From the low peasant to the lord,
The turky smokes on ev'ry board."

At one time it was thought, in England at least, that the turkey had its origin in the country called Turkey—the land of the Turks—but this was a grave mistake, the *naturalists* of England knew better. In a "Perfect description of Virginia," written two hundred and forty years ago, it is recorded that the colonists had "wilde turkies" weighing sixty pounds," and Ray refers to America as the origin of the species from which we derive our domestic bird. It is easy enough to perceive how the name *Turkey* should have been applied to the land of the Turks by "outside barbarians," but it is not so easy to perceive why, or how, it was first applied to an American fowl.

The *Melcagris* of the ancients was not a turkey at all; it was a "Guinea fowl." Linneus, however, has given this as the generic name of our turkey, a bird which was altogether unknown to the ancients, and ornithologists have continued it down to the present time. But this is now of little consequence; for it has been proven, and is generally conceded, that the Europeans only became acquainted with this bird after the discovery of America, and from which it has been spread, in a domesticated state, over the greater part of the civilized world. The wild turkey at one period had a much wider geographical range than it has now, extending from the northwestern states down to the Isthmus of Darien, but civilization, public improvements, and general progress are fast circumscribing that range, and probably the present rising generation may see its entire extinction as a wild bird. Moreover, they do not seem to increase as rapidly in a wild state as they do in a domestic state. The adult males are very hostile toward the young, and kill them whenever they can get an opportunity, and that opportunity is more frequent in a wild state than it is under human intervention in a domestic state.

In the wild bird there is a general uniformity of coloration, but in the domestic bird, there is great variation, from pure white to almost pure black, including almost as many varieties as there are in the genus *Gallus*, which includes our common "chickens." Attempts have been made to demonstrate that there are two distinct species among the wild birds, but it never has had a universal following. It is supposed the Mexican and farther southern bird is specifically different from that which inhabits the United States. It had also been alleged that the tailfeathers of the Mexican bird were not tipped with white, or whitish, presenting the light-colored margin when the tail is expanded; but that distinction can certainly mean nothing. In one of our country strolls about the first of October, we came upon a family of turkeys containing an adult "gobbler," two adult hens, and nine "adolescents" almost as large as the hens aforesaid. Among these were two that lacked the white tips of the tailfeathers; three were entirely white; one was buff, and the remaining three, dark colored with the white tips very conspicuous. The adult gobbler was dark—almost black—and bronzed, and the females were brownish. These were probably the progeny of the two hens, or may have been a single family, and we have only introduced the phenomena here to illustrate the tendencies in the bird to vary from the wild type, in its plumage.

Of course, these different varieties have their different designations, but still they all belong to the Linnæan genus *Melcagris*—"White Holland," "Bronze," "Buff," "Brown," "Mottled," "Grey," etc., etc.

The "Honduras Turkey" (*Melcagris ocellata*) is nearly the size of the common turkey, and is supposed—as the name implies—to be a distinct species. The distinction is based upon the less developed tail, and the fact that the bird has never been known to spread it. The "Brush Turkey" is an Australian spe-

cies. It is the *Tulegalla lathami* of Gould, and inhabits various districts in New South Wales, where it is found in large flocks. It is, however, not a true *Melcagris*, and hence was not derived from our aboriginal stock. As before stated, the "Pintado" or Guinea Fowl, was the *Melcagris* of the ancients, but is now referred to the genus *Numida*, *Melcagris* being retained as a specific name. Both the turkey and the guinea-fowl are related to the "Peacock," and this is recognized not only in physical characteristics, but also in nomenclature. The naked heads, the horizontal necks, and the convexed backs of the turkey and the guinea-fowl show a relation. The characteristic strut, and the erection and expansion of the tail show a relation between the turkey and the peacock. But the relation is still more conspicuous in their names. The generic and specific *Melcagris* of the turkey and the guinea-fowl is a recognition of their affinities.

The Peacock is technically *Pavo cristatus*—the crested *pavo*—and the turkey is specifically *gallo pavo*, which is about equivalent to "chicken-peacock," a name that not only implies that the turkey and the peacock are related, but also that both are related to *Gallus*, or the common fowl. Systematically considered, they all belong to the PHASIANIDÆ, or pheasant family.

Perhaps no other bird in the civilized world has attained to a greater popularity as a "table bird," than the turkey, and none suffer a greater victimization about Thanksgiving and Christmas festivals; and the abrogation of these birds on these occasions would almost be an abrogation of those festivals themselves. If the poor man can only afford to partake of turkey once or twice in a whole year, it will be on one or both of those festivals, in nine cases out of every ten. The turkey has, therefore, a most fearful gauntlet to run in its mission through civilization.

To visit the poultry markets of any of our great cities during the week preceding either of the festivals named, a most formidable scene would be revealed, and the novice not only would be wrapped in wonder as to where all these fowls come from, but also as to where they all go to. True, there are also ducks, geese, and the common fowl in goodly numbers, but none of them garnish the festive board to the extent the turkey does on those annual occasions, and the paraphrastic bard may well have written:

"Who would be a turkey hen,
Fed and fattened in a pen,
Killed and eat by hungry men,
Upon a Christmas-day."

KITCHEN GARDEN FOR DECEMBER.

The care of hot-beds, etc., is nearly all that demands attention; true, other things may be done, but quite as well at a future day, unless the season is over. The annexed hints may, however, prove useful : Compost prepare; dung prepare for hot-beds; hot-beds attend to; radishes and salad sow in frames; trench and drain vacant ground; transplanting trees may still be done.—*Landreth's Rural Register*.

As long as we can recollect anything about garden seeds, except those raised, gathered, sewed up in little bags, and stowed away for future use by our mother, the name of Landreth has been associated in our memory with this business. About sixty years ago a party visiting Philadelphia (on foot) brought home some seeds called by them "Chocolate corn," or "Chinese Chocolate" used as a substitute for coffee, and we have a faint impression that these seeds were obtained from Landreth. The country had not yet recovered from the great financial crisis of 1817, and coffee was so dear that most people were compelled to use "browned rye" instead of coffee. Our mother obtained some of these seeds and cultivated the corn to fruition. It grew something like "broom corn," but the head was compact and so heavy that it bent over and hung with the apex downward. When fully ripe the seeds were large, and had a purple color, and the leaves and stalk were streaked with purple. It was roasted the same as coffee, and to our juvenile taste it was as good, and tasted like chocolate—at any rate, far superior to rye coffee. After coffee "came down" in price, chocolate corn "went under." For fifty years we heard nothing more of it, but within the last five years we saw several varieties of it sent in from California, one of which strongly resembled it. It belongs to the Sorghum family.

When the Landreth seed farm was first established, there were "only thirteen sparsely populated States" in the Union (that was about 1789—93 years ago). An establishment which has sustained itself so long with constantly increasing facilities and reputation, must surely be worthy the patronage of the country; and we can freely allow them the privilege of "blowing their own trumpet," without subjecting them to the charge of egotism, or self-laudation.

EXCERPTS.

HEALTH HINTS.—Try popcorn for nausea.
Try cranberries for malaria.
Try a sun-bath for rheumatism.
Try ginger ale for stomach cramps.
Try clam broth for a weak stomach.
Try cranberry poultice for erysipelas.
Try eating fresh radishes and yellow turnips for gravel.
Try swallowing saliva when troubled with sour stomach.
Try a wet towel to the back of the neck when sleepless.
Try buttermilk for removal of freckles, tan and butternut stains.
Try eating onions and horseradish to relieve dropsical swellings.
Try to cultivate an equable temper and don't borrow trouble ahead.
Try taking your codliver oil in tomato catsup, if you want to make it palatable.
Try breathing the fumes of turpentine or carbolic acid to relieve whooping cough.
Try taking a nap in the afternoon if you are going to be out late in the evening.
Try a cloth wrung out from cold water put about the neck at night for sore throat.—*Dr. Foote.*

DANIEL MURPHY, the noted pioneer who went to California in 1844, died recently at San Jose. He was the owner of immense herds of cattle and thousands of acres of land. He owned 200,000 acres in Nevada, some 6,000,000 acres in Mexico and large tracts of land in Arizona.

MUCH of the sugar sold in English markets is from the beet. It is not an uncommon event in Europe to gain a yield of twelve tons of beets from an acre of ground, and from twelve tons of beets about one and a fifth tons of sugar is extracted.

THE French Minister of Agriculture has placed at M. Pasteur's disposal a further sum of $10,000 to enable him to continue his investigations into the nature, cause, and prevention of contagious diseases among animals.

IN England and Scotland where there are many steam ploughs at work, the most popular sorts are those drawn by stationary engines at each side of the field.

EXTENSIVE lumber fires are becoming alarmingly numerous of late, and a large amount of lumber has been destroyed in this manner.

IT is estimated that the California fruit and vegetable pack this year will amount to about 20 per cent. more than that of 1881.

PROF. BEAL says that all our species of bats are not only harmless, but positively useful, as they are great insect destroyers.

SHEEP-GROWERS of Los Angeles county, Cal., report heavy losses from a poisonous weed on which the sheep fed.

THE quality of the corn crop throughout the South is superior, and most of the Southern States report large yields.

A MISSOURI sheep-grower, after some years of experience, advises breeding from polled rams.

DURING the past year agricultural implements to the amount of $08,000,000 have been made in this country.

WE are informed by old farmers, and they are not far from correct, that next year's wheat crop will be more than double that of any previous year in Umatilla county. The increase of acreage is astonishing, and the amount of land that was summer-fallowed is immense.—*Pendleton, Oregon, Tribune.*

THE oxygen of the air aids and facilitates the germination of seeds, and seeds buried so deeply in the ground as to be out of reach of the atmospheric air will exhibit no signs of life.

ANIMALS when first confined, and supplied with fattening food, always increase largely in weight during the first few weeks, after which the rate of increase diminishes to a considerable extent.

TEXAS has five million head of horned cattle and a superabundance of mast and corn, and thousands thoroughly educated men and women, yet she imports butter, lard and school teachers from Kansas City.

D. BRIGGS, of Davisville, Tolo Co., Cal., has a plantation of 460 acres of graperies from four to eight years, on which he has raised forty-six car loads of raisins, most of which were sent East.

IF those farmers whose farms are soils underlaid with clay would sell one-half of their land and put the proceeds into the judicious tile drainage of the rest, they would make more money from the one-half of the farm under improvement than they now do from the whole area.

JUTE SEED.—The Florida *Times* says that about a year ago Mr. Hamilton Disston sent to India for a supply of jute seed, but the difficulties attending the export of seed prevented the obtaining of a larger quantity than 1,200 pounds. This amount of seed was distributed throughout the State by Mr. Disston, with the offer of liberal premiums for the best exhibit of prepared jute. The competition under the terms of this offer will shortly take place at Jacksonville, and the *Times* says that the planters of the seed promise some choice samples, that will no doubt attract sufficient attention to this industry to warrant the business being taken hold of by capitalists on a large scale.

A QUEER INDUSTRY.—One of the queer industries of New York, says the *United States Miller*, is gathering the stale bread from large hotels and restaurants, and grinding it up into food for poultry and pigs. The Astor House sells its stale bread for $800 annually. The contractor has $100,000 invested in the business, and keeps nine teams at work. We are not posted on the system of reduction employed, whether stones or rolls. Certainly a purifier would be essential.

OUR VARIED INDUSTRIES.—According to the census report there were in the United States, in 1880, 2,486 wool establishments, employing 161,480 hands, and bringing out annually products to the value of $267,182,914; 1,005 cotton establishments, employing 185,-472 hands, and turning out products of the value of $210,950,383; 1,005 iron and steel establishments, employing 140,978 hands, and turning out products of the value of $296,-557,085.

THE honey market is assuming greater importance every year. Now that the foreign trade is clearly established, the demand is almost unlimited, and no fears are entertained of glutting the markets. At present the home markets are fully supplied, but the foreign demand will soon reduce them and increase prices.

NOW is a good time to lay in a stock of vegetables to feed fowls during the winter months. Such food promotes their health, and will induce hens to lay much earlier in the spring than when grain is their only food. Cabbages, turnips, onions, and such vegetables, which need not be of the best quality, are the best for this purpose. Do not forget that bones are of great value to fowls, especially if the poultry is kept closely confined.

THE cellar for roots and apples should be kept cool and rather close and damp to prevent wilting; the temperature should be as near freezing as may be without actual frost, and in warm weather the cellar should be kept close to prevent it from getting too warm. These conditions are more easily obtained in a cellar under the barn or carriage-house than under the dwelling, and moreover the disagreeable, not to say dangerous, smells arising from neglect of the vegetable cellar in spring, and summer point out some other spot as more proper place than our dwelling-house cellar.—*Concord (New Hampshire) Patriot.*

THE CANNED FRUITS.—The canned fruit product of California has largely increased within the last decade. The product of 1875 aggregated in value about $500,000. In 1878 it had reached $1,250,000. In 1880 $1,500,-

000, and in 1882 the product is set down with a value equal to $2,600,000. For the future we have every reason to believe that the rate of increase will be even greater than for the past. And there is no question but that California is destined to become the largest and finest fruit-producing country in the world.--*San Francisco, California, Patriot.*

FROZEN CANE.—" My cane that was frozen all winter I worked up until it got mouldy. It did not sour for two weeks after the frost left it, but it began to have a heavy coat of mould within forty-eight hours after the frost was all cut. Some that was cut from the field a day or two before, froze up and was in good shape when it did freeze up, soured in a few days after the frost left it, while that which was in piles in the yard for three weeks did not sour at all. I think that its drying out kept it from souring, for when I worked it this spring it tested 13¼ B, while the same cane tested 11 to 12 B last fall. The syrup is of a darker color than it was last fall. It made fair syrup, however. I worked part with lime and part without. I liked that the lime was used in best."—*J. A. Jones, Fillmore county, Neb.*

WIRE WORMS.—"I planted my corn in the usual way, but in one row I put wood ashes, in a second sand plaster, and in the third common salt. There was little difference, if any, in the first two and those I had done nothing with, but where I put the salt about nine-tenths of the corn came up well. So I concluded to doctor the whole piece, and sowed a good coat of salt and stirred the ground well, and then sowed another coat not quite so heavy. I then marked out and planted my corn, and it gave me a good stand, something that I never had on that piece before. I made an average yield of corn, and put an end to the wire worms."— *L., Miami county, Ohio.*

A SMALL bantam and a big black hen both began laying eggs together in the same nest. When a sufficient number had accumulated the bantam was given sole possession and began to set. This lasted for over a week, when she came off and the brown hen took her turn, staying until the eggs hatched, the result being only three chickens. Since that time the partnership has been continued, both doing equal duty in obtaining food for the chickens, which can be seen running first to the one and then to the other, according to their success in scratching. At night all the little ones sometimes stay with one mother, sometimes with the other, and sometimes they are divided. This story is told by the Hartford *Evening Mail.*

ONE of the largest silos in Europe is in France on the property of M. Vicompte Arthur de Cheselles, in the Department of the Oise. In this is deposited the produce of 170 acres. The silo is described as an oblong shed, roofed with tiles 72 yards in length, 6¼ yards wide and 4½ yards high, forming an admirable Dutch barn, under which a great portion of the cereal produce of the farm is stored at time of harvest. The floor, instead of being level with the ground, is sunk about twelve feet and is paved and drained. In this great pit is stored the ensilage.

DR. MITTENDORF read a paper on near-sightedness before the New York County Medical Society a few evenings ago, in the course of which he described the singular case of a fine horse in Berlin that had become intractable and which proved on examination to be suffering from myopia. The owner had a pair of glasses for the animal, and on putting them on it became as tractable as ever.

THE cultivation of bamboo in the Southern States is being advised. It is believed that it will thrive well in marshy regions such as fringe the South Atlantic and Gulf States.

A FRENCH farmer writes that he has run out couch-grass on his farm by the cultivation of buckwheat.

UPLAND RICE.—" First, it requires close land to hold the moisture—black gum or gaulberry lands preferable. Perfect drainage is necessary. I have planted both, and prefer the gum to sandy lands. With 75 pounds superphosphate to the acre I made an increase of 14 bushels per acre on a field of twenty acres over four acres that had none on it. The unfertilized land made 18 bushels and the fertilized 32 bushels per acre. One hundred pounds is the outside limit, in my opinion, for rice; any more is injurious—makes too much straw."—*W. A. Jones, Liberty County, Ga.*

GAS TAR ON TIN ROOFS.—"My experience is that while gas tar is a first-rate sort of paint for many things its place is not on a tin roof. I painted over a tin roof with it. The sun and rain together gradually cracked the paint, rendering it worse than worthless, for it affected the rain water that flowed from the roof into the cistern. I also thought it had the effect of attracting the sun more than a light-colored paint would have done, for the tin curled and twisted in places."—*F. M. G., Clark county, Ind.*

NITROGEN ON THE FARM.—" Nitrogen applied on the New Hampshire Agricultural College farm was destructive to corn in successive applications; had but little effect on potatoes, and increased the crop of barley."—*Professor J. W. Sanborn.*

CONTRIBUTIONS.

For the LANCASTER FARMER.

"THE FARMER'S FRIEND."

I have before me a periodical bearing the above title published in Parkesburg, Chester co., Pa., and dated February, 1838. This little Journal, one half the size of the LANCASTER FARMER, but containing less than one-half the reading matter in the Farmer, was "devoted to Horticulture, Agriculture, Botany and Rural Economy." It was edited and published by one Jason M. Mahan, a "Yankee Schoolmaster," who is still remembered by some of the older inhabitants of Salisbury, who attended "Baker's School " nearly have a century ago.

This impecunious pedagogue, who taught school for $18 per month, and for the motto of his paper, "The public good our only aim," More than half the paper is given to the mulberry and sugar beet business. In fact the *Farmers' Friend* lived its brief life during the great Morus Multicaulis boom, and probably gave up the ghost when the mulberry trees were grubbed and piled for bonfires. It is a curious fact that the silk-worm and the sugar beet craze ran their course together, and now, after nearly half a century, are brought prominently before the public again at the same time.

Hear what Editor Mahan says about the silk business: "Having been about ten years engaged in the culture of the mulberry and silk-worm, we flatter ourself that our knowledge of the business is such as will enable us to furnish the necessary information to enable the farmer to raise and prepare silk for market without further knowledge or assistance. * * * * We shall, therefore, sing speed to the plough, wish health and prosperity to the farmer, and rejoice that he is entirely free from such perplexities as disturb the printer."

Of the Morus Multicaulis he says: "Of all the species of mulberry yet introduced into this country, for the purpose of feeding the silk-worm the Morus Multicaulis decidely has the preference, and will speedily be substituted in place of all others in every region of the globe. We would advise all farmers, by all means, to lose no time in supplying themselves with that most invaluable species."

Again he says: "The culture of silk in America succeeds so well in every respect, there is no longer room to doubt of its being eventually very extensively and profitably followed as a pursuit. Heretofore the greater portion of the specimens of this valuable product have been the result of experiments by individuals, on a small scale, but at present larger quantities prepared in factories for sale are beginning to make their appearance in market. The silk thus offered has everything to recommend it in point of lustre, smoothness and strength, and will, it is said, stand a comparison with the Italian. Then why be tributary to foreign nations for this article ?"

Jason Mahan was the author of a work on arithmetic called *Mahan's Instructor*, hence, with his mathematical turn of mind we need not be surprised at the following :

"Mr. I. B. Gray, of Fredericksburg, Va., in April, 1835, at an expense of only $17.50 cost and labor, set out 75 Chinese mulberry trees. In October, 1836, he writes to the editor of the *Silk Culturist* that he had in 18 months multiplied those 75 trees into 5,000 additional trees, and, to crown all, the editor of the *Culturist* asserts those 5,000 trees of the size and height described by Mr. Gray, would be purchased in New England at 50 cents each as soon as offered ! And this enormous profit of $2,500 realized out of an investment of $17.50 in eighteen months required only one-fourth of an acre of ground."

Think of that, ye tobacco growers who "rush into print " with your reports of $500 or $600 per acre for tobacco. Here is a man who makes $10,000 per acre growing mulberry trees! But what about the poor fellows who bought the trees ? and what of the poor fellows who smoke the tobacco ?

In the venerable paper before me there are five articles on the culture of the mulberry and silk-worm and the manufacture of silk, and a long and exhaustive essay on the manufacture of beet-root sugar ; one on the management of horses, and a partial report of the Silver Spring Farmers' Lyceum. This is followed by an article on deep ploughing and a nice essay on "The Importance of Cultivating

Good Fruit." Lastly we have the following on peach trees:

"Mr. William Phillips, of Pennsylvania, has derived great benefit from the application of air slaked, old effete lime to peach trees, the effects of which, according to his own account, are very great. He puts about a peck of lime to each tree ; he thinks it useful as a preservative against the insect so fatal to these trees. We have then two applications recommended, unleached ashes and lime, and from our own experience are able to recommend both. We are not sure which has the preference. The lime and ashes should both be dug up every spring. Washing the trunk with soapsuds will also be serviceable to the tree."—*J. C. L.*

For The Lancaster Farmer.
A SURE PREVENTIVE OF CHICKEN CHOLERA.

Several experiments has been made during the past five years by different parties for the purpose of preventing the spread of chicken cholera by inoculation or vaccination. We have during the past two years vaccinated the fowls in nineteen different yards where the cholera was prevailing badly, and in each yard left some common fowls not vaccinated and they all died, but of the two thousand vaccinated only eleven died, although they were in the same yard with those not vaccinated that were dying daily by the score. We have every reason to believe this chicken vaccination will be as effective in preventing cholera among fowls as vaccination is in preventing smallpox among the human family. Vaccinate a hen and in eight days her system will be thoroughly inoculated; then cut off her head and catch all the blood in some vessel, then pour the blood out on paper to dry; a half drop of this dried blood is sufficient to vaccinate a fowl, and the blood of one hen will vaccinate your whole flock. Catch the fowl you wish to vaccinate, and with a pin or knife make a little scratch on the thigh (just enough to draw blood), then moisten a little piece of the paper with the dried blood on and stick it on the chicken's leg where you scratched it, then let the fowl run and you need have no fear of chicken cholera. As the result of my many experiments I now have enough dried blood to vaccinate, I should suppose, ten thousand fowls, for which I have no use, as I do not sell patent medicines. If any of your readers are enough interested in poultry to try this preventive, by writing to me I will send them them free of any charge enough dried blood to start with; all I ask is that they send immediately, before the blood loses its strength, and report the result of their experiment to your many readers.—*H. H. Griffith, Zanesville, Ohio.*

For The Lancaster Farmer.
THE BALANCE OF TRADE DELUSION.

Editor of The Farmer: I notice, in the last number of your journal, that a correspondent, S. P., of Lincoln, Del., undertakes to discuss the "Balance of Trade" question, and appears particularly desirous of a controversy with me on the subject. According to him my communications to the Farmer have been "false in statistics" and only deserving of "ridicule."

I do not feel under obligation to enter into discussion with one who comes at me in that meat-axe style on his first appearance ; but lest some of your readers might be led to believe from his confident and more or less plausible assertions that they cannot be answered, I will reply to one or two of the most plausible ; but I am not going here to repeat the arguments adduced in my former communications ; they may stand or fall on their own strength or weakness, and in respect to them I will only now say that, in my opinion, they have not been confuted, and cannot be. I have no reason to believe, as alleged, that there was an important error in the figures as I gave them in 1879, though, in adding up the long columns of figures, it is possible that I made a mistake. (I have lost the Report from which they were derived.) If I gave truly the summary of official statistics as furnished by the then Chief of the Bureau of Statistics, even if erroneous, and S. P. gives figures furnished by Mr. Nimmo, or some one else, and the two do not correspond, does that give him a warrant to charge falsification of statistics? Sound logic and good manners alike will answer, No.

But admitting, for argument's sake, that his figures are the right ones, and that our imports were greater than the exports after 1860, as is admitted to be the case in every decade *before* that time, and how does the matter then stand ? It proves that not only part of the time, but all the time, ever since the United States was a nation, we have been importing enormously greater value than we have exported—been thus losing immensely by our foreign trade—been going headlong down the road to commercial ruin uninterruptedly for almost a hundred years ; and most "astounding" fact (to balance-of-trade theorists) we are not yet ruined, but even more wealthy and prosperous than when we set out !

Another fact equally astounding, no doubt, to economists who think it is ruinous to receive more value than we part with, is that England and other European countries of which we have the statistics, show that each one of them imports a great deal more than it exports. And thus it has been going on, decade after decade, each country being impoverished by its foreign commerce in the same way with ours, yet their governments, with a reckless and criminal disregard of their country's welfare, making no effort to prohibit a business so disastrous to their people !

Your Delaware correspondent bases his argument on what he regards as an "axiomatic" proposition, viz : "that all production is gain, and all consumption is loss," from which he argues that an excess of imports over exports must be loss. Elsewhere in his article he argues that it is because so many of the imports are useless luxuries that we are the losers by foreign commerce. It now appears that he regards all imports, except specie perhaps, no matter how useful and valuable, as injurious, and a loss, if in excess of our exports. Thus, if we export $100 worth of corn or tobacco, and get in return $120 worth of cloth or salt, then the balance is against us, and the country loses $20 by the trade, because the cloth and salt are for con-

sumption, "and all consumption is loss." A word about that.

All grain, fruit, &c., is raised for the very purpose of consumption, and in fact is consumed, one way or another. Is it all lost? If consumed by fire or sunk in the sea by shipwreck it is lost undoubtedly ; but if a farmer feeds corn to his cattle and hogs, though the corn is consumed, its value reappears in the form of beef and pork. If he and his family eat it, its value is restored to him and them by life conserved and bodily strength imparted and increased. Is not that as valuable as the money it could be sold for? Is not the very opposite of this alleged "axiomatic" proposition nearer the truth, viz : All the productions of the earth are or will be lost if they are *not* consumed? Were it not for the beneficent effects of their consumption, they would be of no more value than the dirt in the road, and it is only by and through consumption that mankind and all the animal kingdom are kept alive.

' Exports represent consumption, the same as corn fed to the hogs. We get back the value of the exports by our imports and in no other way. Were it not that in place of the exports we could import something of greater value, we would never export anything, for it would be a losing business. Were it not for the imports the exports might as well be thrown in the fire for all the good we would derive from them.

I have not claimed, as insinuated by S. P., that under all circumstances excess of imports must be a gain, but that such excess is not a proof of loss.

A few words about luxuries, of which S. P. alleges we import and consume hundreds of millions worth every year, that amount to nothing of value. When a farmer, say, has supplied himself and family with the essentials of life—plain food and clothing, is out of debt, and has a surplus of grain or wool, or tobacco, and he thinks proper to dispose of part of the surplus in exchange for unessentials or luxuries, such for instance as tea or coffee for his breakfast, silk dresses or jewelry for his wife and daughters, a piano for his parlor, pictures for his rooms, a pleasure carriage for the family, toys for his children, and many other articles of luxury, does the satisfaction derived from the possession of those things "amount to nothing" of value. If he prefers them to the money they cost, is he not entitled to have his choice, and would it would not be a great impertinence for S. P. to come and tell him—"*I* don't care for those things—*I* don't value them a cent, and if you got them from abroad in exchange for your grain, you are a foolish man and a bad citizen, wasting your means and impoverishing your country !" Is nothing but coarse food and clothing of any value ? Is all decoration and ornamentation nothing but criminal waste ? Is the wild Indian whose food is only corn-bread and bear meat, and whose clothing is but a single blanket, the model we should pattern after?

It is said that instead of luxuries we should only import things of real utility, and above all, money. Now, we do not buy luxuries because they are imported. We import them because we want them, and because we can procure them more advantageously abroad than at home. The fault, if there is any, is in our injudicious wants; for, if we *will* have luxuries, it makes no difference to us where they are produced. Our only concern is to get them with as little expenditure of our means as possible; for whether they are made at home or abroad, we know that it is alike the product of our own labor that pays for them, and the fact that they are imported can be no good ground of complaint. Importing money instead of them would not help the matter, for so long as we want luxuries, and so long as there is no prohibitory law enforced against them, the money imported will surely be parted with to procure them. The $12,000,000 expended for imported wines would be laid out in home-made substitutes, or perhaps some of it in cheap whisky, which would not be much improvement. Besides, the hard money itself may be, and a large share of it daily is, melted and fashioned into jewelry and other gold and silver ornaments, which are as much luxuries as anything else.

As I said at the beginning, I will not repeat the general argument in proof that the Balance of Trade theory is a delusion, but in order to show that the views I entertain are sustained by high authority, I ask you to publish the following extract from a speech of Daniel Webster, delivered in the U. S. House of Representatives, April 2, 1824, and which was pointed out to me by a friend *since* my former communications were published in the FARMER. So far as appears, no one in Congress at that day ventured to controvert his argument, and luckily for him, your Delaware correspondent was not there to set him down by telling him that "his arguments were not dangerous," and "only needed statement to meet with ridicule."—J. P., *Lancaster*, *Dec.* 6, *1882*.

EXTRACT FROM D. WEBSTER'S SPEECH.

* * * "Let us inquire, then sir, what is meant by an unfavorable balance of trade. * * * By an unfavorable balance of trade, I understand is meant the state of things in which importation exceeds exportation. To apply it to our own case, if the value of the goods imported exceed the value of those exported, then the balance of trade is said to be against us, inasmuch as we have run in debt to the amount of the difference. Therefore it is said that if a nation continue long in a commerce like this, it must be rendered absolutely bankrupt. It is in the condition of a man that buys more than he sells, and how can such a traffic be maintained without ruin ? Now, sir, the whole fallacy of this argument consists in supposing that whenever the value of imports exceeds that of exports, a debt is necessarily created to the extent of the difference, whereas ordinarily, the import is no more than the result of the export, augmented in value by the labor of transportation. The excess of imports over exports, in truth, usually shows the gains, not the losses of trade. * * * If the value of commodities imported in a given instance did not exceed the value of the outward cargo with which they were purchased, then it would be clear to every man's common sense that the voyage had not been profitable. If such commodities fell far short in value of the cost of the outward cargo, then the voyage would be a very losing one ; and yet it would present exactly that state of things which, according to the notion of a balance of trade can alone indicate a prosperous commerce. On the other hand, if the outward cargo were found to be worth much more than the outward cargo, while the merchant having paid for the goods exported, and all the expenses of the voyage, finds a handsome sum yet in his hands, which he calls profit, the balance of trade is still against him, and whatever he may think of it, he is in a very bad way. Although one individual or all individuals gain, the nation loses. While all its citizens grow rich, the country grows poor! This is the doctrine of the balance of trade."

He then illustrates by instances, and in the course of his remarks says: "There are no shallower reasoners than those political and commercial writers who would represent it to be the only true and gainful end of commerce to accumulate the precious metals," and says that a country at one time may have too much money, as well as too little at another time ; and that when there is too much it is as advantageous to export it as to import it at another time, adding : "We need no more repine when the dollars which have been brought here from South America are despatched to other countries that when coffee and sugar take the same direction."

SELECTIONS.

THE VIRTUES OF COFFEE.

It is getting to be the fashion now for people to say that coffee is injurious to health, and many persons are giving it up regretfully. Perhaps coffee is very injurious in some cases, but of all beverages it is contended that it is the least injurious. Coffee-drinkers are generally cheerful, strong and persevering. The eminent Dr. Bock, of Leipsic, says : " The nervousness and peevishness of the times are chiefly attributable to tea and coffee." He says that "the digestive organs of confirmed coffee-drinkers are in a state of chronic derangement, which reacts on the brain, producing fretful and lachrymose moods." "I cannot agree," says Dr. Henry Segur, of Paris, "that the nervousness and peevishness of the present time are to be attributed to the use of coffee. If people are more nervous or in worse humor now than formerly, we may find other causes arising from the customs and habits of society much more likely to produce a state of things than the use of this particular article of diet."

Let us examine the effects of coffee on the economy. Taken in moderation it is a mental and body stimulant of a most agreeable nature, and followed by no harmful reaction, it produces contentment of mind, allays hunger and bodily weakness, increases the incentive and capacity for work, makes man forget his misfortunes and enables those who use it to remain a long time without food or sleep, to endure unusual fatigue and preserve their cheerfulness and contentment. Jomard says : " An infusion made with ten ounces of coffee enables me to live without other food for five consecutive days without lessening my ordinary occupations and to use more and more prolonged muscular exercise than I was accustomed to without any other physical injury than a slight degree of fatigue and a little loss of flesh."

The mental exhilaration, physical activity and wakefulness it causes explains the fondness for it which has been shown by so many men of science, poets, scholars and others devoted to thinking. It has, indeed, been called the intellectual beverage. It supported the old age of Voltaire and enabled Fontenelle to pass his hundred years.

The action of coffee is directed chiefly to the nervous system. It produces a warming, cordial impression on the stomach, quickly followed by a diffused, agreeable and nervous excitement, which extends itself to the cerebral functions, giving rise to increased vigor of imagination and intellect, without any subsequent confusion or stupor, such as are characteristic of narcotics. Coffee contains essential principles of nutrition far exceeding in importance its exhilarating properties, and is one of the most desirable articles for sustaining the system in certain prostrating diseases. As compared with the nutrition to be derived from the best of soups, coffee has decidedly the advantage and is to be preferred in many instances. The medicinal effects of coffee are very great. In intermittent fever it has been used by eminent physicians, with the happiest effect in cutting short the attack, and if properly managed is better in many cases than the sulphate of quinine. In that low state of intermittent, as found on the banks of the Mississippi river and other malarial districts, accompanied with enlarged spleen and torpid liver, when judiciously administered it is one of the surest remedies. In yellow fever it has been used by physicians, and with some it is their main reliance after other necessary remedies have been administered ; it retains tissue change, and thus becomes a conservator of force: in that state in which the nervous system tends to collapse, because the blood has become impure ; it sustains the nervous power until the depuration and reorganization of the blood are accomplished, and has the advantage over other stimulants in inducing no injurious secondary effects. In spasmodic asthma its utility is well established, as in whooping cough, stupor, lethargy and such troubles. In hysterical attacks, for which in many cases a physician can form no diagnosis, coffee is a great help.

Coffee is opposed to malaria, to all noxious vapors. As a disinfectant it has wonderful powers. As an instantaneous deodorizer it has no equal for the sickroom, as all exhalations are immediately neutralized by simply passing a chafing dish with burning coffee grains through the room. It may be urged that an article possessing such powers and capacity for such energetic action must be injurious as an article of diet of habitual employment, and not without deleterious properties ; but no corresponding nervous disarrangements have been observed after its effects have disappeared, as are seen in narcotics and other stimulants. The action imparted to the nerves is natural and healthy. Habitual coffee drinkers generally enjoy good health. Some of the oldest people have used coffee from earliest infancy without feeling any depressing reaction, such as is produced by alcoholic stimulants.

FEEDING STOCK IN WINTER.

As the season draws near when our domestic animals are to be fed upon artificial forms of food for nearly half a year, it may not be out of place to devote a little thought to the subject of winter feeding. There is no doubt whatever that in years gone by, if not at the present time, many cattle have been kept through the winter with little aim on the part of the feeder, beyond barely carrying the animal through alive; and where such a course has been pursued, there has always been more or less loss of life as well as loss of flesh, and an absence of all forms of profit whatever.

When the country was first settled, and there were no mowing fields of good, sweet hay from which to secure a winter supply; when the only winter fodder was the straw of ripened grain and the interior grasses of our wet meadows, and when the profit from stock husbandry was necessarily pretty much confined to summer pasturage, there was a better excuse for such a practice as was this winter system of partial starvation.

On a large proportion of the farms of New England at the present time, the cattle are kept in quite as good condition in winter as during the summer season. This is particularly true of the herds kept upon milk farms where there is a daily sale of milk to go to the cities or villages. Farmers who are receiving a daily income from their stock are less fearful of daily expenditure for good, nourishing food. Compared with the past, there is now little to complain of regarding the treatment of most of the cattle in the country. The farmers generally feed well, as they understand that term. But feeding animals well, in such a manner that they will produce abundantly of milk, flesh, growth, work, or fat, is something that requires a good deal of thought, study, and considerable practice.

The chemists are informing us something of the relative proportions in which the several food elements should be mixed, how much grain it will be found most profitable to feed with certain amounts of coarser fodders, and they are giving us tables showing the relative chemical and food values of most of the common forms of food used, but they cannot, with their experiments or tables, make a good feeder of one who has no more idea of feeding than to merely stuff an animal's manger full of food one or more times per day. We knew of a barn full of cattle that were fed almost nothing the past winter, but good, merchantable upland hay, grown by high culture and liberal manuring. The cattle were kept warm, were nicely bedded, the stables were cleaned often, and water was freely provided, yet the cattle came out thin in the spring, and made but little growth. The difficulty in this case, as in many others which readers of the Farmer may be familiar with, was, that the good hay was given far too freely, or certainly too much at a time. There was plenty of hay in the barn, and the attendant wanted to make a good showing of his skill in stock feeding, so he filled the racks and mangers full at each feeding. At first the cattle coming in from a short pasture would eat heartily, but with little or no exercise there was less food called for, and the quantity given was greater than the system required. Of course a portion would be left uneaten after the whole had been picked over, and the choicest portions taken out. The rest was breathed over till nothing nothing would eat it, when it was hauled under foot, trodden upon and wasted. The fact is, good English hay used as bedding for idle animals, will al-

most surely spoil them if they can get free access to it with their mouths.

"Under-feeding" is one of the charges brought by the agent of the Society for the Prevention of Cruelty to Animals, against owners or attendants of animals, but there are a great many animals seriously injured, and of course abused, from overfeeding, or, at least, from very injudicious feeding, which the agents are hardly likely to note. We have always found it more necessary to caution hired help against over-feeding than against under-feeding. Hired help do not have the bills to pay, and so have no pecuniary interest in an economical system of feeding. Then it is less labor to feed bountifully two or three times a day, than to give a little at a time, and then to notice how the animals seem to feel. A good observer will know, the moment he steps inside the feeding-room, whether his animals are sufficiently fed or whether they are still hungry, by their appearance. Cattle should have enough, and should then be left by themselves. They should have regular hours for feeding, and then they well know what to depend upon.

Animals that are fed well, and at regular intervals, will rarely call for food except at usual feeding hours. Animals very readily acquire habits, and they will adapt themselves in a considerable degree to the customs of their keepers. It would be difficult to determine from the practice of different feeders whether cattle will do better upon two or three regular meals per day, for there might be many herds instanced that have done well by either system. We have for many years made it a practice to feed cattle but two meals per day, one in the morning, the other in the afternoon, aiming to divide the twenty-four, hours as nearly as convenient, into two equal periods, though the time between night and morning is usually a little longer than the time between morning and evening. A cow's stomach is so constructed that she can easily take enough good food into it to last her twelve hours, and we have long been of the opinion that food is more thoroughly digested when but two meals are given.

It is certainly a great convenience on a dairy-farm, especially in winter, to have the feeding all done at the two ends of a day, so that the middle of the day, while the sun shines, can be used for other purposes. Many families in the country have but two regular meals per day during the short days of winter, and cattle, with their large stomachs designed specially for laying away large quantities of food to be masticated at their leisure, can certainly accommodate themselves to two meals per day as easily as can human beings with their relatively smaller digestive organs. In winter, when farm teams cannot work much more than six or seven hours per day, they can be changed off from three to two meals per day, and will do quite as well as if fed the noon meal. With but two meals per day there is less danger from over-feeding than if digestion be disturbed by a midday feed before the morning meal has been properly disposed of by the digestive organs.

But we would not have our readers understand that we recommend the practice of giving all the food of one meal at a single feeding; on the contrary, we would give it at

three or four different times, say twenty minutes or half an hour apart. The idea is to have the feeding continuous, till the meal is finished, then give no more till the next meal, some eight or ten hours later. A cow or an ox will occupy from one to two hours in eating a breakfast or supper of coarse, dry fodder. Watering, like feeding, may become somewhat a matter of habit as to the number of times and quantity taken. We prefer watering after each meal, but, in practice, find that many cattle will drink heartily but once per day. A good feeder will watch his animals and learn their wants, and endeavor to supply them, but never to over-supply.—*Boston New England Farmer.*

THE RATIONAL METHOD OF TREE-PRUNING.

In this new world little attention has been given to the pruning of trees. With our ample domain, giving space for all the trees, we have left them to grow as they might. Our pruning, where it has not been of that heroic and decisive sort which lays the axe at the root of the trees and cuts them to the ground, has been of a hap-hazard kind, based upon no system and directed by no science. Each one has cut and trimmed according to his own notion or whim. If the limb of a tree has been in the way, becomes an obstruction to the walk, or intercepted some desirable look-out, it has been lopped off, usually by whatever instrument convenience would supply and in a manner to require the least exertion. Tree-growers and the better class of farmers have been somewhat more painstaking in their methods. They have removed or shortened limbs with some study of after effects, and have so performed their work as to secure, if possible, the proper healing of the wounds which they have made. But in most cases the trees have been left to themselves or have been lopped with reckless carelessness. It seems to have been generally thought that they would bear any amount of mishap and the utmost severity of treatment. They have been regarded rather as dead than living matter, and their delicate and sensitive organizations, instead of being guarded and protected with sympathizing care, have been left to be the prey of neglect and violence.

The result has been that many of our forest trees have fallen victims to decay when proper pruning would have insured their healthy growth to full stature, and many trees planted by the roadside or the dwelling for the purpose of shade or ornament, have become deformed and short-lived in consequence of improper pruning.

With the rising interest in trees and tree-planting in America the importance of pruning and its proper method ought to receive attention as a branch of forestry. The object of pruning trees, forest trees especially, is to secure the largest and healthiest, and, therefore, the most profitable growth of timber upon any given area of ground, and experience has shown that by a rational system of pruning a forest may be made to yield a much larger product than when left to itself as is ordinarily the case with us.

Hitherto there has been no adequate treatise on the subject of tree pruning in the English language, and we have been excusable for our ignorance of the subject. The Germans and the French are in advance of us in other departments of forestry, and also in this. The advantages of pruning forest trees, as promoting an increase of timber, was recognized in Germany two hundred and fifty years ago. But the practice of pruning fell into disuse after a time, until it was revived during the present century by the writings of De Courval and Des Cars, who recommend a system of pruning based on the fundamental law of vegetable physiology, and which is now adopted in all the continental forests. The work of Des Cars, entitled, "A Treatise on Pruning Forests and Ornamental Trees," has recently been translated from the French of the seventh edition, by Professor Sargent, of Harvard University, and published by the Massachusetts Society for the promotion of Agriculture. It is essentially a reproduction of a larger treatise of De Courval in a brief and more popular form, it being a duodecimo volume of less than one hundred pages.

The system of these writers is based on the fact that "as wood is alone formed by descending, elaborated sap, a wound made on a tree can only be recovered with healthy, new wood, where its entire circumference is brought into direct communication with the leaves by means of the layer of young and growing cells formed between the wood and the bark. To make this connection it is necessary to prune in such a manner that no portion of an amputated or dead branch shall be left on the trunk. The cut should always be made close to and perfectly even with the outline of the trunk, without regard to the size of the wound thus made. This is the essential rule in all pruning, and on its observance the success of the operation depends. "A tree left entirely to itself," says Des Cars, "generally develops in one of two directions. It does not grow upwards, but assumes the low round form common to the apple-tree; the lower branches grow disproportionately large and absorb too much sap to the detriment of the top of the tree; and these long, heavy branches are often broken by the wind, or by snow and ice, leaving hideous stumps. Trees of this form are very common; they generally decay at the top before reaching maturity, and have little commercial value.

On the other hand, many vigorous trees grow disproportionately at the top, the lower branches die from insufficient nourishment, fall off, and leave, when large, bare decayed spots, which gradually penetrate to the heart of the tree, and ruin also its commercial value. Wounds caused by the breaking off of large branches by wind or snow produce the same results. There is no remedy for the dangerous effects of such accidents except pruning; it is a simple question of surgery. Without pruning, the tree must sooner or later decay; with pruning, its value may be preserved. The secret of obtaining a complete cure in all operations requiring the removal of a branch, either living or dead, consists in cutting close to and perfectly even with the trunk. And it matters not how large the cut may be. This is a universal rule of action; and it is based on the fact that new wood and bark are formed by the descending sap, which passes down between the old wood and the bark and cannot deposit the new woody substance upon the scar of the pruned branch if it is left projecting at all from the line of the sap-vessels in the trunk; but where the cut is made even with the trunk it is soon covered with new woody fibre and bark, and the tree grows on to maturity with unimpaired vigor and soundness. If the limb amputated is large the wound will not heal over completely in a single season. The new wood will form first around the top and the sides of the wound, which will soon be completely surrounded by the new growth. Meantime, to prevent decay taking hold of any portion of the wound, it has been found well to cover the wound with something which will protect it. For this purpose coal-tar, a waste product of gas-works, has been found superior to the many other preparations which have been used. It has remarkable preservative properties, and may be used with equal advantage on living and dead wood. A single application forms an impervious coating to the wood-cells. It produces a sort of instantaneous stoppage of decay, which would otherwise be the case, thus adding to their value, as timber, while more room for a renumerative undergrowth of coppice is thus given, and the total product of the forest greatly increased. In the practical application of the system it is held that the class of young forest trees, that is, those less than forty years old, should be so pruned of their lower branches that the trunk will equal one-third of the entire height of the tree, and the head should be elongated ovoid in form, the lower branches left being more or less shortened in for this purpose. Middle-aged trees, or those between forty and eighty years of age, should have thin trunks, equal to about two-fifths of the total height, and the head should be made to assume a somewhat rounder form than that given to the younger trees. In the old trees, eighty years and upwards, the trunk should be nearly equal to one-half the total height, and the head be still more rounded, and at all times decaying and dead branches should be carefully removed.

Such, without undertaking to go into the minute details of operation, is the system of De Courval and Des Cars. It commends itself at once as a rational system, and ample experience in Europe proves its great value. It is simple and intelligible, and may be put in practice successfully by any one. The Massachusetts society has made a very important contribution to practical forestry in securing the translation and publication of Des Cars' treatise.—*Mr. N. H. Egleston, Williamstown, Mass.*

LETTER FROM THE MOTHER OF BAYARD TAYLOR TO PROF. E. V. RILEY.

KENNETT SQUARE, Nov. 6, 1882.

PROF. RILEY—*Dear Sir:* I send a few silk worm eggs, by Mr. Davis. In 1880, a friend brought some eggs from California from two or three of his friends. I got about 100. I enjoyed feeding and taking care of them very much. I am crippled with the rheumatism, and can neither sew nor knit, but could feed and tend the worms when the leaves were brought to me. When they were done spinning I missed them so much that I thought of trying to raise a second crop, and not knowing anything about their nature, beyond feeding and keeping them clean, I kept some of the eggs that had just been laid, in hope that they would hatch, but after watching them for about four weeks, was told that they were animals, and would not hatch until the next summer. In 1881 I gave many eggs to all who wished to have them. One little girl let some of hers lay eggs, and put them in the garret, thinking it was the coldest place in the house. They were all right, for the winter, but in the spring they commenced hatching before the mulberry put out, and she fed them on lettuce leaves, and kept them alive until the mulberry came; consequently they were much earlier than any others. The little girl got tired of feeding them, as she had to go quite a distance for the leaves, and wanted to sell them. I bought a few. They spun in due time, and as I knew they were the same as my own, was not in a hurry to put them in a cool place, never dreaming they would hatch; but to my utter surprise, in about four or five weeks they commenced hatching, and I had to hurry them off into the vault. I took particular pains with them, and they grew and seemed healthy; but they were not. There did not appear to be any particular disease among them, but every few days one or two would look wilted, and get soft and die in a few hours. There were about 100 hatched, and only 30 lived to spin. The cocoons were not so large as the others I had, but still of a fair size.

I have been intending to write for some time and ask if you could explain why animals can produce two-crop worms. I hope you will pardon me for troubling you with it, but hearing that you were very much interested in the silk culture, I thought perhaps you would excuse the liberty taken by an old woman in her 84th year, and who is very much interested in the same, but unable to do much.

The Japanese eggs didn't hatch well. They were gummed so tight to the card that the poor little worms had a hard struggle to make their way out of the shells.—*Rebecca Taylor.*

SOILING CATTLE.

The necessities of the time demand a modification in methods of husbandry in the older settled parts of the country. The most prominent and obvious of these at the present is that relating to the subsistence of stock. It is necessary that more be kept than formerly, and, to do this, new methods of sustaining the animals must be adopted. The people are ready for a new departure, and, in a few short papers, it will be the aim of the writer to discuss this subject under the general head of soiling cattle. The careful attention of the farmers is invited to the points presented, because they are in full sympathy with the recognized needs of intelligent husbandmen.

This subject is often referred to in a general way, but, so far as the writer knows, no systematic discussion of it has appeared in our local press. References to soiling, when made, convey no well-defined ideas of what is actually meant by the system; nor have directions or processes by which the method is or can be applied in practice, been stated. Many farmers cultivate fodder-corn for their cows, and cut and deliver it to them in the pasture fields, and suppose they are thus practicing the soiling system. At best this is but partial soiling, and is not what is meant by the soiling system in the hands of those who practice it methodically.

Much of what I may say on this subject will appear like quite elementary teaching to those who are acquainted with it. It will be that sort of teaching, because that is the kind needed. Only a few know what the soiling system is. Systematically applied, it is a new and better method than the old, and that it may be understood, the elementary principles must be first taught, as in developing any other methodical system.

What should be understood by soiling cattle is keeping them in the stable all the year round, with short daily liberty in a yard similar to that which dairymen in our vicinity give the cows in winter when not allowed the liberty of the fields. Soiling cattle does not contemplate or allow pasturage in the fields at any season. There are a few dairymen in Bucks county who practice the system in its strict sense. But the number who practice it in the partial way alluded to is large in this and the adjoining counties.

In the first place it will be proper to state some of the principal attained facts developed in the experience of practical dairymen who have tried the system, a few of which are as follows:

1. It saves land.
2. It saves fencing.
3. It economizes food.
4. The cattle are more comfortable and in better condition.
5. They give more milk.
6. A large increase of manure.

The only offset to these and other advantages is the labor of raising and cutting the food, and feeding and taking care of the stock. This additional labor is of the bugbear that frightens farmers, and contemplation of it makes cowards of us all. Farmers generally admit the advantages without argument, and are deterred from giving the system a fair trial through fear of the extra labor it involves. But if the farmer can be reasonably assured that such advantages as are claimed can be realized, the enterprising dairyman should no longer hesitate.

The bare statement of the advantages without argument or elaboration will not convince doubters, or lead even those who admit them to put them in practice. Let us see, then, what can be said to substantiate the claims of the advocates of soiling.

It saves land. In relation to this claim all experience proves it. The only difference found is as to the amount of saving which results. These differences probably arise from variations in the quality of the land and its ability to sustain heavy pasturage. Again, lands used for soiling may be differently cultivated. It is evident that the amount of food raised on an acre of enriched and mowed pasture, and an acre devoted to cultivated crops, must be quite different, and will account for the differences in estimates given of the saving of land. In any case the economy is sufficiently great and should be decisive to the mind of any reasonable man.

Properly enriched and cultivated land will produce two or more soiling crops in a year, and thus greatly increase the amount of food it will yield. Those who practice but the partial soiling method see at once what an increased amount of food can be raised on a small surface. Some assert that the saving is as three to one; others as five to one, and some go even farther and assert that one acre kept for soiling will go as far as three, five, or more kept for pasture. For all practical purposes the testimony is sufficient that on all farms where the land is arable, the economy of surface gained by soiling is very great.

The saving of land will allow more stock to be kept or more feed to be raised and less to be bought. If no more stock be kept, they may be kept on half the product of the farm, and the other half may be sold. The farmer therefore has his choice to increase his stock, or sell half his crops. The farmer who now keeps, say ten cows, and pastures his land to death, can, by the soiling system, keep the ten cows better on ten acres than he now keeps them on his whole farm, and raise better crops on the remaining land. Farmers who think they have not land enough would find that the economy of land secured by the soiling system would practically treble or quadruple the size of their holdings. This is the experience of all who have tried the methods, and this kind of experience has been so uniform that it cannot be doubted. The average farmer, therefore, has it in his power at his option to increase the area and productiveness of his farm by the adoption of the soiling system, and by thus economizing his surface, save land.—A. M. D., Doylestown, Pa., in Weekly Press.

SMOKE-HOUSE AT SMALL EXPENSE.

Every farm should count among its outhouses a good smoke-house. The necessity for such a house is too obvious to call for argument in its favor. When the farm is a small one, and the meat produced thereon is for home consumption only, a large and elaborate smoke-house is, however, not required; in fact, a cheap one serves every purpose, and when meats are to be smoked in a small way an expensive building is a needless extravagance.

The object in smoking meat is to expose the meats to the action of creosote and the vapors resulting from smouldering wood. This is done not only to gain sundry flavors imparted by the smoke, but to gain the preservative principle given by the creosote. All that is necessary to bring this about is space enough in which to hang meat, that can be filled with smoke and shut up tight, with conveniences for suspending the pieces to be cured. In some smoke-houses the fire is made in the centre of the house on a stone slab; in others the fire is placed in a pit in the ground about one foot deep; again the fire oven is built outside the smoke-house.

The very cheapest form of smoke-house is what is termed the hogshead or cask-house. This is made, as the name suggests, of a hogshead or large cask. It is familiar to old readers, but is again described for the benefit of beginners who have no dollars to spend on the construction of a regular house. First, dig a small pit; place a flat stone or a brick across it, upon which the edge of the cask can rest. This pit ought to be about one foot deep and nearly one foot wide, and say three feet long. Remove both head and bottom of the cask. Pass two cross-bars through holes bored in the sides of the cask near the top; upon these rest cross sticks from which the hams are suspended. Then replace the head of the cask and cover with sacks to confine the smoke. Set the cask so that half the pit will be beneath it and half of it outside. Place some live coals in that portion of the pit outside of the cask and feed this fire with damp corn cobs or hardwood chips. The pit must now be covered with a flat stone, by which the fire may be regulated and may be removed when necessary to add more fuel. This fire must, of course, burn slowly so as to produce smoke and not flame.

When a larger house is required then a cask affords, this may be constructed of wood or bricks, as best suits the convenience of the builder. It is a wise plan to build the fireplace of bricks, then there is no danger from fire. A favorite plan is to have fire ovens of brick built on each side of the house; these are constructed upon the outside, but space left between the bricks on the inside, through which the smoke escapes. The outer part of the oven is open at the front, but may be closed by an iron door or a piece of flat stone. When the fire is kindled in these ovens the doors are closed, and the smoke has no means of escape except through the inside spaces. Being so confined, the fire of necessity slowly smoulders, making a steady smoke. Smoke-houses with these outside fire ovens are very clean, there being no ashes inside. The doors to such a house may be of cement or of hard brick laid in cement or mortar. These outside ovens, by the way, can be fitted to any kind of a smoke-house by cutting the necessary openings at the bottom of the walls and protecting the wood work with strips of sheet iron around the bricks.

Meat, to be perfectly smoked, must be continually surrounded by smoke produced from material that imparts a pleasant odor. Corn cobs and good hickory wood furnish admirable material. While the smoke ought to be continuous the smoking process should not be hastened to such a degree as to raise the temperature sufficiently to make the fat ooze out of the meat or prevent the creosote in the smoke from thoroughly permeating it. In a word, the fire must neither be permitted to die out nor to blaze up. It is the slow combustion of the wood that permits the escape of most of the wood acids which impart their flavor and antiseptic properties to the meat.

Old smoke-houses should be thoroughly cleaned previous to use, and the contrivances from which meats are suspended looked after and repaired to prevent their breaking down and bringing the meat in contact with the fire and ashes.

THE SUGAR-BEET.

The Department of Agriculture at Washington appears to have ascertained that the prizes of money offered by the Department to stimulate the production of the sugar-beet and Chinese sorghum in this country were made without authority from Congress, and were therefore illegal. Payments have, therefore, been suspended for the present, and in all probability the matter will be acted upon at the ensuing session of Congress and favorably. We do not ourselves entertain any high expectations of valuable results being likely to be reached by the stimulants of that sort. Still, we always regard with favor any and all efforts made by the Department of Agriculture to encourage the production of saccharine crops in the United States. As regards the sugar-beet, it is rather a reproach upon the enterprise of Northern agriculture that the Mormon colony in the territory of Utah, under the rigorous despotism of Brigham Young, succeeded in producing its own supply of domestic sugar from the beet 20 years in advance of any successful effort in the same way in any of the Atlantic States. This was done before the opening of the Pacific railroad, and at a time when the Mormon prophet was despotically bent upon making the Mormon community entirely independent and self-sustaining. The example thus set has since been slowly imitated by voluntary enterprise in the States of Delaware, New Jersey, Indiana, Illinois and Maine, with irregular and varying results. Perhaps the case would have been different but for the unfortunate attempt of the Department of Agriculture to force a sudden development of the sugar production on an immense scale by means of a rapid and widespread cultivation of the Chinese sorghum, which proved a lamentable failure. Since that date the Department has given encouragement to efforts at sugar making from Indian corn and other crops, the general effect of which has been to weaken the culture of the sugar beet and of the tropical sugar cane, respecting the saccharine properties of which staples there has never been any doubt whatever. Our own judgment is that if the attention of the country as regards sugar making could be concentrated upon those two crops—the one for the Gulf States and the other for the north and West—the result would be far more gratifying than could possibly be attained in any other way. The consumption of sugar in the United States is enormous, and as a vast majority of it is imported from foreign co ntries, which take but little merchandise from us in return, this one article has mostly to be paid for by shipments in American gold and silver, a process entirely too one-sided to be at all pleasant or profitable.—*Germantown Telegraph.*

OUR LOCAL ORGANIZATIONS.

LANCASTER COUNTY AGRICULTURAL AND HORTICULTURAL SOCIETY.

The regular meeting of the Lancaster County Agricultural and Horticultural Society was held in their room on Monday afternoon, December 4th, with the following members present : Joseph F. Witmer, Paradise ; J. C. Linville, Gap ; M. D. Kendig, Creswell ; Casper Hiller, Conestoga ; H. M. Engle, Marietta ; J. M. Johnston, city ; John H. Landis, Manor ; F. R. Diffenderffer, city ; C. A. Gast, city ; Johnson Miller, Warwick ; C. L. Hunsecker, Manheim ; W. B. Paxson, Colerain ; James Wood, Little Britain ; Ephraim S. Hoover, Warwick ; Henry Herr, West Hempfield ; Cyrus Neff, Manor ; Wm. T. Clark, Chestnut Level ; S. P. Eby, Esq., city.

On motion the reading of the minutes of the previous meeting was dispensed with.

Crop Reports.

Casper Hiller said the fruit crop had been a very poor one, but the corn crop has been very good—one farmer reporting as much as 100 bushels to the acre. Potatoes were very scarce, notwithstanding a good crop had been reported. Grain looks remarkably well.

Mr. Engle also reported a scarcity of apples. The grain and clover goes into winter quarters in good condition, notwithstanding the fly was found in some localities. He thought the scarcity of potatoes was due to the fact that the usual amount was not planted. The rainfall for November was three-fourths of an inch.

Mr. Paxson reported the wheat and clover in good condition to stand the winter season.

Mr. Landis concurred in the previous remarks in regard to the wheat. He said he did not know when the tobacco market was so dull as at present—not a single crop having been sold.

Mr. Hunsecker reported the wheat and clover to be in good condition. No tobacco has yet been sold. Potato crop was good, but the apple crop did not amount to anything.

Mr. Wood said the wheat looked well, although in some parts the fly was to be found. He reported a peculiar fact in regard to the grass—it looking as green and fresh as it does at any season of the year.

Mr. Miller was another farmer who did not know of any tobacco being sold. He concurred in the remarks of the other gentlemen in regard to the wheat and clover.

Mr. Witmer had not heard of any sales of tobacco, but he knew one man who was stripping. He had a very good set of grass this fall.

Experimenting with Fertilizers.

Casper Hiller read a brief paper on the subject of some experiments with fertilizers on potatoes, which is published below :

April 21st, planted 3 rows, each 40 yards long, with White Elephant potatoes.

Fertilizers used on row No. 1 : Equal parts of nitrate of soda, dissolved bone, acid S. C. rock and sulphate of potash.

Fertilizer used on row No. 2. Two parts of sulphate of potash. 1 part dissolved bone and 1 part acid rock.

Fertilizer used on row No. 3. A good dressing of well rotted stable manure.

The applications of the special fertilizers was liberal —one half ton to the acre; cost on row No. 1, $25 per acre; on row No. 2, $13 per acre; stable manure no fixed valuation. The weather during July and the early part of August was too dry for a promising crop, but by frequent and thorough cultivation the plants were kept healthy. By the time the rains became frequent the elephants were about done growing.

The yield of No. 1 was 400 bushels per acre.
    "   No. 2   320 "   "   "
    "   No. 3   280 "   "   "

This difference from 80 to 120 bushel per acre is worthy of con sideration. The value of special fertilizers for potatoes has usually been attributed to the potash contained therein. In this experiment it appears that the great increase was owing largely to the use of the nitrates. The White Elephant is a very prolific variety, and in quality it comes nearer to the old Mercer than any of the newer varieties lately introduced.

Mr. Linville said he had been using commercial fertilizers for the past few years on potatoes and found them to produce better results than he was able to obtain from from stable manure.

On motion, a vote of thanks was tendered Mr. Hiller for his experiments.

Mr. Engle said it would be a very good plan if more of the members of the society would experiment more than they do with fertilizers. The question of fertilizers was then discussed at some length by several of the members.

Answering Questions.

At what age should stock cattle be put up for feeders ? This question had been referred to Mr. E. S. Hoover, who answered it by saying that you should begin to feed at the age of 2½ years. His reason for so thinking was that at this age, you have a good, hearty bullock, in good condition to grow and increase in fat. He has had some experience in that line, and had come to the conclusion that this was the proper age at which to begin to feed cattle.

Mr. Linville said the idea now prevailed to get cattle that would mature early, and he thought that the cattle of the future would be of that kind. He agreed with Mr. Hoover, provided we have good cattle. In order to get a great weight we must procure cattle which will grow while we are feeding.

This impression was held by other members of the society—Mr. Engle stating that the cattle should be fattened scientifically as soon as they will commence to eat.

Mr. F. R. Diffenderffer, to whom had been referred the question, " Do farmers eat too much ? " reported progress.

Mr. Paxson read a paper on the subject of the proper treatment of milch cows. It was important, he held, that warm and comfortable quarters during the winter should be had. This is the great secret of success, for cows thrive best the less they are exposed to the cold weather. The animals should not be allowed to gorge themselves with water almost at the freezing point. Regular hours of feeding and a frequent change of nutritious food should be observed. Cotton seed meal was a very good food for cows, and this might be fed in conjunction with other articles.

The article was discussed by H. M. Engle and others, all of whom held substantially the same views as those put forward by Mr. Paxson.

The president said he hoped that members would not forget that at the next meeting of the society officers would be elected for the ensuing year, and he would like to see a large attendance

The following questions were referred : Is it advisable to use constantly the so-called cattle powders ? to H. M. Engle. Should creameries be established in Lancaster county ? to Joseph F. Witmer. What crop would be most profitable as a substitute for oatson corn stubbles? to Johnson Miller. What is the most profitable fertilizer, clover, homemade manure or artificial fertilizers ? to W. B. Paxson.

On motion, the secretary was authorized to secure the services of some person to deliver a lecture before the society.

Adjourned to the second Monday in January, 1883.

POULTRY ASSOCIATION.

The Lancaster County Poultry Society met stately, Monday morning, Dec. 4, 1882, in their room over the City Hall, with the following members present : George A. Geyer, Florin ; J. B. Lichty, city ; John E. Schum, city ; Charles Lippold, city ; Joseph R. Trissler, city ; F. R. Diffenderffer, city ; C. A. Gast, city ; H. H. Tshudy, Lititz ; J. B. Schultz, Elizabethtown ; H. D. Shultz, Elizabethtown ; Isaac Brooks, West Willow ; Dr. E. H. Witmer, Neffsville ; J. M. Johnston, city ; Peter Bruner, Mount Joy ; J. W. Bruckhart, Salunga.

The minutes of the preceding meeting were read and approved.

J. L. Bruner, of Mount Joy, August Lang, of Pittsburg, and Simon Tshudy, of West Willow, were nominated and elected to membership of the society.

The following persons were placed in nomination for officers of the society during the ensuing year : President, H. H. Tshudy, Lititz ; J. B. Lichty, city ; George A. Geyer, Florin. Vice Presidents, M. L. Greider, Mount Joy, and T. Frank Evans, Lititz. Corresponding Secretary, Joseph R. Trissler, city. Recording Secretary, J. B. Lichty, city. Treasurer, John E. Schum, city. Executive Committee, Peter Bruner, Mount Joy ; J. A. Stoler, Schoneck ; Chas. Lippold, city; Wm. A. Schoenberger, city; Dr. E. H. Witmer, Neffsville ; A. S. Flowers, Mount Joy ; S. G. Engle, Marietta. The nominations closed, and the candidates will be balloted for at the next meeting.

The secretary stated that he had secured Excelsior Hall for the exhibition, with storage room for the coops, for $40.

Letters had been received from Messrs. Bickell and Becker, the judges elected by the society, stating that they would accept, and be in Lancaster on time.

The secretary also stated that by had completed

arrangements with the Pennsylvania Railroad Company for excursion rates to Lancaster during the exhibition.

On motion, Joseph R. Trissler was appointed superintendent of the exhibition, he to devote his whole time to the show during its continuance.

Adjourned.

LINNÆAN SOCIETY.

The Linnæan Society met in their museum room on Saturday, November 25, at 2 P. M., Prof. J. S. Stahr, President, in the chair. Prof. Rathvon, Prof. Buehrle, S. M. Sener, Mrs. Zell, Miss Lefevre and two visitors present. Minutes of previous meeting were read and dues collected, after which the following donations were made :

Donations to the Museum.

A pint bottle of spirits, containing two large centipedes (*Scolopendra heros*) a palmated cricket (*Sterropalmata talpa*) and a "Camel cricket (Mantus Carolina) from the vicinity of Carson, Mitchell county, Texas, and donated by Mr. H. A. Rathvon, of Carson, Texas. These centipedes are the largest species among the *Myriapoda*, although occasionally larger specimens may be found. In their present contracted form, they measure fully eight inches in length, but if living, and the segments naturally relaxed and expanded they would measure at least ten inches in length. These animals are no doubt poisonous, but it is probable that many of the stories published about their venomous qualities are fabrications or exaggerations. Each foot is terminated by a sharp bent spine, and we have been informed that when they remove any exposed part of the human body, they leave two rows of dimpled spots, which soon became inflamed.

The Palmated cricket, as its name implies, is related to the "Mole cricket," of which we have one or two species in Lancaster county. This insect in California is called the "Potato cricket," and when numerous, is said to be very destructive to the potato tubers in the ground, excavating them and forming large burrows in them and thus depreciating their culinary values. Some years ago a gentleman in Sacramento Valley sent a number for identification and represented them as destroying his potato crop. Their semi-palmated anterior feet seem to indicate that they are burrowing insects, feeding on roots and tubers. They belong to the family *Gryllidæ* in the order *Orthoptera*.

The Camel cricket is called "Devil's horse" in Texas. Prof. Townsend Glover named it "Rear horse" in one of his reports before the war. In Europe it is called the "Praying mantes," but if an "e" was substituted for the "a," it would more nearly express its character, for it as perhaps the only orthopterous insect that preys upon other in sects. Its large and stout anterior reportal legs explain its character. Specimen of the mole cricket by Dr. M. L. Davis; Steatite by Master H. Halabaugh, and two specimens of superannuated currency and box of Italian Darts by Mr. Beates, of Middletown, Pa. To the library was added, LANCASTER FARMER for November; Patent Office Gazette, Nos. 17 and 20, of vol. 22 and Index of Patentees for 1882; Lancaster Industries, a pamphlet, in reference to removal of duty on books; package of Bicentennial records and envelope of thirteen news paper cuttings. Professor S. S. Rathvon read a paper on " Myriapoda." Prof. R. K. Buehrle was present and paid his initiation fee. After an hour or more passed in science gossip, adjourned. Before adjournment it was resolved, "That when we adjourn, we adjourn to meet early in January, instead of December, on account of the holidays and that said meeting be called by the secretary. Our next meeting will be our annual one.

THE use of a public road is for travel—not to pasture animals upon it. The old barbarism of pasturing on the road is done for in the Northern States, and will be in the South as soon as the south "catches on" to Northern enterprise and judicious tastes.

AGRICULTURE.

Ivory Wheat and Millo Maize.

J. T. Henderson, Commissioner of Agriculture of the State of Georgia, in a report for 1881 and 1882 calls attention to the claims of "Ivory wheat" and " Millo maize" to a place on the list of profitable food crops. These are both members of the large family of sorghums, of the class that have for many years been cultivated in Central Africa and other tropical countries for bread purposes. Analyses made to gain the relative theoretical value of these gratus as compared with ordinary standard wheat show that there is scarcely more difference in the proximate analyses of "Ivory wheat," so called, and Dallas or Red May than appeared between the analyses of the latter two varieties of ordinary wheat. The Ivory wheat shows a larger percentage of albuminoids (flesh formers), slightly less of starch and more of fats (fat and heat producers) than either of the true wheats. The Millo maize has considerably less of the albuminoids or flesh forming substances than either of the others, being about equal to the Indian corn in this respect.

"The flour made from the Ivory wheat, when properly ground and bolted, is rather darker than ordinary 'family' flour, but possesses the property of kneading well, and is therefore adapted to the process of 'raising ' with yeast or by similar means. Bread made from it, though not equal in any sensible respect to that from fine wheaten flour, is by no means unpalatable, and, as indicated by analysis, is probably fully equal in nutritiveness to any. For making the forms of bread for which buckwheat flour, rice flour, middlings of wheat, &c., are usually employed, viz.: waffles, griddle-cakes, muffins, &c., the Ivory flour seems to be well adapted." Mr. Henderson does not speak from actual experiment of the bread qualities of the Millo maize, but is of the opinion that in this respect it will be found to resemble Indian corn meal. It is claimed that both of these plants are enormously productive, rather indifferent as to soil and culture, and almost independent of the seasons after the soil has been prepared and the crop started off. Owing to the extraordinary seasons of this year it has not been practical to test their capacity to resist drought, and a sufficient number of reports of experimenters has not yet been received to form any decided conclusions in reference to productiveness under ordinary circumstances. But Mr. Henderson is of the opinion that the reports will show that both are very productive—far more so than any grain crop now grown in this State. The Millo maize is quite late in maturing, requiring favorable culture and the full season from planting time (April) until frost to mature in north Georgia ; but this difficulty will probably soon yield to the acclimatizing effect of planting home-grown seed a few years. This plant appears to be unusually productive of foliage, will bear two or more cuttings, and promises to be very valuable for soiling and general forage purposes.

Economy on the Farm.

On the farm, and in all the various details of rural and domestic life, prudence and a just economy of time and means are incumbent in an eminent degree. The earth itself is composed of atoms, and in the most gigantic fortunes consist of aggregated items, insignificant in themselves individually considered, but majestic when contemplated in unity and as a whole. In the management of a farm, all needless expenditure should be systematically avoided, and the income made to exceed the outlay as far as possible. Pecuniary embarrassment should always be regarded as a contingency of evil boding, and if contended against with energy and perseverling fortitude, it must soon be overcome. Even, with but little hope of its removal, is a millstone dragging us down and crushing the life-blood out of us. Be careful, therefore, in incurring any pecuniary responsibility which does not present a clear deliverance with the advantages which a wise use of it ought always to insure.

A farmer who purchases a good farm and can pay down one third of the price, give a mortgage for the other two thirds, and possesses the heart and resolution to work it faithfully and well, enters upon the true path to success. He will labor with the encouraging knowledge that each day's exertions will lessen his indebtedness and bring him nearer to the goal when he shall be disenthralled and becomes a freeholder in its most cheering sense. But without due economy in every department, in the dwelling, as well as in the barns and in the fields, this gratifying achievement may not be reached until late in life, or may be indefinitely postponed. A prudent oversight, therefore, over all the operations of the farm, in order that everything may be done that ought to be done and nothing be wasted, will exert a powerful influence in placing a family on the high road to an early independence.—*Germantown Telegraph*.

Rules Adopted by the Hay Trade.

Following are the rules adopted by the Hay Trade in New York, Brooklyn, and Jersey City under the leadership of the Manhattan Hay and Produce Exchange :

No. 1 *Prime Hay*—Shall be pure timothy, properly cured, bright, natural color, sound and well baked.

No. 2, *or Good Hay*—Shall be timothy, not more than one quarter mixed with red top and blue grass, properly cured, bright color, sound and well baled.

No. 3, *or Medium*—Shall include all timothy not good enough for No. 2, proportionately mixed with blue grass, red-top and clover, sound and well baled.

No. 4, *or Shipping Hay*—Shall include all hay not good enough for other grades, and may be natural meadow, free from wild or bog, and must not contain over one-third clover, sound and well baled.

Clover Hay—Shall be medium grown, properly cured, good color, sound and well baled.

No Grade, or Rejected Hay—Shall include all hay badly cured, musty, stained or unsound in any way.

Rules for Inspection—All certificates of inspection shall give the number of bales and grade of each bale inspected.

The expenses for inspection shall be 10 cents per ton for grading, and 20 cents per ton for grading, weighing, and unloading cars of hay, the expenses to be paid half by the buyer and half by the seller.

All hay or straw shall be pressed with wood not to exceed three pounds per bale. All hay or straw wooded in excess of three pounds per bale, the total weight of wood will be deducted. This rule will take effect January 1, 1883.

Effects of Broom-Corn on the Soil.

Professor Shelton, of the Kansas State Agricultural College, gives his views concerning the continued culture of broom-corn on the fertility of the soil in the college paper, the *Industrialist*. He says : " Ultimately, the effects of such crops as broomcorn, hemp, flax and perhaps castor-beans, which furnish no stock feed, or but very little, will be seriously felt in Kansas in the loss of fertility sustained by those lands upon which these are cultivated. The fact that broom-corn is a hoed or cultivated crop makes it much less dangerous than is flax, which receives no cultivation during the period of its growth. The general rule for every farmer who has a higher aim than to scourge his lands and then seek newer ones is to grow no crop upon a considerable scale that cannot be used wholly or in good part as stock feed. This has been the rule of really successful farmers the world over, and at a near day will be the rule in Kansas also."

The Agricultural Interests of the Country.

WASHINGTON, November 23.—George B. Loring, Commissioner of Agriculture, has submitted his annual report to the President. Two and a half million packages of seeds have been distributed and 250,000 copies of special reports printed by the

Department. The statistical division estimates the following as the yield of 1882: Corn, 1,635,000,000 bushels; wheat, 410,000,000 bushels; oats, 470,000,000 bushels; barley, 45,000,000 bushels; rye, 20,000,000 bushels; buckwheat, 12,000,000 bushels. The chemical division has devoted its work largely during the year to the investigation of the sugar-producing qualities of sorghum and other plants. The work of the division was submitted to the National Academy of Science for investigation by that body, and a committee was appointed for that purpose. The report of this committee contains a review of the history of the sorghum industry for twenty-five years and will be issued as a special publication. The report of the veterinary division shows less disease among domestic animals (Texas fever excepted) than in many years. Examinations into the fibres of wool and cotton have been made, and two sites, both in Colorado, have been selected for artesian wells. In the forestry division increased activity has been shown, a special agent having been appointed to collect information west of the Mississippi. His report will soon be submitted.

Small Potatoes.

At a recent session of the San Francisco (Cal.) Academy of Sciences, Mr. J. G. Lemmon, who spent six months in the mountains on the Mexican frontier among Apaches and cowboys, announced that he had brought up five boxes of new plants, and that he had found two or three new kinds of native potatoes, some of which were growing on a peak 10,000 feet high. They were about the size of walnuts. This was regarded as an important find that might throw some light on the nativity of that potato, as the real home of the "Murphys," as they are familiarly styled, has not yet been established. They were reported to have been found in Peru in 1500. Sir Walter Raleigh found some and sent them to England, but they have only been known to the poor man on his table for a century.

HORTICULTURE.

Winter Flowers in the Window.

The man or woman who dislikes to see even a single pot or stand or plants or flowers in the window when they can no longer be grown out-of-doors, we would not care to be acquainted with. When we see these thoughtful, modest efforts to please the eye of the passers-by, we involuntarily accord to that household not only our good wishes, but a due share of natural accomplishment that it did not even dream of evoking. While city people should not omit adorning their front windows in this way, we think that country people beyond all others can make their houses beautiful at small expense in such adornments, and would suggest that the material for this cheap and beautiful arrangement is particularly abundant at this time of the year. There are not only living plants in pots for the window sill, but colored leaves, dried ferns and grasses, and skeletonized plants, out of which innumerable tasteful objects may be made. But among the cheapest and most interesting of these room and window adornments is the hanging plants, which may be suspended from hooks and brackets, and by which the whole window can be made to look like a summer scene in the woods, although we are in the midst of winter. In the closely-built cities some such window-gardening attempts are made, but are generally, with rare, beautiful exceptions, utter failures. Some highly ornamental China work, or some highly-painted or polished "rustic" basket, is employed to hold the plants, but always evidently to the plant's misery. Plants do not seem to admire these elaborate preparations for them. They seem to understand that with them they only occupy a very inferior position in the temple of honor, and hence there is no wonder they fade away and die. In most of our country homes we find them in common pots, shells, stumps, boxes or anything that can be improvised, and always with success.

We remember that at our great Centennial exposition an exhibitor made for hanging-plants purposes the husk of the coconut, and we noticed and ascertained in fact that plants thrived wonderfully in this to us unique arrangement. It reminded us of what we had often read in the English horticultural periodicals, that coconaut refuse is an excellent material for the growth of plants. We are all familiar with the coconaut, but all do not know that the nut is enclosed in a heavy, spongy covering or husk of an inch or more in thickness. These husks are cut across in the middle and make two little baskets, and can be easily suspended with wire. As we have already said, there is a great fertility power in this covering, and there is the basket at the same time, thus securing two desirable things where only one to most people was apparent.

Many beautiful plants and flowers can be had all the winter through in this way at very little cost to time or money—at least, so little as not to be worth mentioning with the pleasure and satisfaction afforded to all, alike to those who provide these most desirable natural ornaments and the obliged public who cordially welcome the heartfelt treat as they pass along. Now, too, is the time, and none to spare, when the preparations should be made about securing them.—*Germantown Telegraph.*

Preserving Garden Flowers.

The time has come when we shall have to part with many garden pets which have given us so much pleasure during the growing season. Such partings always bring regret; and in spite of "nowhere to keep them," people will try if something cannot be done at any rate. It is believed that it is not so much the degree of cold which kills usually hardy plants, as it is the drying influences of a very cold atmosphere, and hence many find a very little covering sufficient to save plants, if the covering be such as to keep them from dying out.

We know, for instance, that a raspberry or a grape vine, which would probably be destroyed if left above ground in its natural way, can be safely preserved by being buried just beneath the surface; and it is found that roses bent over to the ground and covered with earth, so shaped as to throw off the water, will enable rather tender kinds to get through the winter unscathed. A friend once told us that verbenas were much hardier than people supposed. He put dry leaves over the bed and then covered the leaves with a board, and they did not injure by the hardest frost.

We should suppose, however, that green succulent matter would rot by confinement, as well as hard wood get injured by frost; and we would suggest to all who may be disposed to preserve anything in this way the importance of cutting away half ripe wood or succulent green foliage before entombing the plants for winter.

Pampas grasses, the ostrich feather like spikes of which are so commonly seen in gardens, cause much discussion as to the best means to protect. Some take them up and put them into a tub of earth and keep them in a cool cellar; but those who succeed in keeping them over winter in the open ground have finer plants and larger and more numerous spikes. Some of our neighbors turn a barrel over the stocks to keep out the water, filling in dry leaves all about the plant; and, though sometimes the plants will be lost treated in this way, generally it is a success. The "rocket" plant does well on either of the plans named for the pampas grass.

Dahlias, tuberoses, gladiolus and such like summer flowering roots, there is no trouble with. All they require is to be taken up as soon as the first frost has injured their flowers and spoiled their blossoming for the season, and, after drying a little, put them in some moderate cool and dry place secure from the frost.—*Germantown Telegraph.*

HOUSEHOLD RECIPES.

ROAST TURKEY, GARNISHED WITH SAUSAGES.—Wash out the turkey carefully. Stuff as usual, adding a little cooked sausage to the dressing. (Salt the giblets, and keep for to-morrow.) Lay the turkey in the dripping-pan, pour a great cupful of boiling water over it, and roast about ten minutes per pound—slowly for the first hour. Baste faithfully and often, dredging with flour, and basting with butter at the last. Dish the turkey, laying boiled sausages around it. Pour the fat from the gravy; thicken with browned flour; salt and pepper. Boil once, and serve in a boat.

MASHED TURNIPS.—Pare, quarter, and cook tender in boiling water, a little salt. Mash and press in a heated colander; work in butter, pepper and salt; heap smoothly in a deep dish, and put "dabs" of pepper on top.

CANNED CORN PUDDING.—Drain, and chop the corn fine, add a tablespoonful of melted butter, four beaten eggs; a large cup of milk, with an even teaspoonful of corn starch stirred in it, with salt and pepper to taste. Bake, covered, in a greased pudding dish one hour; then brown quickly.

CRANBERRY SAUCE.—Cook a quart of cranberries with a very little water, slowly, in a porcelain or tinned saucepan. Stir often, and when they are broken all to pieces, and thick as marmalade, take off, sweeten liberally, and rub through a colander. wet a mold and put them into form.

ORANGE SNOW AND SNOWDRIFT CAKE.—Four large sweet oranges, juice of all, and grated peel of 1 lemon; 1 package of gelatine soaked in 1 cup of cold water; whites of 4 eggs, whipped stiff; 1 large cup of powdered sugar; 2 cups of boiling water.

Mix the juice and peel of the fruit with the soaked gelatine, add the sugar, stir well, and leave them for one hour. Pour on boiling water, and stir until clear. Strain and press through a coarse cloth. When cold, and beginning to congeal, whip a spoonful at a time into the frothed whites. Put into a wet mold.

OYSTER SOUP.—Two quarts of oysters; 1 quart of milk; 2 tablespoonfuls of butter; 1 teacupful hot water; pepper, salt and blade of mace.

Strain all the liquor from the oysters; add the water, and heat. When near the boil, add the seasoning, then the oysters. Cook about five minutes from the time they began to simmer, until they "ruffle." Stir in the butter, cook one minute and pour into the tureen. Stir in the boiling milk, and send to table.

BOILED CHICKEN.—Clean and truss the chickens, but do not stuff them. Sew up each in a piece of mosquito-netting, and boil in plenty of hot salted water. Allow about twelve minutes to the pound. Undo the netting; wipe the chickens, and rub all over with butter. Send up in a boat a cup of melted butter in which have been stirred the pounded yolks of two hard boiled eggs, and some powdered or minced parsley. Pour a few spoonsfuls over the chickens.

BROWNED POTATOES.—Boil with their skin on. Throw off the water; take each potato in a clean towel, and hold it while you strip off the skin. Lay them, when peeled, in a greased baking-pan, and set this in a hot oven. Roast, with good dripping, until they are all well colored.

BAKED SWEET POTATOES.—Wash, and bake soft in a moderate oven. Serve in their "jackets."

SCALLOPED SQUASH.—Pare, slice, and mash. Stir in while it is hot, a good spoonful of butter, pepper and salt to taste, and two beaten eggs. Pour into a buttered dish; strew fine crumbs on the top, and bake, covered, half an hour—then brown slightly.

BAKED CUSTARDS.—One quart of milk; 4 beaten eggs; 5 tablespoonfuls of sugar, beaten with the eggs; nutmeg, and 2 teaspoonfuls of flavoring extract.

Scald the milk; pour upon the other ingredients; stir together well; flavor, and pour into stone china cups. Set these in a pan of hot water; grate nutmeg upon each, and bake until firm. Eat cold from the cups.

SIMPLE WHITE SOUP.—Take the fat from the top of your turkey soup-stock; strain, rubbing the dressing through the colander. Simmer one hour, with half a sliced onion and four tablespoonfuls of soaked rice in it, or until the rice is soft. Be careful that it does not scorch. Strain through the soup sieve into the tureen, add pepper and salt, if needed—finally a cup of hot milk in which has been stirred and cooked for one minute two beaten eggs.

STEWED FILLET OF VEAL.—Lard the fillet on top with strips of fat salt pork; lay a few slices of corned ham in the bottom of a saucepan; on these the veal; cover with sliced ham; season with pepper, salt, and a pinch of mace; pour in a cup of yesterday's soup, weakened with water. Cover closely and stew two hours, turning the meat at the end of the first hour; take up and keep the meat hot over boiling water; add some browned flour and a tablespoonful of soaked gelatine to the gravy when you have strained it, boil fast and hard until it is thick, and of a glossy brown. Pour on the veal, set in the oven, the larded side upward, and shut the door for a few minutes in "glaze" it. Garnish with light and dark green celery tops. Lay the ham about it.

SPINACH.—Boil in plenty of salt water for twenty-five minutes. Drain chop very fine, put back in the saucepan with a teaspoonful of sugar, a little pepper, salt, and mace, and a few spoonfuls of milk or cream. Beat and toss until it is like a thick green custard, and pour out upon slices of fried bread.

BOILED BEANS.—Soak all night. In the morning, put on in cold water, and pour over them, when dished, a little good drawn butter.

MASHED POTATOES.—Prepare as usual, without browning.

QUEEN'S TOAST.—Cut thick slices of stale baker's bread into rounds with a cake-cutter and fry to a nice brown in hot lard. Dip each slice into boiling water to remove the grease; sprinkle with a mixture of powdered sugar and cinnamon, and pile one upon the other. Serve a sauce made of powdered sugar, dissolved in the strained juice of a lemon and thinned with a glass of wine. Put a very little upon each round. Butter sauces are too rich for queen's toast.

BROWN GIBLET SOUP.—Cut each giblet into three pieces, and put on to boil in stock made of the remnant of your mock turtle soup, diluted with water and strained. Simmer all together one half hour. Chop the gizzard fine, pound the liver. Make what is called technically a roux, by putting two tablespoonfuls of butter into a saucepan, and when it bubbles, stirring in a tablespoonful of browned flour, and continuing to stir until they are well mixed and smooth. Add, pouring by spoonful, half a cup of boiling soup, then the pounded liver; the gizzard, juice of half a lemon, and a half glass of brown sherry. Stir all this in'o the soup, and boil up once. Have in the tureen the yolks of four hard boiled eggs, each quartered with a keen knife, and pour the soup over them.

MINCED TURKEY AND EGGS.—Cut all the meat from the skeleton of the turkey. Put the bones, sinews, skin, and stuffing into a pot with three quarts of cold water. Set at the back of the range and let it simmer down to two quarts. Season, and set way in your stock-pot.

Divide the meat intended for to day into inch-long pieces, tearing rather than cutting it. Heat the skimmed gravy; add as much drawn butter; two beaten eggs; pepper and salt; put in the minced turkey; set back over the fire, and stir until very hot. Cover the bottom of a pudding dish with fine crumbs; pour in the mixture; strew crumbs on top, and bake to a light brown in a quick oven. Serve in the bake dish.

STEWED POTATOES.—Pare and cut into small squares. Lay in cold water half an hour; cook tender in hot water; a little salt. When done—or nearly—pour this off, add a cup of cold milk, and when this begins to simmer, a tablespoonful of butter rolled in flour, pepper, salt, and a little minced parsley. Boil gently one minute, and pour into a dish.

CELERY.—Wash, scrape, and cut off the green leaves. Arrange the best stalks in a celery-glass. Put two or three green pieces into to-morrow's soup-stock while boiling; and if you have time cut up the rest into short bits, and put in a jar or wide-mouthed bottle of vinegar to keep for salad-dressing.

A PLAIN RICE PUDDING.—One large cup of rice, 2 quarts of milk, 8 tablespoonfuls of sugar, 1 teaspoonful of salt, 1 great spoonful of melted butter; nutmeg and cinnamon to taste. Soak the rice two hours in a pint of milk. Add, then, the rest of the milk and the other ingredients. Bake, covered, two hours; brown, and eat cold.

LIVE STOCK.

Cotton-Seed Meal as Stock Feed.

The Commissioner of Georgia says that the true policy of the farmers of Georgia is to encourage the manufacture of cotton-seed oil by exchanging the whole seed for cotton-seed meal at such rates as may be satisfactory, and use the meal, as far as practicable, as food for stock. Chemical analysis proves that the meal is exceedingly rich in both flesh-forming and fat-forming constituents. The one defect to be overcome is the fact that this substitute is too concentrated and must therefore be fed in comparatively small quantities and mixed with less concentrated food. The meal alone is more nutritious than either corn meal or wheat flour and is actually worth more as a stock feed, but it must be fed with greater caution.

"Cut wheat straw, corn forage, or any coarse and comparatively cheap and unnutricious forage that stock can eat, all can be brought up to the requisite standard of nutritiveness," says the Commissioner, "by mixing with due proportion of cotton-seed meal. English feeders very early discovered the immense feeding value of the meal, and for a long time consumed nearly the entire product of our mills. They estimate the manure resulting from a ton of meal fed to cattle at a higher value thau the meal brings in our markets. Northern cattle feeders are now using large quantities, and its use is constantly increasing; while the spectacle is presented in the South of using this valuable food material almost entirely and directly for fertilizing purposes, and this too in view of the fact that the manure from cattle fed ou the meal is nearly as valuable for fertilizing purposes as the meal itself."

Dry Food for Hogs.

Many hogs are kept comparatively poor by the high dilution of their food. They take in so much water that there is not room for a good supply of nutriment. Hence the reason that those farmers who carefully feed undiluted sour milk to their hogs have so much finer animals than those who give them slop. The hog has not room for much water; and if food which contains much is fed to him, it makes him big-bellied, but poor. Hogs, as well as all other animals, should be allowed all the water they will drink; but it should not be mixed with their food in excessive quantity; the hog should not be obliged to take more water than he wants in order to the food he requires.—*S. and P. Index*.

Lincoln Sheep.

The Lincoln sheep are comparatively a rare breed in the United States. They are the largest breed known, under exceptional circumstances dressing up to ninety pounds per quarter. At two years old they are recorded to have dressed one hundred and sixty pounds. They require good care and plenty of succulent food. They have been introduced in some sections of the West and into Canada, and are reported as being well liked, but further time is needed to fully establish their complete adaptability to our Western climate. Other long wooled sheep, as the Cotswold and the larger of the Downs, are giving good satisfaction, and there seems no good reason why these will not on our flush pastures with some succulent food in winter do exceedingly well.

In England fourteen pounds of wool average has been sheared as a first clip from a lot of thirty yearling wethers, the same averaging one hundred and forty pounds each, live weight, at fourteen months old. They have been known in the United States since 1833, and their long, lustrous fleeces, measuring nine inches in length, are the perfection of combing wool.

The Lincolns, originally, were large, coarse, and with ragged, oily fleeces and hard feeders. The improved Lincolns were made by judicious crosses of Leicester rams, careful selection and good feeding, and in England their wool has now a separate class at the fairs.

Pasturing and Soiling Hogs.

The hog is a grass eating animal by nature, and its health is therefore promoted by the use of grass as a part of its food. The grass gives bulk and porousness to the contents of the stomach, and thus aids digestion. If the hogs are to be pushed to fattening, finishing them off in the fall, then they may be kept in a dry pen or yard, and the green, succulent grass brought to them each day and given in three small feeds, in small racks over the troughs. In this way they will not get much under foot, and what falls out of the rack will drop into the trough.

Some years since, we found the best plan in feeding clover to hogs in a pen, was to run it through a straw-cutter, and then feed two quarts of the cut clover, mixed with its ration of meal, to each pig three times per day. We adopted the plan of cutting the clover in the morning, and mixing the proportion of meal with it that we desired the hogs to eat per day, and letting it lie in bulk through the day. It would then become so mingled that the grass and meal would be eaten together. It would warm up some, but not to injure its quality. The hogs were extremely fond of it, and gained in weight from twelve to fifteen pounds each per week. We were feeding for rapid growth through the summer, and fed six pounds of corn-meal to each pig, with the clover, per day, and the result was quite satisfactory.

Growth of Colts.

In order to winter a colt well, and have him come out a fine, showy, sturdy animal in the spring, particular attention must be paid to his growth during the first summer and autumn. If the mare's milk is at all deficient to keep the colt in good flesh and thriving steadily, it is best to have recourse at once to cow's milk. Skimmed milk answers very well for this purpose, especially if a little flax-seed jelly, oil or cotton-seed meal, is mixed with it. A heaped tablespoonful, night and morning, is enough to begin with when the colt is a month old. This can be gradually increased to a pint per day, by the time it is six months old, or double this if the colt be of the large farm or cart-horse breed.

Oats, also, may be given as soon as they can be eaten. Begin with a half-pint night and morning, and go on increasing, according to the age and size of the animal, to four quarts per day. These, together with the meal above, should be supplemented with a couple of quarts of wheat-bran, night and morning. The latter is excellent to prevent worms, and helps to keep the bowels in good condition.

Colts should not be permitted to stand on a plank, cement, paved, or any hard floor the first year, as these are liable to injuriously affect the feet and legs. Unless the yard where colts run in the winter has a sandy, or fine, dry, gravelly soil, it should be well littered, so as to keep their feet dry. Mud, or soft, wettish ground, is apt to make tender hoofs, no matter how well bred the colt may be. One reason why the horses in one district grow up superior to those in another in hoof, bone, muscle, and action, is because it has a dry limestone or silicious soil. When the mare is at work, do not let the colt run with her, and if she comes back from her work heated, allow her to get cool before suckling the colt, as her overheated milk is liable to give the foal diarrhœa.—*National Live Stock Journal*.

Sheep.

American shepherds have much yet to learn in regard to the management of their flocks. For example, the sheep in Siberia are never exposed to much rain. Shelter and shade are provided for them. Nor are they exposed to dust, for that is known to be injurious to the fleece. The greatest possible care is taken in the breeding. Men of experience are employed to go from farm to farm to examine the sheep and select the best rams that can be found. The rams are closely examined as to their fleece-bearing properties, and all but the very best are sold off. The whole economy of the sheep farm is so perfect as intelligence and industry can make it. A ton of wool is worth $750 at 35 cents a pound or $300 at 25 cents. A ton of wheat is worth about $32, and of corn about $16. The freight is about the same for each, and is thus 25 times more for wheat and nearly 50 times more for corn than wool. This is worth considering, and shows how much better it is to turn corn into wool than to sell it.

Cattle Range of Wyoming.

The great cattle range of Wyoming, under the military protection of Fort McKinney, is about 800 miles square. In this area are now grazing 500,000 head of cattle, worth $27 per head, amounting to $13,400,000, to which can be added the value of horses and ranches of the cattle men farmers, and the stock of the grangers, making at least $15,000,000 of property under the protection of this post.—*Indianapolis Price Current*.

Training Horns

If it is desirable to straighten a horn, you may frequently scrape with a piece of glass, or a knife, the hollow side, which will cause it to grow faster on that side; but, in that case, it must not be scraped deeply, for then it becomes weaker on that side, and will be turned towards the weaker side. Some scrape the side towards which they wish to turn the horn quite thin, and then scrape the opposite side just enough to make it grow faster, and that will turn it towards the thinly scraped side. If you wish to turn a horn up, scrape on the under side just enough to make it grow faster on that side. A very barbarous way to turn a horn is sometimes practiced, by searing with a hot iron on the side towards which the horn is to be turned. This prevents the growth of horn on that side, and the growth upon the other side turns the horn.

The horns may be polished by rubbing them with fine sand paper, and then with pumicestone, and oiling them. But this artificial manipulation of horns is seldom necessary. The horns of well-fed cattle will generally grow in comely shape if let alone.

The hair is sometimes oiled to give it a glossy appearance, but the best gloss is put upon the hair by rich and appropriate feeding. Nature, under proper conditions, does this work best.—*National Live Stock Journal*.

POULTRY.

Poultry Nonsense.

It is safe to say that more silly writing finds its way into print on the subject of the poultry yard and the care of poultry, than upon any other that can be named. As a rule this nonsense is uttered by amateurs who have lately taken up the business of poultry growing and who in three cases out of four are unable to distinguish a chicken from a turkey. But the silliness does not all make its appearance in the country newspapers. Poultry journals, that are supposed to know and ought to know a good deal about domestic fowls and their raising, admit articles into their columns that are simply astounding in their ignorance. We remember, f r instance, one of the best known poultry journals contained a long article which was intended to prove that corn should not be fed to domestic poultry. He neglected to say it ought to be fed exclusively to donkeys like himself. The editor of that journal expressed no opinion of his own about the matter, probably because he knew no better himself.

In a well-known agricultural newspaper, published in a neighboring county, the *Bucks County Intelligencer*, we saw this: "The best way to prevent or cure gapes in chickens is to commence feeding them whole grains of corn as soon as they are old enough to swallow them—say two or three weeks old. The effort made by the chick to swallow the whole grain will kill the little red worms in the throat, which are the cause of the gapes, and it is easier and safer to kill the worms in that way than to attempt to take them from the throat with a bent horsehair, as is sometimes done." The learned poultry editor of the above journal must have unusual luck in growing chicks which at the age of two or three weeks are capable of swallowing whole grains of corn. But even if he ever achieved that feat we inform him he might pour the entire corn crop of Illinois down the throats of his infected chicks without destroying or removing a single one of the "little red worms which are the cause of the gapes," and for the very simple reason that none of these destructive parasites ever find their way into the throat. They are always in the windpipe and there they remain, unless removed artificially, until they cause suffocation and death. So much for this learned essay on gapes in chickens.

The Philadelphia *Record* in its agricultural department had an article on the management of young fowls, in which, after saying "gapes are a disgrace to the poultry yard, and their prevention should be sought instead of their cure," proceeds to tell how this can be done. "The disease can be avoided altogether by feeding the chicks on a board or some other hard, clean surface." Chicks find a portion of their food themselves on the ground, where the danger of infection is always present. It is practically impossible to carry out the suggested plan, and experience has taught us the evil would not be remedied if it could be. The same article says " young turkeys, and in fact all young fowls, should be kept away from wet grass or exposure to dampness until well under feather ....The omission of a single feed is sometimes fatal, for once the young fowl becomes debilitated, its progress receives a check from which it seldom recovers." The only way young turkeys can always be kept dry that we know of is to keep them under roof. That plan would prove more fatal to them than a two weeks wet spell. How do the young of the wild turkey manage to survive ? Are they not exposed to wet grass and all the rain and moisture that falls? Does not the domesticated turkey at once lead her young brood into the fields and remain there for the most part until they are half grown? We have tried the plan suggested above year after year, but our success was always better when we turned the brood adrift, whether the grass was dry or wet. Of course, we do not believe a soaking rain will benefit young turkeys; what we mean to say is that the coddling plan is unnatural and unsatisfactory. These are merely a few specimens of the poultry literature with which poultry authorities abound, and which do far more mere harm than good, to say nothing of their irredeemable nonsense.—*New Era*.

Poultry.

The approach of Christmas suggests to our mind how very careless the major portion of our farmers and suburban poulterers are, when they have every facility to raise turkeys every year for market, but, after all, fail to do so. Ducks and geese of the improved breeds are profitably raised on many farms. If a supply of water can be given them, all the better. Good feed is more important than water to swim in and fish for bugs. The Rouen duck stands pre eminent among ducks where size is the consideration; the Aylesbury drake sometimes attains equal size, but the Aylesbury or any other duck seldom does. The white China geese have their admirers, for they have merits of no mean order, though for size the Toulouse geese are preferred in the former, while the Emblem or Bremen have many enthusiastic friends.—*Western Agriculturist*.

Women as Poultry Raisers.

The custom practiced in France of allowing the wife so many francs a month or year as pin money, to use as she pleases, is one that should be generally adopted, especially in the United States. On the farm the care and profits of some, if not all the poultry, could be very properly transferred to the woman of the household. The care of poultry is a business naturally adapted to woman, as it requires patience and attention, and, at the same, kindness and gentleness, traits too often lacking in the sterner sex. There is no event in connection with poultry raising, during the whole year, which has not its interest for those who care for the innocent creatures of the farm yard. Whether it be feeding grateful biped, gathering eggs, hatching the chickens, or reducing the flock in the fall to suit winter quarters—all have their charm, and excite the interest and sympathy of their faithful attendants. There is much complaint among physiologists that American ladies lose health and beauty earlier than they ought for want of sufficient out of door air and exercise, and this occupation has, among its other benefits, that of sending them daily abroad into the pure, outer air, and inciting a love for rural, natural beauty not found among those whom no such beauty tempts from the fireside.—*Lafayette, Ind., Journal*.

To Fat en Fowls or Chickens in Four or Five Days.

Set rice over the fire with skimmed milk, only as much as will serve one day. Let it boil till the rice is quite swelled out ; you may add a teaspoonful or two of sugar, but it will do well without. Feed them three times a day, in common pans, giving them only as much as will quite fill them at once. When you put in fresh, let the pans be set in water, that no sourness may be conveyed to the fowls, as that prevents them from fattening. Give them clean water, or the milk of rice to drink ; but the less wet the latter is when perfectly soaked, the better. By this method the flesh will give a clear whiteness which no other food gives, and when it is considered how far a pound of rice will go, and how much time is saved by this mode, it will be found to be as cheap as barley meal or more so. The coops should be daily cleaned, and no food given for sixteen hours before poultry are killed.—*S. P. Index*.

Winter Rations for Hens.

Fanny Field, who is famous for her success in making her hens lay in winter, tells the *Prairie Farmer* how she feeds them, as follows :

"My way of feeding fowls in winter—and it works wonderfully well—is to give them a warm breakfast every morning just as soon as they can see to eat, a few handfuls of grain at noon, and a full feed of grain at night. The warm breakfast is made of vegetables, turnips, beets, carrots or potatoes boiled and mashed up with wheat bran, or oat meal scalded with skim milk ; or refuse from the kitchen boiled up and the soup thickened with bran; and when sweet apples are plenty, we boil them, and mix with corn meal—sometimes one thing and some times another; we don't believe in feeding one thing all the time, and the hens don't believe in it either. I don't think that my biddies need the noon feed because they are hungry, but I give it to them to make them scratch for exercise, and to keep them out of mischief. I scatter it around among the litter under the shed and let them dig it out. This 'lunch' is generally oats or buckwheat, and once in a while sunflower seed. At night I generally feed corn, but if I could get wheat cheap enough, I should feed that at least half of the time. My fowls have water or milk by them all the time and green food is supplied by fastening cabbage heads up where the fowls can help themselves. Sometimes, when somebody has time to attend to it, we give them a change of green food in the shape of raw turnips or sweet apples chopped fine.

Two winters ago I took a new departure on the meat question, and now, instead of fussing to cook it and dealing out a little at a time, I just hang out a piece and let the fowls eat all they want. When they have meat within their reach all the time there is not the slighest danger of their eating too much. I get cheap meat from the butcher and I am sure I am paid twice over the outlay. Crushed oyster shells, gravel, charcoal and crushed raw bones are kept in the houses all the time. The raw bone is an excellent thing for fowls, and would be the last article of food that I would think of dropping from my biddies' bill of fare. Where the crushed oyster shell cannot be obtained, lime in some other shape will do just as well. One of my neighbors had two of his rooms plastered this fall, and he saved all the old plaster for his hens. The poultry raisers who neglected to get a supply of gravel under cover before the ground froze up, must do the next best thing—feed their broken dishes to their fowls. Break them into bits of a suitable size, and it will do just as well as gravel. I believe in salting all the soft food, and I used to put in a dash of pepper, sometimes mustard or ginger, once in a while, and I honestly thought the fowls were benefited thereby; but doubts are creeping in, and I am very much inclined to drop everything except the salt."

Pekin Ducks.

There are according to the American standard of excellence, ten varieties of domestic ducks, to wit: The Aylesbury, Call Gray, Call White, Cayuga, Crested White, Black East Indian Colored Muscovy, White Muscovy, Pekin and Rouen. Of these the Rouen is probably the most common, but I consider the Pekin as the most profitable. They were first imported into this country in 1773, since which time they have become very popular. Their color is a pure snowy white which makes them very handsome and attractive for small bodies of water or the lawn. They should have rich deep yellow colored bills and legs and perfectly free from any black spots. They can be raised anywhere that chickens can and do not require much water until they are several months old, and then they will thrive and do well with but a small trough of water, if they have a good grass range. It is a very beautiful sight to see them deploy in long lines through the grass in search of crickets and other animal matter. They mature very early and can be marketed in July and August at high prices. It takes in warm weather about three weeks to hatch them out, at birth they are larger and stronger than other varieties and when developed will weigh about eighteen pounds to the pair. They lay about one hundred and fifty eggs per year. During the summer months they require but very little food, as when they have a good range they will pick up enough to keep them in good condition, especially if they have access to the chicken yard, as they will eat what the chickens waste.—*The American Stockman.*

LITERARY AND PERSONAL.

THIRD QUARTERLY REPORT OF THE KANSAS STATE BOARD OF AGRICULTURE.—The Quarterly Report of the Kansas State Board of Agriculture, for the quarter ending September 30, has just been issued.

The report contains the acres and product of principal crops, by counties, accompanied by market quotations of the Kansas C ty market for each month from January, 1877, to September, 1883, for the crops of wheat and corn.

In connection with the statistics on wheat are given instances of extraordinary yields grown in each county in 1882, and the names of the varieties that have been the most successful this year.

Following the crop tables, is an article from the pen of Prof. E. A. Popenoe, entomologist to the Board, and Professor of Botany and Horticulture in the Kansas State Agricultural College, at Manhattan, on the subject of "The chinch bug and the season," giving the history of the operations of these pests during the past season, and the reasons why they were less destructive than was anticipated.

The synopsis of the reports of correspondents on fruit indicates an unusually heavy apple crop, but that peaches, plums, pears and grapes were injured severely, and made less than average crops.

The second division of the volume pertains to live stock. Tables by counties, showing the numbers of each kind of farm animals for both 1881 and 1882, with increase and decrease, are given; also quotations of the Kansas City market on cattle and hogs for a period covering six years. Mr. J. F. True, of Newman, Jefferson county, contributes an article on the feeding of cattle in Kansas. Mr. True is a well-known breeder and feeder of cattle, and is also a member of the State Board of Agriculture.

A summary by counties of reports of correspondents as to the condition of live stock follows the article of Mr. True. Among the valuable points brought out in this summary are, the amount of open range remaining, cost per head for the grazing of cattle, cost per ton of prairie hay, information as to whether the herd law is in operation or not, and facts concerning the raising of sheep and swine.

Statements of the number of acres of public lands yet undisposed of, corrected to date, follow the live stock This has been a prominent feature of the last several reports, and is of much value to those who are seeking homes in the State.

The population of the State for 1881 and 1882, as taken by township assessors, is given in full, by townships and cities, followed by a summary by counties, showing the increase and decrease in each county during the year.

Brief reports as to the principal features of the two State fairs held this year at Topeka and Lawrence precede the meteorological data of the quarter, which closes the volume.

The report is now ready for distribution, and can be obtained by addressing the Secretary, Wm. Sims, Topeka, Kansas, and inclosing the necessary postage, three cents.

HOWE'S LANCASTER CITY AND COUNTY DIRECTORY, containing the names of the inhabitants of Lancaster City and Columbia, together with a Business Directory of Lancaster City and County, to which is added a large list of the farmers of Lancaster county and an appendix of useful information. Compiled and published by C. E. Howe & Co., Philadelphia, Pa., 831, Arch street. Price, $3.00. This is the legitimate successor of "Boyd's Directory," and, if not thoroughly perfect in all its details, it certainly is, on the whole, the best directory of Lancaster county that has ever been placed before the public, including quality of material, letter press, and general arrangement. The paper, type, and print are far superior to any heretofore used for such a purpose; 520 royal octavo pages, 20 pages extra advertisements—indeed, wherever there is any available space it is filled by one or more advertisements. The back, the sides, the edges, and the face margins of every page contain one or more advertisements, illustrating that, whatever else it may be, or fail to be, it certainly is entitled to the distinction of being the best Lancaster County Directory extant. One of the most interesting and useful features of the work, from an agricultural and domestic standpoint, is its list of the names and post-office address of *four thousand six hundred and seventy-four* farmers of Lancaster county, alphabetically arranged. It also includes the population of the United States according to the tenth census (1880), the population of Lancaster county, and the male voting population of Pennsylvania, as well as the entire population. Also Lancaster City Street Guide: the officers of the General Government and of Pennsylvania, Lancaster city and county officers, including the councils and school board, the new fire department, the public buildings, the churches, cemeteries, educational institutions, post office department, stage lines, county post offices and postmasters, distances and rates of railroad fare, newspapers and periodicals, building associations, beneficial and secret societies, literary, scientific, and miscellaneous societies, together with the latest revised lists of their officers; all brought down and adapted to the years 1882 and 1883, in fact, appearing to be all that is necessary in the form of a directory for the territory it covers at the present time.

LANCASTER REAL ESTATE CATALOGUE.—Allen A. Herr & Co., No. 108 East King street, Lancaster, Pa., issue monthly and send free to any address 40 pp. 8vo, containing properties for sale, and prices of many of them, numbering from 10 to 608. A capital publication for those who desire to invest in real estate, and possess the wherewithal to do it.

EARLY GERMAN HYMNOLOGY OF PENNSYLVANIA. By Rev. J. H. Dubbs, D.D.; 27 pages, octavo.—We have been favored with a complimentary copy of this interesting pamphlet by the author, for which we feel exceedingly grateful. It may be a weakness in us, but we all our life have had a leaning towards hymn and song lore, and, at "three score and ten," we often catch ourself humming over those that came to us traditionally in the days of our early boyhood. We read this little work with more than ordinary interest, and, whilst doing so, we found it impossible to divest our mind of the constantly obtruding thought that the unwritten hymnology of "the sable sons of Africa" inhabiting our country would form a prolific theme for a similar literary contribution. It is true they had not much "rhyme" in their composition, and many of them very little "reason;" but, then, this may only have been an *appearance* to "uncircumcised" outsiders blinded by morbid partialities. Be that as it may, they sang their hymns and songs with a *will* that amply compensated their other defects.

BREEDERS' JOURNAL, published by the "Breeders' Live Stock Association," at Beecher, Ill.; $1 00 per year, single number 10 cents. The professed specialty of this journal is to stimulate "Economy of production, and value of product." This is a very ably conducted magazine, of 62 pp. octavo, in tinted covers, and liberally illustrated. The material and typography are of unexceptionable quality, and it seems to "box the compass" in stock breeding, including all that legitimately comes within the sphere of that specialty.

AMERICAN POULTRY BUREAU, devoted to the various interests of poultry, pigeons and pet stock. 16 pp. quarto, at $1.00 a year. Rossall & Gibson, editors and publishers, Des Moines, Iowa, a new candidate for public patronage, which it certainly deserves.

SPRING GARDEN INSTITUTE—Corner of Broad and Spring Garden streets, Philadelphia, Pa. This institute has been organized for the purpose of imparting practical mechanical instruction to the youths of our country: and, from the list of pupils who have attended the handiwork schools belonging to it, it seems to have been successful. The terms of tuition seem to be liberal, $5.00 for a term of 3 months in *Vise* and *Lathe work*—two nights each week. The same for *Pattern making, with the elements of moulding*, and $10.00 for *Steam Engineering.* Lectures will be delivered to the mechanical handiwork classes. The names of *two hundred pupils*, and their addresses are given, as having attended this experimental school in mechanics, during the last term. Perhaps the greatest obstacle to students from a distance would be the expense of boarding, but to those from Philadelphia the Institute furnishes a rare opportunity to receive instruction in filing, turning, drilling, forging, and other mechanical handiwork.

WE doubt the propriety of promulgating the *effete* atheistical views of Robert J. Ingersoll, through the medium of a *Thoroughbred Stock Journal*, as well as any replies to them. It involves a load too heavy to successfully carry, by any journal seeking the patronage of farmers, mechanics and stock breeders; besides, it is not honest to fulminate or diffuse such a rehash of exploded sentiments, under cover of a journal ostensibly devoted to the diffusion of agricultural lore. It is fifty years since we first listened to the atheistical ravings of those of Ingersoll's faith, and to us at that time they seemed to be more able and unanswerable than anything uttered by Ingersoll at the present period. Neither he nor Judge Black have discussed the true theology of the Bible, for it cannot be discussed on the mere plane of "Buckwheat cakes and sausages." There is a line of argument by which the Bible can be sustained, but it is foolishness to the carnal mind, and must be "spiritually discerned," distasteful as such a "Paulineism" may be to Mr. Ingersoll. A theological proposition may not be false, simply because it cannot be proven true by material testimony, or legal argument.

THE LANCASTER FARMER

EVERY lady should send 25 cents to Strawbridge & Clothier, Philadelphia, and receive their *Fashion Quarterly* for 6 mos. 1,100 illustrations and 4 pages new music each issue.

A MANUAL OF ELOCUTION AND READING, embracing the Principles and Practice of Education. By Edward Brooks, Ph. D., Principal of the State Normal School, Millersville, Pa. Philadelphia: Eldridge & Bro. Price, 1.50. To teachers, for examination, $1.00.

PHILIP SCHUM, SON & CO.,
38 and 40 West King Street.

We keep on hand of our own manufacture,

QUILTS, COVERLETS, COUNTERPANES, CARPETS,

Bureau and Tidy Covers, Ladies' Furnishing Goods, Notions, etc.

Particular attention paid to customer Rag Carpet, and scouring and dyeing of all kinds.

PHILIP SCHUM, SON & CO.,
Nov-1y   Lancaster, Pa.

THE PENN HARROW

BEST IN THE WORLD
IT HAS NO EQUAL

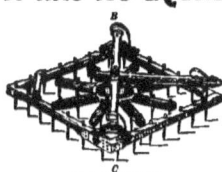

Patented April 13, 1875.

The above cut represents the Penn Harrow complete, with all its combinations of Five Harrows and a sled for each Harrow; and each successive blade is made from this Harrow without the least additional expense. By hooking the team to either end, the center revolves and gives the ground Two Strokes and Two Crosses in passing over it once, making it the most effective pulverizer in the market.

THIS HARROW HAS ONLY TO BE USED TO BE APPRECIATED.

See it before purchasing and you will buy no other.

The Penn Harrow
CHANGED TO A THREE-CORNER ROTARY HARROW.

Indispensable for Orchards, as the revolving wheel harrows right up to and all around the trees without barking them.

The Penn Harrow
CHANGED TO SINGLE "A" HARROW.

DISSOLUTION OF PARTNERSHIP.—

The co-partnership in the merchant tailoring business heretofore existing under the firm of Rathvon & Fisher, is this day dissolved by mutual consent. All persons in any manner indebted to said firm, are respectfully solicited to make immediate payment to S. S. Rathvon, who is hereby authorized to receive the same, and those having claims against said firm, will please present them for settlement.

S. S. RATHVON,
M. FISHER,
101 North Queen Street, Lancaster, Pa.

Until further announcement, the business, without interruption, will be conducted by the undersigned, who solicits a continuation of the patronage heretofore bestowed upon the firm, and which is hereby greatly acknowledged.

S. S. RATHVON,
PRACTICAL TAILOR,

No. 101 North Queen Street,
LANCASTER, PA.

By removing the wing and wheel from the original you have a complete one-horse "A" Harrow.

The Penn Harrow
CHANGED TO DOUBLE "A" HARROW.

Remove the wheel from the original, reverse the wing, and it makes the most complete Double "A" Harrow in the market.

The Penn Harrow
CHANGED TO A SQUARE HARROW.

By removing the wheel from the original you have a Harrow with three seams to hold in. By hooking either of you one harrow in a furrow, and harrow the bottom and both sides, or over a ridge and harrow the top and both sides, or you can lift either out and have three seams, on the ground—something that cannot be done with any other Harrow.

The Penn Harrow
ON ITS SLED.

It has always been a great inconvenience to get the Harrow to and from the field. The Penn Harrow obviates this, as no matter which Harrow you wish to use, by the combination, it has its own sled to haul it on.

The Penn Harrow

is made of the best white oak, with steel teeth, well painted, in every way first-class. Formerly a harrow was the most unhandy implement on the farm, with our improvement it is the most convenient. We have no doubt the work of any one harrow and one farmer and save the farmer half his labor, and is warranted to do all we represent or money refunded. ORDER AT ONCE AND BE CONVINCED.

Price of the light draft Combination Penn Harrow, $30. *Send for a Catalogue and see what farmers say.*

AGENTS WANTED IN EVERY COUNTY.

PENN HARROW MANUFACTURING CO.
CAMDEN, N. J.

TREES
Fruit, Shade and Ornamental Trees.

Plant Trees raised in this county and suited to the climate. Write for prices to

LOUIS C. LYTE
Bird-in-Hand P. O., Lancaster co., Pa.
Nursery at Smoketown, six miles east of Lancaster.
79-1-12

WIDMYER & RICKSECKER,
UPHOLSTERERS,
And Manufacturers of

FURNITURE AND CHAIRS.
WAREROOMS,

102 East King St., Cor. of Duke St.
LANCASTER, PA.
79-1-12]

Special Inducements at the
NEW FURNITURE STORE
OF
W. A. HEINITSH,
No. 16 1-2 E. KING STREET
(over Burck's Grocery Store, Lancaster, Pa.)
A general assortment of Furniture of all kinds constantly on hand. Don't forget the number,
151-2 East King Street,
Nov-1y]   (over Burck's Grocery Store.)

For Good and Cheap Work go to
F. VOLLMER'S
FURNITURE WARE ROOMS,
No. 309 NORTH QUEEN ST.,
(Opposite Northern Market,)
Lancaster, Pa.

Also, all kinds of picture frames.   nov-1y

GREAT BARGAINS.
A large assortment of all kinds of Carpets are still sold at lower rates than ever at the

CARPET HALL OF H. S. SHIRK,
No. 202 West King St.

Call and examine our stock and satisfy yourself that we can show the largest assortment of three Brussels, three ply and Ingrain at all prices—at the lowest Philadelphia prices.

Also on hand a large and complete assortment of Rag Carpet.

Satisfaction guaranteed both as to price and quality. You are invited to call and see my goods. No trouble in showing the articles if you do not want to purchase. Don't forget the number. You can save much here if you want to buy.

Particular attention given to customer work. Also on hand a full assortment of Counterpanes, Oil Cloths and Blankets of every variety.   [nov.-1y

C. R. KLINE,
ATTORNEY-AT-LAW,

OFFICE: 15 NORTH DUKE STREET,
LANCASTER, PA.
Nov-1y

SILK-WORM EGGS.

Amateur Silk-growers can be supplied with superior silk-worm eggs, on reasonable terms, by applying immediately to
GEO. O. HENSEL,
may-30]   No. 278 East Orange Street, Lancaster, Pa.

LIGHT BRAHMA EGGS

For hatching, now ready—from the best strain in the county—at the moderate price of

$1.50 for a setting of 13 Eggs.
L. RATHVON,
No. 9 North Queen st., Examiner Office, Lancaster, Pa.

WANTED.—CANVASSERS for the
LANCASTER WEEKLY EXAMINER
in Every Township in the County. Good Wages can be made. Inquire at
THE EXAMINER OFFICE,
No. 9 North Queen Street, Lancaster, Pa.

Where To Buy Goods
IN
LANCASTER.

BOOTS AND SHOES.

MARSHALL & SON, No. 12 Centre Square, Lancaster, Dealers in Boots, Shoes and Rubbers. Repairing promptly attended to.

M. LEVY, No. 3 East King street. For the best Dollar Shoes in Lancaster go to M. Levy, No. 3 East King street.

BOOKS AND STATIONERY.

JOHN BAER'S SONS', Nos. 15 and 17 North Queen Street, have the largest and best assorted Book and Paper Store in the City.

FURNITURE.

HEINITSH'S, No. 15½ East King st., (over China Hall) is the cheapest place in Lancaster to buy Furniture. Picture Frames a specialty.

CHINA AND GLASSWARE.

HIGH & MARTIN, No. 15 East King st., dealers in China, Glass and Queensware, Fancy Goods, Lamps, Burners, Chimneys, etc.

CLOTHING.

MYERS & RATHFON, Centre Hall, No. 12 East King St. Largest Clothing House in Pennsylvania outside of Philadelphia.

DRUGS AND MEDICINES.

G. W. HULL, Dealer in Pure Drugs and Medicines, Chemicals, Patent Medicines, Trusses, Shoulder Braces, Supporters, &c., 15 West King St., Lancaster, Pa.

JOHN F. LONG & SON, Druggists, No. 12 North Queen St. Drugs, Medicines, Perfumery, Spices, Dye Stuffs, Etc. Prescriptions carefully compounded.

DRY GOODS.

GIVLER, BOWERS & HURST, No. 25 E. King St., Lancaster, Pa., Dealers in Dry Goods, Carpets and Merchant Tailoring. Prices as low as the lowest.

HATS AND CAPS.

C. H. AMER, No. 30 West King Street, Dealer in Hats, Caps, Furs, Robes, etc. Assortment Large. Prices Low.

JEWELRY AND WATCHES.

H. Z. RHOADS & BRO, No. 4 West King St. Watches, Clock and Musical Boxes. Watches and Jewelry Manufactured to order.

PRINTING.

JOHN A. HIESTAND, 9 North Queen st., Sale Bills, Circulars, Posters, Cards, Invitations, Letter and Bill Heads and Envelopes neatly printed. Prices low.

FARMING FOR PROFIT.

It is conceded that this large and comprehensive book, (advertised in another column by J. C. McCurdy & Co., of Philadelphia, the well-known publishers of Standard works,) is not only the newest and handsomest, but altogether the BEST work of the kind which has ever been published. Thoroughly treating the great subjects of General Agriculture, Live-Stock, Fruit-Growing, Business Principles, and Home Life; telling just what the farmer and the farmer's boys want to know, combining Science and Practice, stimulating thought, awakening inquiry, and interesting every member of the family, this book must exert a mighty influence for good. It is highly recommended by the best agricultural writers and the leading papers, and is destined to have an extensive sale. Agents are wanted everywhere. jan-lt

ERISMAN

GLOVES, SHIRTS, UNDERWEAR.
SHIRTS MADE TO ORDER,
AND WARRANTED TO FIT.

E. J. ERISMAN.
56 North Queen St., Lancaster, Pa.
79-1-12]

Thirty-Six Varieties of Cabbage; 26 of Corn; 28 of Cucumber; 11 of Melon; 33 of Peas; 28 of Beans; 17 of Squash; 21 of Beet and 40 of Tomato, with other varieties in proportion, a large portion of which were grown on my five seed farms, will be found in my Vegetable and Flower Seed Catalogue for 1882. Sent free to all who apply. Customers of last Season need not write for it. All Seed sold from my establishment warranted to be fresh and true to name, so far, that should it prove otherwise, I will refill the order gratis. The original introducer of Early Ohio and Burbank Potatoes, Marblehead, Early Corn, the Hubbard Squash, Marblehead Cabbage, Phinney's Melon, and a score of other New Vegetables, I invite the patronage of the public. New Vegetables a specialty.

JAMES J. H. GREGORY,
Marblehead, Mass.

Nov-6mo)

EVAPORATE YOUR FRUIT.
ILLUSTRATED CATALOGUE
FREE TO ALL.

AMERICAN DRIER COMPANY,
Chambersburg, Pa.
Ap1-tf

THE COOLEY CREAMER

Raises all the cream between the milkings. Saves two-thirds the labor. Increases yield of butter. Improves the quality. Quadruples the value of skim milk. Will pay for itself twice or more every season. The Cooley System is the only uniform dairy method in existence. Requires no spring house, or milk room. May be placed in a shed, cellar, or any place that cold water is handy.

The Best Hired Girl.

In the fall of 1879 I bought a Cooley Creamer. I have used it ever since with entire satisfaction. It makes more butter, of better quality, without ice, and half the labor, than the old process. A lady friend who has used one for about six months says it is "the best hired girl" she ever had. I have also used the Davis Swing Churn for the last 16 months, and am highly pleased with it. It churns the cream at a higher temperature and brings the butter in a better condition than any other churn. I have given the Eureka Butter Worker a fair trial, and am happy to recommend it to others. I can work twenty pounds of butter with it in five minutes, and thus save a half hour's work.

Yours truly,   SAMUEL S. OGLKITT.
Mt. Holley, Burlington County, N. J., August 22, 1881.

☞ Send for Circular free to

D. LANDRETH & SONS,
Sole Agents, Philadelphia Pa.

A HOME ORGAN FOR FARMERS.

THE LANCASTER FARMER,

A MONTHLY JOURNAL,

Devoted to Agriculture, Horticulture, Domestic Economy and Miscellany.

Founded Under the Auspices of the Lancaster County Agricultural and Horticultural Society.

EDITED BY DR. S. S. RATHVON.

TERMS OF SUBSCRIPTION:

ONE DOLLAR PER ANNUM,

POSTAGE PREPAID BY THE PROPRIETOR.

All subscriptions will commence with the January number, unless otherwise ordered.

Dr. S. S. Rathvon, who has so ably managed the editorial department in the past, will continue in the position of editor. His contributions on subjects connected with the science of farming, and particularly that specialty of which he is so thoroughly a master—entomological science—some knowledge of which has become a necessity to the successful farmer, are alone worth much more than the price of this publication. He is determined to make "The Farmer" a necessity to all households.

A county that has so wide a reputation as Lancaster county for its agricultural products should certainly be able to support an agricultural paper of its own, for the exchange of the opinions of farmers interested in this matter. We ask the co-operation of all farmers interested in this matter. Work among your friends. The "Farmer" is only one dollar per year. Show them your copy. Try and induce them to subscribe. It is not much for each subscriber to do but it will greatly assist us.

All communications in regard to the editorial management should be addressed to Dr. S. S. Rathvon, Lancaster, Pa., and all business letters in regard to subscriptions and advertising should be addressed to the publisher. Rates of advertising can be had on application at the office.

JOHN A. HIESTAND,
No. 9 North Queen St., Lancaster, Pa.

$72 A WEEK. $12 a day at home easily made. Costly Outfit free. Address TRUE & Co., Augusta, Maine